MICROPROPAGATION OF WOODY TREES AND FRUITS

FORESTRY SCIENCES

Volume 75

The titles published in this series are listed at the end of this volume.

Micropropagation of Woody Trees and Fruits

edited by

S. Mohan Jain

International Atomic Agency,
FAO/IAEA Joint Division,
Vienna, Austria

and

Katsuaki Ishii

Forestry and Forest Products Research Institute,
Tsukuba, Japan

SPRINGER-SCIENCE+BUSINESS MEDIA, B.V.

A C.I.P. Catalogue record for this book is available from the Library of Congress.

ISBN 978-1-4020-1135-1 ISBN 978-94-010-0125-0 (eBook)
DOI 10.1007/978-94-010-0125-0

Printed on acid-free paper

CONTENTS

Section C

Section D

Preface

Global warming, environmental changes, water shortage, and sustainable development are the most up-to-date issues, which have challenged mankind. Researchers worldwide are engaged in addressing some of these problems, including reduction in carbon dioxide accumulation, and enrichment of perennial woody species on the terrestrial ecosystem. About 12 million hectares of the world's forests disappear every year. By 2025, the world population will reach 7.5 billion, and the forest area will be reduced to well below 50 % of the current area. Reforestation is an important to prevent the loss of forest resources including timber, biodiversity and water resources. Therefore, subsequent volume of reforestation over the deforested land should be followed to safeguard the forests and maintain its size, which will require a continuous supply of planting material. Similarly fruit trees, including tropical and subtropical fruit trees, are consumed both as fresh and in the processed form including juices, beverages, and dried fruits. They are an important source of nutrition *e.g.* rich in vitamins, sugars, aromas and flavour compounds, and raw material for food processing industries. The production, cultivation and maintenance of tree species provide highly sustainable production systems that conserve soils, microenvironment and biodiversity. Fruit trees have long juvenile periods and large tree size. In many fruit trees *e.g.* avocado and others controlled crosses are difficult to make due to massive fruit drop. Moreover, fruit trees are faced with agronomic and horticultural problems in terms of propagation, yield, appearance, quality, diseases and pests control, abiotic stresses and poor shelf life. The available genetic information in fruit crops is very limited and their genetic improvement has relied heavily upon vegetative propagation techniques and classical breeding. The key technology for biomass production of trees is propagation via micropropagation. However, there are many recalcitrant species among useful forest and fruit trees to propagate in large numbers. Recent developments of *in vitro* culture techniques have made it possible to commercially propagate useful trees both forest and fruit trees.

In this book, comprehensive information is provided on micropropagation of economically important forest and fruit trees, which is usually available in scattered literatures. Topics cover a wide range, from tropical forest and fruit trees for paper or food supply to *Prunus* species for local craft bark production. This book is divided into four sections. Section A contains five chapters on different parameters influencing micropropagation, which are quality of light, arbuscular mycorrhizal (AM) fungi, pathogen elimination and contamination management, and molecular markers for genetic fidelity assessment.

Section B has five chapters on micropropagation of tropical woody species, *Fagus* spp., *Euycalyptus*, *Acacia* species, and *Aegle marmelos* medicinal tree. Section C includes 14 chapters, which are dealing mainly with micropropagation of fruit trees such as grapevine, bananas, arid zone fruit trees of India, apples, small fruits, Prunus, avocado, papaya, pistachio, olive, kiwifruit, tea, and litchi. Section D has four chapters mainly on bioreactor, photoautotrophic, thin cell layer and *in vitro* germplasm conservation. Even though few papers are published on litchi micropropagation, we have deliberately included this chapter just to remind the readers that plenty of work is still needed in most of the commercially important tropical fruits, including litchi.

The invited authors were selected world wide from east to west. The editors are grateful to all the contributing authors for timely submission of their manuscripts, which made it possible to realise this book project. We appreciate it very much that our colleagues took some time off to review these manuscripts. We sincerely hope that our hard work will benefit all engaged in research on forest and fruit trees.

S. Mohan Jain

K. Ishii

SECTION A

1. EFFECTS OF LIGHT QUALITY ON MICRO-PROPAGATION OF WOODY SPECIES

Morini Stefano[1] and Muleo Rosario[2]
[1]Università degli Studi di Pisa, Dipartimento di Coltivazione e Difesa delle Specie Legnose "G. Scaramuzzi", Sezione Coltivazioni Arboree, Via del Boghetto 80, 56124 Pisa, Italy.
[2]Università della Tuscia, Dipartimento di Produzione Vegetale, Sezione Ortoflorofrutticoltura, Via S. C. De Lellis, 01100 Viterbo, Italy.

CONTENTS

1. INTRODUCTION

Light is a fundamental environmental cue in the life of plants, playing a crucial role, directly or indirectly, in the regulation of plant development and growth. It represents the signal to be elaborated in the photoperception processes by the photoreceptors. Plants have evolved exquisite sensory systems for monitoring their environment and initiating appropriate strategies in their lifetime. In order to optimise the acquisition of light energy for

S.M. Jain and K. Ishii (eds.), Micropropagation of Woody Trees and Fruits, 3–35.

photosynthesis, plants have developed a series of signal-transducing photoreceptors that regulate their growth and development in relation to the presence, amount, direction, duration and quality of incident light radiation. Several families of distinct photoreceptors, sensitive to different regions of the light spectrum, mediate the developmental responses of plants to light signals: phytochromes (red/far-red light-absorbing photoreceptor), cryptochromes (blue/UV-A light-absorbing photoreceptor), phototropin, and UV-B photoreceptor (Briggs and Olney, 2001). Among these regulating signal-transducing photoreceptors, the best characterised, at genetic and physiological levels, is the phytochrome family. In higher plants, the phytochrome family may have as many as five discrete members (Clack et al., 1994; Hauser et al., 1997; Pratt et al., 1997). Different members of the phytochrome family share and have specific biochemical and physiological properties, are differentially expressed and have different functional roles in controlling plant photomorphogenesis (Whitelam and Halliday, 1999; Chory and Wu, 2001). Photomorphogenetic studies at molecular and physiological levels indicate that PHYA is responsible for continuous, monochromatic, far-red light (FRL) driven photomorphogenesis through the FR high-irradiance response (FR-HIR), whereas PHYB is predominantly responsible for continuous, monochromatic, red light (RL) driven photomorphogenesis through the R-HIR (Whitelam and Devlin, 1997; Quail, 1998). Collectively, under *in vivo* conditions, phytochromes regulate all the aspects of plant development from seed germination and plantlet establishment, through the detection of neighbouring vegetation and the initiation of shade avoidance reactions, to the measurement of day length and the control of reproductive development (Smith, 1995). Photoreceptors and signal transduction mutants of *Arabidopsis thaliana* are proving to be valuable tools in the molecular dissection of photomorphogenesis. Mutants deficient in four of the five phytochromes have now been isolated. Their analysis indicates considerable overlap in the physiological functions of different phytochromes. In addition, mutants for components acting downstream from the phytochromes have indicated that different members of the family use different signalling pathways (Casal, 2002).

Blue light (BL) also plays a pivotal role in plant development and gene expression. Recently, molecular genetic studies have identified three BL receptors: cryptochrome 1, cryptochrome 2 and phototropin, which regulate plant growth, flowering time and phototropism. These photoreceptors are similar in structure and chromophore composition to the prokaryotic DNA photolyases, but lacking in photolyase activity. For comprehensive descriptions of the photoreceptors properties and some of the downstream events activated by them, ample reviews have already been published (Cashmore et al., 1999; Briggs and Huala, 1999).

Pure BL responses are rare, whereas more common is the BL receptor action together with phytochrome, which co-act in the regulatory action that modulates the phytochromes downstream response. Genetic studies indicate that phytochromes and BL signal transduction pathways likely converge on common intermediates (Casal, 2002). CRY1 is supposed to be the cryptochrome responsible for BL-HIR, inhibiting stem plant growth and reducing internode elongation, whereas CRY2 should be responsible for the inhibition due to the BL-LIR (Lin, 2000). In particular, in transgenic tomato plants, both CRY1 and CRY2 are implicated in numerous BL-dependent responses, including inhibition of hypocotyl elongation, leaf and cotyledon expansion, pigment biosynthesis, stem growth and internode elongation, control of flowering time and phototropism, all depending on BL intensity (Ninu et al., 1999).

2. Physiological implications of micropropagation

Two fundamental steps can influence the micropropagation efficiency during *in vitro* stages: the first concerns the production of a quantity as large as possible of new axillary shoots (shoot proliferation), while the second is based on the possibility of inducing adventitious root formation at the shoot base. The two growth processes involve different dynamics: while in the case of rooting the final objective can be considered reached once root induction and differentiation have taken place, shoot proliferation is based on two processes, i.e. the formation of lateral meristems from the original shoot tip and the subsequent development of axillary buds into shoots.

It has been widely shown that *in vitro* microenvironmental culture conditions such as light, temperature, and moisture, can modify culture response. While studies on light intensity and photoperiod were extensively performed even on woody species cultures, light quality did inexplicably not attract much attention from researchers. One cause may be the lower culture reactiveness of woody species compared to that of herbaceous species.

In the experiments carried out to date, light quality has mainly been studied with a double objective: the first to improve knowledge on photoreceptors role in the growth processes that take place in *in vitro* microenvironmental conditions; the second to find out possible light spectra photomorphogenetically more active than white light (WL), to apply for increasing micropropagation efficiency. Numerous bibliographic data on the effect of light quality on *in vitro* herbaceous species are available, while much less are those concerning woody species.

The knowledge acquired to date, although fairly extensive, should not yet be considered exhaustive since the results are sometimes conflicting. A reason could be the different response to light quality observed among different species and, for the same species, among different genotypes. Other reasons could be the different explant type and physiological stage, the donor plant age, the methodology and light sources used in the various experiments that are not always homogeneous, particularly in light spectral composition, even if the lights are termed or defined with the same colour. Another cause of variability could be the procedure adopted to measure the radiation, i.e. if inside or outside the culture vessel.

To date, shoot proliferantion is the most studied micropropagation stage in respect of the light quality effects. The objective is to identify optimal ratios among different light spectral bands in order to adapt the radiation system to the apical dominance reduction strategies of the different species. Shoot rooting also attracted the researcher's attention but scientific information on the physiological role of light quality applied during this stage is less comprehensive.

3. Effects of light quality during shoot proliferation stage

Research performed during last years showed that light quality has influence on the biological effictiveness of the growth regulators added to the growth medium, as well as to affect endogenous hormonal balance in the tissues. Therefore, light quality appears as a tool to be manipulated for inducing physiological balances favourable to specific growth responses that might be maintained as long as wanted.

Among the various growth responses, shoot proliferation is particularly important in micropropagation; it is based on cytokinin-induced release from apical dominance of axillary buds. That apical dominance might be affected by light conditions has been accepted for many years (Hillman, 1984). Two light qualities mainly, RL and BL, are able to influence shoot proliferation; their efficacy, as already stated, is related to the presence of specific pigments in the tissues. Various species have been shown to respond better to RL and others to BL. There are also plants that give similar responses to both light qualities but it probably depend on the experimental procedures used.

3.1 EFFECTS ON AXILLARY SHOOT PRODUCTION AND CULTURE MORPHOLOGY

As early as 1977, Thimann showed that light constitutes an environmental factor that is also involved in the regulation of apical dominance. This process is defined as a physiological control exerted by the shoot apex over axillary bud growth. Among the endogenous factors regulating apical dominance, the hormones most involved are auxins and cytokinins (Cline, 1994; Tamas, 1995). In the classical model of auxin involvement in apical dominance, it is hypothesised that this hormone is synthesised by the apex and transported downwards into axillary buds, with subsequent direct down-regulation of outgrowth, or indirect regulation *via* other mechanisms, including activation of secondary growth substances, nutrient diversion, or modification of the auxin/cytokinin ratio (Hillman, 1984; Martin, 1987; Cline, 1991; Napoli et al., 1999). Surgical removal of the shoot apex stimulates outgrowth of axillary buds. Thus, a cytokinin increase in the tissue leads to the growth of axillary buds, opposing the auxin action. This is in line with results of studies on transgenic plants, which have shown that the apical dominance regulation by growth hormones is not determined by the absolute concentration of hormones but by the ratio between them (Klee and Lanahan, 1995).

Shoot proliferation is strongly dependent both on apical dominance exerted by the shoot apex over axillary buds below it, and on the formation of lateral meristems that will origin new axillary buds along the axis of the growing shoot. Under *in vitro* conditions, a number of different factors may exert a marked influence on shoot development and growth and on the inhibitory correlation between the apex and the axillary buds (Murashige, 1974; Zimmerman, 1983). Light quality is one of these factors and has sometimes shown a pronounced influence on *in vitro* culture growth.

3.1.1. EFFECTS OF RED AND FAR RED LIGHTS

RL firstly showed the capacity to influence shoot proliferation. Phytochrome is the active photoreceptor sensitive to RL and FRL. In its active form, it seems to alter the endogenous hormonal balance in favour of reducing apical dominance and increasing lateral shoot development (Jensen et al., 1998; Morelli and Ruberti, 2000; Muday and Murphy, 2002). The earliest work on this subject was carried out by Tucker (1976) who showed that five minutes of FRL after 16 h of fluorescent light inhibited the opening of axillary buds in *in vivo* tomato plants. The formation of abscisic acid in or near the buds, as a consequence of increased auxin synthesis in the apex and young leaves, was indicated as the cause of reduced bud development.

Similar results were obtained *in vitro* by continuous irradiation with FRL. Trials carried out to study the effect of phytochrome on *in vitro* shoot proliferation of GF 655/2 plum rootstock (Baraldi et al., 1988) did not show any increase in the proliferation rate with or without BA under FRL and the result was similar to that detected in darkness (D). However, WL, RL and BL treatments at the intensity of 37 $\mu mol\ m^{-2}\ s^{-1}$ displayed higher and very similar promoting effects on shoot proliferation. If, on the other hand, the light intensity was 9 μmol m^{-2} s^{-1}, the proliferation rate was greater under RL, decreased moving to BL and was reduced even more under FRL. In the presence of BL, FRL and WL, shoot production increased with increasing the photon fluence rates, while for RL this process appeared to be independent from light intensity. It is worth noting that phytochrome seemed to react to HIR and LIR in a trial carried out on microcuttings of peach x almond rootstock GF 677 (Muleo and Morini, unpublished data); WL induced a higher production of shorter shoots compared to those produced under BL and RL. When RL was applied at two levels, 15 and 40 μmol m^{-2} sec^{-1}, the number of shoots produced was 5.2 and 3.5, respectively, indicating that the performance of the explants changed with light quality intensity. Even on *Rhododendron, Potentilla* and *Spiraea* cultures (Norton et al., 1988b), RL increased the number of axillary shoots, while BL induced a lower response in comparison with WL. The authors (Norton and Norton, 1988, Norton et al., 1988b) observed an interaction between cytokinin and RL. Subculturing over a long period of time under WL at both high and low cytokinin concentrations caused a reduction in the proliferation rate. This negative trend was reversed when RL was applied to the cultures, which also showed a lower cytokinin requirement. The authors ascribed this result to the promoting effect of RL on cytokinin biosynthesis (Letham et al., 1978). However, this effect seems to vary in relation to RL treatment duration and should be reduced with the application of prolonged treatments (Van Staden and Wareing, 1972; Wareing and Thompson, 1976). Nevertheless, the physiological effect of RL in promoting cytokinin production could be species dependent since the results of other researches would demonstrate a greater effect of BL with respect to RL (Dörfler and Goring, 1978; Köhler et al., 1980). The cytokinin synthesis in tissues grown under the natural conditions appears to be positively influenced by prolonging the photoperiod through additional illumination provided by tungsten lamps (Lercari and Micheli, 1981). The results were interpreted by various authors as the action that phytochrome and its active form performed in the synthesis of the cytokinins and/or on their action in removing apical dominance. It has recently been observed that the interaction between cytokinins and phytochrome induces in the latter a maintenance of the active form (P_{fr}) even under conditions of D and of FRL. In fact, the photoconversion times should be slowed down by the action of the ARR4 proteins, induced by cytokinins, which, binding to the active form of PHYB, would preserve it from photoreversibility (Fankhauser, 2002). The data could be newly interpreted on the basis that the cytokinins co-act to keep the phytochrome in the active form, without excluding the hypothesis previously mentioned and the possibility that P_{fr} exerts a direct action on the regulation of apical dominance. Furthermore, differences found in experiments where the explants are kept under continuous or photoperiodic light conditions could be attributed to the different effects of phytochrome active form reversibility. In fact, in experiments where the amount of light normally applied with the photoperiod 16 h light/8 h dark was subdivided into light regimes of 8/4 and 4/2, an increase in the axillary shoot formation and a reduction in apical dominance were observed (Morini et al., 1990b, 1991, 1993). Under these conditions, especially with the 4/2 light regime, the rate of permanence of the phytochrome active form probably lasted longer. Furthermore, we cannot exclude an action of the circadian rhythms, which in turn would be regulated by the reversibility of phytochrome (Yanovsky and Kay, 2001).

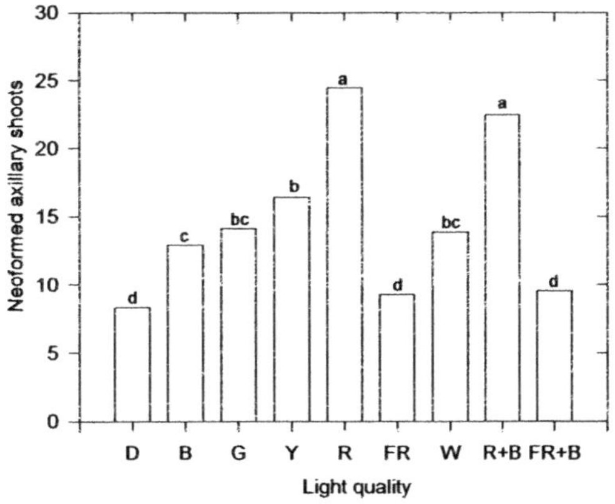

Figure 1. Number of axillary shoots produced by M9 apple rootstock cultures treated for 45 days with different light qualities. Different letters indicate statistically different values per P= 0.05. D = darkness; B = Blue; G = Green; Y = Yellow; R = Red; FR = Far Red; W = White; R+B = Red + Blue; FR+B = Far Red + Blue.

The use of high-pressure sodium vapour lamps on *Potentilla, Rhododendron* and *Spiraea* cultures (Norton et al., 1988b) raised the proliferation rate and increased shoot length compared to fluorescent tubes. RL application produced a similar effect on *Spiraea.* The authors suggested that the difference between the results obtained using sodium and fluorescent lamps might be explained by the different light spectra, since fluorescent lamp has more BL than a sodium lamp, which emits, on the contrary, more RL.

Genetic characteristic effects also exist in response to light quality as shown in a study carried out by Marks and Simpson (1999) on *Rhododendron, Crataegus* and *Disanthus*, where both apical dominance and apical control were differently modified, not only by the BL and RL spectral bands but also by the green bands, all emitted by fluorescent lamps. A high RL/FRL ratio or a low BL/RL ratio, both characterised by a high incidence of RL, promoted axillary buds outgrowth in *Azorina vidalii* (Moreira da Silva and Debergh, 1997) and *Vaccinium corymbosum* (Noè et al., 1998), respectively. In a study carried out in our laboratory on M9 apple rootstock cultures (D'Abramo, 1998), the effects of different light qualities between 400 and 780 nm, and of two combinations of RL+BL and FRL+BL, were investigated. Also in these experiments, RL showed an inhibitory action on apical dominance (Fig. 1), in contrast to FRL. BL, even if its inhibitory action was lower than FRL, did not show important effects on the release from apical dominance. The lower number of new formed shoots would indicate that BL and its photoreceptor promoted apical dominance, as opposed to phytochrome in the active form that reduced it. This interpretation is also supported by the results obtained with the FRL+BL and RL+BL combinations.

Phytochrome seemed to react to HIR and LIR in experiments carried out and Mr.S. 2/5 plum microcuttings (Muleo et al., 2001). Very similar results were observed on GF 677 peach x almond cultures (Table 1) where the neoformed axillary shoots under two RL intensities were less in number, but longer, compared to WL. Moreover, RL at the lower intensity (15 μmol m^{-2} sec^{-1}) induced a higher axillary shoot production than at higher value, but the explant performance did not differed between the two light intensities. Shoot proliferation in the two species was differently influenced by RL increasing intensity: it was positively related in Mr.S. 2/5 cultures and negatively in the GF 677. This result could be due to the different genetic characteristics of the two species, which might differ in behaviour and growth strategy in response to phytochrome stimulus. Finally, attempts to modify phytochrome activity by 15 min treatments of low-intensity RL or FRL after 16 hours of WL treatment had no influence on the proliferation rate of azalea (Economou and Read, 1986) and Mr.S. 2/5 plum (Fortuna, 1990). In the latter species, two interruptions of the D with RL caused an increase in the dry matter content (Table 2).

Table 1: Effect of RL intensity on some shoot proliferation parameters, determined on GF 677 cultures. Different letters within column indicate statistically different values per P = 0.05. (Muleo and Morini, unpublished data)

Light quality	No. of neoformed shoots per explant	% of shoots longer than 1 cm	Internode length (mm)
$White_{(a)}$	7.1 a	28.0 a	1.5 a
$Red_{(a)}$	3.5 b	40.1 b	2.7 b
$Red_{(b)}$	5.2 ab	43.2 b	2.6 b

(a) = 40 μmol m^{-2} s^{-1}; (b) = 15 μmol m^{-2} s^{-1}

Table 2: Fresh weight and dry matter of Mr.S. 2/5 shoot clusters as affected by one or two RL interruptions of the D period under 16 h photoperiod of WL. Different letters within column indicate statistically different values per P = 0.05. (Muleo and Morini, unpublished data)

Light treatment	Fresh weight (g)	Dry matter (%)
16h / 8h	2.4 ns	5.7 a
16h / 8h + 1 RL break	2.1 ns	6.3 ab
16h / 8h + 2 RL break	2.1 ns	7.1 b

3.1.2. EFFECTS OF BLUE LIGHT

Not always a species respond positively to RL during shoot proliferation. In *Vitis* (Chée, 1986; Chée and Pool, 1989), BL, besides enhancing shoot size, increased the proliferation rate by about 50 % over RL as well. These results led the authors to suggest that a B photoreceptor, not phytochrome, was involved, hence determining a BL-induced inhibition of apical dominance. BL, with respect to RL, significantly increased the number of shoots/explant and the number of nodes per shoot produced. The authors attributed this result to BL effect on the release of the axillary buds from IAA inhibition, thus reducing apical dominance.

Based on the knowledge acquired to date, the role of BL and RL on axillary shoot formation would seem contradictory. In fact, in some studies the two light qualities apparently determined different effects on bud induction and apical dominance. The intensity of these development processes likely depend on the higher or lower sensitivity of a species to one or to the other light type mentioned above. Considering again the results reported by Chée (1986) on *Vitis* from this point of view, it emerges that RL had an inhibitory action on bud induction compared with the positive effect of BL. Nevertheless, calculating the frequency of outgrowing buds, to evaluate the degree of apical dominance release under the two light qualities, it clearly appeared that RL reduced apical dominance, compared to BL, while the shoot proliferation rate remained lower. The latter response should be mainly attributed to a lower node formation under RL and thus to a lower amount of potential lateral meristems. Also in other studies performed with intact microcuttings of Mr.S. 2/5 plum (Morini and Muleo, unpublished data; Muleo and Thomas, 1997; Muleo et al., 2001), MM106 (Casano, 1995) and M9 (D'Abramo, 1998) apple and GF 677 peach x almond (Morini and Muleo, unpublished data) cultures, BL induced a higher number of nodes with shorter internodes compared to RL or D. It is worth noticing that the proportion of nodes producing outgrowing lateral shoots recorded in the above mentioned studies, was higher under RL than BL. On M9, the percentage of outgrowing buds under RL was doubled compared to that of BL treated cultures (Table 3). A very important aspect highlighted in these experiments is that axillary bud outgrowth occurred up to the apical portion of the original shoot, while with BL, axillary shoot formation occurred only from nodes located on the main shoot basal portion. These results clearly demonstrated the reduction of apical dominance performed by RL. In decapitated microcuttings, apical dominance was fully released in all the light treatments but not in D (Muleo and Thomas, 1997).

On the basis of the above mentioned finding, shoot proliferation rate appeare to be the final result of two events: bud induction and bud release from apical dominance, which are differentially affected by light quality. The much lower percentage of outgrowing buds observed under BL, as compared to RL, indicates a specific effect of BL photoreceptor, which would act in the opposite direction from the active form (FR absorbing form, P_r) of phytochrome on apical dominance. BL would increase the number of axillary buds differentiated by the apical meristem, without releasing them from apical dominance. The effect of BL appeared to be dependent on the photon fluence rate. On the other hand, RL would remove apical dominance while reducing the formation of axillary buds; both events should be influenced by the putative amount of phytochrome active form, and the effect of light photon fluence rate should be species-dependent. BL-RL treatments related to photomorphogenic events fit an antagonistic model for branching regulated by light *via* cryptochrome and phytochromes interactions. In conclusion, the results obtained in the researches mentioned above confirm that the shoot cluster development depends on the formation of new axillary buds and on their release from apical dominance. Both the processes would interact between them and with light quality in accordance with a dinamic model.

Table 3: Effect of BL and RL treatments applied to M9 apple cultures on some growth parameters. Different letters within column indicate significant differences per P = 0.05 (D'Abramo, 1998).

Light quality	Neoformed nodes per cm	Percentage of outgrowing buds
Darkness	2.7 c	50.4 b
Blue	4.0 a	36.4 c
Red	3.2 b	82.6 a

3.1.3. EFFECTS OF OTHER LIGHT QUALITIES

Research carried out on *in vitro* cultures of some fruit tree rootstocks would indicate that other light qualities, diffrent from RL and BL, are also able to influence shoot proliferation through the regulation of apical dominance. Reduction of apical dominance was in fact detected in Mr.S. 2/5 and in GF 677 cultures subjected to yellow light (YL) (Loreti et al., 1990). Even with M9 cultures, the same light stimulated a higher production of axillary shoots than that detected with BL and FRL, but lower than that of RL (Fig. 1) (D'Abramo, 1998). On MM106 cultures, YL confirmed its favourable effect on shoot production and, above all, green light (GL) induced the highest shoot proliferation rate (Casano, 1995). As observed for RL, these spectra also induced a reduction in apical dominance and promoted a higher percentage of outgrowing buds up to the original shoot apical portion. The distance between the apical meristem and the first outgrowing bud was even shorter in YL and GL compared to RL (Casano, 1995; D'Abramo, 1998). This effect was also detected on the three 2nd order basal shoots in M9 and MM 106 cultures. Furthermore, these two lights had the same effect as RL in increasing the internode length on the original shoot and induced a higher elongation of the neoformed axillary shoots: YL promoted shoot length in Mr.S. 2/5 clusters (Casano, 1995), while GL did it on GF 677 (Table 4).

Table 4: Total number of GF 677 shoots and distribution based on the length, as determined after 45 days of culture under different light qualities, 16 h photoperiod and 40±2 μmol m^{-2} s^{-1}. (Morini and Muleo, unpublished data). Different letters in column indicate values statistically different per P = 0.05.

Light treatment	Total number of shoots	% of shoots shorter than 1.5 cm	% of shoots among 1.5 – 3 cm	% of shoots longer than 3.0 cm
White	13.8 cd	39.1 b	27.2 b	33.7 b
Red	18.9 b	51.2 a	29.3 ab	19.5 c
Yellow	23.2 a	47.3 a	30.8 ab	21.9 c
Green	12.2 cd	26.7 c	14.7 c	58.6 a
Blue	9.8 de	35.6 b	35.3 a	29.1 b
UV-A	7.2 e	51.8 a	17.2 c	31.0 b

Further studies are necessary to better understand whether the wavelengths around 550 nm were, in both light qualities, the active region of the light spectrum or if there are other specific wavelengths nearby that are photobiologically active. It has recently been observed that, in lettuce, besides BL and RL, the YL also acted directly on plant growth and development (Dougher and Bugbee, 2001). Moreover, other experimental evidence would support some specific actions of GL on plant physiology as, for example, the chloroplasts orientation in *Mougeotia* (Lechowski and Bialczyk, 1987), and the regulation of stomatal aperture (Frechilla et al., 2000). Based on our experience, some behaviours of *in vitro* cultures subjected to YL and GL do not appear immediately ascribable to the action of the RL/FRL and of the BL receptors.

3.1.4. HORMONE AND LIGHT QUALITY INTERACTION

It is known that modifications in the production of cytokinin-like compounds in tissues can occur following treatment with some light qualities (Wareing and Thompson, 1976; Köhler et al., 1980; Lercari and Micheli, 1981). Furthermore, exogenous cytokinin applications seem to mime some responses mediated by the light in the development of some substructures such as, for example, chloroplasts (Feierabend and de Boer, 1978; Parthier et al., 1981). Nevertheless, the effect of light quality on cytokinin synthesis and transport has not yet been definitively demonstrated. On *Prunus insititia* Schneider (Baraldi et al., 1988), the effect of benzyladenine on axillary shoots formation was not evident in the D and under FRL but only in the presence of WL, BL or RL. In the absence of cytokinin, shoot proliferation in cultures treated with the various light qualities was practically suppressed, demonstrating that these two factors, light and cytokinin, are indispensable for the above-mentioned process. Similar responses were obtained on Mr.S. 2/5 plum cultures (Muleo et al., 1996) where the various light qualities, in particular RL, had a positive effect on the number of outgrowing buds. These results were partially obtained also on *Potentilla* and *Rhododendron* cultures (Norton et al., 1988a); the average number of shoots formed in these species did not differ statistically under WL, RL and BL. Light effect differed, however, as a function of the benzyladenine concentration in the culture medium: at the optimal concentration, WL gave better responses than those of BL and RL. If, on the other hand, the cytokinin concentration was below the optimal level, axillary shoot production was higher in the RL treated cultures. The interaction between light quality and the exogenous application of benzyladenine should be complex as resulted from *Spirea nipponica* tests carried out by the same authors. These tests highlighted the necessary presence of the hormone, which induced an increase in shoot proliferation related to the increase of its concentration only under BL and WL. Under RL, the optimal cytokinin concentration was 0.25 mgL^{-1} but at higher concentration the proliferation rate decreased. The authors related this response to the promoting effect of RL on the synthesis of cytokinin (Letham et al., 1978), which concentration become too high and exerted a negative effect on shoot proliferation. These results further support the hypothesis of a strong interaction between cytokinin concentration and light quality and that specific wavelength irradiance may mimic or partially replace cytokinin dependency. Recent knowledge would demonstrate that cytokinin can act directly on the structure of PHYB dimer to keep it in the P_{fr} stable form (Fankhauser, 2002), modifying the sensitivity to light quality and to photoperiod. However, in other studies, the incorporation of 2ip cytokinin into the growth medium suppressed the promoting effect of RL (Read and Economou, 1982). Since shoot proliferation is mediated by the effect of the apical dominance on the axillary bud formation and outgrowth, the influence of the different light qualities may be related to both auxin synthesis, thus changing the auxin/cytokinin ratio, and the sensitivity of the hormone receptors. However, there are new evidences from *A. thaliana* that polar transport of both auxin (basipetal direction) and cytokinin (acropetal direction) are equal important to increase or to inhibit respectively, the apical dominance (Chatfield et al., 2000). Following this finding, we can't exclude light quality plays a role into the polar transport of hormones by some photoreceptors like phototropin.

3.2. EFFECTS ON SHOOT ELONGATION

The application of RL stimulated axillary shoots led shoot elongation in a number of woody species (Norton et al., 1988a; Noè et al., 1998; Marks and Simpson, 1999; Muleo and Morini, unpublished data), while BL exerted an inhibitory effect in spite the number of nodes was not influenced by the light quality applied (Moreira da Silva and Debergh, 1997). The favourable effect of RL on shoot elongation was also detected on the three 2nd order basal shoots in MM106 cultures (Casano, 1995) (Fig. 2). From this study it also emerged that YL had a greater effect on shoot elongation than BL. This was particularly evident on the 2nd and 3rd secondary shoot on which shoot elongation increased gradually, moving from BL to GL and YL and, finally, to RL. From studies conducted on decapitated 2-node Mr.S. 2/5 microcuttings, it was confirmed that RL, more than FRL, induced shoot elongation. This result would indicate that RL and FRL effects are mediated by phytochrome, with a stable phytochrome for RL, while FRL effectiveness appears to require prolonged exposure and to be dependent on photon fluence rate: a typical HIR type probably mediated by PHYA (Muleo and Thomas, 1997). Moreover, tissues western blot analysis showed that PHYA is accumulated in D and FRL, but not in RL, WL or BL (Muleo, unpublished data). Under high photosynthetic active radiation (PAR) of 50 μmol m^{-2} s^{-1}, with high and low RL/FRL ratio, 1.1 and 0.6, respectively, the greatest elongation of *Azorina vidalii* stem occurred under low RL/FRL ratio (Moreira da Silva and Debergh, 1997). The BL inhibition on shoot elongation was also observed in experiments conducted *in vivo* on *Prunus cerasus* (Baraldi et al., 1992). This effect has been attributed to the activation of cryptochrome, which, though acting independently of phytochrome, would be more active in the presence of low, rather than high, P_{fr} values. Finally, a different response to light quality exposure would also depends on the genotype and on the species as, for example, observed in *Vaccinium corymbosum*, where the presence of UV bands in the light source differently affected stem elongation among different cultivars (Noé et al., 1998). This behaviour was also observed in light treatments enriched or not enriched with BL.

3.3. EFFECTS ON LEAF DEVELOPMENT

Continuous exposure to D and to FRL has shown to induce a culture development referable to the skotomorphogenic type, with slightly expanded leaf blades, long internodes and fleshy stems, and with chlorophyll, carotenoids and anthocyans synthesis, in most cases inhibited. Although leaf expansion usually occurs under all other light qualities, the response mainly depends on the species analysed. Birch leaves developed in the presence of BL (Sæbø et al. 1995a) had a greater area than those treated with RL, probably due to larger leaf cells. In *Azorina vidalii* (Moreira da Silva and Debergh, 1997), on the contrary, the larger leaves were those subjected to treatments with low RL/FRL ratio. Similar results were obtained even *in vivo* with *Prunus cerasus* leaves (Baraldi et al., 1992), where a greater area was detected when the light induced low P_{fr} levels. From the latter study, phytochrome seemed to control the leaf area development without the involvement of the cryptochrome, which instead would play a role in controlling stem elongation.

3.4. EFFECTS ON LEAF ANATOMY

Besides leaf expansion, light quality has also shown to affect thikness and structure of the leaf cross-section. GL and WL increased leaf thickness in Mr.S. 2/5 cultures (Morini and Muleo, unpublished), compared to other light qualities among which no differences were detected. The ratio between thikness of each leaf layer and total leaf thickness did not show statistical variations among the light treatments. The leaf upper epidermis cells, under all light qualities, showed a larger area compared to leaves exposed to natural radiation. Palisade cell shape was rather circular regardless light treatments,

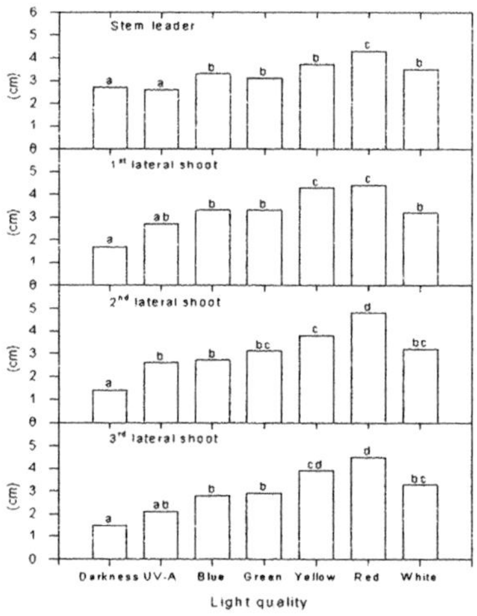

Figure 2. Effect of different light qualities on stem leader and the three basal axillary shoot length, in MM106 apple rootstock cultures. Different letters indicate statistically different values per P = 0.05 (Muleo and Morini, unpublished data).

showing a remarkable difference compared to that of *in vivo* grown leaves. The intercellular space in palisade tissue was apparent, on the contrary of the results obtained in *Liquidambar styraciflua* leaves (Wetzstein and Sommer, 1983). Spongy mesophyll cell size did not differ among the light qualities, and some differences were only related to the free space present in this specific tissue. Free space in spongy mesophyll tissue should be independent of species and light quality treatment, as reported by several authors in woody and herbaceous species cultured under different lamps (Brainerd et al., 1981; Reuther, 1988).

In *Betula pendula* Roth leaves (Sæbø et al., 1995a), treatment with BL determined instead larger palisade and epidermis cells, chloroplasts larger and similar to those of *in vivo* leaves, a lower starch accumulation in the chloroplasts and more densely packed thylakoid membranes. According to the authors, this leaf structure together with the greater photosynthetic capacity induced by BL, reflect a better chloroplast and leaf tissue organisation, which are similar to those of *in vivo* grown leaves. These characters would account for a better quality of the plantlets produced under BL treatment, compared to those subjected to RL, which likely would respond better to acclimatation phase.

3.5. EFFECTS ON CHLOROPHYLL CONTENT

In studies carried out in our laboratory (Morini and Muleo, unpublished data) on *in vitro* Mr.S. 2/5 cultures subjected to different UV and visible region spectra, the production of chlorophyll *a*, *b* and of protochlorophyll in the presence of BL was statistically the same as that under WL treatment, while in the presence of RL the quantity of these pigments in the tissues was statistically smaller. Chlorophyll a/b ratio was higher under BL and GL conditions and significantly lower under UV light (Tab. 5). Similar behaviour was detected for the total chlorophyll content.

These results indicate that RL, in the absence of BL, is not able to fully activate the pathway of chlorophyll synthesis; thus under conditions of RL, the amount of pigment produced by the tissues is smaller than those induced by BL and WL. Similar results were obtained in cultures of *Betula pendula* Roth (Sæbø et al. 1995a), where the total amount of chlorophyll in the presence of BL was double compared to that detected in cultures treated with RL. Greater synthesis of chlorophyll *a* and *b* and carotenoids was also detected in the presence of BL in black pine (Milivojević and Eskins, 1990) when the light intensity was less than 3 μmol m^{-2} s^{-1}; at intensity values lower than 2 μmol m^{-2} s^{-1}, BL was more effective in the light harvesting complex 2 synthesis, while RL promoted light harvesting complex 1. A smaller quantity of chlorophyll *a* and carotenoids was also detected in cultures of *Azorina vidalii* (Moreira da Silva and Debergh, 1997) subjected to a low RL/FRL ratio. In *Rhododendrum*, nevertheless, BL, at an intensity of 22-24 μmol m^{-2} s^{-1}, would have caused a reduction in the total chlorophyll content compared to RL (Marks and Simpson, 1999).

Table 5: Amount of total chlorophyll, chlorophyll *a*, chlorophyll *b*, protochlorophyll, and *a*/*b* chlorophyll ratio, determined in expanded leaflets of Mr.S. 2/5 collected from shoots grown under different light qualities. The values are expressed as mg of chlorophyll per g of fresh weight. Different letters within column indicate statistically different values per P = 0.05 (Morini and Muleo, unpublished data).

Light Quality	Chlorophyll *a*	Chlorophyll *b*	Proto-chlorophyll	Total chlorophyll	*a*/*b* ratio
UV-A	0.952 a	0.342 a	0.029 a	1.325 a	2.78 c
Blue	0.909 a	0.287 a	0.030 a	1.227 a	3.16 a
Green	0.956 a	0.308 a	0.022 ab	1.287 a	3.10 a
Yellow	0.817 a	0.271 a	0.024 ab	1.119 a	2.95 b
Red	0.414 b	0.138 b	0.015 b	0.568 b	2.98 b
White	0.843 a	0.286 a	0.024 ab	1.153 a	2.97 b

3.6. EFFECTS ON STOMATA

Light quality has shown to influence both stomata morphology and functionality. As it is known, stomata allow the exchange of CO_2, O_2, and water vapour between leaf tissue and atmosphere. Guard cell turgor regulates stomatal aperture by inducing changes in the shape of the cells, which, in turn, affect and regulate gas exchange across the leaf surface. Under natural conditions, stomatal aperture is reduced, for example, during drought stress, and, under severe stress, the stomata can close completely. Gas exchange can also be influenced by stomatal density (number of stomata per unit of leaf area) or by the stomatal index (proportion of stomata to total epidermal cell number). Many environmental factors

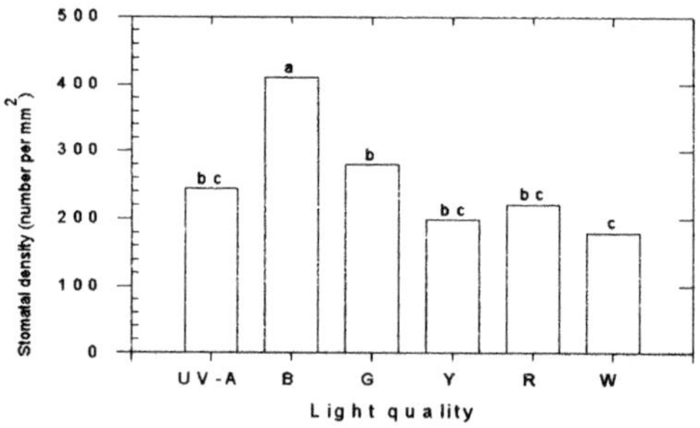

Figure 3. Number of stomata per mm ^{2} detected on Mr. S. 2/5 plum leaves grown under different light qualities. Different letters indicate statistically different values per P = 0.05. UV-A = ultra violet A; B = blue; G = green; Y = yellow; R = red; W = white light.

regulate stomatal functionality, density and size. The concentration of atmospheric CO_2, relative humidity, and temperature are very remarkable environmental cues that are already known to influence stomata properties. In particular, ABA is thought to play a pivotal role in triggering stomatal closure under the plethora of environmental stresses (Blatt and Grabov, 1997; Grill and Himmelbach, 1998).

Fluorescent cool-WL induced a relatively small number of stomata on plum leaflets, compared to GL and BL (Fig. 3) (Morini and Muleo, unpublished data). This response might depend on YL and RL bands present in the WL emission spectrum. In fact, BL promoted stomata differentiation and the density value was double under this light treatment compared to WL, where the value did not significantly differ from that of RL.

Stomatal functionality is strongly affected by microenvironmental conditions and, in some cases, may influence the post-vitro adaptation of the new plantlets. As observed in our laboratory, stomatal density and functionality under WL change with the photoperiod and with leaf age (Zacchini et al., 1997; Zacchini and Morini, 1998). In particular, the stomatal closure response was active under conditions of D but only in the youngest leaf, indicating that stomata shape is correlated to stomata functionality. In fact, in the youngest leaf, stomata were found to be elliptical as opposed to the circular shape found in older leaves. Recent evidence in photobiology and molecular biology of stomatal functionality (Frechilla et al., 2000), would demonstrated that GL plays a pivotal role in stomatal closure, indicating that the stimulatory action of BL can be reversed. In other experiments, we observed that stomata shape in leaves grown under GL was elliptical (Fig. 4), whereas under RL or BL conditions, the circular shape was predominant (Fig. 5). Finally, under WL, the elliptical stomata shape appears to be the most widespread. Therefore, our hypothesis is that, under *in vitro* culture conditions, as well as under other environmental conditions, the ratio among the different light qualities can play an important regulatory effect on stomatal development. Light spectral composition would also assume pronounced importance in the ratios between GL/BL and GL/RL, which would play a decisive role in guard cells shape and functionality.

BL and RL, through electro-membrane variations (Kinoshita and Shimazaki, 1999; Spalding, 2000) would induce stomata aperture, while GL would determine its closure by means of zeoxanthyne (Frechilla et al., 1999). It is noteworthy that the environmental conditions that induce stomatal closure as, for example, water stress, are absent in *in vitro* conditions, thus the photoperiod and the spectral bands ratio can assume a decisive role in evoking the rhythmicity of stomatal functionality.

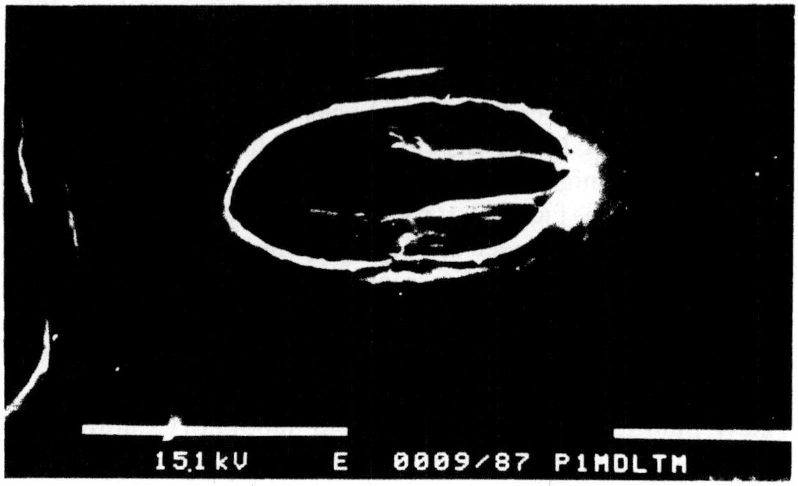

Figure 4. Stomata appearance on a Mr.S. 2/5 leaf grown *in vitro* under GL conditions.

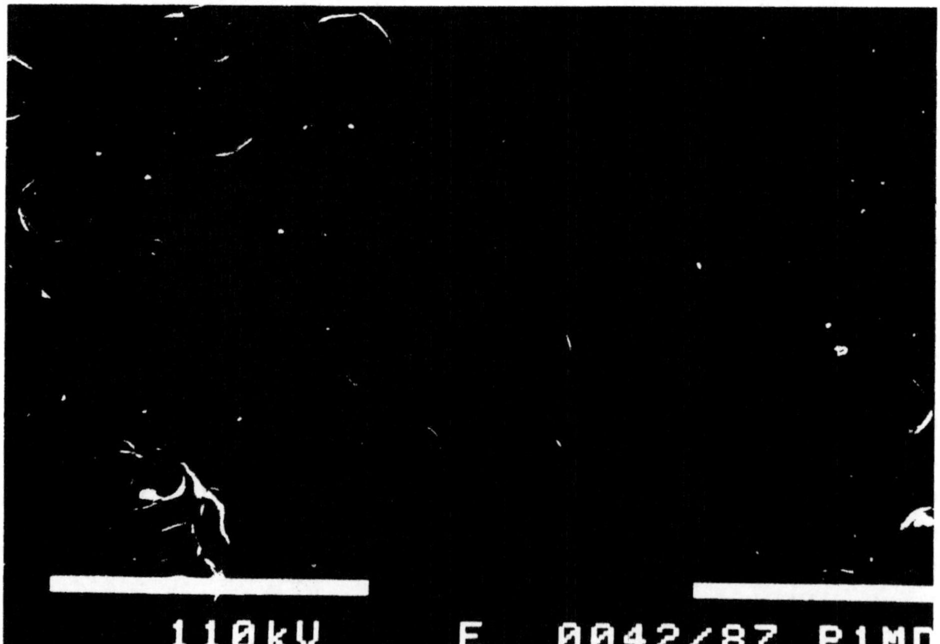

Figure 5. Stomata appearence on a Mr.S. 2/5 leaf grown *in vitro* under RL or BL conditions.

3.7. EFFECTS ON PHOTOSYNTHETIC ACTIVITY

Knowledge in this field are very few. Photosynthetic activity per leaf area unit in *Betula pendula* Roth cultures (Sæbø et al. 1995a) was higher in the presence of BL and cool WL, than in the presence of RL or incandescent light. Nevertheless, photosynthetic activity expressed on chlorophyll unit basis was very similar under BL and RL treatments, leading the authors to presume that in cultures containing a smaller amount of chlorophyll, i.e. those subjected to RL, pigment effectiveness was greater compared to that of cultures treated with BL, characterised by a higher amount of chlorophyll. According to the authors, the effects of RL and BL on photosynthetic activity are multiple and complex. BL would determine a low ratio between carbohydrates and proteins (Voskresenskaya, 1972) and would promote the protein synthesis of the photosynthetic system (Richter et al., 1987). RL would inhibit the carbohydrate translocation process, thus favouring starch accumulation in the chloroplasts (Britz and Sager, 1990; Maas, 1992) and inhibition the photosynthetic process. It is likely that, as in *in vivo* conditions, the source/sink relationship could also regulate *in vitro* photosynthetic activity (Hansen, 1975) in shoot clusters during active proliferation, influencing directly the regulation of RuBisCo activity (Van Oosten and Besford, 1994).

To evaluate the photosynthetic activity in *in vitro* conditions some factors, known to exert a marked influence on cultures growth, are to be taken into consideration. As indicated in the literature, the light intensity used in the laboratories is mostly too low to stimulate a high photosynthetic activity (Kozai, 1991). Furthermore, gaseous exchanges in the culture vessels are strongly limited by the vessel closure system. Just a few hours after the light has been turned on, the CO_2 concentration inside the vessels decreases considerably, remaining at low levels until the light is turned off. During D period the CO_2 concentration increases again, following dark respiration (Morini et al., 1993). If the 16-h light/8-h dark regime is fractionated in 4-h light/2-h dark, the amount of CO_2 taken up by the cultures over a number of light/dark cycles increased, and shoot proliferation and shoot cluster growth were found to increase as well. Thus the regulative role of light quality could be masked by the limitation of the above-mentioned environmental factors. It should be necessary, in order to better understand the regulatory action of the different light spectral qualities on *in vitro* photosynthetic activity, to perform experiments under conditions of exogenous CO_2 enrichment and higher light intensity, for a long culturing period.

4. Effects of light quality on adventitious root induction.

In the early experiments carried out on *Prunus mahaleb* (Hedtrich, 1977), the application of FRL showed negative effects on rhizogenesis. BL also inhibited rooting in birch shoots cultivated *in vitro* (Pinker et al., 1989). Similar responses of BL were detected in *Prunus serotina* cultures (Fuernkranz et al., 1990). Other studies, however, pointed out that BL, compared to RL, favourably affected rhizogenesis. In *Vitis* (Chée, 1986), the positive effect of BL was manifested through a higher percentage of rooted shoots and longer roots. A significant interaction was detected between light quality and manganese sulphate concentration as regards root number/explant. According to the author, the lower rooting response of RL treated cultures in the presence of an optimal growth medium NAA concentration, depended on an excess of auxin accumulated in the shoot tissues as a consequence of the lower IAA destruction performed by RL. Nevertheless, on the same *Vitis* genotype, the author detected in other experiments, after a long day of RL application, a doubling of the root number and a five times increase of their length compared to BL (Chée and Pool, 1989). The greatest leaf area detected on the RL treated shoots was correlated to the highest production of roots. Moreover, RL induced an increase in the number of roots per shoot and gave a greater total root length per shoot compared to BL. Similar trend was observed under cool WL lamps compared to daylight lamps and it was possibly due to the higher proportion of red irradiance present in the former lamps.

In *Prunus* GF 655/2 shoots subjected to RL, rooting was higher than under BL (Rossi et al., 1993). The effects of RL and BL on root induction appeared independent of NAA exogenous application. Under both light spectra, rooting responses without auxin were as high as with the addition of NAA to the culture medium but, in the presence of RL, rooting reached 100 %. By contrast, under FRL without NAA, rooting fell to about 10 % but rose up to 100 % with the addition of auxin to the growth medium. At the end of the culturing period, no effect was visible due to the presence or absence of exogenous NAA under RL and BL, while a dramatic response reduction was observed under FRL and D.

In MM106 shoots (Casano, 1995), the presence of IBA in the growth medium induced rooting responses rather high in all light treatments but rhizogenesis was particularly pronounced

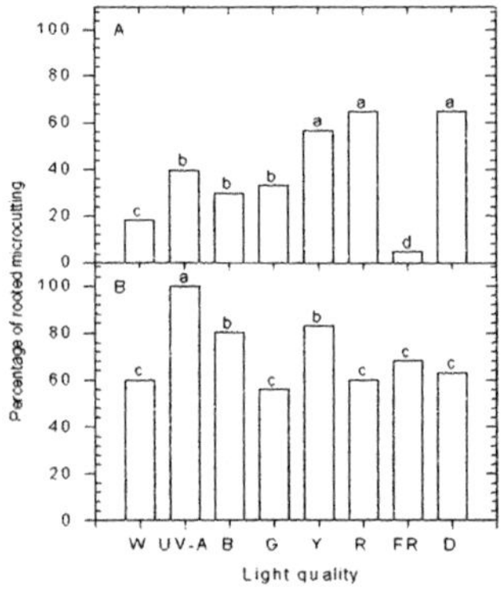

Figure 6. Effect of different light qualities on MM106 apple microcutting rooting in the absence (A) or in the presence (B) of 1 mgL^{-1} IBA into the medium. Different letters indicate statistically different values per P = 0.05.

with the simultaneous application of UV-A light, followed by YL and BL (Fig. 6). On an auxin free growth medium, shoot rooting showed the highest values in the presence of RL, YL and D. RL and D gave similar values both in the presence and in the absence of auxin. Moreover, while there was a strong inhibition of rhizogenesis in the presence of FRL without IBA, rather high rooting values were detected when this light was applied simultaneously to auxin. Possibly RL and FRL interact with the endogenous IAA in particular interfering with its transport system (Kraepiel et al., 2001).

Enhancement of rooting by RL in comparison to WL and BL was also observed in Mr.S. 2/5 plum shoots (Casano, 1995; Muleo and Morini unpublished data). In the absence of auxin, rooting was quite high particularly in the presence of YL, whose values were higher than those obtained with RL. The absolute lowest values were detected with WL and D while FRL completely inhibited rhizogenesis (Casano, 1995). Similar effects were observed with RL and FRL in the absence of auxin in the rooting medium of *Myrtus communis* shoots (Rolli et al., 2002). In this species, simultaneous treatment with IBA and RL showed that the positive interaction between these two factors was genotype dependent. A genotype dependent light quality effect was also observed in pear (Bertazza et al., 1995). In easy to root genotype, cv Conference, the phytochrome role was evident and root formation without IBA application occurred under RL, BL, and WL but not under FRL and D. On IBA enriched medium, a positive co-action with low light quality intensity was detected regardless the genotype. On M9 apple microcuttings dipped in a 400 ppm IBA solution, RL and YL increased the rooting response compared to BL and WL, and positively affected root length and number (D'Abramo, 1998).

In other studies on birch (Sæbø et al., 1995b), nevertheless, different light qualities, among which BL and RL, applied to shoots during *in vivo* rooting, had no effect. However, if BL was applied *in vitro* to shoot donor cultures, rooting response was quicker and a greater number of roots was obtained compared to treatment with RL. According to the authors, this result was attributable to the higher photosynthetic rate of the shoots produced under BL, while the negative effect of RL would be mediated by the inhibition of carbohydrate translocation (Britz and Sager, 1990). Finally, when plantlets treated with BL were planted in the field, they were shorter compared to those treated with RL but, at the end of the first and second year of growth, they were statistically taller and had a greater number of longer axillary shoots.

5. Effects of light quality on other culture growth characteristics

Besides the effects of light quality on axillary shoot production and on adventitious root formation, there are also some qualitative aspects that should be taken into consideration since they may considerably influence the success deriving from the application of certain light qualities in the micropropagation technique. Observations on the morphology of Mr.S. 2/5 and MM106 (Casano, 1995) shoots treated with RL and YL during proliferation phase, pointed out the presense of basically etiolated cultures, with small leaves and low chlorophyll content. On the contrary, with BL shoot clusters appeared very similar to those grown under WL, with well shaped leaves rich in chlorophyll. Responses caused by GL were intermediate between WL and RL. Under UV-A radiation, plum culture growth appeared negatively influenced and after some days, leaf and shoot stem apical necroses, followed by total cluster degeneration, were often observed. Apple cultures, on the other hand, did not show so negative effects to UV-A radiation but only a reduction in leaf area and internode length was recorded, indicating that this species might be more resistant than plum to UV-A light.

Shoots of both Mr.S. 2/5 and MM106 species, treated with UV-A during rooting, exibited a good morphology with normally shaped and intensively green coloured leaves. Thinner roots and absence of callus at the shoot base characterized the UV-A treated cultures compared to those submitted to WL. With GL, RL and WL, not consistent variations in the shoot and root appearance were observed while under BL the leaves were smaller and the shoots and roots shorter. With YL, the rooting response was high but the shoot apical portions were etiolated. Pronounced differences were highlighted under FRL: in the presence of IBA in the growth medium, the roots were well differentiated, the leaves were well shaped and intensively green coloured up to the shoot apical portion, but the shoot apex growth appeared blocked since the beginning of the rooting stage. On auxin free medium, rhizogenesis under FRL was completely absent and the shoot apical portion was elongated, etiolated and exhibited very small leaves. The different culture behaviour observed among the light qualities applied during shoot proliferation or rooting, is a further evidence of light-hormone interaction.

6. Alternative propagation systems in *in vitro* culture

Recently, quite a high number of research programmes aimed to study organogenesis and somatic embryogenesis from differentiated and undifferentiated tissue. The interest in these processes arose from the important fields of application in which regenerated somatic embryos and/or adventitious shoots could be utilised. As a propagation technique, somatic embryogenesis has proven to have, in terms of potentially producible plantlets, a much higher efficiency than micropropagation, further increased by synthetic seed production. The genetic variability possibly present in the new plantlets would be a problem for woody species cultivated for fruit production but it would be less important for others as, for example, forest species, where a certain amount of variability may be, on the contrary, a factor of interest. For a number of fruit tree species there is not serious problems of adaptation to *in vitro* culture, while many forest, shrubbery and ornamental woody species, perhaps due to the donor plant age, are recalcitrant to *in vitro* propagation systems. Furthermore, in some fields of application, such as reforestation, propagation by cutting is not effective and economic, while seed distribution may allow covering large areas at reduced costs.

Genetic engineering is another field of application for somatic embryogenesis and caulogenesis. In this case, more than real propagation techniques, these processes represent the only course to obtained a complete plantelet from genetically modified cells. The poor regenerating capacity observed in *in vitro* cultures of many woody species is presently a bottleneck to the advancement of knowledge in this sector. The scientific and pratical importance of these application fields increased the interest to extend knowledge on the effects of light quality on somatic embryogenesis and caulogenesis as well.

6.1 EFFECTS OF LIGHT QUALITY ON SOMATIC EMBRYOGENESIS

Data available on the effect of light quality on somatic embryogenesis are rather scarce, above all for woody species. On leaf explants of apple seedlings, embryo induction occurred only if the explants were placed in the dark (Liu et al., 1983a). The same authors (Liu et al., 1983b) observed that embryo induction and development from cotyledons explants or from seedling leaves were induced by exposure to D and by exposure for 5 minutes to FRL. Induction was inhibited if the treatment with FRL was followed by treatment with RL, thus indicating that phytochrome played an active role in the induction of the embryogenic process. On quince (*Cydonia oblonga* Mill.) leaves, a positive correlation was found between somatic embryo regeneration and phytochrome photoequilibrium: the majority of the somatic embryos was obtained with RL ($\varphi = 0.86$), followed by RL+BL ($\varphi = 0.82$) and by WL ($\varphi = 0.76$) (D'Onofrio et al., 1998). These treatments proved effective both on the percentage of embryogenic leaves and on the number of embryos per leaf. Furthermore, light effect varied with the age of the *in vitro* donor cultures under WL. In long-term cultures, the leaf embryogenic capacity increased gradually in those treatments with lower phytochrome photoequilibrium values, in which the embryogenic potential was completely similar to that of cultures subjected to lights inducing higher photoequilibrium values (Bellocchi et al., 2001). In the experiments on quince, the formation of adventitious roots at the same time as somatic embryos was detected. In this case, root regeneration was higher under WL, followed by RL and RL+BL (Morini et al., 2000). Brief RL treatment (15 min) carried out at the end of the photoperiod with WL had a favourable effect on the formation of somatic embryos on *Araujia sericifera* petals (Torné et al., 1996, 2001), and FRL (5 min) had a negative effect. These lights, however, showed different effects depending on day length (long day or short day) and on room temperature. Previously, on asparagus callus, RL was observed to have a stimulating action on post-torpedo embryo development that was accentuated even more when fructose was present in the culture medium (Muleo, 1989). In *Picea abies* (Latkowska et al., 2000), the growth response of embryogenic callus cultures to different light qualities differed as a function of the genotypes tested. BL had strongly negative effects on the growth of the callus in only one of all the genotypes tested, while the responses to the other lights, among which RL, were no different than those obtained under D or WL. On the same species (Kvaalen and Appelgren, 1999), RL improved germination and root growth in somatic embryos compared to BL and WL, but hypocotyl elongation was lower than that observed under the latter two light qualities. Finally, root and hypocotyl browning was less pronounced in somatic embryos cultivated under RL.

6.2 EFFECTS OF LIGHT QUALITY ON SHOOT REGENERATION

Most studies were conducted to evaluate the effects of WL and D on the production of adventitious shoots. It has been shown that the presence of light in the caulogenic process is important in various species. In *Pinus radiata* (Villalobos et al., 1984), the interaction between light and cytokinin was determinant for the development of meristematic tissue and the subsequent development into shoot. In *Rubus* (Fiola et al., 1990), caulogenesis varied with the genotype but under WL it was not much higher than that observed under D. In callus cultures from *Pyrus serotina* seeds, D and a short photoperiod improved the regeneration of adventitious shoots (Hiratsuka and Katagiri, 1988). An interaction between light-dark conditions and growth regulators was found on *Paulownia tomentosa* Steud. (Marcotrigiano and Stimart, 1981).

Hedtrich (1977), on discs of *Prunus mahaleb* leaf, observed that D had a totally negative effect and that a light source containing near-UV had a positive effect on the formation of adventitious roots. Light sources lacking in FRL induced a much higher rhizogenic response and the FRL+UV-AL combination showed a negative effect. The author interpreted rhizogenesis inhibition by FRL as a promoting effect of RL. This hypothesis was subsequently confirmed on *Pseudotsuga menziesii* callus (Kadkade and Jopson, 1978), where a positive effect of RL on adventitious shoot production was detected when it was used at 0.42 mW/cm^2 intensity, whereas BL and UV-AL induced the same effect as under D. Subsequently, on cultures of *Actinidia deliciosa* callus, substantial shoot regeneration occurred with RL at 39 μmol m^{-2} s^{-1} light intensity while root regeneration was observed in all treatments but occurred earlier under RL (Muleo and Morini, 1990).

7. Technical aspects and lamp types

As it is known, the photosynthetic activity of *in vitro* cultures is limited by the reduced light intensity and by the rapid depletion of CO_2. In these conditions, the cultures are characterised by photomixotrophic growth and the enrichment of the medium with exogenous sucrose represents the prevalent energy resource for tissue growth. Due to this, the growth chamber of a micropropagation laboratory does not necessarily require high photon flux density like that commonly used for the growth of *in vivo* plants into cabinets. Light, therefore, assumes predominant importance for photomorphogenic processes, as we stated earlier. Thus the choice of the lamp type to use represents a relevant aspect for the regulation of culture growth and development. In general, cool fluorescent lamps have found widespread distribution in the commercial laboratories since, besides a quality of the spectrum that simulates solar radiation, they have the property of emitting a limited amount of heat, whose accumulation could cause an increase in growth chamber temperature.

Thank to the new technology, an ample range of lamps, with different spectral characteristics (Cathey and Campbell, 1980) are now available. Regardless lamp technical aspects, which can be easily found in the company's catalogues or directly in Internet sites, the most common types available, besides cool fluorescent lamps, are represented by High Pressure and Low Pressure Sodium Lamps, Metal Halide Lamps, Mercury Lamps, and Incandescent lamps. The use of special filters allows the selection of the emission spectra of these lamps in order to modify the radiation quality applied to the cultures. A comprehensive and wide-ranging review of the above-mentioned materials can be found on the site: http://cc.joensuu.fi/photobio/photobio.html edited by Aphalo. Recently, light-emitting diodes (LEDs) have been tested as a light source for *in vitro* culture. Their characteristics were evaluated about ten years ago (Bula et al., 1991), and some interesting advantages were pointed out. These include the BL and RL emission peaks which would coincide closely with the absorption peaks of photoreceptor proteins and chlorophylls, the long life and, therefore, the need to change less frequently, the very small amount of heat generated, thereby minimising cooling system costs and, finally, their compactness, which is another attractive feature (Tanaka et al., 1998). However, due to their small size, a very high number should be required to light a surface as large as that of a micropropagation laboratory growth chamber.

Some important aspects to be taken into consideration when selecting lamp type concern: I) the best combination of light spectra depending on the phases of the *in vitro* culture; II) the light intensity and the photoperiod, set up in order to obtain the maximum biological responses and avoid energy wastes;

III) the heat emitted by the lamps, which must be minimal in order not to affect the temperature and the costs of the growth chamber thermal conditioning system.

8. Concluding remarks

The knowledge acquired to date on the behaviour of *in vitro* cultures subjected to different light qualities has evidenced that some important morphogenic processes are regulated by the action of photoreceptors which, even in the particular microenvironmental conditions of the *in vitro* culture, play a primary role in promoting the culture growth and development. Although a wide range of information is available, the applicability of the knowledge in this field appears however quite limited at the present, especially for two orders of reasons. The first is the lack of response uniformity to the same light spectrum observed among different species and genotypes, which hampers the utilisation of a technical protocol to be applied for a large number of species. As plants, under natural conditions, differ also in relation to the higher (heliophilous plants) or lower (sciophilous plants) light requirement, it is likely that even *in vitro* they retain this peculiar behaviour, adapting the response to the different light qualities and to the ratios among the different spectral light qualities. The second order of reasons is represented, at the moment, by the limited supply of ready-to-use lamps characterised by the emission of spectral light qualities biologically effective in the activation of specific photoreceptors. The lamps used in the experimental research performed up to date were often set up resorting to special devices and/or materials (incandescent lamps, special filters, etc.), and only a few of them should be technically suitable to be used in the growth chamber of a commercial laboratory. Several evidences indicate that a single photomorphogenic event needs a specific light quality balance, which in turn activates different signal transduction sensitivity regulated by both different levels and ratios of photoreceptors. Therefore, only a greater effort towards a better understanding of the relationship between light quality and specific photomorphogenic processes in *in vitro* grown cultures, will provide wider knowledge on the technical characteristics that the lamps should have and that the light companies should consider in setting up new light sources. Finally, an important point to be emphasised is the uniformity of the experimental design and the detection of more stringent physiological parameters that should be utilized by the researchers in the aim of comparing the scientific informations obtained and set up more suitable protocols for each phase of the culture system.

9. References

Baraldi R., Rossi F., Lercari B. *In vitro* shoot development of *Prunus* GF 655-2: interaction between light and benzyladenine. Physiol. Plant. 1988;74: 440.443

Baraldi R., Cristoferi G., Facini O., Lercari B. The effect of light quality in *Prunus cerasus*. I. Photoreceptors involved in internode elongation and leaf expansion in juvenile plants. Photochem. and Photobiol. 1992; 4: 541-544

Bellocchi G., D'Onofrio C., Morini S. Effect of long-term *in vitro* shoot culture on somatic embryogenesis of quince leaves treated with different light qualities. In Vitro Cell. Dev. Biol. - Plant 2001; 37: 767-772

Bertazza G., Baraldi R., Predieri S. Light effects on *in vitro* rooting of pear cultivars of different rhizogenic ability. Plant Cell. Tiss. Org. Cult. 1995; 41: 139-143

Blatt M.R., Grabov A. Signalling gates in abscisic acid-mediated control of guard cell ion channels. Physiol. Plant. 1997; 100: 481-490

Brainerd K.E., Fuchigami L.H., Kwiatowski S., Clark C.S. Leaf anatomy and water stress of aseptically cultured pixy plum grown under different environments. HortSci 1981; 16: 173-175

Briggs W.R., Huala E. Blue-light photoreceptors in higher plants. Annu. Rev. Cell Dev. Biol. 1999; 15: 33-62

Briggs WR and Onley M.A. Photoreceptors in plant photomorphogenesis to date. Five phytochromes, two cryptochromes, one phototropin and one superchrome. Plant Physiol. 2001; 125: 85-88

Britz S.J., Sager J.S. Photomorphogenesis and photoassimilation in soybean and sorghum grown under broad spectrum or blue-deficient light sources. Plant Physiol. 1990; 94: 448-454

Bula R. J., Morrow T. W., Tibbitts T. W., Barta D. J., Ignatius R. W., Martin T. S. Light-emitting diodes as a radiation source for plants. HortSci. 1991, 26: 203-205

Casal J.J. Environmental cues affecting development. Cur. Opin. Plant Biol. 2002; 5: 37-42

Casano S. Effetti della qualità della luce sulla morfogenesi di specie arboree da frutto coltivate *in vitro*. PhD thesis, 1995,University of Pisa, Italy

Cashmore AR., Jarillo J.A., Wu Y-J, Liu D. Crypthocrome: blue light receptors for plants and animals. Science 1999; 284: 760-765

Cathey H. M., Campbell L. E. Light and lighting systems for horticultural plants. Hortic. Rev. 1980; 2: 491-537

Chatfield S.P., Stirnberg P., Forde B.G., Leyser O. The hormonal regulation of axillary bud growth in *Arabidopsis*. Plant J. 2000; 24(2): 159-170

Chée R. *In vitro* culture of Vitis: the effect of light spectrum, manganese sulfate and potassium iodide on morphogenesis. Plant Cell Tiss. Org. Cult. 1986; 7: 121-134

Chée R., Pool R.M. Morphogenic responses to propagule trimming, spectral irradiance, and photoperiod of grapevine shoots recultured *in vitro*. J. Amer. Soc. Hort. Sci. 1989; 114(2): 350-354

Chory J., Wu D. Weaving the complex Web of signal transduction. Plant Physiol. 2001; 125: 77-80

Clack E., Mathews S., Aharrock R. A. The phytochrome apoprotein family in *Arabidopsis* is encoded by five genes: the sequences and expression of *phyD* and *phyE*. Plant Mol. Biol. 1994; 25: 413-427

Cline M.G. Apical dominance. Bot. Rev. 1991; 57: 318-359

Cline M. G. The role of hormones in apical dominance. New approaches to an old problem in plant development. Physiol. Plant. 1994; 90: 230-237

D'Abramo A. Effetto della qualità della luce sulla proliferazione e sulla radicazione di germogli coltivati *in vitro* del portinnesto di melo M9. Dissertation thesis, 1998, University of Pisa, Italy

D'Onofrio C., Morini S., Bellocchi G. Effect of light quality on somatic embryogenesis of quince leaves. Plant Cell Tiss. Org. Cult. 1998; 53: 91-98

Dörfler M., Göring H. Der Einfluss verschiedener Lichtqualität (Blau und Rotlicht) auf den Cytokiningehalt von Kürbisjungpflanzen. Biol. Rundschau 1978; 16: 186-188

Dougher T. A. O., Bugbee B. Evidence for yellow light expression of lettuce growth. Photochem. Photobiol. 2001; 73(2): 208-212

Economou A.S., Read P.E. Effect of red and far-red light on azalea microcutting production *in vitro* and rooting *in vivo*. VI Int. Congr. Plant Tissue Cell Cult. 1986; abstract n. 431

Fankhauser C. Light perception in plants: cytokinins and red light join forces to keep phytochrome B active. Trends in Plant Sci. 2002; 7(4): 143-145

Feierabend J., de Boer J. Comparative analysis of the actions of cytokinin and light on the formation of ribulose-bisphosphate carboxylase and plastid biogenesis. Planta 1978, 142: 75-82

Fiola J.A., Hassan M.A., Swartz H.J., Bors R.H., McNicols R. Effect of thidiazuron, light fluence rates and kanamycin on *in vitro* shoot organogenesis from excised *Rubus* cotyledons and leaves. Plant Cell Tiss. Org. Cult. 1990; 20: 223-228

Fortuna P. Effetto della diversa distribuzione del fotoperiodo sull'accrescimento *in vitro* di specie arboree da frutto. Dissertation thesis, 1990; University of Pisa, Italy

Frechilla S., Talbott L.D., Bogomolni R.A., Zeiger E. Reversal of blue light-stimulated stomatal opening by green light. Plant Cell Physiol. 2000; 41(2): 171-176

Frechilla S., Zhu J. Talbott L.D., Zeiger E. Stomata from npq1, a zeaxanthin-less Arabidopsis mutant, lack a specific response to blue light. Plant Cell Physiol. 1999; 40(9): 949-54

Fuernkranz H.A., Novak C.A., Maynard C.A. Light effects on *in vitro* adventitious root formation in axillary roots of mature *Prunus serotina*. Physiol. Plant. 1990; 80: 337-341

Grill E., Himmelbach A. ABA signal transduction. Cur. Opin. Plant Biol. 1998; 1(5): 412-418

Hansen P. Produktion, fordling og udnyttelse af fotosyntater i æbletræer. 1204. Beretning fra statens forsøgsvirksomhed i plantekultur. Tidsskrift for planteavl 1975; 79: 133-170

Hauser B.A., Pratt L.H., Cordonnier-Pratt. M.M. Absolute quantification of five phytochrome transcripts in seedlings and mature plants of tomato (*Solanum lycopersicum* L.). Planta 1997; 201: 379-387

Hedtrich C. M. Differentiation of leaf discs of *Prunus mahaleb*. Acta Hortic. 1977; 78: 177-183

Hillman J.R. Apical dominance. In: *Advanced Plant Physiology,* M.B. Wilkins, ed. London, Pitman Press, 1984

Hiratsuka S., Katagiri M. Shoot and callus formation in explants from immature seeds of Japanese pear. Sci. Hort. 1988; 34: 193-199

Jensen P.j., Hangarter R.P, Estelle M. Auxin transport is required for hypocotyl eleongation in light-grown but not dark-grown arabidopsis. Plant Physiol. 1998; 116: 455-462

Kadkade P.G., Jopson H. Influence of light quality on organogenesis from embryo-derived callus of Douglas fir (*Pseudotsuga menziesii*). Plant Sci. Lett. 1978; 13: 67-73

Kinoshita T., Shimazaki K. Blue light activates the plasma membrane H^+-ATPase by phosphorylation of the C-terminus in stomatal guard cells. EMBO J. 1999; 20: 5548-5558.

Klee H.J., Lanahan M.B. Transgenic plants in hormone biology. In: *Plant hormones, physiology, biochemistry and molecular biology,* 2nd edn, P.J. Davies, ed. Dordrecht, Kluwer Academic Publishers, 1995

Köhler K.H., Dorfler M., Gohring H. The influence of light on the cytokinin content of *Amaranthus* seedlings. Biol. Plant. 1980; 22(2): 128-134

Kozai T. Micropropagation under photoautotrophic conditions. In *Micropropagation. Technology and application,* P.C. Debergh, R.Z. Zimmerman, eds Dordrecht, Kluwer Academic Publishers, 1991

Kraepiel Y., Agnès C., Thiery L., Maldiney R., Miginiac E., Delarue M. The growth of tomato (*Lycopersicon esculentum* Mill.) hypocotyls in the light and in darkness differentially involves auxin. Plant Sci. 2001; 161:1067-1074

Kvaalen H., Appelgren M. Light quality influences germination, root growth and hypocotyl elongation in somatic embryos but not in seedlings of Norway spruce. In Vitro Cell. Dev. Bio. - Plant 1999; 35: 437-441

Latkowska M.J., Kvaalen H., Appelgren M. Genotype dependent blue and red light inhibition of the proliferation of the embryogenic tissue of Norway spruce. In Vitro Cell. Dev. Biol. - Plant 2000; 36: 57-60

Lechowski Z. and Bialczyk J. Interaction between green and far-red light on the low fluence rate chloroplast orientation in *Mougeotia.* Plant Physiol. 1987; 85: 581-584

Lercari B., Micheli P. Photoperiodic regulation of cytokinin levels in leaf blades of *Allium cepa* L. Plant & Cell Physiol. 1981; 22: 501-505

Letham D.S., Higgings T.J.V., Goodwin P.B., Jacobsen J.V. Phytohormones in retrospect. In: *Phytohormones and related compounds – a comprehensive treatise,* D.S. Letham, P.B. Goodwinand, T.J.V Higgings, eds. Elsevier North Holland Press, Amsterdam, 1978

Lin C. Plant Blue-light receptors. Trends in Plant Sci. 2000; 5(8): 337-342

Liu J.R., Sink K.C., Dennis F.G.jr. Adventive embryogenesis from leaf expalnts of apple seedlings. HortSci. 1983a; 18: 871-873

Liu J.R., Sink K.C., Dennis F.G.jr. Plant regeneration from apple seedlings explants and callus cultures. Plant Cell Tiss.Org. Cult. 1983b; 2: 293-304

Loreti F., Muleo R., Morini S. Effect of light quality on growth of *in vitro* cultured organs and tissues. Inter. Plant Prop. Soc. Proc. 1990; 40: 615-623

Maas F.M. Photomorphogenesis in roses. Thermo- and photomorphogenesis. Acta Hort. 1992; 305: 109-112

Marcotrigiano M., Stimart D.P. *In vitro* organogenesis and shoot proliferation of *Paulownia tomentosa* Steud. (Empress tree). HortSci. 1981; 16: 405 (abstract)

Marks T.R., Simpson S.E. Effect of irradiance on shoot development *in vitro*. Plant Growth Regul. 1999; 28: 133-142

Martin G. C. Apical dominance. HortSci. 1987; 22: 1071-1080

Milivojević D., Eskins K. Effect of light quality (blue, red) and fluence rate on the synthesis of pigments and pigment-proteins in maize and black pine. Physiol. Plant. 1990; 80: 624-628

Moreira da Silva M.H., Debergh P.C. The effect of light quality on the morphogenesis of *in vitro* culture of *Azorina vidalii.* Plant Cell Tiss Org. Cult. 1997; 51: 187-193

Morelli G., Ruberti I. Shade avoidance responses. Driving auxin along lateral routes. Plant Physiol. 2000; 122: 621-626

Morini S., D'Onofrio C., Bellocchi G., Fisichella M. Effect of 2,4-D and light quality on callus production and differentiation from *in vitro* cultured quince leaves. Plant Cell Tiss. Org. Cult. 1990a; 63: 47-55

Morini S., Fortuna P., Sciutti R., Muleo R. Effect of different light-dark cycles on growth of fruit tree shoots cultured *in vitro*. Adv. Hort. Sci. 1990b; 4: 163-166

Morini S., Muleo R., Sciutti R., Fortuna P. Relationship between evolution of CO_2 and growth of plum shoot tips cultured *in vitro* under different ligh/dark regimes. Physiol. Plant. 1993; 87: 286-290

Morini.S., Sciutti R., Muleo R., Fortuna P. Growth patterns of *in vitro* cultured shoot tips as influenced by different light-dark regimes. Acta Hortic. 1991; 289: 137-139

Muleo R. Ricerche sull'interazione tra irradianza ed attività fotosintetica, accrescimento e morfogenesi *in vitro*. PhD dissertation thesis, 1989, University of Pisa, Italy

Muleo R., Morini S. Effect of light quality on regeneration from callus of *Actinidia deliciosa*. Acta Hortic. 1990; 280: 155-158

Muleo R., Thomas B. Effect of light quality on shoot proliferation of *Prunus cerasifera* are the result of differential effects on bud induction and apical dominance. J. Hortic. Sci. 1997; 72: 483-491

Muleo R., Morini S., Casano S. Photoregulation of growth and branching of plum shoots: physiological action of two photosystems. In Vitro Cell. Dev. Biol.-Plant 2001; 37: 609-617

Muleo R., Casano S., Bigozzi L., Morini S. Ruolo dei fotorecettori nella proliferazione e radicazione *in vitro* di germogli di susino Mr.S. 2/5, III Giornate Scientifiche SOI, Proc. 1996; 341-342

Murashige T. Plant propagation through tissue culture. Ann. Rev. Plant Physiol. 1974; 25:135-166

Muday G.K., Murphy A.S. An emerging model of auxin transport regulation. Plant Cell 2002; 14: 293-299

Napoli A. C., Beveridge A.C., Snowden C. K. Reevaluating concepts of apical dominace and the control of axillary bud outgrowth. Cur. Top. Devel. Biol. 1999; 44: 127-169

Ninu L., Ahmad M., Miarelli C., Cashmore A.R., Giuliano G. Cryptochrome 1 controls tomato development in response to blue light. Plant Journal 1999; 18(5): 551-556

Noè N., Eccher T., Del Signore E., Montoldi A. Growth and proliferation *in vitro* of *Vaccinium corymbosum* under different irradiance and radiation spectral composition. Biol. Plant. 1998; 41(2): 161-167

Noh B., Murphy A.S., and Spalding E.P. Multidrog resistance-like genes of Arabidopsis required for auxin transport and auxin-mediated development. Plant Cell 2001; 13: 2441-2454

Norton R.C., Norton E.M. Manipulation of differentiation in *Spirea in vitro.* Acta Hortic. 1988; 226: 191-194

Norton C.R., Norton M.E., Herrington T. Light quality and the control of shoot lenght in woody ornamental plants grown *in vitro.* Acta Hortic.1988a; 227: 453-456

Norton C.R., Norton M.E., Herrington T., Phillips D. Light quality and light pipe in the micropropagation of woody ornamental plants. Acta Hortic. 1988b; 226: 413-416

Parthier B., Lerbs S., Klyachko N.L. Plastogenesis and cytokinin action. Cytokinin and light interactions in plastid enzim formation of detached *Cucurbita* cotyledons. In *Metabolism and Molecular Activity of Cytokinins,* J. Guern, C. Peaud-Lenoel, eds Springer-Verlag, Berlin, 1981

Pinker I, Zoglauer K., Göring H. Influence of light on adventitiuos root formation in birch shoot cultures *in vitro.* Biol. Plant. 1989; 31: 254-260

Pratt L. H., Cordonnier-Pratt M-M., Kelmenson P. M., Lazarova G. I., Alba R. M. The phytochrome gene family in tomato (Solanum lycopersicum L.). Plant Cell Env.1997; 20: 672-677

Quail P.H. The phytochrome family: dissection of the functional roles and signalling pathways among family members. Phil. Trans. R. Soc. Lond. B 1998, 353: 1399-1403

Read P.E., Economou A.S. Effect of red and far-red light on microcutting production and rooting of hardy deciduous azaleas. XXI Int. Hort.Congr. 1982; abstract n. 1784

Reuther G. Comparative anatomical and physiological studies with hornamental plants *in vitro* and greenhouse conditions. Acta Hort. 1988; 226: 91-98

Richter G., Dudel A., Einspanier R., Danhauer I., Hüsemann W. Blue-light control of mRNA level and trascription during chloroplat differentation in photomixotrophic and photoautotrophic cell cultures (*Chenopodium rubrum* L.). Planta 1987; 172: 79-87

Rolli S., Predieri S., Baraldi R. Ruolo della luce e del metabolismo delle auxine durante la fase di radicazione *in vitro* di *Myrtus communis* L. VI Giornate Scientifiche SOI, Proc. 2002; 73-74

Rossi F., Baraldi R., Facini O., Lercari B. Photomorphogenic effects on *in vitro* rooting of *Prunus* rootstoc GF 655-2. Plant Cell Tiss. Org. Cult. 1993; 32: 145-151

Sæbø A., Krekling T., Appelgren M. Light quality affects photosynthesis and leaf anatomy of birch plantlets *in vitro.* Plant Cell Tiss. Org. Cult. 1995a; 41: 177-185

Sæbø A., Skjeseth G., Appelgren M. Light quality of the *in vitro* stage affects the subsequent rooting and field performance of *Betula pendula* (Roth). Scand. J. For. Sci. 1995b; 10: 155-160

Smith H. Physiological and ecological function within the phytochrome family. Ann. Rev. Plant Physiol. Plant Mol. Biol. 1995; 46: 289-315

Spalding E.P. Ion channels and the transduction of light signals. Plant Cell. Env. 2000; 23: 665-674

Tamas I.A. Hormonal regulation of apical dominance. In: *Plant hormones, physiology, biochemistry and molecular biology.* 2nd edn, P.J. Davies, ed Dordrecht, Kluwer Academic Publishers, 1995

Tanaka M., Takamura T., Watanabe H., Endo M., Yanagi T. *In vitro* growth of *Cymbidium* plantlets cultured under superbright red and blue light-emitting diodes (LEDs). J. Hort. Sci.& Biotec. 1998; 73: 39-44

Thimann K.V. Hormone action is the whole life of plants. Amherst: University of Massachusetts Press, 1977

Torné J.M., Moysset L., Claparols I., Simon E. Photocontrol of somatic embryogenesis and polyamine content in *Araujia sericifera* petals. Physiol. Plant. 1996; 98: 413-418

Torné J.M., Moysset L., Santos M., Simon E. Effect of light quality on somatic embryogenesis in *Araujia sericifera.* Physiol. Plant. 2001; 111: 405 -411

Tucker D.J. Effect of far red light on the ormonal control of side shoot growth in the tomato. Ann. Bot. 1976; 40: 1033-1042

Van Oosten J.J., Besford R.T. Sugar feeding mimics effect of acclimatation to high CO_2-rapid down regulation of RuBisCo small subunit trascripts but not of the large subunit transcripts. J. Plant Physiol. 1994; 143: 306-312

Van Staden J., Wareing P.F. The effect of phoperiod on levels of endogenous cytockinens in *Xanthium strumarium.* Physiol. Plant. 1972; 27: 331-337

Villalobos V., Leung D.W.M., Thorpe T.A. Light-cytokinin interaction in shoot formation in cultured cotyledon explants of radiata pine. Physiol. Plant. 1984; 61: 497-504

Voskresenskaya N.P. Blue light and carbon metabolism. Annu Rev. Plant Physiol. 1972; 23: 219-234

Wareing P.F., Thompson A.G. Rapid effects of red light on hormone levels. In *Light and Plant Development,* H. Smith, ed. Butterworths Publ. Co., London, 1976

Wetzstein H.Y., Sommer H.E. Scanning electron microscopy of *in vitro*-cultured *Liquidambar styraciflua* plantlets during acclimatization. J. Amer. Soc. Hort. Sci. 1983; 108: 475-480

Whitelam G.C., Devlin P.F. Role of different phytochromes in *Arabidopsis* photomorphogenesis. Plant Cell Env. 1997; 20: 752-758

Whitelam G.C, Halliday K.J. Phytochrome takes a partner! Curr. Biol. 1999; 9: 225-227

Yanovsky M. J., Kay S. A. Signalling networks in the plant circadian system. Cur. Opin. Plant Biol. 2001; 4: 429-435

Zacchini M., Morini S. Stomatal funtioning in relation to leaf age in in-vitro-grown plum shoots. Plant Cell Rep. 1998; 18: 292-296

Zacchini M., Morini S., Vitagliano C. Effect of photoperiod on some stomatal characteristics of *in vitro* cultured fruit tree shoots. Plant Cell Tiss. Org. Cult. 1997; 49: 195-200

Zimmerman R.H. Factors affecting *in vitro* propagation of apple cultivars. Acta Hort. 1983; 131: 171-178

2. PROPAGATION AND DNA MARKERS CHARACTERIZATION *OF POPULUS TREMULA* L. AND *POPULUS ALBA* L.

M. Angeles Bueno[1], Aránzazu Gomez[1] and José Antonio Manzanera[2]
[1]INIA-CIFOR .Ctra de la Coruña km 7. E-28040 Madrid. Spain.
[2]IMIA. Apdo 127. E-28800. Alcala de Henares, Madrid. Spain

Chapter contents

S.M. Jain and K. Ishii (eds.), Micropropagation of Woody Trees and Fruits, 37–74.

1. INTRODUCTION

The genus *Populus* is a part of the family Salicaceae which appeared in the early Cretaceous and expanded in the Tertiary, presently consisting of about 40 species of temperate and cold regions of the Northern Hemisphere, America, Europe and Asia, but also reaching Eastern Africa (Tanzania, Uganda). *Populus* spp. are dioecious, deciduous trees with simple, glabrous leaves, scale-covered buds and flowers born in hanging catkins. Fertilisation is wind-driven, producing abundant seeds inside dry fruits (capsules).

The reproduction of these trees is easy in many ways, e.g., by seed, root suckers, stump sprouts, adventitious budding, cuttings, layering and grafting. This, together with the easiness of interspecific hybridisation and their rapid growth and good development, has led to their use in extensive plantations of fast growing clones and hybrids, rendering *Populus* spp. a major economical importance in many countries all around the world. Only in Europe, about one million hectares are planted with *Populus* clones (Douglas GC 1989). The wood is light, soft and homogeneous, without well-differentiated heartwood, being used in different industrial applications. The interest in the culture of fast growing trees such as *Populus* species for wood, fuel and fibre plantations has been developed in the last few decades (Zsuffa *et al.* 1984). According to Kleinschmit (2000), poplars come in fourth place after pine, spruce and oak regarding breeding activities. The reason for this is the high economic value of poplar plantations in Europe.

Ecologically, *Populus* are water-demanding trees, living near rivers or lakes, in fertile, moist and even wet soils, indicating the presence of a high phreatic layer. They can also be used to fix and restore riversides and shores.

The genus *Populus* comprises five sections, of which Leuce, Aigeiros and Tacamahaca are of major economic importance. Section Leuce includes some abundant and interesting species frequently exploited and planted, such as quacking aspen (*P. tremuloides* L.) from North America, and two of them are the object of this study, namely, European aspen (*P. tremula* L.) and White poplar (*P. alba* L.).

1.1 European aspen

1.1.1 The species

Populus tremula is a short-living tree, seldom reaching more than 120 years. It is medium-sized, about 20 m tall, less frequently up to 30 m height. The trunk is straight, cylindrical, up to one metre of diameter at the base. The crown is usually conic-shaped when young, later open, round and broad, with scarce foliage. The bark is smooth, greenish-grey, becoming dark and rough when old, with suckers from the base. The buds are oval, glabrous and sticky, yellow or brown-coloured (Figure 1).

Figure 1: *Populus tremula* L. stand.

The leaves are hanging, 3 to 8 cm long, bigger in young suckers and saplings, with a long thin petiole, favouring the quaking movement of the leaf to which the specific name refers. The catkins of male trees are eight to 12 cm long, with five to 12 anthers per flower, normally eight. Female catkins are shorter during flowering but elongate more at maturity. Flowering takes place in March, and seed disseminates in April or May.

1.1.2 Geographic distribution

Aspen prefers moist and wet soils of slopes and valleys, growing in open spaces without competition, frequently as a pioneer species after fires, preparing the site for other trees, such as Scots pine.

Aspen occupies a broad area, Central and Northern Europe, reaching Scandinavia, the Baltic and Russia. In the North, it lives in plains while in Southern Europe prefers the mountains, up to 1700 m height in the Alps and 1800 m in the Pyrenees, appearing in open patches between Scots pine, birch, beech and fir forests. Aspen is also present in Northern and Western Asia, Siberia, Kamtchatka and Northern Japan. It is scarcely found in Northern Africa, in Algerian mountains between 1500 and 2000 m altitude. In the extremely continental climate of Siberia, aspen appears where the summer lasts at least four months (mean temperature above 10°C), accompanying other small-leaved hardwoods, such as birches and mountain ash (*Sorbus aucuparia*), playing an important ecological role in the transition zone between the coniferous forest and the steppe (Walter, 1973).

1.1.3. Economic importance

Populus tremula has a white, soft and light wood, the finest and more homogeneous of the genus. Traditionally used for matches, toothpicks, toys, cabinet-making and furniture interiors, it is also used as sown wood and lumber in long rotations, and for pulp, fibreboard and particle board in the lowest grades of wood, obtained with shorter rotations. It is economically very important in Northern Europe, where it is abundant. Northern races are usually of better trunk shape and wood. Also in Scandinavia, the hybridisation with the American aspen (*P. tremuloides*) provides very fast-growing clones, which furthermore are resistant to common cryptogamic diseases of the European aspen.

Aspen is also used as an ornamental, taking golden and purple colours in the autumn. There are weeping, upright and other garden forms of economic interest.

1.2 White poplar

1.2.1. The species

Populus alba is a medium-sized tree, up to 30 m high, with a stem of up to 3 m diameter and a deep and broadly extended root system. The crown is broad, round, with horizontal or weeping, twisted branches. The growth is rapid, loosing vigour and ageing at 60 or 70 years. The bark is smooth, white or grey-green in young trees and dark or even black and rough in old ones. The buds are small, oval, brown-orange, covered with white hairs.

There are two leaf types. Long branches have palm-like tooth-edged leaves with three or five lobes, the face being dark green and the back hairy white. Short shoots have smaller, narrower oval leaves with wavy or toothed edge and Grey back. The petiole is hairy white (Figure 2).

The catkins are four to eight centimetres long in the flower stage and reach up to 14 cm during fructification. Male catkins are grey with red anthers, while female catkins are green. Flowering takes place in March-April, and seeds are dispersed by wind in April-May.

Figure 2: *Populus alba* L. plantation.

1.2.2. Geographic distribution

White poplar is commonly found in wet soils of valleys, plains and plateaus near rivers, up to 1000 m altitude in Southern Europe. It may also be found in sandy coasts, due to its capacity to resist salinity and wind.

White poplar is native to Southern and Central Europe, Russia, The Caucasus, Central Asia (Anatolia, Iran, Himalayas) and Northern Africa (Algeria and Morocco, up to 2000 m altitude), growing well in a wide range of temperatures, from –15 °C to 50 °C, provided the water supply is guaranteed. In some countries, like Chile, where it has been introduced, has become invasive.

Populus alba is accompanied by other riparian species, such as ash (*Fraxinus angustifolia*), *Tamarix gallica, Nerium oleander,* elm (*Ulmus minor*), *Celtis australis,* willows (*Salix alba*, *S. pedicellata*) and alder (*Alnus glutinosa*), needing light and space for a good growth.

1.2.3. Economic importance

The wood is white, with pinkish heartwood, with density between 0.45 and 0.7, depending on the growth rate. It has been used for carpentry, boxes, chairs, toys and traditional handicrafts. The industry uses it as pulpwood.

In ecological works, this species has been used for riverside and shore protection and restoration. White poplar has also ornamental value, with many forms, clones, varieties and subspecies, such as the abundant "bolleana", of Asiatic origin.

A related taxon, grey poplar (*P. canescens* (Aiton) Sm.), commonly appears in the same areas of Europe as white poplar, and has some features, such as oval tooth-edged leaves with grey hairy back and hairy buds and shoots, which are intermediate between aspen and white poplar. Some botanists have classified grey poplar as a different species or a subspecies of white poplar, while others suspect that grey poplar is actually a naturally occurring hybrid between *P. tremula* and *P. alba* (Ruiz de la Torre and Ceballos, 1979).

2. PROPAGATION

2.1. Conventional methods of vegetative propagation

Vegetative propagation is the traditional method of reproducing poplars and aspen, specially *P.* x *euramericana* hybrids. Nevertheless, there are some species of this genus of present or future interest that are difficult to propagate by cuttings, such as *P. tremula* and *P. euphratica*, and even in the putatively "easy" species, such as *P. alba*, *P. deltoides*, *P* x *canescens* and *P* x *euramericana*, some interesting clones have a medium to low rooting capacity (Grau, 1991). Actually, the capacity for vegetative propagation depends not only on the species but also on the genotype (table 1), hampering the clonal propagation of some autochthonous individuals adapted to ecosystems where *P. x euramericana* presents problems of adaptation, mainly in mountainous areas.

2.2 Alternative propagation methods: tissue culture

Aging is a problem associated to the genus *Populus* which hampers vegetative propagation of old trees by traditional methods. Sometimes it may be solved by using root suckers, grafting or seeds. Nevertheless, these solutions are more expensive, incompatibility between rootstock and scion, is present or the propagation rate is low. Tissue culture can solve most of these difficulties, providing a rapid production of a large amount of propagules from selected phenotypes of adult *Populus* (Bueno *et al.*, 1992c). Furthermore, this technique may be combined with the propagation from adventitious buds induced from root suckers, providing a rejuvenated plant material (Hartmann *et al.,* 1990).

Populus has been used for germplasm preservation. Dormant buds of *P. tremula, P. tremuloides* and their hybrid were stored in liquid nitrogen (-196 °C) for 24 h by slow-freezing previously to –40 °C. The dormant buds remained viable. Shoots differentiated on bud explants of all clones retrieved from liquid nitrogen, although a genotype effect was observed. The microshoots were rooted and plantlets were regenerated. On the other hand, quick-freezing methods were also assayed but all explants died (Ahuja, 1993).

Table 1. Capacity of vegetative propagation in the genus *Populus*

SECTION / species	Propagation
TURANGA	
P. euphratica Oliv.	Rooting difficult
LEUCE	
P. alba L.	Propagation by cuttings. Very good sprouting from roots. Poor sprouting from stumps
P. tremula L.	Propagation difficult by cuttings. Sprouting from roots very easily and from stump very bad
P. x canescens Smith	Propagation by cuttings with a previous treatment
AIGEIROS	
P. nigra L.	Propagation by cuttings, stick and seed very easy
P. deltoides March.	Rooting of cuttings is variable, depending on the clone
P. deltoides ssp *angulata*	Difficult by cuttings
P. deltoides ssp *missouriensis monilifera*	Difficult by cuttings
P. *x euramericana* (Dode) Guinier	Depending on the clone, rooting of cuttings may be good or excellent
TACAMAHACA	
P. trichocarpa	Excellent propagation by cutting, and stick

Source: González Antoñanzas (1990).

Somaclonal variation has been studied in *Populus* by Antonetti and Pinon (1993). They suggested that somaclonal induction might occur more frequently in species from the Leuce section than from Aigeiros or Aigeiros x Tacamahaca hybrids. Although most somaclones were tetraploid, the more pronounced Leuce variants were diploid. This suggests that Leuce somaclones might be induced without undergoing major chromosome change. Evidence from herbaceous species suggests that somaclones are more likely in callus cultures. Also, culture duration or repeated subculturing contribute to instability (Fry *et al.*, 1997).

Somatic embryogenesis also has been studied in *Populus* (Park and Son, 1997). Isolated leaf pieces from tissue cultured plantlets were cultured on MS medium supplemented with BA and 2,4-D. Leaves were punctured with a pin or sliced to stimulate embryogenic callus production. Both direct and indirect somatic embryogenesis was reported. To maintain somatic embryos or embryogenic callus for further development, they must be selected at an early stage. Embryogenic callus provided somatic embryos by altering the 2,4-D level. Overall, the culture media for maintenance of embryogenic cell lines are similar to induction media except that plant growth regulator levels are lower. Maturation media included BA, alone or combined with NAA. Increased levels of osmoticum stimulated maturation.

2.3. Biotechnological propagation of *Populus alba* L.

2.3.1. Tissue culture of *Populus alba* L.

Although *Populus alba* is supposedly easy to propagate, certain clones fail to root. Tissue culture has been developed of these recalcitrant clones (Bueno *et al.*, 1992a). Also the conservation of these clones in a germplasm bank will facilitate the utilisation in genetic improvement. For the establishment in culture, catkins were collected during the opening of floral buds of clone "Siberia Extremeña". Catkins were aseptically immersed in 70% ethanol for 30 s, then in 0.8% NaOCl with a few drops Tween 20 for 15 min, followed by three rinses in distilled sterile water 5 min each.

The catkins were cut in segments 3 to 5 mm long and cultured on MS (Murashige and Skoog, 1962) with sucrose (30 g/l), 8 g/l agar, pH 5.8 and plant growth regulators for the initiation, 2,4-dichlorophenoxiacetic acid (2,4-D) 2 mg/l and Kinetin (1 mg/l). The medium was autoclaved at 1.2 atmospheres (120°C) during 20 min.

Approximately a month later, a callus was formed and transferred to MS medium without growth regulators. The dividing callus was then passed to MS medium with thidiazuron (TDZ) and α-naphthaleneacetic acid (NAA) (Chalupa, 1987). The best results were obtained with 1 mg/l TDZ and 0.1 mg/l NAA respectively (table 2). In *Populus alba,* organogenesis from callus (Figure 3) was obtained in MS medium with 0.7 mg/l 6-benzyl adenine (BA) and 0.002 mg/l NAA (Kapusta and Skibinska, 1985) and in *Populus deltoides* with medium WPM plus 0.05 mg/l BAP and 0.2 mg/l ANA (Uddin *et al.* 1988) but these treatments were ineffective in *P. alba* "Siberia Extremeña" catkin explants.

Figure 3: *Populus alba* adventitios bud induced in callus.

Adventitious buds (Figure 3) were cultured on a modified MS medium supplemented with 40 mg/l of Sequestrene 138 FE G-100 (Ciba Geigy) as chelating agent against chlorosis (Rutledge and

Douglas, 1988), plus 0.1mg/l BA and 0.002 mg/l NAA. Multiplication rate was close to 4 shoots per explant.

Table 2. Bud induction by TDZ and NAA, in callus from P. alba "Siberia Extremeña" catkins.

	0.1 mg/l TDZ		1 mg/l TDZ	
	Control	0.1 mg/l NAA	Control	0.1 mg/l NAA
No. of buds	0	334	0	424

2.4. Biotechnological propagation of *Populus tremula* L.

Plant material was collected from six Spanish provinces (Fig. 4) with a sample size chosen proportionally to the abundance of European aspen in each region (table 3). Furthermore, collected individuals were micropropagated for reforestation purposes. Genetic diversity was analysed in the collection of micropropagated clones.

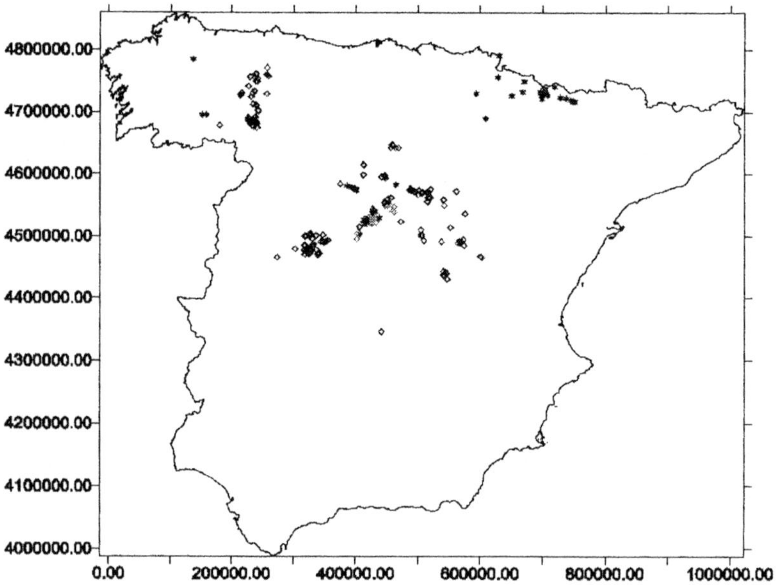

Figure 4. Location of *P. tremula* samples collected for tissue culture

Table 3. Number of samples of *P. tremula* from each location.

Province	Stand location	Number of samples
Segovia	Valsaín	11
	Cuellar	4
	Navafría	4
	Riaza	2
	Saldaña	1
Madrid	Canencia	2
	La Jarosa	1
	Somosierra	2
Navarra	Burgui	1
	Isaba	5
	Biguezal	2
	Puente de la Reina de Pamplona	3
Huesca	Villanua	6
	Borau	7
	Aisa	3
	Sallent de Gallego	1
	Canfranc	4
	Fanlo	3
Lugo	Lugo	1
Orense	Orense	2
	Viana del Bollo	2
Zamora	Justel	2
	Pedralba de la Pedrería	1
	Porto	1
León	Cabrillanes	2
	Santibañez	2
	Coloma de Somoza	2
	Villagatón	3
	Quintana del Castillo	2
	Magaz de Cepeda	1
	Astorga	1
	Luyego	1
	Quintana y Congosto	1
	Castrocontrigo	3

2.4.1. Meristem culture of *Populus tremula* L.

Meristem cultures of selected *Populus tremula* trees were established (Bueno *et al.*, 1992d). As initial plant material, root fragments were collected from 20 to 70 year-old-trees, transported to the laboratory, injured and sprayed with 130 mg/l BA and covered with a sterile mixture of peat and perlite (2:1). After approximately seven weeks, suckers were induced and cut for culture *in vitro*.

Nodal segments and apical buds of the root suckers were immersed in 70% ethanol for 30 s, then in 1% NaOCl with Tween 20 for 20 min and followed by three rinses in distilled sterile water, 5 min each.

Tissue cultures were initiated in the "Aspen Culture Medium" (ACM, Ahuja, 1983), a modification of "Woody Plant Medium" (WPM, Lloyd and McCown, 1980), with sucrose (20 or 30 g/l), 8 g/l agar, pH 5.6, added with 0.5 mg/l BA and 0.02 mg/l NAA. Bud proliferation was maintained in the same medium but elongation was achieved with 0.1 mg/l BA and 0.02 mg/l NAA (Figure 5). Bud proliferation rate was genetically determined.

2.4.2. Endogenous contamination

A problem of endogenous contamination by bacteria is frequent due to the root origin of the explants. This can be solved by adding antibiotics (Cefotaxime sodium salt, Tetracycline chlorhidrate and Rifampicine) at concentrations of 25, 25 and 6 mg/l respectively (Young *et al.* 1984) for three subcultures during the proliferation phase.

2.4.3 Leaf culture in *Populus tremula* L.

Adventitious buds induced on leaves from tissue cultured *Populus tremula* micro-shoots can be used as an adequate explant for the initiation of new cultures and genetic transformation (Bueno *et al.*, 1995). The petiole is removed and incisions are practised to stimulate bud formation. The medium used was WPM with 0.5 mg/l TDZ and

0.1 mg/l NAA. Bud induction was also genotype-dependent. Adventitious bud proliferation was obtained in WPM with 0.5 mg/l BA and 0.02 mg/l NAA (Figure 5).

Figure 5: Populus tremula microshoot.

2.5. Rooting, acclimation and transfer to soil of the micropropagated plants.

Regeneration of plantlets by micropropagation is relatively easy in this genus in comparison to other groups (Ahuja, 1987). The whole process duration varies with genotype and the degree of juvenility, lasting between 4 and 5 months. Economic evaluation is not easy, but the conservation of genetic resources and the production at industrial scale may assure the profitability of the initial investment for reforestation purposes (Figure 6).

2.5.1. *Populus alba*

Rooting of shoots induced in catkin-derived callus of *P. alba* "Siberia Extremeña" was obtained in WPM with 0.2 mg/l AIB, in the same environmental conditions as in *P. tremula*. The rooting rate obtained is 95%, three weeks later. Acclimation in the greenhouse in the same conditions as described above provides a survival rate of 96% (Bueno *et al.* 1992b) with good growth.

2.5.2. *Populus tremula*

Rooting of shoots from proliferating buds of *P. tremula* previously induced in meristem culture can be stimulated by 0.5 mg/l indole-3 butyric acid (IBA) and 0.1 mg/l ANA, under a 16 hour light/ 8 hour obscurity photoperiod and a photon flux density of 70 $\mu E/m^2/s$ provided by fluorescent "cool white" lamps. Temperature is 25°C by day and 20°C by night. Rooting averages between 75 and 95%, depending on the genotype.

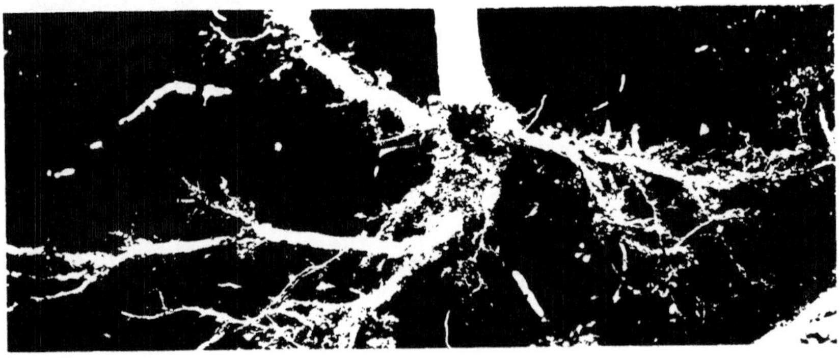

Figure 6: In vitro plantlet transferred to soil.

Regenerated plantlets are acclimated in a mist tunnel in the greenhouse. The substrate used is a sterile mixture of peat and perlite 1:1. In approximately two weeks, plantlets are transferred to the nursery. Survival rate surpasses 95%.

Protocols

Species	*P. alba*		
Explant	Catkins		
	Initiation*	Callus	Adv. buds
Basal medium (g/l)	MS 30 sucrose 8 agar	MS 30. sucrose 8. agar	MS mod. 0.04. Sequestrene 30. sucrose 8. agar
Optimal hormone concentration (mg/l)	2. 2,4D 1. Kn	1. TDZ 2. 0.1 NAA	0.1 BA 0.002 NAA
Flux density	70 μE7m^2/s		
Photoperiod	16 h light/ 8 h obscurity		

*Initiation step is followed by one month in MS medium without growth regulators.

Species	*P. tremula*			
Explant	Meristems		Leaves*	
	Bud	Elongation	Initiation	Adv. buds
Basal medium (g/l)	ACM 30. sucrose 8. agar	ACM 30. sucrose 8. agar	WPM 30. sucrose 8. agar	WPM 30. sucrose 8. agar
Optimal hormone concentration (mg/l)	0.5 BA 0.02 NAA	0.1 BA 0.02 NAA	0.5 TDZ 0.1 NAA	0.5 BA 0.02 NAA
Flux density	70 μE7m^2/s			
Photoperiod	16 h light/ 8 h obscurity			

* *In vitro* culture neoformed buds.

Rooting	Basal medium (g/l)	Optimal hormone concentration (mg/l)	Acclimation
P. alba	WPM 30. sucrose 8. agar	0.2 IBA	Mist tunnel and substrate of peat-perlite 1:1
P. tremula	WPM 30. sucrose 8. agar	0.5 IBA 0.1 NAA	

Experimental plots with vitroplants of two to four years from different clones and origins were established in the field for evaluation of their behaviour in natural conditions. Plot locations were "Monte de Valsaín", "Torremocha del Jarama" and "Riaño" forests. Survival rate was close to 90%. The successful micropropagation of *Populus tremula* from adult trees overcomes the difficulties of the classical methods of vegetative propagation (Bueno *et al.,* 1996).

3. GENETIC CHARACTERISATION

3.1. Characterisation of genetic diversity

The strategy for the conservation of genetic resources and for large scale production for reforestation purposes, is in part supported by the creation of *in vitro* culture collections of material with a natural origin. Collections via "*in vitro*" culture of forest trees are repositories of the biodiversity available for each species. They are also a valuable source of useful plants for preservation of biodiversity and genetic resources programs specially with species, which produce recalcitrant seeds, and those that are vegetatively propagated (Bueno *et al.* 2001; Thorpe and Malhotra 1996).

The characterisation of genetic diversity is a prerequisite for conservation and breeding purposes. Both objectives have to be based on available information about the genetic diversity in natural populations of the species. In this sense, the use of molecular markers for differentiation between species, namely of autochthonous species, for a species genetic diversity studies and for clones identification is especially interesting.

The importance of a proper identification of poplar clones was appreciated long ago, based on the observation of morphological and phenotypic traits (UPOV, 1981). Isozyme analysis has previously been used to distinguish poplar clones (Rajora, 1989). However, its resolving power is limited by the number of enzymatic systems that can be conveniently analysed. This information can be obtained using DNA molecular markers which afford many benefits for the identification of variation and the estimation of biological diversity.

The development of the polymerase chain reaction (PCR) for amplification of DNA (Saiki *et al.*, 1988) has been important for the detection of DNA polymorphisms. One of the PCR-based applications, RAPDs, uses single DNA primers of arbitrary nucleotide sequence (Williams *et al.,* 1990). Amplification of genomic DNA sequences occurs wherever the primers find sufficient sequence identity at a favourable distance and in a converging orientation. RAPD may be used to detect DNA variability at different levels, from single base changes to deletions and insertions (Williams *et al.*, 1990). RAPDs have several advantages over other techniques, e.g. the use of restriction fragment length polymorphisms because it requires only a few nanograms of DNA, it is rapid, and relatively economical. These markers have been used for DNA fingerprinting to determine genetic and taxonomic relationships among *Populus* clones (Castiglione *et al.*, 1993; Sigurdsson *et al.*, 1995; Rani *et al.*, 1995).

Other DNA markers like microsatellites have become the genetic markers of choice in mammalian and many plant systems because of their abundance, high degree of polymorphism and amenability to automation (Weber and May, 1989). Due to their co-dominant inheritance, simple sequence repeats (SSRs) have become the preferred tool for investigations of critical importance for germplasm managers, such as the establishment of unique genetic identities or fingerprints, determination of genetic relatedness between accessions, and the assessment of genetic diversity within a collection (Hokanson *et al.*, 1998). For this purpose, SSR markers have been analysed in a number of crop species: soybean (Rongwen *et al.,* 1995); grape

(Lamboy and Alpha, 1998); apple (Guilford *et al.*, 1997); sorghum (Brown *et al.*, 1996); barley (Russell *et al.*, 1997); avocado (Lavi *et al.*,1994), etc.

3.2 Species identification by RAPD markers in the genus *Populus*

The molecular identification of forest species is important for the assessment of differences and relationships among them, specially in *Populus*. species which readily hybridise. Twenty five poplar clones (table 4): 5 of *Populus nigra,* 5 of *P. alba,* 5 of *P. tremula,* 1 of *P. trichocarpa,* 3 of *P. x canescens* and 1 of clone *"Platero" (P. tremula x P. alba "Bolleana")* were screened for random amplified polymorphic DNA (RAPD) markers for distinction of the cited species (Sánchez *et al.*, 1998).

Clones *P. x euramericana* I-214, *P. deltoides x P. alba* 7/32 B and the other cultivated clones (Gonzalez and Domingo, 1987) used in this work of several origins (Turkey, The Netherlands, USA), and originated from different geographic regions. The rest of them were of Spanish origin. DNA was extracted from leaf, branch or cutting samples. For some clones of *P. alba* and *P. tremula,* DNA was extracted from leaves of "*in vitro*" explants (Bueno *et al.*, 1993; Sánchez *et al.*, 1997).

Genomic DNA was extracted following protocol developed by Doyle and Doyle (1990) and stored at 4°C before amplification. The primers tested in the reactions were Kit-N and OPA 1 to 5 from Operon Technologies, Inc. USA. PCRs were performed in a "personal cycler" (Biometra) thermal controller in a final reaction volume of 12.5μl, containing 10 mM of Tris HCl pH 8.3, 1.5 mM of $MgCl_2$, 200μM each of dATP, dCTP, dGTP and dTTP (Boehringer Mannheim, Germany), 0.2 μM primer (Operon Technologies, Inc.), 0.5 ng/μl to 1 ng/μl of poplar DNA and 0.04 units/μl of Taq DNA polymerase (Boehringer Mannheim, Germany). The DNA was initially denatured at 95°C for 5s, followed by 1 min, 55s at 92°C. The amplification profile consisted of 45 cycles at 95°C for 5s, 92°C for 55s (denaturation); 34°C for 1min (annealing); and 72°C for 2 min (extension). A final extension was performed at 72°C for 7 min. RAPD reactions were performed several times on each DNA sample to verify the reproducibility of the data.

Amplification products were analysed by electrophoresis in 1% agarose gel with 40 mM of Tris[hydroxymethyl]aminometane-Acetic acid and 1mM of Ethylenediaminetetraacetic acid (TAE) buffer and run in the same buffer for 2-3h at 80V. The gels were stained with 1µg/ml ethidium bromide and photographed under UV light.

The PCR-based DNA amplification from leaf, branch and cutting material from the same clone gave similar results. Extraction from both branches and cuttings are useful for collection of plant material throughout the year.

Amplifications resulted in products for 11 primers ,and 8 of them gave reproducible and polymorphic patterns. From the eight selected primers, one of them (OPN06, 5′-GAGACGCACA-3′) produced polymorphic bands that were informative for each of the 5 species. The amplification of DNA with the OPN06 primer showed the same banding pattern for clones of the same species but a different one for individuals of different species-. Hybrid clones *P. deltoides x P. alba* 7/32B and *P. deltoides x P. nigra* could be distinguished from their parent species. In these two cases, the banding pattern was different from both parents although containing some bands from either one or the other parent species.

For *P. x canescens* (a *P. tremula x P. alba* natural hybrid) and *"Platero" (P. tremula x P. alba "Bolleana"),* the banding pattern was similar to *P. alba* or *P. tremula,* probably because all bands obtained in *P. tremula* samples were also found in *P. alba.* Therefore, the hybrids could not be distinguished from the parents. These results indicated RAPD-based markers dominant. Alternatively, it is possible that the hybrids are genetically contaminated as a result of back-crossing to one of the parents in wild. Finally, it is possible that some individuals may have been phenotypically wrougly typed as hybrids.

Table 4: Populus clones for RAPDs analysis.

CLONE	ORIGIN (and COUNTRY)	TREE MATERIAL
P.deltoides "N. Dakota"	Populetum M Central (USA)	cutting
P. deltoides "Ohio"	" (USA)	cutting
P. deltoides "Missouri"	" (USA)	cutting
P. deltoides "Harvard"	" (USA)	cutting
P. deltoides "77/55"	" (USA)	cutting
P. nigra "NA2"	Navarra (Spain)	leaf
P. nigra "NA8"	Navarra (Spain)	leaf
P. nigra "( )"	Populetum M. Central (Spain)	Cutting
P. nigra "Covat"	" Guadalajara (Spain)	Cutting
P. nigra "Tr.56/75"	" (Turkey)	Cutting
P. alba "5-41"	Jaén (Spain)	Leaf
P. alba "T"	Torremocha (Spain)	Leaf
P. alba "FA"	Almería (Spain)	Leaf
P. alba "Bolleana"	(The Netherlands)	Branch
P. alba "Raket"	Populetum M. Central (Belgium)	Cutting
P. x canescens "Bet"	Cuenca (Spain)	leaf
P. x canescens "LPG5"	Granada (Spain)	Leaf
P. x canescens "7"	Teruel-Alfambra (Spain)	Leaf
"Platero" *P. tremula x P. alba* "Bolleana"	Populetum M. Central (Spain)	Cutting

CLONE	ORIGIN (and COUNTRY)	TREE MATERIAL
P. tremula "5C"	Canencia (Spain)	Leaf
P. tremula "3B"	Balsaín (Spain)	Leaf
P. tremula "3"	Teruel (Spain)	Leaf
P. tremula "4J"	La Jarosa (Spain)	Leaf
P. trichocarpa "F. Pauley".	Populetum M. Central(USA)	Cutting
P. nigra "Bordils"	" Gerona (Spain)	Cutting

Nevertheless, the reliability of markers based on the OPN06 primer is demonstrated by the fact that clones of the same species from different geographic areas have a common RAPD banding pattern.

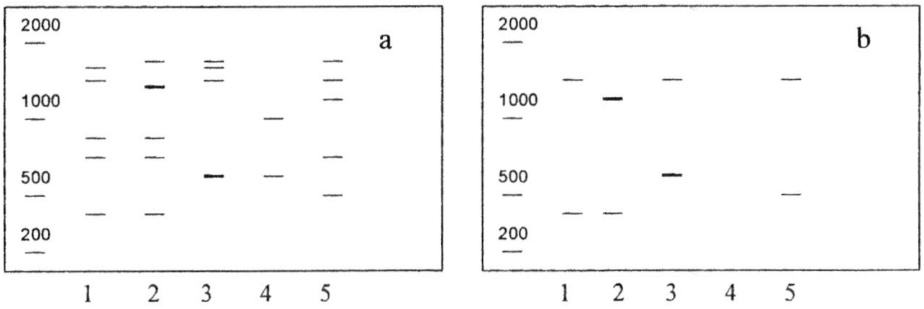

Figure 7. a) Banding patterns obtained mingling OPN06 amplification products of five clones and b) identificative banding patterns in each species (1: *P. deltoides*, 2: *P. nigra*, 3: *P.alba*, 4: *P. tremula*, 5: *P. trichocarpa*).

In order to identify bands patterns for each analysed species, a software derived schematics (Aldea *et al.*, 1988) was used (Figure 7). A scheme of the banding pattern characterising each species was defined by the bands that are common to all clones of the same species. Other scheme was established by indicative bands for each species, (bands that are present in a species but absent in all others). In *P. tremula* there were no identical bands because the two characteristics bands were present in some clones of *P. alba,* for instance in *"Raket"* and *"Bolleana"*. However, *P. tremula* was identified by the absence of the *P. alba* bands. Three different groups of *P. alba* were indicator of them had the of *P. alba,* indicator band but at the same time have other bands that identify the group.

Discrimination between *Populus* species may be done using RAPD markers. One of the markers revealed different banding patterns between species and similar banding patterns for clones of the same species. For hybrids such as *P. x canescens* and *"Platero"*, the banding pattern was the same as either *P. alba* or *P. tremula.* On the other hand, for other hybrids analysed, such as *P. x euramericana* I-214 and *P. deltoides x P. alba* 7/32 B, the banding pattern differed from both parents. The mentioned marker showed characteristic bands for each species, and in the particular case of *P. alba* three different groups could be distinguished.

3.3. RAPD analysis of regions of origin of *Populus tremula*

Random Amplified Polymorphic DNA (RAPD) were used to identify the collection of 65 *Populus tremula* micropropagated trees sampled from different Spanish origins (Sánchez *et al.,* 1997). The combination of four primers provided a set of four banding patterns that permitted the identification of all genotypes with a high discrimination capacity. Some trees located in the same stand or in near places showed the same fingerprint, probably due to clonal propagation.

DNA extraction from branches and leaves of adult trees was done by Doyle & Doyle (1990) protocol. Twenty six oligonucleotide primers from Operon Technologies Inc. USA were used for PCR amplification (table 5). Amplifications were repeated in the same conditions with Taq DNA polymerase from both Pharmacia and Boehringer Mannheim brands, corroborating that banding patterns obtained with both enzymes were identical, RAPD reactions were performed several times on each DNA sample to verify the reproducibility of the data. The amplification products were size-separated by gel electrophoresis in 1%

agarose with Tris-Alcohol-EDTA (TAE) buffer and run at 80 V for 2h. Gels were stained with 1 µg/ml of ethidium bromide.

Table 5. Oligonucleotide primers from Operon Technologies Inc. USA used for PCR amplifications of *P. tremula* DNA. (+: amplification, -: no amplification, P: polymorphism).

Primer	Amplification	Primer	Amplification
OPA01	P	OPN04	+
OPA02	P	OPN05	-
OPA03	+	OPN06	+
OPA04	-	OPN07	+
OPA05	+	OPN08	-
OPA10	P	OPN09	-
OPP01	+	OPN10	-
OPP02	+	OPN11	-
OPP03	P	OPN12	-
OPP04	+	OPN13	-
OPN01	+	OPN14	-
OPN02	+	OPN15	-
OPN03	+	OPN16	+

Four primers (OPA01, OPA02, OPA10 and OPP03) were chosen, out of the 26 tested (table 5), on the basis of the number and frequency of polymorphisms produced among the samples. It is noteworthy that the primer OPN6, which is successfully used to discriminate between species, was not polymorphic in aspen.

The characterisation of European aspen genotypes was obtained by the combination of diverse banding patterns each one for each primer amplification (Figure 7). Each fingerprint has been allocated with a letter. Eight clones have been identified with a single primer: **"V"**, "**Y**", and "**N**" for primer OPA01; with a **"K"**, "**T**" and a "**X**" for primer OPA10; with a **"n"** and a "**q**" for primer OPP03 (table 6). Since, other individuals can only be distinguished by the combination of all four banding patterns corresponding to the four primers used.

A special approach was followed with the primer OPP03 where the absence of bands has been considered a different pattern, named **0**. The explanation was: first, because patterns **l**, **m** and **n** of this primer have only one or two characteristic bands but not several, as it is the case in other primers studied. Second, the relatively high frequency (0.21) of pattern **0** (no band) cannot be attributed exclusively to a technical failure. Therefore, it can be assumed that, at least in this case, the non-amplification pattern is due to a variation in the sequence where other individuals show amplification.

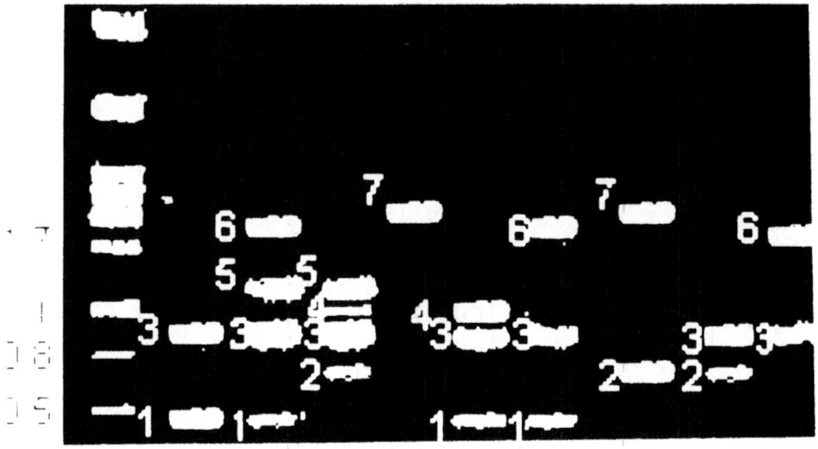

Figure 7. Banding patterns detected by RAPD primer OPA02 in *P. tremula* analysis.

The high discrimination capacity of RAPD markers was proved by the power of discrimination (PD, Kloosterman et al. 1993) defined as: PD = $1-\Sigma(P_i)^2$; where P_i represents the frequency of each genotype. In this case the indices i represent different and independent fingerprints of a certain primer. $\Sigma(Pi)^2$ for primer OPA01 is 0.32, for OPA10 is 0.2, for OPA02 is 0.14 and for OPP03 is 0.43. The probability of two individuals having the same four primers is: p = 0.32 x 0.2 x 0.14 x 0.43 = 0.004. Therefore, the PD of this primer combination is: 1- 0.004 = 0.996. This means that 249 individuals in a set of 250 may be identified.

Thirty six per cent of the trees from the same stand have the same banding pattern for all the primers tested (Table 6). This suggests that these trees from the same stand are actually ramets of a single clone propagated by root suckers that have been extended through the stand (Heybroek 1984). The other 64% showed RAPD pattern differences between individuals of the same stand, due to sexual reproduction (Sanchez *et al.*, 2000).

Table 6. Banding patterns detected by RAPD primers in *P. tremula* analysis.

Primer	Pattern
OPA01	A,B,C,M,V,I, Y, O, N
OPA10	D,E,F,G,K,R,T,X,L
OPA02	a,b,c,e,f,d,h,g,i
OPP03	m,l,n,0,p,s,q

Relatively short distances (1 km) between stands suggested a high degree of consanguinity if similarity in banding patterns is considered. An extreme situation was detected for Balsain stands 5 and 6 that showed identical patterns. Other explanation was given for Riaza stands 31 and 32, where equal banding patterns were attributed to former intensive artificial plantations.

For statistical purposes individuals lacking at least one of the four patterns were excluded. Also, individuals from the same origin with the same banding pattern for all four primers were considered ramets of a single clone, thus being managed as only one individual. The analysis were performed using the software packages, Statistical for Windows Release 4.5 A. Statsoft, inc. 1993; and Population genetic analysis (Pop gene Version 1.2, 1997).

Four subgroups at 65-70% distance were distinguished by cluster analysis (Figure 8) based in Nei's genetic distance matrix (Nei 1978). One group was formed by only two individuals from Huesca and Navarra, a second one by individuals from Segovia; a third one from Huesca, Orense and Lugo; and a fourth one from Segovia, Huesca, Navarra and Madrid. Similar results were also obtained by Multidimensional Scaling analysis of Nei's distance, with three groups well delimited. The group integrated by trees from Huesca, Madrid, Segovia and Navarra origins showed small genetic distances,

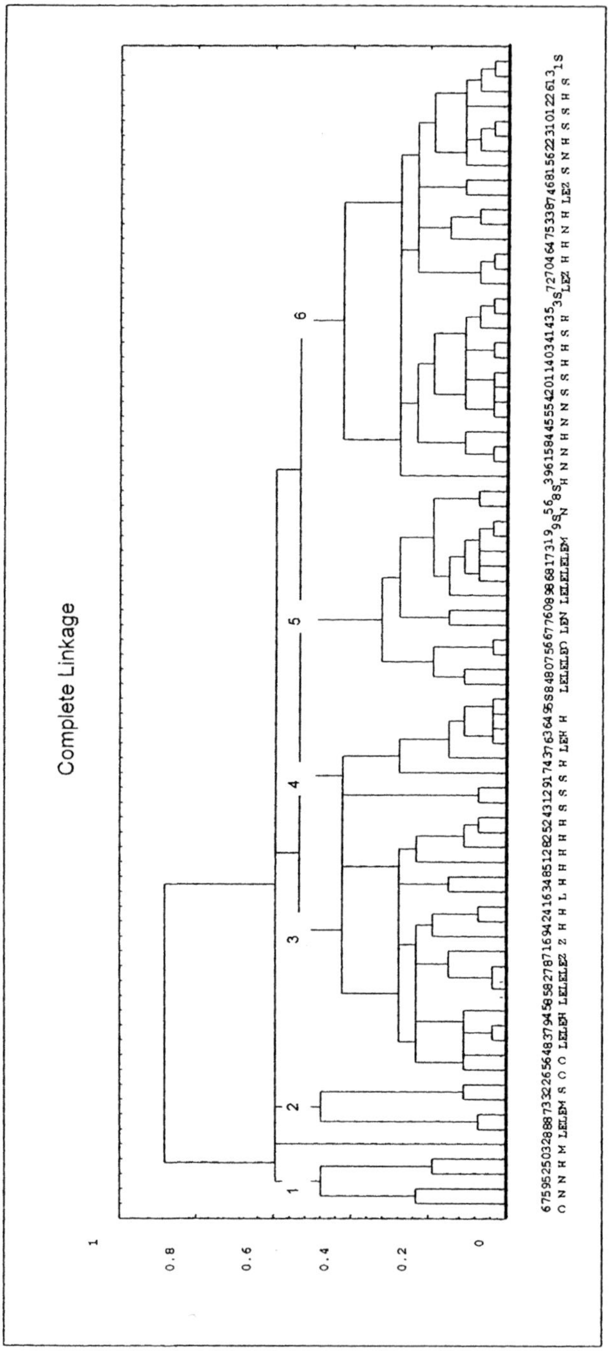

Figure 8. UPGMA Phenogram of *Populus tremula* by RAPD markers.

in contrast to the other groups. Local populations from Huesca showed the highest variability.

Two main trends of genotypic relationship deriving from Huesca have been observed in the statistical analysis. A close relationship between North-western populations and those from the Pyrenees was found, and also between populations from the Central mountain range and those from the Pyrenees but not between North-western and Central populations.

The analysis of the RAPD fingerprints obtained with *P. tremula* from different Spanish origins showed that Pyrenean populations showed the highest variability and the greatest number of banding patterns in common with the other studied locations, suggesting a genetic flow from the Northeast to the rest of the Iberian Peninsula.

3.4. Microsatellite analysis of *Populus tremula* L.

Eighty-two individuals of *Populus tremula* selected from eight Spanish provinces representing the main area of distribution were analysed: One located in the Central Mountain Range (Segovia); one in the Iberian Mountains (Cuenca); two near the Pyrenees (Northeast: Huesca and Navarra); and four in the Northwest (Lugo, Orense, Zamora and León). Leaf samples from 51 trees were collected from the *in vitro* collection of *P. tremula*. Branch material from the remaining 31 clones were collected directly from the trees. DNA was extracted following the method described in Doyle & Doyle (1990).

Six microsatellites were amplified in a final volume of 25 μl amplification reaction contained: 20 ng of total DNA, 0.2 μM of fluorescently labelled forward primer and unlabelled reverse primer (Progenetic), 200 μM each dNTPs, 50 mM KCl, 10 mM Tris-HCl (pH 9), 2.5 mM $MgCl_2$ and 0.5 U of Taq-DNA polymerase (Ecogen). Fluorescently labelled PCR products were separated and analysed on a semiautomatic sequencer (ABI-Prism, Perkin-Elmer). Standards were used for length determination of alleles. Oligonucleotide primers from *P. tremuloides* genome were designed by Dayanandan *et al.* (1998) and Rahman *et al.* (2000). Amplifications were carried out in a Perkin

Elmer 9600 thermocycler, with two profiles: for SSR1, SSR2, SSR3 and SSR4 loci 1 min at 94°C, 5 cycles 1 min at 94°C (denaturation), 1 min 55°C (annealing) and 1 min 72°C (extension) followed by 30 cycles 30 s at 94°C, 30 s at 60°C and 30 s at 72°C. For SSR5 and SSR6 amplification was carried out following Rahman *et al.* (2000).

Amplification products were analysed with an ABI-Prism automatic sequencer (Perkin Elmer). From each sample three SSRs were loaded and run together with internal molecular weight standards (Genescan/350 TAMRA). Resulting fragment sizes were calculated with the Genescan software (Perkin Elmer) by comparison with the standards.

All six SSRs from *P. tremuloides* have resolution in *P. tremula* probably due to the close relationship between both species during evolution (Primmer *et al.,* 1996; Sun and Kirkpatrick, 1996; Fields and Scribner, 1997; Gómez *et al.*, 2001). In fact, both quaking and European aspen have been placed in the subsection *Trepidae* of the section *Leuce* and hybridise very easily.

Although, all SSR markers are polymorphic differences were found between both species, so the mean number of alleles detected per *locus* in *P. tremula* ranged from 2 to 5, which is lower than the mean number of alleles (3 to 11) detected by the same SSRs in *P. tremuloides* (Dayanandan *et al.,* 1998; Rahman *et al.,* 2000). The explanation may be that North American plant species are more abundant and show a higher degree of variability than their European relatives (Takhtajan, 1986). In North America, the North-South-oriented mountains facilitated migrations during the glaciations, while in Europe, the East-West-oriented mountains would have formed a barrier that caused population extinction.

A total of 65 genotypes have been detected (table 7) by the combination of the 6 SSRs amplified fragments. In general, when different trees have the same genotype, they are located at closer distances between them than other sampled trees.

Heterozygosity was calculated using Nei's formula: $H=(1-\Sigma p_i^2)* n/(n-1)$, where p_i is the frequency of allele i in the n analysed trees. The heterozygosities for individual *loci* ranged from 0.115 to 0.659 for the six loci with a mean value for all loci of 0.475 (table 8).

Power of discrimination for each *locus* was calculated using the formula by Kloosterman *et al.* (1993) and gave a value of 99.2% (table 8) similar to the result obtained through RAPD markers (99.6%).

Table 7. Highest frequency *P. tremula* genotypes detected by 6 SSR markers.

Genotype (sizes in bp)	Frequency
254 204/210 216/238 202 250/260 202	0.024
254 204 216 200 242 202	0.049
254 204 216 200 260 196/202	0.024
254 204 216 202 250 200	0.061
254 204 216 202 250 196/202	0.037
254 210 216/230 200 250 190/202	0.024
254 204/210 216 200 260 202	0.024
254 204/210 216 200/202 260 196/202	0.024
260 204/214 218 200 250/260 196	0.024
254/260 204/214 216/230 200 250 190/202	0.024
254/260 204/214 216/230 200/202 260 202	0.024

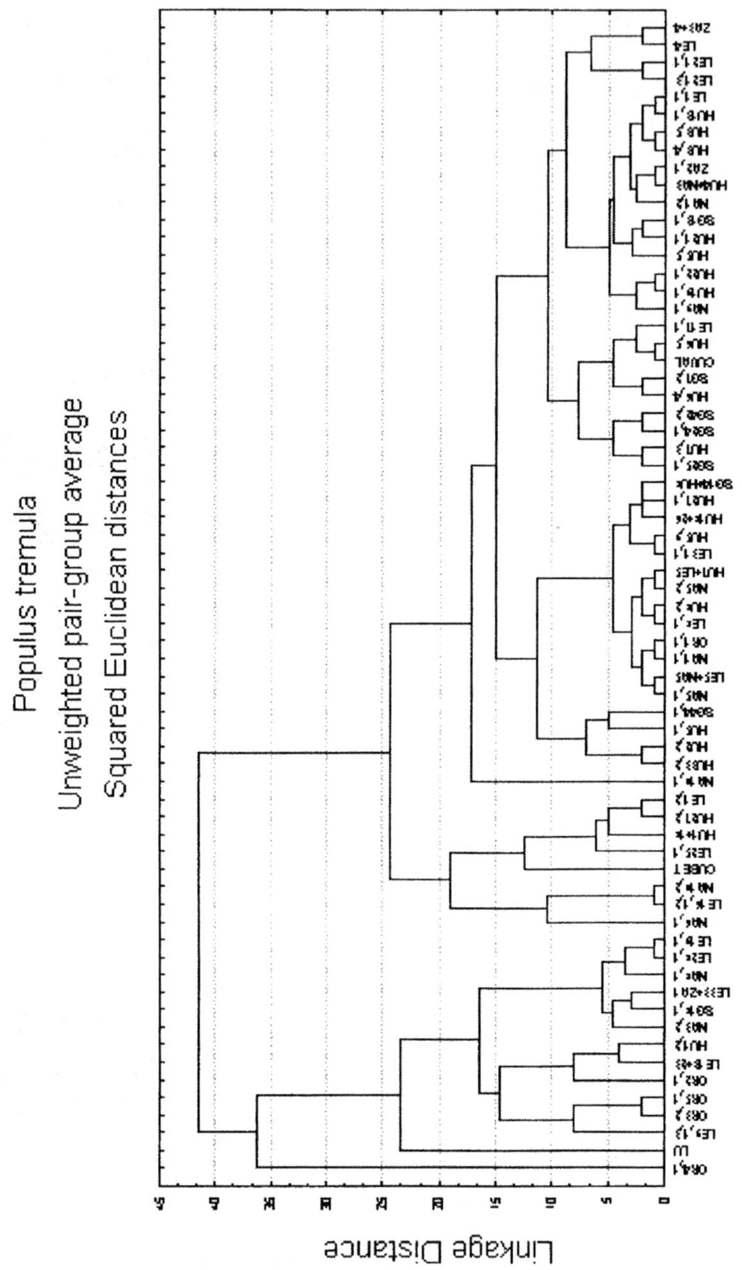

Figure 9. UPGMA phenogram for *P. tremula* by SSR markers.

Table 8. Heterozygosity (H) and Discrimination Power (DP) of 6 SSR loci in *Populus tremula* from Spanish origin.

	SSR1	SSR2	SSR3	SSR4	SSR5	SSR6	Mean	Total
H	0.115	0.535	0.497	0.468	0.578	0.659	0.475	
DP	0.182	0.718	0.612	0.647	0.667	0.808	0.606	0.992

Genetic relationships among trees were investigated using an unweighted pair-group method (UPGMA) cluster analysis computed with the program NTSYS-pc (Rholf 1994). The phenogram generated resulted in several groupings which appear to be related with the geographical origin (Figure 9). Samples from Huesca and Navarra may be found in all groups of the phenogram. The first group are in general trees collected in the Northwest of the Iberian Peninsula. Whereas group 2 has two smaller groups, called A and B, formed by samples from Segovia and from León-Zamora respectively.

SSR markers have resulted a good tool for genetic diversity knowledge approach in *P. tremula* and increasing the number of markers they will offer a possibility for clone identification that not only can help to clarify the genetic variability for improvement programs, but also will be useful for the recognition of patent rights.

4. CONCLUSIONS AND PROSPECTS

Clonal propagation of both *Populus tremula* and *P. alba* through tissue culture has been successfully obtained. Catkin explants provided a rejuvenated material for the establishment in culture of an adult clone ("Siberia Extremeña", *P. alba*). A similar result has been reached with root suckers of *P. tremula*. Juvenility and vigour were critical factors for the propagation of adult trees.

The whole process of micropropagation, from the establishment *in vitro* to acclimation varies with genotype but was estimated between four and five months. Response to field conditions has permitted the establishment of experimental plantations with clones of both species

(*P. tremula* y *P. alba*) in the frame of programmes for the conservation of genetic resources and reforestation.

Other important point is the genetic composition of the *in vitro* collection. The information provided by DNA markers is valuable for the assessment of the sampling strategies, specially when the trees of a stand are both seed- and vegetatively propagated. Both RAPD and SSR markers were robust enough to prove that aspen stands were composed of both seedlings and ramets of clones propagated by root suckers through the stand.

The spatial distribution of genetic diversity in *P. tremula* detected by RAPDs and microsatellites showed that Pyrenean populations contain the highest variability and the greatest number of genotypes in common with the other studied locations, suggesting a genetic flow from the Northeast to the rest of the Iberian Peninsula.

Concerning the low diversity detected compared with other forest species, it has been suggested that the seedling reproduction is low at the present time. Nevertheless, all the results suggest that, in addition to this low variability in aspen, the glaciation bottle-neck has further reduced the genetic variability of the Spanish *P. tremula* populations. The conservation of the remaining genetic diversity is an urgent task for the preservation of these species.

5. Acknowledgements

This study has been funded by projects INIA SC 94-138; CC 96-105; and CDTI P-970027. A. Gomez had a scholarship of INIA.

6. REFERENCES

Ahuja, M. R. 1983. Somatic cell differentiation and rapid clonal propagation of aspen. Silvae Genet. 32: 131-135.

Ahuja, M. R., 1987. In vitro propagation of poplar and aspen. In Cell and tissue culture in forestry. Vol. 3. Case Histories: Gymnosperms, Angiosperms and Palms. Bonga, J.M.; Durzan Don J. Eds. Martinus Nijhoff Publishers. 416 pp.

Ahuja MR, 1993. Regeneration and germplasm preservation in aspen-*Populus*. In: Ahuja MR (ed.): Micropropagation of woody plants. Kluwer Acad Publ., 187-194.

Aldea, M., Vicente, M. and Kushner, S.R. 1988 Cloning Version 2.0. For the IBM PC/TX/AT. (Aldea, M. and Kushner, S.R.: Cloning: a microcomputer program for cloning simulations. Gene 65: 111-116 .

Antonetti PLE, Pinon J, 1993. Somaclonal variation within polar. Plant Cell, Tissue and Organ Culture 35:99-106.

Brown, S.M.; Hopkins, M. S.; Mitchell, S. E.; Senior, M. L.; Wang, T.Y.; Duncan, R.R.; González-Candelas, F. and Kresovich, S. 1996. Multiple methods for the identification of polymorphic simple sequence repeats (SSRs) in sorghum (*Sorghum bicolor* (L.) Moench). Theor. Appl. Genet. 93:190-198.

Bueno, M.A.; Astorga, R.; Manzanera, J.A., 1992a. Micropropagación de *Populus alba* "Siberia Extremeña" a partir de amentos. Investigación Agraria. Serie Sistemas y Recursos Forestales. Vol 1(2) pp 163-171.

Bueno, M.A.; Alba, N.; Astorga, R.; Manzanera, J.A. 1992b. Conservación de *Populus alba* "Siberia Extremeña". Regeneración por organogénesis adventicia. Montes 28: 76-77.

Bueno, M.A.; Astorga, R.; García de los Ríos, M.D.; Manzanera, J.A. 1992c. Nuevas Técnicas de propagación: Cultivo in vitro del género *Populus*. Cap. 9° de "Tecnología Selvícola e Industrial del Chopo". 19ª Sesión de la Comisión Internacional del Álamo. Zaragoza. Vol II. pp 171-189.

Bueno M.A. Astorga R. Manzanera J.A. Garcia de los Rios M.D. 1992d. Propagación clonal de árboles adultos de *Populus tremula* de la Sierra de Madrid por cultivo de tejidos. Libro de Resúmenes de la 19° Sesión de la Comisión Internacional del Alamo. Zaragoza. Volumen I. pp 523-531

Bueno, M.A.; Grau, J.M.; García de los Ríos, M.D. 1993. Micropropagación de árboles adultos de *P. tremula* e identificación de clones en rodales mediante electroforesis. I Congreso Forestal. Tomo II pp 177-182.

Bueno, M.A.; Grau, J.M.; Sánchez, N.; Manzanera, J.A. 1995. Obtención de yemas adventicias de *P. tremula* a partir de hojas neoformadas *in vitro*. XI Reunión Nacional de la Sociedad Española de Fisiología Vegetal. IV Congreso Luso-Español de Fisiología Vegetal. Libro de Resúmenes pp 250.

Bueno, M.A.; Sánchez, N.; Manzanera, J.A.; Grau, J.M. 1996. Propagación in vitro de *Populus tremula* L.: Una alternativa para la reforestación. ITEA Volumen extra n° 17 pp 341-350.

Bueno MA, Manzanera JA, Grau JM, Sanchez N, Gomez A 2001. Propagación *in vitro* de *Populus tremula* y *Populus* alba. In: Populicultura en Castilla y Leon. Junta de Castilla y Leon. S-599-2001. 53-81 pp.

Castiglione, S., Wang, G., Damiani, G., Bandi, C., Bissofi, S. and Sala, F 1993. RAPD Fingerprinting for identification and for taxonomic studies of elite poplar *(Populus spp.)* clones. Theor. Appl. Genet. 87: 54-59 .

Chalupa, V. 1987. Effect of benzylaminopurine and thidiazuron on *in vitro* shoot proliferation of *Tilia cordata, Sorbus aucuparia* L. and *Robinia pseudoacacia* L. Biol. Plant. 29 (6): 425-429.

Dayanandan S, Rajora OP, Bawa 1998. Isolation and characterization of microsatellites in Trembling aspen (*Populus tremuloides*). Theor. Appl. Genet. 96:950-956.

Douglas GC, 1989. Poplar (*Populus spp.*). In: Bajaj YPS (ed.) Biotechnology in Agriculture and Forestry 5, Trees II, Springer, ISBN 3-540-19158-5, 300-323.

Doyle, J. and Doyle, J 1990. Isolation of DNA from fresh tissue. Focus 12: 13-15.

Fields RL, Scribner KT 1997. Isolation and characterisation of novel waterfowl microsatellite *loci*: cross-species comparisons and research application. Molecular Ecology 6:199-202.

Fry, DJ, Douglas GC, Sieed NT, 1997. Somaclonal variation in *Populus*: an evaluation. In: Klopfenstein NB, Chun YW, Kim MS nad Ahuja MR (eds.) Micropropagation, genetic engineering and molecular biology of *Populus*. Gen Tech Rep RM-GTR-297, Fort Collins, CO, USDA Forest Service, Rocky Mountain Research Station, 33-43.

Gomez A, Pintos B, Aguiriano E, Manzanera JA, Bueno MA 2001. SSR markers for *Quercus suber* tree identification and embryo analysis. Journal of Heredity 92: 292-295.

Gonzalez, F. and Domingo, G 1987. Primeros resultados obtenidos en la comparación de clones de chopo en los populetum del INIA en la Meseta Central. Comunicaciones INIA 45: 1-51.

González Antoñanzas, F. 1990. Principales especies variedades y/o clones de chopos más cultivados en España y en algunos países de nuestro entorno europeo. INIA-EUITF. 100 pps.

Grau, J.M., 1991. El cultivo del chopo. Monografías universitarias. Universidad Internacional Alfonso VIII. Soria, 135 pp.

Guilford P, Prakash S, Zhu JM, Rikkerink E, Gardiner S, Bassett H, Forster R 1997. Microsatellites in *Malus* x *domestica* (apple): abundance, polymorphism and cultivar identification. Theor Appl Genet 94: 249-254.

Hartmann, H.T.; Kester, d.E.; Davies, F.T. 1990. Plant propagation: principles and practices. Englewood Cliffs, N.J. U.S.A.: Prentice Hall. 647 pp.

Heybroek, H.M 1984 Clones in Forestry and in Nature. Arboricultural Journal 8: 275-286.

Hokanson SC; Szewc-McFaddden AK; Lamboy WF; McFerson JR. (1998) Microsatellites (SSR) markers reveal genetic identities, genetic diversity and relationships in a *Malus domestica* borkh. Core subset collection Theor Appl Genet 97:671-683.

Kapusta, J.; Skibinska, A. 1985. Induction of morphogenesis and regeneration in the callus of *Populus alba* L. and *Populus nigra* L. Journal of Tree Sciences 4(2): 34-38.

Kleinschmit, J., 2000. Tree breeding and plantations in Europe: A regional situation report. IUFRO WP 2.02.00 and 2.08.00 meeting in Kuala Lumpur, Malaysia, 7. August 2000 (manuscript)

Kloosterman AD, Budowle B, Daselaar, P 1993. PCR-amplification and detection of the human D1S80 VNTR *locus*. Amplification conditions, population genetics and application in forensic analysis. Legal Medicine 105:257-264.

Lamboy WF, Alpha CG 1998. The utility of simple sequence repeats (SSRs) for DNA fingerprinting germplasm accessions of grape (*Vitis* L.) species. J Am Soc Hortic Sci 123:182-188.

Lavi U, Akkaya M, Bhagwat A, Lahav E, Cregan PB 1994. Methodology of generation and characteristics of simple sequence repeat DNA markers in avocado (*Persea americana* M.). Euphytica 80:171-177.

Lloyd, McCown, 1980. Commercially feasible micropropagation of mountain laurel, *Kalmia latifolia*, by shoot-tip culture. Comb. Proc. Int. Plant Prop. Soc. 30: 421-427.

Murashige, T.; Skoog, F. 1962. A revised medium for rapid growth and bioassays with tobacco tissue culture. Physiologia Plantarum 15: 473-497.

Nei, M 1978. Estimation of average heterozygosity and genetic distance from a small number of individuals. Genetics 89:583-590 .

Park YG and Son SH, 1997. Somatic embryogenesis from poplar leaf tissue. In: Klopfenstein NB, Chun YW, Kim MS nad Ahuja MR (eds.) Micropropagation, genetic engineering and molecular biology of *Populus*. Gen Tech Rep RM-GTR-297, Fort Collins, CO, USDA Forest Service, Rocky Mountain Research Station, 10-12.

Primmer CR, Moller AP, Ellegren H 1996 Polymorphisms revealed by simple sequence repeats. Trends in Plant Science 1:215-222.

Rahman MH, Dayanandan S, Rajora OP 2000. Microsatellite DNA markers in *Populus tremuloides*. Genome 43:293-297.

Rajora, O.P.1989. Characterisation of 43 *Populus nigra* L. clones representing selection, cultivar and botanical varieties based on their allozyme genotypes. Euphytica 43: 197-206.

Rani, V., Parida, A. and Raina, S.N. 1995 Random amplified polymorphic DNA (RAPD) markers for genetic analysis in micropropagated plants of *Populus deltoides* Marsh. Plant Cell Reports. 14 : 459-462 .

Rholf FJ (1994) NTSYS-pc, version 1.8. Exeter Software, Setauket, New York.

Rongwen J, Akkaya MS, Bhagwat AA, Lavi U, Cregan PB 1995. The use of microsatellite DNA makers for soybean genotype identification. Theor Appl Genet 90: 43-48.

Ruiz de la Torre, J.; Ceballos, L. 1979. Arboles y Arbustos de la España Peninsular. ETSI de Montes. Madrid. 512 pp. , ISBN 84-600-1541-6,

Russell J, Fuller J, Young G, Thomas B, Tamarino G, Macaulay M, Waugh R, Powell W 1997. Discriminating between barley genotypes using microsatellite markers. Genome 40:442-450.

Rutledge, C.B.; Douglas, G.C. 1988. Culture of meristem tips and micropropagation of commercial clones of poplar *in vitro*. Physiologia Plantarum 72:367-373.

Saiki, R.K., Gelfand, S., Stoffel, S., Scharf, S.T., Higuchi, R., Horn, G.T., Millis, K.B., Erlich, H.A.: Primer-detected enzymatic amplification of DNA with a thermostable DNA polymerase. Science 239: 487-491 (1988)

Sánchez, N.; Manzanera, J.A.; Grau, J.M.; Bueno, M.A. 1997. RAPDs para la identificación y determinación de la estabilidad genómica en clones de *Populus tremula* obtenidos in vitro. Cuadernos de la SECF pp 247-251.

Sanchez N, Grau, J.M., Manzanera, J.A. and Bueno, M. A. 1998: RAPD markers for the identification of *Populus* species. Silvae Genetica 47: 67-71.

Sanchez N, Grau, J.M., Manzanera, J.A. and Bueno, M. A. 2000: Genetic characterisation of *Populus tremula* regions of origin in Spain using RAPD fingerprints. Silvae Genetica 49 (2): 66-71 .

Sigurdsson, V., Ananthawat-Jonsson, K. and Sigurgeirsson A. 1995: DNA fingerprinting of *Populus trichocarpa* clones using RAPD markers. New Forests 10: 197-206.

Sun HS, Kirkpatrick BW 1996. Exploiting dinucleotide microsatellites conserved among mammalian species. Mammalian Genome 7:128-132.

Takhtajan A 1986 Floristic regions of the world, Univ. Calif. Press, Berkeley

Thorpe R, Malhotra A 1996 Molecular and morphological evolution within small islands. Phil Trans R Soc B 351: 815-822.

Uddin, M. R.; Meyer, M.M.; Jokela, J.J. 1988. Plantlet production from anthers of Eastern cottonwood (*Populus deltoides*) Can. J. For. Res. 18: 937-941.

UPOV. 1981. Guidelines for the conduct of test for distinctness, homogeneity and stability of poplar (*Populus* L.). TG/21/7. UPOV

Young, P.; Hutchins, A.S.; Canfield, M.L. 1984. Use of antibiotics to control bacteria in shoot cultures of woody plants. Plant Science Letters 34: 203-209

Walter H, 1973. Vegetation of the Earth in relation to climate and the Eco-Physiological conditions, London, NY.

Weber JL, May PE 1989. Abundant class of human DNA polymorphisms which can be typed using the polymerase chain reaction. Am J Hum Genet 44: 388-396.

Williams, J.G., Kubelik, A,R,, Livak, K.J., Rafalski, J.A. and Tingey, S.V. 1990 DNA polymorphisms amplified by arbitrary primers are useful as genetic markers. Nucleic. Acids. Res. 18: 6531- 6535 .

Zsuffa, .L, Mosseler, A. and Raj, Y. 1984. Prospects for interspecific hybridisation in willows for biomass production. In: Perttu, K. (ed.). Ecology and Management of Forest Biomass Production Systems. Dept. Ecol. and Environ. Res, Swed. Univ. Agric Rep. 15: 261-281 .

3. MOLECULAR DNA MARKER ANALYSIS TO ASSESS THE GENETIC FIDELITY OF MICROPROPAGATED WOODY PLANTS

Vijay Rani and S. N. Raina

Laboratory of Cellular and Molecular Cytogenetics, Department of Botany, University of Delhi, Delhi 110007, India. E-mail: vijay15@bol.net.in; snraina@satyam.net.in

CONTENTS

S.M. Jain and K. Ishii (eds.), Micropropagation of Woody Trees and Fruits, 75–101.

1. INTRODUCTION

1.1. Woody plants micropropagation and maintenance of genetic fidelity

Successful micropropagation of woody plants is relatively a recent phenomenon (Thorpe, 1990; Bajaj, 1997). Several woody species such as poplars, wild cherry, eucalypts, red wood, radiata pine and teak are at present commercially micropropagated (Thorpe, 1990; Bajaj, 1997), while for others, such as sandal wood, birch, loblolly pine, *Shorea* and *Dalbergia,* the protocols are being standardized for mass multiplication. Very soon, standardized and efficient protocols for mass multiplication of a large number of diverse woody plant species would be available. From the practical standpoint, however, the maintenance of genetic integrity among micropropagated plants vis-à-vis explant source(s) will be one of the most crucial concerns for the reasons that advantages like high yield, uniform quality of commercial product and shorter rotation period, in the use of elite genotype(s) over natural seedlings are maintained, and no untoward mutation(s) creep in (Larkin and Scowcroft, 1983; D'Amato 1985; Scowcroft, 1985; Larkin, 1987; Karp and Bright, 1985; Peschke and Phillips, 1992; Cloutier and Landry, 1994; Karp, 1991,1995; Phillips et al., 1990, 1994; Kaeppler et al., 1998; Jain et al., 1998a,b; Jain, 2001). This is all the more important for tree species and other perennial crops where the life span is long and the performance of *in vitro*-derived plants can be ascertained only after their long juvenile stage (Brown and Sommer, 1982; Bonga and Durzan, 1987; Schoofs, 1992; Gamborg, 1993). After such a long period, it could be disastrous if the plants turn out to be abnormal. In this context, Dunstan and Thorpe (1986) have suggested that the commercial application of tissue culture to perennial crops must await adequate quality checks and field testing with proper controls to prevent large-scale losses, as experienced with oil palm and poplars. Therefore, it is very clear that the assessment of genetic integrity of micropropagated plants of woody plant species is essential before a particular protocol for a specific genotype is released for mass multiplication.

1.2. Methods of micropropagation and factors affecting genetic fidelity of micropropagated plants

Micropropagation using tissue culture can be achieved using five different approaches i) single node culture; ii) Regeneration of adventitious buds/shoots; iii) regeneration of plants from callus; iv) enhanced axillary branching, and v) somatic embryogenesis (Pierik, 1987; Thorpe, 1990; see Rani and Raina, 2000). The first four approaches lead to plantlet production via organogenesis through the production of unipolar shoots, which must then be rooted in a multistaged process. In contrast, somatic embryogenesis leads to the formation of a bipolar embryo, capable of producing a complete plantlet (Thorpe, 1990). Somatic embryogenesis can be induced directly or indirectly with an intermediate callus phase. Primary somatic embryos can be multiplied through secondary embryogenesis.

Of the various methods described above, somatic embryogenesis and enhanced axillary branching are most extensively utilized in commercial micropropagation system and have been considered to be principal pathways of plant micropropagation (Vasil, 1994). Of these two methods, plant micropropagation through somatic embryogenesis is preferred, wherever possible, because i) of the time required and difficulty involved in the rooting of the shoots derived through enhanced axillary branching cultures; ii) rapid propagation of a large number of plants is possible using somatic embryogenesis and iii) the technology is amenable to mechanization thus making possible the use of bioreactors for large scale production of somatic embryos and their delivery through encapsulation into artificial seeds (Ammirato and Styer, 1985; Lutz et al., 1985; Redenbaugh et al., 1987). However, the standard protocol for application of somatic embryogenesis is available for only a few plant species (Thorpe, 1990). Enhanced axillary branching method involves the abolition of apical dominance resulting in de-repression and multiplication of axillary buds, and has become a very important micropropagation method due to simplicity of approach and a faster propagation rate (Debergh and Read, 1990; Wang and Charles, 1991). Of the remaining three methods, single node culture is the simplest and most natural method but the rate of propagation is very low and is strongly dependent on the number of buds formed within a given time period. Serious problems may arise while micropropagating shrubs and trees due to dormancy of buds and failure of stem elongation. Micropropagation through regeneration of adventitious buds and shoots is mostly limited to herbaceous plants and propagation of plants through callus has many disadvantages, the most important being the associated genetic instability.

Various kinds of changes such as numerical and structural chromosomal changes, deamplification and amplification of genes, transposable element activation, alteration in DNA methylation have been observed to be associated with somaclonal variation (Scowcroft, 1985; Phillips et al., 1994, see Rani and Raina, 2000). In general, tissue culture conditions, for example, the presence of various hormones and the released bye-products act as a stress environment which induces a programmed loss of cellular controls (Phillips et al., 1994; Kaeppler et al., 1998). In addition to the culture method and environment (Pierik, 1987, 1991; Thorpe, 1988, 1990; Debergh and Read, 1990; Wang and Charles, 1991), genotype (Zehr et al., 1987; Carver and Johnson, 1989; Mohmand et al., 1990), explant source (Morere-Le Paven et al., 1992), ploidy level and *in vitro* culture age (Henry et al., 1996; Lee and Phillips, 1987; DeBuyser et al. 1988) are also known to be associated with the occurrence of somaclonal variation.

2. METHODS FOR THE ASSESSMENT OF GENETIC FIDELITY OF MICROPROPAGATED PLANTS

Before the advent of molecular biological techniques, several strategies that were used to assess the genetic integrity of *in vitro* raised plants were beset with many serious limitations and most of such studies were used primarily to answer various issues concerning tissue culture rather than to test genetic integrity with authenticity of at least the bare minimum number of plants required for such studies. The detection of off-types among micropropagated plants, especially tree taxa, for example, by phenotypic identification based on morphological traits requires extensive observations of plants until maturity and in many cases lacks definition and objectivity (Orton, 1983 a, b, 1985). Despite all these constraints and in the absence of better and fast approach(es) till recently this has remained the most widely used method for off-type detection. Similarly, chromosome research particularly with respect to the determination of rearrangements and/or numerical variation in chromosomes in the tissue culture plants has serious limitations in, for example, most of the economically important hard wood tree species wherein, besides severe difficulty in obtaining good number of analyzable cells at metaphase, the chromosomes are small in size and/or zygotic number is high due to polyploidy (D'Amato, 1985; Bhojwani et al., 1986). Moreover, in almost all the cases, the chromosome analysis is done in regenerated root tips. The root and shoot tips meristems in *in vitro* arise from different type of cells cultured in different hormonal conditions, and they may not always have the same genetic constitution. In other words, it will be more prudent to analyse the chromosome preparation made from shoot tip rather than root tip cells for the simple reason that it is the genetic integrity of shoot tip meristem about which we are more concerned with.

The determination of nuclear DNA amounts in the shoot tip cells, on the other hand, serves dual purpose. It gives rapid appraisal of DNA deviation, if any, independent of a change in the chromosome number and/or chromosome number dependent DNA change (Bohanec et al., 1995; Rani et al., 2001). The assessment of genetic integrity of micropropagated plants by biochemical markers such as isozymes, proteins and flavanoides etc. has also several disadvantages. The major drawbacks of this technique lie in the limited number of informative markers from only transcribed region of the genome, and its proneness to environmental and developmental conditions (Brown et al., 1993). Besides these limitations, the above approaches are not fully suitable for detecting polymorphism at DNA sequence level, if any, in the micropropagated plants.

The advent of molecular DNA markers has opened up new vistas for an easy and precise detection and better understanding of somaclonal variation. It is in this context that in the recent past there has been an upsurge in molecular DNA markers based assessment of the genetic integrity of micropropagated plants of many plant species (Shimron-Abarbanell and Breiman, 1991; Shirzadigan et al., 1991; Chowdhury et al., 1994; Isabel et al., 1993, 1996; DeVerno et al., 1994; Hartmann et al., 1994; Kawata et al., 1995; Wallner et al., 1996; Henry, 1998; Brar and Jain, 1998; Rani, 1995; Rani et al., 1995, 2000, 2001; Rani and Raina, 1998a, b, 2000; Jaligot et al., 2000; Rival et al., 2000; Matthes et al., 2001; Tang, 2001; Gimenez et al., 2001). The advantages of molecular DNA markers include an unlimited number of probes/primers which can scan both coding as well as non-coding regions of the genome. The incidence of somaclonal variation can be specified to particular genome(s) since this is the only approach which detects variation, if any, in each of the nuclear, mitochondrial and chloroplast genome separately. The molecular DNA markers can be used at various developmental stages in tissue culture, and thus help in identifying the specific culture condition(s) that induce variation. However, molecular DNA marker based assessment also has some limitations. Since, the plant genome is believed to contain 100 000 or more genes and repetitive sequences, the number of enzyme-probe combinations and primers required to be used in RFLPs and PCR reactions is very important in assessing genetic fidelity. In such circumstances, RFLP analysis with more restriction endonucleases with many probes and more primers for PCR amplifications should be screened to identify an adequate number of polymorphic markers. Even then, it is difficult to visualize as to how large number of enzyme-probe combinations or primers can appropriately cover the whole genome in different species.

Several molecular DNA marker techniques are available now and new ones continue to be developed. Different techniques may lend themselves to help in ascertaining genetic stability/instability related problems in the micropropagated plants. In the following section, different types of most popular molecular DNA markers will be outlined. These different markers will then be briefly discussed with respect to their application in the assessment of genetic fidelity of micropropagated woody plant species.

2.1. Fingerprints that are probe-hybridization based

These markers include Restriction Fragment Length Polymorphism (RFLP) analysis in which total, nuclear, chloroplast or mitochondrial genomic DNA is digested with restriction endonucleases, the resulting fragments separated by gel electrophoresis, transferred to nylon membrane by blotting and then probes are hybridized to the membrane. The probes could be short random or genomic DNA fragments, characterized or uncharacterized cDNAs, synthetic oligonucleotide fragments and homologous or heterologous gene probes originating from nuclear, mitochondrial and chloroplast genome. Variation in RFLP patterns can arise if mutation(s) alter restriction sites or the fragment length between them. Changes in DNA methylation can be detected by RFLPs by using methylation sensitive restriction endonucleases. Since change(s) in DNA methylation has been proposed as a mechanism that may in part explain the breakdown of normal developmental pathway leading to genetic instability in *in vitro* conditions, this aspect has recently, aroused much interest in the studies dealing with somaclonal variation (Karp, 1991, 1995; Peschke and Phillips, 1992; Kaeppler and Phillips, 1993; Phillips et al., 1994; Kaeppler et al., 1998; Shimron-Abarbanell and Breiman 1991; Brown and Lorz, 1986; Brown, 1989; Rani et al., 2000; Jaligot et al., 2000; Rival et al., 2000; Matthes et al., 2001). The technical complexity of performing RFLP analysis is associated with mainly high cost and wide spread use of radioisotopes in the detection method and prevents its routine application especially, in commercial micropropagation system.

2.2. Fingerprints that are based on PCR with arbitrary or generic primers

Since the development of polymerase chain reaction (PCR) (Saiki et al., 1988) many standard molecular biological procedures have been revolutionized. The PCR technique basically involves many fold amplification of the specific DNA sequence(s) from the particular genome. The resulting amplified band(s) is complementary to the template DNA, and hence serves as a genetic marker(s). Originally, sequence information at the ends of the target sequence was a prerequisite for designing the two primers that amplified it but later on, many modifications of the original procedure were designed to suit a number of diverse needs of whole gamut of molecular biology. The modifications have been primarily categorized into two classes. The first class of markers (RAPD, DAF, AP-PCR, ISSR and SSRP) are amplified by using a single arbitrary primer in a PCR reaction.

2.2.1. Randomly Amplified Polymorphic DNA (RAPD) markers

RAPD technique was developed by Williams et al., (1990) to detect nucleotide sequence polymorphism in DNA by using a single, usually 10-mer primer, of arbitrary nucleotide sequence. This primer anneals to complementary sequences at many places on the genomic DNA, and initiates the amplification process. The amplified products are either separated by Agarose or Polyacrylamide Gel Electrophoresis (AGE or PAGE) and stained with ethidium bromide or silver staining, respectively. RAPD markers are generally dominant markers and the polymorphisms between individuals result from sequence difference in one or both primer binding sites and are visible as the presence or absence of a band(s). RAPD fingerprinting has the advantage of being technically simple. It requires very small amount of DNA with no need for prior information of the sequence of primers and there are no blotting or hybridization steps involved. Due to these factors, RAPD analysis has become one of the most commonly used marker systems in tissue cultures. It is, however, necessary that reaction conditions must be strictly standardized to achieve reproducible profiles.

2.2.2. DNA Amplification Fingerprinting (DAF) markers

Instead of using a 10-mer primer as in RAPDs, Caetano Annollès et al., (1991) employed single arbitrary primers as short as five nucleotide sequence to amplify random DNA sequences using PCR. The amplified fragments are separated on PAGE and visualized with silver staining. DAF requires careful optimization of reaction conditions. The technique is amenable to automation with fluorescent tagging of primers for an early and easy detection of amplified fragments.

2.2.3. Arbitrary Primed Polymerase Chain Reaction (AP-PCR) markers

AP-PCR was developed by Welsh and McClelland (1990), wherein the single arbitrary primer used is 10-15 nucleotides in length, and the primer-template annealing is kept non-stringent in the first two cycles. Initially, the technique involved the use of radioisotopes but now separation and visualization of the amplified products is brought about by using agarose gel elecrophoresis and ethidium bromide staining.

2.2.4. Inter Simple Sequence Repeats (ISSRs) markers

Microsatellite repeats also known as simple sequence repeats (SSRs) have recently been shown to be abundant as well as polymorphic in eukaryotic genomes (Weber and May, 1989; Litt and Lutty, 1989). In ISSR amplification technique, introduced by Zietkiewicz et al., (1994), primers constitute of di-, tri-, tetra-, and penta-nucleotides motifs. Most of them are 10-20 mers. Further, various SSR primers are anchored at 3′ to increase their specificity. The polymorphism is based on the difference in the number of repeat units present between the two corresponding SSRs, which changes the size of the amplified products. The amplified products are size-separated by AGE and stained with ethidium bromide. The ISSR markers are mostly dominant, although, occasionally a few might exhibit co-dominance. Like RAPDs, ISSR fingerprinting is technically simple and easy to perform.

2.3. Fingerprints that combine the above two techniques

This second class of modifications is technically more demanding but gives greater reproducibility. It involves both PCR and RFLP techniques combined in one methodology. The procedures have been modified to yield various types of markers and are discussed in the following text.

2.3.1. Cleaved Amplified Polymorphic Sequences (CAPS) markers

This assay captures some of the advantages of RFLP analysis while avoiding the disadvantages of Southern blot analysis. CAPS also known as PCR-RFLP is a codominant marker system. The PCR amplified products are digested with a number of restriction endonucleases to reveal RFLPs within the amplified bands between the analysed samples (Konieczny and Ausubel, 1993; Jarvis et al., 1994). Despite the limited size of the amplified fragments, capturing RFLP polymorphism in the amplified band can be little more difficult than the traditional RFLP analysis, however, many restriction endonucleases can easily be tested on the amplified DNAs with this technique.

2.3.2. Sequence Characterized Amplified Region (SCAR) markers

By this technique (Michelmore et al., 1991; Martin et al., 1991), genetic markers are derived from specific PCR amplification of individual RAPD marker(s), by using longer (22-24 nucleotides), specific markers based on termini sequences of the RAPD band of interest. SCARs are usually dominant but can be converted to co-dominant markers by digesting the amplified products with restriction endonucleases. The digested DNA is analysed on denaturing gradient gel electrophoresis (DGGE) which resolves partially denatured double stranded DNA in precisely defined conditions of temperature and denaturations.

2.3.3. Amplified Fragment Length Polymorphism (AFLP) markers

AFLP technology is based on the selective PCR amplification of a subset of restricted fragments of genomic DNA (Vos et al., 1995). Total genomic DNA is digested with restriction endonucleases and double stranded DNA adapters are ligated to the ends of the DNA fragments to generate template for DNA amplification. The sequence of the adapters and the adjacent restriction site serve as primer binding sites for the subsequent amplification of restriction fragments by PCR. The primers are designed to contain the sequences that are complementary to the adapters and the restriction sites, alongwith one to three selective base(s) added to the their 3′ ends. The use of selective bases results in PCR amplification of a subset of restriction fragments in which the nucleotides flanking the restriction site match the selective nucleotides. The subset of the amplified fragments are analyzed by denaturing polyacrylamide gel electrophoresis to generate the fingerprint. The use of methylation sensitive restriction endonucleases in AFLP analysis represents an excellent system to identify polymorphisms resulting due to methylation changes, if any (Xiong et al., 1999). The polymorphism in the DNA fingerprints are represented in terms of presence and/or absence of a band which can result from alterations in the DNA sequences including mutations abolishing or creating a restriction site, and insertions, deletions, or inversions between two restriction sites. The AFLP generated polymorphisms are inherited in a Mendelian fashion, and the selected AFLP bands can be converted into SCAR markers.

2.3.4. Selective Amplification of Microsatellite Polymorphic Loci (SAMPL) markers

SAMPL technology has been developed by Morgante and Vogel (1994) as a microsatellite based modification of AFLP. The construction of pre-amplified AFLP library is based on the selective amplification of restriction fragments generated by a rare and a frequent cutter restriction enzyme. However, the selective amplification utilizes an AFLP primer with three selective nucleotides in combination with a microsatellite based SAMPL primer that is in part complementary to the microsatellite sequences. The SAMPL primer (18-20 nucleotides long) comprises of two adjoining repeat sequences (with an intervening sequence), as it has been shown that SSRs are present in plant systems as two juxtaposed SSR sequences (Vogel and Scolnik, 1998). SAMPL primers with a single SSR (with or without a small non-microsatellite sequence at 5' end for anchoring) have also been used, but amplification patterns obtained were not very clear and scorable. Since the SAMPL markers target hypervariable loci and produce large number of genetic loci, they can be very useful to detect somaclonal variation in *in vitro* plants.

2.4. Fingerprints that are gene-expression based

2.4.1. Differential Display

Differential display technique (Liang and Pardee, 1992) can be used to study the genetic differences in terms of differential RNA expression. The information on the broad scale surveys of transcribed gene expression patterns, directly or indirectly related to the modification of transcribed DNA including DNA hypo/hyper methylation, will show induction or repression in compared mRNA populations of micropropagated plants, if any. The technique basically involves cDNA synthesis and a differential display PCR. A single or multiple cDNA(s) is synthesized using either single or different oligo(dT) primers to subdivide the pool of total RNA. The differential display PCR amplifies sequences on chance homology to the arbitrary primers. The amplified products are examined on PAGE using autoradiography. No prior information regarding primer sequences is required for producing fingerprints of transcribed DNA.

Although, a number of other molecular DNA markers like Single Sequence Repeat Polymorphisms (SSRPs), Sequence Tagged Sites (STSs), Sequence Tagged Microsatellites (STMs), Single Strand Conformational Polymorphism (SSCPs) have been developed, and are being routinely used for the genetic analysis of plant species for various purposes; the genetic analysis of microproapgated plants is generally carried out by RFLPs of nuclear, mitochondrial and chloroplast genomes, and PCR based assays like RAPD, and ISSR fingerprinting. Of late, though, techniques like AFLP fingerprinting with standard and methylation sensitive enzymes, and differential display have also been used in the characterization of somaclonal variation in tissue culture systems. Further, the random amplification products that are derived from various arbitrarily primed assays are being sequenced and converted into SCAR markers for better and precise studies.

3. ASSESSMENT OF GENETIC INTEGRITY OF MICROPROPAGETED WOODY TREE SPECIES BY MOLECULAR DNA MARKERS

3.1 Angiosperm (Hardwood) tree species

The last decade has witnessed the use of various molecular markers in the assessment of genetic fidelity of micropropagated plants of various angiosperm tree species. Table 1 gives a summarized account of these analyses in various tree taxa micropropagated through somatic embryogenesis and/or enhanced axillary branching cultures.

3.1.1. Oilpalm

Application of molecular markers such as RAPDs have been helpful in the detection of somaclonal variation in oil palm (Shah et al., 1994; Rival et al., 1998a). But, it has not been able to reveal any specific polymorphism associated with the mantled variants which can be easily identified by their phenotype (Corley et al., 1986; Shaw et al., 1994; Rival et al., 1998a). Recently, Jaligot et al., (2000) and Rival et al., (2000) investigated the relationship between the mantled somaclonal variants and alterations in the DNA methylation, and demonstrated a correlation between the DNA hypomethylation and mantled phenotypes. The studies on the change in methylation status were further investigated by Matthes et al., (2001) by using AFLPs with methylation sensitive enzymes. No somaclonal variant was revealed among the regenerants when standard AFLPs were used. However, AFLPs with methylation sensitive enzymes revealed genetic variation in the form of additional bands showing a reduction in methylation in a few regenerants. Nine of the polymorphic AFLP bands were converted into probes and used in Southern hybridization. The RFLPs showed that seven fragments were partially methylated single copy sequences while the remaining two showed homology to 25S ribosomal RNA and chalcone synthase gene family. However, no specific polymorphism was consistent between normal and mantled clones. All these investigations indicate that the appearance of *in vitro* induced mantled phenotype in oil palm is a complex phenomenon and a comprehensive analysis combining different approaches would be required to understand factors responsible for its origin in tissue culture. Rival et al., (1998b), have recently used differential display (ddRT-PCR) technique to identify the mRNAs that are different in normal and abnormal tissue cultured-derived plants of oil palm, followed by confirmation of their expression pattern by Northern blotting. Of these, one has been confirmed to be characteristic of the normal culture.

3.1.2. Banana

In banana (considered as a fruit tree, but botanically a perennial herb) micropropagation system, the somaclonal variants such as dwarf plants bearing smaller hairy fruit bunches of inferior quality are detected at about mature stage of development, and hence brings about reduced profit margins to the farmers. In this context, it is worthwhile to mention that RAPD fingerprinting has been able to discriminate the above mutants from the normal plants (Damasco et al., 1996; Vuylsteke et al., 1998; Shoseyov et al., 1998) and have helped identify specific primer sequences that amplify characteristic amplification bands only in the dwarfs (Damasco et al., 1996). This has enabled to detect the mutants very early in the multiplication cycle, and thus minimize the loss. The distinguishing marker was associated to chloroplast genome. Sequencing of this marker and detailed analysis of chloroplast DNA of off-types may define the exact molecular basis of variation. In yet another study, RAPD markers were able to detect somaclonal variation among micropropagated plants of banana cv. CIEN BTA-03 resistant to Yellow Sigatoka disease (Gimenez et al., 2001). Damasco et al., (1998) have recently sequenced a specific RAPD band and used it as a Sequence Characterized Amplified Region (SCAR) marker for detecting dwarf off-types in micropropagated Cavendish bananas. They emphasized that SCAR allows a more robust and reliable method of detecting somaclonal variation in banana than RAPDs (Damasco et al., 1998). Using SCAR marker, they detected dwarf off-types as early as 4th subculture after inoculation of shoot tips. The results indicated that among the various factors that induced somaclonal variation in banana, a high phytohormonal level especially of benzylaminopurine (BAP) is responsible for a shift of the micropropagation protocol from shoot tip culture to adventitious branching. The inherent instability of particular cultivar(s) during micropropagation was another factor for induction of somaclonal variation. New Guinea Cavendish bananas, for instance, produced dwarf variants at higher frequency than the cv. Williams, and so on.

3.1.3. *Populus* clones

To assess the genetic integrity of meristem culture-derived micropropagated plants of *Populus deltoides* clones, multiple molecular DNA markers (RFLP, RAPD, ISSR) were used for screening various regions of nuclear, mitochondrial and chloroplast genomes (Rani, 1995; Rani et al., 1995, 2001; Rani and Raina, 1998a, 2000). 125 micropropagated plants raised by enhanced axillary branching cultures and the corresponding mother plant(s) of five poplar clones were subjected to RAPD analysis with eleven primers. While, the micropropagated plants of three clones D121, G48, S7C15 revealed genetic uniformity at the analyzed genetic loci, distinct RAPD polymorphism was revealed in the micropropagated plants of the remaining two clones G3 and L34 (Rani, 1995; Rani et al., 1995, 2001). In the clone L34, five primers distinguished a total of 13 polymorphic bands common across 6 out of the 23 analysed plants. The same number of primers revealed differences across 11 common micropropagated plants in clone G3. The incidence of polymorphism in same set of micropropagated plants in both clones indicates that variation was induced due to a single somatic mutation event each early in the multiplication cycles and was then carried forward. The observed polymorphism in the micropropagated plants of clones L34 and G3 was further substantiated with nuclear RFLPs of ribosomal DNA (rDNA) (Rani, 1995). Rahman and Rajora (2001) used microsatellite markers derived from 10 simple sequence repeat (SSR) loci to examine somaclonal variation, if any, in morphologically similar micropropagated plants of three genotypes of *Populus tremuloides*. In two genotypes, two out of the ten SSR loci revealed new variation in one plant each. The variation was manifested in terms of either appearance of new allele(s) of increased size due to either an addition in repeat units or duplication of a chromosome and/or a chromosomal segment (Rahman and Rajora, 2001).

3.1.4. *Eucalyptus* species

In two species of *Eucalyptus*, 46 enhanced axillary branching-derived micropropagated and the corresponding mother plant(s) digested with 8 restriction enzymes hybridized to 13 nuclear, mitochondrial and oligonucleotide probes revealed minor variation in the nuclear rDNA profiles of micropropagated plants of *E. tereticornis* alone. Further, RAPD and SSR primers produced completely monomorphic profiles across micropropagated plants of both *E. tereticornis* and *E. camaldulensis*. The results demonstrated that enhanced axillary branching culture of mature trees could be profitably utilized for mass commercial clonal propagation of these two important *Eucalyptus* species (Rani and Raina, 1998). Earlier, Haque et al., (1992) demonstrated by RAPD analysis that somatic embryogenesis-derived plants of *Eucalyptus* lines produces genetically stable regenerants.

3.1.5. Coffee

In *Coffea arabica* cv. Cauvery (S-4347), novel nuclear and mitochondrial genome organizations were detected among the high frequency somatic embryogenesis (HFSE) derived micropropagated plants (Rani et al., 2000). Out of total 480 genetic loci scored by hybridization of 16 nuclear, mitochondrial and chloroplast gene probes with 9 restriction endonuclease digests, 38 RAPD and 17 SSR primers, as many as 44 genetic loci (9.16% of the total) were observed to be polymorphic among the 27 micropropagated plants and the single control. A relatively low level of polymorphism (4.36%) was found in the nuclear genome while polymorphism in the mitochondrial genome (41%) was much higher. No polymorphism was detected in the chloroplast genome. In the mitochondrial RFLPs, the polymorphism was detected in only four plants while such selective polymorphism was not true for the nuclear genome. The mitochondrial genome has been considered to be a model system to study the effects of *in vitro* induced changes (Hanson, 1984). The *in vitro* induced changes in mitochondria are thought to be due to infrequent homologous recombination between the short dispersed repeat elements accompanied by preferential amplification and rearrangement of these minor pre-existing forms (Shirzadegan et al., 1991). It has also been suggested that changes in the organization of mtDNA might be controlled by nuclear rather than mitochondrial information alone (Hartmann et al., 1994).

3.1.6. Other plant species

Hashmi et al., (1997) on the basis of RAPD analysis distinguished two embryo lines (30 and 156) of peach (*Prunus persica*) cv. Sunhigh with respect to the relative frequency of somaclonal variation. They emphasized that frequency of *in vitro* induced variation is genotype dependent in peach. The number of polymorphic RAPD bands in the regenerants of embryo line 156 was much higher to those observed in the embryo line 30 regenerants. RAPD fingerprinting revealed complete fidelity among the somatic embryo-derived plants of two mango cultivars (Jayasankar et al., 1998). The true-to-type nature of somatic embryo-derived plants of the two cultivars Barhi and Mejdool of date palm (*Phoenix dactylifera*) was confirmed by the occurrence of monomorphic RAPD fingerprints among the regenerants of each cultivar (Javouhey et al., 2000).

Vendrame et al., (1999) employed AFLP markers to assess genetic integrity of four embryogenic lines of pecan (*Carya illinoinensis* (Wangenh.) E. Koch. They found that the four lines could be easily distinguished on the basis of AFLP fingerprints. The interesting observation was the occurrence of polymorphism within the line as well. In

Piper longum, on the other hand, 56% RAPD markers were observed to be polymorphic across the mother and micropropagated plants (Parani et al., 1997) Not even a single micropropagated plant was shown to possess completely identical RAPD fingerprints with the mother plant.

Merkle et al., (1998) analyzed embryogenic cell lines of yellow poplar (*Liridendron tulipifera* L.) by RFLP fingerprinting using homologous probes of highly repeated DNA sequences. No polymorphism was found. Even the sublines of embryogenic suspension cultures and plantlets derived from the same clonal line lacked any variation in the restriction fragments. Thakur et al., (1999) used RAPD markers to reveal genetic stability in somatic embryogenesis derived regenerated plantlets, somatic embryos, and embryogenic callii in *Quercus* serrata. In another *Quercus* species, somatic embryogenesis results in the production of normal and anomalous embryos. AFLP and RAPD markers could not distinguish the mutant types, while flow cytometry revealed tetraploidy in some cell lines after several years of continuous culture (Wilhelm, 2000).

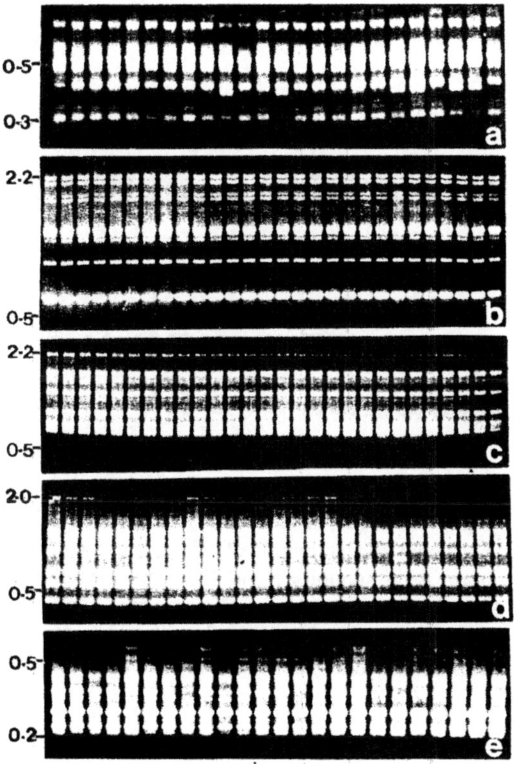

Figure 1. Gel electrophoresis of amplification products obtained in field-grown mother plant (lane 1) and enhanced axillary branching and/or somatic embryo-derived plants (remaining lanes) of *Populus deltoides* clone L34 (a), *Eucalyptus tereticornis* (b), *E. camaldulensis* (c) and *Coffea arabica* (d, e) with RAPD and SSR primers OPC09 (a), OPI08 (b), $(GATA)_4$ T(c), OPA04 (d) and $(GT)_8$C (e).

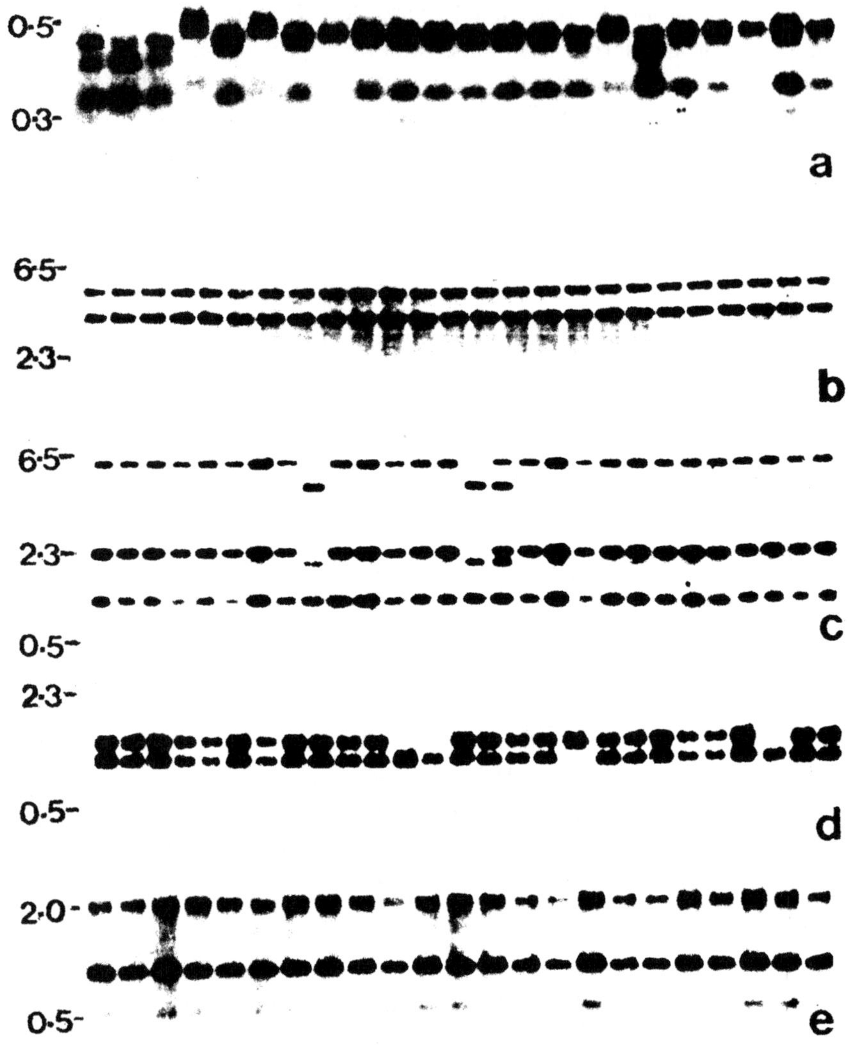

Figure 2. *Taq*I (a), *Eco*RI (b), *Hind*III (c, d) and *Bam*HI (e) restriction fragments that hybridized to ribosomal repeat length of wheat : probe pTA71 (a, b), mitochondrial cytochrome oxidase gene subunits I and III from Oenothera : probes CoxI (c) and CoxIII (d) and part of tobacco chloroplast gene rbcL : probe B29-22c (e) in the field grown mother plant (lane 1) and enhanced axillary branching and/or somatic embryo-derived plants (remaining lanes) of *P. deltoides* clone G3 (a), *E. tereticornis* (b) and *C. arabica* (c-e).

Table 1. Summary of molecular marker(s) based analysis of genetic fidelity in micropropagated angiosperm woody plants

Plant Species	Markers	Somaclonal Variation	References
Oil Palm	DNA methylation	Present	Jaligot et al., 2000 Rival et al., 2000
	AFLPs and DNA Methylation	Present	Matthes et al., 2001
Banana	RAPD	Present	Damasco et al., 1996 Vuylsteke et al., 1998 Shosryov et al., 1998 Gimenez et al., 2001
	SCAR	Present	Damasco et al., 1998
Populus deltoides	RFLP	Present	Rani, 1995, Rani et al.,
Clones L34, G3	RAPD	Present	1995, 2000, Rani and
	ISSR	Present	Raina, 1998a, 2001
Clones D121,	RFLP	Absent	Rani 1995, Rani et al.,
S7C15, G48	RAPD	Absent	1995, 2000, Rani and
	ISSR	Absent	Raina, 1998a, 2001
Populus tremuloides	SSR	Present	Rahman and Rajora, 1992
Eucalyptus tereticornis	RFLP	Present	Rani, 1995,
	RAPD	Absent	Rani and Raina,
	ISSR	Absent	1998a, b, 2000
Eucalyptus camaldulensis	RFLP	Absent	Rani, 1995,
	RAPD	Absent	Rani and Raina,
	ISSR	Absent	1998a,b, 2000
Coffea arabica	RFLP	Present	Rani, 1995,
cv. Cauvery S-4347	RAPD	Present	Rani and Raina, 2000
	ISSR	Present	Rani et al., 2000
Prunus persica cv. Sunhigh	RAPD	Present	Hashmi et al. 1997
Mangifera indica	RAPD	Absent	Jayasankar et al., 1998
Phoenix dactylifera cv. Barhi, Mejdool	RAPD	Absent	Javouhey et al., 2000
Carya illinoinensis	AFLP	Present	Vendrame et al., 1999
Piper longum	RAPD	Present	Parani et al., 1997
Liriodendron tulipifera	RFLP	Absent	Merkle et al., 1998
Quercus serrata	RAPD	Absent	Thakur et al., 1999
Quercus sps.	RAPD	Present	Wilhelm, 2000
	AFLP	Present	

3.2. Gymnosperm (Softwood) tree species

Micropropagation system using mainly somatic embryogenesis has been well established in softwood gymnosperm tree species of *Picea, Pinus, Larix* and others. The molecular analyses of genetic fidelity in micropropagated plants of various softwood tree taxa has been detailed in Table 2.

3.2.1. *Picea* species

The RAPD technique has been used to assess genetic stability of somatic seedlings of *Picea mariana* (Isabel et al., 1993), *P. glauca* (Isabel et al., 1996; DeVerno et al., 1999) and *P. abies* (Heinz and Schmidt, 1995; Fourre et al., 1997; Chandelier et al., 1999). In *P. mariana,* ten RAPD primers were used to evaluate the genetic stability of somatic embryos derived from three embryogenic cell lines. No variation was detected within the lines suggesting that the somatic embryos of *P. mariana* were genetically stable (Isabel et al., 1993). In *P. glauca* somatic embryogenesis-derived plants, Isabel et al., (1998) identified four different variegated phenotypes among 2270 micropropagated plants. One RAPD profile with a specific 10-mer primer was correlated with white needles of the variegated plants. DeVerno et al., (1999) employed RAPD analysis to assess the genetic integrity of *P. glauca* plants regenerated from six embryogenic clones after cryopreservation for 3 and 4 years. Their results showed that different clones could be distinguished by the presence of diagnostic RAPD fragments. Somatic embryogenic tissue may be composed of a mixture of genetically altered and unaltered cells present in different proportions at different stages of *in vitro* culture. They classified the somatic embryos into : (1) normally maturing and germinating somatic embryos, and (2) embryos showing some phenotypic variation either at maturation and/or germination stage. The RAPD profiles of the trees regenerated in successive years from the former type of somatic embryos maintained their genetic integrity while the plants regenerated from phenotypically aberrant somatic embryos revealed amplification of additional RAPD bands which were absent in the mother and normally germinating somatic embryo-derived plants.

In Norway spruce (*Picea abies*), Heinz and Schmidt (1995) distinguished various cell lines by RAPD markers. The plants derived from the same cell line were, however, genetically fidel. The phenotypic and cytogenetic variation among the somatic embryo-derived regenerants of *P. abies* was not reflected in the differences in RAPD profiles (Fourre et al., 1997). Chandelier et al., (1999) extracted mitochondrial DNA from embryonal suspensor masses of *P. abies* and identified a family of plasmid like DNAs in addition to the master chromosome. The plasmid like DNAs had no evident sequence homology with the master chromosome indicating that they constitute a different component of the mt genome. The variability of this plasmid-like DNA was revealed by Southern blot analysis.It has been suggested that the observed polymorphism might be due to sequence polymorphism between plasmid monomers and intra- and inter- molecular recombinations mediated by small repeats and topoisomeric forms (Chandelier et al., 1999).

Somatic embryogenic cultures of interior spruce (*Picea glauca* x *engelmannii* complex) were stable in the hybridization profiles of genomic DNA using probes that hybridized to the hypervariable regions. No differences in hybridization patterns were observed between parental and the subcloned lines indicating that somatic embryogenic lines of interior spruce were genetically stable in tissue culture (see DeVerno, 1995).

3.2.2. *Larix* species

RFLP analysis with heterologous mitochondrial gene probes of somatic embryogenic cell lines of *Larix leptolepis*, *L. decidua* and their reciprocal hybrids revealed qualitative as well as quantitative differences in the mitochondrial genome in the form of loss of few restriction sites and amplification of certain regions of mitochondrial genome, respectively (DeVerno et al., 1994). Quantitative differences in mtDNA hybridization signals were also observed in a 4-year old somatic embryo-derived *Larix* x *eurolepis* plant. Based on the analysis, it was emphasized that the changes in the mitochondrial genome were either induced after the trees were regenerated or the somatic embryogenic culture might be composed of a mixture of cells containing both normal and altered mitochondrial genome, and the cells with altered genome generally lacked regenerative capacity (DeVerno et al., 1994).

3.2.3. *Pinus* species

In the genus *Pinus*, Goto et al., (1998) used RAPD markers to determine the genetic stability of long term (more than 10 years) micropropagated shoots of Japanese black pine *Pinus thunbergii*. Micropropagated shoots representing three phenotypic variants with short, medium and long needles were analyzed with 30 RAPD primers. The results revealed monomorphic RAPD profiles in the morphologically aberrant regenerated plants of *P. thunbergii*. Similar observations were made in loblolly pine, *P. taeda* plants regenerated from callus derived from mature zygotic embryos. Twenty one RAPD primers produced a total of 107 monomorphic bands across the 21 analyzed plantlets (Tang, 2001).

Table 2. Summary of molecular marker(s) based analysis of genetic fidelity in micropropagated gymnosperm woody plants

Plant Species	Markers	Somaclonal Variation	References
Picea mariana	RAPD	Absent	Isabel et al., 1993
Picea glauca	RAPD	Present	Isabel et al., 1996 DeVerno et al., 1999
Picea abies	RAPD	Absent	Heinz and Schmidt, 1995 Fourre et al., 1997
	RFLP	Present	Chandelier et al., 1999
Picea glauca x *engelmanii* complex	RFLP	Absent	DeVerno, 1995
Larix leptolepis, L.decidua and reciprocal hybrids	RFLP	Present	DeVerno et al., 1994
Larix x *eurolepis*	RFLP	Present	DeVerno et al., 1994
Pinus thunbergii	RAPD	Absent	Goto et al., 1998
Pinus taeda	RAPD	Absent	Tang, 2001

4. CONCLUSION

The advent of molecular DNA markers has revolutionized the genetic studies in plant cell and tissue culture. It has opened up new vistas for addressing the current issues of plant tissue cultures with ease and precision. With the continuous sophistication of molecular techniques, and the development of simpler, reliable and cost effective protocols, the molecular markers are finding wide spread applications in plant production systems in *in vitro* conditions. A growing choice of molecular markers has allowed the selection of precise markers for achieving specific objectives.

One recurrent concern in many-a-tissue cultures has been the maintenance of genetic integrity among plant cells/whole plants, which is vital in procedures like micropropagation, plant transformation and *in vitro* conservation experiments, and in studies involving the understanding of biochemical and molecular pathways to various cellular processes. Of the various classes of markers developed, the molecular DNA markers have shown great potential in detecting somaclonal variation at very early stages in tissue cultures so as to help modify the specific culture stage(s) that induces variation. An early detection of somaclonal variation is especially relevant in perennial crops and woody plant species which have long juvenile phases. Application of molecular DNA markers for the assessment of genetic fidelity of the micropropagated plants of various hardwood and softwood plant taxa, raised by a particular protocol, have revealed mixed results with some taxa being able to retain the genetic integrity while other show the induction of discrete somaclonal variation. The differences in the response could be influenced by the genotype, explant source, culture medium, age of donor plant and duration of culture process. A better understanding of the factors influencing somaclonal variation by molecular techniques may allow more efficient ways to eliminate somaclonal variation , where it is not desired.

5. ACKNOWLEDGEMENTS

This research was funded by a grant from the Department of Biotechnology, Ministry of Science and Technology, Government of India.

6. REFERENCES

Ammirato, P.V. and D.J. Styer. 1985. Strategies for large scale manipulation of somatic embryos in suspension cultures. In: *Biotechnology in Plant Science:Relevance to agriculture in the Eighties.* pp. 161-178 (eds. Zaitlin, M., P. Day and A. Hollaender) New York:Academic Press.

Bajaj, Y.P.S. 1997. *Biotechnology in Agriculture and Forestry.* vol. 39. New York: Springer Verlag.

Bohanec, B., M. Jakse, A. Ihan and B. Juornik. 1995. Studies of gynogenesis in onion (*Allium cepa* L.) : induction procedures and genetic analysis of regenerants. *Plant Sci.* 104:215-224.

Bonga, J.M. and D.J. Durzan. 1987. *Cell and Tissue Culture in Forestry.* vol 1. Dordrecht: Martinus Nijhoff.

Brar, D.S. and S.M. Jain.. 1998. Somaclonal variation : mechanism and applications in crop improvement. In: *Somaclonal Variation and Induced Mutations in Crop Impovement.* pp. 15-37 (eds. Jain, S..M.., D.S. Brar and B.S. Ahloowalia) Dordrecht: Kluwer.

Brettell, R.I.S., M.A. Palota, and J.P. Gustafson. 1986. Variation at the Nor loci in *triticale* derived from tissue culture. *Theor Appl Genet.* 71:637-643.

Brown, P.T.H. 1989. DNA methylation in plants and its role in tissue culture. *Genome.*31:717-729.

Brown, P.T.H., F.D. Lange, E. Kranz and H. Lorz. 1993. Analysis of single protoplasts and and regenerated plants by PCR and RAPD technology. *Mol Gen Genet.* 237:311-317.

Brown, P.T.H. and H. Lorz. 1986. Molecular changes and possible origin of somaclonal variation. In: *Somaclonal variation and crop improvement,* pp. 138-159 (ed. J. Semal). Dordrecht: Martinus.

Brown, C.L. and H.E. Sommer. 1982. Vegetative propagation of dicotyledonous trees. In: *Tissue culture in forestry*, pp. 109-149 (eds. Bonga, J.M. and D.J. Durzan). Dordrecht: Martinus.

Caetano-Annolles, G., B.J Bassam and P.M. Gresshoff. 1991. DNA amplification fingerprinting using very short arbitrary oligonucleotide primers. *Biotechnology*. 9:553-557.

Carver, B.E and B.B. Johnson. 1989. Partitioning of variation derived from tissue culture of winter wheat. *Theor Appl Genet*. 78:405-410.

Chandelier, A.P., D.U. Jardin, C. Avril and M. Paques. 1999. Identification of mitochondrial plasmid-like DNAs in *Picea abies* (L.) Karst. *Plant Cell Rep*. 18:841-847.

Chowdhury, M.K.U., V. Vasil and I.K. Vasil. 1994. Molecular analysis of plants regenerated from embryogenic cultures of wheat (*Triticum aestivum* L.). *Theor Appl Genet*. 87:821-828.

Cloutier, S and B.S. Landry. 1994. Molecular markers applied to plant tissue culture. *In Vitro Cell Dev Biol-Plant*. 30 P:32-39.

Corley, R.H.V., C.H. Lee, Law, I.H. and C.Y. Wong. 1986. Abnormal flower development in oil palm clones. *Planter*. 62:233-240.

D'Amato, F. 1985. Cytogenetics of plant cell and tissue cultures and their regenerants. *CRC Crit Rev Plant Sci*. 3:73-112.

Damasco, O.P., G.C. Graham, R.J. Henry, S.W. Adkins and I.D. Godwin. 1996. Random amplified polymorphic DNA (RAPD) detection of dwarf off-types in micropropagated Cavendish (Musa spp. AAA) bananas. *Plant Cell Rep*. 16:118-123.

Damasco, O.P., S.W. Adkins and I.D.Godwin. 2000. Use of a SCAR-based marker for the early detection of dwarf off-types in micropropagated Cavendish Bananas. *Acta Hort*. 461:157-164.

Debergh, P.C. and P.E. Read. 1990. Micropropagation. In: *Micropropagtion, Technology and Application*. pp 1-13 (eds. Debergh, P.C. and R.H. Zimmerman) Dordrecht:Kluwer.

De Buyser, J., C. hartmann, Y. Henry and A. Rode. 1988. Variation in long-term wheat somatic tissue culture:chromosomal number, plant morphologym, and mitochondrial DNA. *Can J Bot*. 66:1891-1895.

DeVerno, L.L., P.J. Charest and L. Bonen. 1994. Mitochondrial DNA variation in somatic embryogenic cultures of *Larix*. *Theor Appl Genet*. 88:727-732.

DeVerno, L.L. 1995. An evaluation of somaclonal variation during somatic embryogenesis. In: *Somatic Embryogenesis in Woody Plants*. Vol 1, pp. 361-377 (eds. Jain, S.M., P. Gupta and R. Newton) Dordrecht : Kluwer.

DeVerno, L.L., Y.S. Park, J.M. Bonga and J.D. Barrett. 1999. Somaclonal variation in cryopreserved embryogenic clones of white spruce (*Picea glauca* (Moench) Voss.). *Plant Cell Rep*. 18:948-953.

Dunstan, D.I. and T.A. Thorpe. 1986 Regeneration in forest trees. In: *Cell Culture and Somatic Cell Genetics of Plants*. Vol 3, pp. 223-241 (ed. Vasil, I. K) Orlando: Academic Press.

Fourre, J.L., P. Berger, I. Niquet and P. Andre. 1997. Somatic embryogenesis and somaclonal variation in Norway spruce: morphogenetic, cytogenetic and molecular approaches. *Theor Appl Genet*. 94:159-169.

Gamborg, O.L. 1993. Potentials of biotechnology in micropropagation and genetic improvement of woody species. pp. 8-11, In: *Proc Intl Symp Appl Plant In Vitro Tech*, University of Agriculture Malaysia, Malaysia.

Goto, S., R.C. Thakur and K. Ishii. 1998. Determination of genetic stability in long-term micropropagated shoots of *Pinus thunbergii* Parl. using RAPD markers. *Plant Cell Rep*. 18:193-197.

Gimenez, C., E.D. Garcia, N.X.D. Enrech and I. Blanca. 2001. Somaclonal variation in banana: cytogenetic and molecular characterization of somaclonal variant CIEN BTA-03. *In Vitro Cell Dev Biol–Plant.* 37:217-222.

Hanson, M.R. 1984. Stability, variation and recombination in plant mitochondrial genomes via tissue culture and somatic hybridization. *Oxf Surv Plant Mol Cell Biol.* 1:33-52.

Hartmann, C., H. Recipin, M. Jubier, C. Valor, B. Delcher, Y. Henry, J. DeBuyser, B. Lejeune and A. Rode. 1994. Mitochondrial DNA variability in a single wheat regenerant involves a rare recombination event across a short repeat. *Curr Genet.* 25:456-464.

Hashmi, G., R. Huettel, R. Meyer, L. Krusberg and P. Hammerschlag. 1997. RAPD analysis of somaclonal variants derived frm embryo callus cultures of peach. *Plant Cell Rep.* 6:624-627.

Haque, N.S., N.W. Fish and M. Keil. 1992. Assessment of somaclonal variation in *Eucalyptus* using random amplified polymorphic DNA markers. Proceedings : Fifth Wkshp. IUFRO Wk. Party S2.04.06. Carcans-Maubisson, June 15-18, 1992. INRA, France.

Heinz, B. and J. Schmidt. 1995. Monitoring genetic fidelity vs somaclonal variation in Norway Spruce (*Picea abies*) somatic embryogenesis by RAPD analysis. *Euphytica.* 85:341-345.

Henry, R.J. 1998. Molecular and biochemical characterization of somaclonal variation. In: *Somaclonal Variation and Induced Mutations in Crop Impovement.* pp. 485-499 (eds. Jain, S.M., D.S. Brar and B.S. Ahloowalia) Dordrecht: Kluwer.

Henry, Y., J.L. Marcotte and J De Buyser. 1996. The effect of aneuploidy on karyotype abnormality in wheat plants regenerated from short and long-erm somatic embryogenesis. *Plant Sci.* 114:101-109.

Isabel, N., L. Tremblay, M. Michaud, F..M. Tremblay and J. Bousquet. 1993. RAPDs as an aid to evaluate the genetic intergrity of somatic embryogenesis derived populations of *Picea mariana* (Mill.) B.S.P. *Theor Appl Genet.* 86:81-87.

Isabel, N., R. Bovin, C. Levasseur, P.M. Charest, J. Bousquet and F.M. Tremblay. 1996. Occurrence of somaclonal variation among somatic embryo-derived white spruces (*Picea glauca*, Pinaceae). *Am J Bot.* 83:1121-1130.

Jain, S..M. 2001. Tissue-culture derived variation in crop improvement. *Euphytica* 118:153-166.

Jain, S.M., B.S. Ahloowalia and R.E. Vielleux. 1998a. Somaclonal variation and crop improvement. In: *Somaclonal Variation and Induced Mutations in Crop Impovement.* pp. 203-218 (eds. Jain, S.M., D.S. Brar and B.S. Ahloowalia) Dordrecht: Kluwer.

Jain, S.M., D.S. Brar and B.S. Ahloowalia. 1998b. *Somaclonal Variation and Induced Mutations in Crop Improvement.* Dordrecht : Kluwer.

Jaligot, E., A. Rival, T. Beule, S. Dussert and J.L. Verdeil. 2000. Somaclonal variation in oil palm (*Elaeis guineensis* Jacq.) : the DNA methylation hypothesis. *Plant Cell Rep.* 19:684-690.

Jarvis, P., C. Lister, V. Szabo and C. Dean. 1994. Integration of CAPs markers into RFLP map generated using recombinant inbred lines of *Arabidopsis thaliana. Pl Mol Biol.* 24:685-687.

Javouhey, M., F. Daguin and R. Letouze. 2000. Somatic embryogenesis, an efficient tool for date palm (*Phoenix dactylifera* L.) industrial micropropagation, characterization and genetic stability of original offshoots and regenerated plantlets by RAPD markers. *Acta Hort.* 530:237-241.

Jayasankar, S., R.E. Litz, R.J. Schenell and A.C. Hernandez 1998. Embryogenic mango cultures selected for resistance to *Colletotrichum gloeosporiodes* cultures filtrare show variation in random amplified polymorphic DNA markers. *In Vitro Cell Dev Biol-Plant.* 34:112-116.

Kaeppler, S.M. and R.L. Phillips. 1993. Tissue culture-induced DNA methylation in maize. *Proc Natl Acad Sci.* 90:8773-8776.

Kaeppler, S.M., R.L. Phillips and P. Olhoft. 1998. Molecular basis of heritable tissue culture-induced variation in plants. In: *Somaclonal variation and induced mutations in crop improvement.* pp. 465-484 (eds. Jain, S.M., P.S. Brar and D.S. Ahloowalia) Dordrecht:Kluwer.

Karp, A. 1991. On the current understanding of somaclonal variation. *Oxf Surv Plant Mol Cell Biol.* 7:1-58.

Karp, A. 1995. Somaclonal variation as a tool for crop improvement. *Euphytica* 85:295-302.

Karp, A. and S.W.J. Bright. 1985. On the causes and origins of somaclonal variations. *Oxf Surv Plant Mol Cell Biol.* 2:199-234.

Kawata, M., A. Ohmiya, Y. Shimamoto, K. Oono and F. Takaiwa. 1995. Structural changes in the plastid DNA of rice (*Oryza sativa* L.) during tissue culture. *Theor Appl Genet.* 90:364-371.

Konieczny, A. and M.A. Ausubel. 1993. A procedure for mapping *Arabidopsis* mutations using codominant ecotype-specific PCR-based markers. *Plant J.* 4:403-410.

Larkin, P.J. 1987. Somaclonal variation:history, method and meaning. *Iowa State J.* 61:393-434.

Larkin, P.J. and W.R. Scowcroft. 1983. Somaclonal variation and crop improvement. In: *Genetic engineering of plants.* pp 289-314 (eds Konsuge, T., C. Meredith and A, Holloender) New York:Plenum Press.

Lee, M and R.L. Phillips. 1997. Genetic variants in the progeny of regenerated maize. *Genome.* 29:834-838.

Litt, M and J. A. Lutty. 1989. A hypervariable microsatellite revealed by *in vitro* amplification of a di-nucleotide repeat within the cardiac muscle action genes. *Am J Hum Genet.* 44:397-401.

Liang, P and A.B. Pardee. 1992. Differential display of eukaryotic messenger RNA by means of the polymerase chain reaction. *Science.* 257:967-971.

Lutz, J.D., J.R. Wong, J. Rowe, D..M. Tricoli, R.H.Jr. Lawrence. !985. Somatic embryogenesis for mass cloning of crop plants. In: *Tissue Culture in Forestry and Agriculture*, pp.105-116 (eds. Henke, R.R., K.W. Hughes, M.P. Constantin and A. Hollaender). New York:Plenum.

Martin, G. B., J.G.K. Williams and S.D. Tanksley. 1991. Rapid identification of markers linked to a *Pseudomonas* resistance gene in tomato by using random primers and near isogenic lines. *Proc Natl Acad Sci USA.* 88:2336-2340.

Matthes, M., R. Singh, S.C. Cheah and A. Karp. 2001. Variation in oil palm (*Elaeis guineensis* Jacq.) tissue culture-derived regenerants revealed by AFLPs with methylation-sensitive enzymes. *Theor Appl Genet.* 102:971-979.

Merkle, S.A., P.L. Chou and H.E. Sommer. 1998. Stability of highly repeated sequences in the DNA of embryogenic cultures of yellow poplar. In : *Molecular Genetics of Forest Trees,* pp. 85-88 (eds. Cheliak, W.M. and A.C. Yapa). Petwawa National Forestry Institute Information Report PI-X-80.

Michelmore, R. W., I. Paran and R.V. Kesseli. 1991. Identification of markers linked to disease-resistance genes by bulked segregant analysis:a rapid method to detect the markers in specific genomic regions by using segregating populations. *Proc Natl Acad Sci USA.* 88:9828-9832.

Mohmand, A.S and M.W. Nabors. 1990. Somaclonal variant plants of wheat derived from mature embryo eplants of three genotypes. *Plant Cell Rep.* 8:558-560.

Morere-Le paven, M.C., J. De Buyser, Y. Henry, F. Corre, C. Harmann and A. Rode. 1992. A multiple pattern of mtDNA reorganization in plants regenerated from different *in vitro* cultured explants of a single wheat variety. *Theor Appl Genet.* 85:9-14.

Morgante, M and J, Vogel. 1994. Compound microsatellite primers for the detection of genetic polymorphisms. US Patent Appl 08/326456.

Parani, M., A. Anand and A. Parida. 1997. Application of RAPD fingerprinting in selection of micropropagated plants of *Piper longum* for conservation. *Current Sci.* 73:81-83.

Peschke, V.M. and R.L. Phillips. 1992. Genetic implications of somaclonal variation in plants. *Adv Genet.* 30:41-75.

Phillips, R.L., S.M. Kaeppler and V.M. Peschke. 1990. Do we understand somaclonal variation? In: *Progress in Plant Cellular and Molecular Biology*, pp. 131-141 (eds Nijkamp, H.J.J., L.H.W. VanderPlas and J. Van Aartrijk). Proceedings, Sixth International Congress on Plant Tissue and Cell Culture. Dordrecht:Kluwer.

Phillips, R.L., S.M. Kaeppler and P. Olhoft. 1994. Genetic instability of plant tissue cultures : breakdown of normal controls. *Proc Natl Acad Sci USA.* 91:5222-5226.

Pierik, R.L.M. 1987. *In Vitro* Culture of Higher Plants. Dordrecht:Kluwer.

Pierik, R.L.M. 1991. Commercial aspects of micropropagation. In: *Horticulture-new technologies and applications*, pp.141-153 (eds Prakash, J. and R.L.M. Pierik). Dordrecht:Kluwer.

Rahman, M.H. and O.P. Rajora. 2001. Microsatellie DNA somaclonal variation in micropropagated trembling aspen (*Populus tremuloides*) *Plant Cell Rep.* 20:531-536.

Rani, V. 1995. Potential use of molecular markers in the genetic analysis of *in vitro* propagated tree systems. Ph. D. thesis, Department of Botany, Univ. of Delhi, Delhi, India.

Rani, V., A. Parida and S.N. Raina. 1995. Random amplified polymorphic DNA (RAPD) markers for genetic analysis in micropropagated plants of *Populus deltoides* Marsh. *Plant Cell Rep.* 14:459-462.

Rani, V and S.N. Raina. 1998a. Genetic fidelity and molecular diagnostics in micropropagation system. In : *Genetics and Biotechnology in Crop Improvement.* pp. 370-388 (eds. Gupta, P. K., S. P. Singh, H. S. Balyan, P. C. Sharma and B. Ramesh) India : Rastogi Publications.

Rani, V and S.N. Raina. 1998b. Genetic analysis of enhanced axillary branching derived *Eucalyptus tereticornis* and *E. camaldulensis* plants. *Plant Cell Rep.* 17:236-242.

Rani, V. and S. N. Raina. 2000. Genetic fidelity of organized meristem-derived micropropagated plants: A critical reappraisal., *In Vitro Cell Dev Biol-Plant.* 36:319-330.

Rani, V., K.P. Singh, M. Shiran, S. Nandy, S. Goel, R.M. Devarumath, H. Sreenath and S.N. Raina. 2000. Evidence for new nuclear and mitochondrial genome organizations among high frequency somatic embryogenesis-derived plants of allotetraploid *Coffea arabica* L. (Rubiaceae). *Plant Cell Rep.* 19:1013-1020.

Rani, V., A. Parida and S.N. Raina. 2001. Chromosome number dependent genome size and RAPD fingerprinting diagnostics for genetic integrity of enhanced axillary branching-derived plants of ten forest tree species. *Acta Hort.* 560:531-534.

Redenbaugh, K., P. Viss, D. Slade and J.A. Fujii. 1987. Scale up:artificial seeds. In: *Plant Cell and Tissue Culture.* pp 473-493 (eds. Green, C.E., D.A. Somers, W.P. Hackett and D.D. Bisboer). New York : A.R.Liss.

Rival, A., J. Tregear, J.L. Verdeil, F. Richard, T. Beale, C. Hartmann, and A. Rode A. 1998a. Molecular search for mRNA and genomic markers of the oil palm "mantled" somaclonal variants. *Acta Hort.* 461:165-171.

Rival, A., L. Bertrand, T. Beale, M.C. Combes, P. Trouslot and P. Leshermes. 1998b. Similarity of RAPD analysis for detection of somaclonal variation in oil palm (*Elaeis guineensis* Jacq.) *Plant Breed.* 117:73-76.

Rival, A., E. Jaligot, T. Beale, J.L. Verdeil and J. Tregear. 2000. DNA methylation and somaclonal variation in oil palm. *Acta Hort.* 530:447-454.

Saiki, R., S. Scharf, F. Faloona, K.B. Mullis, G.T. Horn, H.A. Erlich and N. Arnheim. 1988. Enzymatic amplification of β-globin genomic sequences and restriction site analysis for diagnosis of sickle cell anemia. *Science.* 230-1350-1354.

Schoofs, J. 1992. Rapid clonal propagation, storage and exchange of *Musa* species. In: *Rapid Propagation of Fast Growing Woody Species.* pp. 29-41 (ed. Baker, F.W.G.) Wallingford: CAB International.,

Scowcroft, W. R. 1985.Somaclonal variation: the myth of clonal uniformity. In: *Genetic Flux in Plants.* pp. 217-245 (eds. Hohn, B., E.S. Dennis) Berlin: Springer Verlag.

Shaw, F.H., O. Rashid, A.J. Simons and A. Dunson. 1994. The utility of RAPD markers for the determination of genetic variation in oil palm (*Elaeis guineensis*). *Theor Appl Genet.*89:713-718.

Shimron-Abarbanell, D. and A. Breiman. 1991. Comprehensive molecular characterization of tissue-culture-derived *Hordeum marinum* plants. *Theor Appl Genet.* 83:71-80.

Shirzadegan, M., J.D. Palmer, M. Christey and E.D. Earle. 1991. Patterns of mitochondrial DNA instability in *Brassica campestris* cultured cells. *Plant Mol Biol.* 16:21-37.

Shoseyov, O., G. Tsabary and O. Reuveni. 1998. Detection of dwarf somaclones of banana cultivars (*Musa*) by RAPD markers. In: *Somaclonal Variation and Induced Mutation in Crop Improvement.* pp. 595-601 (eds. Jain, S.M., D.S. Brar and B.S. Ahloowalia) Kluwer: Dordrecht.

Tang, W. 2001. *In vitro* regeneration of loblolly pine and random amplified polymorphic DNA analyses of regenerated plantlets. *Plant Cell Rep.* 20:163-168.

Thakur, R.C., S. Goto, K. Ishii and S. M. Jain. 1999. Monitoring genetic stability in *Quercus serrata* Thunb. somatic embryogenesis using RAPD markers. *J For Res.* 4:157-160.

Thorpe, T.A. 1988. *In vitro* somatic embryogenesis. *ISI Atlas of Science-Animal and Plant Sciences.* 1:81-88.

Thorpe, T.A. 1990. The current status of plant tissue culture. In: *Plant tissue culture: applications and limitations.* pp.1-33 (ed. Bhojwani, S.S.) Elsevier:Amsterdam.

Vasil, I.K. 1994. Automation of plant propagation. *Plant Cell Tissue Organ Cult.39:105-108.*

Vogel, J.M. and P.A. Scolnik. 1998. Direct amplification from microsatellites : detection of simple repeat-based polymorphisms without cloning. In: *DNA markers : Protocols, Applications and overviews.* pp. 33-150 (eds. Caetano Annolles, G and P. M. Gresshoff) New York: Wiley-Liss Inc.

Vos, P., R. Hogers, M. Reijnans, T. vande Lee, M. Hornes, A. Friters, J. Pot, J. Peleman, M. Kupier and M. Zebeau. 1995. AFLP: a new technique for DNA fingerprinting. *Nucleic Acids Res.* 23:4407-4414.

Vuylsteke, D., R. Swennen, G.F. Wilson, and E. DeLanghe. 1998. Phenotypic variation among *in vitro* propagated plantain (*Musa* sp cultivar AAB). *Sci Hort.* 36:79-88.

Wang, P.J. and Charles A. 1991. Micropropagation through meristem culture. In: *Biotechnology in Agriculture and Forestry-Hightech and Micropropagation* I. vol 17, pp32-52 (ed Bajaj, Y.P.S.) NewYork: Springer.

Welsh, J and M, McClelland. 1990. Figerprinting genomes using PCR with arbitrary primers. *Nucleic Acids Res.* 18:7213-7218.

Weber, J.L. and P.E. May. 1989. Abundant class of human DNA polymorphisms which can be typed using the polymerase chain reaction. *Am J Hum Genet.* 44:388-396.

Wilhelm, E. 2000. Somatic embryogenesis in oak. *In Vitro Cell Dev Biol- Plant.* 36:349-353.

Williams, J.G.K., A.R. Kubelik, K.J. Livak, J.A. Rafalski and S.V. Tingey. 1990. DNA polymorphisms amplified by arbitrary primers are useful as genetic markers. *Nucleic Acids Res.* 18:6531-6535.

Wallner, E., K. Weising, R. Rompf, G. Kahl and B. Kopp. 1996. Oligonucleotide and RAPD analysis of *Achillea* species: characterization and long-term monitoring of micropropagated clones. *Plant Cell Rep.* 15: 647-652.

Xiong, L.Z., C.G. Xu, M.A. Saghai-Maroof and Q, Zhang. 1999. Patterns of cytosine methylation in an elite rice hybrid and its parental line, detection by a methylation sensitive amplification polymorphism technique. *Mol Gen Genet.* 261:431-446.

Zehr, B.E., M.E. Williams and R.D. Widholm . 1987. Somaclonal variation among the progeny of plants regenerated from callus cultures of seven inbred lines of maize. *Can J Bot.* 65:491-499.

Zietkiwicz, E., A. Rafalski and D. Labvada. 1994. Genome fingerprinting by simple sequence repeats (SSR)-anchored polymerase chain reaction. *Genomics.* 20:176-183.

4. MICROPROPAGATION OF WOODY TREES AND FRUITS: PATHOGEN ELIMINATION AND CONTAMINATION MANAGEMENT

Alan C Cassells and Eileen A O'Herlihy,
Department of Plant Science, National University of Ireland Cork, Ireland
(a.cassells@ucc.ie)

Table of contents

S.M. Jain and K. Ishii (eds.), Micropropagation of Woody Trees and Fruits, 103–128.

1. INTRODUCTION

Losses due to bacterial contamination of cultures and the risk of the release of pathogens in planting material produced by micropropagation are serious treats to the industry (Cassells, 2000a; Leifert and Cassells, 2001). In early conferences on contamination in tissue culture (*e.g.* Cassells, 1988), health status was discussed without reference to microplant quality, however, it was evident from the losses to pathogens at weaning (Williamson *et al.*, 1997) that microplant epigenetic and physiological quality are important in the process. It is now recognised that aseptic microplants are a 'biological vacuum' lacking the biological buffering of a natural microflora against opportunistic pathogens (Cassells, 2000b). This leads to efforts both to improve the physiological quality of microplants *in vitro* and to biotization or inoculation *in vitro* and *post vitrum* with beneficial micro-organisms (Vestberg *et al.*, 2001). Both the health status and genetic, epigenetic, and physiological quality of microplants have been the subject of several conferences in recent years (Cassells 1997; Cassells *et al.*, 2000). An emerging consensus on pathogen and contamination management and protection against damping-off diseases at weaning is that these may be best regarded as three separate phases in the micropropagation process.

In the first phase, pathogen indexing requires knowledge of the crop pathogens and the availability of diagnostics with appropriate sensitivity and specificity (O'Donnell, 2000). For many economically important crops, certain diseases are of unknown etiology; where the causal agent is known, there are no commercially available diagnostics for the micropropagator. With regard to laboratory contamination, the risk is primarily that cultivable contaminants may over-run the cultures. Even organisms not pathogenic to plants in nature may affect plants *in vitro* (so called 'vitropaths') (Herman, 1990a,b) *e.g.*, the human pathogen *E. coli* has been reported to be pathogenic to cabbage microplants *in vitro* (Rafferty *et al.*, 2000). In general, pathogen-contaminated clones pose the greatest threat to the growing-on nursery customer through the release of clonally infected material; where not only the crop itself may be affected, it may also be a source of inoculum for wider dissemination of the disease. Where pathogen-infected microplants cross quarantine zones, this may result in the serious consequence of the introduction of alien pathogens to a region.

It is recognized that, due to uneven distribution of pathogens, sampling for pathogen indexing in *in vivo* material poses technical problems. An associated issue is cyclical variation in virus titer in plant tissues. A further problem, particularly in perennial plants, is the presence of pathogen complexes, usually in the form of multiple infections with viruses or co-infections with virus, viroids and prokaryote pathogens. In many cases, symptoms may be expressed which reflect interactions among pathogens, with positive or negative interactions. The latter is termed 'cross protection'. Cross protection is used practically in citrus, grape etc. (Lecoq, 1998). Suppression of virus titer in complexes poses the risk of detection failure. These issues are discussed in detail in Matthews (1991).

Given the above issue, of critical importance is the availability of diagnostics with appropriate sensitivity and specificity. The general belief is that the higher the sensitivity of diagnostics the better for detection. This has been questioned when the application is for diagnosis of pathogens in the field where environmental influences may be critical in determining whether a disease outbreak will occur (Scott and Bainbridge, 1978), but not when the host is systemically infected *in vitro* with viruses, viroids or fastidious bacteria which may be clonally propagated in a latent form. Arguably, in micropropagation, there should be zero tolerance of pathogens as they may be clonally propagated to proliferate during the

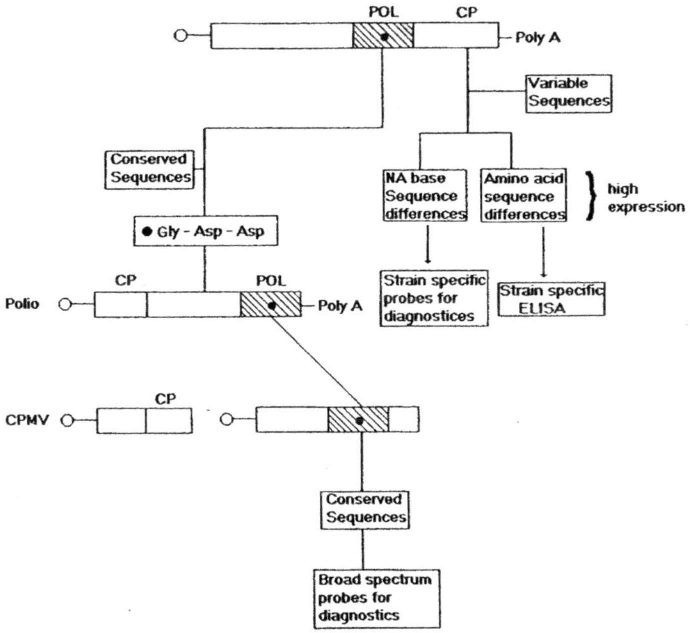

Fig. 1. Diagrammatic representation of Tobacco mosaic virus (TMV) showing the location of the coat protein gene (CP) and the RNA polymerase gene (POL). (*left*) The hatching of POL and the black spot indicate conserved regions that are found in related polio and cowpea mosaic virus (CPMV) that may be used to construct broad spectrum probes for use as diagnostics. (*right*) The CP contains variable sequences expressed in strain differences in coat proteins which can be used to develop strain specific ELISAs; alternatively these differences may be used to develop strain-specific probes (after Matthews, 1991).

subsequent cropping cycle. The question of specificity of diagnostics is also one that micropropagators view in a perspective different from that of field crop producers. In the field, pathogen isolates tend to be 'local' or characterized. It is unlikely that a pathogen will escape detection during vegetative propagation therefore roguing can be used. In micropropagation, pathogen symptoms may not be expressed or may go undetected, ascribed to *in vitro* variation and consequently, the pathogen spreads clonally, in latent form. Also, micropropagators may be working with poorly characterized 'quarantine' pathogens or uncommon pathogen strains (Frison and Diekmann, 1998). Consequently, micropropagators may have a dual need for pathogen strain-specific diagnostics for detection of certified crop pathogens and broad-spectrum diagnostics for the detection of uncharacterized pathogens strains along the lines of quarantine testing (Fig. 1).

Here, disease and contamination elimination and monitoring, and measures to protect against biotic stress at weaning will be discussed in relation to the stages in micropropagation as defined by Murashige (1974) and revised by Debergh and Maene (1981). Sampling for pathogen testing, new approaches to *in vitro* testing and the development of broad-spectrum diagnostics will be discussed with emphasis on woody plants and fruit crops. Guidelines for good laboratory practice based on HACCP (Hazard Analysis Critical Control Points) principles will be introduced. Finally, *in vitro* weaning and biotization with arbuscular mycorrhizal fungi and beneficial bacteria, singly and in combination, to protect against damping-off diseases will be reviewed.

2. GENERAL GUIDELINES FOR PATHOGEN ELIMINATION AND CONTAMINATION MANAGEMENT

- *The first phase is the establishment of pathogen-free, aseptic cultures.* Pathogen detection and elimination should be carried out prior to establishment of cultures. If meristem culture is used to produce pathogen-free cultures, then, progeny plants from these cultures should be established. After pathogen indexed *in vivo, in vitro* multiplication can commence from these indexed plants, or from the clones maintained *in vitro.* The reason for this approach is that the standard diagnostics and diagnostic tests have been validated only for use on *in vivo* material. Their reliability for indexing of *in vitro* material has yet to be confirmed (Cassells, 2000a; O'Herlihy *et al.*, 2001). The procedure of *in vivo* indexing must also be applied where thermo- or chemotherapy has been used *in vitro* for pathogen elimination.
- *The second phase is concerned with maintaining contaminant-free cultures by using good working practices.* This can be

achieved by applying HACCP principles to production (Kvenberg *et al.*, 2000). In addition to providing training and equipment for workers, this involves the routine servicing of laboratory equipment to the manufacturers' specifications and regular inspection of the cultures for signs of bacterial and fungal infection and infestation by micro arthropods. Cultures should be routinely indexed for latent bacterial infection. Contaminant organisms that indicate specific problem areas have been reported (Leifert and Waites, 1994; Leifert and Cassells, 2001).

- *The third phase is concerned with development of the protection of microplants against damping-off disease and can extend to bioprotection against diseases of the crop.* This is based on optimisation of the physiological status of the microplants (Cassells, 2000b) and biotization with beneficial organisms singly, or in combination to establish a beneficial rhizosphere (Vestberg *et al.*, 2002). This may also involve applications of elicitor fungicides to the microplants *in vitro* or at weaning to protect against airborne pathogens (Cassells, 2000b).

These guidelines are summarized in Table 1.

Table 1. The phases of pathogen and contaminant management in micropropagation.

Phase 1. Establishment of an aseptic culture

Stage 0
- Selection of vigorous stock plants
- Seek pathogen escape; index for pathogens
- Thermotherapy followed by re-indexing as necessary

Stage 1
- Meristem culture; index for bacterial contamination
- Meristem culture for pathogen elimination; regenerate plants for indexing *in vivo*
- Antiviral, antibacterial chemotherapy; regenerate plants and index for pathogens *in vivo*

Phase 2. Aseptic cloning

Stage 2
- Monitor for contamination using HACCP guidelines

Phase 3. *In vitro* weaning and biotization.

Stage 3
- Transfer cultures to vessels with vented lids or bottom cooling
- Transfer cultures to autotrophic culture; inoculate with beneficial organisms

Stage 4
- Apply inoculum of beneficial organisms to the microplant roots or put in plant hole in growth substrate

3. PHASE 1 - INDEXING OF STOCK PLANTS

3.1 Training and resources

Assuming a basic training in plant pathology, the first step in pathogen indexing is crop intelligence. Electronic databases have revolutionized access to information, including information on pests and diseases of crops. Important websites, which usually have search facilities, are those of international publishers of books and journals in plant pathology, *e.g.*, Kluwer (www.wkap.nl), internet marketing companies, *e.g.*, Barnes & Noble (www.bn.com), and international organizations such as the FAO (www.fao.org) and the American Phytopathological Society (www.apsnet.org). Details of crop certification schemes and approved indexing procedures can be obtained from the European and Mediterranean Plant Protection Organization (www.eppo.org). Links from the latter sites and literature searches on the ISI Web of Science (www.isi.com) and other databases also provide relevant information and usually contact addresses for specialists actively working on specific or related diseases. On-line electronic databases, *e.g.*, Science Direct (www.elseviersciencedirect.com) can be rapid and very efficient but, unfortunately, access to many of the services is expensive for micropropagation companies. These may find it more cost effective to seek advice from university departments or state agricultural extension services.

3.2 Diagnostic procedures

Biological testing by mechanical, graft and vector transmission to indicator plants remains important in the detection and indexing of plant pathogens (Agrios, 2001; EPPO, 1998). Where cultivable organisms are concerned, Koch's postulates should be confirmed. In the case of confirmation of known-pathogen identity, diagnostic features should be confirmed within the recognized constraint of the cost effectiveness of the screening. While serological tests based on ELISA are widely used, it is good practice to confirm negative ELISA results with biological testing because of uneven pathogen distribution in the test tissue, seasonality and *in vitro* suppression, and possible pathogen titer/counts below the threshold for ELISA detection (Fig. 2; O'Herlihy *et al.*, 2001). Transmission, particularly by grafting, allows biological amplification of the pathogen to occur.

A number of serological methods exist such as precipitin test, latex aggregation and ELISA (Dijkstra and de Jager, 1998). The latter is widely used. The procedure can be partially automated and the results read optically. A key feature in ELISA testing is the setting of the positive/negative threshold (Fig. 2). This is dependent on having appropriate positive and negative controls. Serological electron microscopy can be a useful technique in research but as with *in situ* examination of tissues, it involves high capital cost and skilled labor (Hari and Das, 1998).

In addition to their contribution to classification and thus clarification of the relations between organisms, DNA techniques, supported by bioinformatics, have the potential to resolve all the issues in diagnosis although this may take some time to be effective. The problems that exist relate to sample preparation, design of probes, and automation. These factors and intellectual property costs of patent fees restrict application of the technology (O'Donnell, 2000). The potential of DNA diagnostics is in: 1) ability to design either broad spectrum probes against conserved genomic sequences, or specific probes that can be used to identify quarantine pathogens (Fig. 1) or, in the case of bacteria and viruses, pathogenic isolates; and 2) ability to carry out multiplex assays for a number of pathogens in a single test (Candresse *et al.*, 1998). In the future biosensors may overcome the above difficulties but, currently, although limited in practical application, research has confirmed the potential of DNA diagnostics.

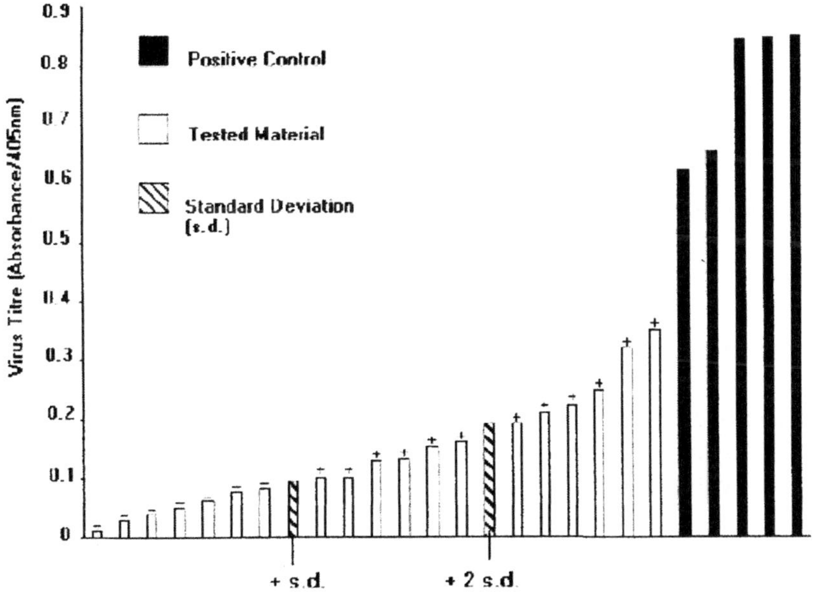

Fig. 2. Setting the positive/negative threshold in ELISA. The data are for the ELISA of apple chlorotic leaf spot virus (ACLSV) in *Malus* microplants *in vitro*. The results were validated by PCR. The data show the importance of setting an appropriate threshold or cut-off point. Values greater than negative mean + 1 standard deviation were tested positive and were confirmed as positive for virus by PCR. Values below 1 s.d. were all negative (from O'Herlihy *et al.*, 2001).

3.3 The influence of in vitro conditions on virus titer

A major factor in pathogen testing is the prejudice of the certification authorities that ELISA testing, and especially testing of *in vitro* material, may not be reliable (see EPPO, 1998). The time taken to index plants *in vivo* (that is to establish and grow into microplants) and the isolation facilities required are prohibitively expensive for many micropropagators. Part of the problem lies in the myth that virus titer may be low and its distribution uneven in *in vitro* tissues. Certainly, media hormones and the type of culture (heterotrophic versus autotrophic) may influence virus titer *in vitro.* Virus titer may be higher and its distribution less uneven in *in vitro* culture than in *in vivo* tissue as indicated by recent results (Fig 3; O'Herlihy *et al.*, 2001).

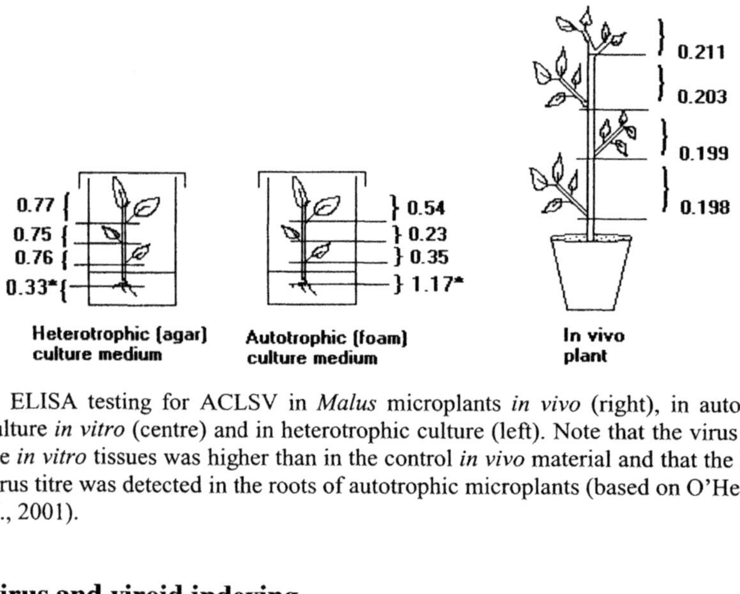

Fig. 3. ELISA testing for ACLSV in *Malus* microplants *in vivo* (right), in autotrophic culture *in vitro* (centre) and in heterotrophic culture (left). Note that the virus titre in the *in vitro* tissues was higher than in the control *in vivo* material and that the highest virus titre was detected in the roots of autotrophic microplants (based on O'Herlihy *et al.*, 2001).

3.4 Virus and viroid indexing

In approaching the problems of indexing plants for viruses, it is important to make the distinction between known disease indexing for which diagnostics are available and indexing for quarantine pathogens for which diagnostics may lack specificity and for viruses present in complexes for which titer of one virus may be affected by the presence of others (Matthews, 991). There is, also, the roblem of uneven virus distribution in tissues and cyclic variation in virus titer and in symptom expression (Matthews, *loc. cit.*). For decades, plant virologists were dependent on indicator species for virus identification. These gave distinctive symptoms upon inoculation with specific virus/virus strains. Transmission was either by sap inoculation (mechanical inoculation) or grafting or dodder transmission. Such methods were also valuable in separating virus complexes in cases in which a host plant was immune to one virus but susceptible to other(s). Propagation species (hosts in which a specific virus replicate to a high titer) were also important in building up virus material for antiserum production or biochemical and biophysical analyses. Indicator plants still have an important role in support of modern techniques in which they may be used to confirm negative results. Virus indexing has been reviewed by Noordam (1973); Matthews (1991); Dijkstra and de Jager (1998) and Hadidi *et al.* (1998).

Detection of known viruses is now usually based on ELISA (enzyme linked immunosorbent assay) (Clark and Adams, 1977). ELISA has had wide acceptance and is relatively inexpensive, requiring little capital investment apart from a plate reader and a micro-pippeting system. It has the disadvantage of being relatively strain specific and, as some suggest, of relatively low sensitivity compared to DNA-based diagnostics (O'Herlihy *et al.*, 2001). It also can suffer the interference of sap components such as tannins. There are many suppliers of ELISA kits, *e.g.*, Agdia (www.agdia.com). Many plant pathologists construct ELISA kits from in-house raised antibodies and bought-in enzyme-linked secondary antibody. Suppliers of the latter include Amersham (www.amersham.com).

DNA-based diagnostics offer several advantages over the ELISA detection method. Firstly, the probes can be against conserved or variable regions of the virus genome (Fig. 1). This offers the potential of strain specific or non-strain specific detection, *i.e.*, of certification strains or possible members of the virus family (Fig. 1). The latter 'quarantine' diagnostics are very valuable when virus strains are poorly characterised, in order to avoid the problem of virus escape. Unlike ELISA in which each test has to be carried out independently, it is possible to carry out 'mutiplex' or multiple assays on the same sample by using several virus-specific probes (Candresse *et al.*, 1998). The use of amplification steps can give a higher sensitivity as compared to ELISA. The disadvantages of DNA diagnostics are the high cost (both high consumables and labour skills) and the interference of sap components. This may require sample purification and the use of a reverse transcriptase step because most plant viral genome are of RNA (O'Donnell, 2000; Candresse *et al.*, 1998).

Like viruses, viroids were indexed by transmission to indicator species. This was superseded by polyacrylamide gel electrophoresis. The principle of this method is that nucleic acids are extracted from the infected tissues and digested with nucleases. The viroid, which has a double stranded RNA genome, is resistant to enzymatic digestion and is detected as a band on the gels. This method is relatively insensitive and was rapidly replaced by DNA probe technology (Singh and Dhar, 1998). Bioinformatics has transformed detection and identification by making available sequence data on a wide range of viruses and viroids that can be used in probe design (see *e.g.* www.embl-heidelberg.de).

Serological electron microscopy (ISEM - immunosorbent electron microscopy) is also used in virus testing but requires high capital investment and, consequently, is not generally available to micropropagators except as a service (Hari and Das, 1998; Dijkstra and de Jager, 1998).

3.5 Bacteria including phytoplasma indexing

While viruses are intracellular, bacteria may be intercellular and/or intracellular (endophytic). An important aspect is whether the bacteria are cultivable or fastidious (Lelliot and Stead, 1987). If cultivable, they can usually be detected by plating the sap produced on common bacteriological media (Schaad *et al.*, 2001). Leifert and Waites (1994) have proposed a broad-spectrum medium for this purpose, which is commercially available. However, not all potentially cultivable bacteria are detected in sap, possibly due to a combination of low colony forming units (cfu) in the tissues and/or because of inhibition by tannins or other compounds in the sap. It may be more reliable to place explants for testing on bacteriological media. In this way, penetration of nutrients into the tissue may produce an enrichment effect and bacteria number may increase to be expressed and grown on the medium.

If required, *e.g.*, for certification purposes, bacteria may be characterised by conventional keying out on bacterial media and grown on selective media, followed by pathogenicity testing by inoculation on indicator species as for viruses and viroids (Lelliot and Stead, 1987). Serological testing, *e.g.*, by ELISA, for bacteria is less common and arguably less reliable than for viruses, due to difficulty in obtaining specific antiserum and to the occurrence of many non-pathogenic isolates in the environment. Miniaturised biochemical tests, *e.g.*, API kits (www.biomerieux.fr), provide relatively rapid testing but are limited by the lack of reference data for plant pathogenic isolates in databases. They require a preliminary test to identify the appropriate test kit and inoculation of indicator species to confirm pathogenicity (Cassells, 2000a). Another identification method, available also as a service, is fatty acid profiling. This method has an advantage over biochemical test kits in that specific fatty acids may be taxonomic markers and hence, even if the organism is not in the data base, taxonomic clues may be provided to guide further screening (Stead *et al.*, 2000).

One of the greatest successes of genomics has been in the analysis of bacterial genomes (Brown and Koretke, 2000). The relationships between fastidious bacterial like organisms (molicutes) and Eubacteria have been confirmed and probes are now available to detect and identify fastidious bacteria (Bove and Garnier, 1997). There remains the problem, in some cases, of confirming pathogenicity due to difficulties in transmission of the organisms to indicator species. Other techniques available for bacterial detection and identification include *in planta* localization by serological specific staining or detection by non-specific nucleic acid staining. In plant, localization may also be carried out using transmission electron microscopy (TEM). These techniques are not used in commercial practice (Hari and Das, 1998). Bacterial indexing has been reviewed by Lelliott and Stead, 1987; Stead *et al.*, 2000; Schaad *et al.*, 2001.

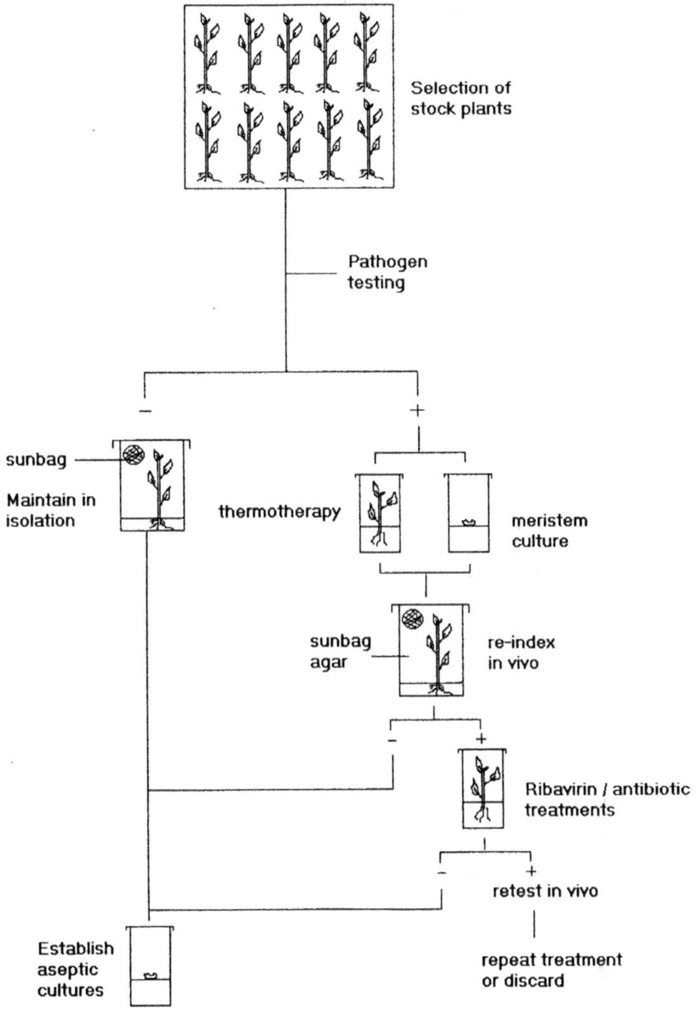

Fig. 4. Scheme for production of pathogen tested plants via micropropagation for certification standards. Note: the key step is testing plants for pathogens *in vivo*, following thermotherapy, meristem culture or *in vitro* thermo- or chemo-therapy. While ELISA may be used as a preliminary screen, certification requires specified *in vivo* tests. (See EPPO, 1998).

3.6 Fungal detection and identification

Usually, fungal contamination is not a serious problem in plant tissue culture. Fungal symptoms are expressed on the stock plant and, usually, do not extend to the plant apex. Therefore fungi are eliminated when only meristem explants are used to initiate the cultures. Fungi can occur during phase 2 as laboratory contamination. The fungal contaminants most frequently encountered are yeasts (red yeasts) (Leifert *et al.*, 1990) and common environmental fungi including *Penicillium, Aspergillus* spp. (blue moulds), and *Fusarium* (pink mould) (Agrios, 2001). Fungal endophytes occur widely in nature (Saikkonen *et al.*, 1998) and have been isolated from *Musa* sp. and re-introduced into 'axenic' plants *in vitro* (Pereira *et al.*, 1999). The possibility of latent endophytic fungal contamination of *in vitro* cultures warrants investigation and should be taken into consideration by micropropagators.

3.7 Pathogen elimination

Three strategies are employed to obtain pathogen free *in vitro* cultures; namely, seeking escapes, applying thermotherapy to the stock plant *in vitro*, meristem culture and by thermo- and chemo-therapy *in vitro* (Fig. 4). For woody plants, thermotherapy is widely used. For reviews see heat treatment and meristem culture (Dijkstra and de Jager, 1998; Faccioli and Marani, 1998; Mink, 1998) and antibacterial and antiviral chemotherapy (Leifert *et al.*, 1991a; Barrett and Cassells, 1994 and Cassells, 1998).

3.7.1 Strategy 1 – Thermotherapy of stock plants

Thermotherapy involves exposing plant propagules, (cuttings, budwood etc.) to temperatures of approx. 50°C for a few minutes to hours or exposing growing plants to air temperatures of 35-40°C for periods of weeks (Dijkstra and de Jager, 1998; Mink, 1998). Thermotherapy may also be applied to *in vitro* cultures. This method is, principally, used to eliminate viruses. It has been assumed that virus elimination results from the effects of elevated temperature on virus particle stability and/or effects on viral replication. However, caution should be exercised in accepting the results of virus testing of the progeny plants. There are indications from *in vitro* thermotherapy that virus may be suppressed (virus titer may be reduced) below the threshold for detection in standard ELISA tests, giving false negatives for *in vitro* tested material. This virus can re-emerge in the grown-on material (Fig. 4; Roistacher, 1998). This has serious consequences where mass clonal propagation *in vitro* step is involved.

3.7.2 Strategy 2 - Pathogen elimination by meristem tip culture

Meristem culture was applied for the elimination of pathogens from plants, subliminally, also eliminating endophtyic contaminants (Fig. 4). In general, the smallest apical explant (the apex plus first pair of leaf primordial) compatible with explant survival is used. This is preferably placed on a bacterial expression medium such that bacterial contaminants may also be detected at this stage (Menard *et al.*, 1985). In the case of woody plants and some fruit crops, meristem grafting has been used to improve the establishment of the excised meristem (for reviews, see George, 1993; 1996; Dijkstra and de Jager, 1998).

3.7.3 Strategy 3 – In vitro therapy

Both chemo-and thermo-therapy (see Strategy 1 above) can be used with *in vitro* cultures. In thermotherapy the treatment temperatures and temperature cycles are determined empirically. When using antimicrobial compounds it is important to recognise that most are biostatic rather than biocidal and that elimination may depend on excising new tissue growth and on serial subculture on the chemical. Protocols for using the antiviral chemical Ribavirin are discussed in Cassells (1998). Falkiner (1998) has discussed the selection criteria for antibiotics for *in vitro* use. Barrett and Cassells (1994) have described the procedure for screening antibiotics and for determining inhibitory concentrations. Falkiner (2000) has warned about the risks of developing antibiotic resistance bacterial isolates when using antibiotics in tissue culture.

3.8 Pathogen testing following thermotherapy or meristem culture

When stock plant tissues have been exposed to *in vivo* thermotherapy, the plants can be tested by using conventional tests as recommended e.g. by the EPPO (1998). When pathogens have been eliminated by meristem culture or/and *in vitro* thermotherapy, caution should be exercised. Diagnostic tests have not been validated for use on *in vitro* tissues and may give false positives for a number of reasons. Firstly, the hormones used in the culture medium may suppress virus/pathogen titer (Quak, 1977; O'Herlihy *et al.*, 2001). Secondly, microplants *in vitro* may be stressed and express induced systemic resistance or the effect of other mechanism that suppress virus titer (Cassells and Curry, 2000). For these reasons, it has been argued that samples of production should either be grown on and tested under standard *in vivo* conditions or, as microplants, grown under standardized *in vitro* conditions *e.g.* under autotrophic conditions (Fig. 3; O'Herlihy *et al.*, 2001).

Table 2. Critical Control Points (CCPs), indicator microorganisms, equipment and methods used for microbiological contamination control in micropropagation.

Stages of Plant Tissue Culture	Parameters (CCPs) to be monitored	Indicator microorganisms	Equipment needed	References for methods
Stage 0 Stock plant treatment	Symptoms of viroid, viral, bacterial, and fungal pathogens. Detection and identification of viroids, viruses and bacteria and fungi		Hand lens, light microscope Phase contrast microscope; selective bacteriological media; serological tests/DNA probes; indicator plants	**Viruses**: 1, 2 **Bacteria**: 3, 4 **Fungi**: 5, 6
Stage 1-3				
Surface sterilization	Active chlorine content in hypochlorite solutions	Gram-negative bacteria: *Pseudomonas fluorescens* *Erwinia* spp., *Klebsiella* spp., *Serratia* spp., *Agrobacterium* spp.	Whatman indicator paper, arsenite solution, titration column	7
	Activity of other biocides		Depending on compound	8-11
Indexing	Presence of viruses, bacteria, and fungi on explants after surface disinfected		Indexing media	3, 4, 12
			Serological tests	1-4
			Nuclei acid hybridization tests	1-4

Table 2 cont.

Stages of Plant Tissue Culture	Parameters (CCPs) to be monitored	Indicator microorganisms	Equipment needed	References for methods
Media preparation	3.1 Autoclaving	*Bacillus* spp. (e.g. *B. circulans, B. subtilis, B. pumilus*)	Physical methods (autoclave tape, temperature indicators)	13,14
			Biological methods (e.g. spore strips)	
	3.2 Media pouring	*Penicillium* spp.	"Sterility" testing of media	
		Candida spp.	Testing of laboratory air and laminar flow (as described below)	15,16
		Staphylococcus spp.		
Subculturing of plants in the laboratory	4.1 Air "sterility"	*Rhodotorula* spp. (pink yeasts)	Air sampler and/or particle counter	16,17
	4.2 Laminar flow cabinets	*Penicillium* spp.	Particle counter	17
	4.3 Instrument sterilisation	*Bacillus* spp.	Indexing media	17
	4.4 Aseptic technique of operators	*Staphylococcus* spp., *Streptococcus* spp., *Candida albicans*	Recording of individual contamination rates	
Storage of media and plants	As 4.1	As 4.1	As 4.1	As 4.1
	5.1 Micro arthropods	'spore trails' (see Fig. 6)	Light microscope	18
Weanings of plants	As in stage 0	As in stage 0	As in stage 0	As in stage 0

Note: (1) Hill (1984); (2) Matthews (1991); (3) Lelliott and Stead (1987); (4) Schaad *et al.* (2001); (5); Smith *et al.* (1986); (6) Hoffman and Schmutterer (1983); (7) Hoffman *et al* (1981); (8) Collins *et al.* (1981); (9) Hugo (1971); (10) Russell *et al.* (1982); (11) Sykes (1965); (12) Leifert *et al.* (1991); (13) Kelsey (1961); (14) Anonymous (1982); (15) Danby *et al.* (1994); (16) Gregory (1973); (17) Leifert *et al.* (1991). (18) George (1993). Based on Leifert and Waites, 1994; redrawn from Leifert and Cassells (2001)

3.9 Certification protocols

Phytosanitary legislation exists to regulate and facilitate trade in plant propagation material both nationally and internationally. An important aspect of certification is that standard validated tests be carried out. Models of such tests can be found in the literature published by the European and Mediterranean Plant Protection Organization (www.eppo.org). These schemes involve the following stages:

- Selection of elite individual plants
- Selection for virus-escapes among plants using approved testing protocols
- Elimination of viruses and other pathogens from plants by heat therapy or meristem culture: testing of the progeny plants using approved tests (*e.g.* EPPO, 1998).
- Maintenance of nuclear stock under conditions designed to prevent re-infection, with re-testing as appropriate
- Multiplication of nuclear stock (propagation stock) under conditions ensuring freedom from re-infection, with pathogen testing as appropriate
- Production of pathogen-tested plants (certified stock under strict official control)
- Release of certified plants from pathogen-tested plants. These plants should satisfy national and international quarantine regulations.

The key elements are that approved pathogen testing schemes be used and, where micropropagation is used for mass propagation, that the cloning protocol maintains genetic stability. Details of such testing methods are published by the EPPO in the Bulletin OEPP/EPPO Bulletin *e.g.* the certification scheme for strawberry and woody plants (EPPO, 1998). The recommended testing may involve grafting to indicator plants and allow ELISA testing. For example, production of fruit trees for certification allows screening of candidate material by ELISA (assay result in hours) and grafting to herbaceous material (approx. 20 days for

symptom development) but testing on woody indicator plants *in vivo* (average time 12 weeks) is necessary for certification of nuclear stock (Fig. 5). Clearly, the long-term nature of *in vivo* testing present the micropropagator with logistical problems. Currently, molecular diagnostic tests are not generally recommended for other purposes than screening of candidate material. EPPO guidelines are updated, based on recommendations from the International Society for Horticultural Science working group on Fruit Tree Viruses. This group meets every three years and its proceedings are published in Acta Horticulturae (see www.ishs.org).

4. PHASE 2 - LABORATORY CONTAMINATION MANAGEMENT

4.1 Good laboratory practice

Leifert and Waites (1994) are to be credited with introducing the concept of hazard and critical control point (HACCP) analysis to micropropagation. HACCP is used in support of good manufacturing practice in industry and helps to focus micropropagators on issues related to maintaining contaminant-free production. The following are the critical control points to be monitored: in Stage 0, the stock-plant health (see above) and, in Stages 1-3, autoclaving, media preparation, media pouring, air quality, laminar flow cabinet efficiency, instrument sterilisation and aseptic technique of the operators. Leifert and Waites (1994) identified indicator organisms whose presence as contaminants indicates failures in the latter procedures. The indicator organisms and equipment needed are given in Table 2 for each of the stages in micropropagation.

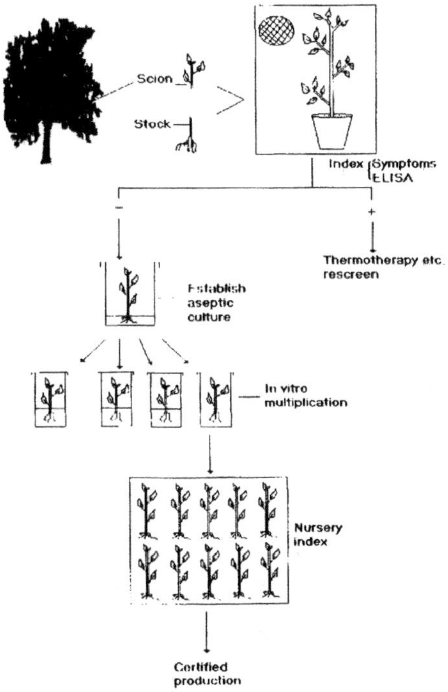

Fig. 5. Pathogen testing of woody plants. Buds or bud wood is grafted to indicator species and tested for virus. Symptomatology or ELISA may be used in pre-screening to reduce the throughput of infected material. Material testing negative may then be micropropagated but is returned to vector-proof isolation facilities for further approved testing and then released for certified production.

4.2 Monitoring of cultures for biological contamination

An important aspect of monitoring micropropagation production is sampling for contamination. The indicator organisms in Table 2 are expressed and their identity confirmed. However, there are contaminants that are not clearly expressed on plant tissue culture media, because of inhibition either by the medium or the medium components (Cassells *et al.*, 1988). These organisms may be clonally propagated at subculture to over-run the cultures when culture conditions are changed e.g. on transfer to a rooting medium. To detect these organisms, routine indexing on bacteriological media is recommended (Lelliot and Stead, 1987). When sampling production for bacterial indexing, it should be remembered that Poisson distribution does not apply. Production should be sequentially arranged on the growthroom shelves. Random sampling may be applied but, when a contaminated culture is detected, production up and down stream of the contaminated vessel should be screened to detect the limits of contamination. This is because all of a production batch may be contaminated after a failure of good operator practice (Cassells, 2000a).

4.3 Culture vessel design and contamination

Plant tissue culture is carried out in a range of culture vessel types. These range from inexpensive glass and plastic food containers (George, 1993) to partial submersion systems (Teisson and Alvard, 1995) and bioreactors (Ziv et al., 1998). Open vessels, which allow free gaseous exchange with the atmosphere, also provide entry points for spores and spore-bearing micro arthropods from the environment (Cassells, 2000a). Fluctuations in day and night temperatures (i.e. lights on/off) in culture vessels may cause a vacuum to develop in the vessel, which can suck in air with contamination from the growthroom. Micro arthropods are a special problem in growth rooms as they migrate and can spread contamination widely (Fig 6; Pype *et al.*, 1997). Equally, it can be

Fig. 6. Spore trails on the medium indicating the presence of contamination with micro arthropods

difficult to maintain aseptic conditions in large vessels as, once a contaminant enters the system, it can spread unhindered (Cassells, 1991). Several culture vessel design strategies are used to control laboratory contamination. Firstly, to maintain good microplant physiological quality in sealed vessels, these cultures may be kept on shelves with bottom cooling or in vessels either lidded with vented closures or wrapped in gas permeable membranes e.g. polyvinyl chloride ('cling wrap') with the appropriate permeability characteristics (Cassells, 2000b). Secondly, plants can be grown in bioreactors with bacteria filters on line in circulating the nutrient medium (Ziv *et al.,* 1998). Finally, plants can be grown in autotrophic culture ('aseptic microhydroponics') (Cassells, 2000b). When environmental microorganisms enter the latter systems, growth is limited by the availability of nutrients, which are mainly derived from root leakage and so the contaminants do not build up to inhibitory levels. Beneficial contamination has also been reported in heterotrophic cultures but this is exceptional as most environmental contaminants tend to over grow the cultures (Cassells, 2000b).

5. PHASE 3 - MICROPLANT WEANING AND BIOTIZATION

5.1 Microplant physiological quality and disease susceptibility

Losses to damping-off diseases at weaning are unquantified but have been reported (Williamson *et al.*, 1997). The latter disease is expressed in micropropagation. The greater the economic consequences and so considerable effort is made to prevent such losses. A complicating factor is that microplants may respond adversely to common pesticides used to prevent damping-off in seedlings (Werbrouck *et al.*, 1996). Several strategies have emerged to prevent losses due to disease at weaning namely, attempts to improve the physiological quality of microplants *in vitro* and the establishment of special nurseries for microplant weaning. Improvement of microplant quality is based on strategies of both media and environmental manipulations to improve the physiological functioning of the microplants or growth and transfer to autotrophic culture for *in vitro* weaning (Cassells, 2000b). The rational behind the latter approach is based on the development of plant horizontal (polygenic) resistance with ontogeny. Young plants show low resistance to pathogens, especially damping-off. A characteristic of *in vitro* plants is that they may have undergone 'rejuvenation' or 'reinvigoration' and may be especially vulnerable to damping-off (George 1993; Joyce *et al.*, 1999; Joyce and Cassells, 2001).

5.2 Biotization in vitro and in vivo

Biotization with arbuscular mycorrhizal fungi (AMF) and/or plant growth promoting or biocontrol bacteria has become the subject of considerable research in recent years. The aim is to block potential niches for colonization by facultative pathogens and to protect against weaning stresses and antagonist pathogens. Biotization may be carried out *in vitro*

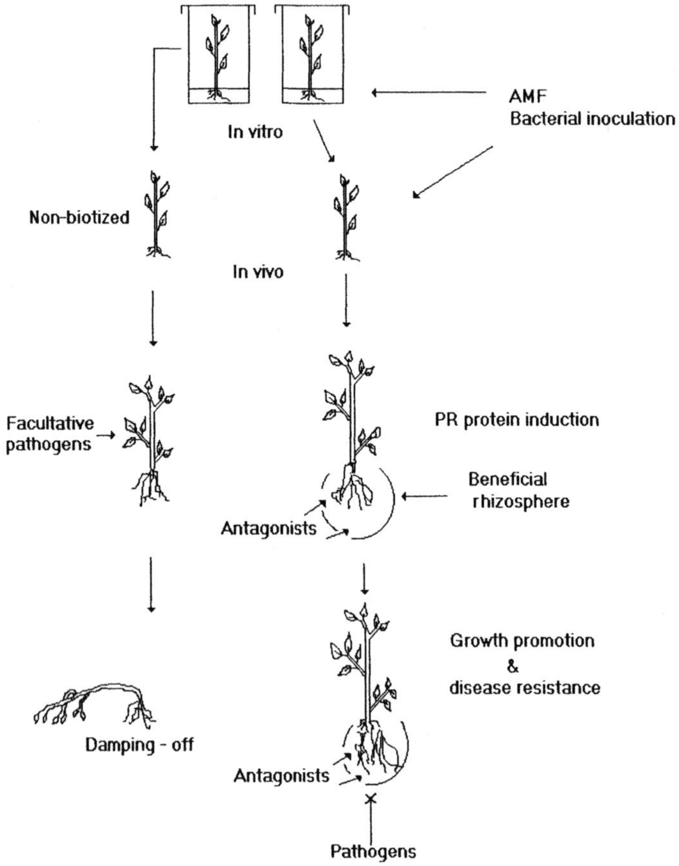

Fig. 7. Establishment of beneficial rhizosphere. Aseptic microplants may be inoculated *in vitro* or at establishment *post vitrum* with beneficial bacteria and/or fungi, singly or in combinations. The objective is to promote microplant growth and to establish a pathogen antagonistic rhizosphere with the objective of preventing losses due to damping-off pathogens. A secondary objective may be to carry over these beneficial effects after transplantation.

(Cassells *et al.*, 1996) but, more commonly, inoculation is done at establishment (Fig. 7; see review by Vestberg *et al.* 2001). Inoculation may be with a single microorganism but, there is an increasing interest in establishing a beneficial rhizosphere by co-inoculating with several microorganisms, usually a combination of fungi (AMF and/or *Trichoderma* sp.) and bacteria (commonly *Pseudomonas* and/or *Bacillus* and/or *Agrobacterium* species) (Fig. 7). The beneficial rhizosphere was the subject of a recent conference (Atkinson, 2000) and has been reviewed by Cordier *et al.* (2000).

The mechanism of action of microbial inoculants has been the subject of considerable research and may consist of one or more elements, namely, competitive exclusion based on niche colonization on the plant, elicitation of host plant defences, modification of plant root signalling compounds, production of inhibitory compounds, and lytic enzymes.

6. CONCLUSIONS

Inevitably, in reviews, generalizations are presented. The reality can sometimes be more difficult to cope with. Here, the emphasis has been on the importance of trying to resolve the problem of contamination in three discrete phases. The key issue in phase 1 is the confirmation of the establishment of aseptic redundant cultures as when using micropropagation there should be zero tolerance to the presence of pathogens that may be undetected *in vitro* and clonally propagated. It has been emphasized that progress in *in vitro* pathogen testing may result from standardization of the *in vitro* material *e.g.* by culture under autotrophic conditions which avoid many of the *in vitro* media and environmental variables (O'Herlihy *et al.*, 2001), and by the development of more sensitive DNA diagnostics which may allow multiplex screening. Laboratory contamination management is based on good laboratory practice and the implementation of HACCP guidelines. The justification for biotization in micropropagation, on theoretical grounds, is convincing but economic cost has to be absorbed. If the consumer can be convinced of continuing benefit during the growth of the crop, then, he/she may be willing to accept the additional propagule costs.

One underlying factor that may lead to false negatives in pathogen testing *in vitro* and/or after thermotherapy stress the induction of host resistance to pathogens and contaminants by stress (Cassells and Curry, 2000). This may also explain how microplants of weak appearance survive transplantation and damping-off diseases in general. This area warrants further investigation.

7. REFERENCES

Agrios, G.N. (2001). Plant Pathology. Academic Press. London, UK.

Anonymous (1982). The Oxoid Manual, Turnergraphic Ltd, Basingstoke, UK

Atkinson, D. (ed.) (2000). The establishment of the beneficial rhizosphere. Appl. Soil Ecol. 15: 99-240

Barrett, C. and Cassells, A.C. (1994). An evaluation of antibiotics for the elimination of *Xanthomonas campestris* pv. *pelargonii* (Brown) from *Pelargonium* x *domesticum* cv. Grand Slam explants *in vitro*. Plant Cell Tiss. Org. Cult. 36:169-175

Bove, J.M. and Garnier, M. (1997). Walled and wall-less Eubacteria from plants: sieve-tube-restricted plant pathogens. In: Pathogen and Microbial Contamination Management (A. C. Cassells, Ed.) Kluwer, Dordrecht, The Netherlands pp. 45-60

Brown, J.R. and Koretke, K.K. (2000). Universal trees: discovering the archaeal and bacterial legacies. In: Applied Microbial Systematics (Priest, F.G. and Goodfellow, M., Eds.). Kluwer, Dordrecht, The Netherlands pp. 19 – 56.

Candresse, T., Hammond R.W. and Hadidi, A. (1998). Detection and identification of plant viruses and viroids using polymerase chain reaction. In: Plant Virus Disease Control (Hadidi, A., Khetarpal, R.K., Koganezawa, H., Eds.). APS Press, St. Paul, USA pp. 399-416

Cassells, A.C. (ed.), (1988). Bacterial and Bacteria-like Contaminants of Plant Tissue Cultures. Acta Hortic. 225

Cassells, A.C. (1991). Control of contamination in automated plant propagation. In: Cell Culture and Somatic Cell Genetics of Plants. Vol. 8 (ed. Vasil, I.K.), Academic Press, New York, USA pp. 197-212

Cassells, A.C. (ed.), (1997) Pathogen and Microbial Contamination Management in Micropropagation. Kluwer Academic Publishers, Dordrecht, The Netherlands.

Cassells, A. C. (1998). Factors relevant to the use of Ribavirin to eliminate viruses *in vitro*. In: Reuther, G. (ed.) Physiology and Control of Plant Propagation *in Vitro*. CEC, Luxembourg, pp. 122-126

Cassells, A.C. (2000a). Contamination detection and elimination. Encyclopedia of Plant Cell Biology. Wiley, Chichester, pp. 577-586

Cassells, A.C. (2000b). Aseptic microhydroponics: a strategy to advance microplant development. Acta Hortic. 530: 187-194

Cassells, A.C. and Curry, R.F. (2000). Oxidative stress and physiological, epigenetic and genetic variability in plant tissue culture: implications for micropropagators and genetic engineers. Plant Cell Tiss. Org. Cult. 64: 145-156

Cassells, A. C. Mark, G. L. and Periappuram, C. (1996). Establishment of arbuscular mycorrhizal fungi in autotrophic strawberry culture, *in vitro*: Comparison with modulation of microplants *in vivo*. Agronomie 16: 625-632

Cassells, A.C., Harmey, M.A., Carney, B.F., McCarthy, E. and McHugh, A. (1988). Problems posed by cultivable bacterial endophytes in the establishment of axenic cultures of *Pelargonium* x *domesticum*: the use of *Xanthomonas pelargonii*- specific ELISA, DNA probes and culture indexing in the screening of antibiotic treated and untreated donor plants. Acta Hortic. 225: 153-161

Cassells, A.C., Doyle, B.M. and Curry, R.F. (Eds.) (2000). Methods and Markers for Quality Assurance in Micropropagation. Acta Hortic. 530

Clark, M.F. and Adams, A.N. (1977). Characteristics of the microplate method of enzyme-linked immunosorbent assay for the detection of plant viruses. J. Gen. Virol. 34: 475-483

Collins, C.H., Allwood, M.C., Bloomfield, S.F. and Fox, A. (1981). Disinfectants; Their use and evaluation of effectiveness. Academic Press, London, UK.

Cordier, C., Lemoine, M.C., Lemanceau, P., Gianinazzi-Pearson, V. and Gianinazzi, S. (2000). The beneficial rhizosphere: a necessary strategy for microplant production. Acta Hortic. 530: 259-268

Danby, S., Berger, F., Howitt, D.J., Wilson, A.R., Dawson, S. and Leifert, C. (1994). Fungal contaminants of *Primula, Coffea, Musa* and *Iris* tissue cultures. In: Lumsden, P.J., Nicholas, J.R. & Davies, W.J. (eds.). Physiology, Growth and Development of Plants in Culture. Kluwer Academic Publishers, Dordrecht, The Netherlands pp. 397-403

Debergh, P.C. and Maene, L.J. (1981). A scheme for commercial micropropagation of ornamental plants by tissue culture. Sci. Hort. 14: 335-345

Dijkstra, J. and de Jager, C.P. (1998). Practical Plant Virology. Springer, Berlin, Germany.

EPPO (1998). Certification Schemes PM4/1-26. European and Mediterranean Plant Protection Organization, Paris, France.

Faccioli, G and Marani, F. (1998). Virus elimination by meristem tip culture and tip grafting. In: Plant Virus Disease Control (Hadidi, A., Khetarpal, R.K., Koganezawa, H., Eds.). APS Press, St. Paul, USA pp. 346-380

Falkiner, F.R. (1988). Strategy for the selection of antibiotics for use against common bacterial pathogens and endophytes of plants. Acta Hortic. 225: 53 – 56.

Falkiner, F. R. (2000). Antibiotics and Antibiotic resistance associated with plants, fruits and vegetables. Acta Hortic. 530: 83 – 86

Frison, E. and Diekmann, M. (1998). IPGRI's role in controlling virus diseases in germplasm. In: Plant Virus Disease Control (Hadidi, A.,

Khetarpal, R.K., Koganezawa, H., Eds.). APS Press, St. Pau, USA pp. 230-236

George, E.F. (1993) Plant Propagation by Tissue Culture: Part 1 – The Technology Exegetics, Basingstoke, UK.

George, E.F. (1996) Plant Propagation by Tissue Culture: Part 2 – In Practice Exegetics, Basingstoke

Gregory, P.H. (1973). The microbiology of atmosphere. Leonard Hill Books, Aylesbury, UK.

Hadidi, A., Khetarpal, R.K., Koganezawa, H. (1998). Plant Virus Disease Control (eds.). APS Press, St. Paul, USA.

Hari, V and Das, P. (1998). Ultra microscopic detection of plant viruses and their gene products. In: Plant Virus Disease Control (Hadidi, A., Khetarpal, R.K., Koganezawa, H., Eds.). APS Press, St. Paul, USA pp. 417-427

Herman, E.B. (1990a). Non-axenic plant tissue culture: possibilities and opportunities. Acta Hortic. 280:112-117

Herman, E.B. (1990b). Bacterial contamination in micropropagation. Agricell Rep. 14:41-43

Hill, S.A. (1984). Methods in plant virology. Blackwell Scientific Publications, Oxford, UK.

Hoffmann, G.M. and Schmutterer, H. (1983). Parasitare Krankheiten und Schadlinge an landwirtschaftlichen Kulturpflanzen. Eugen Ulmer Verlag, Stuttgart, Germany.

Hoffman, P.N., Death, J.E. and Coates, D. (1981). The stability of sodium hypochlorite solutions. In: Disinfectants; Their use and evaluation of effectiveness [Collins, C.H., Allwood, M.C., Bloomfield, S.J. and Fox .A.(eds.)]. Academic Press, London, UK pp 77-83

Hugo, W.B. (1971). Inhibition and destruction of the microbial cell. Academic Press, London, UK.

Joyce, S.M. and Cassells, A.C. (2001). Variation in potato microplant morphology *in vitro* and DNA methylation. Plant Cell Tiss. Org. Cult. (in press).

Joyce, S. M., Vaughan, P., McCarthy T. F. and Cassells, A. C. (1999). PCR analysis of bisulphite-modified DNA to investigate developmental variation in micropropagated potatoes (*Solanum tuberosum* L.). Potato Res. 549-558

Kelsey, J.C. (1961). The testing of sterilizers 2 thermophilic spore papers. J. Clin. Path 14:313-319

Kvenberg, J., Syolfa, P., Stringfellow, D. and Garrett, E.S. (2000). HACCP development and regulatory assessment in the United States of America. Food Control, 11, 387-401

Lecoq, H. (1998). Control of plant vitus diseases by cross protection. In: Plant Virus Disease Control (Hadidi, A., Khetarpal, R.K., Koganezawa, H., Eds.). APS Press, St. Paul, USA pp. 33 - 40

Leifert, C. and Waites, W.M. (1994). Dealing with microbial contaminants in plant tissue culture; hazard analysis and critical control points. In: Physiology, Growth and Development of Plants in

Culture [Lumsden, P.J., Nicholas, J.R. and Davies, W.J. (eds.)]. Kluwer Academic Publishers, Dordrecht, The Netherlands pp. 363-378.

Leifert, C., Nicholas, J.R. and Waites, W.M. (1990). Yeast contaminants of micropropagated plant cultures. J. Appl. Bacteriol. 69:471-476

Leifert, C., Ritchie, J. and Waites, W.M. (1991). Contaminants of plant tissue and cell cultures. World J.Microbiol. Biotechnol. 7:452-469

Leifert, C. and Cassells, A. C. (2001). Microbial hazards in plant tissue and cell cultures. In Vitro - Plant 37: 133-138

Lelliot, R.A. and Stead, D.E. (1987). Methods for the diagnosis of bacterial diseases in plants. Blackwell Scientific Publications, Oxford, UK.

Matthews, R.E.F. (1991). Plant Virology. Academic Press, New York, USA.

Menard, D., Coumans, M. and Gaspar, T.H. (1985). Micropropagation du *Pelargonium* a partir de meristems. Mededelingen van de Faculteit Landbouwettenschappen Rijksuniversiteit Gent. 50:327-331

Mink, G.I. (1998). Virus certification of deciduous trees in the United States and Canada. In: Plant Virus Disease Control (Hadidi, A., Khetarpal, R.K., Koganezawa, H., Eds.). APS Press, St. Paul, USA pp. 294-300

Murashige, T. (1974). Plant propagation through tissues. Annu. Rev. Plant Physiol. 25: 135-166

Noordam, D. (1973). Identification of Plant Viruses. Methods and Experiments. PUDOC, Wageningen, The Netherlands.

O'Donnell, K.J. (2000). Molecular Diagnostics: Technical and Economic Issues. Acta Hortic: 530: 39-44

O'Herlihy, E.A., Croke, J.T. and Cassells, A.C. (2001). Influence of cytokinin and sucrose on the titre of ACLSV in *Malus* and PRNSV in *Prunus*: interaction between cytokinin and Ribavirin on virus elimination. Plant Cell Tiss. Org. Cult. (submitted)

Pereira, J.O., Vieira, M.L.C. and Azevedo, J.L. (1999). Endophytic fungi from *Musa acuminata* and their reintroduction into axenic plants. World J. Microbiol. Biotechnol. 15: 43-46

Pype, J., Everaert, K. and Debergh, P.C. (1997). Contamination by micro-arthropods. In: Cassells, A.C. (ed.) Pathogen and Microbial Contamination Management in Micropropagation. Kluwer Academic Publishers, Dordrecht, The Netherlands pp. 259-266

Quak, F. (1977). Meristem culture and virus-free plants. In: Applied and Fundamental Aspects of Plant Cell, Tissue and Organ Culture [Reinert, J and Bajaj, Y.P.S., (eds.)]. Springer, Berlin, Germany pp. 598-615

Rafferty, S.M., Williams, S., Falkiner, F.R. and Cassells, A.C. (2000). Persistence in *in vitro* cultures of cabbage (*Brassica oleracea* var. *capitata* L.) of human food poisoning pathogens: *Escherichia coli* and *Serratia marcescens*. Acta Hortic. 530: 145- 152

Roistacher, C.N. (1998). Indexing for virus in citrus. In: Plant Virus Disease Control (Hadidi, A., Khetarpal, R.K., Koganezawa, H., Eds.). APS Press, St. Paul, USA pp. 301-319

Russell, A.D., Hugo, W.B. and Ayliffe, G.A.J. (1982). Principles and practice of disinfection, preservation and sterilisation. Blackwell Scientific Publications, Oxford, UK.

Saikkonen, K., Faeth, S.H., Helander, M. and Sullivan, T.J. (1998). Fungal endophytes: a continuum of interactions with host plants. Annu. Rev. Ecol. Systemat. 29: 319-343

Schaad, N.W., Schaad, J.B., Jones, J.B. and Chun, W. (2001). Laboratory Guide for Identification of Plant Pathogenic Bacteria (3rd Ed.). APS Press, St Paul, USA.

Scott, P.R. and Bainbridge, A (Eds.) (1978). Plant Disease Epidemiology. Blackwell, London, UK.

Singh, R.P. and Dhar, A.K. (1998). Detection and management of viroids. In: Plant Virus Disease Control (Hadidi, A., Khetarpal, R.K., Koganezawa, H., Eds.). APS Press, St. Paul, USA pp. 428-447.

Smith, I.M.; Dunez, J.; Lelliot, R.A.; Phillips, D.H.; Archer, S.A. (1986). European handbook of plant diseases. Blackwell Scientific Publications, Oxford, UK.

Stead, D.E., Elphinstone, J.G., Weller, S., Smith, N. and Hennessy, J. (2000) Modern methods for characterizing, identifying and detecting bacteria associated with plants. Acta Hortic. 530: 45-60

Sykes, G. (1965). Disinfection and sterilisation. Spon, London, UK.

Teisson, C. and Alvard, D (1995). A new concept of plant *in vitro* cultivation in liquid medium: temporary immersion. In: Current Issues in Plant Molecular and Cellular Biology (Terzi, M., Cella, R. and Falavigna, A., Eds.). Kluwer, Dordrecht, The Netherland pp. 105-109

Vestberg, M., Cassells, A. C., Schubert, A., Cordier, C. and Gianinazzi, S. (2002). AMF and micropropagation of high value crops. In: Mycorrhizal Technology: From Genes to Bioproducts – Achievements and Hurdles in Arbuscular Mycorrhizal Research. Chapter 16 Birkhäuser Verlag, Basel, Switzerland (in press)

Werbrouck, S.P.O., Redig. P., VanOnckelen, H.A. and Debergh, P.C. (1996). *Gibberellins* play a role in the interaction between imidazole fungicides and cytokinins in Araceae. J. Plant Growth Regul. 15: 87-93

Williamson, B., Cooke, D.E.L., Duncan, J.M., Leifert, C., Breese, W.A. and Shattock, R.C. (1997). Fungal infections of micropropagated plants at weaning. In: Pathogen and Microbial Contamination Management (A. C. Cassells, Ed.) Kluwer, Dordrecht, The Netherlands pp 309-320.

Ziv, M., Ronen, G. and Raviv, M. (1998). Proliferation of meristematic clusters in disposable presterilized plastic bioreactors for the large-scale micropropagation of plants. In Vitro Cell. Devel. Biol.- Plant 34: 152-158

5. BENEFICIAL INFLUENCES OF ARBUSCULAR MYCORRHIZAL (AM) FUNGI ON THE MICROPROPAGATION OF WOODY AND FRUIT TREES

Jeanette Taylor and Lucy A. Harrier
Plant Science Division, Scottish Agricultural College, West Mains Road, Edinburgh, Scotland, UK.

CONTENTS

ABSTRACT

Arbuscular mycorrhizal (AM) fungi are ancient soil microbes that form symbiotic interactions with plant root systems of over 80% of all terrestrial flowering plants. In colonising the plant root systems the fungus confers benefits to the plant in the majority of cases. Benefits to the plant range from enhanced mineral nutrition to increased tolerance to biotic and abiotic stresses. The fungus benefits from this interaction through the acquisition of a source of carbon from the host plant that permits it to fulfil its lifecycle. Many woody and fruit trees are dependent on the formation of the AM symbiosis for natural plant growth and development. Therefore, AM fungal inoculation of micropropagated woody and fruit trees may help to overcome problems associated with micropropagation of these plants. In this chapter

S.M. Jain and K. Ishii (eds.), Micropropagation of Woody Trees and Fruits, 129–150.

the recent developments and insights into the use of (AM) fungal inoculation strategies for micropropagated woody and fruit trees is presented.

KEYWORDS

Arbuscular mycorrhizal (AM) fungi, biocontrol, benefits, nutrition, rooting, stress and symbiosis.

INTRODUCTION

The term mycorrhiza, literally means *fungus-root,* was first applied to fungal-tree associations in 1885 by the German forest pathologist Frank (1885). He stated that the association was mutualistic due to the covering of the root by the host fungus and the lack of detrimental responses within the host tree. Since then we have learned that the vast majority of land plants form symbiotic associations with fungi and an estimated 95 % of all plant species belong to genera that characteristically form mycorrhiza (Smith & Read, 1997).

Mycorrhizal associations can vary in structure and function, but the most common structure is the arbuscular mycorrhizal (AM) association. This association is formed between the root systems of most higher plants and zygomycete fungi belonging to the Order Glomales. The order encompasses five recognised families: the Glomaceae (*Glomus* and *Sclerocystis*), Acaulosporaceae (*Acaulospora* and *Entrophosphora*), Gigasporaceae (*Gigaspora* and *Scutellospora*), Archaeosporaceae (*Archaeospora*) and Paraglomaceae (*Paraglomus*) into which 150 species have been classified (Morton & Benny, 1990; Morton & Redecker, 2001). It is estimated that more than 80% of all terrestrial plants form this type of association with these fungi. This includes many agriculturally and horticulturally important crop species (Smith & Read, 1997).

The formation of a AM symbiosis is a complex process and is characterised by distinct developmental stages in the growth of the AM fungus: spore germination, hyphal differentiation, appressorium formation, root penetration, intercellular growth, intracellular arbuscule formation and nutrient exchange. The developmental stages are represented schematically in Figure 1 and for discussions on aspects of the development of the arbuscular mycorrhizal symbiosis, readers are referred to a number of comprehensive reviews (Smith & Gianinazzi-Pearson, 1988; Koide & Schreiner, 1992; Bonfante & Perotto, 1995; Harrison, 1997, 1998 & 1999; Hirsch & Kalpulnik, 1988; Albrecht *et al.*, 1999; Harrier, 2000).

The interaction between the plant and the fungus begins before the fungus makes physical contact with the host. Metabolites exuded by the host root specifically enhance spore germination and hyphal differentiation. Upon contact with the root epidermis the AM fungal hypha swell and form appressoria (Figure 1). This is followed by penetration of the outer root tissues, this can occur either intercellularly or intracellularly with the formation of a simple or large coiled hyphae (Figure 1) within the root cells. On reaching the inner cortex of the root system the hyphae penetrate the cortical cell walls, differentiate within the cells to form highly branched, terminal structures known as arbuscules (Figure 1). The arbuscule is assumed to be the site of nutrient exchange between the two partners. Some species of AM fungi also form vesicles within the root system. Vesicles are thin walled lipid containing bodies produced terminally or intercalarily from hyphae in the root cortex by species of the fungi in the suborder Glomineae.

AM fungi also develop an extensive network of external hyphae (extraradical hyphae) that ramifies into the surrounding soil matrix and exploits an area beyond the limits of the root system into the bulk soil. The hyphal network contributes greatly to the stability of soils and to better aeration and water percolation. The role and function of AM fungi extraradical hyphae is still not well understood due to the problems associated with observation, collection and quantification. Initial studies have demonstrated that the extraradical system consists of different types of hyphae, runner hyphae which divide dichotomously and extend the fungal colony radially and finer absorptive hyphae which play a role in nutrient uptake (Bago *et al.*,1998 a, b).

AM fungi and the symbiosis is an ancient interaction, as shown by fossils in the Devonian Rhynie Chert (Remy *et al.*, 1994: Taylor *et al.*, 1995), the Ordovacian of Wisconsin (Redecker *et al.*, 2000) and by molecular phylogeny of fungi (Simon *et al.*, 1993). Compared to other plant-microbe interactions, AM fungi have extremely low host specificity and this may reflect both their mutualistic biotrophic nature and the very long period of co-evolution with plant species. It is believed that during evolution, the AM symbiosis has been lost in about 10% of plants, and is completely absent in families such as Chenopodiaceae and Brassicaceae (Tester *et al.*, 1987).

The majority of woody and fruit trees, often exhibit a high dependency on the AM symbiosis for normal growth and development (Nemec, 1986). Due to the high mortality rates and poor rooting for many micropropagated woody and fruit tree species, it is an important pre-requisite to develop strategies that reduce these losses. Furthermore, in micropropagation the plants will not have the benefits of the symbiosis, as these fungi are removed along with other beneficial/pathogenic microbes and only by re-inoculation of the AM fungi will the benefits be obtained. Therefore, the integration of AM fungi into a micropropagation strategy is a key development. There are three potential time points during the micropropagation process that is suitable for inoculation by AM fungi. These include, 1) *in vitro*, during the rooting phase; 2) *ex vitro*, at the beginning acclimatisation (weaning) phase and 3) *ex vitro*, after the acclimatisation phase but before starting the post acclimatisation period under greenhouse conditions (Vestberg & Estaún, 1994; Azcon-Aguilar & Barea, 1997).

Inoculation of AM fungi within micropropagated crop plants, such as bamboo (Verma & Arya, 1998), cassava (Azcon-Aguilar *et al.*, 1997), strawberry (Hrselova *et al.,* 1988; Niemi & Vestberg, 1992; Vestberg, 1992; Williams *et al.*, 1992; Varma & Schüepp, 1994; Vestberg *et al.*, 1994; Cassels *et al.*, 1996; Hernandez-Sebastia *et al.*, 1996, 2000: Taylor & Harrier, 2001) and raspberry (Varma & Schüepp, 1994; Taylor & Harrier, 2000) has proven to be successful. However, the application of AM fungi to micropropagated woody and fruit trees has been demonstrated to be particularly beneficial to many plant species. Table 1 highlights the range of micropropagated woody and fruit trees that have been inoculated with species of AM fungi.

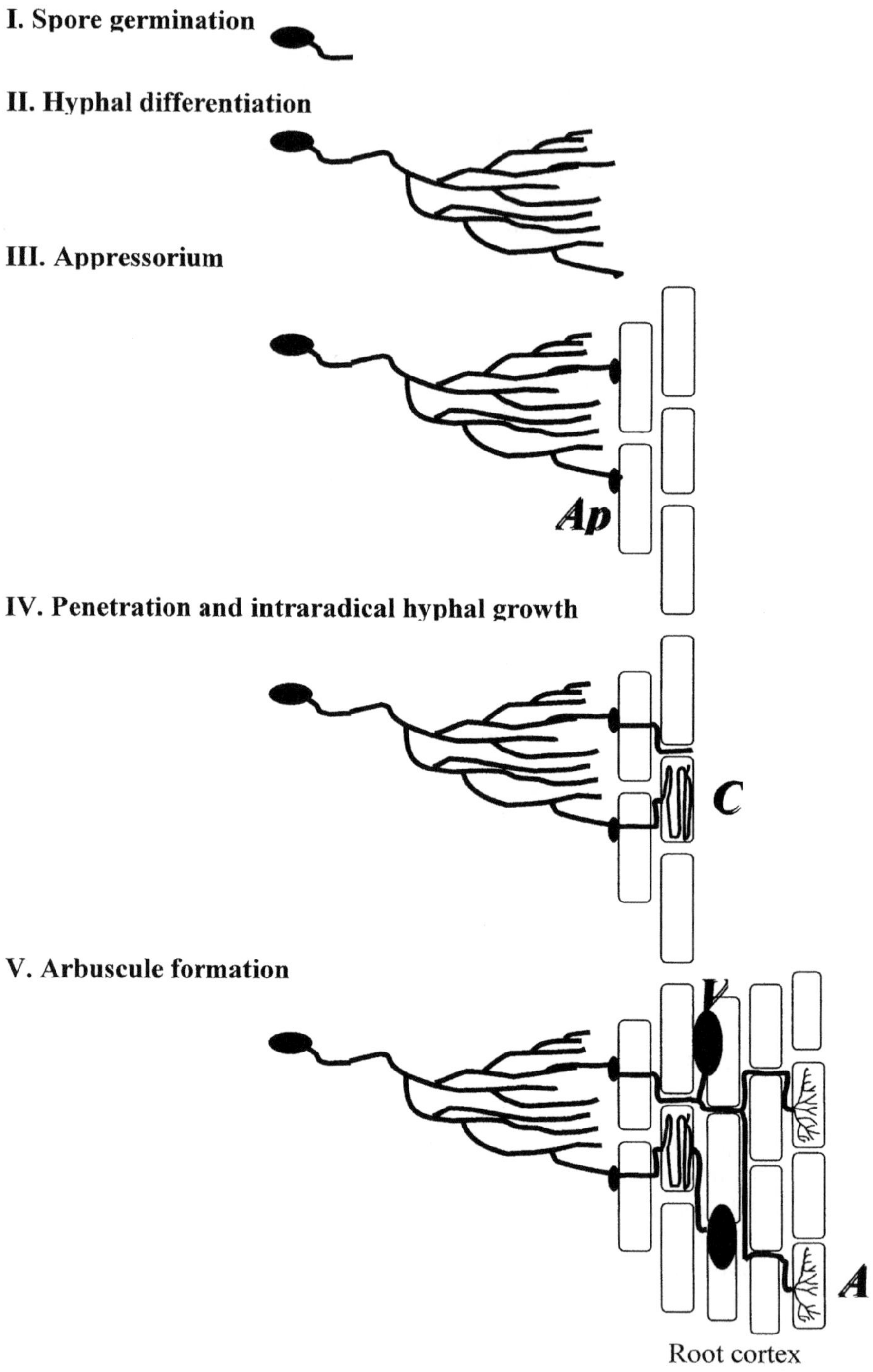

Figure 1:- Developmental stages in the growth and development of an arbuscular mycorrhiza. (Ap- appressoria; C - hyphal coils; A- arbusucle)

BENEFITS OF INOCULATION OF AM FUNGI

AM fungi are generally beneficial to host plant growth by increasing nutrient uptake (Marschner & Dell, 1994; Clark & Zeto, 2000), by protecting roots from soil pathogens (Newsham *et al.*, 1995), by enhancing the ability of plants to survive in the presence of heavy metals at toxic levels (Ricken & Hofner, 1996) and by improving water relations (Subramanian *et al.*, 1995). AM fungi also develop an extensive external hyphal network which makes a significant contribution to the improvement of soil texture (Bethlenfalvay & Schüepp, 1994). Moreover, AM fungi are known to have a positive effects on the growth and development of many micropropagated fruit species as many micropropagated species are AM fungal dependent. Furthermore, inoculation of micropropagated fruit plantlets with AM fungi improves survival and development of the plants, thus providing plants which are larger and healthier (Sbrana *et al.*, 1994). Many articles have been published on the influence of AM fungi on micropropagated woody and fruit trees of 37 host species representing 18 genera.

Four genera of AM fungi, including twenty species of *Glomus* (not including un-identified *Glomus* sp.) have been studied (Table 1).

Improvement to mineral nutrition

One of the main problems for the *in vivo* establishment of the micropropagated fruit plants is the survival and growth rate. AM fungal inoculation has been shown to not only enhance the growth of micropropagated fruit plants, but also improve survival and growth after removal from the *in vitro* system (Ravolanirina *et al.*, 1989; Guillemin *et al.*, 1994; Branzanti *et al.*, 1992; Fortuna *et al.*, 1992; Vidal *et al.*, 1992: Hooker *et al.*, 1994; Sbrana *et al.*, 1994; Fortuna *et al.*, 1996; Subhan *et al.*, 1998; Schubert & Lubraco, 2000). This increase in growth and survival in micropropagated fruit plants is often associated with enhanced mineral nutrition in particular improvements to the phosphorous (P) supply (Graham & Syvertsen, 1984; Fortuna *et al.* 1996; Lopez *et al.*, 1997; Schubert & Lubraco, 2000). One of the major benefits associated with an AM symbiosis is the enhanced P status of mycorrhizal plants. The enhancement is due to the uptake of P from the surrounding matrix by the AM fungus and then its subsequent translocation and transfer to the host plant (Smith & Smith, 1989). The phosphate flux has been estimated at 13 nmol $m^{-2}s^{-1}$ across the arbuscule in the mycorrhiza, although this rate can increase if exogenous phosphate is supplied (Cox & Tinker, 1976; Smith *et al.*, 1994). However, different AM fungi have varying capacities to take up P and give it to the host plant. In the study by Taylor & Harrier (2000, 2001) which compared P shoot concentrations of micropropagated raspberry and strawberry plants demonstrated that the same AM fungal isolates have different effects on host plants. Furthermore, different species of AM fungi have different effects within the same host plant in terms up P uptake. This observation also held true for the 11 other mineral nutrient concentrations analysed in the raspberry and/or strawberry plants analysed. Thus suggesting that individual isolates of AM fungi have unique effects on the mineral status of the host plants. Enhanced mineral

Table 1: Range of micropropagated woody and fruit trees inoculated with species of arbuscular mycorrhizal fungi.

Plant Species	**AM fungal Species**	**Reference**
Actinidia deliciosa	*Glomus* sp. isolate E3	Schubert *et al.*, 1992
Actinidia deliciosa cv. Hayward	*Glomus mosseae* isolate BEG12	Gribaudo *et al.*, 1996
Ananas comosus	*Glomus fasiculatum*	Jaizme Vega & Azcon, 1991
	Glomus mosseae	
	Acaulospora sp.	
Ananas comosus	*Glomus clarum* isolate LPA16	Guillemin *et al.*, 1992
	Glomus sp. isolate LPA21	
	Glomus sp. isolate LPA22	
	Glomus sp. isolate LPA25	
	Scutellospora pellucida isolate LPA20	
Ananas comosus	*Glomus* sp. isolate LPA21	Guillemin *et al.*, 1994
Ananas comosus	*Glomus fasiculatum*	Jaizme-Vega & Azcon, 1995
	Glomus mosseae	
	Acaulospora laevis	
Ananas sp.	Inoculant from AGC (UK)	Lovato *et al.*, 1992
	Inoculant from Phytotec (Belgium)	
	Glomus sp. isolate LPA4	
	G. intraradices isolate LPA8	
Annona cherimola	*Glomus deserticola*	Azcon-Aguilar *et al.*, 1994 & 1996
	Glomus fasiculatum	
	Glomus mosseae	
Anthyllis cytisoides	*Glomus fasiculatum*	Salamanca *et al.*, 1992
Cydonia oblonga	*Glomus intraradices*	Calvet *et al.*, 1995
Elaeis guineensis	*Glomus* sp. isolate LPA21	Blal *et al.*, 1990
	Glomus sp. isolate LPA22	
Fraxinus excelsior clone 71	*Glomus deserticola* isolate LPA27	Lovato *et al.*, 1994
	Glomus intraradices isolate LPA8	
	Gigaspora rosea isolate LPA23	

Table 1 continued

Plant Species	AM fungal Species	Reference
Jugulans regia	*Glomus mosseae* isolate BEG12	Dolcet-Sanjuan *et al.*, 1996
	Glomus intraradices isolate BEG72	
Leucaena leucocephala	*Glomus fasiculatum*	Puthur *et al.*, 1998
	Glomus macrocarpum	
Malus baccata	*Glomus claroidum*	Uosukainen & Vestberg, 1994
	Glomus mosseae	
	Glomus hoi	
	Glomus fistulosum	
Malus pumila	*Glomus* sp. isolate A6	Sbrana *et al.*,1994
Malus pumila	*Glomus intraradices* PI-HMZ-9	Fortuna *et al.*, 1996
	Glomus mosseae PI-HMZ-4	
	Glomus viscosum PI-HMZ-10	
Malus pumila	*Glomus mosseae* isolate BEG12	Gribaudo *et al.*, 1996
Malus pumila	*Glomus mosseae* isolate BEG12	Schubert & Lubraco, 2000
Malus silvestris	*Glomus mosseae*	Pinochet *et al.*, 1993
Malus sp.	*Glomus epigaeum*	Granger *et al.*, 1983
Malus sp.	*Glomus fasiculatum*	Branzati *et al.*, 1992
	Glomus intraradices	
Malus sp.	*Glomus mosseae*	Avanzato *et al.*, 1993
Musa acuminata	*Glomus intraradices*	Jaizme-Vega & Azcon, 1995
	Glomus proliferum	
	Glomus versiforme	
Musa acuminata	*Glomus fasiculatum*	Jaizme-Vega *et al.*, 2001
	Glomus mosseae	
	Scutellospora heterogama	
Musa sp.	*Acaulospora scrobiculata*	Yano-Melo *et al.*, 1999
	Glomus clarum	
	Glomus etunicatum	
Persea americana	*Glomus fasiculatum*	Vidal *et al.*,1992

Table 1 continued

Plant Species	AM fungal Species	Reference
Persea americana	*Glomus deserticola*	Azcon-Aguilar *et al.*,1992
	Glomus mosseae	
	Glomus sp	
Pistacia integerrima	*Glomus caledonium*	Schubert & Martinelli, 1988
	Glomus occultum	
Prunus amygdalus	*Glomus spp.* isolate A6	Sbrana *et al.*,1994
Prunus avium cv. F12/1	*Glomus deserticola* isolate LPA27	Lovato *et al.*, 1994
	Glomus intraradices isolate LPA8	
	Gigaspora rosea isolate LPA23	
Prunus avium	*Glomus intraradices* isolate LPA8	Cordier *et al.*, 1996
	Glomus caledonium isolate BEG20	
Prunus cerasifera	*Glomus caledonium*	Fortuna *et al.*, 1992
	Glomus coronatum	
	Glomus mosseae	
	Glomus sp. isolate A6	
Prunus cerasifera	*Glomus* sp. isolate A6	Sbrana *et al.*,1994
Prunus cerasifera	*Glomus intraradices* isolate LPA8	Berta *et al.*, 1995
	Glomus mosseae isolate BEG12	
Prunus cerasifera	*Glomus intraradices* PI-HMZ-9	Fortuna *et al.*, 1996
	Glomus mosseae PI-HMZ-4	
	Glomus viscosum PI-HMZ-10	
Prunus cerasifera	*Glomus intraradices*	Fortuna *et al.*, 1998
	Glomus mosseae	
Prunus cerasifera	*Glomus mosseae*	Camprubi *et al.*, 1993
Prunus cerasifera X P. munsoniana	*Glomus mosseae*	
Prunus insititia	*Glomus mosseae*	
Prunus cerasifera X P. munsoniana	*Glomus intraradices*	Pinochet *et al.*, 1998
	Glomus mosseae	

Table 1 continued

Plant Species	AM fungal Species	Reference
Prunus dulcis	*Glomus mosseae*	Monticelli *et al.*,2000
Prunus persica X Prunus dulcis	*Scutellospora calospora*	
Prunus domesticus X Prunus persica		
Prunus persica X Prunus amygdalus	*Glomus mosseae*	Estaún *et al.*, 1994
	Glomus sp. isolate E3	
	Acaulospora laevis	
Prunus persica X Prunus amygdalus	*Glomus* sp.	Rapparini *et al.*, 1994
Prunus persica X Prunus amygdalus	*Glomus intraradices* isolate BEG72	Estaún *et al.*, 1999
Prunus persica	*Glomus* sp. isolate A6	Sbrana *et al.*,1994
Prunus persica	*Glomus mosseae*	Pinochet *et al.*,1995
Pyrus communis.	*Glomus* sp.	Rapparini *et al.*, 1994
Pyrus communis.	*Glomus* sp.	Rapparini *et al.*, 1996
Pyrus communis.	*Glomus mosseae*	Lopez *et al.*, 1997
	Glomus intraradices	
Sesbania sesban	*Glomus fasiculatum*	Subhan *et al.*, 1998
Spatium junceum	*Glomus fasiculatum*	Salamanca *et al.*, 1992
Vitis berlandieri X Vitis rupestris	*Glomus caledonicum*	Schubert *et al.*, 1990
	Glomus constrictum	
	Glomus fasiculatum	
	Glomus monosporum	
	Glomus occultum	
	Glomus versiforme	
	Glomus sp. isolate E3	
Vitis berlandieri X Vitis riparia	*Glomus mosseae* isolate BEG12	Gribaudo *et al.*, 1996
Vitis vinifera	*Gigaspora margarita* Isolate LPA2	Ravolanirana *et al.*, 1989
	Glomus caledonium. Isolate LPA12	
	Glomus fasiculatum. Isolate LPA7	
	Glomus mosseae. Isolate LPA 5	

Table 1: continued

Plant Species	**AM fungal Species**	**Reference**
Vitis vinifera	*Glomus fasiculatum* isolate LPA 7	Schellenbaum *et al.*, 1991
Vitis vinifera	*Glomus agreggatum*	Aguin *et al.*, 2001
Vitis vinifera	Inoculant from AGC (UK)	Lovato *et al.*, 1992
	Inoculant from Phytotec (Belgium)	

nutrition within micropropagated plants is key to their survival, growth and development and a greater understanding of the effects of different AM fungi on mineral nutrition is required. Therefore, specific isolates of AM fungi need to be screened for their ability to promote growth, development and mineral nutrition prior to be utilised on a commercial basis.

Induction of root system modifications

One of the major problems associated with micropropagated woody and fruit trees is the development of functioning root systems (Hooker *et al.*, 1994; Azcon - Aguilar & Barea, 1997). AM fungi have been shown to modify root architecture and development within the host plant root system (Berta *et al.*, 1993; Atkinson *et al.*, 1994). In general, AM fungal colonised plants have more numerous and branched roots, this is particularly true for several herbaceous and woody species (Amijee *et al.*, 1989; Berta *et al.*, 1990, 1993; Schellenbaum *et al.*, 1991; Tisserant *et al.*, 1992). Moreover, different AM fungal species can induce different effects on root system morphology (Berta *et al.*, 1995). The effects of inoculation of AM fungi has been investigated in micropropagated woody and fruit trees with beneficial effects being observed during the rooting phase and morphological changes in the plant root system observed for *in vivo* rooted shoots during the acclimatisation stage. (Schellenbaum *et al.*, 1991; Uosukainen & Vestberg, 1994; Berta *et al.*, 1995; Fortuna *et al.*, 1998). During the *in vivo* rooting phase, AM fungi have been shown to enhance the development of a more branched root system, irrespective of nutrient supply (Fortuna *et al.*, 1998). The more branched root system was achieved through enhancing branching of the first order lateral roots; increasing the number of second order lateral roots and by increasing total root length (Fortuna *et al.*, 1998). Conversely, inoculation with AM fungi has been shown to have negative effects in some species of *Malus* (Avanzato *et al.*, 1993; Lovato *et al.*, 1994; Uosukainen & Vestberg, 1994). For example, *Glomus hoi* inoculation of crab apple 'Hanna' caused severe rooting of microcuttings while micropropagated common ash plants inoculated with *G. intraradices*, growth and development was depressed during the weaning phase (Lovato *et al.*, 1994).

In studies investigating root system morphogenesis of micropropagated transplants, in particular *Prunus cerasifera*, Berta *et al.*, (1995) demonstrated that inoculation by AM fungi increased the intensity of branching in all root orders. This effect was most pronounced on first order lateral roots where the number of branches increased from under 100 to over 300 branches m^{-1}. Anatomical studies on the root apices also revealed that there were changes in the diameter of the root cap, cell numbers and cell sizes (Berta *et al.*, 1995).

The mechanisms for root system modification are complex as there is a range of additional physiological changes occurring within a mycorrhizal plant such as altered mineral nutrition status, altered hormone balance that may account for such changes. However, AM fungal inoculation influences processes in the root system at different levels, and not all effects are due to improved P nutrition (Berta *et al.*, 1995).

Increased tolerance to pathogenic organisms

Fungal pathogens

Micropropagated plants are particularly susceptible to foliar and root fungal pathogen attack. AM fungi are known for their bioprotection against fungal pathogens in particular soil pathogens such as *Phytophthora, Gauemannomyces, Fusarium, Chalara, Pythium, Rhizoctonia, Sclerotium, Verticillium* and *Aphanomyces* (Rosendahl, 1985; Cordier *et al.*, 1996a: Kjoller & Rosendahl, 1997; Slezack *et al.*, 1999, 2000). However, the protection is not effective for all plant pathogenic fungi and the level of biological control conferred by the AM fungal colonisation is plant and AM fungal species specific. AM fungi have also been shown to bioprotect micropropagated plants against root pathogens (Bartschi *et al.*, 1981; Bethlenfalvay & Linderman, 1982; Guillemin *et al.*, 1994; Cordier *et al.*, 1996b; Azcon -Aguilar & Barea, 1997). Disease reduction within host plants is the outcome of complex interactions between plant, pathogen and AM fungi. The mechanisms proposed to explain the bioprotection vary and include:- Improvement of mineral nutrition; Root damage compensation (Trotta *et al.*, 1996; Cordier *et al.*, 1996a); Competition for photosynthesis and/or colonisation and infection sites (Caron *et al.*, 1985); Production of anatomical and morphological changes in the root system (Atkinson *et al.*, 1994); Changes in the microbial composition of the mycorrhizosphere and activation of plant defense mechansisms (Cordier *et al.*, 1998; Pozo *et al.*, 1996, 1999)

Root lesion nematodes

Pratylenchus vulnus Allen and Jensen is considered one of the most important lesion nematode attacking fruit crops in nurseries and established orchards within both Europe and America (Day & Serr, 1953, Camprubi *et al.*, 1993;Pinochet *et al.*, 1995; Lopez *et al.*, 1997). In Europe it has been found to attack apple (*Malus silvestris* L.) rose (*Rosa multiflora* L.) quince (*Cydonia oblonga* Miller), pear (*Pyrus communis* L.) and plum (*Prunus insititia* L.) (Camprubi *et al.*, 1993). While in America it is known to damage peach, almond, walnut (*Juglans regia* L.), cherry (*Prunus avium* L.), fig (*Ficus carica* L.) and grape (*Vitis* spp.). Damage can seriously affect early stages of plant development within the nursery or when rootstocks are transplanted into nematode infested fields.

Root colonisation by AM fungi has been shown to have different effects on *P vulnus* within different host plants (Camprubi *et al.*, 1993; Calvet *et al.*, 1995; Pinochet *et al.*, 1993, 1995; Lopez *et al.*, 1997). Early AM fungal colonisation increases tolerance to *P. vulnus* in micropropagated apple (Pinochet *et al.*, 1993); quince (Calvet *et al.*, 1995), plum (Camprubi *et al.*, 1993), cherry (Pinochet *et al.*, 1995) and pear (Lopez *et al.*, 1997).

The lower levels of nematode parasitism observed within some AM fungal colonised plants, suggests a suppressive effect from the AM fungus on root lesion nematode reproduction (Lopez *et al.*, 1997). The mechanisms of nematode suppression are currently unknown. However, mechanisms could include the competition for space and food source and the production and accumulation of compounds within mycorrhizal root tissue that affect nematode feeding. Nematode migration may be impaired through increased lignification and suberization in mycorrhizal tissue (Hayman, 1982).

CONSIDERATIONS FOR MYCORRHIZATION

Choice of AM fungal species

The choice of AM fungal isolate is critical for the successful establishment of a functional symbiosis. One of the most important factors which influences the positive effects conferred by AM fungal colonisation on micropropagated plants, is the choice of AM fungus (Guillemin *et al.* 1992; Fortuna *et al.* 1992). Therefore, species and isolates of AM fungi require to be screened for their potential to confer positive effects on the growth, development and mineral nutrition of the host plant prior to being utilised on a commercial basis.

Problems associated with screening different species of AM fungi originates from the fact that many AM fungal species have preferences for soils with different properties. For example, some species of AM fungi may be ineffective in colonising their hosts under certain environmental conditions such as soil pH conditions (Abbott & Robson, 1985).

Mycorrhizal dependency varies greatly among taxa and varieties of plants (Lambert & Cole, 1980; Graham & Eissenstat, 1994). Many fruit and woody trees represent plants that have been bred and selected for enhanced fruit production, tolerance to pathogens etc. These plants were not selected on the basis of mycorrhizal functioning. Therefore, there was no mechanism to link their gene pool with mycorrhizal functioning. Consequently these management practices have possibly selected for mycorrhizal fungi that are 'cheaters' (Soberon & Martinez del Rio, 1985; Smith & Smith, 1996; Johnson *et al.*, 1997). An AM fungal 'cheater' would represent an isolate that receives the benefits from the host plant (i.e. C source) but does not reciprocate by transferring mineral nutrient (P, etc.) (Smith & Smith, 1996). The AM fungal species utilised in the experiments by Taylor & Harrier (2000, 2001) with micropropagated raspberry and strawberry plants caused growth depressions may represent cheating fungi as they did not enhance the mineral nutrition of the host plants. However, the underlying reasons as to why particular AM fungi cause growth depressions in host plants is complex (Smith & Smith, 1996) but one reason is that they may be too great a carbon drain on the host plant and giving no nutritional benefits in return.

Quality and quantity of inoculum

One of the limiting factors in using AM fungi is the inability to culture them in the absence of the host plants and this has hampered the mass production of inoculum and their utilisation within crop systems (Jarstfer & Sylvia, 1992). However, it is possible to grow AM fungi in sterile culture with plant root explants (for example, Mosse & Hepper, 1975; Miller-Wideman & Watrud, 1984; Diop *et al.*, 1994) and/or with hairy roots transformed with *Agrobacterium rhizogenes* (for example:- Mugnier & Mosse, 1987; Becard & Fortin, 1988; Diop *et al.*, 1992; Declerck *et al.*, 1996 & 1998; Pawlowska *et al.*, 1999). Furthermore, *in vitro* culture of AM fungi provides an abundant source of contaminant free AM fungal hyphal and spore tissue. The provision of 'sterile inoculum' is suitable for the use in *in vivo* systems. However, it should be noted that not all species of AM fungi can grow well and sporulate in these *in vitro* systems and only a relatively few species have been established in these systems to date. Therefore, other AM fungal species require traditional pot culture production.

Although the production of sterile AM fungal inoculum is important, the loss of additional beneficial microbes may be important in the observable beneficial effects observed upon inoculation of plant by AM fungi. The AM fungus also harbours structures called bacterium-like organisms (blos), which have been detected ultrastructurally in spores and germinating and/or symbiotic mycelium (Bianciotto*et al.*, 1996). These bacteria have been shown to be group II Pseudomonads (genus *Burkholde*ria) by DNA sequence analysis of small subunit rRNA genes. However, the significance of these bacteria remains unknown, but indicates that mycorrhizal systems can include plant, fungal and bacterial cells (Bianciotto *et al.*, 1996). However, with the development of new micropropagation systems such as phototrophic culture systems, it may help overcome the need to use sterile AM fungal inoculum (Cassels *et al.*, 1996). This type of system would facilitate inoculation with AM fungi and beneficial helper and/or biocontrol bacteria. Normally such bacteria would be likely to overgrow the culture causing microplant death. Furthermore, phototrophic culture is more inoculum efficient, with AM fungal colonisation occurring with a minimum of two viable spores (Cassels *et al.*, 1996).

An another factor to consider is the type of inoculum i.e. spores versus colonised root pieces. In the suborder Glomineae, all fungal parts appear to be highly infective: intraradical or extraradical spores, intraradical vesicles and attached or detached extraradical hyphae (Jasper *et al.*, 1989). In the Gigasporineae, the spores are the only infective units. Species of the Gigasporineae do not form vesicles, however they form auxiliary cells and no evidence has ever been presented to indicate that auxiliary cells are infective. Furthermore, extraradical hyphae are not infective once mycorrhizal development has ceased (Pearson & Schweiger, 1994). In addition to the type of inoculum, the amount of inoculum that can affect AM fungal root colonisation is also important (Schubert *et al.*, 1992). The volume ratio of soil/inoculum used should be sufficient for colonisation, as too little will reduce the capacity for colonisation and any benefits will not be observed.

Timing of AM fungal inoculation

The timing of AM fungal inoculation is also critical to the successful development of a functional AM symbiosis. Inoculation can occur at either of the three time points during the micropropagation process as described previously. However, the majority of AM fungal inoculation has been conducted either after the rooting phase, at the very beginning of the weaning stage or after acclimatisation. The technique to inoculate at the beginning of the acclimatisation stage is normally employed due to its easier feasibility (Ravolanirina*et al.*, 1989) and in some cases a better plant response is produced when the inoculation is carried out at this stage

(Vidal *et al.*, 1992). The AM fungal colonisation of the roots takes place only in young secondary roots and these can be produced later during the *ex vitro* development (Azcon-Aguilar & Barea, 1997). The low AM fungal colonisation rates observed by Monticelli *et al.* (2000) demonstrated how important the duration of time is available for AM fungal colonisation to occur.

Inoculation of the AM fungi at the time of rooting is applicable especially since sterile spores of specific AM fungi can be produced by *in vitro* culture techniques (see above). However, some problems are encountered such as particular rooting culture media may be inhibitory to AM fungal spore germination and growth (Azcon-Aguilar & Barea, 1997).

Inoculation at the beginning of weaning and or after, in which the axenically propagated plantlets are transplanted appears to be beneficial. Studies have highlighted that although AM fungal colonisation is possible at both stages, a better plant response was produced when inoculation was carried out at the beginning of weaning (Vidal *et al.*, 1992). Moreover, inoculation at these time points for a nursery is more feasible (Azcon-Aguilar & Barea, 1997).

Substrates for mycorrhization during acclimatisation

The effects of AM fungal inoculation on the growth and development of micropropagated woody and fruit trees may be promoted and/or negated by the choice of substrate which is used during acclimatisation (Azcon-Aguilar & Barea, 1997). Commercial fruit plant micropropagation normally uses peat based substrates, which contain no soil and are enriched in nutrients such N, P and K (Preece and Sutter, 1991). This enrichment, particularly of P will have negative effects on the growth and development of AM fungi. Studies have been conducted which evaluate the growth and development of micropropagated woody and fruit trees in commercial substrates after inoculation with AM fungi (Schubert & Lubraco, 2000). Therefore, fertilisation regimes, substrates, plant species and AM fungal combinations must be evaluated, to establish the most suitable combination. Chemical treatments are also currently used to kill pathogenic organisms present in the propagation substrates however, these may also inhibit the subsequent growth and development of AM fungi and beneficial microbes. These negative chemical effects can often be overcome by a period of aeration prior to inoculation (Azcon-Aguilar & Barea, 1997)

CONCLUSIONS

Despite the successful applications of AM fungal inoculation to micropropagated woody and fruit trees, several considerations should be made when developing an inoculation strategy for each individual woody and fruit tree species. Aspects to be considered include root development rate, length of weaning and acclimatisation periods, the objectives of AM fungal inoculation and the species of AM fungi used for inoculation (Vestberg & Estaún, 1994;Azcon-Aguilar & Barea, 1997). The length of weaning and acclimatisation periods is important for AM fungi as the length of time for particular AM fungi can be considerable for example 4-8 weeks. Therefore, to observe the potential benefits time must be allowed for the symbiosis to establish, but the advantages of AM fungal inoculation are not always evident in the short-term. The long-term performance of micropropagated plants transplanted in the field may respond better. The objectives of AM fungal inoculation should be determined at the onset before any screening of potential AM fungal species. The objectives will consider the end-user requirements, such as enhanced survival rates, tolerance to biotic and or abiotic stresses, plant developmental patterns and lower fertiliser inputs. Such considerations are important for AM fungal species selection, as each isolate and species will on different effect on each host plant and maximal benefits will be obtained the selection of the most appropriate AM fungal species.

Taking into consideration the high production costs of micropropagated woody and fruit trees, the application of AM fungal inoculum may represent only a small contribution to the input costs. In addition, AM fungal inoculation can reduce the length of production cycle, thus saving on energy and chemical imputs (Uosukainen & Vestberg, 1994). Furthermore, in light of the recent developments in the understanding of AM fungal biology, with particular reference to *in vitro* culture for inoculum production, the utilisation of AM fungal inoculants in the micropropagation process as bioprotectors, biostimulators and/or biofertilisers is becoming a reality.

ACKNOWLEDGEMENTS

SAC receives financial support from the Scottish Executive Environmental Rural Affairs Division (SEERAD).

References

Abbott, L.K., Robson, A.D. The effect of soil pH on the formation of VA mycorrhizas by two species of *Glomus*. Australian Journal of Soil Research 1985; 23:253-261.

Aguin, O., Vilarino, A., Mansilla, J.P., Sainz, M.J. Pre-inoculation of grapevine rootstocks increase plant tolerance to white root rot. In; Programs, Abstracts and Papers of the COST 8.38 Action Working Groups 1 and 4 meeting on *Knowledge on population biology of AMF as a tool for mycorrhizal technology*, 2001, Průhonice Czech Republic.

Albrecht, C., Geurts, R., Bisseling, T. Legume nodulation and mycorrhizae formation two extremes in host specificity meet. The EMBO Journal 1999; 18: 281-289.

Amijee, F., Tinker, P.B., Stribley, D.P. Effects of phosphorous on the morphology of VA mycorrhizal root system of leek (*Allium porrum*). Plant & Soil 1989; 119: 334-336.

Atkinson, D., Berta G., Hooker, J.E. Impact of mycorrhizal colonisation on root architecture, root longevity and formation of growth regulators. In ; Impact of Arbuscular Mycorrhizas on Sustainable Agriculture and Natural Ecosystems (S. Gianinazzi and H. Schuepp eds.). Birkhauser Verlag Basel, Switzerland. 1994; pp89-99.

Avanzato, D., Cherubini, S., Jaque, E. Portainjertos de manzano de obtención *in vitro*. Potencialidad de la aclimatación y enraizamiento directo de estaquilas. Hortofruticultura 1993; 1: 67-72.

Azcon-Aguilar, C., Barceló, A., Vidal, MT., Viña, G. de la Further studies on the influence of mycorrhizae on growth and development of micropropagated avocado plants. Agronomie 1992; 12: 837-840.

Azcon-Aguilar, C., Barea, J.M., Applying mycorrhiza biotechnology to horticulture: significance and potentials. Scientia Horticulturae 1997; 68: 1-24.
Azcon-Aguilar, C., Cantos, M., Troncoso, A., Barea, J.M., Beneficial effect of arbuscular mycorrhizas on acclimatisation of micropropagated cassava plantlets. Scientia Horticulturae 1997; 72: 63-71.
Azcon-Aguilar, C., Encina, C.L., Azcon, R., Barea, J.M. Mycotrophy of *Anona cherimola* and the morphology of its mycorrhizae. Mycorrhiza 1994; 4: 161-168.
Azcon-Aguilar, C., Padilla, I.G., Encina, C.L., Azcon, R., Barea, J.M. Arbuscular mycorrhizal fungal inoculation enhances plant growth and changes root system morphology in micropropagated *Annona cherimola* Mill.. Agronomie 1996; 16: 647-652.
Bago, B., Azcon-Aguilar, C., Goulet, A., Piche, Y. Branched absorbing structures (BAS): a feature of the extraradical mycelium of symbiotic arbuscular mycorrhizal fungi. New Phytologist 1998a; 139: 375-388.
Bago, B., Azcon-Aguilar, C., Piche, Y. Architecture and developmental dynamics of the external mycelium of the arbuscular mycorrhizal fungus *Glomus intraradices* grown under monoxenic conditions. Mycologia 1998b; 90: 52-62.
Bartschi, H., Gianinazzi-Pearson, V., Vegh I. Vesicular arbuscular mycorrhiza formation and root rot disease *Phytophthora cinnamomi* development in *Chamaecyparis lawsonia*. Phytopath Z 1981; 102: 213-218.
Becard G., Fortin, JA. Early events of vesicular-arbuscular mycorrhiza formation on Ri T-DNA transformed roots. New Phytologist 1988; 108: 211-218.
Berta, G., Fusconi, A., Trotta, A. VA mycorrhizal infection and the morphology and function of root systems. Environmental Experimental Botany 1993; 33: 159-173.
Berta, G., Fusconi, A., Trotta, A., Scannerini, S. Morphogenic modifications induced by the mycorrhizal fungus *Glomus* E3 in the root system of *Allium porrum* L. New Phytologist 1990; 114: 207-215.
Berta, G., Trotta, A., Hooker, J., Munro, M., Atkinson, D., Giovanetti, M., Marini, S., Loreti, F., Tisserant, B., Gianinazzi-Pearson, V., Gianinazzi, S. The effects of arbuscular mycorrhizal infection on plant growth, root system morphology and soluable protein content in *Prunus cerasifera* L. Tree Physiology 1995; 15: 281-293.
Bethlenfalvay, G.J., Linderman, R.G. *Mycorrhizae in sustainable agriculture*. ASA Spec. Publ. Madison Wisconsin. 1992.
Bethlenfalvay, G.J., Schuepp, H. Arbuscular mycorrhizas and agrosystem stability. In *Impact of Arbuscular Mycorrhizas on Sustainable Agriculture and Natural Ecosystems*, S. Gianinazzi & H. Schuepp eds., Basel, Switzerland: Birkhauser Verlag. pp117-131; 1994.
Bianciotto, V., Bandi, C., Minerdi, D., Sironi, M., Volker-Tichy, H., Bonfante, P. An obligately endosymbiotic mycorrhizal fungus itself harbours obligately intracellular bacteria. Applied and Environmental Microbiology 1996; 62: 3005-3010.
Blal, B., Gianinazzi-Pearson, V., Fardeau, J.C., Gianinazzi, S. Influence of vesicular arbuscular mycorrhizae on phosphate fertiliser efficiency in two tropical acid soils planted with micropropagated oil palm (*Elaeis guineensis* jacq.). Biology and Fertility of Soils 1990; 9: 43-48.
Bonfante, P., Perotto, S. Strategies of Arbuscular Mycorrhizal Fungi when infecting host plants. New Phytologist 1995; 130: 3-21
Brazanti, G., Gianinazzi-Pearson, V., Gianinazzi, S. Influence of phosphate fertilisation on the growth and nutrient status of micropropagated apple infected with endomycorrhizal fungi during weaning stage. Agronomie 1992; 12: 841-845.
Calvet, C., Pinochet, J., Camprubi, A. Fernandez, C. Increased tolerance to the root lesion nematode *Pratylenchus vulnus* in micropropagated BA-29 quince rootstock. Mycorrhiza 1995; 5: 253-258.
Camprubi, A., Pinochet, J., Calvet, C., Estaun, V. Effects of the root lesion nematode *Pratylenchus vulnus* and the vesicular arbuscular mycorrhizal fungus *Glomus mosseae* on the growth of three plum rootstocks. Plant & Soil 1993; 153: 223-229.
Caron, M., Fortin J.A., Richard, C. Influence of substrate on the interaction of *Glomus intraradices* and *Fusarium oxysporum* f. sp. *radicis-lycopersici* on tomatoes. Plant and Soil 1985; 87: 233-239.
Cassels, A.C., Mark, G.L., Periappuram, C. Establishment of arbuscular mycorrhizal fungi in autotrophic strawberry cultures *in vitro*. Comparison with inoculation of microplants *in vivo*. Agronomie 1996; 16: 625-632.
Clark, R.B., Zeto, S.K. Mineral acquisition by arbuscular mycorrhizal plants. Journal of Plant Nutrition 2000; 23: 867-902.
Cordier, C., Gianinazzi, S., Gianinazzi-Peaarson, V. Colonisation patterns of root tissues by *Phytophthora nicotiania* var *parasitica* related to reduced diseases in mycorrhizal tomato. Plant & Soil 1996a; 185: 223-232.

Cordier,C., Trouvelot, A., Gianinazzi, S., Gianninazzi-Pearson, V. Arbuscular mycorrhiza technology applied to micropropagated *Prunus avium* and to protection against *Phytophthora cinnamomi*. Agronomie 1996b; 16: 679-688.

Cordier, C., Pozo, M.J., Gianinazzi, S., Gianinazzi-Pearson, V. Cell defense responses associated with localised and systemic resistance to *Phytophthora parasitica* induced in tomato by an arbuscular mycorrhizal fungus. Molecular Plant -Microbe Interactions 1998; 11: 1017-1028.

Cox, G., Tinker, P.B. Translocation and transfer of nutrients in vesicular arbuscular mycorrhizas. I. The arbuscule and phosphorus transfer: a quantitative ultrastructural study. New Phytologist 1976; 77: 371-78.

Day, L.H., Serr, E.F. Comparative resistance of rootstocks of fruit and nut trees to attack by a root lesion nematode. American Society of Horticultural Science 1953; 114: 30-35.

Declerck, S., Strullu, D.G., Plenchette, C. *In vitro* mass-production of the arbuscular mycorrhizal fungus *Glomus versiforme*, associated with Ri T-DNA transformed carrot roots. Mycological Research 1996.; 100: 1237-1242.

Declerck, S., Strullu, D.G., Plenchette, C. Monoxenic culture of the intraradical forms of *Glomus* sp. isolated from a tropical ecosystem: a proposed methodology for germplasm collection. Mycologia 1998; 90: 579-585.

Diop, T.A., Becard, G., Piche, Y. Long term *in vitro* culture of an endomycorrhizal fungus *Gigaspora margarita* on Ri T-DNA transformed roots of carrot. Symbiosis 1992; 12: 249-259.

Diop, T.A., Plenchette, C., Strullu, D.G. Dual axenic culture of sheared root inocula of vesicular-arbuscular mycorrhizal fungi associated with tomato roots. Mycorrhiza 1994; 5: 17-22.

Dolcet-Sanjuan, R., Claveria, E., Camprubi, A., Estaun, V., Calvet, C. Micropropagation of walnut trees (*Juglans regia* L.) and response to arbuscular mycorrhizal inoculation. Agronomie 1996; 16: 639-645.

Estaún, V., Calvet, C., Camprubi, A. Arbuscular mycorrhizae and growth enhancement of micropropagated *Prunus* rootstock in different soilless potting mixes. Agricultural Science in Finland 1994; 3: 263-267.

Estaún, V., Calvet, C., Camprubi, A., Pinochet, J. Longterm effects of nursery starter substrate and AM inoculation of micropropagated peach X almond hybrid rootstock GF677. Agronomie 1999; 19: 483-489.

Fortuna, P., Citernesi, A., Morini, S., Giovannetti, M., Loreti, F. Infectivity and effectiveness of different species of arbuscular mycorrhizal fungi in micropropagated plants of Mr S2/5 plum root stock. Agronomie 1992; 12: 825-829.

Fortuna, P., Citernesi, A., Morini, S., Vitagliano, C., Giovannetti, M. Influence of arbuscular mycorrhizas and phosphate fertilisation on shoot apical growth of micropropagated apple and plum rootstocks. Tree Physiology 1996: 16: 757-763.

Fortuna, P., Morini, S., Giovannetti, M. Effects of arbuscular mycorrhizal fungi on *in vivo* root initiation and development of micropropagated plum shoots. Journal of Horticultural Science and Biotechnology 1998; 73: 19-28.

Frank, B. Ueber die auf Wurzelsymbiose beruhende Ernahrung gewiser Baume durch unterirdische Pilze. Berichte der Deutsch Botanische Gesellschaft 1885; 3: 128-145.

Graham, J.H., Eissenstat, D.M. Host genotype and formation and function of VA mycorrhizae. Plant & Soil 1994; 159: 179-185.

Graham, J.H., Syvertsen, J.P. Influence of vesicular arbuscular mycorrhizae on the hydraulic conductivity of roots of two citrus rootstocks. New Phytologist 1984; 97: 277-284.

Granger, R.L., Plenchette, C., Fortin, J.A. Effect of a vesicular-arbuscular (VA) endomycorrhizal fungus (*Glomus epigaeum*) on the growth and leaf mineral content of two apple clones propagated *in vitro*. Can. Journal of Plant Science 1983; 63: 551-555.

Gribaudo, I., Zanetti, R., Morte, M.A., Previati, A., Schubert, A. Development of mycorrhizal infection in *in vitro* and *in vivo* formed roots of woody fruit plants. Agronomie 1996; 16: 621-624.

Guillemin, J.P., Gianinazzi, S., Trouvelot, A. Screening of arbuscular endomycorrhizal fungi for establishment of micropropagated pineapple plants. Agronomie 1992; 12: 831-836.

Guillemin, J.P., Gianinazzi, S., Gianinazzi-Pearson, V., Marchal, J. Contribution of arbuscular mycorrhizas to biological protection of micropropagated pineapple (*Ananas comosus* (L.) Merr.) against *Phytophthora cinnamomi* Rands. Agricultural Science of Finland 1994; 3: 241-251.

Harrier, L.A. Arbuscular Mycorrhizal (AM) Symbiosis: A Review of Signalling and Molecular Aspects of Root Colonisation. Botanical Journal of Scotland 2000; 52: 159-169.

Harrison, M.J. The arbuscular mycorrhizal symbiosis: an underground association. Trends in Plant Science 1997; 2: 54-60.

Harrison, M.J. Development of the arbuscular mycorrhizal symbiosis. Current Opinion in Plant Biology 1998; 1: 360-365.

Harrison, M.J. Molecular and Cellular Aspects of the Arbuscular Mycorrhizal symbiosis. Annual Review of Plant Physiology and Plant Molecular Biology 1999; 50; 361-389.
Hayman, D.S. The physiology of vesicular-arbuscular endomycorrhizal symbiosis. Canadian Journal of Botany 1982; 6: 944-963.
Hernandez-Sebastia, C., Poulin, M.J, Piche, Y., Desjardins, Y. Effect of vesicular arbusuclar mycorrhizal fungi on water relations of micropropagated strawberry plantlets during the rooting phase. In: *Mycorrhizas in integrated systems from genes to plant development*, C., Azcon-Aguilar, J.M., Barea, Eds.. Office for official publications of the European Communities, Luxembourg, 1996; pp440-443.
Hernandez-Sebastia, C., Samson, G., Bernier, P., Piche, Y., Desjardins, Y. *Glomus intraradices* causes differential changes in amino acid and starch concentrations of *in vitro* strawberry subjected to water stress. New Phytologist 2000; 148: 177-186.
Hirsch, A.M., Kalpulnik, Y. Signal transduction pathways in mycorrhizal associations: comparisons with the *Rhizobium*-legume symbiosis. Fungal Genetics and Biology 1998; 23: 205-212.
Hooker, J.E., Gianinazzi, S., Vestberg, V., Barea, J.M., Atkinson, D. The application of arbuscular mycorrhizal fungi to micropropagated systems and opportunity to reduce chemical inputs. Agricultural Science in Finland 1994; 3: 227-231.
Hrselova, H., Vejsadova, H., Prikryl, Z., Vachova J., Vancura, V., Vit, A., Effect of inoculation with vesicular-arbuscular mycorrhizal fungi on growth of strawberries. In: *Interrelationships between plants and micro-organisms in soil*, V., Vancura, F., Kunc, Eds.. Academia, Prague, 1988; pp109-114.
Jaizme-Vega, M.C., Azcon, R. Effect of vesicular arbuscular mycorrhizal fungi on pineapple (*Ananas comosus* L.) in the canary islands. Fruits 1991; 46: 47-50
Jaizme-Vega, M.C. , Azcon, R. Response of some tropical and sub-tropical cultures to endomycorrhizal fungi. Mycorrhiza 1995; 5: 213-217
Jaizme-Vega, M.C.,Declerck, S., Romero, R.A.S., Hermoso, M.C. Effect of *in vitro* monoxenically produced arbuscular mycorrhizal fungi on the first stages of development of micropropagated banana plantlets. In; Programs, Abstracts and Papers of the COST 8.38 Action Working Groups 1 and 4 meeting on *Knowledge on population biology of AMF as a tool for mycorrhizal technology*, 2001, Průhonice Czech Republic.
Jarstfer, A.G., Sylvia, D.M. Inoculum production and inoculation technologies of vesicular arbuscular mycorrhizal fungi. In: *Soil Technologies: Applications in Agriculture, Forestry and Environmental Management*, B., Metting ed., Marcel Dekker Inc., New York, USA. 1992; pp349-377.
Jasper, D.A., Abbott, L.K., Robson, A.D. Soil disturbance reduces the infectivity of external hyphae of vesicular arbuscular mycorrhizal hyphae. New Phytologist 1989; 112: 93-99.
Johnson, N.C., Graham, J.H., Smith, F.A. Functioning of mycorrhizal associations along the mutualism-parasitism continuum. New Phytologist 1997; 135: 575-585.
Kjoller, R., Rosendahl, S. The presence of the arbuscular mycorrhizal fungus *Glomus intraradices* influences enzymatic activities of the root pathogen *Aphanomyces euteiches* in pea roots. Mycorrhiza 1997; 6: 487-491.
Koide, R.T., Schreiner, R.P. Regulation of the vesicular-arbuscular mycorrhizal symbiosis. Annual Review of Plant Physiology and Plant Molecular Biology 1992; 43: 557-581.
Lambert, D.H., Cole, H. Effects of mycorrhizae on establishment and performance of forage species in mine spoil. Agronomy Journal 1980; 72: 257-260.
Lopez, A., Pinochet, J., Fernandez, C., Calvet, C., Camprubi, A. Growth response of OHF-333 pear rootstock to arbuscular mycorrhizal fungi, phosphorous nutrition and *Pratylenchus vulnus* infection. Fundamentals in Applied Nematology 1997; 20: 87-93
Lovato, P., Guillemin, J.P., Gianinazzi, S. Application of commercial arbuscular endomycorrhizal fungal inoculants to the establishment of micropropagated grapevine rootstock and pineapple plants. Agronomie 1992; 12: 873-880.
Lovato, P., Hammatt, N., Gianinazzi-Pearson, V., Gianinazzi, S. Mycorrhization of micropropagated mature wild cherry (*Prunus avium* L.) and common ash (*Fraxinus excelsior* L.). Agricultural Science of Finland 1994; 3: 297-301.
Marschner, H., Dell, B. Nutrient uptake in mycorrhizal symbiosis. Plant & Soil 1994; 159: 89-102.
Miller-Wideman, M.A., Watrud, L.S. Sporulation of *Gigaspora margarita* on root cultures of tomato. Canadian Journal of Microbiology 1984; 30: 642-646.
Monticelli, S., Puppi, G., Damiano, C. Effects of *in vivo* mycorrhization of micropropagated fruit tree rootstocks. Applied Soil Ecology 2000; 15: 105-111.
Morton, J.B., Benny, G.L. Revised classification of arbuscular mycorrhizal fungi (Zygomycetes); A new order, Glomales, 2 new suborders, Glomineae and Gigasporineae, and 2 new families, Acaulosporaceae and Gigasporaceae with an recommendation of Glomaceae. Mycotaxon 1990; 37: 471-491.

Morton, J.B., Redecker, D. Two new families of Glomales, Archaesporaceae and Paraglomaceae, with two new genera *Archaeospora* and *Paraglomus*, based on concordant molecular and morphological characters. New Phytologist 2001; 93: 181-195.
Mosse, B., Hepper, C. Vesicular-arbuscular mycorrhizal infections in root organ cultures. Physiological Plant Pathology 1975; 5: 215-223.
Mugnier, J., Mosse, B. Vesicular-arbuscular mycorrhizal infections in transformed root inducing T-DNA roots grown axenically. Phytopathology 1987; 77: 1045-1050.
Nemec, S. VA mycorrhiza in horticultural systems. In: *Ecophysiology of VA mycorrhizal plants*, G.R. Safir ed.. CRC Boca Raton, FL. USA. 1986; pp59-69.
Newsham, K.K., Fitter, A.H., Watterson, A.R. Arbuscular mycorrhiza protect an annual grass from root pathogenic fungi in the field. Journal of Ecology 1995; 83: 991-1000.
Niemi, M., Vestberg, M. Inoculation of commercially grown strawberry with VA mycorrhizal fungi. Plant and Soil 1992; 144: 133-142.
Pawlowska, T.E., Douds, D.D., Charvat, I. *In vitro* propagation and life cycle of the arbuscular mycorrhizal fungus *Glomus etunicatum*. Mycological Research 1999; 103: 1549-1556.
Pearson, J.N., Schweiger, P. *Scutellospora calospora* (Nicol. & Gerd.) Walker and Sanders associated with subterranean clover produces non-infective hyphae during sporulation. New Phytologist 1994; 127: 697-701.
Pinochet, J., Calvet, C., Camprubi, A., Fernandez, C. Growth and nutritional response of Nemared peach rootstock infected with *Pratylenchus vulnus* and the mycorrhizal fungus *Glomus mosseae*. Fundamentals in Applied Nematology 1995; 18: 205-210.
Pinochet, J., Camprubi, A., Calvet, C. Effects of the root lesion nematode *Pratylenchus vulnus* and the mycorrhizal fungus *Glomus mosseae* on the growth of EMLA-26 apple rootstocks. Mycorrhiza 1993; 4: 79-83.
Pinochet, J., Camprubi, A., Calvet, C., Fernandez, C., Kabana, R.R. Inducing tolerance to the root lesion nematode *Pratylenchus vulnus* by early mycorrhizal inoculation of micropropagated myrobalan 29C plum rootstock. Journal of American Society Horticultural Science 1998; 123: 342-347.
Pinochet, J., Fernandez, C., Jiazme, M.D., Tenoury, P. Micropropagated banana infected with *Meloidogyne javanica* responds to *Glomus intraradices* and phosphorous. HortScience 1997; 32: 101-103.
Pozo, M.J., Azcon-Aguilar, C., Dumas-Gaudot, E., Barea, J.M. Beta-1-3 glucanase activities in tomato roots inoculated with the arbuscular mycorrhizal fungi and/or *Phytophthora parasitica* and their possible involvement in bioprotection. Plant Science 1999; 141: 149-157.
Pozo, M.J., Dumas-Gaudot, E., Slezack, S., Cordier, C., Asselin, A., Gianiazzi, S., Gianinazzi-Pearson, V., Azcon-Aguilar, C., Barea, J.M. Induction of new chitinase isoforms in tomato roots during interactions with *Glomus mosseae* and/or *Phytophthora nicotianae* var *parasitica*. Agronomie 1996; 16: 689-697.
Preece, J.E., Sutter, E.G. Acclimatisation of micropropagated plants to the greenhouse and the field. In: *Micropropagation* PC Debergh, RH Zimmerman eds.. Kluwer Academic publishers, Dordrecht 1991, pp71-93.
Puthur, J.T., Prasad, k.V.S.K., Sharmila, P., Saradhi, P.P. Vesicular arbuscular mycorrhizal fungi improves establishment of micropropagated *Leucaena leucocephala*. Plant Cell, Tissue and Organ Culture 1998; 53: 41-47.
Rapparini, F., Baraldi, R., Bertazza, G. Growth and carbohydrate status of *Pyrus communis* L. plantlets inoculated with *Glomus* sp. Agronomie 1996; 16: 653-661.
Rapparini, F., Baraldi, R., Bertazza, G., Brazanti, B., Predieri, S. Vesicular arbuscular mycorrhizal inoculation of micropropagated fruit trees. Journal of Horticultural Science 1994; 69: 1101-1109.
Ravolanirina, F., Gianinazzi, S., Trouvelot, A., Carre, M. Production of endomycorrhizal explants of micropropagated grapevine rootstock. Agricultural Ecosystems and Environment 1989; 29: 323-327.
Redecker, D., Kodner, R., Graham, L.E. Glomalean fungi from the Ordovacian. Science 2000; 289: 1920-1921.
Remy, W., Taylor, T.N., Hass, H., Kerp, H. Four hundred million year old vesicular arbuscular mycorrhizae. Proceedings of the National Academy of Science in the USA 1994; 91: 11841-11843.
Ricken, B., Hofner, W. Effect of Arbuscular Mycorrhizal fungi (AMF) on heavy-metal tolerance of Alfafa (*Medicago sativa* L.) and Oat (*Avena sativa* L.) on a sewage sludge treated soil. *Zeitschrift fur Planzenernahrung und Bokenkunde* 1996; 159: 189-194.
Rosendahl, S. Interactions between the vesicular-arbuscular mycorrhizal fungus *Glomus fasiculatum* and *Aphanomyces euteiches* root rot of pea. Phytopathology Z 1985; 114: 31-40.

Salamanca, C.P., Herrera, M.A., Barea, J.M. Mycorrhizal inoculation of micropropagated woody legumes used in revegetation programmes for desertified Mediterranean ecosystems. Agronomie 1992; 12: 869-872.
Sbrana, C., Giovannetti, M., Vitagliano, C. The effect of mycorrhizal infection on survival and growth renewal of micropropagated fruit rootstocks. Mycorrhiza 1994; 5: 153-156.
Schellenbaum, L., Berta, G., Ravolanirana, F., Tisserant, B., Gianinazzi, S., Fitter, A.H. Influence of endomycorrhizal infection on root morphology in micropropagated woody plant species (*Vitis vinifera* L.). Annals of Botany 1991; 68: 136-141.
Schubert, A., Bodrino, C., Gribaudo, I. Vesicular-arbuscular mycorrhizal inoculation of kiwi fruit (*Actinidia delicosa*) micropropagated plants. Agronomie 1992; 12: 847-850.
Schubert, A., Lubraco, G. Mycorrhizal inoculation enhances growth and nutrient uptake of micropropagated apple rootstocks during weaning in commercial substrates of high nutrient availability. Applied Soil Ecology 2000; 15: 113-118.
Schubert, A., Martinelli, A. Effects of vesicular-arbuscular mycorrhizae on growth *in vitro* propagated *Pistacia integerrima* Acta Hortic., 1988 441-443.
Schubert, A., Mazzitelli, M., Ariusso, O., Eynard, I. Effects of vesicular arbuscular mycorrhizal fungi on micropropagated grapevines. Influence of endophyte strain, P fertilisation and growth medium. 1990 Vitis; 29: 5-13.
Simon, L., Bousquet, J., Levesque, R.C., Lalonde, M. Origin and diversification of endomycorrhizal fungi and coincidence with vascular plants. Nature 1993; 363: 67-69.
Slezack, S., Dumas-Gaudot, E., Paynot, M., Gianinazzi, S. Is a fully established arbusuclar mycorrhizal symbiosis required for bioprotection of *Pisum sativum* roots against *Aphanomyces euteiches*? Molecular Plant-Microbe Interactions 2000; 13, 238-241.
Slezack, S., Dumas-Gaudot, E., Rosendahl, S., Kjoller, R., Paynot, M., Negrel, J., Gianinazzi, S. Endoproteolytic activities in pea roots inoculated with the arbuscular mycorrhizal fungus Glomus mosseae and/or *Aphanomyces euteiches* in relation to bioprotection. New Phytologist 1999; 142: 517-529.
Smith, F.A., Smith, S.E. Membrane transport at the biotrophic interface: an overview. Australian Journal of Plant Physiology 1989; 16: 33-43.
Smith, F.A., Smith, S.E. Mutualism and parasitism: diversity in function and structure in the " Arbuscular" (VA) mycorrhizal symbiosis. Advances in Botanical Research 1996; 22: 1-43.
Smith, S.E., Dickson, S., Morris, C., Smith, F.A. Transfer of phosphate from fungus to plant in VA mycorrhizas: calculation of the area of symbiotic interface and fluxes from two different fungi to *Allium porrum* L. New Phytologist 1994; 127: 93-99.
Smith, S.E., Gianinnazzi-Pearson, V. Physiological interactions between symbionts in vesicular arbuscular mycorrhizal plants. Annual Review of Plant Physiology and Plant Molecular Biology 1988; 39: 221-244.
Smith, S.E., Read, D.J. *Mycorrhizal symbiosis*. Academic Press, New York. 1997.
Soberon, J.M., Martinez, del Rio C. Cheating and taking advantage in mutualistic associations. In The biology of mutualism ecology and evolution, D.H. Boucher Ed.. Croom Helm, London. 1985; pp 192-216.
Subhan, S., Sharmila, P., Saradhi, P. *Glomus fasciculatum* alleviates transplantation shock of micropropagated *Sesbania sesban*. Plant Cell Reports 1998; 17: 268-272.
Subramanian, K.S., Charest, C., Dwyer, L.M., Hamilton, R.I. Arbuscular mycorrhizas and water relations in Maize under drought stress at Tasseling. New Phytologist 1995; 129: 643-650.
Taylor, J., Harrier, L.A. A comparison of nine species of arbuscular mycorrhizal fungi on the development and nutrition of micropropagated *Rubus idaeus* L. cv. Glen Prosen (Red Raspberry). Plant & Soil, 2000; 225: 53-61.
Taylor, J. & Harrier, L.A. A comparison of development and mineral nutrition of micropropagated *Fragaria x ananassa* cv. Elvira (Strawberry) when colonised by nine species of arbuscular mycorrhizal fungi. Applied Soil Ecology 2001; In Press.
Taylor, T.N., Remy, W., Hass, H., Kerp, H. Fossil arbuscular mycorrhizae from the early Devonian. Mycologia 1995; 87: 560-573.
Tester, M., Smith, S.E., Smith, F.A. The phenomenon of non-mycorrhizal plants. Canadian Journal of Botany 1987; 65: 419-431.
Tisserant, B., Schellenbaum, L., Gianinazzi-Pearson, V. Gianinazzi, S., Berta, G. Influence of infection by endomycorrhizal fungus on root development and architecture in *Platanus acerifolia*. Allionia 1992; 30: 171-182.

Trotta A., Varese, G.C., Gnavi, E., Fusconi, A., Sampo, S., Berta, G. Interactions between the soil-borne root pathogen *Phytopthora nicotianae* var. *parasitica* and the arbuscular mycorrhizal fungus *Glomus mosseae* in tomato plants. Plant & Soil 1996; 185: 199-209.

Uosukainen, M., Vestberg, M. Effect of inoculation with arbuscular mycorrhizas on rooting, weaning and subsequent growth of micropropagated Malus (L.) Moench. Agricultural Science Finland 1994; 3: 269-279

Varma, A., Schüepp, H. Infectivity and effectiveness of *Glomus intraradices* on micropropagated plants. Mycorrhiza 1994; 5: 29-37.

Verma, R.K., Arya, I.D. Effect of arbuscular mycorrhizal fungal isolates and organic manure on growth and mycorrhization of micropropagated *Dendrocalamus asper* plantlets and on spore production in their rhizosphere. Mycorrhiza 1998; 8: 113-116.

Vestberg, M. The effect of vesicular arbuscular mycorrhizal inoculation on the growth and root colonisation of ten strawberry cultivars. Agricultural Science of Finland 1992; 1: 527-535.

Vestberg, M., Estaun, V. Micropropagated plants, an opportunity to positively manage mycorrhizal activities. In: *Impact of arbuscular mycorrhizas on sustainable Agriculture and Natural Ecosystems*, S. Gianinazzi, H. Schüepp, Eds., Birhäuser, Basel. 1984; pp217-226

Vestberg, M., Palmujoki, H., Parikka, P., Uosukainen, M. Effect of arbuscular mycorrhizas on crown rot (*Phytophthora cactorum*) in micropropagated strawberry plants. Agricultural Science of Finland 1994; 3: 289-295.

Vidal, M.T., Azcon-Aguilar, C., Barea, J.M., Pliego-Alfaro, F. Mycorrhizal inoculation enhances growth and development of micropropagated plants of avocado. Horticultural Science 1992; 27: 785-787.

Williams, S.C.K., Vestberg. M., Uosukainen, M., Dodd, J.C., Jefferies, P. Effects of fertilisers and arbuscular mycorrhizal fungi on the post-*vitro* growth of micropropagated strawberry. Agronomie 1992: 12: 851-857.

Yano-Melo, A.M., Saggin, O.J., Lima-Filho, J.M., Melo, N.F., Maia, L.C. Effect of arbuscular mycorrhizal fungi on the acclimatisation of micropropagated banana plantlets. Mycorrhiza 1999; 9:119-123.

SECTION B

6. Micropropagation of Tropical Woody Species

Adriana Pinheiro Martinelli Rodriguez
Laboratório de Biotecnologia Vegetal
Centro de Energia Nuclear na Agricultura
Universidade de São Paulo
Av. Centenário 303
Piracicaba, SP, 13416-970, Brazil

Wagner Aparecido Vendrame
Tropical Research and Education Center
Institute of Food and Agricultural Sciences
University of Florida
18905 S.W. 280th Street
Homestead, FL 33031, USA

1. INTRODUCTION

During the last thirty years plant micropropagation has been one of the most significant contributions of biotechnology to the horticultural industry (Giles and Friesen, 1994). The possibility of mass propagation of elite, endangered, or difficult-to-propagate genotypes, in relatively small areas and time period, is the most important characteristic of micropropagation. In association with breeding programs micropropagation can have a significant impact not only in the multiplication of important genetic material, but also for the *in vitro* maintenance of germplasm.

In vitro multiplication of forest species has many advantages including the possibility of clonal mass propagation, reversion to juvenility, and uniformity of stands. All of them are important characteristics for the establishment of new areas including productivity of stands and harvest of large areas such as tropical forest plantations.

S.M. Jain and K. Ishii (eds.), Micropropagation of Woody Trees and Fruits, 153–179.

The selection or definition of a culture medium is vital for the success of an *in vitro* process. The inorganic salt formulations can vary, but in most cases MS medium (Murashige & Skoog, 1962) is the most routinely used culture medium in plant tissue culture (Smith, 1992). Other culture media commonly used for the micropropagation of woody species are WPM (Lloyd and Mc Cown, 1980), de Fossard (1974), Quoirin and Lepoivre (1977) and Gonçalves (1980), which are considered as having a medium total ionic concentration, and are commonly used for *Eucalyptus* spp. The use of diluted concentrations of MS salts is also very common due to the high total salt concentration in MS medium, which was originally defined for tobacco callus cultures. The optimization of culture conditions results in better productivity of cultures not only in terms of number of plants produced, but also in quality, which in turn facilitates their acclimatization and growth *ex-vitro*.

Several factors are responsible for the success of micropropagation and they need to be well defined through experimentation. Changes in culture medium components such as nutrients, growth regulators and other additives, and also in the environmental conditions of culture vessels are essential for the development of micropropagation protocols. Among various factors influencing plant micropropagation, type of growth regulator is the most frequently studied. Different types of growth regulators may affect different stages of the process: culture establishment, rate and efficiency of multiplication, plant elongation, and rooting. Other factors studied are type of explant, physical state of the culture medium (liquid or solidified with different gelling agents), culture conditions (temperature, and light intensity and duration), addition of various substances (casein hydrolysate, coconut water, amino acids, carbohydrates, etc.), and the levels of nutrients in the culture medium.

This chapter will cover three aspects related to micropropagation: firstly, a description on recent advances in the micropropagation of *Eucalyptus* and passion fruit (*Passiflora* spp), and new strategies for the formulation of improved culture media for various species; secondly, a review of micropropagation of different tropical woody species including forest, medicinal and fruits; and finally, a brief review on somaclonal variation and the use of molecular techniques to approach this type of variation.

2. STRATEGIES FOR IMPROVING MICROPROPAGATION PROCEDURES FOR EUCALYPT AND PASSION FRUIT

2.1. Micropropagation of *Eucalyptus* spp

Clonal propagation is an efficient technique for increasing the productivity and uniformity in forest species. The rooted cuttings and *in vitro* propagation through axillary and adventitious bud induction are commercially used for the propagation of *Eucalyptus*. Hybridization is a common practice in Eucalyptus breeding followed by selection. Micropropagation can greatly shorten the breeding cycle, and the selection and multiplication of interspecific and intraspecific hybrids of commercial value.

Most of the protocols for eucalypt micropropagation use BA alone, or in combination with a lower concentration of an auxin.. There are few other reports on the use of cytokinins and thidiazuron (TDZ), probably due to the positive results obtained with BA. Tibok *et al.* (1995) obtained callus from hypocotyls of *E. urophylla* by having BA or TDZ in combination with a low concentration of auxin. The use of BA promoted regeneration through organogenesis, while TDZ promoted not only the formation of buds from callus, but also the development of somatic embryos. Lima and Gonçalves (1998) used clones of an hybrid of *E. grandis* x *E. tereticornis* and tested different concentrations of TDZ for the multiplication of bud aggregates maintained *in vitro*. Very low doses of TDZ (0.01 $mg.L^{-1}$) promoted the best results whereas at higher concentrations (0.1 or 1.0 $mg.L^{-1}$), shoot proliferation occurred, but the shoots were shorter and frequently associated with callus formation and hyperhydricity.

Although micropropagation has many advantages, in many cases it is an expensive method of propagation. For some genotypes of eucalypts, *ex vitro* rooting is a feasible strategy that reduces costs. When thousands of plants are produced commercially, the presence of roots in plants transferred to acclimatization is undesirable, as it is difficult to outplant them without causing injury to the roots. *Ex vitro* rooting of *Eucalyptus* micropropagated plants was successfully obtained by Xavier and Comério (1997). Grattapaglia and Machado (1990) cited several advantages of *ex vitro* rooting in a commercial set up in order to cut down the cost of *in vitro* produced plants. Rooting fo shoots *in vitro* represents one extra tissue culture step, implying that cultures remain for a longer period of time in the culture room, extra consumption of electric energy and culture media, and more specialized personnel as compared to *ex vitro* rooting. The root system developed directly in the substrate also shows a higher quality, with the formation of more secondary roots and absence of callus. The term microcuttings, conventionally used to define small cuttings harvested from hedges and greenhouse-grown parent plants, can also be used for shoots produced *in vitro* and rooted *ex vitro* .

Xavier & Comério (1997) tested *ex vitro* rooting of three micropropagated clones of hybrids of *Eucalyptus grandis* x *E. urophylla*. Four concentrations of IBA (indol butyric acid) – 0, 600, 1500, 3000 mg.L^{-1} - were compared. The microcuttings, which consisted of elongated shoots higher than 1.5 cm, were maintained in the greenhouse under high humidity for 10, 15, 20 or 25 days, followed by 10 days in a shaded greenhouse and 90 days under full sunlight. Differences in response were observed among clones, with one of them (H13) being tolerant to the hormonal treatments. The other two clones (H5 and H12) tested showed more pronounced losses in parameters such as percent rooting, plant height and rate of plants obtained after 90 days, with increasing IBA concentration. Despite differences among clones, the best results for *ex vitro* rooting of all three clones were obtained without IBA under high humidity in 10 to 15 days in the greenhouse. This confirms that, at least for some eucalypt genotypes, micropropagation is an effective method for the reversion to juvenility (Gonçalves 1980), since no auxin treatment was needed for rooting of the micropropagated shoots.

Large-scale production of *Eucalyptus* plants for pulp and paper industry plantations is still done, in many cases, by rooting of cuttings for the production of clonal material from selected mother plants. Higashi *et al.* (2000 a, b) described the evolution of the clonal gardens for the production of eucalypt plants by Brazilian pulp and paper companies. In the early years, clonal gardens, commonly used by these companies, needed large areas for the production of planting material (Campinhos and Ikemori 1983). More recently, taking advantage of the rejuvenation ability of *in vitro* propagation (Gonçalves 1980), new systems of clonal gardens were developed and are being employed by 'Champion Papel e Celulose Ltda.' (SP, Brazil) for *Eucalyptus* spp. The use of micropropagated material as a source of propagules for clonal gardens has improved the plant propagation programs (Higashi *et al.* 2000). In this case, clonal material, micropropagated from nodal segments through several subcultures, is outplanted, rooted in the greenhouse under mist and transferred to a microclonal garden (Iannelli *et al.* 1996). Microcuttings (3-4 cm-long shoot apices) are obtained every 15 to 20 days and rooted in the greenhouse. Iannelli *et al.* (1996) cited several advantages of this method, when compared with the 'conventional' propagation of cuttings harvested from plants (*e.g.* hedge) not originally produced through *in vitro* culture. Among the advantages recognized by 'Champion Papel e Celulose Ltda.' (SP, Brazil) are the shortening of the breeding program due to a faster clonal recommendation; better rooting in the greenhouse when compared with rooting of other types of cuttings; reduction in the propagation period from

90-120 days to 75-85 days; reduction in greenhouse investment; and higher rate of survival of plants in the field due to high vigor, and other improved characteristics.

2.2 Micropropagation of passion fruit

Passiflora edulis f. *flavicarpa* (yellow passion fruit) and *Passiflora edulis* (purple passion fruit) are the two most important species of *Passiflora* spp. The fruits are commercially used mainly for concentrated and fresh juice. According to Vaz (1991), passion fruit can benefit greatly from biotechnology via micropropagation of superior clones directly or for rootstocks, and virus elimination through meristem culture. However, programs employing micropropagation must take into account the high auto-incompatibility of these species, hence, clonal orchards must have several clones to assure cross pollination (Grattapaglia *et al.* 1991).

Most of the *in vitro* work done in *Passiflora edulis* and *P. edulis* f. *flavicarpa* was related to shoot multiplication from nodal segments, internode, or leaf discs. Moran Robles (1978, 1979) used axillary buds and internodal segments as explants in medium containing kinetin, followed by transfer to medium with IAA. Drew (1991) micropropagated *P. edulis* f. *flavicarpa* in medium containing BA, kinetin and IAA. Kantharajah and Dodd (1990) micropropagated *P. edulis* from nodal segments on MS medium supplemented with 2 $mg.L^{-1}$ BA, followed by medium devoid of growth regulators and then on 1 $mg.L^{-1}$ NAA for rooting. The efficiency of coconut water favoring bud formation was also observed in *P. edulis* (Kantharajah and Dodd 1990), and other *Passiflora* species, including *P. edulis* f. *flavicarpa* (Dornelas and Vieira 1994).

Passiflora edulis f. *flavicarpa* has shown sensibility to high levels of ethylene in the culture vessel. The use of silver thiosulfate (STS) was attempted by Faria and Segura (1997 a,b), as an ethylene inhibitor. The use of 8.8 μM STS favored the formation of higher number of adventitious buds from leaf discs as a result of reduction in ethylene levels.

Although most of the research articles on micropropagation of *Passiflora* spp report the use of full- or half-strength MS medium (Kantharajah and Dodd 1990, Dornelas and Vieira 1994, Manders *et al.* 1994, Kawata *et al.* 1995, Otoni *et al.* 1995); changes in the nutrient composition improved yields and plant quality. Kantharajah and Dodd (1990) reported mineral deficiency in micropropagation of *P. edulis* from nodal segments in MS medium; plants showied chlorosis and eventually died. Moran Robles (1978) tested different media formulations for *P. edulis* f. *flavicarpa* and *P. mollisima*, obtaining differences in growth and rooting *in vitro*. Different forms and levels of inorganic nitrogen (NO_3^- and NH_4^+) were used for the induction of adventitious buds, from hypocotyl and leaf discs (Faria and Segura 1997a), and shoot apices (Faria and Segura 1997b) in *P. edulis* f. *flavicarpa*. Faria and Segura (1997a) also compared MS and WPM (Lloyd and McCown, 1980) and observed that MS was more effective than WPM for the differentiation of adventitious buds. The omission of either source of nitrogen in MS medium, or a disequilibrium of the 2:1 ratio (NO_3^-:NH_4^+) reduced the ability of adventitious bud formation (Faria and Segura, 1997a).

2.3. Adequation of the mineral composition of the culture medium

Micropropagation efficiency is highly influenced by the ionic equilibrium of the culture medium, diffusion and assimilation of nutrients by the explant, for an active homogeneous and effective growth in the shortest period of time (George and Sherrington 1984). The research group leaded by Dr. A. Natal Gonçalves, at the Department of Forest Sciences, University of São Paulo, Brazil, has been engaged in defining specific media formulations for *Eucalyptus* species, hybrids and clones. Based on the information obtained in the literature on the nutritional characteristics of young plants of *E. grandis*, an optimal concentration range was defined for each major inorganic element. A specific culture medium was then formulated, named JADS (Table 1), with adequate nutrient levels for the production of 150 mg of dry matter per explant per 10 ml of culture medium (Correia, 1993; Correia *et al.* 1995).

Following the same approach, Grossi (1995) collected nutritional information from the literature on *E. saligna* to select a suitable medium for this species. Considering that *E. saligna* requires higher soil fertility than *E. grandis*, initially MSA medium was developed, and tested *in vitro* with *E. saligna* clones. This medium was modified based on the preliminary results. MSA-1 medium was then defined and compared with MS medium, which has higher levels of PO_4^{3-}, Ca^{2+}, Mg^{2+}, SO_4^{2-}, Fe^{2+}, Cu^+, Co^{2+}, lower levels of NO_3^-,NH_4^+, K^+, BO_3^-, Zn^+, Mn^+, the same amount of MoO_4^{2-}, and no Cl^-. Seven *in vitro* micropropagated clones were tested. MSA-1 medium was better than MS and MSA media in producing higher dry matter after 35 days in culture, and lower occurrence of hyperhydricity. No significant differences were observed in the multiplication rate of adventitious buds and further elongation.

These studies were also extended to cultures of passion fruit, which had shown severe mineral deficiency when cultured on MS medium (Monteiro 2000, Monteiro *et al.* 2000). Since other researchers had also reported similar observations, an adequation of the culture medium for *Passiflora edulis* f. *flavicarpa* was attempted (Monteiro 2000, Monteiro *et al.* 2000) using a similar, albeit more precise approach to that used for the formulation of JADS or MSA-1 media for *E. grandis* and *E. saligna*, respectively.

Leaf mineral analysis data of young healthy plants were collected from the literature (Malavolta *et al.* 1997, Reuter and Robinson 1997), and a medium was formulated according to the nutrient balance in the leaves. The chemical equilibrium of the new medium was then adjusted using the software Geochem (Sposito and Mattigod, 1980). For that, the concentration of each element was inserted, the pH set to 5.8 and the software calculated the chemical equilibrium. Adjustments were then made to ensure high availability of the elements and low or no precipitation (Monteiro 2000, Monteiro *et al.* 2000).

The newly defined medium was named MSM (Table 1). A comparison between MS and MSM medium was done by culturing nodal segments in both media supplemented with 3.0 $mg.L^{-1}$ BA. After three subcultures, mineral analysis was done in leaves of both treatments. At this stage, mineral deficiency symptoms, such as chlorosis and reduced growth, were observed in shoots cultured on MS medium, while those on MSM medium had green leaves and were more elongated. Chlorosis on MS-produced shoots was reversed after transfer of shoots to MSM medium. Leaf mineral analysis showed that shoots on MSM had higher Ca, Cu, Fe, Mg and S, and lower B and Cl levels, than those on MS. The nutrient levels were also compared with information collected from the literature for healthy greenhouse-grown plants (Malavolta *et al.* 1997, Reuter and Robinson 1997). Levels of N, P, K, S, Cu, Mn and Zn in plants cultivated in both media (MS and MSM) were in accordance with the recommended levels for this species by Malavolta *et al.* 1997 and Reuter and Robinson 1997. Levels of Fe and Mg were adequate in plants cultivated in MSM (109 $mg.kg^{-1}$ and 2.96 $g.kg^{-1}$, respectively) and low in those cultivated in MS (62 $mg.kg^{-1}$ and 1.92 $g.kg^{-1}$, respectively), while B levels were low in MSM (22 $mg.kg^{-1}$) and adequate in MS (47 $mg.kg^{-1}$). The Ca level in plants cultured in MSM (9.38 $g.kg^{-1}$) was twice as much when compared to that in plants in MS (4.64 $g.kg^{-1}$). Ca level in MSM was still low according to the range suggested in the literature (15-27.5 $g.kg^{-1}$). It should be considered that the high humidity in the tissue culture vessels may interfere in Ca absorption. Finally, Cl level in MS was excessively high (7100 $mg.kg^{-1}$), while in MSM (300 $mg.kg^{-1}$) although still above the recommended level (< 20 $mg.kg^{-1}$), it was much lower (Monteiro 2000, Monteiro *et al.* 2000).

Besides the adequation of media for eucalypt and passion fruit, this approach has also been tested for banana micropropagation with positive results (data not published), indicating that it should be useful for the definition of the mineral composition of media specific to each species.

Table 1: Composition of JADS (Correia 1993, Correia *et al.* 1995) and MSM (Monteiro 2000, Monteiro *et al.* 2000) media, defined for the micropropagation of *Eucalyptus grandis* and *Passiflora edulis* f. *flavicarpa*, respectively, compared to MS (Murashige and Skoog, 1962).

Components	JADS	MSM	MS
		$g.L^{-1}$	
NH_4NO_3	324.00	1000.00	1650.00
KNO_3	809.00	1520.00	1900.00
KH_2PO_4	408.00	272.00	170.00
$Ca(NO_3)_2.4H_2O$	1181.00	1300.00	----
$CaCl_2.2H_2O$	----	----	440.00
$MgSO_4.7H_2O$	739.50	616.00	370.00
$FeSO_4.7H_2O$	55.60	55.60	27.80
Na_2EDTA	74.50	74.84	33.60
$MnSO_4.4H_2O$	16.90	----	22.30
$MnSO_4.H_2O$	----	28.00	----
$ZnSO_4.7H_2O$	4.32	2.90	8.60
$Na_2MoO_4.2H_2O$	0.15	0.25	0.25
$CuSO_4.5H_2O$	1.25	1.25	0.025
$CoCl_2.6H_2O$	0.25	0.025	0.025
H_3BO_3	3.10	4.50	6.20
KI	----	----	0.83
Nicotinic acid	0.50	0.50	0.50
Pyridoxine	0.50	0.50	0.50
Thiamine	5.00	0.10	0.10
Glycine	----	2.00	2.00
Myo-inositol	100.00	100.00	100.00
L-arginine	7.00	----	----
L-glutamine	146.00	----	----
L-cysteine	2.50	----	----
Ca pantotenate	2.40	----	----
Sucrose	30000.00	30000.00	30000.00

3. MICROPROPAGATION OF TROPICAL WOODY SPECIES

3.1. Micropropagation of tropical pines (*Pinus* spp)

A great number of studies on *in vitro* propagation of angiosperm and gymnosperm woody trees has been reported, usually emphasizing somatic embryogenesis as the most commonly used tissue culture technique for pine species (Jain *et al.* 1998). Here we selected a few publications on micropropagation of pines, in particular for some tropical and sub-tropical species, such as: *Pinus elliotii* Engelm, *Pinus caribaea* Morelet, and *Pinus kesiya* Royle ex Gord. *P. caribaea* is a forest species of economical importance for its wood quality and for its reforestation potential. This pine species is native to South America: Belize, Guatemala, Honduras, Mexico - Quintana Roo, and Nicaragua. *P. elliotii* 's economical importance includes fiber, gum/resin, wood (for wood pulp and for miscellaneous purposes). Its native range extends through most of the Southern US, including Alabama, Florida, Georgia, Louisiana, Mississippi, and South Carolina. *P. kesiya* is generally used for its gum/resin and wood. It is native to Asia, being distributed from Temperate-Asia (China) to Tropical-Asia, including Bhutan, India, Laos, Myanmar, Philippines (Luzon), Thailand, and Vietnam (USDA 2001).

Meyer (1998) reported *in vitro* formation of adventitious buds on mature embryos of pine hybrids (*P. elliotii* Engelm x *P. caribaea* Morelet). Gresshoff and Doy (1972) (GD) medium was used for adventitious bud formation, which performed better than modified Murashige and Skoog (MS), Risser and White (RW), and Schenk and Hildebrandt (SH). Several growth regulators at different concentrations were tested, and a combination of 1.2 $mg.L^{-1}$ benzyladenine (BA) and 0.2 $mg.L^{-1}$ indolebutyric acid (IBA) was most effective for adventitious bud formation. Elongation of buds was obtained on half-strength Gresshof and Doy (GD) medium supplemented with 6.4 $mg.L^{-1}$ IBA. Plants were transferred to soil and hardened off. Different hybrid seed families resulted in differences in morphogenic potential and in growth and rooting of adventitious shoots, suggesting a genotypic effect.

Recently, Nadwani *et al.* (2001) studied the regeneration of *Pinus kesiya* Royle ex Gord from shoot tips of 14- 20-day-old seedling. Highest shoot formation and growth occurred on MS medium containing 5 $mg.L^{-1}$ kinetin. When transferred from MS to SH or Leproivre media, a high frequency of multiple shoot formation was obtained. The best multiple shoot induction was achieved by maintaining the cultures on a medium with cytokinin for four weeks. Following induction, shoot elongation and multiplication were obtained by transferring cultures to a cytokinin-free medium. For rooting of shoots, GD medium was used and shoots were exposed to naphtaleneacetic acid (NAA) or IBA followed by transfer to a medium without growth regulators.

3.2. Micropropagation of rubber tree (*Hevea brasiliensis* (Willd. ex A. Juss.) Müll. Arg.)

The rubber tree, *Hevea brasiliensis* (Willd. ex A. Juss.) Müll. Arg, is the most economical important species for the production of rubber or latex. It is native to Southern America, ranging from Bolivia (Beni), Brazil (Amapá, Amazonas, Mato Grosso, Pará), Colombia (Amazonia), to Peru (Huanuco, Loreto, Madre de Dios, Pasco, San Martin) (USDA 2001).

In the past ten years some articles have reported on micropropagation and related factors affecting the *in vitro* culture of rubber tree. Here we selected some of the most relevant work. Although somatic embryogenesis has played a major role in tissue culture of rubber tree, the *in vitro* organogenesis ability of somatic embryogenesis-derived clones has become of greater interest, as shown by Perrin *et al.* (1996). Shoots were used as explants for micropropagation of rubber tree. Repeated grafting onto juvenile rootstocks and somatic embryogenesis were used as intermediate steps, followed by *in vitro* axillary shoot and root formation. Comparisons were made between mature and rejuvenated clones regarding endogenous levels of selected plant growth regulators, such as zeatin, zeatin riboside, isopentenyladenine, and isopentenyladenosine. Clones whose shoots were micrografted *in vitro* showed higher levels of both zeatin forms when compared with mature non-grafted material. The high levels of zeatin were related to the ability for *in vitro* organogenesis (production of axillary shoots and roots). This shows that both forms of this growth regulator can be used for *in vitro* organogenesis after rejuvenating treatments in rubber tree clones.

Nodal cuttings were used to initiate microcutting cultures of rubber tree, with explants taken from branches showing different physiological growth stages (Lardet *et al.* 1998). This considerably affected the *in vitro* culture initiation and the organogenic potential. Buds of nodal segments obtained from dormant branches showed higher ability for *in vitro* shoot formation and elongation than those removed from branches with actively growing leaves. When *in vitro*-developed shoots were used as the source of secondary nodal explants, the branch growth stage persisted into the multiplication phase. However, this effect was temporary, disappearing after subculture of primary nodal explants. A positional effect of the buds on organogenesis was also observed, with buds from nodes located furthest from the apex showing higher *in vitro* competence for shoot morphogenesis. In addition, the organogenic potential was also dependent on the type and combination of growth regulators, their appropriate concentration and inhibitors of the growth regulators present in the buds (which negatively affected the organogenic potential).

Axillary bud formation has also been studied. Mendanha *et al.* (1998) used newly expanded leaves and axillary buds of rubber trees as explants for shoot organogenesis. MS medium was used for shoot formation from axillary buds, supplemented with kinetin (1.0 $mg.L^{-1}$), 2,4-D (1.0 $mg.L^{-1}$), sucrose (20.0 $g.L^{-1}$), and Difco agar (4.0 $g.L^{-1}$). The use of either axillary buds or leaf explants resulted in prolific callus growth in medium supplemented with kinetin, BA , 2,4-D, IBA., NAA, and IAA. No regeneration was obtained after transferring cultures to MS medium without plant growth regulators. However, shoots developed from axillary buds when cultured on MS medium with a combination of NAA, IAA, and IBA. Vigorous rooting was also observed.

Recently, Carron *et al.* (2000) propagated a variety of juvenile and mature genotypes either by somatic embryogenesis or *in vitro* microcuttings. The main purpose of the research was to compare the architecture of the root system developed in field conditions with that from trees derived from different propagation methods, such as seedlings, plants from *in vitro* microcuttings, and somatic embryogenesis-derived plants. This particular work faced the challenge of overcoming the problems encountered in rooting cuttings, such as the poor structure of the root system and aimed to check their anchoring ability. Studies of the root system included both static and dynamic components observed at different stages of growth up to 3 years. Comparisons were made to the root systems of seedlings of the same age, and vigor and balance of the root systems were evaluated. The development of a strong taproot and lateral root system was observed in *in vitro* plants. The root architecture resembled that of seedlings, although differences were noted in taproot and lateral root vigor among selected clones.

3.3. Micropropagation of endangered tropical tree species

During the last decade, there has been increased interest in alternative propagation methods as a means for the conservation of endangered tropical tree species, and a number of articles have been published. Tissue culture techniques (*e.g.* micropropagation) are valuable tools for the rapid multiplication of difficult-to-propagate, rare and endangered tropical trees, both for commercial production and germplasm conservation.

Albizia guachapele*, *Cedrella odorata and ***Swietenia macrophylla***, are three important timber species of the Meliaceae family, commonly known as "Guayaquil", "Spanish Cedar" and "Honduran Mahogany", respectively. Hypocotyl (*A. guachapele*, *C. odorata*) and epicotyl (*S. macrophylla*) segments were used by Valverde-Cerdas *et al.* (1998) as explants for induction of adventitious buds in half strength MS medium supplemented with four concentrations of BA. In addition, a combination of BA and NAA was also tested for *C. odorata*. Direct organogenesis was obtained for all three species, indicating their potential for *in vitro* propagation. *A. guachapele* and *S. macrophylla* responded to BA, with a maximum of 20 (4 mg.L^{-1} BA) and 15% (6, 8 or 10 mg.L^{-1} BA) of responsive explants, respectively. *C. odorata* hypocotys produced adventitious buds only in the presence of BA and NAA, with the highest total number of adventitious buds obtained with 4 mg.L^{-1} BA and 1 mg.L^{-1} NAA

Miconia sp. ("Jacatirão") is an endangered tropical woody species from the Brazilian Atlantic Forest that can be used for timber, furniture or as an ornamental species (USDA 2001). Micropropagation of "jacatirão" (*Miconia* sp., Melastomaceae, probably belonging to the species *Miconia cinnamomifolia* (DC) Naud.) has been obtained from explants of leaf and seedling nodal and internodal segments (Barrueto Cid *et al.* 1997). Shoot regeneration and multiplication was obtained from all three explants in MS modified medium containing TDZ. Leaf and nodal segments responded with higher rates of bud formation (60-90%), compared to internodal segments (20-30%). For leaf explants, increasing regeneration was observed with an increase in TDZ concentration, from 0.1 to 0.4 mg.L^{-1} TDZ, while for nodal segments, an increase in TDZ concentration (0.08 to 0.9 mg.L^{-1}) caused a decrease in shoot regeneration. Shoots were elongated in medium containing 0.01 mg.L^{-1} NAA, 0.2 mg.L^{-1} BA and 0.7 mg.L^{-1} GA_3. Rooting of the adventitious shoots was obtained in media containing 0.25, 0.5 or 1.0 mg.L^{-1} IBA, with no significant difference among the three concentrations.

Syzigium travancoricum Gamble is a highly endangered plant species restricted to the Western ghats. Its essential oils are of important medicinal use (USDA 2001). A micropropagation protocol was developed for the rapid multiplication of this species by Anand *et al.* (1999), involving nodal explants cultured on MS medium containing macronutrients and vitamins, WPM micronutrients and plant growth regulators. Using a basal MS medium supplemented with 4 mg.L^{-1} BA and 0.25 mg.L^{-1} NAA, multiple shoots were induced from non-woody branches from wild trees and from 1 to 2-yr-old seedlings under mist conditions (70% relative humidity). An average of 12 multiple shoots produced in rosettes per nodal explant (70% response) was obtained on multiplication medium. The nodular callus showed the best regeneration under 0.05 mg.L^{-1} BA and 2 mg.L^{-1} NAA. By the third subculture, their number increased to over 25 shoots per node on further incubation in fresh MS medium. Microshoots were regenerated *in vitro* from the nodular callus in a half-strength MS basal medium. Rooting was obtained in half-strength medium with 0.2 mg.L^{-1} IAA. The field survival rate was 40% of rooted plants.

3.4. Micropropagation of various tropical tree species

Aegle marmelos (L.) Corr. (Bael) is a woody species native to India. The tree has medicinal importance and its fruit, mildly astringent, is used for the cure of dysentery, diarrhea and stomach aches. Leaves and roots are also used for medicinal purposes (USDA 2001).

Ajithkumar and Seeni (1998) studied clonal propagation of bael via *in vitro* enhanced axillary bud proliferation. Young nodal segments from a 25-year-old tree were cultured on MS medium to induce bud break. Cytokinin supplement was essential for induction of bud break. The combination of 2.5 $mg.L^{-1}$ BA and 1 $mg.L^{-1}$ IAA provided a synergistic effect with the induction of about 12 shoots after 7 weeks in culture. In the same study, the shoots grown *in vitro* were used as the source of explants (node, whole leaf, shoot tip, and internode). Explants were subcultured in MS medium supplemented with 0.05 - 2.5 $mg.L^{-1}$ BA. After 7 weeks in culture, all the explant types produced from 3 to 11 shoots and shoot buds at frequencies varying from 40 to 100%. Regeneration ability was similar between nodal segments and leaves. Adventitious organogenesis from leaves was confined to the petiolar ends. The best responses were obtained with nodal segments in the presence of 0.05 - 0.1 $mg.L^{-1}$ BA., and uniform shoots were produced between 3.8 - 5.3 cm, facilitating the simultaneous harvest for rooting. Cultures were repeatedly subcultured with a continuous production of healthy shoots, without presence of callus or decline in multiplication for at least five cultivation cycles. Rooting of shoot cuttings was obtained in half-strength MS medium supplemented with 0.5 $mg.L^{-1}$ IAA or 10.0 $mg.L^{-1}$ IBA. A high rate of establishment was observed among the rooted plants raised in the presence of IAA (90%) and IBA (86%) compared to NAA (35%) after hardening. The post-transplantation growth of the micropropagated plants was uniform as observed visually. Further visual observations confirmed that stem-node-derived plants had normal growth and morphology, with no abnormalities. Isozyme analysis confirmed the uniformity of plants derived from axillary bud proliferation.

Azadirachta excelsa L. (Sentang) is an important tropical timber tree native to South East Asia and is being increasingly promoted and planted in Malaysia as a plantation species for timber (USDA 2001). Liew and Teo (1998) used axillary buds from 5-month-old seedlings of sentang to initiate *in vitro* cultures in MS medium supplemented with BA and NAA. The highest multiplication rate (number of axillary buds elongated) was obtained using a combination of 0.1 $mg.L^{-1}$ BA with 0.01 $mg.L^{-1}$ NAA, with an average of 8 shoots per explant. Normal shoot production was achieved subsequently in MS medium supplemented with 0.8 $mg.L^{-1}$ BA and 0.05 $mg.L^{-1}$ NAA, at an average of 4 shoots with normal appearance per explant. Continuous multiplication of elongated shoots resulted in abnormal shoots, when using the same medium and plant growth regulators. The combination of two different culture media for the different phases of sentang *in vitro* culture is essential for normal shoot production.

Bixa orellana L. ("Annatto") is a small tropical American tree, that belongs to Bixaceae family, grown in the hottest parts of India for its medicinal value and production of food coloring (carotenoids) from mature seeds (USDA 2001). The use of seeds for its propagation is limited due to low seed availability and poor germination rate. Rooting of cuttings also presents difficulties due to the production of a gummy substance and phenolics on the cut ends. Micropropagation was attempted by D'Souza and Sharon (2001) as an alternative method to obtain large number of planting material. Shoot apex and nodal explants from 6-week-old seedlings were cultured in B5 medium supplemented with various concentrations and combinations of growth regulators. Plants were regenerated in medium containing 1.0 $mg.L^{-1}$ 2-isopentenyl adenine (2iP), at a rate of 5 and 9 shoots per explant in shoot apex and nodal explants, respectively. Rooting of the regenerated shoots was obtained in medium supplemented with 0.01 $mg.L^{-1}$ NAA, for shoot apex-derived plants, or 0.5 $mg.L^{-1}$ NAA, for nodal-segment-derived plants. After 10 months (8 subcultures), 3350 rooted plants were obtained from one shoot with one apical shoot and four nodal segments, proving the commercial value of this protocol.

Ficus spp are tropical trees of interest either for their fruit production (*i.e.*, *F. carica* L.) or for their ornamental characteristics (*i.e.*, *F. benjamina* L.) (USDA 2001). Gabryszewska *et al.* (1997) studied *in vitro* organogenesis in *F. benjamina* L. cv. Golden King. Growth and development of roots and shoots were evaluated under different light spectra (white, blue, green, and red) and under the presence of 2iP and IAA. Using red light and 15 $mg.L^{-1}$ 2iP, a high number of shoots were obtained. The combination of red light with 0.4 $mg.L^{-1}$ IAA resulted in high number of roots, although all light treatments produced roots and regenerated rooted plants. The presence of 2iP in the medium further inhibited leaf senescence, when using in combination with red light.

Kumar *et al* (1998) developed a protocol for multiple *in vitro* shoot induction of *F. carica* L. cv. Gular. Apical buds were used as explants, collected from 7- to 8-year-old trees, and cultured on MS medium supplemented with 2.0 $mg.L^{-1}$ BA and 0.2 $mg.L^{-1}$ NAA. Culture establishment was achieved with 90% success. Shoots regenerated *in vitro* were multiplied on MS medium with the same supplements used for culture initiation. After 8 weeks, 1 to 4 shoots per explant were obtained. An average multiplication rate of 4 shoots per subculture was reported. Rooting of excised shoots was obtained in liquid half strength MS medium supplemented with 2.0 $mg.L^{-1}$ IBA. and 2.0 $g.L^{-1}$ activated charcoal. Regeneration was achieved with 68% of plantlets showing well-formed root systems. Plantlets were successfully hardened-off and acclimatized

***Garcinia mangostana* L.** (Mangosteen) is a woody species native to Tropical Asia, widely cultivated in Malaysia. Normah *et al.* (1995) reported shoot multiplication in mangosteen, by culturing seed explants on MS medium supplemented with 9.2 $mg.L^{-1}$ BA and 0.5 $mg.L^{-1}$ NAA. Cultures were maintained under an 8-hour light regime (25 $\mu mol\ m^{-2}\ s^{-1}$ cool white fluorescent light), at 30° C, forming an average of 16.8 shoots per explant. The addition of 2 $g.L^{-1}$ activated charcoal to the same medium reduced the number of shoots produced. However, those shoots showed a more organized growth and 75% rooting. WPM medium was not suitable for shoot multiplication. Rooted plantlets were established on substrate containing 3:2:1 sand, soil and organic matter, respectively.

An improved micropropagation system for this species was reported by Huang *et al.* (2000). Plant growth regulator treatments were monitored through successive subcultures, and could only be observed after two 8-week periods between subcultures. Photoperiod and temperature did not affect shoot proliferation and effects due to those parameters were not evident until after the fifth subculture. Bud primordium differentiation in cotyledon explants or proliferation of regenerated shoots were obtained without the presence of auxins. Highest primordium differentiation and shoot proliferation were obtained with 3.0 $mg.L^{-1}$ BA, and 1.0 - 3.0 $mg.L^{-1}$ BA, respectively. Shoot regeneration was obtained with successive subcultures under a 16-h daily exposure to 25 $\mu mol\ m^{-2}\ s^{-1}$ cool white fluorescent light and at a constant 26° ± 2° C. Rooting rates of 85% were achieved after cultures were maintained for 3 days on a medium supplemented with 98 $mg.L^{-1}$ IBA, or 10 days with 98.0 $mg.L^{-1}$ IBA. Two subsequent subcultures at 8-week intervals onto a medium without plant growth regulators were essential for a high rooting yield frequency. Shoots were acclimatized in a growth chamber and transferred to greenhouse conditions. A survival rate of 95% was obtained.

Hancornia speciosa Gom. ("Mangabeira") is a tropical fruit tree of the Apocynaceae family. Native to Brazil, *H. speciosa* is important for its tasty and nutritive fruits, used fresh or in ice creams, juices, wines, and vinegars (Aguiar Filho and Bosco, 1995).

Pereira-Netto (1996) first reported the *in vitro* propagation of *H. speciosa*, and Pereira-Netto and McCown (1999) developed a shoot microculture protocol for this tropical fruit tree. High temperature stimulation of axillary branching was obtained by increasing the culture temperature from 25 to 31° C for a period of 4 weeks. This caused the stimulation of elongation in newly subcultured microshoots. Subsequent increases in temperature resulted in different effects. Increase to 35° C inhibited elongation, but induced shoot branching. Further increase to 37° C caused inhibition of both shoot elongation and branching. When cultures were maintained at a temperature regime of 35° C for 2 weeks, followed by growth at a constant temperature of 31° C, vigorous branching was obtained. The authors raise the hypothesis that branching induced by temperature regimes is mediated by reduced ethylene biosynthesis. Ethylene was monitored in cultures of nodal segments maintained at 31° C, with peaks observed at 12 to 16 days after subculture. After that period, a decrease in ethylene was observed until the end of the culture cycle. In cultures maintained at 35° C, an earlier peak was detected, with a reduced rate in ethylene evolution throughout the culture cycle.

Melia azedarach **L.** (Chinaberry or China tree) is a woody species native to Asia and of medicinal importance (USDA 2001).In addition, the seeds are used for semi-drying oil, and its wide range of uses and adaptability makes it a suitable woody species for reforestation purposes. Sujatha and Chandran (1997) compared plants derived from seedlings and from micropropagation. *In vitro*-derived plants showed higher values for the characteristics analyzed, such as plant height, stem girth and plant canopy when compared to seedling-derived plants. Evaluations started 6 months after establishment in the field and were recorded at 3-month intervals for 15 months. Shoot organogenesis was obtained using MS medium with half-strength macroelements and 4.0 $g.L^{-1}$ sucrose. China grass was used as the gelling agent and fertilizer-grade urea as the nitrate source. This allowed a significant reduction (up to 7-fold) in the cost of micropropagation. The medium used was effective for shoot growth, shoot multiplication and elongation. Further cost reduction was achieved by adopting a microcutting method by which rootless microshoots were directly transferred to soil (*i.e. ex vitro* rooting).

Punica granatum **L.** (Pomegranate) distributional range includes the Temperate Asia (Afghanistan, Armenia, Azerbaijan, Georgia, Iran, Iraq, Russian Federation - Dagestan, Tajikistan, Turkey, Turkmenistan) and Tropical Asia (Pakistan). The tree has environmental importance, used as boundary, barrier, support, and as an ornamental. It is also used as food source (fruit) and as a beverage base (USDA 2001).

Naik *et al.* (1999) reported a procedure for *in vitro* clonal propagation of an elite cultivar of pomegranate, 'Ganesh', using nodal stem segments of mature trees as explants. MS medium supplemented with benzyladenine (BA), zeatin riboside (ZR) and thidiazuron (TDZ) was used for induction of bud break and shoot proliferation. The use of 20 $mg.L^{-1}$ of ZR resulted in the highest number of shoots, although elongation was not obtained. Further subculture of shoot clumps to a medium with lower level of ZR promoted shoot elongation. Half-strength MS supplemented with 1.0 $mg.L^{-1}$ IBA promoted 86 % rooting of regenerated shoots. Transferring rooted shoots onto half strength MS medium without auxins promoted root elongation. In addition, new roots were developed. In 2000, Naik and collaborators, developed a new protocol for *in vitro* regeneration of pomegranate. It entails using cotyledonary nodes derived from axenic seedlings as explants and MS medium with 0.5 - 5 $mg.L^{-1}$ BA or kinetin for shoot induction and development. An average of 9.8 shoots per explant was obtained with 2 $mg.L^{-1}$ BA. This protocol also adds the establishment of shoot cultures by repetitive subcultures of the original explant onto fresh medium, after newly formed shoots are removed. Basically, elongated shoots are divided into nodal segments and cultured onto a fresh medium for shoot multiplication. A single cotyledonary explant produced about 30-35 shoots in a period of 60 days. Elongated shoots were rooted on half strength MS medium with 0.01 - 1 $mg.L^{-1}$ NAA. The highest rooting percentage and number of roots were obtained on the same medium supplemented with 0.1 $mg.L^{-1}$ NAA. In both cases, using the 1999 and 2000 protocols, plantlets were acclimatized and successfully established in soil conditions.

Sterculia urens Roxb. is a tree species, known commonly as gum karaya, native to India. The exudate or gum is used as a fiber-deflocculating agent for the long-fibered paper pulps (USDA 2001). Protocols for adventitious shoot organogenesis and somatic embryogenesis were developed by Sunnichan *et al.* (1998). Selected adult trees were used as explant sources, from which nodal segments were removed and cultured on MS medium supplemented with 1.5 $mg.L^{-1}$ BA. An average of six adventitious shoots per explant was produced in 30 days. *In vitro* rooting was achieved by using ¼ strength MS medium with 2.0 $mg.L^{-1}$ IBA. For somatic embryogenesis, hypocotyl explants produced nodulated callus on MS medium supplemented with 1.0 $mg.L^{-1}$ 2,4-D and 2.0 $mg.L^{-1}$ BA. When nodulated callus was transferred to ½ strength MS medium supplemented with 0.1 $mg.L^{-1}$ TDZ, globular, embryo-like structures formed after 3 weeks in culture, followed by heart-shaped and cotyledonary-stage somatic embryos. Culturing of nodulated callus for 2 days in medium with TDZ followed by transfer to basal medium provided the best results for somatic embryo formation. Nearly 40% of the somatic embryos transferred individually to ¼ strength MS medium without plant growth regulators developed into plantlets within 14 days. Sixty plantlets were acclimatized and transferred to soil, of which 47 survived.

***Sesbania grandiflora* (L.)** Persoon is a leguminous tree of multipurpose use. Native to Indo-Malaysia, it is widely cultivated as an ornamental species and also an important nitrogen-fixing tree. Common names used include "hummingbird tree", "sesban", and "scarlet wisteria tree" (USDA 2001). *In vitro* regeneration of sesban has been reported by Detrez *et al.* (1994) using cotyledon segments as explants, cultured on medium with BA and NAA. Some seedling characteristics were fundamental in determining bud induction and elongation abilities, including seedling age and lighting conditions for cultivation. Seedlings grown in the dark for 24 to 36 hours post imbibition yielded up to 96% of explants with more than 30 buds each, after one week in culture. Direct organogenesis allowed for bud development. Histological analysis showed that bud development occurred from the proximal and adaxial cut surfaces of the explant. Regeneration was also obtained from wounded epidermal surfaces, indicating that the regeneration ability was distributed among cells along the explant surface. Bud isolation and subsequent subculture into liquid medium for one week promoted and improved bud elongation, with stem differentiation and leaf growth being observed. Shoots were rooted on medium containing auxin after 3 weeks in culture. Rooted plantlets were established in soil with a 92% survival rate.

4. SOMACLONAL VARIATION IN *IN VITRO* CULTIVATED PLANTS

Clonal fidelity is one of the main concerns in micropropagated plants. True-to-type propagules are the expected product of clonal propagation whether from organogenic or somatic embryogenic regeneration. Genetic stability is thus a prerequisite for the application of micropropagation and transformation systems.

The occurrence of variation in plants regenerated from *in vitro* cultures was first reported by Heinz *et al.* (1969) and Heinz and Mee (1971). The variation observed among plants regenerated from *in vitro* culture was named somaclonal variation by Larkin and Scowcroft (1981). Somaclonal variation has been reported in several tissue culture systems, such as morphological and yield variation in micropropagated plantain (Vuylsteke *et al.* 1988); dwarfism, mosaic-like leaves and reddish leaf petioles in banana micropropagated plants (Reuveni and Israeli 1990); in poplar, variation occurred in plants regenerated from callus cultures (Antonetti and Pinon 1993), among others (reviewed by Bajaj 1990, Deverno 1995). Biochemical analysis have shown that *in vitro*-regenerated plants may not be genetically identical to plants conventionally propagated (Warner 1995).

Karp (1994) suggested that cellular organization is a critical feature in some tissue culture systems and that somaclonal variation is related to unorganized growth. The longer the duration of the unorganized phase (i.e., callus), the greater the chances of somaclonal variation. The frequency of somaclonal variants may increase with length of time in culture (George 1993, Taylor *et al.* 1995). According to Larkin (1987), the factors that can affect the frequency of somaclonal variation are explant (cotyledons, embryos, leaves, etc.), culture mode (solid vs. liquid suspension), culture conditions (nutrients, growth regulators, temperature, light, etc.) and duration (young vs. old cultures). Nehra *et al.* (1992) obtained strawberry regenerants showing dwarfism under high concentrations of BA and malformed and yellowish leaves in older cultures.

George (1993) reported that in addition to ploidy changes, structural changes may occur during *in vitro* cultivation, such as translocation, deletions, and breakages. Transposable elements may be involved in such genetic changes and induction may be a result of DNA methylation, lack of nucleic acid precursors, growth regulators, medium composition, and cultural conditions (pro-oxidant states, temperature, and method of culture).

Somaclonal variation may not occur in all systems, or may not affect the phenotype of the regenerants. Hammerschlag (1992) cited examples of phenotypic stability in regenerated plants of peach and thornless loganberry, and little or no variation in apple, barley, cherry, daylily and pear, suggesting that genotype and the nature of the explant are important determining factors for the occurrence of somaclonal variation. Therefore, thorough studies of such factors may allow the control of somaclonal variation in tissue culture-derived plants.

Studies evaluating somaclonal variation in somatic embryogenic systems have been limited. One major difficulty in conducting such studies has been the need to develop a method whereby material can be easily and rapidly screened to reveal genetic differences from non tissue culture-derived controls. Additionally, it must be determined whether changes are heritable (true somaclonal variation) or temporary (epigenetic or physiological variation). Some methods for evaluation of somaclonal variation commonly used are: phenotypic analysis, chromosome counts and structure, protein analysis, or the evaluation of DNA restriction fragments (De Klerk 1990).

Despite the long-term field evaluations, phenotypical analysis is still largely applied to evaluation of variation, based on the percentage of plants showing mutations for one or more defined characteristics. A number of studies have compared the field performance of tissue-cultured versus conventionally-propagated fruit crops (Webster *et al.* 1985, Liverani *et al.* 1989, Hammerschlag and Scorza 1991, Wetzstein and Myers 1994, Isabel *et al.* 1996, Zimmerman 1997, Tetsumura and Yukinaga 1998) and forestry species (Ritchie and Long 1986, Warrag *et al.* 1989a, Warrag *et al.* 1989b, Gupta *et al.* 1991). However, fewer studies have attempted to address the extent of genetic diversity in tissue culture-derived plants.

Chromosome analysis reported for a number of tissue cultured species may show karyological changes such as amplification or loss of DNA, and re-arrangements of nuclear DNA (Karp *et al.* 1987, Cecchini *et al.* 1992). However, chromosome structural alterations may also be present in regenerated plants with normal chromosome numbers. The detection of such variation is very laborious and not routinely practical (De Klerk 1990). In addition, karyological analysis cannot reveal alterations in specific genes or small chromosomal rearrangements (Isabel *et al.* 1993).

Isozyme analysis, which relies on gene expression, has been used to assess genetic stability of tissue culture-regenerated shoots and plants (Ishii *et al.* 1987, Karp *et al.* 1987, Sabir *et al.* 1992, Khavkin and Zabrodina 1994, and Richard *et al.* 1995). However, this technique is limited by the number of available markers and subject to environmental and developmental alterations. It may also lead to incorrect interpretations if gene expression is not stable during the development of a plant (Isabel *et al.* 1993).

Molecular markers, brought about by the great development in plant molecular genetics, are very useful tools now being applied for several research purposes. Uses include pedigree analysis, linkage mapping and identification and discrimination of closely related cultivars. Molecular markers are unlimited in number and are not affected by the phenotype. At the nuclear level, several markers have been used with more or less success (Cloutier and Landry 1994).

RFLP (Restriction Fragment Length Polymorphism), a hybridization-based technique, can determine precise changes in a particular gene sequence, but RFLP analysis is time demanding and the result is limited only to the gene sequence used as probe, which may not be directly responsible for somaclonal variation (Brown *et al.* 1993). It also requires a relatively large amount of DNA (10-15 μg). Automation is difficult and the cost per data point is high.

RAPD (Random Amplified Polymorphism DNA), a PCR-based molecular technique, enables discrimination not only between tissues at different culture stages, but also among regenerated plants (Brown *et al.* 1993, Isabel *et al.* 1993). Advantages include the small amounts of DNA (10-25 ng) required, amenability to automation and rapidity of results (Ragot and Hoisington1993, Cloutier and Landry 1994). However, there are still difficulties with reproducibility (Rafalski and Tingey 1993), and some data for RAPD analysis are contradictory. Hashmi *et al.* (1997) reported RAPD as a feasible technique for evaluation of somaclonal variants of peach, and showed evidence for genetic differences among them. Isabel *et al.* (1996) observed only one, out of 250 RAPD markers correlated with white needles of variegated *Picea glauca*, while Fourré *et al.* (1997), despite detecting genotypic variation by cytogenetic analysis in Norway spruce clones, were not able to detect somaclonal variation by

RAPD analysis, even using a large sample of DNA and many primers. Additional studies have been unsuccessful in detecting variation in different species by either RFLP or RAPD (Isabel 1993, Vallés *et al.* 1993).

A novel PCR-based assay for plant DNA fingerprinting, AFLP (Amplified Fragment Length Polymorphism), has been reported as a relatively easy assay to perform, which is able to identify significant polymorphisms (Lin and Kuo 1995). It is reportedly more efficient than RFLP and RAPD (Faye *et al.* 1996), reliable and reproducible (Cato and Richardson 1996), requires small amounts of DNA and can detect a great number of polymorphisms with only one or two sets of primers (Hill *et al.* 1996). The cost per data point is lower than RFLP or RAPD, it is automatable, and recent fluorescence techniques eliminate the use of radioisotopes for detection.

Vendrame *et al.* (1999, 2000) used AFLP to analyze pecan (*Carya illinoinensis*) somatic embryos and 4-yr-old field-grown trees derived from somatic embryos. AFLP could readily distinguish differences among tissue culture lines even with a single primer combination, providing between 121-127 scorable markers per primer combination. Individual embryos derived from the same culture line exhibited high similarity and generally grouped together. However, within a culture line, some embryo-to-embryo differences were also observed with some divergent embryos exhibiting higher polymorphisms (Vendrame *et al.* 1999). Phenotypical and AFLP analysis were performed between and within two lines of pecan trees derived from somatic embryogenesis (Vendrame *et al.* 2000). Somatic embryogenic cultures produced trees that appeared to be phenotypically stable and maintained the characteristics of the line. In contrast, AFLP analysis showed polymorphisms that were not reflected in phenotypic variation. The usefulness of AFLP to evaluate somatic embryo-derived plants for clonal fidelity will be dependent on how closely the number of polymorphic loci correlates with economically important phenotypic characteristics.

5. CONCLUSIONS

Micropropagation is a valuable technique for the rapid multiplication of difficult-to-propagate plants, both for commercial production and germplasm conservation. Depending on the cost of production and the commercial value of the species, micropropagation can be the method of choice for multiplication of plants, or part of the process. Cultures can be started from different types of explants, showing different multiplication rates, which can be improved by manipulating culture medium components and culture conditions. The commercial use of micropropagation, as with any other commercial protocol, can benefit from the optimization of the process. Several parameters can be studied aiming for high rate of multiplication and a low cost of production, maintaining the high quality of the plants obtained. As in any *in vitro* process, care should be taken to avoid somaclonal variation and more studies should be done to determine the stability of cultures *in vitro* and rate of somaclonal variation in a given *in vitro* process.

6. REFERENCES

Aguiar Filho S.P., Bosco J.A 1995 Mangabeira e sua importância para o tabuleiro costeiro paraibano. Informativo SBF, Itajaí, v. 14, nº 4. 26p.

Ajithkumar D., Seeni S. 1998 Rapid clonal multiplication through *in vitro* axillary shoot proliferation of *Aegle marmelos* (L.) Corr., a medicinal tree. Plant Cell Rep 17:422-426.

Anand A., Rao C.S., Balakrishna P. 1999 *In vitro* propagation of *Syzygium travancoricum* Gamble – an endangered tree species. Plant Cell Tiss Org Cult 56:59-63.

Antonetti P.L.E., Pinon J. 1993 Somaclonal variation within poplar. Plant Cell Tiss Org Cult 35:99-106.

Bajaj Y.P.S. 1990 Somaclonal variation - origin, induction, cryopreservation, and implications in plant breeding. In: Bajaj Y.P.S. (ed) Somaclonal variation in crop improvement I. Springer-Verlag, Berlin. pp. 3-48.

Barrueto Cid L.P., Gomes A.C.M., Costa S.B.R., Teixeira, J.B. 1997 Micropropagation of *Miconia* sp, a woody Melastomaceae from Brazil, using thidiazuron as plant growth regulator. R Bras Fisiol Veg 9:21-25.

Brown P.T.H., Lange F.D. Kranz E, Lorz H. 1993 Analysis of single protoplasts and regenerated plants by PCR and RAPD technology. Mol Gen Genet 237:311-317.

Campinhos E., Ikemori Y.K. 1983 Introdução de novas técnicas na produção de mudas de essências florestais. Silvicultura 8:226-228.

Carron M.P., Le Roux Y., Tison J., Dea B.G., Caussanel V., Clair J., Keli J. 2000 Compared root system architectures in seedlings and *in vitro* plantlets of *Hevea brasiliensis*, in the initial years of growth in the field. Plant and Soil 223:73-85.

Cato S., Richardson T.E. 1996 Development and evaluation of AFLPs in *Pinus radiata*. Plant Genome IV Program, 69. International Plant Genome Conference, January 14-18, 1996. San Diego, CA.

Cecchini E., Natali L., Cavallini A., Durante M. 1992 DNA variations in regenerated plants of pea (*Pisum sativum* L.). Theor Appl Genet 84:874-879.

Cloutier S., Landry B.S. 1994 Molecular markers applied to plant tissue culture. In Vitro Cell Dev Biol - Plant 30P:32-39.

Correia D. 1993 Crescimento e desenvolvimento de gemas na multiplicação de Eucalyptus spp *in vitro* em meio de cultura líquido e sólido. MS Thesis, ESALQ, University of São Paulo, Piracicaba, 113p.

Correia D., Gonçalves A.N., Couto, H.T.Z. do, Ribeiro, M.C. 1995 Efeito do meio de cultura líquido e sólido no crescimento e desenvolvimento de gemas de *Eucalyptus grandis* x *Eucalyptus urophylla* na multiplicação *in vitro*. IPEF, 48/49:107-116.

De Klerk G.-J. 1990 How to measure somaclonal variation. Acta Bot Neerl 39:129-144.

De Fossard R.A. 1974 Tissue culture of *Eucalyptus*. Australian Forestry 37:43-54.

Detrez C., Ndiaye S., Dreyfus B. 1994 *In vitro* regeneration of the tropical multipurpose leguminous tree *Sesbania grandiflora* from cotyledon explants. Plant Cell Rep 14:87-93.

Deverno L.L. 1995 An evaluation of somaclonal variation during somatic embryogenesis. In: Jain S.M., Gupta P., Newton R. (eds.) Somatic embryogenesis in woody plants. Kluwer Academic Publishers, Dordrecht. Vol. 1, pp. 361-377.

Dornelas M.C., Vieira M.L. 1994 Tissue culture studies on species of *Passiflora*. Plant Cell Tiss Org Cult 36:211-217.

Drew R.A. 1991 *In vitro* culture of adult and juvenile explants of *Passiflora* species. Plant Cell Tiss Org Cult 26:23-27.

D'Souza M.C., Sharon, M. 2001 *In vitro* clonal propagation of annatto (*Bixa orellana* L.). *In Vitro* Cell Dev Biol Plant 37:168-172.

Faria J.L.C., Segura J. 1997a *In vitro* control of adventitious bud differentiation by inorganic medium components and silver thiosulfate in explants of *Passiflora edulis* f. *flavicarpa*. *In Vitro* Cell Dev Biol - Plant 33:209-212.

Faria J.L.C., Segura J. 1997b Micropropagation of yellow passion fruit by axillary bud proliferation. HortSci 32:1276-1277.

Faye C., Murigneux A., Bonjean A., Lacaze P., Peleman J. 1996 AFLP™ mapping in maize:distribution and stability of the bands. Plant Genome IV, 73. International Plant Genome Conference, January 14-18, 1996. San Diego, CA.

Fourré J.-L., Berger P., Niquet L., André P. 1997 Somatic embryogenesis and somaclonal variation in Norway spruce: morphogenetic, cytogenetic and molecular approaches. Theor Appl Genet 94:159-169.

Gabryszewska E., Rudnicki R.M., Blacquiere T., Gude H. 1997 The effects of light quality on the growth and development of shoots and roots of *Ficus benjamina in vitro*. Acta Hort 418:163-167.

George E.F. 1993 Plant propagation by tissue culture, part 1 - The technology. Exegetics Ltd, Edington, Wilts, England. pp. 67-87; 261-263.

George E.F., Sherrington P.D. 1984 Plant propagation by tissue culture: handbook and directory of commercial laboratories. Exegetics, Eversley. 709p.

Giles K.L., Friesen K.R.D. 1994. Micropropagation. In. Shargool, P.D and Ngo, T.T. (eds.) Biotechnologial applications of plant cell cultures. CRC Press. Boca Raton. p. 111-128.

Gonçalves A.N. 1980. Reversion to juvenility and cloning of *Eucalyptus urophylla* S. T. Blake in cell and tissue culture system. In. Simpósio IUFRO em melhoramento genético e produtividade de espécies florestais de rápido crescimento. Águas de São Pedro, SP. Brazil. p. 15-30.

Grattapaglia D., Machado M.A. 1990 Micropropagação. In: Torres, A.C.; Caldas, L.S. Técnicas e aplicações da cultura de tecidos de plantas. ABCTP/EMBRAPA - CNPH. 433p.

Grattapaglia D., Caldas, L.S., Silva J.R. da, Machado M.A. 1991 Cultura de tecidos de maracujá. In. San José A.R., Ferreira, F.R., Vaz, R.L. (eds.). A cultura do maracujá no Brasil. FUNEP, Jaboticabal. p. 61-77.

Gresshoff P.M., Doy C.H. 1972 Development of differentiation of haploid *Lycopersicon esculentum* (tomato). Planta 107: 161-170.

Grossi F. 1995 Adequação nutricional do meio de cultura para crescimento e desenvolvimento de gemas de *Eucalyptus saligna* Smith *in vitro*, procedência Itatinga. MS thesis. ESALQ, University of São Paulo, Piracicaba. 125p.

Gupta P.K., Timmis R., Mascarenhas A.F. 1991 Field performance of micropropagated forestry species. *In Vitro* Cell Dev Biol - Plant 27P:159-164.

Hammerschlag, F.A. 1992 Somaclonal variation. In: Hammerschlag F.A., Litz R.E. (eds). Biotechnology of perennial fruit crops. CAB International, Wallingford, Vol 8 pp. 35-36.

Hammerschlag F.A., Scorza R. 1991 Field performance of micropropagated, own-rooted peach trees. J Amer Soc Hort Sci 116:1089-1091.

Hashmi G., Huettel R., Meyer R., Krusberg L., Hammerschlag F. 1997 RAPD analysis of somaclonal variants derived from embryo callus cultures of peach. Plant Cell Rep 16:624-627.

Heinz D.J., Mee G.W.P., Nickell L.G. 1969 Chromosome number of some *Saccharum* species hybrids and their cell suspension cultures. Am J Bot 56:450-456.

Heinz D.J., Mee G.W.P. 1971 Morphologic, cytogenetic, and enzymatic variation in *Saccharum* species hybrid clones derived from callus tissue. Am J Bot 58:257-262.

Higashi E.N., Silveira R.L.V.A., Gonçalves, A.N. 2000 Evolução do jardim clonal de eucalipto para a produção de mudas. IPEF Notícias 24:4-6.

Hill M., Witsenboer H., Zabeau M., Vos P., Kesseli R., Michelmore R. 1996 PCR-based fingerprinting using AFLPs as a tool for studying genetic relationships in *Lactuca* spp. Theor Appl Genet 93:1202-1210.

Huang L.C., Huang B.L., Wang C.H., Kuo C.I., Murashige T. 2000 Developing an improved *in vitro* propagation system for slow-growing species using *Garcinia mangostana* L. (mangosteen). *In Vitro* Cell Dev Biol - Plant 36:501-504.

Iannelli C., Xavier A., Comério J. 1996 Micropropagação de *Eucalyptus* spp na Champion. Silvicultura 17:33-35.

Isabel N., Tremblay L., Michaud M., F.M. Tremblay F.M., Bousquet J. 1993 RAPDs as an aid to evaluate the genetic integrity of somatic embryo-derived populations of *Picea mariana* (Mill.) B.S.P. Theor Appl Genet 86:81-87.

Isabel N., Boivin R., Levasseur C., Charest P.-M, Bousquet J., Tremblay F.M. 1996 Occurrence of somaclonal variation among somatic embryo-derived white spruces (*Picea glauca*, Pinaceae). Amer J Bot 83:1121-1130.

Ishii K., Moran G.F., Bell J.C., Hartney V. 1987 Genetic stability examination of micropropagated radiata pine (*Pinus radiata*) using isozyme assays. J Jap For Soc 69:487-488.

Jain S.M., Ishii K., Bruns, S. 1998 Recent advances in somatic embryogenesis in forest trees. In: Proceedings of an IFS Workshop, Florianopolis, Brazil, 15-19 Sept. 1997. International Foundation os Sciences, Stockholm. pp. 214-231.

Kantharajah A.S., Dodd W.A. 1990 *In vitro* micropropagation of *Passiflora edulis* (purple passionfruit). Ann. Bot. 63:337-339.

Karp A. 1994 Origins, causes and uses of variation in plant tissue cultures. In: Vasil, I.K. and T.A. Thorpe (eds). Plant Cell and Tissue Culture. Kluwer Academic, Dordrecht. pp. 139-151.

Karp A., Steele S.H., Parmar S., Jones M.G.K., Shewry P.R. 1987 Relative stability among barley plants regenerated from cultured immature embryos. Genome 29:405-412.

Kawata K., Ushida C., Kawai F., Kanamori M., Kuriyama A. 1995 Micropropagation of passionfruit from subcultured multiple shoot primordia. J Plant Physiol 147:281-284.

Khavkin E.E., Zabrodina M.V. 1994 Heritable variations in peroxidase and esterase isozyme patterns of maize somaclones. Russ J Plant Phys 41:859-867.

Kumar V., Radha A., Kumar Chitta S. 1998 *In vitro* plant regeneration of fig (*Ficus carica* L. cv. Gular) using apical buds from mature trees. Plant Cell Rep 17(9): 717-720.

Lardet L., Aguilar M.E., Michaux-Ferriere N., Berthouly M. 1998 Effect of strictly plant-related factors on the response of *Hevea brasiliensis* and *Theobroma cacao* nodal explants cultured *in vitro*. *In Vitro* Cell Dev Biol - Plant 34:34-40.

Larkin P.J. 1987 Somaclonal variation: history, method, and meaning. Iowa St. J. Res., 61:393-434.

Larkin P.J., Scowcroft W.R. 1981 Somaclonal variation-a novel source of variability from cell culture for plant improvement. Theor Appl Genet 60:197-214.

Liew T.K., Teo C.K.H. 1998. Multiple shoot production *in vitro* of the tropical timber tree, sentang (*Azadirachta excelsa* L.). HortSci 33:1073-1075.

Lima M.M. de, Gonçalves A.N. 1998 Efeito do thidiazuron na multiplicação *in vitro* de gemas de um clone de *Eucalyptus grandis* x *E. tereticornis*. Sci Forestalis 53:49-56.

Lin J.-J. Kuo J. 1995 AFLP™: a novel PCR-based assay for plant and bacterial DNA fingerprinting. Focus 17:66-70.

Liverani A., Cobianchi D., De Salvador F.R., Insero O., Minguzzi A. 1989 Field performance of two peach and nectarine cultivars propagated *in vitro*. Acta Hort 254:41-46.

Lloyd G., McCown B. 1980 Commercially-feasible micropropagation of Mountain Laurel, *Kalmia latifolia*, by use of shoot-tip culture. Comb. Proc. Intl. Plant Prop. Soc. 30:421-427.

Malavolta E., Vitti G.C., Oliveira S.A. 1997 Avaliação do estado nutricional das plantas: princípios e aplicações. 2 ed. POTAFOS, Piracicaba. 319p.

Manders G., Otoni W.C., D'utra Vaz F.B., Blackhall N.W., Power J.B., Davey M.R. 1994 Transformation of passionfruit (*Passiflora edulis* f. *flavicarpa* Degener.) using *Agrobacterium tumefaciens*. Plant Cell Rep. 13:697-702.

Mendanha A.B.L., Torres R.A.D., Freire A.D. 1998 Micropropagation of rubber tree (*Hevea brasiliensis* Muell. Arg.). Genetics and Molecular Biol 21:395-398.

Meyer H.J. 1998 *In vitro* formation of adventitious buds on mature embryos of *Pinus elliottii* Engelm. x *P. caribaea* Morelet hybrids. South African J Bot 64:220-225.

Monteiro A.C.B.A. 2000 Cultivo *in vitro* de três espécies do gênero *Passiflora*. MS thesis. ESALQ, University of São Paulo, Piracicaba. 82 p.

Monteiro A.C.B.A., Higashi E.N., Gonçalves A.N., Rodriguez A.P.M. 2000 A novel approach for the definition of the inorganic medium components for micropropagation of yellow passionfruit (*Passiflora edulis* Sims. f. *flavicarpa* Deg.). *In Vitro* Cell Dev Biol - Plant 36:527-531.

Moran Robles M. J. 1978 Multiplication végétative, *in vitro*, des bourgeons axillares de *Passiflora edulis* var. *flavicarpa* Degener et de *P. mollissima* Bailey. Fruits 33:693-699.

Moran Robles M. J. 1979 Potentiel morphogénétique des entrenoeuds de *Passiflora edulis* var. *flavicarpa* Degener et *P. mollissima* Bailey en culture *in vitro*. Turrialba 29:224-228.

Murashige T., Skoog F. 1962 A revised medium for rapid growth and bioassays with tabacco tissue culture. Physiologia Plant 15:473-497.

Naik S.K., Pattnaik S., Chand P.K. 1999 *In vitro* propagation of pomegranate (*Punica granatum* L. cv. Ganesh) through axillary shoot proliferation from nodal segments of mature trees. Sci Hort 79:175-183.

Naik S.K., Pattnaik S., Chand P.K. 2000 High frequency axillary shoot proliferation and plant regeneration from cotyledonary nodes of pomegranate (*Punica granatum* L.). Sci Hort 85:261-270.

Nadwani D., Kumaria S., Tandon P. 2001 Micropropagation of *Pinus kesiya* Royle ex Gord (Khasi pine). Gartenbauwissenschaft 66:68-71.

Nehra N.S., Kartha K.K., Stushnoff C., Giles K.L. 1992 The influence of plant growth regulator concentrations and callus age on somaclonal variation in callus culture regenerants of strawberry. Plant Cell Tiss Org Cult 29:257-268.

Normah M.N., Nor-Azza A.B., Aliudin R. 1995 Factors affecting *in vitro* shoot proliferation and *ex vitro* establishment of mangosteen. Plant Cell Tiss Org Cult 43:291-294.

Otoni W.C., Casali V.W.D., Cecon P.R., Davey M.R., Power J.B. 1995 Regeneração de plantas de maracujazeiro (*Passiflora coccinea* Aubl.) a partir de protoplastos derivados de mesófilo. Rev Ceres 42:461-468.

Pereira-Netto A.B. 1996 *In vitro* propagation of *Hancornia speciosa*, a tropical fruit tree. *In Vitro* Cell Dev Biol Plant 32:253-256.

Pereira-Netto A.B., McCown B.H. 1999 Thermally induced changes in shoot morphology of *Hancornia speciosa* microcultures: evidence of mediation by ethylene. Tree Phys 19:733-740.

Perrin Y., Doumas P., Lardet L., Carron M-P. 1996 Endogenous cytokinins as biochemical markers of rubbertree (*Hevea brasiliensis*) clone rejuvenation. Plant Cell Tiss Org Cult 47:239-245.

Quoirin M., Lepoivre P. 1977 Etude de milieux adaptes aux cultures *in vitro* de *Prunus*. Acta Hort 78:437-442.

Rafalski J.A., Tingey S.V. 1993 Genetic diagnosis in plant breeding: RAPDs, microsatellites and machines. Trends in Genetics 9:275-280.

Ragot M., Hoisington D.A. 1993 Molecular markers for plant breeding: comparisons of RFLP and RAPD genotyping costs. Theor Appl Genet 86:975-984.

Reuter D.J., Robinson J.B. (eds.) 1997 Plant analysis: an interpretation manual. 2 ed. CSIRO, Collingwood. 572p.

Reuveni O., Israeli Y. 1990 Measures to reduce somaclonal variation in *in vitro* propagated bananas. Acta Hort 275:307-313.

Richard S., Gauthier S., Laliberté S.1995 Isozymes assessment of the genetic stability of micropropagated hybrid larch (*Larix* X *eurolepsis* Henry). Can J For Res 25:1103-1112.

Ritchie G.A., Long A.J. 1986 Field performance of micropropagated Douglas Fir. New Zeal J For Sci 16:343-356.

Sabir A., Newbury H.J., Todd G., Catty J., Ford-Lloyd B.V. 1992 Determination of genetic stability using isozymes and RFLPs in beet plants regenerated *in vitro*. Theor Appl Genet 84:113-117.

Smith R. 1992 Plant tissue culture: techniques and experiments. Academic Press, San Diego. 171p.

Sposito G., Mattigod S.V. 1980 Geochem: a computer program for the calculation of chemical equilibria in soil solutions and other material water systems. Kearney Foundation of Soil Science, University of California, Riverside. 107p.

Sujatha M., Chandran K.. 1997 A commercially feasible micropropagation method for *Melia azedarach* L. Indian Journ of Experimental Biology 35:787-791.

Sunnichan V.G., Shivanna K.R., Ram H.Y.M. 1998 Micropropagation of gum karaya (*Sterculia urens*) by adventitious shoot formation and somatic embryogenesis. Plant Cell Rep 17:951-956.

Taylor P.W.J., Geijskes J.R., Ko H.-L., Fraser T.A., Henry R.J., Birch R.G. 1995 Sensitivity of random amplified polymorphic DNA analysis to detect genetic change in sugarcane during tissue culture. Theor Appl Genet 90:1169-1173.

Tetsumura T., Yukinaga H. 1998 Early field performance of micropropagated Japanese persimmon trees. HortSci 33:751-753.

Tibok A., Blackhall N.W., Power, J.B., Davey, M.R. 1995 Optimized plant regeneration from callus derived from seedling hypocotyls of *Eucalyptus urophylla*. Plant Science 110:139-145.

USDA, ARS, National Genetic Resources Program. 2001 *Germplasm Resources Information Network - (GRIN)*. [Online Database] National Germplasm Resources Laboratory, Beltsville, Maryland.

Vallés M.P., Wang Z.Y., Montavon P., Potrykus I., Spangenberg G. 1993 Analysis of genetic stability of plants regenerated from suspension cultures and protoplasts of meadow fescue (*Festuca pratensis* Huds.). Plant Cell Rep 12:101-106.

Valverde-Cerdas L., Dufour, M., Villalobos V. 1998 *In vitro* organogenesis in *Albizia guachapele*, *Cedrella odorata* and *Swietenia macrophylla* (Fabaceae, Meliaceae). Rev. Biol Trop 46:225-228.

Vaz R.L. 1991 A biotecnologia aplicada à fruticultura: necessidades e perspectivas potenciais. In. San José A.R., Ferreira, F.R., Vaz, R.L. (eds.). A cultura do maracujá no Brasil. FUNEP, Jaboticabal. p. 241-247.

Vendrame W.A., Kochert G.D., Wetzstein H.Y. 1999 AFLP analysis of variation in pecan somatic embryos. Plant Cell Rep 18:853-857.

Vendrame W.A., Kochert G.D., Sparks D., Wetzstein H.Y. 2000 Field performance and molecular evaluations of pecan trees regenerated from somatic embryogenic cultures. J Amer Soc Hort Sci 125:542-546.

Vuylsteke D., Swennen R., Wilson G.F., De Langhe E. 1988 Phenotypic variation among in-vitro propagated plantain (*Musa* sp. cultivar AAB). Sci Hort 36:79-88.

Warner G. 1995 Identity query halts rootstock production. Good Fruit Grower, 46(16):18-23.

Warrag E.I., Lesney M.S. Rockwood D.L. 1989a Comparative greenhouse study of *Eucalyptus grandis in vitro* plantlets and half-sib seedlings, I. Net photosynthesis. Plant Cell Rep., 8:497-499.

Warrag E.I., Lesney M.S., Rockwood D.L. 1989b Comparative greenhouse study of *Eucalyptus grandis in vitro* plantlets and half-sib seedlings, II. Dry matter accumulation and relative distribution. Plant Cell Rep 8:500-503.

Webster A.D., Oehl V.H., Jackson J.E., Jones O.P. 1985 The orchard establishment, growth and precocity of four micropropagated apple scion cultivars. J Hort Sci 60:169-180.

Wetzstein H.Y., Myers S.C. 1994 Vegetative and yield component characteristics of micropropagated muscadine grape (*Vitis rotundifolia* Michx.). J Hort Sci 69:747-753.

Xavier, A., Comério, J. 1997 Enraizamento *ex vitro* de gemas de *Eucalyptus* spp. multiplicadas e alongadas *in vitro*. Sci Forestalis 51:29-36.

Zimmerman R.H. 1997 Orchard variation in micropropagated trees of 'Redspur Delicious' apple. HortSci 32:935-936.

7. MICROPROPAGATION OF *Fagus* spp.

A.M. Viéitez, M.C. San-José, M.C. Sánchez & A. Ballester

Instituto de Investigaciones Agrobiológicas de Galicia, CSIC, Apartado 122, 15080 Santiago de Compostela, Spain

Contents

1. INTRODUCTION

The genus *Fagus*, a member of the Fagaceae family, comprises ten species of monœcious trees native to the Northern Hemisphere Temperate Zone regions of Eurasia and Eastern North America. The name *Fagus* (related to Greek *phagein*, to eat) is a reference to the distinctive triangular nuts of these species, which are eaten by both wildlife and humans. The chief members of this genus are *Fagus sylvatica* L. (European beech), which is one of the economically most important deciduous trees of Central Europe and together with oaks defines the climax vegetation of this region; *F. orientalis* Lipski (Oriental beech), a native of the temperate regions of Eastern Europe, the Balkan peninsula, the Caucasus and Asia Minor; *F. grandiflora* Ehrh. (American beech), a native of Eastern North America that is both of great

S.M. Jain and K. Ishii (eds.), Micropropagation of Woody Trees and Fruits, 181–215.

ornamental value and highly prized as timber; and *F. japonica* Max. and *F. sieboldi* Engl., both of which are Japanese species that are very similar to European beech (Chalupa 1996). The closely related genus *Nothofagus* (Southland beech, Antarctic beech) comprises 45 tree or shrub species native to the Southern Hemisphere that are of great importance in the mesophile forests and high-altitude and high-latitude communities of the Southern Andes, New Guinea, Southern Australia and New Zealand. Beech trees are thus important both as major components of ecosystems and economically. Their fine-grained wood is used for flooring, furniture and veneers, and is also excellent firewood because of its high density and good burning qualities. Ornamental varieties are also of economic importance: ornamental varieties of European beech include purple beech (var. *atropunicea* or var. *purpureae* Ait.), which has purplish bronze foliage; a variety with pendant branches (var. *pendula*); a cut-leaved form (var. *laciniata*); and var. *asplenifolia* Lodd (Ceballos and Ruiz de la Torre, 1979).

Beeches are slow-growing, shade-tolerant monœcious deciduous trees with an average life-span of 150-200 years, although 300-year-old specimens have been recorded. They have a short trunk and a broad, rounded crown, and can reach a height of 20-40 m. They grow best in soils with ample surface moisture because their root system is wide-spreading but shallow. Their bark is smooth and grey, their elliptical leaves have prominent veins and wide-spaced marginal teeth, and their long, spindle-shaped, scaly buds protrude at a sharp angle from slender twigs. Their flowers appear with the leaves, the staminate yellowish green in long-stemmed heads and the anemophilously pollinated pistillate flowers in two- to four-flowered spikes. Their triangular, edible nuts are borne in twos or threes in a husk covered in weak, unbranched spines. Reproductive maturity is attained at about 50 years in forest stands and 35-40 years in the case of isolated trees.

A number of ornamental beech species are currently propagated by bench grafting, but conventional vegetative propagation of beech is generally difficult (Meier and Reuther, 1994; Chalupa, 1996; Cuenca and Viéitez, 2000) and has not been successful on a commercial scale. In particular, beeches are extremely difficult to propagate by cuttings taken from mature trees, especially if the cuttings are taken from crown branches (Ahuja, 1984a). Cornu *et al.* (1977) achieved a rooting rate of 80-90% using cuttings from stocks under 5 years old or from stump sprouts taken from mature trees after felling. Chalupa (1982) reported a rate of 60-80% for softwood cuttings taken from 1-3—year-old plants, but the survival rate of rooted beech cuttings is low. Beech is accordingly propagated mainly by seeds, which between collection and sowing in the following spring are often stored at 5°C with a moisture content of 20-25% to maintain viability. However, good harvests occur only every 4-6 years, and storage for long periods is problematic. Furthermore, reliance on seed means that tree improvement depends on conventional breeding techniques and is therefore seriously hindered by its taking 35-50 years for trees to attain reproductive maturity (a drawback shared by propagation via cuttings taken from seedlings). Hibridization is very limited and the genetic basis for selection of elite types is thus restricted (Nadel *et al.* 1991a). New methods of rapid clonal propagation of beech are therefore needed due to the lack of methods for producing large quantities of superior plant material by vegetative propagation of proven mature trees (Meier and Reuther, 1994).

The need for means of rapid clonal propagation of material from proven mature trees is made especially acute by the decline of American beech forests under the ravages of beech bark disease, a major pathology in which elimination of the outer bark by the beech scale insect, *Cryptococcus fagisuga* Lind., renders the tree susceptible to invasion by several species of fungi (Barker *et al.* 1997). Viable rapid propagation methods would allow forests to be restocked with beeches derived from the few resistant trees that have been identified in endemically affected stands.

Given this panorama, attention has naturally turned to the development of *in vitro* methods for the propagation of selected seeds or somatic material obtained from proven mature trees. As we shall see, both somatic embryogenesis and shoot induction have been explored. An incentive in these endeavours has been the fact that some kind of viable micropropagation method is essential for exploitation of genetic engineering for beech improvement (Viéitez and San-José, 1996). It is safe to say that *in vitro* techniques will increasingly be employed for improvement and propagation of *Fagus*. In the remainder of this chapter we review the current situation of *Fagus* micropropagation, and also briefly summarize some other aspects of *in vitro* manipulation of *Fagus*.

2. WORK ON TISSUE CULTURE

In vitro studies of *Fagus* have involved micropropagation, protoplast culture, and hormone research.

The use of *in vitro* tissue culture for rapid clonal propagation of *Fagus* species was first suggested in the mid-eighties. Its early proponents included Ahuja (1984a), who achieved limited shoot formation by initial cultures of *F. sylvatica* embryo explants, and Chalupa (1985, 1987a), whose work is described below in Sections 3.1 and 3.2. Although progress has been slow, substantial advances have been made since Chalupa's review of 1996. Most work has concerned *F. sylvatica*, *F. orientalis* and *F. grandiflora* (Table 1), and has concerned three micropropagation methods: somatic embryogenesis, the proliferation of shoots by axillary branching, and the induction of adventitious buds. Most studies have used material of juvenile origin, although the micropropagation of mature genotypes has also recently been reported (Table 1). The greatest advances have been made in proliferation by axillary branching (Viéitez *et al.* 1993; Meier and Reuther, 1994; Barker *et al.* 1997) and in adventitious shoot induction (Viéitez and San-José, 1996; Cuenca *et al.* 2000).

Some work has also been done on culture of European beech protoplasts. Ahuja (1984b) isolated protoplasts from leaves of mature trees, but was unable to achieve sustained division and colony formation in cultures. However, Lang and Kohlenbach (1988) isolated viable protoplasts from young leaves of 3-5-year-old and 40-year-old trees. Isolation in saline solution containing 0.5% (w/v) Pectinol and 2% (w/v) Cellulase R-10 yielded 3×10^7 protoplasts per gram fresh weight). They were able to culture them successfully in modified Kao and Michayluk (1975) medium containing 4.8 μM 1-naphthaleneacetic acid (NAA), which induced cell wall regeneration and sustained cell division with the formation of colonies and microcalli.

In vitro methods have also been employed in research on hormone activity in European beech buds (Nadel *et al.* 1991a,b). Optimal culture conditions broke the dormancy of buds taken from 10-12—year-old trees, giving rise to sprouting, shoot elongation and leaf expansion. *In vitro* growth was associated with changes in polypeptide patterns, the decreased intensity of many polypeptide bands suggesting that they might correspond to storage proteins (Nadel *et al.* 1991a). Growth and polypeptide changes were modified by exogenous indole-3-acetic acid (IAA), gibberellic acid (GA_3) and zeatin, the effects of these growth regulators depending on the degree of innate, seasonal dormancy of the buds (Nadel *et al.* 1991b). Although the aim of these studies was not the micropropagation of beech, Nadel *et al.* (1991a) suggested that application of exogenous growth regulators might induce the growth of otherwise dormant buds, and that bud culture might be useful for detecting and monitoring hormonal processes controlling forest tree growth.

Table 1. Summary of *in vitro* culture studies in *Fagus*

Species	Explant source	Culture medium*	Growth response	Reference
F. sylvatica	Shoot tips and axillary buds of seedlings	WPM, MS, DKW+PBA or BA (0.44-2.7) + IBA (0.49-0.98)	Shoot multiplication, rooting, plants in soil	Chalupa, 1979; 1985; 1987a; 1990; 1996
	Buds of mature trees	WPM+ BA (2.2) + NAA (0.1) + 2,4D (0.09)	Limited shoot differentiation	Ahuja, 1984a
	Embryonal explants	WPM + BA (2.2) + NAA (0.1)	Shoot formation, occasionally plantlets	Ahuja, 1984a
	Leaves of mature trees	AC	Protoplast isolation, cell wall regeneration	Ahuja, 1984b
	Immature zygotic embryos	MS + BA (2.2) + 2,4D (2.3-4.5)	Somatic embryos	Chalupa, 1987b
	Leaves of juvenile and mature trees	Kao and Michayluk + NAA (4.8)	Protoplast isolation, microcalli formation	Lang and Kohlenbach, 1988, 1991
	Anthers	MS + BA (0.5-10) + 2,4D (0.5-10)	Somatic/androgenic embryo	Jörgensen, 1988, 1991
	Buds of mature trees	MS, AC + IAA (5) + GA_3 (5) + Z or BA (5)	Bud sprouting, shoot elongation	Nadel *et al.*, 1991a, b
	Immature zygotic embryo	WPM + BA (2.2) + 2,4D (0.45-4.5)	Somatic embryos, plantlet regeneration	Viéitez *et al.*, 1992
	Embryonic axes	WPM + BA (4.4)	Plantlet development	Viéitez *et al.*, 1993
	Shoot tips, nodal segments of two-month-old plantlets	WPM + BA (2.2) +Z (9.1) + NAA (1.08)	Shoot multiplication, rooting, plants in soil	Viéitez *et al.*, 1993
	Cotyledon, hypocotyl explants	WPM + BA (2.2) + NAA (0.54)	Adventitious shoots	Viéitez *et al.*, 1993
	Winter buds from mature and juvenile trees, graftings	WPM + BA (4.5)	Shoot multiplication; rooting	Meier and Reuther, 1994
	Leaves from shoot cultures	WPM + IAA (2.9) + BA (8.9) or TDZ (2.3)	Adventitious shoots; rooting	Viéitez and San-José, 1996
	Internode segments from shoot cultures	WPM + IAA (2.9) + TDZ (2.3-4.5)	Adventitious shoots, rooting	Cuenca *et al.*, 2000
	Shoot tips from shoot cultures	WPM + BA (2.2) + Z (9.1) + IAA (2.9) and different carbon sources	Shoot multiplication	Cuenca and Viéitez, 2000
F. orientalis	Leaves from shoot cultures	WPM + IAA (2.9) + TDZ (4.5)	Adventitious shoots, histological study	Cuenca and Viéitez, 1999
	Internode segments from shoot cultures	WPM + IAA (2.9) + TDZ (4.5)	Adventitious shoots, rooting	Cuenca *et al.*, 2000
	Shoot tips from shoot cultures	WPM + BA (2.2) + Z (9.1) + IAA (2.9) and different carbon sources	Shoot multiplication	Cuenca and Viéitez, 2000
	Leaf and internode segments from shoot cultures	WPM + IAA (2.9) + TDZ (4.5) and different carbon sources	Adventitious shoots	Cuenca and Viéitez, 2000
F. grandiflora	Embryonic axes	MS + BA (2 $mg.L^{-1}$) + IAPhe (2 $mg.L^{-1}$)	Root elongation, shoot greenning and callus following cryostorage	Pence, 1990
	Shoot tips from seedlings, root sprouts and branches of mature trees	AC + BA (0.89) + NAA (0.27)	Shoot multiplication, rooting, plants in soil	Barker *et al.*., 1997

*Quantities in brackets are expressed in μM, unless otherwise stated

Mineral media: AC – Aspen Culture medium (Ahuja, 1983); DKW – Driver and Kuniyuki (1984); MS – Murashige and Skoog (1962); WPM – Woody Plant Medium (Lloyd and McCown, 1980).

Supplements: BA – benzylaminopurine; IAPhe – indoleacetylphenylalanine; IBA – indole-3-butyric acid; NAA – 1-naphtaleneacetic acid; PBA – N-(phenylmethyl)-9-(tetrahydro-2H-pyran-2-yl)-9H-purin-6-amine; TDZ – thidiazuron; Z – zeatin.

3. MICROPROPAGATION

3.1. Micropropagation by Somatic Embryogenesis

Most attempts at micropropagation of a *Fagus* species by somatic embryogenesis have used immature *F. sylvatica* zygotic embryos as starting material. Success has been very limited.

Chalupa (1987b) collected immature zygotic embryos in the first fortnight of July and found that repeated subculture through fresh Murashige and Skoog medium (MS; Murashige and Skoog, 1962) containing 500 mg L^{-1} glutamine, 500 mg L^{-1} casein hydrolysate, 2.22 µM benzyladenine (BA) and 2.26-4.52 µM 2,4-dichlorophenoxyacetic acid (2,4-D) led to their developing embryogenic tissue that produced somatic embryos when transferred to MS with lower hormone content. However, regeneration of plantlets was not reported.

F.J. Viéitez *et al.* (1992) collected immature zygotic embryos at the end of July, cultured them for 5 weeks on semi-solid Woody Plant Medium (WPM; Lloyd and McCown, 1980) containing 1 g L^{-1} L-glutamine, 50 mg L^{-1} glycine, 50 mg L^{-1} arginine, 0.45-4.52 µM 2,4-D and 2.22 µM BA, and then transferred them to medium with a lower BA concentration (0.44-0.89 µM) and 0.27-0.54 µM NAA instead of 2,4-D. Although a few somatic embryos had developed to the cotyledonary stage 4 weeks after transfer (Fig. 1), none developed into plantlets. However, after 4-5 months, several cultures had formed areas of yellowish, translucent mucilaginous callus that proved to be embryogenic. Using callus developed in this way, it was possible to establish embryogenic cell suspensions of five genotypes in liquid 1/2MS medium with 1/3-strength nitrates containing 0.22-2.26 µM 2,4-D and 0.45 µM BA. These cultures were kept in a shaker at 90 rpm in the dark. The origin of somatic embryos was traced from single cells. Development progressed no further than the globular stage if the embryos remained in liquid medium, but continued to the cotyledonary stage when aggregates of globular embryos were plated on MS with half-strength major salts supplemented with glutamine and low levels of growth regulators. The success rate at this stage depended on genotype more than on the growth regulator treatment. It seems that genotype may be a more significant factor than culture medium in affecting the behaviour of beech embryogenic cultures. When isolated cotyledonary embryos were cultured on MS supplemented with low concentrations of BA, GA_3 and indole-3-butyric acid (IBA), 40% of embryos underwent partial germination (radicle elongation). Best conversion rates (10%) were obtained if the isolated cotyledonary embryos were stored in the cold for up to 7 months before transfer to conversion medium.

It may be mentioned here that Jörgensen (1988, 1991) attempted androgenesis of *F. sylvatica* by culturing anthers on MS containing 500 mg L^{-1} glutamine, 30 mg L^{-1} serine and various combinations of BA and 2,4-D. After 2 months of culture, one embryo developed in anthers in the one nucleus stage of the pollen grain during the first four days. It was not established whether this regenerant was diploid or haploid.

3.2. Micropropagation by Axillary Shoot Proliferation

The particular current importance of axillary branching for *in vitro* culture of *Fagus* arises from its being the only form of culture that has allowed the micropropagation of genotypes derived from mature trees. However, it has also been applied to material of juvenile origin, and it is this latter (and more successful) application that will be discussed first.

3.2.1. Material from Juvenile Origin

The first reported attempts to regenerate plants of a *Fagus* species by axillary branching were made by Chalupa (1979, 1985), who took shoot tips and nodal segments from *F. sylvatica* seedlings and cultured them on a range of mineral media, including MS, WPM and Driver and Kuniyuki medium (DKW; Driver and Kuniyuki, 1984), that were supplemented with various concentrations of the cytokinins N-(phenylmethyl)-9-(tetrahydro-2H-pyran-2-yl)-9H-purin-6-amine (PBA), BA and thidiazuron (TDZ). On the most successful medium (WPM supplemented with 0.1-0.6 mg L^{-1} PBA or BA and with 0.02-0.10 mg L^{-1} IBA), new axillary shoots appeared within 4-6 weeks. Shoot proliferation by subculture was limited by the number of shoots per explant being usually only one or two, although exceptionally as many as eight were obtained. Thidiazuron stimulated axillary shoot formation at low concentrations (0.001-0.005 mg L^{-1}), but high concentrations promoted callus formation and inhibited shoot elongation (Chalupa 1985). Shoots excised from cultures rooted within 3-4 weeks when cultured on WPM supplemented with 0.3 mg L^{-1} IBA and 0.1 mg L^{-1} NAA, the success rate ranging from 22% to 76%, depending on genotype (Chalupa 1987a). It was also found possible to root well-developed shoots from fast-growing cultures *ex vitro* in a potting mixture following a root induction treatment consisting in dipping the bases of the shoots in auxin powder (Chalupa 1996); within 6-8 weeks, 18-64% of the shoots had formed roots that were suitable for growth in soil. After 2 months of acclimatization of these rooted plantlets in a soil mixture under continuous light and high air humidity, the formation of adapted leaves made it possible to grow the plants in a greenhouse in partial shade. The survival rate among those planted out in the nursery was 87%, and the growth of those surviving in the field was similar to that of control seedlings, a height of 40-65 cm being reached by the end of the fourth growing season.

An *in vitro* shoot multiplication and plant regeneration system for European beech that starts from juvenile tissues and is based on axillary branching has also been established by Viéitez *et al.* (1993). Embryonic axes excised from beech seeds were germinated *in vitro* on a BA-supplemented medium to obtain *in vitro* plantlets from which axillary shoots were excised to establish shoot proliferation cultures. To establish the initial embryonic axis cultures, seeds were surface-sterilized by successive immersion in 70% (v/v) ethanol (1 min) and 4% active chlorine solution (20 min), and were then rinsed three times in sterile distilled water. The embryonic axes were dissected out and cultured on WPM with or without BA (0.1, 1.0 or 2.5 mg L^{-1}). On BA-free medium only 27% of the embryonic axes developed into plantlets; after 8 weeks, the plantlets that had developed had a 1-3 cm main stem with two or three leaves, a 3-4 cm hypocotyl and a slender root, but no axillary shoots. Addition of 0.1-1.0 mg L^{-1} BA increased germination rates (maximum 61%) and induced the formation of axillary shoots, mainly at the cotyledonary nodes, although it resulted in shorter hypocotyls and roots. The greatest average number of separable axillary shoots per germinated embryonic axis was afforded by the 1.0 mg L^{-1} BA treatment, and it was therefore shoots from plantlets obtained with this treatment that were used to develop a stabilized shoot multiplication system. On the basis of experiments with a variety of mineral media and growth regulators (0.1-1.0 mg L^{-1} BA, 1-2 mg L^{-1} zeatin, 0-0.2 mg L^{-1} NAA), the combination adopted as shoot multiplication medium was WPM supplemented with 0.5 mg L^{-1} BA, 2 mg L^{-1} zeatin and 0.2 mg L^{-1} NAA. In subsequent experiments the NAA was replaced by 0.5 mg L^{-1} IAA (Fig. 3A). The standard growth chamber conditions consisted in a 16-hour photoperiod in which cool-white fluorescent lamps supplied a photosynthetic density of 30 µmol $m^{-2} s^{-1}$, with light and dark temperatures of 25°C and 20°C, respectively. Six weeks of culture in multiplication medium afforded 1.8-4.0 shoots per explant, depending on clone.

In the experiments carried out to optimize the multiplication medium, repeated subculture on DKW medium was found to be detrimental (Vieitez *et al.* 1993), which contrasts with Chalupa's (1985) finding that, WPM and DKW were the mineral media that best stimulated shoot growth. As in Chalupa's experiments, however, it was found that *in vitro* proliferation of beech shoots appears to require the presence of both cytokinin and auxin in the medium, whereas other Fagaceae, including chestnut (Viéitez *et al.* 1983) and oak (Viéitez *et al.* 1985), require only BA.

To improve shoot proliferation further, shoots from clones established *in vitro* were cut into 8-10 mm apical and basal segments and their multiplication rates were compared (Viéitez *et al.* 1993). At the same time, for each explant type, two subculture cycles were compared: maintenance in the same medium for 6 weeks, and transfer to fresh medium every 2 weeks during the 6-week of multiplication cycle. Regardless of culture regime, the production of both shoots and culturable segments was greater for basal explants than for apical ($p < 0.005$), but multiplication rates were improved by transfer to fresh medium every 2 weeks ($p < 0.005$). This latter finding is interesting in that it appears to echo the fact that polycyclic growth patterns showing both annual and intra-annual growth rhythms are a particularly complex feature of European beech that seems to be more frequent when the soil provides an abundant supply of water and nutrients (Thiébaut *et al.* 1985, 1990). Similarly, the transfer to fresh medium every two weeks could supply sufficient nutrients favouring the *in vitro* development of beech shoots.

In rooting experiments (Viéitez *et al.* 1993), shoots 15-20 mm long were isolated from multiplication cultures and transferred to rooting medium consisting of WPM with half-strength macronutrients and no growth regulators. Rooting was induced either by previously dipping the basal end of shoots in 1 $g L^{-1}$ IBA solution for 1 min, or by the first 10 days being spent in medium supplemented with 3 $mg L^{-1}$ IBA or 3 $mg L^{-1}$ IBA + 0.1 $mg L^{-1}$ NAA. Addition of both IBA and NAA to the rooting medium had previously been used by Chalupa (1985, 1987a). The best rooting rate (88.8%) was afforded by the dipping treatment, but even more important was the application of an initial 7-day dark period, which significantly improved rooting capacity ($p < 0.0001$) and accelerated root emergence (Fig. 2), making the choice of induction treatment less important than for non-dark-treated shoots. After one month in rooting medium, and after having their roots cut back to a length of 10-15 mm, 60 plantlets were transferred to pots containing a 2:1 mixture of commercial peat substrate and perlite, and placed under a 80% air humidity in the greenhouse. After 8 weeks in the humidity tunnel the survival rate was 72%, and the surviving plants had initiated new growth (Fig. 3B).

The 6-week subculture regimen described in the previous paragraphs has also allowed the establishment of *F. orientalis* micropropagation cultures of juvenile origin, the shoot-tip initial explants being taken from 2-month-old seedlings and from a 4-year-old plant. Finally, American beech seedlings aged between 3 months and 2 years have been propagated *in vitro* using Aspen Culture medium (AC; Ahuja, 1983) containing 0.89 μM BA and 0.27 μM NAA for establishing and multiplying cultures and a 1-min dip in 12.3 mM IBA for root induction (Barker *et al.* 1997).

3.2.2. Material from Adult Origin

Explants taken from mature beech trees are in general difficult to grow and differentiate *in vitro*. For example, when Ahuja (1984a) attempted to culture bud explants from three 70-year-old *F. sylvatica* trees on WPM with low levels of phytohormones, only those of one, a specimen of var. *purpurea*, produced any shoots at all, and no success in multiplying or rooting these shoots was reported. Successful micropropagation of adult beech has only been achieved using rejuvenated tissues such as graftings (Meier and Reuther, 1994) or tissues that retain juvenile characteristics, such as epicormic shoots, stump sprouts (Gebbardt and Weisgerber, 1990) or root sprouts (Barker *et al.* 1997). These results are summarized below.

Meier and Reuther (1994) attempted the *in vitro* micropropagation of *F. sylvatica* genotypes obtained from trees of 10 and more than 100 years. They used two kinds of initial explant: buds taken directly from the mature trees, and buds taken from shoots that had sprouted from grafts implanted on seedling rootstocks 1 year previously. In both cases, dormant buds longer than 20 mm were taken from the apical region of shoots in February or March. They found that tissue culture could be started only with dormant buds disinfected by dipping in ethanol and subsequent flaming without damage to the tissues if the bud scales were entirely closed. For establishment *in vitro*, the best of five media tried was WPM supplemented with 4.5 μM BA, 1 $g L^{-1}$ casein hydrolysate and 15 $g L^{-1}$ fructose. Depending on genotype, shoots formed on 2.6-12.8% of the cultured buds. For shoot subculture the best medium was in general WPM followed by those of Shenk and Hildebrandt (1972; SH) and Gresshoff and Doy (1972; GD), although the clones from the 10-year-old tree multiplied best on GD. The general superiority of WPM agrees with the results obtained by Chalupa (1985, 1996) and Viéitez *et al.* (1993) with juvenile beech, and WPM has in fact been used in most studies of *in vitro* culture of beech tis-sues (Table 1). In the clones derived directly from bud explants of mature trees, multiplication rates declined progressively in successive subcultures until the total loss of the *in vitro* material. In the clones derived following grafting onto seeding rootstocks, the physiological activation so induced allowed successful *in vitro* establishment and maintenance in some cases but not all: among 51 grafted genotypes from 35-year-old trees, only seven afforded stable *in vitro* shoot cultures. Shoots from these seven were rooted with 65-75% success rates by dipping their basal ends in 4.92 mM IBA solution for 30 s, a procedure similar to that applied to shoots of juvenile origin by Viéitez *et al.* (1993).

Micropropagation of mature *F. grandiflora* trees resistant to beech bark disease has also been reported (Barker *et al.* 1997). Again, the performance of explants from mature tissues (buds from branches) was compared with that of explants from rejuvenated tissue (shoot tips from root sprouts induced on root segments). In all cases the explants were disinfected in 0.5% (v/v) sodium hypochlorite for 5-15 min and then cultured and multiplied on AC medium supplemented with 0.89 μM BA, 0.27 μM NAA and 20 g L^{-1} sucrose, the same medium as was used by the same authors for explants from seedlings of this species (see above, Section 3.2.1). Of 33 mature trees from which root segments were taken, 24 (73%) afforded segments that responded to sprout induction treatment, and 13 of these (39% of the original 33) were successfully micropropagated. By contrast, of the 41 trees from which buds from mature branches were taken, only 6 (15%) were amenable to micropropagation, and these cultures grew more slowly than those established via root sprouts. Rooting of shoots was achieved using a 1-min dip in 12.3 mM IBA solution followed by culture in a Horticube satured with half-strength AC medium containing 20 g L^{-1} sucrose. Rooting rates ranged from 48% to 97% among shoots derived from root sprouts, and from 50% to 97% among shoots derived from mature buds. As in the case of European beech (Viéitez *et al.* 1993; Meier and Reuther, 1994), brief dipping in relatively concentrated IBA solution was found to be superior as a rooting treatment than inclusion of lower IBA concentrations in the rooting medium.

3.3. Micropropagation by Regeneration of Adventitious Buds

While germinating embryonic axes of juvenile *F. sylvatica* as initial explants (see Section 3.2.1), Viéitez *et al.* (1993) observed the differentiation of adventitious buds on the intact hypocotyls of the resulting plantlets. In particular, adventitious buds developed on 29% of plantlets cultured on medium containing 4.44 μM BA. Capacity for adventitious bud formation was also shown by segments of cotyledon and hypocotyl isolated from 3-week-old seedlings produced by *in vitro* culture of zygotic embryos: 20% of such segments produced adventitious buds when cultured on medium supplemented with 2.22 μM BA plus 0.54 μM NAA. These observations led to studies of the development of adventitious buds on leaves or internode segments explanted from *in vitro* shoot multiplication cultures. These tissues are more abundant than the hypocotyl or cotyledon sections used as starting material, and the development of adventitious buds on them is a procedure that is well suited to genetic engineering approaches to genotype improvement.

3.3.1. Adventitious Buds Induced on Leaf Explants

Viéitez and San-José (1996) took leaves from shoot cultures established from 2-month-old and 3-year-old *F. sylvatica* seedlings, and investigated their ability to develop adventitious buds. The source cultures had been maintained on a shoot multiplication medium as described by Viéitez *et al.* (1993) (see Section 3.2.1 above). The two uppermost unfolded leaves were excised from the shoot apices of 6-week-old cultures and were cut transversally across the midvein into distal and proximal (basal) halves that were cultured for 8 weeks, abaxial face down, in bud induction medium consisting of WPM containing 30 $g\,L^{-1}$ sucrose, 7 $g\,L^{-1}$ agar, 2.9 µM IAA and either BA (4.4, 8.9 or 17.8 µM) or TDZ (0.2, 2.3, 4.5 or 9.1 µM). Adventitious buds generally developed within 5-7 weeks on the petiole stub callus of proximal explants(Fig. 3C) or on the callus closing the midvein wound of distal explants, although some buds appeared to emerge directly from the adaxial leaf blade in association with the midrib or major vasculature (Fig. 3D). Best results were obtained by culturing proximal leaf sections on induction medium including 8.9 µM BA or 2.3 µM TDZ (Table 2). Regardless of the induction medium, the percentage of explants producing buds, and their productivity, were significantly greater for proximal leaf halves than for distal halves. The growth regulator treatment had a significant effect on the number of buds in both the clones evaluated, but only in one did it affect the percentage of explants that developed buds. Bud formation capacity also depended on genotype of the stock shoot culture, and was enhanced by an initial 10 days in darkness but was reduced by longer periods of darkness. The genotype of the explants is one of the greatest determinants of organogenic response (Brown and Thorpe, 1986).

Table 2. Effect of cytokinin treatment and explant type (distal or proximal) on bud regeneration frequency (%) and the number of buds per regenerating explant, for two *Fagus sylvatica* clones cultured on regeneration medium containing 2.9 µM indole-3-acetic acid (IAA)[a].(Vieitez and San-José, 1996).

	Regeneration %		Number of buds[b]	
Treatment (µM)	Distal	Proximal	Distal	Proximal
Clone 5B				
BA[c] 4.4	6.3±3.0	43.8±9.9	1.0±0.4	1.5±0.2
BA 8.9	9.0±4.6	36.0±6.2	1.6±0.7	2.1±0.2
BA 17.8	4.0±1.4	28.0±3.5	1.3±0.6	1.7±0.2
TDZ 0.2	13.2±4.5	35.7±7.6	3.6±0.7	3.4±0.5
TDZ 2.3	24.3±5.4	49.4±8.9	2.7±0.5	3.4±0.5
TDZ 4.5	13.6±2.6	54.5±6.9	2.2±0.2	2.5±0.3
TDZ 9.1	8.8±3.0	47.0±10.3	1.0±0.4	2.5±0.1
Clone FS-104				
BA 4.4	1.7±1.4	17.0±1.3	0.3±0.3	2.6±0.4
BA 8.9	10.0±4.1	27.2±8.5	1.3±0.6	1.7±0.3
BA 17.8	0	2.2±0.9	0	0.7±0.3
TDZ 2.3	4.4±2.4	35.6±6.6	3.1±1.4	5.6±0.7

Analysis of Variance	Clone 5B				Clone FS-104			
Source of variation	Regeneration %		N° of buds		Regeneration %		N° of buds	
	F-test	LSD 5%	F-test	LSD 5%	F-test	LSD 5%	F-test	LSD 5%
Treatment (A)	2.262^{ns}	----	6.899^{***}	1.34	6.684^{**}	15.04	9.822^{**}	2.30
Explant (B)	69.434^{***}	6.27	4.795*	0.54	23.431^{**}	8.68	7.124^{ns}	----
A X B	0.541^{ns}	----	0.621^{ns}	----	1.369^{ns}	----	0.965^{ns}	----

[a]Means ± standard errors in four experiments, each with 30 replicates per treatment/explant/clone combination.
[b]In thidiazuron (TDZ) treatments of clone 5B, the number of bud values actually represent the number of bud clusters. F-test: ns, not significant; * P = 0.05; ** P = 0.005; *** P = 0.0001
[c]BA = benzyladenine

With a view to promoting the elongation of the adventitious shoots, the effect of the duration of exposure to TDZ on bud induction and shoot elongation was investigated. Basal leaf halves were incubated on induction medium with 2.3 μM TDZ for 1-8 weeks. After these times, the explants were transferred to multiplication medium to give a total experimental time of 8 weeks. Thidiazuron exposure time significantly affected caulogenic competence: the percentage of explants generating buds was greatest after 4 weeks of TDZ treatment, and the average number of shoots per explant after 3 weeks of treatment, while shoot length fell with increasing exposure time, possibly because of the high cytokinin activity of TDZ. Bud clusters produced in response to continuous exposure to this compound were usually stunted, compact and more difficult to convert into elongated shoots than buds induced by BA medium. Overall, three weeks culture with TDZ was the most efficient treatment for shoot production and elongation, in accordance with the two-stage culture procedure proposed by Huetteman and Preece (1993).

Thidiazuron also induced shoot buds on leaf explants from Oriental beech cultures derived from 4-year-old or 2-month-old seedlings of several genotypes (Cuenca and Viéitez 1999). The percentage of explants developing adventitious buds ranged from 70% to 90%, with 13-19 buds per responsive explant, regardless of the age of the clone-founding seedling, but as in the case of *F. sylvatica* was significantly affected by genotype, at least in the case of the clones derived from 2-month-old plants.

Taken together, the above findings suggest that beech tissue is more amenable to adventitious bud induction than some other Fagaceae such as chestnut or oak, although similarly good adventitious bud induction results were obtained by Jordan *et al.* (1996) with *Nothofagus alpina*, a member of the closely related genus *Nothofagus*. In these latter experiments, *N. alpina* leaf explants were cultured on medium containing 0.45 μM thidiazuron and 0.005 μM IBA.

3.3.2. Adventitious Buds Induced on Internode Explants

Cuenca *et al.* (2000) evaluated the adventitious bud formation capacity of internode segments obtained from shoots of two clones of European beech and five clones of Oriental beech (the same as clones had been used in the experiments with leaf explants described in Section 3.3.1 above). Donor shoots were collected from *in vitro* cultures, and the two uppermost internodes of each were excised and divided in segments 2-3 mm long that were then cultured on induction medium consisting of WPM supplemented with 2.9 μM IAA and either 0.4-17.8 μM thidiazuron or 0.4-

17.8 μM BA. After 4 weeks of culture on induction medium, the explants were transferred to the multiplication medium used for maintenance of stock shoot cultures, where they spent a further 8 weeks. Adventitious buds formed on callus that developed on the internode segments (Fig. 3E) of four of the five *F. orientalis* genotypes and both the *F. sylvatica* genotypes (Table 3). Callus in fact seemed to be an essential intermediate in bud formation, since the only clone (FO-9) that developed no buds also failed to develop callus. As in the case of leaf explants this study provides further evidence of the importance of explant genotype in *in vitro* shoot regeneration. The mean number of buds produced was affected by both species and clone, and three genotypes performed significantly better on 4.5 μM TDZ than on other concentrations of this compound, which emphasizes the need to optimize growth regulator levels for each individual clone if maximum bud production is desired. Previously, interclonal variation in sensitivity to growth regulators had been put forward by Lainé and David (1994) as a possible cause of the dependence of capacity for adventitious bud formation on genotype, which has also been reported for other woody species (Coleman and Ernst 1989; Bergmann and Stomp 1994).

As in the case of leaf explants (Viéitez and San-José, 1996), TDZ was much more efficient than BA in inducing adventitious buds from internode segments, the optimal TDZ concentration being 4.5 μM as against 17.8 μM for BA (Fig.4), although higher thidiazuron concentrations resulted in the formation of dense clusters of many minute buds which failed to develop individually. These findings are in agreement with Murthy *et al.*'s (1998) assertion that TDZ is more effective for bud induction than purine-type cytokinins. The high cytokinin activity of TDZ has been attributed to its modulating the biosynthesis of endogenous cytokinins (Huetteman and Preece 1993) and/or inhibiting cytokinin oxidase activity (Eapen *et al.* 1998). It has also been suggested that the TDZ morphogenic effect is a stress response (Murch *et al.* 1997) and the occurrence of regenerants in TDZ-treated tissues may be an adaptative reproductive mechanism for overcoming imposed stress (Murthy *et al.* 1998).

Thidiazuron induction medium supplemented with a range of individual auxins was also investigated, and we have found that IAA or IBA at 2.9 μM enhanced the bud forming capacity of internode explants (Cuenca *et al.* 2000)

The percentage of explants developing adventitious buds, and the number of buds developed, were greater among the more apical than among the more basal internodes. Such a polarity in bud regeneration was also shown in the experiments with leaf explants, proximal leaf halves were more productive than distal halves (see Section 3.3.1), and in both cases buds

Table 3. Effect of genotype and TDZ concentration on adventitious bud induction and callus fresh weight from *Fagus orientalis* and *F. sylvatica* internode explants obtained from two-month-old plants (clones FO-3, FO-9, FO-23, FO-34, FS-5B and FS-104) and from a four-year-old plant (clone FO-BS). Data recorded after 12 weeks of culture.(Cuenca *et al.* 2000).

	Bud regeneration frequency (%)			Bud number per regenerating explant			Fresh weight per explant (mg)		
	TDZ (μM)			TDZ (μM)			TDZ (μM)		
	0.4	2.3	4.5	0.4	2.3	4.5	0.4	2.3	4.5
F. orientalis									
FO-3	82.5 ± 5.4	100.0 ± 0.0	90.0 ± 3.5	11.6 ± 2.1	33.2 ± 3.9	43.8 ± 2.5	236.4 ± 31.6	229.9 ± 6.2	210.0 ± 26.7
FO-9	0.0 ± 0.0	0.0 ± 0.0	0.0 ± 0.0	0.0 ± 0.0	0.0 ± 0.0	0.0 ± 0.0	4.7 ± 0.8	5.4 ± 1.1	6.5 ± 1.1
FO-23	43.2 ± 8.7	44.1 ± 3.0	51.4 ± 4.4	5.7 ± 0.7	12.7 ± 2.5	15.7 ± 5.6	80.9 ± 7.8	119.5 ± 7.0	116.5 ± 15.2
FO-34	10.0 ± 6.1	6.8 ± 2.0	15.0 ± 10.3	1.4 ± 0.7	1.3 ± 0.4	1.1 ± 0.6	111.3 ± 12.8	113.5 ± 4.4	122.3 ± 17.4
FO-BS	7.5 ± 4.1	15.0 ± 4.3	25.0 ± 5.6	4.5 ± 2.3	5.1 ± 1.9	11.3 ± 3.4	47.6 ± 24.3	58.7 ± 10.2	66.3 ± 11.9
F. sylvatica									
FS-5B	85.4 ± 5.1	81.3 ± 3.9	82.3 ± 7.2	11.2 ± 1.9	22.5 ± 2.8	30.6 ± 5.6	186.8 ± 6.5	146.7 ± 3.1	167.3 ± 12.6
FS-104	5.3 ± 2.6	15.8 ± 5.7	30.3 ± 12.1	1.1 ± 0.6	1.4 ± 0.5	3.1 ± 1.3	157.6 ± 33.4	181.0 ± 22.6	239.0 ± 12.5

Source of Variation	df	Bud regeneration frequency (%)			Bud number per regenerating explant			Fresh weight per explant (mg)		
		MS	F	LSD (5%)	MS	F	LSD (5%)	MS	F	LSD (5%)
TDZ concentration	2	414.354	2.974*	6.81	718.239	21.898 ***	3.30	1593.716	1.188^{NS}	
Genotype	6	10110.050	72.572***	11.80	1537.720	46.882 ***	5.72	71399.294	53.228 ***	36.61
TDZ conc. x Genotype	12	165.663	1.189^{NS}		156.257	4.764 ***	8.10	1687.976	1.258^{NS}	
Error	63	139.312			32.800			1341.394		

Values are means ± SE for four replications (10 explants for each replication)
NS - not significant; * - significant at 0.05; *** - significant at 0.001

generally arose either from callus that had developed on proximal wounds or in the vicinity of leaf veins. These results with leaves suggest that the sites at which meristemoids are initiated may be determined by nutrient and growth regulator gradients mediated by vascular tissues, including endogenous auxin gradients set up in the whole explant (the whole leaf or whole stem) prior to its division in proximal and distal segments (Brown and Thorpe, 1986). Similar considerations would seem to apply to the results on internode explants, growth regulator polar transport processes appearing to result in the balance between auxins and cytokinins being more favourable for adventitious bud formation in the more apical internodes than in the more basal.

3.3.3. Anatomical Study

The identity of the cells originating adventitious buds on *F. orientalis* leaf explants, and the anatomical events involved in bud development, were studied by Cuenca and Viéitez (1999). Samples were collected every 2-3 days between explantation on day 0 and day 28, inclusive. On each occasion, transverse 8-10 μm sections were cut from the petiole stub region (where most adventitious buds developed) and from the leaf lamina and were stained with safranin-fast green. Suberized tissues were detected by contrasting sections stained with PAS (periodic acid-Schiff stain) with Sudan Black B.

Observation of the sections showed petioles to consist of epidermis, subepidermis, inner parenchyma and vascular bundles. The latter were mostly arranged in a continuous arc open towards the adaxial side of the petiole, where there were also additional bundles. The first visible change was observed as early as day 3, when the nuclei and nucleoli of a number of subepidermal cells and cells adjacent to vascular bundles began to enlarge. Between days 3 and 6 these cells underwent mitosis; predominantly periclinal cell divisions in subepidermal layers produced clusters of cells sharing a common outer wall (Fig. 5A). By day 9, continued predominantly periclinal cell divisions in subepidermal and outermost parenchyma cells had given rise to cell files that caused bulging outgrowths. Cell proliferation in the inner parenchyma, and the development of cambium-like zones from perivascular petiole tissue, were also observed at this stage. These areas were destined to form a substantial callus tissue differentiating tracheary elements.

Between days 11 and 13, radial cell files perpendicular to the explant surface were further extended by extensive periclinal division of the new cells derived from the subepidermal layer. Following Gautheret (1959), this region may be described as an undefined secondary meristem or diffuse cambium, which is the result of *in vitro* proliferation in a preferential direction induced by contact with the culture medium. In cross-sections its component cells were rectangular, vacuolated, radially flattened and smaller than subepidermal cells, and they had a prominent nucleus in a lateral or central position (Fig. 5B). Tissue resembling diffuse cambium developed much more intensely on the abaxial side of the explant which was in contact with the culture medium. In the adaxial region, the division of parenchyma and subepidermal cells in various directions gave rise to vascular cambium-like zones.

After 11-13 days in culture, some of the outermost cells of the newly formed radial files seemed to de-differentiate, acquiring a meristematic appearance. These cells had a large, prominent, centrally located nucleus and a high nucleus-to-cell area ratio. They were easily distinguished from the surrounding larger, flattened, more vacuolated radial file cells. The dedifferentiation process was generally associated with anticlinal division of cells originated by periclinal division (Figs. 5B,C). Meristematic feature were also exhibited by some epidermal cells, which started to undergo anticlinal mitotic division. De-differentiation into meristematic cells progressed in succeeding days, giving rise to small meristematic centres that continued to divide. By day 16, meristemoids had thus formed in the outermost regions of tissues that had derived primarily from the subepidermis (Fig. 5D), although the formation of some meristemoids had also involved overlying epidermal cells.

By days 19-21 the meristemoids had increased in both size and number. Most were polarized structures that were broader towards the surface. They were located between radial cell files in which phenolic compounds were found to have been deposited (Fig. 5E). At the same time, extensive cell division in the parenchyma and in the vascular cambium-like zones gave rise to an increase in callus tissue. Organized cell divisions in the meristemoids originated bud primordia that emerged from the explant surface and differentiated a protoderm (Fig. 5F).

By days 24-26 the bud primordia had developed procambial strands that had established vascular connections with the subjacent vascular tissue originated within the callus. The progressive structural differentiation of the shoot apical meristem and leaf primordia led to the formation of well-developed shoot buds by the end of the fourth week of culture (Fig. 5G). During these last few days, a phellogen appeared to differentiate at the outer boundary of the proliferating regions, and the activity of the diffuse cambium appeared to be replaced by that of this phellogenic cambium. Also, some bud primordia were observed to emerge through the complementary or filling tissue of lenticels, and some meristemoids were differentiated from the lenticel phellogen (Fig. 5H).

Most of the shoot buds that differentiated on leaf lamina originated in the abaxial region of the midvein, which underwent modifications similar to those of the petiole stub. In the adaxial part of the lamina, callus formation was limited to a small protuberance on the midvein or on lateral leaf veins, where adventitious buds occasionally arose directly without extensive intervening cell division.

In this study, both direct and indirect development of meristemoids was observed. Most meristemoids originated indirectly from cell files that had been formed by superficial, periclinally dividing secondary meristems (areas of diffuse cambium or that presented the appearance of phellogenic cambium), which provided the bulk of cells for development of bud primordia (Fig. 5I). Cell division patterns therefore seem to play an important role in determining the location and characteristics of cells that will become meristematic and give rise to the organization of meristemoids. Similar observations have been reported for the differentiation of promeristemoids in certain conifer species (Flinn *et al.* 1988). The finding that meristemoid precursors originated in surface layers of the explant is in keeping with reports that shoot buds induced on cultured explants of many other species also originated directly or indirectly in epidermis, subepidermis or both (Thorpe and Kumar 1993; Pawlicki and Welander 1994; Torregrosa and Bouquet 1996; Azmi *et al.* 1997), although none of these other reports mentioned the involvement of a phellogenic cambium-like zone.

An histological study of adventitious bud induction on beech internode segments was performed by Cuenca *et al.* (2000). In this case, too, it was found that adventitious buds arose from surface layers of callus which had in turn originated from the subepidermal and epidermal layers of the explants, and that the buds had vascular connections with vascular tissues that had differentiated within the callus.

3.3.4. Shoot Proliferation and Rooting

The utility of adventitious bud induction for micropropagation depends not only on the efficiency of the initiation and differentiation processes discussed in the preceding sections, but also on the proliferation and rooting of *in vitro* clones derived from the induced buds. To study these aspects, leaf and internode explants bearing adventitious buds were transferred to fresh multiplication medium (the same as used for stock shoot cultures) 8 weeks (for leaves) or 12 weeks (for internodes) after their culture initiation on adventitious bud induction medium. Shoot elongation was promoted by transferring the explants to fresh medium every 2 weeks. After 2-3 months, when the adventitious shoots were about 1.5-2.0 cm long (Fig. 3F), they were excised from the leaf or internode explant, sectioned, and subcultured on multiplication medium, so establishing shoot culture lines that were identified by the adventitious shoot of origin (Viéitez and San-José, 1996; Cuenca *et al.* 2000). The number of elongated shoots on internode explants was found to benefit from 2 months of storage under dim light at 3-4°C. Following transfer to standard culture conditions, 70% of these explants developed elongated shoots, as against 40% of those that had not been subjected to cold storage. The importance of chilling for elongation of shoots induced by TDZ had been reported by Huetteman and Preece (1993).

Shoots from clones originated by adventitious shoots were rooted using the same procedure as for the stock shoot cultures, i.e. by dipping the basal ends in 4.9 mM IBA solution for 2 min and subsequent transfer to medium with half-strength WPM macronutrients and no auxin, where they spent the first 8 days in darkness (Fig. 3G). Shoot cultures of adventitious origin became morphologically indistinguishable from the stock cultures that had provided the explants on which the adventitious buds had been induced, and likewise exhibited very similar *in vitro* behaviour as regards proliferation rate and rooting capacity. For example, the rooting frequencies of two shoot lines originated from adventitious shoots on leaf explants, 80% and 88%, did not differ significantly from that of the stock culture from which the leaf explants had been taken, 83% (Viéitez and San-José 1996).

3.4. Clonal Fidelity

In an unpublished study carried out in our laboratory we used random amplified polymorphic DNA (RAPD) analysis to investigate the genetic stability of shoot culture lines established from adventitious buds induced on leaves taken from two *in vitro* clones of *F. sylvatica* (clones FS-104 and 5B), on internode from one *in vitro* clone of *F. orientalis* (clone 3), and on hypocotyl excised from cultured embryonic axis of another *F. orientalis* genotype (clone 11). Six lines originated from *F. orientalis* internode were used (lines 3.1, 3.2, 3.3, 3.5, 3.7 and 3.8), one originated from *F. orientalis* hypocotyl (line 11.1), one originated from leaves of one of the *F. sylvatica* clones (line FS-104.1), and two originated from leaves of the other *F. sylvatica* clone (lines 5B.1 and 5B.2); each of these lines derived from a single adventitious bud and was subcultured separately by axillary shoot proliferation.

DNA amplification was carried out by PCR using 100 ng of genomic DNA and 30 arbitrary 10-base primers included in Operon Technologies

Inc.'s kits A and S. RAPD profiles were obtained for the ten shoot lines of adventitious origin and the four axillary shoot lines from which they had been derived, making a total of 14 DNAs analysed. A total of 293 amplified sequences were obtained from these 14 DNAs, 57.7% of which were involved in polymorphism. The total number of sequences amplified by each primer ranged from 6 to 14 (mean 9.8), and there were certain individual primers, and several sets of two primers, that sufficed to detect polymorphism allowing both discrimination between species and also discrimination between the four axillary shoot lines (Fig. 6). For example, two fragments of 600 (Fig. 6A) and 1100 (Fig. 6B) base pairs are present in the *F. sylvatica* clones but not in the *F. orientalis* clones. Similar results were obtained when other *F. sylvatica* clones were included in the analysis. However, for none of the axillary shoot lines did any of the primers detect RAPD differences either among the lines of adventitious origin derived from that axillary shoot line or between any of these lines and the axillary shoot line itself. There therefore seems to be no somaclonal variation associated with *in vitro* adventitious bud induction in these *Fagus* species. Organogenesis of loblolly pine from callus has likewise been shown by RAPD analysis to cause no somaclonal variation (Tang 2001), whereas a similar analysis of sugar beet plantlets regenerated by organogenesis from leaves did find low levels of intraclonal polymorphism (Munthali *et al.* 1996).

3.5. Effect of Carbon Source on Shoot Production

The chemical nature of the sugar added to a culture medium may be critical for morphogenic reactions. Cuenca and Viéitez (2000) performed experiments to determine the effects of carbon source and concentration on both the *in vitro* multiplication of *F. sylvatica* and *F. orientalis* shoots by axillary branching and on the induction of adventitious buds on leaf and internode explants of *F. orientalis*.

In the axillary shoot multiplication experiments, the mean number of shoots, the longest shoot length and the percentage of shoots that were hyperhydric were all significantly affected by both the type of carbohydrate (sucrose, glucose or fructose) and its concentration (1-4%). More shoots were obtained with glucose than with the other sugars, and glucose also afforded the greatest longest shoot length in two of the three clones used; in the other, the greatest longest shoot length was obtained with 4% sucrose. Both the number and the length of shoots grown on glucose medium were greater with a glucose concentration of 3-4% than with lower concentrations. The percentage of shoots that were hyperhydric was significantly affected by carbohydrate type (ranging from 0-1.5% for sucrose to 8-15% for fructose), but not by concentration. Fructose had an adverse effect on both the length of shoots and their quality, and the use of filter-sterilized rather than autoclaved fructose neither stimulated growth nor reduced the incidence of hyperhydricity in all three clones. Meier and Reuther (1994), who included 15 $g L^{-1}$ fructose in the shoot multiplication medium they used for micropropagation of mature European beech, did not investigate the suitability of other carbohydrates.

In the adventitious bud regeneration experiments with leaf explants, the percentage of explants responding to bud induction treatment depended on sugar type ($p < 0.05$) but not on sugar concentration; glucose-containing media were superior to sucrose-containing media (Table 4). The number of buds per responsive explant was affected by both the type and concentration of sugar, with significant interaction between the two factors ($p < 0.001$). Whereas sucrose concentration did not affect the number of buds, increasing glucose concentration raised bud production, with 3% and 4% glucose giving the highest values (Fig. 3H). The numbers of buds originating from leaf lamina and petiole callus showed trends similar to those observed for the total number of buds per leaf explant (lamina plus petiole), although glucose level had a more significant influence on lamina than on petiole in this respect.

Table 4. Effect of carbohydrate type and concentration on bud regeneration frequency (%) and the number of buds per responsive explant (leaf section or internode segment) in *Fagus orientalis* clone BS. Data were recorded after a total of 8 weeks culture for leaf explants or 12 weeks culture for internode explants (Cuenca and Vieitez, 2000).

Carbohydrate concentration (%)	Regeneration %			Mean bud number		
	Sucrose	Glucose	Mean	Sucrose	Glucose	Mean
Leaf explant						
1%	76.7±6.7	100.0±0.0		7.17±1.3a	26.1±3.3b	
2%	66.7±18.6	93.3±3.3		6.9±2.8a	50.6±8.2c	
3%	86.7±6.7	93.3±6.7		18.7±3.5ab	76.7±9.0d	
4%	86.7±6.7	90.0±6.4		16.7±1.6ab	91.0±5.5d	
Mean	79.2±4.8a	94.2±2.0b				
Internode explant						
1%	13.3±6.7	46.7±8.8	30.0±16.7a	1.0±0.0	2.9±0.6	1.8±1.1a
2%	36.7±3.3	80.0±10.0	58.3±21.7b	1.67±0.2	6.4±1.4	4.0±2.4ab
3%	66.7±14.5	83.3±6.7	75.0±8.3b	8.81±3.4	12.4±0.7	10.6±1.8b
4%	60.0±5.8	56.7±6.7	58.3±1.7b	4.9±2.5	19.2±6.8	12.1±7.1b
Mean	41.2±12.1a	66.7±8.9b		4.0±1.8a	10.2±3.6b	

In the adventitious bud induction experiments with internode explants, both the percentage of explants responding to treatment and the number of buds per responsive explant were significantly affected by both the type and concentration of sugar, albeit without significant interaction between the two factors (Table 4). As in the case of leaf explants, glucose stimulated the caulogenic response more than sucrose in terms of both response frequency and the number of buds.

In general, a 3% concentration of glucose was the best carbon source for both axillary branching and adventitious shoot induction, although the responses of shoot cultures to the different sugars appeared to depend to some extent on genotype. Thus although most research on the micropropagation of woody species has used sucrose, it is not necessarily the most effective carbon source for this purpose. Other studies of the influence of carbon source on shoot formation by *in vitro* cultures of Fagaceae have reached the same conclusion: as in Cuenca and Viéitez's (2000) study, sucrose was found to be inferior to glucose by Belaizi and Boxus (1995) and by Romano *et al.* (1995), and inferior to fructose by Chauvin and Salesses (1988). Although non-reducing carbohydrates such as sucrose constitute the main translocated form of carbon in the Fagaceae whole plant (Zimmermann and Ziegler 1975), reducing sugars such as glucose can readily diffuse across the cut tissue surfaces and are easily metabolized by *in vitro* tissue cultures, for which they can therefore be a better carbon source.

The positive effect of sugar concentration found in this study (Cuenca and Viéitez, 2000) may have been due to its increasing the osmotic pressure of the medium, which stimulates mitochondrial activity and hence, presumably, production of the energy required for shoot initiation (Bonga and von Aderkas 1992).

4. COLD STORAGE AND CRYOPRESERVATION

The application of *in vitro* techniques to germplasm storage is of particular interest for the conservation of species with recalcitrant seeds, a class that includes beech species. The particular techniques that may be applicable depend on the desired storage time. For medium-term storage (1-12 months) the aim is to reduce growth rate and increase the period between subcultures, which can be achieved by simply lowering the culture temperature. However, for long-term storage of germplasm the only method currently available is cryopreservation (Engelmann 1997).

In an unpublished study carried out in our laboratory, the survival and proliferation capacity of beech shoot cultures stored for up to 14 months at 2-4°C was studied. The sources of starting materials were 6 clones of European beech and 42 clones of Oriental beech that had been maintained by subculturing 1.0-1.5 mm shoot tips at 6-week intervals on WPM supplemented with 2.2 μM BA, 9.1 μM zeatin, 2.9 μM IAA, 30 g L^{-1} sucrose and 7 g L^{-1} agar. Fifteen days after the last subculture, the 500 mL jars containing the shoot tip cultures were transferred from standard culture conditions to cool cabinets (340 L Sanyo Medicool MPR 311D), where they were kept for up to 14 months at 2-4°C with 12 h per day of dim light (8 μmol $m^{-2} s^{-1}$). In general, survival rates of 70-90% were recorded 1 month after transfer from cold storage to normal culture, the only exceptions being three *F. orientalis* clones that could not be profitably stored for more than 6 months. The proliferation capacity and growth of the shoots increased in the first subculture after storage, although this effect was not a permanent one, but returned to control levels (non stored cultures) in subsequent subcultures. It is thus clearly possible to keep beech cultures in cold storage for at least a year without subculturing.

As noted above, storage of beech seeds is problematic because of their recalcitrance, which results from their sensitivity to desiccation. However, there have been reports that for some species with recalcitrant seeds cryopreservation can be applied to embryonic axes excised from the seeds and then dried before freezing. Pence (1990) investigated the feasibility of this approach for several temperate species with recalcitrant seeds, including *F. grandiflora*. Excised embryonic axes were surface-sterilized by immersion in a 5% solution of commercial sodium hypochlorite for 10 min, and were then dried overnight in open sterile dishes under the air stream of a laminar flow hood, after which half were frozen by immersion in liquid nitrogen. Both dried and thawed axes were rehydrated and grown *in vitro* on a medium containing 2 mg L^{-1} BA and the same concentration of indoleacetylphenylalanine. Regardless of treatment (control, dried, and dried + frozen), the American beech embryos showed some elongation of the hypocotyl, the beginning of shoot growth and some axillary shoot development, together with the development of callus. It seems likely that absence of further development may have been due to the concentrations of auxin and cytokinin in the medium, since high levels of these compounds tend to inhibit radicle elongation and stimulate the development of callus. As pointed out by Viéitez *et al.* (1993) the germination of European beech embryonic axes was favoured by media with less BA (0.1-1.0 mg L^{-1}) and no auxin (see Section 3.2.1). Therefore, as Pence (1990) mentioned, it seems quite likely that better results would be forthcoming following optimization of growth media and other conditions for *in vitro* germination of cryopreserved embryonic axes.

Jörgensen (1990) also demonstrated that immature embryos of *F. sylvatica* (globular stage and 1-2 mm in diameter) can be frozen in liquid nitrogen. The best results were obtained when using 10% DMSO and 5% sucrose as cryoprotectants and a slow cooling technique. More than 90% of the embryos survived with this method, although subsequent growth and development was delayed due to 'cryoshock'.

5. CONCLUDING REMARKS

Conventional propagation methods are unable to supply the large quantities of genetically superior individuals of *Fagus* species that are required for reforestation programs. There is therefore considerable interest in the possibility of regenerating plantlets from cultures multiplied *in vitro* following initiation from embryos, shoot tips and other tissues explanted from juvenile or adult trees. As described above, *in vitro* proliferation and regeneration systems have in fact been developed that use somatic embryogenesis, axillary shoot multiplication, and adventitious bud induction.

Culture of axillary shoots allows successful propagation of genotypes that have been established *in vitro* using material from either juvenile or mature plants. Although it is difficult to induce morphogenesis directly in mature tissues, good results have been obtained using tissues that have been rejuvenated by grafting on juvenile rootstock or by induction of root sprouts. The rooting and acclimatization rates of shoots obtained from *in vitro* axillary shoot multiplication cultures are generally acceptable, although further studies are required on the transfer of rooted plantlets to soil and to the field.

Plantlets can also be regenerated from cultures established from adventitious buds induced by BA or TDZ on leaves or internodes excised from *in vitro* shoot cultures, at least in the cases of *F. sylvatica* or *F. orientalis*. The response depends on explant type and polarity, on the cytokinin in the medium, and on genotype. Histological monitoring shows that adventitious buds arise directly or indirectly from the more superficial layers of the explant. A system of this kind, in which a large number of surface cells can be converted into adventitious buds, would be of great value for improvement of *Fagus* species by genetic engineering, for which it is now necessary to develop the appropriate efficient transformation protocols.

The performance of beech shoot cultures is affected by both the type and concentration of sugar in the culture medium. In general, glucose is better than sucrose or fructose for both axillary shoot multiplication and adventitious bud induction, and this should be taken into account in trials designed to allow further improvement of *Fagus* micropropagation conditions.

The development of somatic embryogenesis of beech has not progressed as far as other methods of micropropagation, and further studies are required for optimization of culture conditions for the induction, development and germination of somatic embryos of these species. In view of the enormous potential of somatic embryogenesis for production of genetically improved plants, these studies should be put in motion.

Though not surprising, it must be mentioned that the efficiency of all three of the above approaches to micropropagation of beech (axillary branching, adventitious bud induction and somatic embryogenesis) is greatly influenced by genotype.

Beech germplasm can be preserved for at least a year, without loss of viability, by storage at 2-4°. For long-term storage of beech germplasm, cryopreservation of embryonic axes seems to be promising, but more research needs to be done to optimize this procedure, and for long-term storage of clonal material (bud cultures or embryogenic cultures) it will also be necessary to investigate the cryopreservation of these tissues.

Finally, a crucial question in all programs of *in vitro* propagation of improved genotypes is the genetic stability of the *in vitro* cultures. In general this depends on the micropropagation method used and must be investigated separately for each. In what is to date the only study of this issue in regard to beech, no somaclonal variation was found, within the limits of RAPD analysis, among shoots derived from adventitious buds induced on leaf or internode explants.

6. ACKNOWLEDGEMENTS

The financial support from CICYT, Spain (AGF96-0453) and Xunta de Galicia, Spain (PGIDT 00BIO40001) is gratefully acknowledged.

7. REFERENCES

Ahuja M.R. 1983. Somatic cell differentiation and rapid clonal propagation of aspen. Silv Genet 32: 131-135.

Ahuja M.R. 1984a. *In vitro* induction of organogenesis in juvenile and mature beech. Silv Genet 33: 241-242.

Ahuja M.R. 1984b. Isolation and culture of mesophyll protoplasts from mature beech trees. Silv Genet 33: 37-39.

Azmi A., Noin M., Landré P., Prouteau M., Boudet A.M. and Chriqui D. 1997. High frequency plant regeneration from *Eucalyptus globulus* Labill. hypocotyls: Ontogenesis and ploidy level of the regenerants. Plant Cell Tiss Org Cult 51: 9-16.

Barker M.J., Pijut P.M., Ostry M.E. and Houston D.R. 1997. Micropropagation of juvenile and mature American beech. Plant Cell Tiss Org Cult 51: 209-213.

Belaizi M. and Boxus P. 1995. *In vitro* shoot multiplication of cork oak (*Quercus suber* L.). Influence of different carbohydrates. Bull Rech Agron Gembloux 30: 39-46.

Bergmann B.A. and Stomp A.M. 1994. Effect of genotype on *in vitro* adventitious shoot formation in *Pinus radiata* and correlations between pairs of phenotypic traits during *in vitro* shoot development. Plant Cell Tiss Org Cult 39: 185-194.

Bonga J.M. and Von Aderkas P. 1992. In Vitro Culture of Trees. Dordrecht: Kluwer Academic Publishers.

Brown D.C.W. and Thorpe T.A. 1986. Plant regeneration by organogenesis. In: Vasil I.K. (Ed.), Cell Culture and Somatic Cell Genetics of Plants, Vol 3, pp 49-65. Academic Press, Orlando, FL.

Ceballos L. and Ruiz de la Torre J. 1979. Arboles y Arbustos de la España Peninsular. Madrid: Escuela Técnica Superior de Ingenieros de Montes.

Chalupa V. 1979. *In vitro* propagation of some broad-leaved forest trees. Commun Inst For Cech 11: 159-170.

Chalupa V. 1982. Vegetative propagation of broadleaved trees by cuttings. Lesnictvi (Forestry) 28: 21-30.

Chalupa V. 1985. *In vitro* propagation of *Larix, Picea, Pinus, Quercus, Fagus* and other species using adenine-type cytokinins and thidiazuron. Commun Inst For Cech 14: 65-90.

Chalupa V. 1987a. European hardwoods. In: Bonga J.M. and Durzan D.J. (Eds.), Cell and Tissue Culture in Forestry, Vol 3, pp. 224-246. Martinus Nijhoff, Dordrecht.

Chalupa V. 1987b. Somatic embryogenesis and plant regeneration in *Picea, Quercus, Betula, Tilia, Robinia, Fagus* and *Aesculus*. Commun Inst For Cech 15: 133-148.

Chalupa V. 1996. *Fagus sylvatica* L. (European Beech). In: Bajaj Y.P.S. (Ed.), Biotechnology in Agriculture and Forestry, Vol 35, Trees IV, pp. 138-154. Springer-Verlag, Berlin, Heidelberg.

Chauvin J.E. and Salesses G. 1988. Effect du fructose sur la micropropagation du châtaignier *Castanea* sp. C R Acad Sci Paris 306 Série III: 207-212.

Coleman G.D. and Ernst S.G. 1989. *In vitro* shoot regeneration of *Populus deltoides*: effect of cytokinin and genotype. Plant Cell Rep 8: 459-462.

Cornu D., Delran S., Garbaye J. and Le Tacon F. 1977. Recherche des meilleures conditions d'enracinement des boutures herbaceés de chêne rouvre (*Quercus petraea* (M.) Liebl.) et de hêtre (*Fagus sylvatica* L.). Ann Sci For (Paris) 34: 1-16.

Cuenca B. and Viéitez A.M. 1999. Histological study of *in vitro* development of adventitious buds on leaf explants of Oriental beech (*Fagus orientalis* Lipsky). In Vitro Cell Dev Biol Plant 35: 326-332.

Cuenca B. and Viéitez A.M. 2000. Influence of carbon source on shoot multiplication and adventitious bud regeneration in *in vitro* beech cultures. Plant Growth Regul 32: 1-12.

Cuenca B., Ballester A. and Viéitez A.M. 2000. *In vitro* adventitious bud regeneration from internode segments of beech. Plant Cell Tiss Org Cult 60: 213-220.

Driver J.A. and Kuniyuki A.H. 1984. *In vitro* propagation of paradox walnut rootstock. HortScience 19: 507-509.

Eapen S., Tivarekar S. and George L. 1998. Thidiazuron-induced shoot regeneration in pigeonpea (*Cajanus cajan* L.). Plant Cell Tiss Org Cult 53: 217-220.

Engelman F. 1997. *In vitro* conservation research activities at the International Plant Genetic Resources Institute (IPGRI). Plant Tissue Cult Biotech 3: 46-52.

Flinn B.S., Webb D.T. and Newcomb W. 1988. The role of cell clusters and promeristemoids in determination and competence for caulogenesis by *Pinus strobus* cotyledons *in vitro*. Can J Bot 66: 1556-1565.

Gautheret R.J. 1959. La Culture des Tissues Végétaux. Paris: Masson et Cie.

Gebbardt K. and Weisgerber H. 1990. Viability and *in vitro* growth of shoot tips from adult beech. Abstr VIIth Int Congr Plant Tissue Cell Cult, IAPTC, Amsterdam, pp 101.

Gresshoff P.M. and Doy C.H. 1972. Development and differentiation of haploid *Lycopersicon esculentum*. Planta 107: 161-170.

Huetteman C.A. and Preece J.E. 1993. Thidiazuron: a potent cytokinin for woody plant tissue culture. Plant Cell Tiss Org Cult 33: 105-119.

Jordan M., Velozo J. and Sabja A.M. 1996. Organogenesis *in vitro* of *Nothofagus alpina* (P. et E.) Oerst., Fagaceae. Plant Cell Rep 15: 795-798.

Jörgensen J. 1988. Embryogenesis in *Quercus petraea* and *Fagus sylvatica*. J Plant Physiol 132: 638-640.

Jörgensen J. 1990. Conservation of valuable gene resources by cryopreservation in some forest tree species. J Plant Physiol 136: 373-376.

Jörgensen J. 1991. Androgenesis in *Quercus petraea, Fagus sylvatica* and *Aesculus hippocastanum*. In: Ahuja M.R. (Ed.), Woody Plant Biotechnology, pp 353-354. Plenum Press, New York.

Kao K.N. and Michayluk M.R. 1975. Nutritional requirements for growth of *Vicia hajastana* cells and protoplasts at very low population density in liquid media. Planta 126: 105-110.

Lainé E. and David A. 1994. Regeneration of plants from leaf explants of micropropagated clonal *Eucalyptus grandis*. Plant Cell Rep 13: 473-476.

Lang H. and Kohlenbach H.W. 1988. Callus formation from mesophyll protoplasts of *Fagus sylvatica* L. Plant Cell Rep 7: 485-488.

Lloyd G. and McCown B.H. 1980. Commercially-feasible micropropagation of mountain laurel, *Kalmia latifolia*, by use of shoot tip culture. Comb Proc Int Plant Propagators' Society 30: 421-427.

Meier K. and Reuther G. 1994. Factors controlling micropropagation of mature *Fagus sylvatica*. Plant Cell Tiss Org Cult 39: 231-238.

Munthali M.T., Newbury H.J. and Ford-Lloyd B.V. 1996. The detection of somaclonal variants of beet using RAPD. Plant Cell Rep 15: 474-478.

Murashige T. and Skoog F. 1962. A revised medium for rapid growth and bioassays with tobacco tissue cultures. Physiol Plant 15: 473-497.

Murch S.J., Krishnaraj S. and Saxena P.K. 1997. TDZ-induced morphogenesis of Regal Geranium (*Pelargonium domesticum*): a potential stress response. Physiol Plant 101: 183-191.

Murthy B.N.S., Murch S.J. and Saxena P.K. 1998. Thidiazuron: a potent regulator of *in vitro* plant morphogenesis. In Vitro Cell Dev Biol Plant 34: 267-275.

Nadel B.L., Altman A., Pleban S. and Hüttermann A. 1991a. *In vitro* development of mature *Fagus sylvatica* L. buds. I. The effect of medium and plant growth regulators on bud growth and protein profiles. J Plant Physiol 138: 596-601.

Nadel B.L., Altman A., Pleban S., Kocks R. and Hüttermann A. 1991b. *In vitro* development of mature *Fagus sylvatica* L. buds. II. Seasonal changes in the response to plant growth regulators. J Plant Physiol 138: 136-141.

Pawlicki N. and Welander M. 1994. Adventitious shoot regeneration from leaf segments of *in vitro* cultured shoots of the apple rootstock Jork 9. J Hortic Sci 69: 687-696.

Pence V.C. 1990. Cryostorage of embryo axes of several large-seeded temperate tree species. Cryobiology 27: 212-218.

Romano A., Noronha C. and Martins-Loução M.A. 1995. Role of carbohydrates in micropropagation of cork oak. Plant Cell Tiss Org Cult 40: 159-167.

Schenk R.U. and Hildebrandt A.C. 1972. Medium and techniques for induction and growth of monocotyledonous and dicotyledonous plant cell cultures. Can J Bot 50: 199-204.

Tang W. 2001. *In vitro* regeneration of loblolly pine and random amplified polymorphic DNA analyses of regenerated plantlets. Plant Cell Rep 20: 163-168.

Thiébaut B., Cuguen J. and Dupré S. 1985. Architecture des jeunes hêtres (*Fagus sylvatica* L.). Can J Bot 63: 2100-2110.

Thiébaut B., Comps B. and Teissier du Cros E. 1990. Dévelopement des axes des arbres: pousse annuelle, syllepsie et prolepsie chez le Hêtre (*Fagus sylvatica*). Can J Bot 68: 202-211.

Thorpe T.A. and Kumar P.P. 1993. Cellular control of morphogenesis. In: Ahuja M.R. (Ed.), Micropropagation of Woody Plants, pp. 11-29. Kluwer Academic Publishers, Dordrecht.

Torregrosa L. and Bouquet A. 1996. Adventitious bud formation and shoot development from *in vitro* leaves of *Vitis* x *Muscadinia* hybrids. Plant Cell Tiss Org Cult 45: 245-252.

Viéitez A.M. and San-José M.C. 1996. Adventitious shoot regeneration from *Fagus sylvatica* leaf explants *in vitro*. In Vitro Cell Dev Biol Plant 32: 140-147.

Viéitez A.M., San-José M.C. and Viéitez E. 1985. *In vitro* plantlet regeneration from juvenile and mature *Quercus robur* L. J Hortic Sci 60: 99-106.

Viéitez A.M., Ferro E.M. and Ballester A. 1993. Micropropagation of *Fagus sylvatica* L. In Vitro Cell Dev Biol 29P: 183-188.

Viéitez A.M., Ballester A., Viéitez M.L. and Viéitez E. 1983. *In vitro* plantlet regeneration of mature chestnut. J Hortic Sci 58: 457-463.

Viéitez F.J., Ballester A. and Viéitez A.M. 1992. Somatic embryogenesis and plantlet regeneration from cell suspension cultures of *Fagus sylvatica* L. Plant Cell Rep 11: 609-613.

Zimmermann M.H. and Ziegler H. 1975. List of sugars and sugar alcohols in sieve-tube exudates. In: Zimmermann M.H. and Milburn J.A. (Eds.), Encyclopedia of Plant Physiology. New Series Vol 1, Transport in Plants I, pp. 480-503. Springer-Verlag, Berlin, Heidelberg, New York.

FIGURE LEGEND

Figure 1. Development of somatic embryos in a culture derived from a cotyledon of a *Fagus sylvatica* immature zygotic embryo. F.J. Vieitez *et al.* (1992).

Figure 2. Rooting ability of beech shoots relative to incubation time. Shoots were cultured in rooting medium with or without an initial 7-day period of darkness. Rooting was induced either by previous dipping in 1 g l^{-1} IBA solution for 1 min or by the first 10 days being spent in medium supplemented with 3 mg l^{-1} IBA or 3 mg l^{-1} IBA + 0.1 mg l^{-1} NAA. Vieitez *et al.* (1993).

Figure 3. A-B, Micropropagation of *Fagus sylvatica* by axillary shoot development. A, Shoot proliferation after culture for six weeks in multiplication medium. B, *In vitro* regenerated plants established in soil two years after acclimatization. C-H, Plantlet regeneration through adventitious shoot induction in beeches. C, Shoot-bud regeneration on petiole stub callus of a *F. sylvatica* proximal leaf explant cultured on medium with 4.5 μM TDZ and 2.9 μM IAA (clone 5B). D, Buds formed directly on the midrib of a distal explant cultured on medium with 4.5 μM TDZ and 2.9 μM IAA. E, Adventitious shoot buds from an internodal explant of *F. orientalis*, after culture on 2.3 μM TDZ and 2.9 μM IAA induction medium. F, Shoot development on multiplication medium following two-month subculture of regenerating *F. orientalis* internodal explants. G, Plantlets produced from adventitious shoots of *F. orientalis* one month after root induction treatment. H, Adventitious shoot-bud regeneration from *F. orientalis* clone BS leaf explant cultured with 3% glucose-containing medium.

Figure 4. Effects of five concentrations of TDZ and BA on bud induction frequency (A) and number of buds per responding internodal explant (B) of *Fagus orientalis* clone FO-3. Means with the same letter are not significantly different at the P = 0.05 level (LSD test). Bars represent standard errors of the means. Cuenca *et al.* (2000).

Figure 5. Development of adventitious buds in petiole tissues of *Fagus orientalis* leaf explants. A, Early activation of cells and periclinal divisions in the subepidermal layers after 6 days of culture.(X200). B-C, Explants after 13 days in culture, showing radial cell file proliferation derived from subepidermal layers. Dedifferentiation into meristematic cells is evident in the surface layers of the proliferated petiole; the arrangement of cells suggests that meristematic cells are a result of anticlinal cell divisions (arrows) (X300). D, Few-celled meristematic group, or meristemoid (m), exhibiting different planes of division apparent after 16 days of culture (X250). E, Transverse section of petiole tissues after 19 days, showing four different-sized meristemoids (arrows) located between radial cell files of the diffuse cambium (X200). F, Explant after 24 days in culture showing bud primordia emerging from the proliferated petiole surface (X250). G, Organized shoot apex and leaf primordia after 28 days of culture (X100). H, Bud primordia originated from lenticel phellogen cells (X200). I, Petiole callus showing multiple bud primordia and shoot-buds as induced after 35 days of culture (X20).

Figure 6. Results of agarose gel electrophoresis of PCR amplification products obtained using A) primer OPS-7 and B) primer OPS-11. DNA was extracted from axillary shoot culture lines of *Fagus orientalis* (3, 11) and *F. sylvatica* (FS-104 and F5B) and from adventitious shoot culture lines (3.1, 3.2, 3.3, 3.5, 3.7, 3.8, 11.1, FS-104.1 and F5B.1). M: DNA sizing ladder.

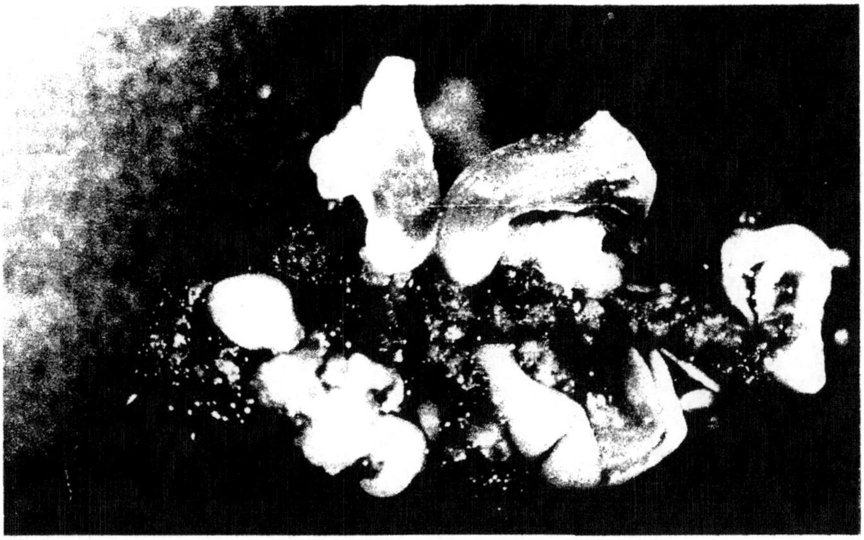

Figure 1.

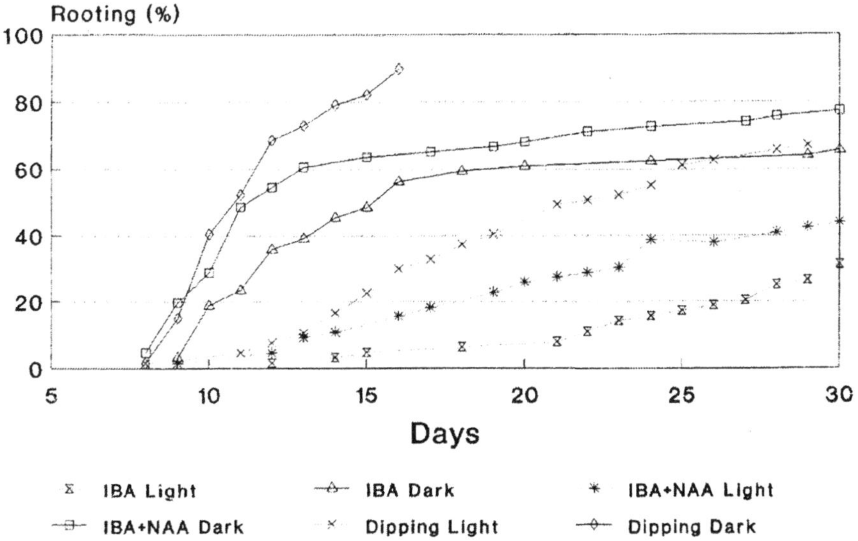

Figure 2.

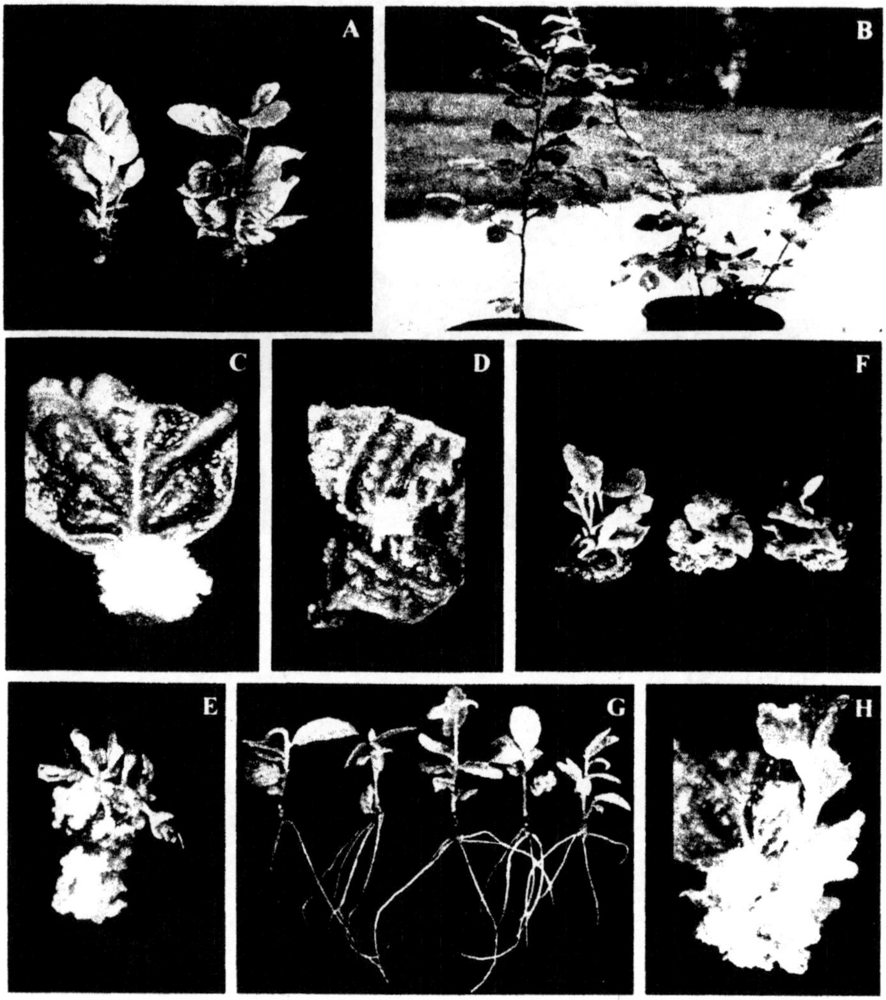

Figure 3.

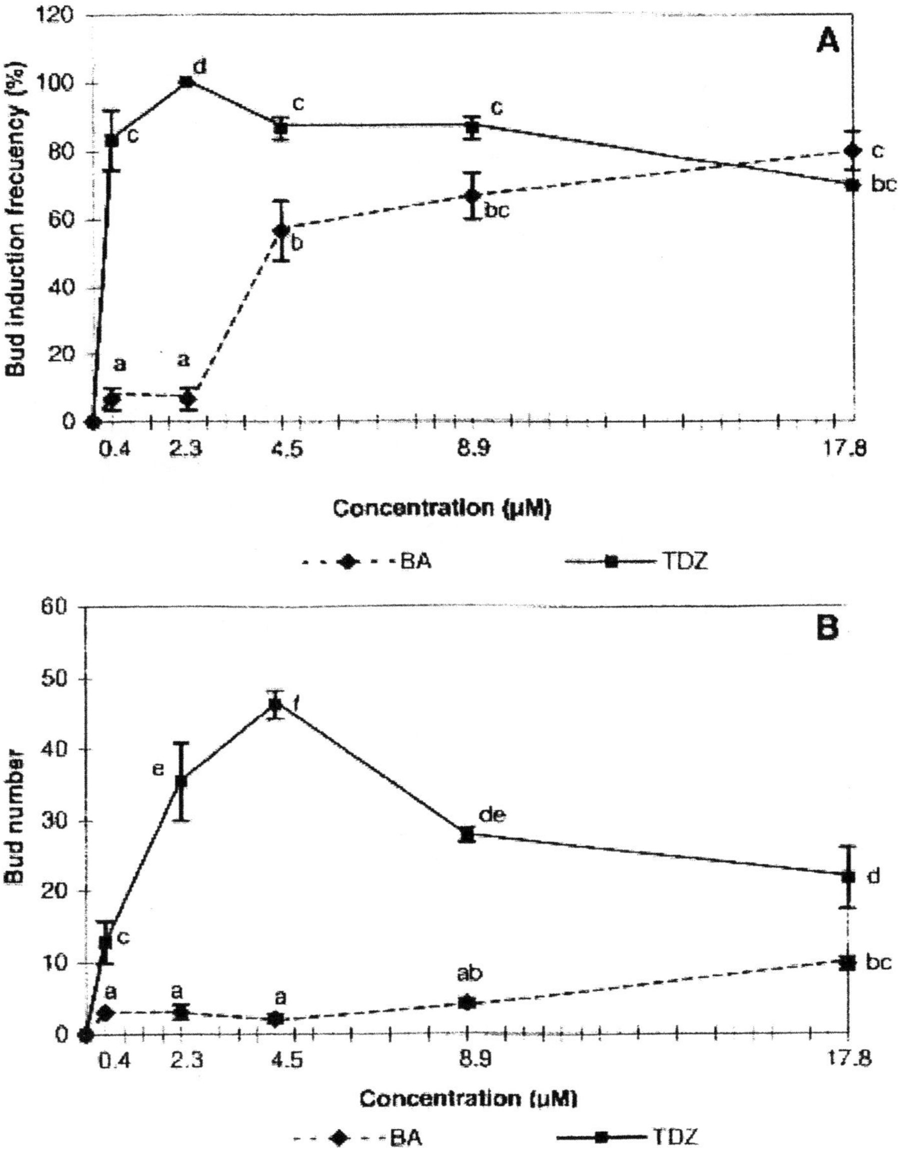
A
Bud induction frecuency (%)
120
100
80
60
40
20
0
0.4
2.3
4.5
8.9
17.8
Concentration (µM)
BA
TDZ
d
c
c
c
c
c
bc
bc
b
a
a
B
Bud number
60
50
40
30
20
10
0
0.4
2.3
4.5
8.9
17.8
Concentration (µM)
BA
TDZ
f
e
de
d
c
bc
ab
a
a
a

Figure 4.

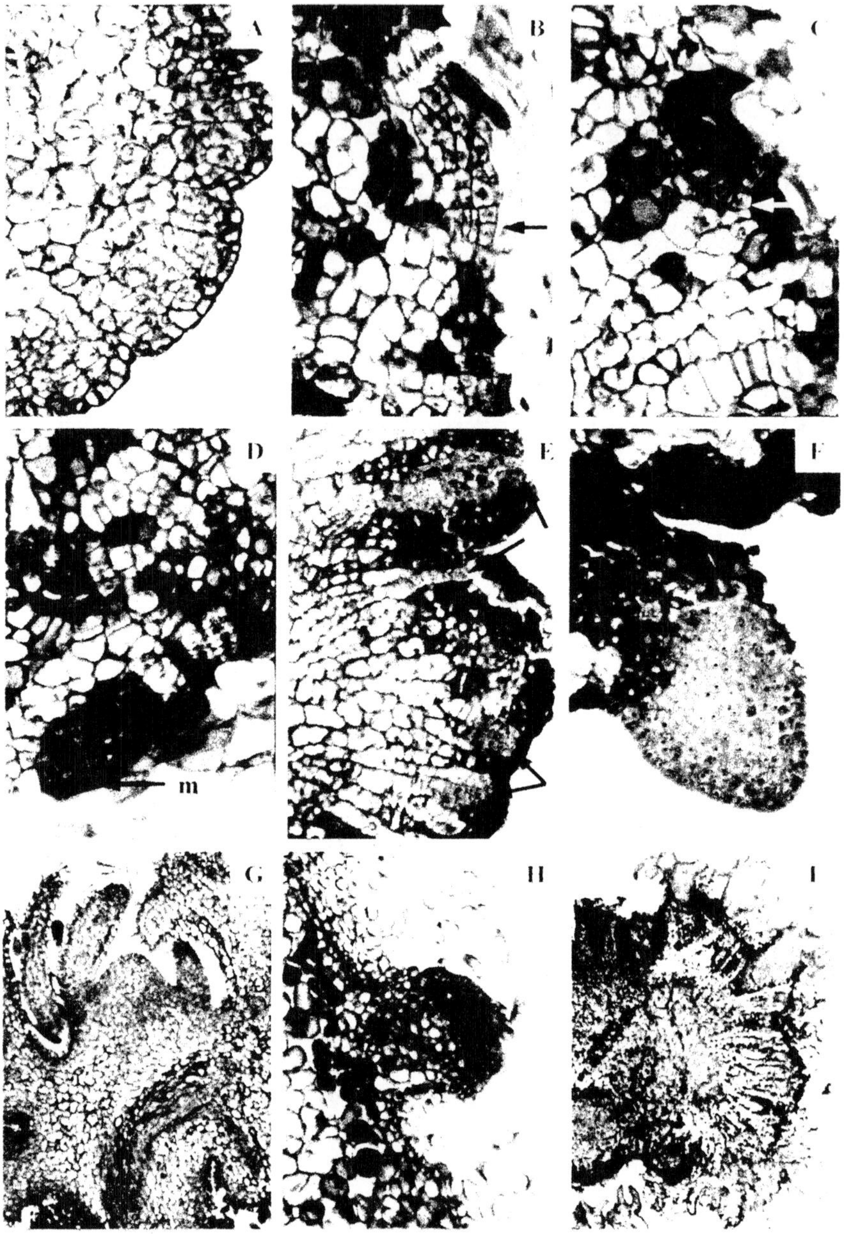

Figure 5.

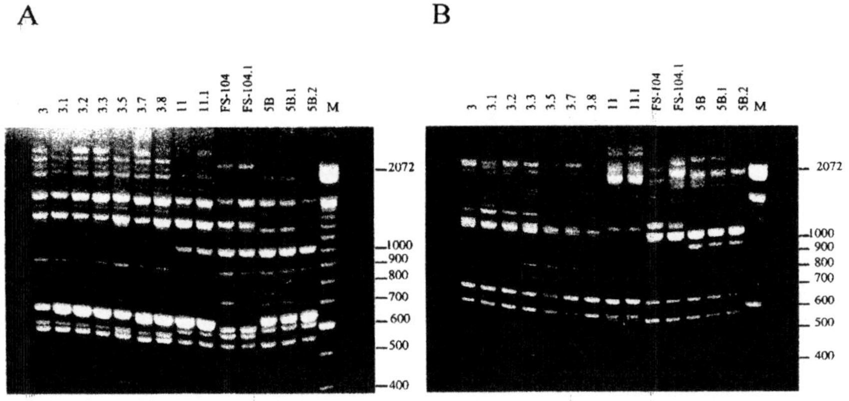
A
B
3
3.1
3.2
3.3
3.5
3.7
3.8
11
11.1
FS-104
FS-104.1
5B
5B.1
5B.2
M
2072
1000
900
800
700
600
500
400

Figure 6.

8. MICROPROPAGATION OF *EUCALYPTUS*

M.P. Watt[1], F.C. Blakeway[2], M.E.O. Mokotedi[1] and S. M. Jain[3]
[1]*School of Life and Environmental Sciences, University of Natal, Durban 4041, South Africa;* [2]*Mondi Forests, PO Box 12, Hilton, South Africa;* [3] *International Atomic Energy Agency, FAO/IAEA Joint Division, Box - 100, A - 1400, Wagramerstrasse 5, Vienna, Austria*

Chapter contents

1. THE GENUS *EUCALYPTUS*: UTILIZATION AND ECONOMIC IMPORTANCE

The genus *Eucalyptus* L'Hér. (Myrtaceae) is indigenous to Australia. A few species are found in the Philippines, West Timor and New Guinea. Many *Eucalyptus* spp. and hybrids are extensively used as exotic plantation species in many temperate and subtropical areas of the world including Argentina, Australia, Brazil, Morocco, Portugal, South Africa, Spain, USA and Uruguay. This genus contains a wide range of species. No other tree genus has been propagated so widely throughout the world in terms of adaptation to sites, types of management systems and variety of uses (Eldridge *et al.*, 1994). The species of choice in each country varies according to the suitability to the particular geographical and climatic conditions, as well as to the final end-product requirements. These include ornamentals, wind breaks, fuel wood for

S.M. Jain and K. Ishii (eds.), Micropropagation of Woody Trees and Fruits, 217–244.

locomotives, saw timber, mine props, poles, firewood, charcoal, honey, essential oils, tannin, pulp and paper (Turnbull, 1999). The versatility of the genus, together with its fast growth rate, have maintained a sustained interest (MacRae and van Staden, 2000) as the worldwide demand for pulp, fuel and construction wood increases.

Turnbull (1999) suggested that 12-14 million hectares are globally planted with Eucalypts (Table 1). The top-ranking species in terms of mean annual increment of wood production are *E. grandis*, *E. camaldulensis*, *E. tereticornis*, and *E. globulus*; *E. urophylla*, *E. viminalis*, *E. saligna*, *E. deglupta*, *E. exserta*, *E. citriodora*, *E. paniculata* and *E. robusta* (Eldridge *et al.*, 1994).

Table 1: **Estimated *Eucalyptus plantation* areas**

country	ha
Argentina	240 000
Australia	300 000
Brazil	3 000 000
Chile	320 000
China	600 000
India	> 1 000 000
Indonesia	80 000
Morocco	200 000
New Zealand	45 000
Papua Guinea	25 000
Portugal	550 000
South Africa	600 000
Spain	500 000
Thailand	80 000
Uruguay	300 000
Vietnam	250 000

2. VEGETATIVE PROPAGATION PRACTICES

Eucalyptus spp. can be propagated from seed with different degree of successes; and there is always a choice of seed source. The selection and maintenance of seed production areas are recognized as the first stage of eucalypt tree improvement, after selection of the best species and provenances (Eldridge et al, 1994). Seed orchards enable the production of large quantities of genetically improved seed.

The breeding of *Eucalyptus* requires maximizing genetic gain whilst minimising genetic erosion (Burley, 1989). The ideal approach to capturing desired gene combinations, and thus maximising genetic gain, is through selection followed by vegetative propagation. In addition, the vegetative route allows advanced material to be commercially produced more rapidly than through seed orchards (Eldridge et al, 1994). Breeding strategies utilising vegetative propagation to produce improved planting stock (clonal forestry) are promoted as an alternative to conventional seed-orchard procedures (Mullin and Park, 1992). Accelerated orchard systems are dependent on reliable vegetative propagation systems (Barbour and Butcher, 1995).

In 1950's, a French forester, Bouvier, working in Morocco, explored for the first time vegetative propagation of juvenile eucalypts. By the mid-1950s thousands of cuttings from seedlings of *E. camaldulensis*, *E. gomphocephala* and hybrids of *E. robusta* were produced (Eldridge et al, 1994). By 1954, trials were undertaken by using the cuttings of mature eucalypts, and since then research in this area has continued with great intensity. The utilisation of cuttings for producing elite genotypes has become an integral part of plantation species production (Zobel, 1993). Successful clonal propagation programmes employing cuttings for the commercial propagation of homogenous stands of select material have been developed at Celulosa Beira Industrial (CELBI) in Portugal (Cotterill and Brindbergs, 1997), Brazil (Bertolucci and Penchel, 1993; Zobel, 1993; Ikemori *et al.* 1994), South Africa (Denison and Quaile, 1987; Denison and Kietzka, 1993), Southern France (Eldridge *et al*, 1994), and Pointe Noire in the Congo (Eldridge *et al*, 1994).

Plantations, established from vegetatively propagated planting stocks, are characterized by uniformity, adaptation and homogeneity of wood properties (Eldridge *et al*, 1994). Uniformity of diameter, height, form and wood properties have extremely economic importance. Vegetative propagation provides an opportunity to match selected clones very closely to particular sites (Eldridge *et al*, 1994) and reduces the time and potentially the cost of deployment of improved material.

While conventional vegetative propagation through rooted macrocuttings has enabled the development of highly successful clonal programmes, it has been necessary to look at alternative vegetative propagation methods, for example smaller rooted cuttings (microcuttings), hydroponic and *in vitro* culture systems. The vegetative propagation of *Eucalyptus* species using cuttings is limited by the variability of rooting ability among clones and the gradual decrease of rooting potential with aging of parent plants. Frequently, only the clones that propagate readily are acceptable for large-scale production (Wilson, 1998), and quite often they represent a small number of clones within a species. However, selected genotypes that prove difficult to root through cuttings may be propagated and rooted with greater success using *in vitro* methods (Yasodha *et al*., 1997). Furthermore, *in vitro* plants appear to have root systems that resemble more closely those of a seedling than do plants derived from macrocuttings. Consequently, when used together with cutting production methods, *in vitro* propagation contributes significantly to tree improvement and clonal production programmes (Watt *et al*, 1995; Wilson, 1995) (see chapter by Rodriguez and Vendrame in this book). In the later part of this chapter, we will discuss more the indispensable nature of tissue culture propagation systems for the production of transgenic trees in the future.

3. MICROPROPAGATION PROTOCOLS

3.1. Historical perspective (1968-1991)

Some of the earliest reports on *in vitro* cultures of *Eucalyptus* spp. dates back to the 1960s (Sussex, 1965; Aneja and Atal, 1969). They were followed by the prolific work of De Fossard's (De Fossard et al., 1974; 1977) and Durand-Cresswell's (Durand-Cresswell *et al.*, 1977; 1982) research groups. Since then, reports describing *in vitro* culture of a number of species regularly appeared in the literature. In a comprehensive review on the micropropagation of *Eucalyptus*, Le Roux and van Staden (1991a) cited 204 reports (including 51 species and five hybrids), 30 of them had complete regeneration protocols. Among them, the most extensively studied species were *E. grandis* (23 reports), *E. gunnii* and *E. citriodora* (each 15 reports) and *E. camaldulensis* (13 reports). For 28 species, the direct organogenesis route via axillary bud proliferation was followed, whereas only eight species were regenerated indirectly from callus. Further, by 1991, plantlet regeneration via somatic embryogenesis had only been reported for *E. citriodora* (Muralidharan and Mascarenhas, 1987; Muralidharan *et al.*, 1989) and *E. grandis* (Watt *et al.*, 1991).

3.2. The past decade (1991 – 2001)

Over the last decade some progress has been made in developing complete plant regeneration protocols for the micropropagation of *Eucalyptus* species, *i.e.* those that yield plantable units. Whereas, during 1968 and 1991, only 30 out of 204 publications (see review Le Roux and van Staden 1991a) included protocols for plantlet production, twenty-nine out of 65 reviewed reports published since then present complete methodology (Table 2). They encompass 20 species and six hybrids, including the first studies on *E. microcorys*, *E. occidentalis*, *E. sargentii*, *E. smithii* (one report each) and *E. radiata* (two reports). Among the 15 species, such as *E. grandis* and *E. globulus*, for which regeneration work has been ongoing for the last 20 years, only eight have complete regeneration protocols (Table 2).

Table 2. Summary of yield of *Eucalyptus* species micropropagated during the past decade. The figures shown were derived from the information provided in literature. Symbols used: DSE = direct somatic embryogenesis, ISE = indirect somatic embryogenesis, DO = direct organogenesis, IDO = indirect organogenesis, NR = not reported. II = insufficient information. Unless otherwise indicated, the geographical source of all species listed is Australia and solid support media were used (*liquid and solid support media, **geographical source = Brazil; ***liquid support medium).

Species	Explant type	Route	Protocol final product	Culture period (weeks)	Hardening-off details	Plants/explant		Reference
						before hardening-off	after hardening-off	
*E. botryoides**	Shoot tips from seedlings & field-grown plants	IDO	Plantlets	56 – 69	Yes	NR	NR	Ito *et al.*, 1996
E. camaldulensis	Shoot tips from seedlings & field-grown plants	IDO	Plantlets	56 – 69	Yes	NR	NR	Ito *et al.*, 1996
	Nodal segments with single axillary buds	DO	Plantlets	21 – 30	Yes	[1]3 -4	2.25 – 3.2	Yasodha *et al.*, 1997
	In vitro leaves	IDO	Shoots	10 – 11	No	NR	NR	Mullins *et al.*, 1997
E. deglupta	Shoot tips from seedlings & field-grown plants	IDO	Plantlets	56 – 69	Yes	NR	NR	Ito *et al.*, 1996
*E. dunnii***	Hypocotyls	DSE	Somatic embryos	5.4 – 6.3	No	NR	NR	Termignoni *et al.*, 1996

Table 2. Continued

Species	Explant type	Route	Protocol final product	Culture period (weeks)	Hardening-off details	Plants/explant		Reference
						before hardening-off	after hardening-off	
E. globulus	Axillary buds from 4 – 5 year old trees	DO	Plantlets	58 – 63	Yes	0.85 – 3.2	0.83 – 3	Bennett *et al.*, 1992
provenance 1	Micropropagated shoots of seedlings	DO	Plantlets	10.2	No	0.5 – 0.85	NR	Willyams *et al.*, 1992
provenance 2	Micropropagated shoots of seedlings	DO	Plantlets	10.2	No	0.7 – 1.1	NR	
provenance 3	Micropropagated shoots of seedlings	DO	Plantlets	10.2	No	1 – 1.6	NR	
	Axillary buds from 4-5 year old trees	DO	Plantlets	66	Yes	2.4 – 9.3	2.4 – 8.8	Bennett *et al.*, 1994
	Cotyledons & hypocotyls	IDO	Plantlets	5 – 6	Yes	7 & 14	II	Azmi *et al.*, 1997a
	Leaves & internodes of micropropagated plantlets	IDO	Plantlets	54 – 55	No	NR	NR	Azmi *et al.*, 1997b
	Epicormic shoots from 10 – 30 year old trees	DO	Plantlets	58 – 60	Yes	[1]64 – 76	58 – 72	Trindade and Pais, 1997

Table 2. Continued

Species	Explant type	Route	Protocol final product	Culture period (weeks)	Hardening-off details	Plants/explant		Reference
						before hardening-off	after hardening-off	
	Cotyledons	IDO	Plantlets	20 – 27	Yes	2.2	NR	Bandyopadhyay *et al.*, 1999
	Hypocotyls	IDO	Plantlets	20 - 27	Yes	1.3	NR	Bandyopadhyay *et al.*, 1999
E. grandis	Micropropagated seedling shoots	DO	Plantlets	23	Yes	1.2 – 2	1 – 1.6	Lubrano, 1991
	Seedling leaves	ISE	Plantlets	19 – 20	Yes	4 –6	3.6 – 5.4	Watt *et al.*, 1991
	In vitro leaves from coppice shoots	IDO	Plantlets	27 - 29	No	II	NR	Lainé and David, 1994
	Leaves from micropropagated plantlets	IDO	Roots and nodules	15	No	NR	NR	Quoirin and Vieira, 1995
	Shoot tips from seedlings & field-grown plants	IDO	Plantlets	56 – 69	Yes	NR	NR	Ito *et al.*, 1996
	Seedling shoots	DO	Plantlets	21 – 23	Yes	2	2	Watt *et al.*, 1996

Table 2. Continued

Species	Explant type	Route	Protocol final product	Culture period (weeks)	Hardening-off details	Plants/explant		Reference
						before hardening-off	after hardening-off	
	Stump nodal explants from 25 year-old trees	DO	Plantlets	26 - 27	Yes	14	NR	Wachira, 1997
E. grandis[*]	Mature zygotic embryos & hypocotyls	DSE	Bipolar embryo-like 'structures'	18	No	NR	NR	Major *et al.*, 1997
E. macarthurii	Axillary buds from seedlings & clonal hedges	DO	Plantlets	~13 (II)	Yes	0 – 2	0 – 1.6	Le Roux and van Staden, 1991b
E. marginata	Axillary buds from 4 – 5 year old trees	DO	Plantlets	58 – 63	No	[1]8.8 - 36	NR	Bennett *et al.*, 1992
E. microcorys	Shoot apices from newly germinated and 9 month old seedlings	DO	Plantlets	~12 (II)	No	4	NR	Niccol *et al.*, 1994
E. nitens	Shoots from basal coppice and seedlings	DO	Plantlets	NR	No	[1]0.4 - 9	NR	Rasmussen, 1991

Table 2. Continued

Species	Explant type	Route	Protocol final product	Culture period (weeks)	Hardening-off details	Plants/explant		Reference
						before hardening-off	after hardening-off	
provenance 1	Micropropagated shoots of seedlings	DO	Plantlets	10.2	No	0.5 – 0.8	NR	Willyams *et al.*, 1992
provenance 2	Micropropagated shoots of seedlings	DO	Plantlets	10.2	No	0.3 – 0.5	NR	
provenance 3	Micropropagated shoots of seedlings	DO	Plantlets	10.2	No	1 – 1.6	NR	
*E. nitens****	Hypocotyls, leaves cotyledons,& epicotyls	ISE	Somatic embryos	12.3	No	NR	NR	Ruaud *et al.*, 1997
	Hypocotyls	IDO	Plantlets	20 – 27	Yes	0.7	NR	Bandyopadhy-ay *et al.*, 1999
	Cotyledons	IDO	Plantlets	20 - 27	Yes	2	NR	Bandyopadhy-ay *et al.*, 1999
E. occidentalis	Axillary buds from 4 – 5 year old trees	DO	Plantlets	58 – 63	No	[1]40	NR	Bennett *et al.*, 1992
E. radiata	Epicormic shoots from mature trees	DO	Plantlets	24	No	NR	NR	Donald and Newton, 1991

Table 2. Continued

Species	Explant type	Route	Protocol final product	Culture period (weeks)	Hardening-off details	Plants/explant before hardening-off	Plants/explant after hardening-off	Reference
E. radiata ssp *radiata*[*] (*E. australiana*)	Axillary buds from coppice	DO	Plantlets	20 – 21	No	0.45	NR	Chang *et al.*, 1992
	Axillary buds from mature branches	DO	Plantlets	20	No	0.46	NR	Chang *et al.*, 1992
E. regnans	Seedling microcuttings	DO	Plantlets	36 – 48	Yes	2.3 - 4	2 – 3.5	Blomstedt *et al.*, 1991
E. saligna	Axillary buds from seedlings & clonal hedges (both 1 year old)	DO	Plantlets	~13 (II)	Yes	0.7 – 4	0.6 – 2.7	Le Roux and van Staden, 1991b
E. sargentii	Axillary buds from 4 – 5 year old trees	DO	Plantlets	58 – 63	No	[1]30	NR	Bennett *et al.*, 1992
E. smithii	Axillary buds from seedlings & clonal hedges (both 1 year old)	DO	Plantlets	~13 (II)	Yes	0.05 – 0.75	0.04 – 0.6	Le Roux and van Staden, 1991b

Table 2. Continued

Species	Explant type	Route	Protocol final product	Culture period (weeks)	Hardening-off details	Plants/explant before hardening-off	Plants/explant after hardening-off	Reference
*E. tereticornis**	Nodal segments from coppices shoots of mature trees	DO	Plantlets	20 – 25	Yes	2.6 – 3.4 (liquid medium)	2.1 – 2.8	Gill and Gosal, 1996
	Nodal segments from coppices shoots of mature trees	DO	Plantlets	20 - 25	Yes	NR (solid medium)	II	Gill and Gosal, 1996
	Nodal segments with single axillary buds	DO	Plantlets	21 – 30	Yes	[1]3 - 4	2.2 – 3.2	Yasodha *et al.*, 1997
	Epicormic shoots from 15 – 20 year old trees	DO	Plantlets	11 – 14	Yes	0.87 – 1.45	0.4 – 0.6	Patil and Kuruvinashetti, 1998
	Coppice shoots from 15 – 20 years old trees	DO	Plantlets	11 – 14	Yes	2 – 3.3	1.3 – 2.0	Patil and Kuruvinashetti, 1998
	Axillary buds from 4-year-old trees	DO	Plantlets	58 - 60	Yes	4 - 5	3.2 - 4	Sharma and Ramamurthy, 2000
E. urophylla	Hypocotyls	IDO	Plantlets	22	Yes	NR	II	Tibok *et al.*, 1995

Table 2. Continued

Species	Explant type	Route	Protocol final product	Culture period (weeks)	Hardening-off details	Plants/explant		Reference
						before hardening-off	after hardening-off	
E. grandis hybrids	Hypocotyls	IDO	Plantlets	~ 9 (II)	Yes	2 - 19	2 - 18	Warrag *et al.*, 1991
E. grandis x *macarthurii*	Axillary buds from seedlings and clonal hedges (both 1 year old)	DO	Plantlets	~13 (II)	Yes	0.3 – 2.4	0.2 – 2	Le Roux and van Staden, 1991b
E. grandis x *nitens*								
clone 1	Axillary buds from 1- year- old hedges	DO	Plantlets	20 – 21	Yes	7	4	Mokotedi *et al.*, 2000
clone 2	Axillary buds from 1- year- old hedges	DO	Plantlets	20 - 21	Yes	8	6	
E. grandis x *urophylla*								
clone 1	Axillary buds from hedges	DO	Plantlets	23 - 27	Yes	5 – 8	3.5 – 5.6	Jones and van Staden, 1994
clone 2	Axillary buds from hedges	DO	Plantlets	23 – 27	Yes	3.2 – 5.4	1.3 – 2.3	
clone 3	Axillary buds from hedges	DO	Plantlets	23 – 27	Yes	3 – 5	0.7 – 1.1	

Table 2. Continued

Species	Explant type	Route	Protocol final product	Culture period (weeks)	Hardening-off details	Plants/explant before hardening-off	Plants/explant after hardening-off	Reference
	Leaves from micropropagated plantlets	IDO	Roots and nodules	15	No	NR	NR	Quoirin and Vieira, 1995
	Apical and axillary buds from 3-year - old trees	DO	Plantlets	18.4	Yes	13	11	Yang *et al.*, 1995
	Hypocotyls, cotyledons & cotyledonary nodes	IDO	Plantlets	35 - 42	Yes	24 - 40	24 - 40	Cid *et al.*, 1999
	Primary leaves	IDO	Plantlets	41 - 50	Yes	24 - 40	24 - 40	Cid *et al.*, 1999
E. nitens x *globulus*	Shoots from basal coppice and seedlings	DO	Plantlets	NR	No	[1]0.2 – 9.6	NR	Rasmussen, 1991
E. tereticornis x *camaldulensis*	Seedling nodal segments	DO	Plantlets	23	Yes	28 - 218	18 - 196	Bisht *et al.*, 1999

[1]Number of shoots produced per initial explant was not reported, hence, the number of plantlets shown was calculated from the number of shoots transferred to rooting medium.

Out of the 65 studies reviewed since 1991 (Table 2), 40 were conducted on seedling material, 13 on axillary buds, six on explants from mature trees, and six on *in vitro* material. Of all these studies, 32 were on direct organogenesis, 25 on indirect organogenesis, and only eight focused on somatic embryogenesis. At least 12 of the 19 studies conducted with axillary buds and explants from mature trees (epicormic and coppice shoots) resulted in complete plant regeneration protocols, compared to 14 out of 32 studies (both direct and indirect organogenesis) on seedling material.

During the past decade, out of 25 reports on indirect organogenesis, the details of complete plantlet regeneration protocols have been reported only for *E. globulus* (Azmi *et al.*, 1997a), *E. urophylla* (Tibok *et al.*, 1995), *E. grandis* hybrids (Warrag *et al.*, 1991) and *E. grandis* x *urophylla* (Cid *et al.*, 1999). Plant regeneration via somatic embryogenesis has been achieved only with *E. grandis* (Watt *et al.*, 1991), *E. citriodora* (Muralidharan and Mascarenhas, 1995), *E. dunnii* and *E. saligna* (Termignoni *et al.*, 1998). However, in those studies, plantlet production rate from seedling explants (Watt *et al.*, 1995; Muralidharan and Mascarenhas, 1995) was low (up to one and five plants per explant, respectively). Termignoni *et al.* (1998) improved yields of plantlet regeneration from aerial parts of seedlings and micropropagules of *E. dunnii* and *E. saligna*, but complete details of the somatic embryogenesis protocol were not furnished (Patent No. PI 9801485-4 INPI) (Table 2). In other *Eucalyptus* species, only the production of embryo-like structures and somatic embryos has been reported (Nugent *et al.*, 1997; Major *et al.*, 1997; Ruad *et al.*, 1997). The limiting step in the indirect organogenesis route appears to be the hardening-off stage (Azmi *et al.*, 1997a, b; Bandyopadhyay *et al.*, 1999), whereas in somatic embryogenesis, it is poor somatic embryo conversion rate (Watt *et al.*, 1999). Overall lack of progress in increasing the frequency rate of plant production, through both morphogenic routes, is of great concern due to the negative impact on genetic improvement programmes.

The most commonly used explants for the micropropagation of eucalypts are seedlings and axillary buds from field-grown plants (61% and 21% of reports, respectively) (Table 2). This may be attributed to difficulties experienced in propagating certain species such as *E. nitens* by other means, such as cuttings (Rasmussen, 1991; Willyams *et al.*, 1992; Maile and Nieuwenhuis, 1996; Tibbits *et al.*, 1997). A few high yielding protocols have been reported for both seedlings (Warrag *et al.*, 1991; Bisht *et al.*, 1999; Cid *et al.*, 1999) and axillary buds (Yang *et al.*, 1995; Bennett *et al.*, 1994; Mokotedi *et al.*, 2000), with the former resulting in the highest plantlet production However, to our knowledge, no study has been undertaken on a direct comparison of yields between the two explants types from the same genotype. As, in *Eucalyptus*, the usefulness of micropropagation of seedlings is limited to seedling orchards and initial selection stages, for large-scale commercial production of superior selected trees, axillary bud explants predominates micropropagation. In the literature, the number of plantlets produced via a micropropagation protocol has been quantified on the basis of the number of shoots cultured on rooting medium. The yield is expressed on the basis of the total number of plantlets obtained from all the explants after the (variable) number of subcultures. With notable exceptions (Le Roux and van Staden, 1991b; Bisht *et al.*, 1999; Mokotedi *et al.*, 2000) few publications indicated the success of a protocol on the basis of rooted or hardened-off plantlets per original explant. The latter was, however, the approach used in this review to assess performance of published protocols. Inevitably, such yield data (Table 2) are lower than those recorded in the literature. .

While compiling Table 2, sufficient information lacked in published reports on a variety of essential parameters, viz. number of regenerated shoots per initial explant (Yasodha *et al.*, 1997), exact number of shoots in rooting medium (Cid *et al.*, 1999), number of shoot subcultures (Ito *et al.*, 1996), length of rooting period (Bennett *et al.*, 1992) and number of plantlets that survived the hardening-off phase (Azmi *et al.*, 1997a). For these reasons, results on Table 2 were often estimated and presented as a range. For example, plantlet yield, when researchers used combined results from several clones to express shoot multiplication (Table 2) and/or shoots from different sources (explants, media) for the rooting experiments (Cid *et al.*, 1999). Hardening-off details were often confined to statements such as "regenerants were transferred to potting mix without significant loss" (Azmi *et al.*, 1997a). When made available, the original reports present hardening-off data as percentages of rooted plantlets only (Yang *et al.*, 1997) and not in relation to yield per explant. Applications for patents and confidentiality problems may be some of the factors preventing publication of details of the extensive regeneration protocols.

Plantlet yield is affected by interplay of several factors including the type of explants, plant genotype, duration of culture period, geographical location, appropriate hardening facilities and labour skills. They all contribute to the cost of micropropagated plants, which is considered very high when compared with macropropagated plants (Wilson, 1996, 1998). By assessing plantlet yield from rooting of cuttings and *in vitro* propagation via axillary bud proliferation of *E. globulus* seedlings, Wilson (1996) reported no significant differences over similar culture periods (30 plantable plants in 54.9 and 56.9 weeks, respectively). Furthermore, he concluded that high shoot multiplication rate via micropropagation had not increased yields due to high plantlet losses during hardening-off stage, probably because "*in vitro* plants are more fragile than rooted cuttings" (Wilson, 1998). Although most of Wilson's conclusions are valid, many of the comparisons made were between yields of axillary bud and cutting experiments undertaken at separate times and on different genotypes.

As discussed earlier, hardening-off of micropropagated plants is a crucial step as it is the determinant of final total number of plantlet production. Out of the 65 publications reviewed (Table 2), 36 either omitted or failed to report the hardening-off data. In those studies, difficulties may have not been encountered at that stage or they were addressed, but not reported

Among the 16 studies with seedling explants that present hardening-off data, there are at least three high-yielding, commercially viable protocols related to *E. grandis* hybrids (Warrag *et al.*, 1991), *E. tereticornis* x *camaldulensis* (Bisht *et al.*, 1999), and *E. grandis* x *urophylla* (Cid *et al.* (1999). They produced (depending on genotype) up to 18, 196 and 40 plantlets per explant, respectively (Table 2). When considering propagation via axillary buds from field-grown trees, protocols of *E. grandis* x *urophylla* (Yang *et al.*, 1995) and *E. globulus* (Bennett *et al.*, 1994) were developed and are commercially viable. They resulted in 12 and nine plantlets (per initial explant), respectively (Table 2). However, in most other studies with axillary buds, less than five plantable plantlets per explant were obtained. In *E. tereticornis*, Sharma and Ramamurthy (2000) and Yasodha *et al.* (1997) obtained up to four and three plantlets per explant, respectively; and Le Roux and van Staden (1991b) reported approximately three plantlets per explant of *E. saligna*. The question remains as to the yield and cost of propagation of such genotypes through conventional ways (*i.e.* rooting of cuttings) as compared to published *in vitro* protocols. *In vitro* methods may be deemed cost effective in cases where rooting of cuttings is variable and/or low for certain genotypes.

In addition to yield, the quality of micropropagated plants is highly critical while determining the survival frequency of plants in the field. Furthermore, plant quality also determines the appropriate commercial applications of the material obtained from plants at the end of the rotation period. However, very little is known about the effect of a propagation system on plant quality, and whether rooted cuttings and *in vitro*-produced plants are similar or different with respect to growth rates, physiological age, wood quality, etc. In terms of growth (tree height and diameter at breast height), micropropagated plants grow initially faster (depending on the genotype) than their macropropagated counterparts in the field. Watt *et al.* (1995) found significant growth differences for some genotypes of the *E. grandis* x *camaldulensis* hybrid during the first 36 months in the field. However, for some species and/or hybrids such differences may be insignificant at any time or after one year of field growth, as reported for *E. grandis* (Rockwood and Warrag, 1994) and *E. grandis* x *urophylla* (Yang *et al.*, 1995), respectively. However, in all studies, tissue culture-derived plants were highly uniform (Rockwood *et al.*, 1994; Yang *et al.*, 1995; Watt *et al.*, 1995).

In conclusion, replicated field-testing of *in vitro*-produced plants is essential to determine the realization of expected potential gains from micropropagation. However, such information is scarce, despite the fact that micropropagation of *Eucalyptus* species and hybrids via axillary bud proliferation are used routinely in numerous research and commercial laboratories, e.g. Mondi Forests, South Africa. The few available reports in the literature (Gupta *et al.*, 1991; Rockwood and Warrag, 1994; Watt *et al.*, 1995) are insufficient to formulate significant conclusions. Nevertheless, those studies indicate that growth rate and uniformity of micropropagated plants are at least comparable, if not higher than plants belonging to the same genotype produced through cuttings. Further, to our knowledge, no published reports are available on field-testing of material produced via different propagation techniques, under different climatic conditions.

Currently, we are investigating the physiological efficiency of roots of micropropagated versus macropropagated *E. grandis* x *nitens* clones at juvenile stages (four and nine months). In these studies, roots of micropropagated plants resembled those of seedlings, with one thick root originating directly from the base of the stem. In contrast, in macropropagated plants, more than two roots of similar size were produced from the side of the stem above the stem base. Despite such morphological differences between the two root systems, no significant differences have been detected in specific root hydraulic conductivity, *i.e.* both types of roots were equally efficient in transporting water to the shoot.

4. APPLICATIONS OF *IN VITRO* PROPAGATION

The current applications of the *in vitro* propagation of *Eucalyptus* are: i) mass production of selected genotypes, ii) conservation of germplasm, and iii) as a vehicle for molecular genetics research.

i) 4.1. Mass production of selected genotypes

The ability to mass propagate improved genetic material produced through breeding and clonal activities is highly desirable. Large forestry and forest products companies use tissue culture systems for the mass production of selected genotypes for successive breeding generations, bulking up of hybrid genotypes and for clonal plantation (Aitken-Christie, pers. comm., MacRae, pers. comm.). Although it is perceived that this may provide an opportunity for immediate value recovery from the products of tree improvement programmes, there is not a great deal of published information in this arena.

It is apparent that the system most utilised commercially is organogenesis, and that somatic embryogenesis is being investigated primarily at the research level (Aitken-Christie, pers. comm). While somatic embryogenesis systems are not yet commercialized, there is considerable interest in improving existing systems because somatic embryos are highly desired in genetic transformation and for cryopreservation.

In 1997, Nippon Paper Industries (through their affiliated company Volterra S.A.) established a research programme for the clonal micropropagation of superior clones of *E. globulus*. The companies involved were Mininco S.A., Angol S.A. Los Lagos S.A. and Reginales S.A. Following this research, the Nippon Tissue Culture Laboratory, in Adelaide, Australia propagated *E. globulus* using *in vitro* techniques. International Paper currently produces 40 000 tissue culture plants per year as feedstock for their clonal production operations (Brun, pers. comm.). In Mondi Forests (South Africa), axillary shoot multiplication is used for the multiplication of selected clones for clonal trials and production of clonal nursery feedstock (macro- and micro-cuttings). The annual production varies between 20 000 and 100 000 plants per annum, depending on requirements.

In 1997, Shell Forestry Ltd. (United Kingdom) patented the Bulk-Up ® liquid tissue culture system. This system has been tested across more than 150 *Eucalyptus* genotypes, and offers faster deployment of elite germplasm together with rapid access to and multiplication of germplasm. Plantlets can be rooted *ex vitro*. A pilot laboratory attached to the Forestal Yguaza S.A nursery in Paraguay is capable of producing 75 000 plants per year (Griffin, pers. comm). Mondi Forests are currently testing (for research purposes) the RITA temporary immersion system for *E. grandis* and *E. grandis* x *E. urophylla* clonal material.

4.2. ii) Germplasm conservation

In vitro methods are not commonly used to conserve eucalypt germplasm for environmental and biodiversity conservation. With the focus of this chapter being on eucalypts as plantation species, the term 'germplasm conservation' will be regarded as being a procedure used to benefit tree improvement programmes, and thus with perceived commercial value.

In *Eucalyptus* and other plantation trees, genotypes that are no longer directly utilised in the current generation of tree improvement are often discarded, as large arboreta are costly to maintain. In addition, land availability, legislation pertaining to land and water use, labour constraints and maintenance costs sometimes preclude the conservation of many genotypes. However, successful breeding programmes rely on the conservation of broad genetic variation to facilitate gains in future generations and selection for changing objectives, environments and environmental pressures. *In vitro* storage methods offer an alternative to the conservation of potentially useful germplasm on productive landholdings.

The common practices for the *in vitro* multiplication of selected *Eucalyptus* genotypes is to harvest branches from field plants, set cuttings and then use the established plants in the greenhouse or in hedges as stock material for micropropagation. This strategy is presently necessary as shoots collected in the field wilt and endogenous microbial contaminants proliferate during the usually long period between field collection and culture initiation in the laboratory. An alternative approach that would accelerate micropropagation and germplasm storage is *in vitro* collection, i.e. surface sterilisation of material in the field and transport under sterile conditions. Although this approach has proven successful for some species (e.g. coconut; Assy-Bah *et al.*, 1987), it has not been reported for *Eucalyptus*. Preliminary work in our laboratories on various *Eucalyptus* spp. and hybrid clones has indicated that such techniques can be established with relative ease (unpublished).

In vitro germplasm storage, involves the maintenance of material at low temperature for minimal growth, high viability and minimal risk to genetic stability. Further, successful systems also effect significant savings in labour input, materials and commitment of specialised growing facilities (Grout, 1995a, b). At present, two main storage methods may be considered: slow growth and cryopreservation.

a) 'Slow growth', 'growth limitation/inhibition' and 'minimal growth' describe direct ways of restricting *in vitro* growth and development by altering the culture environment, thereby increasing the time interval between sub-cultures. Germplasm is thus maintained as active collections, in controlled sterile environments, from where material can be rapidly recovered and subsequently propagated (Aitken-Christie and Singh, 1987). Shoots of *Eucalyptus* spp. and hybrids have been successfully stored in this way for six to ten months (Mascarenhas and Agrawal, 1991; Watt *et al.*, 2000 a,b).

b) Long-term conservation of biological material can be achieved through cryopreservation, i.e. in liquid nitrogen (-196 °C) where cellular activities are arrested (Engelmann, 1997). However, the optimum conditions for cryostorage have to be determined empirically, on a species basis, taking into consideration a series of variables, e.g. water content, cryoprotectant, tissue desiccation tolerance/sensitivity, etc. As a consequence, and despite the significant advances in the past decade, the technology has been mainly restricted to a few species. Nevertheless, the reports on *E. gunnii* (Poissonier *et al.*, 1992) and, recently, on *E. grandis* x *camaldulensis* axillary buds (Blakesley and Kierman, 2001) have demonstrated the feasibility of this approach to the species. Yet, until such time as high yielding somatic embryogenesis protocols are available, axillary buds are necessarily the most suitable storage material and the focus of studies in our laboratories.

In vitro storage is invaluable in allowing germplasm to be securely maintained in a pest- and disease-free environment. However, as *in vitro* cultures and micropropagated plants are not microbe and disease-free *ipso facto*, they need to be checked for presence of pathogens before distribution and/or use. In addition, materials in germplasm banks need to be well-identified and characterised in terms of genetic diversity and stability (Ashmore, 1997; McFerson, 1998). Procedures to index (Kartha, 1986; George, 1993; Barker and Torrance, 1997), characterise and monitor germplasm collections (Tanksley and McCouch, 1997; McFerson, 1998) are well established for various species and need only to be adapted for *Eucalyptus*.

Plant germplasm can be distributed as *in vitro* cultures and this is routinely carried out for a number of species (Withers, 1988; Ashmore, 1997). In case of trees such as *Eucalyptus*, germplasm is usually exchanged as rooted cuttings to distant nurseries. The main advantages of using *in vitro* plants/cultures are the reduction in volume and weight of the shipment t, and the protection of the shipments against the external environment while in transit.

4.3. iii) Vehicle for genetic modification

The plantation of genetically modified forest trees, such as *Eucalyptus*, has been the topic of public debate (Griffin, 1996; Mathews and Campbell, 2000). In this review, we will highlight the importance of *in vitro* plant regeneration for producing transgenic *Eucalypts*.

The prospect of genetic manipulation of *Eucalyptus* is of great interest to commercial forestry, especially for enhanced adaptation to environmental stresses (*e.g.* disease tolerance/resistance, drought and temperature stresses), more efficient management practices (*e.g.* insect and herbicide tolerance/resistance) and product improvement (*e.g.* lignin and cellulose content) (Bauer, 1997; Pullman *et al.*, 1998; Tzfira *et al.*, 1998; Harcourt et al, 2000; MacRae and van Staden, 2000; Mathews and Campbell, 2000). In clonal forestry, even though clones are selected and strictly matched to plantation sites, production is limited by a variety of environmental challenges, such as insects, pathogens, diseases and water use efficiency. Resistance/tolerance to herbicides such as glyphosate, chlorosulfuron, glufosinate and bromoxynil would permit aerial spraying of such agents, and that would lead to more efficient management of young plantations (Dickson and Walker, 1997; Tzfira *et al.*, 1998; MacRae and van Staden, 2000), and reduction in tillage control of weeds, that would curtail soil erosion (James *et al.*, 1998).

Improvements in lignin composition and/or content by genetic manipulation have immense potential economic benefits to the Pulp and Paper industries. Researchers are aiming to reduce the cost of lignin extraction process (Pullman *et al.*, 1998; MacRae and van Staden, 2000). Several groups are engaged in optimising wood quality (e.g. density), trunk shape and quality of the raw product (Dickson and Walker, 1997; Pullman *et al.*, 1998; Tzfira *et al.*, 1998); and alteration of lignin composition for easy extraction (James *et al.*, 1998). Notwithstanding the perceived value, by the Plantation Forestry and Forest Products Industries, of integrating transgenic eucalypts with conventional breeding programmes, significant progress ha yet to be reported. Further, the limited advances and successes in the production of transgenic *Eucalyptus* have been achieved mainly in commercial laboratories and/or are the subject of patent applications. The recent review article of MacRae and van Staden (2000) has indicated recovery of transgenic plants of seven species and three hybrids of *Eucalyptus,* out of which most information is available on *E. camaldulensis* (Diallo and Duhoux, 1984; Mullins *et. al.*, 1997; (Ho *et. al.*, 1998), *E. citriodora* (Muralidharan and Mascarenhas, 1987), *E. grandis* (Lainé and David, 1994) and *E. grandis* hybrids (Warrag *et. al.* 1991). In addition, transgenic plants of *Eucalyptus globulus* have also been produced (Moralejo *et. al.*, 1998)

Several factors have limited the progress in gene transfer in *Eucalyptus*, *viz.*: (i) poor knowledge on molecular control of traits such as growth, stem and wood quality; (ii) unavailability of genes of interest; (iii) limited susceptibility of the species to *Agrobacterium tumefaciens* and *A. rhizogenes*; and (iv) unavailability of high yielding *in vitro* regeneration methods (MacRae and van Staden, 2000).

Regardless of transformation technique, a highly reproducible protocol for plant regeneration from transformed cells is indispensable. The main obstacle in the successful production of transgenic plants is the lack of appropriate culture methods (indirect organogenesis, somatic embryogenesis) and/or the loss of regenerative ability of culture systems when used in conjunction with *Agrobacterium*-mediated transformation procedures. Notable reported exceptions include the production of transgenic plants of *Eucalyptus globulus* (Moralejo *et. al.*, 1998) and *E. camaldulensis* (Ho *et. al.*, 1998), via *Agrobacterium tumefaciens*-mediated transformation of hypocotyls and plantlet regeneration through indirect organogenesis. Significantly, the regeneration protocol developed by Ho *et al.* (1998) yielded approximately 10 plants/explant, and the transformed callus maintained its organogenic ability for more than a year. However, reports on similar successes are scarce.

Some reports are available on field trials of transgenic *E. grandis* and *E. camaldulensis*. It appears that gross morphology of transgenic plants is normal and that transgene expression is largely stable and can be predicted from *in vitro* culture material (MacRae and van Staden, 2000). Despite such encouraging results, much more long-term research is necessary to assess the added value of transgenic *Eucalyptus* benefiting industry, especially the on tolerance to a broad range of biotic and abiotic stresses, and raw product improvements. This can only be achieved by improving the genetic transformation rate, availability of trait-specific transgene, and a reliable protocol for plant regeneration of different *Eucalyptus* spp. At present, the most viable option for improving yields is to exploit the natural genetic variation, some of which is yet untapped.

5. CONCLUSION

Tree improvement is a process of managing genetic resources through recurrent cycles of selection and breeding. The Plantation Forestry companies have recognized the benefits of incorporating micropropagation into their tree improvement and propagation activities. In *Eucalyptus* breeding and clonal programmes, *in vitro* culture systems are being used for a variety of applications, e.g. the rapid multiplication of selected genotypes, rejuvenation of physiologically old shoots and germplasm conservation. Yet, with notable exceptions, the potential commercial benefits of using *in vitro* technology for mass propagation of selected genotypes have yet to be realised.

There should be more close interaction between industry and academia for mutual benefits in developing innovative technologies in forestry. For example, genetic stability of *in vitro* elite planting material is highly desirable for commercialization of eucalypts - any genetic loss can have adverse economic consequences. However, there is hardly any work on tissue culture-derived or somaclonal variation in *Eucalyptus*, even though there is a plenty of information available on somaclonal variation and molecular marker analysis in relation to genetic fidelity among *in vitro* grown plants (Jain *et al*., 1998; Jain, 2001). In addition to mass multiplication of selected genotypes, *in vitro* plant regeneration protocols can be used for a variety of purposes but, in the past decade, few of them have been exploited. For example, *in vitro* conservation techniques should be the focal point for long- or short-term conservation of elite genetic material, which can be made available to industry and researchers. Furthermore, *in vitro* conservation techniques, in combination with traditional methods, should be practiced to alleviate the pressure on land utilisation.

Transgenic research in forestry has not taken long strides, perhaps, due to long life cycle of forest trees. It is rather difficult to predict the behavior of transgenes during the long life cycle of trees. Moreover, a reliable plant regeneration method is needed for large-scale plant multiplication of transgenic material. The most rate-limiting step in the development of transgenic *Eucalyptus* is the regeneration of transgenic plants (MacRae and van Staden, 2000). Considerable efforts are still needed in the evaluation of stable transgene expression, and on risks and benefits of this technology, before its commercial exploitation in Forestry.

ACKNOWLEDGMENTS

We thank The University of Natal Research Fund, National Research Foundation and Mondi Forests for continued support. We are also deeply grateful to all colleagues in Research Institutions and Industry who generously provided information for and comments on this chapter. In particular, Dr Sharmane MacRae, Dr Tim White, Dr. Rod Griffin, Dr Fabio Brun, Dr Jenny Aitken-Christie and Dr Dale Smith.

6. REFERENCES

Aitken-Christie, J.A. (pers. com) Leader – Partnerships and Innovation, Carter Holt Harvey Forest Genetics, PO Box 2463, Rotorua, New Zealand

Aitken-Christie, J.A., Singh, A.P. 1987. Cold storage of tissue culture. In: *Cell and Tissue Culture in Forestry*, vol. 2: *Specific principles and methods, growth and development,* J.M. Bonga, D.J. Durzan, eds Martinus Nijhoff Publishers, Dordrecht, pp 285-304

Aneja, S., Atal, C.K. 1969. Plantlet formation in tissue cultures from lignotubers of *Eucalyptus citriodora* Hook. Current Sciences 38: 69 –70

Ashmore, S.E. Status Report on the Development and Application of *In Vitro* Techniques for the Conservation and Use of Plant Genetic Resources. International Plant Genetic Resources Institute, Rome, Italy, 1997, p 67

Assy-Bah, B., Durnad-Gasselin, T., Pannetier, C. Use of zygotic embryo culture to collect germplasm of coconut (*Cocos nucifera*). Plant Genetic Resources Newsletter 1987; 71: 4 -10

Azmi, A., Noin, M., Prouteau, M., Boudet, A.M. Chriqui, D. High frequency plant regeneration from *Eucalyptus globulus* Labill. hypocotyls: Ontogenesis and ploidy level of the regenerants. Plant Cell, Tissue and Organ Culture 1997a; 51: 9 - 16

Azmi, A., Dewitte, W., Drevet, C., van Onckelen, H., Landré, P., Boudet, A.M., Jouanin, L., Chriqui, D. Bud regeneration from *Eucalyptus globulus* clones and seedlings through hormonal imbalances induced by *Agrobacterium tumefaciens* strain 82.139. Plant Science 1997b; 127: 81 - 90

Bandyopadhyay, S., Cane, K., Rasmussen, G., Hamill, J.D. Efficient plant regeneration from seedling explants of two commercially important temperate eucalypt species – *Eucalyptus nitens* and *E. globulus*. Plant Science 1999; 140: 189 - 198

Barbour, E.L. and Butcher, T. "An accelerated Clonal Orchard System for *Eucalyptus globulus*". In: *Eucalypt Plantations: Improving Fibre Yield and Quality*. Proceedings of the CRCTHF-IUFRO conference, Hobart Australia, February 1995, pp 281 - 284

Barker, H., Torrance, L. "Importance of Biotechnology for germplasm health and quarantine". In: *Biotechnology and Plant Genetic Resources. Conservation and Use*, J.A. Callows, B.V. Ford-Lloyd, H.J. Newbury, eds CAB International, New York, 1997, pp 235 - 254

Bauer, L.S. Fiber farming with insecticidal tress. Journal of Forestry 1997, 95: 20 - 23

Bennett, I.J., McComb, J.A., Tonkin, C.M., McDavid, D.A.J. Effect of cytokinins on multiplication and rooting of *Eucalyptus globulus* and other *Eucalyptus* species. In: *Mass Production Technology for Genetically Improved Fast Growing Forest Tree Species*. Proceedings of an AFOCEL/IUFRO symposium, Bordeaux, France, September 1992, pp 195 - 202

Bennett, I.J., McComb, J.A., Tonkin, C.M., McDavid, D.A.J. Alternating cytokinis in multiplication media stimulates *in vitro* shoot growth and rooting of *Eucalyptus globulus* Labill. Annals of Botany 1994; 74:53 - 58

Bertolucci, F.l., Penchel, R.M. Clonagem do eucalipto: efeitos sobre a productividade e qualidade da madeira. Ciencia Hoje Supplemento 1993; 16: 16 - 21

Bisht, P., Sharma, V.K., Joshi, I., Kapoor, M.L. Micropropagation of newly produced F_1 hybrid of *Eucalyptus* (*E. tereticornis* SM. x *E. camaldulensis* Dehn. Southern Form) Silvae Genetica 1999; 48: 104 - 108

Blakesley, D., Kiernan, R.J. Cryopreservation of axillary buds of *Eucalyptus grandis* x *Eucalyptus camaldulensis* hybrid. CryoLetters 2001; 22: 13 - 18

Blomstedt, C., Cameron, J., Whiteman, P., Chandler, S.F. Micropropagation and juvenile *Eucalyptus regnans* (Mountain Ash). Australian Journal of Botany 1991; 39: 179 – 186

Brun, F. (pers. com.) International Paper Brazil.

Burley, J. "Application of biotechnology for forestry and rural development". In: *Application of Biotechnology in Forestry and Horticulture*, V. Dhawan, ed Plenum Press New York, 1989, pp 9 - 20.

Chang, S.H., Donald, D.G.M., Jacobs, G. Micropropagation of *Eucalyptus radiata* ssp. *radiata* using explants from mature and coppice material. South African Forestry Journal 1992; 162: 43 - 47

Cid, L.P.B., Machado, A.C.M.G., Carvalheira, S.B.R.C., Brasileiro, A.C.M. Plant regeneration from seedling explants of *Eucalyptus grandis* x *E. urophylla*. Plant Cell, Tissue and Organ Culture 1999; 56: 17 – 23

Cotterill, P.P. and Brindbergs, M.L. Growth of first-and second-generation *Eucalyptus globulus* clonal cuttings and seedlings. In: *IUFRO conference on Silviculture and Improvement of Eucalypts*, Salvador, Brazil. 1997. pp 233- 238.

De Fossard, R.A., Nitsch, C., Cresswell, R.J., Lee, E.C.M. Tissue and organ culture of *Eucalyptus*. New Zealand Journal of Forestry Science 1974; 4: 267-278

De Fossard, R.A., Barker, P.K., Boume, R.A. The organ culture of nodes of four species of *Eucalyptus*. Acta Horticulturae 1977; 73: 157 - 165

Denison, N.P., Kietzka, J.E. The development and utilisation of vegetative propagation in Mondi for commercial afforestation programmes. South African Forestry Journal 1993; 165: 47-54

Denison, N.P., Quaile, D.R. The applied clonal Eucalypt progamme in Mondi Forests. South African Forestry Journal 1987; 142: 60 - 66

Diallo, N. Duhoux, E. Organogenése et multiplication *in vitro* chez l' *Eucalyptus camaldulensis*. Journal of Plant Physiology 1984; 115: 177 – 182

Dickson, R. Walker, J. Pines: growing commodities or designer trees. Forestry Review 1997, 76: 273 - 279

Donald, D.G.M., Newton, D.J. *In vitro* propagation of *Eucalyptus radiata*. South African Forestry Journal 1991; 157: 66 - 68

Durand-Cresswell, R.J., Boulay, M., Franclet, A. "Vegetative propagation of *Eucalyptus*." In: *Tissue Culture in Forestry*, J.M. Bonga, D.J. Durzan, eds The Hague, Nijhoff, 1982, pp 150 - 181

Durand-Cresswell, R.J., Nitsch, C. Factors influencing the regeneration of *Eucalyptus grandis* by organ culture. Acta Horticulturae 1977; 78: 149 - 155

Eldridge, K., Davidson, J., Harwood, C and van Wyk, G. Eucalypt domestication and breeding. Claredon Press, Oxford, 1994.

Engelmann, F. Importance of desiccation for the cryopreservation of recalcitrant seed and vegetatively propagated species. Plant Genetic Resources Newsletter 1997; 112: 9 - 18

George, G.E. *Plant Propagation by Tissue Culture*. Part 1, The Technology, Exegenics Ltd., Edington, 1993, p 574

Gill, R.I.S., Gosal, S.S. Micropropagation of economically important tropical forest trees. In: *Tree Improvement for sustainable tropical forestry*. Proceedings of the QFRI-IUFRO conference. M.J. Dieters, A.C. Matheson, D.G. Nickles, C.E. Harwood, S.M. Walker, eds Caloundra, Queensland, Australia, 27 October - 1 November 1996, pp 230 - 233

Griffin, A. R. (pers. com.) Shell Forestry Technical Services, HR1 East Malling, West Malling Kent ME 196BJ, UK.

Griffin, A.R. Genetically modified tress – the plantations of the future or an expensive distraction? Commonwealth Forestry Review 1996, 75: 169 – 175.

Grout, B.W.W. "Introduction to the *in vitro* preservation of plant cells, tissues and organs". In: *Genetic Preservation of Plant Cells* In Vitro, B.W.W. Grout, ed Springer-Verlag, Berlin, Heidelberg, New York, 1995a, pp 1 – 20

Grout, B.W.W. Minimal growth storage. In: *Genetic Preservation of Plant Cells* In Vitro, B.W.W. Grout, ed Springer-Verlag, Berlin, Heidelberg, New York, 1995b, pp 21 - 28

Gupta, P.K., Timmis, R., Mascarenhas, A.F. Field performance of micropropagated forestry species. In Vitro Cellular and Developmental Biology 1991; 27: 159 - 164

Harcourt, R.L., Kyozuka, J., Floyd-R.B., Bateman, K.S., Tanaka, H., Decroocq, V., Llewellyn, D.J., Zhu, X., Peacock, W.J. and Dennis, E.S. Insect- and herbicide –resistant transgenic eucalypts. Mol. Breed. 2000, 6: 307-315.

Ho, C.K., Chang, S.H., Tsay, J.Y., Tsai, C.J., Chiang, V.L., Chen, Z.Z. *Agrobacterium tumefaciens*-mediated transformation of *Eucalyptus camaldulensis* and production of transgenic plants. Plant Cell Reports 1998; 17: 675 - 680

Ikemori, T.K., Penchel, R.M., Bertolucci, F.L.G. Integrating biotechnology into eucalypt breeding. In: *Proceedings of the International Symposium on Wood Biotechnology*. 31 August - 1 September 1994, Fuchu, Tokyo, Japan

Ito, K., Doi, K., Tatemichi, Y., Shibata, M. Plant regeneration of eucalypts from rotating nodule cultures. Plant Cell Reports 1996; 16: 42 – 45

Jain, S.M. Tissue culture-derived variation in crop improvement Euphytica 2001. 118: 153-166.

Jain, S.M., Ahloowalia, B.S. and Veilleux, R.E. Somaclonal variation in crop plants. In: Somaclonal variation and induced mutations in crop improvement, S.M.Jain, D.S. Brar and B.S. Ahloowalia (eds.). Kluwer Academic Publishers, Dordrecht, The Netherlands. 1998. pp 203 - 218

James, R.R., DiFazio, S.P., Brunner, A.M., Strauss, S.S. Environmental effects of genetically engineered woody biomass crops. Biomass Bioenergy 1998, 14: 403 - 414

Jones, N.B., van Staden, J. Micropropagation and establishment of *Eucalyptus grandis* hybrids. South African Journal of Botany 1994; 60: 123 - 126

Kartha, K.K. "Production and indexing of disease-free plants." In: *Plant Tissue Culture and its Agricultural Applications*, L.A. Withers, P.G. Alderson, eds Butterworths, London, 1986, pp 219 - 238

Lainé, E., David, A. Regeneration of plants from leaf explants of micropropagated clonal *Eucalyptus grandis*. Plant Cell Reports 1994; 13: 473 - 476

Le Roux, J.J., van Staden, J. Micropropagation and tissue culture of *Eucalyptus* - a review. Tree Physiology 1991a; 9: 435 - 477

Le Roux, J.J., van Staden, J. Micropropagation of *Eucalyptus* species. HortScience 1991b; 26: 199 - 200

Lubrano, L. Micropropagation of *Eucalyptus grandis*. Acta Horticulturae 1991; 300: 89 – 94

MacRae, S. (pers. com.) Tree Improvement Manager, State Forests of New South Wales, Research and Development Division, 121 – 131 Oratava Avenue, West Pennant Hills NSW 2125

MacRae, S., van Staden, J. "Transgenic *Eucalyptus*". In: *Biotechnology in Agriculture and Forestry 44: Transgenic Trees*." Y.P.S. Bajaj, ed Springer, 2000, pp 88 - 114

Maile, N., Niewenhuis, M. Vegetative propagation of *Eucalyptus nitens* using stem cuttings. South African Forestry Journal 1996; 175: 29 - 34

Major, G., Krause, M., Ross, S., Sotelo, M. Preliminary studies on the somatic embryogenesis in *Eucalyptus grandis*. In: *Silvilculture and Improvement of Eucalypts*, vol. 2: *Biotechnology applied to genetic improvement of tree species*. Proceedings of the IUFRO conference, Salvador, Brazil, August 1997, pp 123 - 126

Mascarenhas, A.F., Agrawal, D.C. "Prospects for *in vitro* conservation of tree germplasm". In: *Plant Genetic Resources Through* In Vitro *Methods*, A.H. Zakri, N.M. Normah, M.T. Senawi, A.C. Abdul Karim, eds FRIM/MNCPGR, Kuala Lumpur, Malaysia, 1991, pp 231-252

Mathews, J.H., Campbell, M.M. The advantages and disadvantages of the application of genetic engineering to forest trees: a discussion. Forestry 2000, 73 – 371 - 380

McFerson, J.R. From *in situ* to *ex situ* and back: the importance of characterizing germplasm collections. HortScience 1998; 33: 1134 - 1135

Mokotedi, M.E.O., Watt, M.P., Pammenter, N.W., Blakeway, F.C. *In vitro* rooting and subsequent survival of two clones of a cold-tolerant *Eucalyptus grandis* x *E. nitens* hybrid. HortScience 2000; 35 (6): 1163 - 1165

Moralejo, M., Rochange, F., Boudet, A.M., C. Teulieres. Generation of transgenic *Eucalyptus globulus* plantlets through *Agrobacterium*-mediated transformation. Australian Journal of Plant Physiology 1998; 25: 207 - 212

Mullin, T.J., Park, Y.S. Estimating genetic gains from alternative breeding strategies for clonal forestry. Canadian Journal of Forest Research 1992; 22: 14 - 23

Mullins, K.V., Llewellyn, D.J., Hartney, V.J., Strauss, S., Dennis, E.S. Regeneration and transformation of *Eucalyptus camaldulensis*. Plant Cell Reports 1997; 16: 787 - 791

Muralidharan, E.M., Mascarenhas, A.F. "Somatic embryogenesis in *Eucalyptus*." In: *Somatic embryogenesis in woody plants*, vol. 2 – *Angiosperms*, S.M. Jain., P.K. Gupta., R.J. Newton eds Kluwer Academic Publishers, Dordrecht. 1995 pp 23 - 40

Muralidharan, E.N., Gupta, P.K., Mascarenhas, A.F. Plantlet production through high frequency somatic embryogenesis in long-term cultures of *Eucalyptus citriodora*. Plant Cell Reports 1989; 8: 41 - 43

Muralidharan, E.M., Mascarenhas, A.F. *In vitro* plantlet formation by organogenesis in *E. camaldulensis* and by somatic embryogenesis in *E. citriodora*. Plant Cell Reports 1987; 6: 256 - 259

Niccol, R.J., Regan, P.A., de Filippis, F.L. Simplified protocol for the micropropagation of selected *Eucalyptus* and *Banksia* species. Australian Forestry 1994; 57: 143 - 147

Nugent, G., Chandler, S.F., Whiteman, P., Stevenson, T.W. Somatic embryogenesis in *Eucalyptus globulus* Labill. In: *Silvilculture and Improvement of Eucalypts*, vol. 2: *Biotechnology applied to genetic improvement of tree species*. Proceedings of the IUFRO Conference, Salvador, Brazil, August 1997, pp 183 - 185

Patil, V., Kuruvinashetti, M.S. Micropropagation of *Eucalyptus tereticornis* using explants from mature and coppice shoots. International Journal of Experimental Botany 1998; 63: 129 - 132

Poissonnier, M., Monod, V., Paques, M., Dereuddre, J. Cryopreservation in liquid nitrogen of *Eucalyptus gunnii* shoot tips grown *in vitro* following encapsulation and dehydration. Annales de Reserches Sylvicoles, 1992; 5 – 23

Pullman, G., Cariney, J., Peter, G. Clonal forestry and genetic engineering: where we stand, future prospects, and potential impacts on mill operations. TAPPI Journal 1998, 8: 57 - 64

Quoirin, M., Vieira, R.C. Rhizogenesis and nodule formation from callus of *Eucalyptus grandis* and *E. grandis* x *urophylla*. Arquivos de Biologia e Tecnologia 1995; 38: 793 - 798

Rasmussen, G.F. Micropropagation of *E. nitens* and hydrid clones. In: *Forest Genetics*. Proceedings of the 11th Research Working Group 1, C. Dean, ed Coonawara, South Australia 1991, pp 106 – 109

Rockwood, D.I., Warrag, E.I. Field performance of micropropagated, macropropagated and seed-derived propagaules of three *Eucalyptus grandis* ortets. Plant Cell Reports 1994; 13: 628 - 631

Ruaud, J.-N., Churchill, K., Pepper, S. Somatic embryogenesis in *Eucalyptus nitens*. Acta Horticulturae 1997; 447: 185 - 187

Sharma, S.K., Ramamurthy, V. Micropropagation of 4-year-old elite *Eucalyptus tereticornis* trees. Plant Cell Reports 2000; 19: 511 - 518

Smith, A., Jones, D., Wilson, M. Studies of forest trees. Plant Physiology 1990; 9: 543 - 547

Sussex, I.M. The origin and morphogenesis of *Eucalyptus* cell populations. In: *Proceedings of the International Conference on Plant Tissue Culture*. White, R.R., Grove, A.R. eds McCutchan, Berkeley 1965, pp 383 - 391

Tanksley, S.D., McCouch, S.R. Seed banks and molecular maps: unlocking genetic potential from the wild. Science 1997; 277: 1063 - 1066

Termignoni, R.R., Wang, P.-J., Hu, C.-Y. Somatic embryo induction in *Eucalyptus dunnii*. Plant Cell, Tissue and Organ Culture 1996; 45: 129 - 132

Termignoni, R.R., Jobim, C.P., Moraes, L. Somatic embryogenesis in *Eucalyptus* spp.: Regeneration systems from elite adult clones. In: *Plant Biotechnology and* In Vitro *Biology in the 21st century*. Abstracts of the Proceedings of the IX International Congress on Plant Tissue and Cell Culture, Jerusalem, Israel, June 14 - 19, 1998. p 114

Tibbits, W.N., White, T.L., Hofge, G.R., Joyce, K.R. Genetic control of rooting ability of stem cuttings in *Eucalyptus nitens*. Australian Journal of Botany 1997; 45: 203 - 210

Tibok, A., Blackhall, N.W., Power, J.B., Davey, M.R. Optimized plant regeneration from callus derived from seedling hypocotyls of *Eucalyptus urophylla*. Plant Science 1995; 110: 139 - 145

Trindade, H., Pais, M.S. *In vitro* studies of *Eucalyptus globulus* rooting ability. In Vitro Cellular and Developmental Biology – Plant 1997; 110: 139 - 145

Turnbull, J.W. Eucalypt plantations. New Forests 1999; 17: 37 – 52

Tzfira, T., Zuker, A., Altman, A. Forest tree biotechnology: genetic transformation and its application to future forests. Trends in Biotechnology 1998, 16: 439 - 446

Wachira, F. *In vitro* shoot multiplication of *Eucalyptus grandis*. African Crop Science Journal 1997; 5: 239 - 251

Warrag, E., Lesney, M.S., Rockwood, D.J. Nodule culture and regeneration of *Eucalyptus grandis* hybrids. Plant Cell Reports 1991; 9: 586 - 589

Watt, M.P., Blakeway, F.C., Cresswell, C.F., Herman, B. Somatic embryogenesis in *Eucalyptus grandis*. South African Forestry Journal 1991; 157: 59 - 65

Watt, M.P., Duncan, E.A., Ing, M., Blakeway, F.C., Herman, B. Field performance of micropropagated and macropropagated *Eucalyptus* hybrids. South African Forestry Journal 1995; 173: 17 - 21

Watt, M.P., Gauntlett, B.A., Blakeway, F.C. Effect of anti-fungal agents of *in vitro* cultures of *Eucalyptus grandis*. South African Forestry Journal 1996; 175: 23 - 27

Watt, M.P., Blakeway, F.C., Termignoni, Jain, S.M. Somatic Embryogenesis in *Eucalyptus grandis* and *E. dunnii*. In: *Somatic Embryogenesis in Woody Plants*, vol. 5. S.M. Jain,. P.K. Gupta, R.J. Newton, eds Kluwer Academic Publishers, 1999, pp 63 - 78

Watt, M.P., Thokoane, N. L., Mycock, D., Blakeway, F.C. *In vitro* storage of *Eucalyptus grandis* germplasm under minimal growth conditions. Plant Cell, Organ and Tissue Culture 2000a; 61: 161-164

Watt, M.P., Mycock, D.J., Blakeway, F.C., Berjak, P. Application of *in vitro* methods to *Eucalyptus* germplasm conservation. Southern African Forestry Journal 2000b; 187: 3 - 10

Willyams, D., Whiteman, P., Cameron, J., Chandler, S.F. Inter- and intra-family variability for rooting capacity in micropropagated *Eucalyptus globulus* and *Eucalyptus nitens*. In: *Mass Production Technology for Genetically Improved Fast Growing Forest Tree Species*. Proceedings of the AFOCEL/IUFRO symposium, Bordeaux, France, September 1992, pp 177 - 181

Wilson, P.J. Multiplication rates *in vitro* and by stem cutting propagation, and the development of seedling-origin clones of *Eucalyptus globulus*. In: *Eucalypt Plantations: Improving Fibre Yield and Quality*. Proceedings of the CRCTHF-IUFRO conference, Hobart, Australia, February 1995, pp 304 - 310

Wilson, P.J. Multiplication rates *in vitro* and by stem cuttings propagation, and clonal development from *Eucalyptus globulus* seedlings. Forest Science 1996; 42: 415 - 418

Wilson, P.J. *In vitro in vacuo*. Southern African Forestry Journal 1998; 183: 53 – 58

Withers, L.A. "Germplasm Preservation". In: *Applications of Plant Cell and Tissue Culture*, L.A. Withers, ED Wiley and Sons, Sussex, 1988, pp 163-177

Yang, J.-C., Chung, J.-D., Chen, Z.-Z. Vegetative propagation of adult *Eucalyptus grandis* x *urophylla* and comparison of growth between micropropagated plantlets and rooted cuttings. Plant Cell Reports 1995; 15:170 - 173

Yasodha, R., Sumathi, R., Gurumuthi, K. "Micropropagation of difficult to propagate clones of eucalypts". In: *Silvilculture and Improvement of Eucalypts*, vol. 2: *Biotechnology applied to genetic improvement of tree species*. Proceedings of the IUFRO conference, Salvador, Brazil, pp 123 - 126

Zobel, B.J. "Clonal forestry in the Eucalypts". In: *Clonal Forestry II: Conservation and Application*. Springler-Verlag Berlin Heidelberg. Ahuja, M.R. and Libby, W.J. eds 1993. pp 139 – 148

9. MICROPROPAGATION OF *Acacia* SPECIES

Marguerite Quoirin
Departamento de Botânica, Universidade Federal do Paraná, Caixa postal 19031, Curitiba-PR, 81531-990, Brazil. E-mail: quoirin@bio.ufpr.br

Chapter contents

1. INTRODUCTION

1.1. Botany and geographical distribution

The *Acacia* group (*Leguminosae* family, *Mimosaceae* subfamily) comprises more than 1250 known species of shrubs and trees (Duhoux et al., 1998). Mostly they grow in arid and semi-arid regions and are resistant to dry climate due to their highly developed root system. This explains their role in the rural economy of all arid and semi-arid regions. Other species grow in humid tropical zones. All *Acacia* species are important for sustainable development of arid and tropical regions (Duhoux et al., 1998).

According to Dommergues et al. (1999), the genus *Acacia* has been subdivided into sub-genera (Vassal, 1981). This classification was not

S.M. Jain and K. Ishii (eds.), Micropropagation of Woody Trees and Fruits, 245–268.

universally accepted. Based on their morphological characters, the *Acacia* species can be classified into:

- *Acacia sensu stricto*, with stipular thorns, no other kind of thorns, bipinnate and deciduous leaves,
- *Aculeiferum*, without stipular thorns, with bipinnate and deciduous leaves,
- *Heterophyllum*, with few stipular thorns, bipinnate leaves transforming into phyllodes, not deciduous when mature.

Four types of acacia can be distinguished, according to their geographical origin and the presence of phyllodes: 1) African acacias, as *A. nilotica, A. senegal, A. tortilis.* They are thorny and produce pods very appreciated by cattle; 2) Australian acacias with phyllodes, as *A. auriculiformis, A. bivenosa, A. crassicarpa, A. mangium, A. salicina, A. saligna* and *A. victoriae*; 3) Australian acacias with compound leaves, as *A. dealbata, A. decurrens* and *A. mearnsii*; 4) American acacias: for instance *A. caven, A. farnesiana* and *A. koa.* However, *A. albida* (syn. *Faidherbia albida*), which has specific botanical and phytochemical characters, is classified in a separate genus (Dommergues et al., 1999).

1.2. Uses and economic importance

Most *Acacia* species form a natural symbiosis with the soil bacterium *Rhizobium* and fix atmospheric nitrogen. Other species, that do not fix nitrogen, seem to improve their nutritional status by exploiting symbiosis with mycorrhizal fungi. They therefore often survive on land, which is low in nitrogen and organic matter where other tree species fail (National Research Council, 1983). They are also widely used for dune stabilization, erosion control and soil improvement (Dart et al., 1991). *Acacia* species are also important in both arid and moist regions as fuel wood, forage and timber. *A. mangium* wood is used for furniture, door frames, moldings, sliced veneer. It is a light-duty building timber. It is also suited for fuel. The pulp is used in fine papers due to its high-quality fibre (National Research Council 1983; Tsai, 1988). *A. crassicarpa*, along with other Australian acacias, is a promising species for tropical forestry. Its wood properties are similar to those of *A. mangium* and it may also prove to be fast growing. It produces excellent fuel wood and other useful wood products (National Research Council, 1983). *A. auriculiformis* is used for soil improvement trials (Yap, 1987). The hybrid *A. mangium* x *A. auriculiformis* has better stem form and longer clear bole height than *A.*

auriculiformis (Darus, 1991). It has a higher wood density and cellulose content than *A. mangium* and is less prone to heart-rot disease (Galiana et al., 2002).

The bark of *A. mearnsii* contains a tannin of exceptional quality, which is used in the leather industry and in water filtration. Other species such as *A. auriculiformis* produce tannins too, but with red colour, which are less appreciated. *A. mearnsii* also provides charcoal, green manure and cellulose (Resende et al., 1991; National Academy of Sciences, 1980; Nitrogen Fixing Tree Highlights, 1985). The leaves and pods of *A. albida* serve as feed for sheep, goats cattle and camels. It is also used to provide firewood and is important in agroforestry systems (Nitrogen Fixing Tree Highlights, 1985). *A. saligna* is one of the most promising forage crops for Egyptian desert farming systems (Barakat and El-Lakany, 1992). *A. salicina* and *A. saligna* have proved effective for erosion control and fuel wood plantations, particularly in North Africa, and *A. salicina* is important as an agroforestry plantation crop in water catchment basins in Israel (Jones et al., 1990). *A. senegal* yields the true arabic gum known as "Senegal gum", used in food and pharmaceutical industries, in the manufacture of adhesives and in the paint, ink and cosmetic industries. It is a valuable export for India which exports 5,000 tones of gum per year (Shahana and Gupta, 2000). Some acacias have medicinal properties: pods of *A. sinuata* are used in India as diuretic, emetic, detergent, depurative and antihelminthic (Vaidyaratnam, 1994).

Tropical *Acacia* species have been introduced into southeast Asia for afforestation of denudated land resulting from mining activities or excessive burning. *A. mangium* has become the main plantation tree species in this region in the last two decades with estimated planted areas that are now reaching about 600,000 ha (Turnbull et al., 1997). The hybrid *A. mangium* x *A. auriculiformis* is a promising plantation species for the future, since it is more vigorous and more adaptable than both parental species (Chia, 1993, in Galiana et al., 2002).

A. mearnsii is also extensively used worldwide with a total planted area of about 500,000 ha, of which the majority is in Brazil, South Africa and east Africa (Turnbull et al., 1997). In Southern Brazil, it is cultivated for cellulose and tannin production. The planted area is between 120 and 130.000 ha and has been increasing over the last few years. This species is planted by small producers in agroforestry systems (Higa, A. pers. comm.).

On the other hand, in some regions where tropical leguminous species have been introduced, as the islands of the western Indian Ocean, *Acacia* species are now considered as invaders and constitute a threat for biodiversity that need to be controlled. This applies to *A. farnesiana* and *A. mearnsii* (Tassin, 1999).

1.3. Diseases

Many of the acacias are susceptible to insect defoliators. When planted outside their native range and in pure stands, and subjected to unusual stress such as drought, excess moisture, heat or cold, they become susceptible to disease. For instance, *A. koa* is the host of 99 known insect species including 11 seed borers, 12 defoliators, mining sucking and boring insects, 2 mistletoes, 2 rust fungi and a host of root and butt rots (Skolmen, 1986). The seedlings of *A. mangium* suffer from root diseases (*Phellinus noxius* and *Macrophomina*); the adult trees can be defoliated by bagworms and Lepidopterous (Yap, 1987). *A. auriculiformis* roots are infested by fungus, while beetles damage the foliage. Expensive chemicals are required to control these problems.

1.4. Propagation

Acacias are prolific seed setters and are generally propagated from seeds that are long-lived and easy to store. They also coppice well and several species spread by root suckers. As the trees are out breeders, there is a high degree of genetic variability in the seedling progeny. There is therefore a need to develop propagation methods for rapid clonal multiplication of elite individuals (Dhawan, 1993). Some species, like *A. mangium,* can be easily propagated by stem cuttings. This method, as well as micropropagation, are used for production of genetically improved planting stock for large-scale planting program in Malaysia (Darus, 1991a).

2. MICROPROPAGATION

The first report on tissue culture of *Acacia* was from Bonner (1942), who cultured root of *A. melanoxylon* and maintained them in liquid White's medium for over one year (Dhawan, 1993). So far, approximately 26 *Acacia* species and one hybrid have been cultured *in vitro* for micropropagation, bud regeneration or somatic embryogenesis.

Several techniques of micropropagation have already been developed and applied to a great number of plant species. The most widely used consists of increasing the development of axillary buds by suppressing apical dominance by means of tissues isolation and modification of the hormonal balance favoring cytokinins. Other technique, like indirect organogenesis, may lead to somaclonal variation. Somatic embryogenesis is another useful way to multiply plant material.

The principal reports on the micropropagation of *Acacia* species by axillary budding are indicated in table 1. The majority of them refer to the process initiated from juvenile explants. When the micropropagation is initiated from adult material, it has the advantage of cloning genetically superior trees and fixing the desired characteristics. This procedure has been used with coppice material of *A. mearnsii* (Beck et al., 1998b) and *A. melanoxylon* (Meyer and van Staden 1987), with nodal explants or apices from adult trees of *A. catechu* (Kaur et al., 1998), *A. ligulata* (Williams et al., 1985), *A. mangium* (Rajadurai et al., 1989), *A. mangium x A. auriculiformis* (Darus, 1991b), *A. nilotica* (Singh et al., 1993) and *A. saligna* (Barakat and El-Lakany, 1992). In case of *A. albida* (Gassama and Duhoux, 1987; Gassama, 1989), sprouts or suckers from adult trees were used as starting material.

For multiplication, benzylaminopurine (BA) is the most widely used cytokinin, sometimes combined with indol-3-acetic acid (IAA), gibberelic acid (GA_3) or naphthalenacetic acid (NAA). After this phase, there is no need for transfer to a specific elongation medium, as *Acacia* species form shoots on a multiplication medium.

Rooting is obtained *in vitro* on a full strength or diluted mineral medium. The requirement for an auxin and the concentration thereof are species-dependent. Generally, NAA or indol 3-butyric acid (IBA) is used. As for other species, the rooting ability of *Acacia* microcuttings is affected by BA concentration in the multiplication medium (Galiana et al., 1991a). It is therefore recommended that a low concentration of this cytokinin be used during the multiplication phase. Darkness combined with auxin treatment improved rooting of *A. mangium* microshoots from juvenile or mature trees (Monteuuis & Bon, 2000).

Activated charcoal was sometimes added to basal medium at 2.10^3 or 2.5 10^3 mg/L. It avoids leaf chlorosis of *A. mearnsii* shoots and allows rooting without addition of auxin (Quoirin et al., 2001) or reduces leaf abscission and results in more leaves with higher leaflet numbers in *A. melanoxylon* plantlets (Jones, 1986).

Multiplication rate is highly variable depending on the species and, for the same species, of the procedure. For instance, the value reported for *A. mangium* (Table 1) varies between 2.3 (Saito et al., 1993b) and 40 (Galiana et al., 1991a). It is difficult to compare these values, as they can differ greatly between the first and the subsequent subcultures. For instance, Galiana et al. (1991a) indicated that the multiplication rate was 40 in the first subculture, then decreased to 10-20 in the second one.

In the case of the hybrid *A. mangium* x *A. auriculiformis,* Darus (1991b) calculated the number of shoots that could be obtained after one year of multiplication: 6,048,000 from juvenile material (5 subcultures) and 90,720,000 from adult material (6 subcultures). These numbers were obtained on the base of a mean multiplication rate of 30 every month on MS medium supplemented with 2.22 μM.BA. The rooting ability of such material varied between 50 and 73% and decreased with increasing number of subcultures. For the same hybrid, Galiana et al. (2002) obtained a multiplication rate of only 2.8 every two months, which varied greatly according to the clone. In this case the culture medium was not described.

It is difficult to indicate a general micropropagation protocol valid for all *Acacia* species. In most of them, MS medium containing BA is appropriate for multiplication phase and half-strength MS mineral medium with IBA for rooting. In some species, however, mineral MS medium must be modified or substituted by another formula, for instance modified B_5 (Gamborg, 1968) or QL (Quoirin and Lepoivre, 1977) medium.

Field trials were conducted with plantlets of *A. mangium, A. melanoxylon* and *A. mangium x A. auriculiformis.* In the first case, micropropagated plantlets were compared to seedlings. After three years, the growth values of plants from *in vitro* cultures were higher than those obtained from seedlings. It was suggested that this could be due to the development of a better rooting system, with many lateral roots, in plants from *in vitro* culture (Setiawan and Umboh, 1990). A field test with *A. melanoxylon*, observed one year after planting, revealed that tissue cultured planting stock had a similar growth rate to seedlings (Jones et al., 1991). In an experiment with 20,000 micropropagated plantlets of *A. mangium* x *A. auriculiformis*, Galiana and coworkers (2002) recovered 94% when they were maintained one month under misting system and 85% when acclimatized in polybags during three months. Studies using molecular markers are necessary to detect the possible presence of somaclonal variants among the micropropagated plants.

Studies consisting in inoculating micropropagated plantlets with *Bradyrhizobium* revealed that the plants maintain its nodulation and N_2-fixing ability after *in vitro* propagation (Galiana et al., 1991b).

Table 1. Micropropagation of *Acacia* species by axillary budding (concentration of growth regulators in µM).

Species	Explant type	Multiplication medium	Rooting medium	Result	Reference
A. albida	Cotyledonary buds from 3–week-old seedlings	MS salts, NN vitamins, BA 13.3, NAA 2.7	LS salts, NN vit., sucrose 5%	Acclimatized plants	Duhoux & Davies, 1985
A. albida	Shoot apex from seedlings	MS, BA 0.1, NAA 0.1 or 0.01	NAA 0.1	Acclimatized plants	Ruredzo & Hanson, 1993
A. albida	Nodes from mature trees sprouts Nodes from suckers	Modified MS, NN vit., BA 2.2, NAA 0.01	LS, sucrose 5%	*In vitro* plantlets	Gassama & Duhoux, 1987 Gassama, 1989
A. auriculiformis	Axillary buds from one-month-old seedlings	B_5, CM 5%, BA 1, sucrose 3%	B_5, NAA or IBA 1	*In vitro* plantlets	Mittal et al., 1989
A. auriculiformis	Nodes from seedlings	MS, BA 10, IBA 0.5, sucrose 3% MR 2.6 after 2 subcult	White, IBA 2, NAA 1	Acclimatized plantlets	Semsuntud & Nitiwattanachai 1991
A. auriculiformis	Epicotyl segments	1/2MS, BA 6.7 Elong. GA_3 14.4, NAA 0.11	1/2MS alone or + NAA 0.11	Acclimatized plants	Ide et al., 1994
A. bivenosa	Nodal explants from seedlings	MS, BA 3 MR: 8 - 10	1/2MS, IAA 6	*In vitro* plantlets	Dhawan, 1993

Table 1 cont'd

A. catechu	Nodal explants from mature trees	MS, BA 17.7, NAA 2.7, asc. acid 0.11, adenine sulf. 135, glut. 1. Elongation: BA, Kin. MR: 8.7	1/4MS, IAA 17.1, sucrose 1.5%	Acclimatized plantlets	Kaur et al., 1998
A. catechu	Shoot tips from seedlings	MS, BA 6.7, Kin 7 MR: 12	1/4MS, IAA 1.3, sucrose 1.5%	Acclimatized plantlets	Kaur & Kant, 2000
A. crassicarpa	Nodal segments from seedlings	MS; BA 4.4; GA_3 1.4 to 2.9	½ MS; IBA 2	Acclimatized plantlets	Nishikawa & Seido, 2001
A. flava	Cotyledonary buds	MS; Kin 2; sucr. 5%	MS NAA 0 or 1	*In vitro* plantlets	Di Michele & Bray, 1994
A. koa	Shoot apex	Modif. MS, BA 4.4	Modif. MS, IBA 1.5	Acclimatized plants	Skolmen & Mapes, 1978
A. ligulata	Nodal segments from mature trees	Modif. MS, BA 0.5, Kin 0.5	Modif. MS	Acclimatized plants	Williams et al., 1985
A. maconochiena	Nodal explants from seedlings	MS, BA 3, NAA 3 MR: 6	MS, IAA 3, NAA 1	Acclimatized plants	Dhawan, 1993
A. mangium	Epicotyls from 2-week-old seedlings	½ MS salts, BA 1, NAA 1	same	Acclimatized plants	Crawford & Hartney, 1987
A. mangium	Five-day-old seedlings	MS, 2,4-D 9 to 45, BA 44.4		*In vitro* shoots	Hamzah et al., 1989

Table 1 cont'd

A. mangium	Seedling apex & adult tree apex	MS & B_5 BA	IAA or IBA	*In vitro* plants	Rajadurai et al., 1989
A. mangium	One-month-old seedling nodes	MS, sucrose 3%, BA 2.2 MR: 25.4 after 2 months	Seradix 3 (0.8% IBA)	Acclimatized plants	Darus, 1989 and 1991a
A. mangium	Eight-month-old seedling nodes	MS, BA 4.4 MR: 17	Seradix 3 (0.8% IBA)	Acclimatized plants	Darus, 1991a
A. mangium	2-week-old seedling 7-month-old plants	MS, BA 4.4 -8.9 MR: 14 to 41	1/2MS, IBA 0.25	Acclimatized plants	Galiana et al., 1991a
A. mangium	Six-month-old seedling nodes	MS, sucrose 3%, BA 10, IBA 5 MR: 2.3	1/2MS, sucrose 1%, IAA 10	Acclimatized plants	Saito et al, 1993
A. mangium	Three-month-old seedlings nodes	Macron.: 3/4B_5, Micron.: 2/3SH, BA 4.4, IBA 2.5	Macron.: 2/3QL Mícron.: 1/2MS	*In vitro* plantlets	Bon et al., 1998
A. mangium x A. auriculiformis	Nodes from one-month-old seedlings	MS, BA 2.2, sucr. 3% MR 30 after 6 subcult.	Seradix 3 (0.8% IBA)	Plantlets	Darus, 1991a
A. mangium x A. auriculiformis	Nodal explants from mature trees	MS, BA 4.4, sucrose 3%. MR 30 after 6 subcult.	Seradix 3	Plantlets	Darus, 1991b

Table 1 cont'd

A. mangium x A. auriculiformis		Unpublished medium MR 4.4 after 3 subcult.	Unpublished medium	Acclimatized plantlets	Galiana et al., 2002
A. mearnsii	Shoot apex from seedlings	3/4MS, BA 8.9 MR: 3	MS1/2, NAA 3.2	Acclimatized plants	Huang et al., 1994
A. mearnsii	Nine-month-old plant nodes	1/2MS, BA 0.22-13.3, IBA 0.05		*In vitro* shoots	Correia & Graça, 1995
A. mearnsii	Nodes from one-month-old seedlings and from 3-9 month-old plants	MS, BA 8.9	MS, IBA 4.9	Acclimatized plants	Beck et al., 1998a
A. mearnsii	Nodal explants from coppice material (2- to 10- year-old)	MS, BA 8.9, sucrose 3%		Rejuvenated shoots	Beck et al., 1998b
A. mearnsii	Nodes from one-month-old seedlings	3/4MS, AC 0.2%	Same as multipl.	*In vitro* plants	Quoirin et al., 2000
A. melanoxylon	Seedlings	QL, AC 0.25%	QL, IBA 24.6, NAA 13.42, BA 0.88	Preliminary studies	Jones, 1986
A. melanoxylon	Lateral shoots of young plants	AM, BA 4.44 in pulse treatment		*In vitro* shoots	Jones & Smith, 1988

Table 1 cont'd

A. melanoxylon	Juvenile material from stump coppice	MS, sucrose 2%, BA 1, IAA 1-5	IBA 10	Acclimatized plantlets	Meyer & van Staden, 1987
A." mimosa"	Axillary buds	BA 2.2, IAA 2.8, GA_3 1.4	IAA 5.7	Acclimatized plantlets	Ruffoni et al., 1992
A. nilotica	Cotyledonary nodes from 15-day-old seedlings	B_5, sucrose 3%, BA 6.6 MR: 6.3	B_5, IAA 11.4	Acclimatized plantlets	Dewan et al., 1992
A. nilotica	Nodal segments from 20-25-year-old trees	MS or WPM BA 1.1-2.2 or Adenine SO_4 108.6		*In vitro* shoots	Singh et al., 1993
A. salicina	Nodal segments form 7-month-old trees	MS, BA 1, IAA 1, glutamine 20	IBA 8.9	Acclimatized plantlets	Jones et al., 1990
A. saligna	Shoot tips from mature trees	MS, BA 22 to 40 Elongation: BA 1.3, IAA 1.1	MS, IBA 9.8	Acclimatized plantlets	Barakat & El-Lakany, 1992
A. sclerosperma	Nodal explants from seedlings	MS, BA 6, NAA 1 MR: 8-10	1/2MS, IAA 6	*In vitro* plants	Dhawan, 1993
A. senegal	Nodes from 4-year-old plants	MS (1/2 macron.), Zea 50. MR: 3.3	JN (1/2 macron.); NAA 50	Acclimatized plantlets	Badji et al., 1993
A. seyal	Shoot tips from 6-week-old seedlings	MS, BA 8.9, NAA 0.5 or 2.7. MR: 6.5	1/2MS, IBA 19.6	Acclimatized plants	Al-Wasel, 2000

Table 1 cont'd

A. stenophylla	Epicotyls from 2-week-old seedlings	½ MS salts, BA 1, NAA 1	same	Acclimatized plants	Crawford & Hartney, 1987
A. tortilis sb *raddiana*	Cotyledonary nodes from seedlings	MS, BA 22.2, NAA 0.5. MR: 13-15	1/2MS, IBA 14.7	Acclimatized plantlets	Nandwani, 1995
A. victoriae	Nodal explants from seedlings	MS, BA 6 MR: 9	MS, IBA 3 or 1/2MS NAA 1	*In vitro* plants	Dhawan, 1993

Culture media: AM = Acacia medium (Jones & Smith, 1988); B_5 = Gamborg (1968); JN = Jordan medium (Jordan et al., 1978); LS = Lin & Staba (1961); MS = Murashige & Skoog (1962); NN = Nitsch & Nitsch (1965); QL = Quoirin & Lepoivre (1977) ; White, 1943; WPM = Lloyd & McCown, 1980. Others: glut. = glutamine, spm = spermidine, AC = activated charcoal, CM = coconut milk, MR = multiplication rate. * Followed by JN medium alone.

3. ORGANOGENESIS

Methods of direct or indirect shoot regeneration are important for *in vitro* propagation of these species, and for production of transformed plants, as the introduction of DNA into plant cells implies the posterior differentiation of these cells into a whole plant.

The procedure of *in vitro* plant regeneration has been established for four species: *A. albida, A. auriculiformis, A. mangium* and *A. sinuata* (Table 2). For other species (*A. crassicarpa, A. koa* and *A. nilotica*), only buds or shoots were obtained.

In the case of *A. albida* (Ahée and Duhoux, 1994; Gassama-Dia & Duhoux, 1992), the process is particular, as buds were formed from root suckers in the presence of BA and spermidine. In the other species, the explants used to start the cultures were hypocotyls, cotyledons, shoot tips, parts of mature embryos and stem segments.

Shoot regeneration sometimes occurred directly from the plant tissues, without prior callus formation from cotyledons of *A. auriculiformis, A. crassicarpa* and *A. mangium.* The growth regulators used to induce this direct regeneration were Thidiazuron alone (Quoirin et al., 2000) or combined with an auxin (Kitani and Yasutani, 1997), or combinations of auxin and cytokinin (Skolmen, 1986).

In other cases, adventitious shoots were regenerated after callus formation. This callogenesis was induced by the addition of BA to the basal medium (Ranga-Rao and Prasad, 1991), 2,4-D (Skolmen and Mapes, 1978; Xie and Hong, 2001a; Vengadesan et al., 2000). IAA was also used (Mathur and Chandra, 1983) or a combination of IBA and kinetin (Yanxiu et al., 1990). Shoot regeneration was observed when the calli were cultured on a medium containing BA alone (Skolmen and Mapes, 1978), BA and an auxin (Ranga-Rao and Prasad, 1991, Vengadesan et al., 2000; Yanxiu et al, 1990) or Thidiazuron and an auxin (Kitani and Yasutani, 1997; Toda et al., 1995; Douglas and McNamara, 2000). Here too, some antioxidants have been applied in the culture medium: polyvinylpyrrolidone for *A. auriculiformis* (Ranga Rao and Prasad, 1991) or ascorbic acid for *A. catechu* (Das et al., 1996).

4. SOMATIC EMBRYOGENESIS

Somatic embryogenesis is a potentially efficient means of propagation for superior and genetically engineered forest tree genotypes (Gupta et al., 1991; Jain et al., 1999). For both coniferous and hardwood species, somatic embryogenesis has a number of advantages over other micropropagation techniques, which include very high multiplication rates, potential for scale-up in bioreactors and for delivery to the greenhouse or field as artificial seeds (Merkle, 1995; Jain et al., 2000). In addition, somatic embryos are bipolar propagules that can develop into complete plantlets in a single step, thus circumventing the problem of root induction encountered in regeneration procedures, especially with hardwood perennials (Shahana and Gupta, 2000). The root system of trees developed from somatic embryos is formed by a principal-root which is better than adventitious roots in rendering the adult tree more stable.

The first published results on somatic embryogenesis of legume trees were that of Gharyal and Maheshwari (1981) who described the formation of somatic embryos from hypocotyls of *Albizia lebbeck*. To date, this process has been reported in nine species of acacias: *A. auriculiformis, A. catechu, A. farnesiana, A. koa, A. mangium, A. nilotica, A. schaffneri, A. senegal* and *A. tortilis* (Table 3). Immature zygotic embryos or cotyledons were used as starting material, except for *A. nilotica* where somatic embryos developed from immature endosperm (Garg et al., 1996).

For *A. senegal*, somatic embryogenesis was obtained directly in the presence of zeatin (Shahana and Gupta, 2000). In the other cases, the process was indirect and included the previous formation of a callus that developed in media containing an auxin or a combination of an auxin and a cytokinin: for *A. tortilis*, Sané et al. (2000) used 9.05 µM 2,4-D with or without BA. For *A. farnesiana* and *A. schaffneri*, Ortiz et al. (2000) recommended 9.05 µM 2,4-D and 4.65 µM Kin; for *A. nilotica*, 10 µM 2,4-D, 25 µM BA and 1,000 $mg.L^{-1}$ casein hydrolysate gave the best callus proliferation (Garg et al., 1996); for *A. catechu*, it was 4.6 µM NAA and 21.5 µM Kin (Rout et al., 1995), while for *A. mangium* 11.42 µM IAA and 0.25 µM TDZ was the best hormonal balance (Xie and Hong, 2001a). Salts and vitamins formulas were that of MS medium (for *A. farnesiana, A. mangium, A. nilotica, A. schaffneri* and *A. senegal*), WPM (for *A. catechu*), Von Arnold (Von Arnold and Eriksson, 1981) medium (for *A. tortilis*) or Schenk and Hildebrandt (1972) medium (for *A. koa*). In some cases, the embryos developed after three subcultures on the first medium and when transferred from light to dark conditions (*A. nilotica* and *A. tortilis*).

After the development of the embryogenic callus, a second medium was sometimes necessary to induce the formation of the embryos: a mixture of MS and VA (Von Arnold and Eriksson, 1981) salts supplemented with BA (2.22 µM) and IBA (0.245 µM) for *A. tortilis*. For the other species, the calli differentiated into embryos on the first medium. Embryos maturation occurred on MS medium for most of the species (*A. farnesiana, A. koa, A. mangium, A.schaffneri, A. senegal* and *A. tortilis*). For *A. senegal,* the addition of 0.68 µM glutamine to basal medium was recommended and for *A. mangium* 7.2 to 14.4 µM GA_3 and 3% sucrose. The presence of abscisic acid (ABA) (3.78 to 20 µM) or 5% sucrose at this stage was beneficial for maturation of *A. tortilis, A. farnesiana* and *A. schaffneri* embryos.

The conversion of embryos into plantlets constitutes a limiting step in the process of somatic embryogenesis, sometimes due to a lack of maturation of somatic embryos. Such a conversion was possible in the case of *A. catechu, A. farnesiana, A. schaffneri* and *A. tortilis* and 2 triploid plants of *A. nilotica.* In the other cases, the somatic embryos did not develop into plantlets. The maturation treatment consisted of a combination of BA and Kin for *A. senegal,* 217 µM adenine sulfate for *A. farnesiana* and *schaffneri,* 5.10^4 mg/L sucrose and 35.10^2 mg/L Phytagel for *A. mangium,* 75.10^3 mg/L polyethyleneglycol 4000 and 25.10^2 mg/L charcoal for *A. nilotica,* 2.10^4 mg/L sucrose for *A. catechu* and 5.10^4 mg/L for *A. tortilis.* These results show the importance of osmotic conditions of the medium for maturation and germination of the embryos. PEG was used in association with abscisic acid for proembryos formation in *Araucaria angustifolia* (Astarita and Guerra, 1998), for the maturation of conifer somatic embryos (Attree and Fowke, 1993) and of *Aspidosperma polyneuron* somatic embryos (Ribas et al., 2000). ABA and sucrose seem to have a role in the tolerance of the somatic embryos to hydric stress and in accumulation of storage compounds. ABA would also have an effect on embryo dormancy (Sané et al., 2000).

The percentage of conversion of the embryos into plantlets is very variable according to the species and the treatment: the higher value reported was 69% in *A. farnesiana,* 47% in *A. schaffneri* (Ortiz et al., 2000), 23% in *A. tortilis* (Sané et al., 2000), 11% in *A. mangium* (Xie and Hong, 2001a) and only 2% in *A. nilotica* (Garg et al., 1996).

In some cases, secondary somatic embryogenesis was observed: in *A. catechu* (Rout et al., 1995) and *A. mangium* (Xie and Hong, 2001a).

The results obtained so far for *Acacia* species indicates the recovery of entire plantlets of *A. catechu, A. farnesiana, A. mangium, A. schaffneri* and *A. tortilis.* They illustrate the major limitations to forest tree somatic embryogenesis: low multiplication rates and inability to initiate embryogenic cultures from mature trees. In fact, in all the successful cases reported here, the somatic embryos were regenerated from juvenile tissues and not from mature trees. Thus, in *Acacia* species, some efforts are needed in order to overcome the limitations of this technique.

Table 2. Organogenesis in *Acacia* species (growth regulators concentration in μM).

Species	Origin of explant	Callus medium	Regeneration medium	Rooting medium	Result	Reference
A. albida	Root segments from seedlings	No callus	1/5MS		*In vitro* shoots	Ahée & Duhoux, 1994
A. albida	Root fragments from seedlings	No callus	B5 or 1/2MS, BA 44.4, Spm 0.01	1/2MS IBA 5	Acclimatized plants	Gassama-Dia & Duhoux, 1992
A auriculiformis	Hypocotyl/ Callus	MS modified, BA 4.4, NAA 0.5	NAA 2.7, BA 4.4 Glut 150mg/L	IBA 4.9	Plantlets	Ranga Rao & Prasad, 1991
A. auriculiformis	Cotyledons	No callus	MS, TDZ 0.01, NAA, 2,4-D or IAA 0.001		Shoots	Kitani & Yasutani, 1997
A. caven	Embryonic axes	No callus	MS; BA 22 or TDZ 4.5		Shoots	Jordan, 1998
A. crassicarpa	Cotyledons	No callus	MS salts, NN vit., TDZ 2.3		Buds	Quoirin et al., 2000
A. koa	Various	No callus	MS, IAA, NAA, CPA, 2,4-D, BA, 2iP, Kin, Zea		Shoots	Skolmen, 1986
A. koa	Shoot tips	SH, 2,4-D 0.9	MS, BA 4.4	IBA 1.5	Rooted shoots	Skolmen & Mapes, 1978

Table 2 cont'd

A. mangium	Cotyledons		MS, TDZ 0.01 and NAA, 2,4-D or IAA 0.001		*In vitro* shoots	Kitani & Yasutani, 1997
A. mangium	Hypocotyl	MS	NAA, BA and TDZ or CPPU		Acclimatized plants	Toda et al., 1995
A. mangium	Cotyledonary nodes		DKW, 2,4-D and Kin or TDZ and AIA	NAA (10.7) Kin (2.3)	Buds	Douglas & McNamara, 2000
A. mangium	Various types from mature embryos	MS 2,4-D 9.05 and Kin 13.9	Nodules: TDZ 4.5, IAA 1.4 Shoots: TDZ 0.0045, GA_3 7.2		Plantlets	Xie & Hong, 2001b
A. nilotica	Young stem segments	MS IAA 2.8-5.7	IAA or NAA	same	Shoots	Mathur and Chandra, 1983
A. salicina	Cotyledon and hypocotyl	MS, IBA 2.5, Kin 9.4	NAA 2.7, BA 4.5-9.0	IBA 2.5		Yanxiu et al., 1990
A. sinuata	Hypocotyl /callus	MS, 2,4-D 6.8, BA 2.2	1/2MS, CW 10%, BA 13.2, IAA 3.4	IBA 7.4	20 plantlets/ hypocotyl/ 4 months	Vengadesan et al., 2000
A. visco	Embryonic axes	No callus	MS; BA 22 or TDZ 2.3	NAA or IBA 0.1 to 0.5	In vitro plantlets	Jordan, 1998

Culture media: B5 = Gamborg et al. (1968); DKW = Driver & Kuniyuki (1984); LS = Lin & Staba (1961); MS = Murashige & Skoog (1962); NN = Nitsch & Nitsch (1965); QL = Quoirin & Lepoivre (1977); Others: AC = activated charcoal; CW= coconut water ; glut = glutamine; spm = spermidine.

Table 3. Somatic embryogenesis in *Acacia* species.

Species	Origin of explant	First and second media	Germination medium	Result	Reference
A. catechu	Immature cotyledons	WPM, Kin, NAA, proline	1/2MS 2% sucrose	Acclimatized plants	Rout et al., 1995
A. farnesiana	Immature zygotic embryo	MS, 2,4-D and Kin	MS + adenine sulfate	Acclimatized plants	Ortiz et al., 2000
A. koa	Cell suspension	SH, 2,4-D, CPA, Kin	MS, CW 1%	Embryo-like organized structures	Skolmen, 1986
A. mangium	Immature zygotic embryo	MS, TDZ, IAA + amino acids	1/2MS, GA_3, 3% sucrose	Acclimatized plants	Xie & Hong, 2001a
A. nilotica	Immature endosperm	MS, 2,4-D, BAP, CH	Modified MS + PEG and charcoal	Triploid plantlets	Garg et al., 1996
A. schaffneri	Immature zygotic embryo	MS, 2,4-D and Kin	MS, adenine sulfate	Acclimatized plants	Ortiz et al., 2000
A. senegal	Immature cotyledons	MS, ZEA	MS, BA, Kin	Mature embryos	Shahana & Gupta, 2000
A. tortilis	Immature zygotic embryo	(1) VA, 2,4-D or 2,4-D and BA (2) 1/2MS, BA, IBA	1/2MS, sucrose	Plantlets	Sané et al., 2000

Culture media: SH = Schenk and Hildebrandt (1972); VA = salts and vitamins of Von Arnold and Erikson (1981); WPM = Woody Plant Medium, Lloyd and McCown (1980). Others: CH = casein hydrolysate; CPA = p-chlorophenoxyacetic acid; CW = coconut water.

5. CONCLUSION AND PROSPECTS

Acacias, among other woody legume species, are especially important in tropical and semiarid ecosystems. They are multipurpose trees producing food for animals and wood for construction, paper and fuel. They also positively influence soil nitrogen status (Trigiano et al., 1992).

Micropropagation techniques have been used on a large-scale for propagation of elite material of tree species like *Eucalyptus* and *Pinus*. These techniques constitute an important tool for tree breeding programmes. In *Acacia* species, progress has recently been achieved and the micropropagation of fifteen species has been reported. However, further research is needed to develop methods aiming at restoring juvenility to superior trees and establishing high rates of multiplication and rooting. There are few reports on field trials with these species. The best results have been obtained for *A. mangium*. On the other hand, the acacias, like other tree species, show low shoot regeneration ability. Some positive results were obtained in somatic embryogenesis of nine species of *Acacia* , indicating the difficulties encountered to implement this process.

Considering the actual problems of deforestation and need of fuel wood for growing populations of poor countries, further development of *in vitro* techniques for *Acacia* species should greatly be encouraged.

6. References

Ahée J. & Duhoux E. Root culturing of *Faidherbia=Acacia albida* as a source of explants for shoot regeneration. Plant Cell Tiss Org Cult 1994; 36:219-225.

Al-Wasel A. Micropropagation of *Acacia seyal* Del. *in vitro*. J Arid Environm 2000; 46:425-431.

Astarita L.V. & Guerra M.P. Early somatic embryogenesis in *Araucaria angustifolia*. Induction and maintenance of embryonal-suspensor mass cultures. Rev Bras Fisiol Veg 1998; 10:113-118.

Attree S.M. & Fowke I.C. Embryogeny of gymnosperms advances in synthetic seed technology of conifers. Plant Cell Tiss Org Cult 1993; 35:1-35.

Badji S., Mairone Y., Ndiaye I., Merlin G., Danthu P., Neville P. & Colonna J.P. In vitro propagation of the gum Arabic tree (*Acacia senegal* (L.) Willd.). 1. Developing a rapid method for producing plants. Pl Cell Rep 1993; 11:629-633.

Barakat M.N. & El-Lakany M.H. Clonal propagation of *Acacia saligna* by shoot tip culture. Euphytica 1992; 59:103-107.

Beck S.L., Dunlop R. & van Staden J. Micropropagation of *Acacia mearnsii* from *ex vitro* material. Plant Growth Reg 1998a; 26:143-148.

Beck S.L., Dunlop R & van Staden J. Rejuvenation and micropropagation of adult *Acacia mearnsii* using coppice material. Plant Growth Reg 1998b; 26:149-153.

Bon M. C., Bonal D., Goh D.K. & Monteuuis O. Influence of different macronutrient solutions and growth regulators on micropropagation of juvenile *Acacia mangium* and *Paraserianthes falcataria* explants. Plant Cell Tiss Org Cult 1998; 53:171-177.

Correia D. & Graça M.E.C. *In vitro* propagation of black wattle (*Acacia mearnsii* de Wild). IPEF, Piracicaba 1995; 48/49: 117-125.

Crawford D. F. & Hartney V. J. Micropropagation of *Acacia mangium* and *Acacia stenophylla.* In: Australian acacia in developing countries. J.W. Turnbull, ed. Proceedings of international workshop. ACIAR-Proc-Ser., Australian Center for International Agricultural Research, 1987, pp. 64-65, Canberra.

Dart P., Umati-Garcia M. & Almendras A. Role of symbiotic associations in nutrition of tropical acácias. Advances in Tropical Acacia Research, J.W. Turnbull, ed. ACIAR Proceedings, Proc. Int. Workshop. Bangkok, Thailand: 1991 February 11-15, pp. 13-19. Canberra.

Darus H.A. Micropropagation of *Acacia mangium* from aseptically germinated seedlings. J Trop Forest Sc 1989; 3:204-208.

Darus H.A. Multiplication of *Acacia mangium* by stem cuttings and tissue culture techniques. Advances in Tropical Acacia Research, J.W. Turnbull, ed. ACIAR Proceedings, Proc. Int. Workshop. Bangkok, Thailand: 1991a February 11-15, pp. 32-35. Canberra.

Darus H.A. Micropropagation techniques for *Acacia mangium* x *A. auriculiformis.* In: L.T. Carron & K.M. Aken, ed. ACIAR Proceedings, 1991b, No 37, pp. 119-121.

Das P., Samantaray, S. & Rout, G.R. *In vitro* propagation of *Acacia catechu*, a xerophilous tree. Pl Tissue Cult 1996; 6:117-126.

Dewan A., Nanda K. & Gupta S.C. *In vitro* micropropagation of *Acacia nilotica* subsp. *indica* Brenan via cotyledonary nodes. Pl Cell Rep 1992; 12:18-21

Dhawan V. "Micropropagation of N-fixing trees". In *Micropropagation of Woody Plants*, M. R. Ahuja, ed. Dordrecht, The Netherlands: Kluwer Acad. Publishers, 1993.

Di Michele M.N. & Bray L. " Multiplication végétative *in vitro* d'*Acacia flava* syn. *Ehrenbergiana*". In *Quel avenir pour l'amélioration des plantes*? AUPELF-UREF, ed. Pp.195-204. Paris: John Libbey Eurotext: Paris, 1994.

Dommergues Y., Duhoux E. & Diem G.H. "*Acacia*" In *Les Arbres Fixateurs d'Azote,* Montpellier, France: CIRAD, Editions Espaces, Rome, Italy: FAO, Paris, France: IRD, 1999.

Douglas G. C. & McNamara J. Shoot regeneration from seedling explants of *Acacia mangium* Willd. In Vitro Cell Dev Biol - Plant 2000; 36:412-415.

Driver J.A. & Kuniyuki A.H. *In vitro* propagation of paranox walnut rootstock. HortSci 1984; 19:507-509.

Duhoux E. & Davies D. Caulogenèse à partir des bourgeons cotylédonaires *d'Acacia albida* et influence du saccharose sur la rhizogenèse. J Plant Physiol 1985; 121:175-180.

Duhoux E., Galiana A., Ahée J. & Franche C. "Applications des cultures *in vitro* dans le genre *Acacia*". In *L'acacia au Sénégal*. Claudine Campa, Claude Grignon, Mamadou Gueye, Serge Hamon, ed., Paris: Editions de l'ORSTOM, 1998.

Galiana A., Goh D., Chevallier M-H., Gidiman J., Môo H., Hattah M. & Japarudin Y. Micropropagation of *Acacia mangium* x *Acacia auriculiformis* hybrids in Sabah. Bois Forêts Trop 2002 (in press)

Galiana A., Tibok A. & Duhoux E. *In vitro* propagation of the nitrogen-fixing tree-legume *Acacia mangium* Willd. Plant and Soil 1991a; 135:151-159.

Galiana A., Tibok A. & Duhoux E. Nitrogen-fixing potential of micropropagated clones of *Acacia mangium* inoculated with different *Bradyrhizobium* spp. strains. Plant and Soil 1991b; 135:161-166.

Gamborg O.L., Miller R.A. & Ojima K Nutrient requirements of suspension cultures of soybean root cells. Exp Cell Res 1968; 50:151-158 .

Garg L., Bhandari N.N., Rani V. & Bhojwani S.S. Somatic embryogenesis and regeneration of triploid plants in endosperm cultures of *Acacia nilotica*. Plant Cell Rep 1996; 15:855-858.

Gassama Y.K. *In vitro* culture and improvement of symbiosis using mature *Acacia albida*. International Foundation for Science 1989: 286-290.

Gassama Y.K. & Duhoux E. Micropropagation de *l'Acacia albida* Del. (Leguminosae) adulte. Bull. I.F.A.N. 1986-87; 46:314-320.

Gassama-Dia Y.K. & Duhoux E. Régénération de bourgeons à partir de culture de racines d'*Acacia albida*. In: *Interactions Plantes-Microorganismes* IFS (International Foundation for Science) Dakar: 1992

Gharyal P.K. & Maheshwari S.C. *In vitro* differentiation of somatic embryoids in a leguminous tree – *Albizia lebbeck* L. Naturwissenschaften 1981; 68:379-380

Gupta P.K., Timmis R. & Mascarenhas, A. F. Field performance of micropropagated forestry species. In Vitro Cell. Dev. Biol. 1991, 27P:159-164.

Hamzah M.B., Alang Z.C. & Salekan, J. *In vitro* propagation of *Acacia mangium* from young seedlings. Proceedings of the seminar on tissue culture of forest species (Rao, A.N. and Aziah Mohe, Y., eds), 1989.

Huang F. H., Al-Khayri J. M. & Gbur E. E. Micropropagation of *Acacia mearnsii*. In Vitro Cell Dev Biol - Plant 1994; 30P: 70-74.

Ide Y., Watanabe Y. & Ikeda H. Tissue culture of *Acacia auriculiformis* using the aseptically germinated seedlings. J Japan Forestry Soc 1994; 76:576-583.

Jain S. M., Gupta P. K. & Newton R. J. *Somatic embryogenesis in woody plants*. Vol. 4-5. Kluwer Academic Publishers, The Netherlands, 1999.

Jain S. M., Gupta P. K. & Newton R. J. *Somatic embryogenesis in woody plants*. Vol. 6. Kluwer Academic Publishers, The Netherlands, 2000.

Jones C. Getting started in micropropagation of Tasmanian Blackwood (*Acacia melanoxylon*). Combined Proceedings Intern Plant Propagator's Society 1986; 36: 477-481.

Jones C. & Smith D. Effect of 6-benzylaminopurine and 1-naphthylacetic acid on *in vitro* axillary bud development of mature *Acacia melanoxylon*. Combined Proceedings Intern Plant Propagator's Society 1988; 38:389-393.

Jones C., Smith D., Gifford H. & Nicholas I. Field trial results of *Acacia melanoxylon* from tissue culture. Combined Proceedings Intern Plant Propagator's Society 1991; 41: 116-119.

Jones T.C., Batchelor C.A. & Harris P.J.C. *In vitro* culture and propagation of *Acacia* species (*A. bivenosa, A. holosericea, A. salicina, A. saligna,* and *A. sclerosperma*). International Tree Crops J 1990; 6:183-192.

Jordan M. *In vitro* regeneration of some leguminous multipurpose trees from seedling explants. Phyton 1998; 63:249-256.

Jordan M., Apablaza G. & Lippi P. Ciencia e Investigación Agraria 1978; 5:207-211.

Kaur K. & Kant U. Clonal propagation of *Acacia catechu* Willd. by shoot tip culture. Plant Growth Regul 2000; 31:143-145.

Kaur K., Verma B. & Kant U. Plants obtained from the Khair tree (*Acacia catechu* Willd.). Plant Cell Rep. 1998; 17:427-429.

Kitani S. & Yasutani I. Callus cultures of Acacia species: initiation, growth optimisation and organogenesis. Proc. Inter. Workshop BIO-REFOR, IUFRO/SPDC, 1997; Brisbane.

Lin M.L. & Staba J. Peppermint and spearmint tissue culture. I. Callus formation in submerged culture. Lloydia 1961; 24:139-145.

Lloyd G. & McCown B. Commercially-feasible micropropagation of Mountain Plant laurel, *Kalmia latifolia*, by use of shoot tip culture. Proc Int Prop Soc 1980; 30:421-427.

Mathur I. & Chandra N. Induced regeneration in stem explants of *Acacia nilotica*. Current Sci 1983; 52:882-883.

Merkle S.A. Strategies for dealing with limitations of somatic embryogenesis in hardwood trees. Plant Tissue Cult and Biotechn 1995, 1:112-121.

Meyer H.J. & van Staden J. Regeneration of *Acacia melanoxylon* plantlets *in vitro*. South African J Bot 1987; 53:206-209.

Mittal A., Agarwal R. & Gupta S.C. *In vitro* development of plantlets from axillary buds of *Acacia auriculiformis* a leguminous tree. Plant Cell Tiss Org Cult 1989; 19:65-70.

Monteuuis O. & Bon M.-C. Influence of auxin and darkness on *in vitro* rooting of micropropagated shoots from mature and juvenile *Acacia mangium*. Plant Cell Tiss Org Cult 2000; 63:173-177.

Murashige T. & Skoog F. A revised medium for rapid growth and bioassays with tobacco tissue culture. Physiol Plant 1962; 15: 473-497

Nandwani D. In vitro micropropagation of a tree legume adapted to arid lands *Acacia tortilis* subsp. *raddiana*. Ann Sc Forest 1995; 52:183-189.

National Academy of Sciences. Firewood crops: shrubs and tree species for energy production Washington, D.C: US Natl Acad Press, 1980.

National Research Council. *Mangium and other fast growing acacias for the humid tropics*. Washington, DC: US Natl Acad Press, 1983.

Nitrogen Fixing Tree Highlights. 1985, Nitrogen Fixing Tree Association, Waimanelo, Hawaii.

Nishikawa H. & Seido K. *In vitro* mass propagation fron aseptically germinated seedlings of *Acacia crassicarpa*, tropical legume tree. Bull Yamanashi For Res Inst 2001; 22:10.

Nitsch J.P. & Nitsch C. Néoformation de fleurs *in vitro* chez une espèce de jours courts: *Plumbago indica* L. Ann Physiol Vég 1965; 7:251-256.

Ortiz B.O.C., Reyes M.E.P. & Balch E.P.M. Somatic embryogenesis and plant regeneration in *Acacia farnesiana* and *A. schaffneri*. In Vitro Cell Dev Biology- Plant 2000; 36:268-272.

Quoirin M., Galiana A., Goh D.K.S., Limanton A., Gratio V., Ahée J., Rio M., Oliveira D., Duhoux E. & Franche C. "Progress towards the genetic transformation of four tropical *Acacia* species: *Acacia mangium, Acacia crassicarpa, Acacia mearnsii* and *Acacia albida*". In: *Molecular Biology of Woody Plants*.Vol. 2, S. M. Jain and S. C. Minocha, ed. Dordrecht, The Netherlands: Kluwer Acad. Publ., 2000.

Quoirin, M. & Lepoivre, P. Etude de milieux adaptés aux cultures *in vitro* de *Prunus*. Acta Hort 1977; 78:437-442.

Quoirin M., Silva M. da, Martins K. G. & Oliveira D. E. de. Multiplication of juvenile black wattle (*Acacia mearnsii* de Wild) by microcuttings. Plant Cell Tiss Org Cult 2001; 66:199-205.

Rajadurai D., Rao A.N. & Loh C.S. *In vitro* culture studies on two leguminous species. In: Proceedings of the seminar on tissue culture of forest species. A.N. Rao and Y. Aziah Mohe, ed, 1989, 104-128.

Ranga Rao G.V. & Prasad M.N.V. Plantlet regeneration from the hypocotyl callus of *Acacia auriculiformis* – Multipurpose Tree Legume. J Plant Physiol 1991; 137:625-627.

Resende M.D.V., Souza S.M., Higa A.R. & Stein P.P. Estudos da variação genética e métodos de seleção em testes de progênies de *Acácia mearnsii* no Rio Grande do Sul. Bol. Pesq. Fl., Colombo, Brazil,1991; 22/23:45-.

Ribas L.L.F., Guerra M. P., Zanette F. & Kulchetscki L. "Somatic embryogenesis in *Aspidosperma polyneuron* Mull. Arg.". In: *Somatic Embryogenesis in Woody Plants*, Vol. 6, pp. 509-537. S. M. Jain, P. K. Gupta and R. J. Newton, ed. Dordrecht, The Netherlands: Kluwer Acad. Publ., 2000.

Rout G.R., Samantaray S. & Das P. Somatic embryogenesis and plant regeneration from callus culture of *Acacia catechu* – a mutipurpose leguminous tree. Plant Cell Tiss Org Cult 1995; 42:283-285

Ruffoni B., Massabo F., Costantino C., Arena V. & Damiano C. Micropropagation of *Acacia "mimosa"*. Acta Hortic 1992; 300:95-102.

Ruredzo T.J. & Hanson J. Plant recovery from seedling-derived shoot tips of *Faidherbia albida* grown *in vitro*. Agroforest Syst 1993; 22:59-65.

Saito Y., Kojima K., Ide Y. & Sasaki S. In vitro propagation from axillary buds of *Acacia mangium*, a legume tree in the tropics. Plant Tiss Cult Letters 1993; 10:163-168.

Sané D., Borgel A., Verdeil J.L. & Gassama-Dia Y.K. Plantlets regeneration via somatic embryogenesis in immature zygotic embryos calli from a tree species adapted to arid lands: *Acacia tortilis* subsp. *raddiana* (Savi.) Brenan. Acta Bot Gallica 2000; 147:257-266.

Schenk R.V. & Hildebrand A.C. Medium and techniques for induction and growth of monocotyledoneous and dicotyledoneous plant cell cultures. Can J Bot 1972; 50:199-204.

Semsuntud N. & Nitiwattanachai W. Tissue culture of *Acacia auriculiformis*. Advances in Tropical Acacia Research, J.W. Turnbull, ed. ACIAR Proceedings, Proc. Int. Workshop. Bangkok, Thailand: 1991 February 11-15, pp. 13-19. Canberra.

Setiawan I. & Umboh M.I. Growth and rooting system of *Acacia mangium* obtained by tissue culture. Biotropia; 1990; 4:1-8.

Shahana S. & Gupta S.C. "Somatic embryogenesis in a leguminous tree - *Acacia senegal* (L) Willd". In: *Somatic Embryogenesis in Woody Plants*, Vol. 6, pp. 539-552. S. M. Jain, P. K. Gupta and R. J. Newton, ed. Dordrecht, The Netherlands: Kluwer Acad. Publ., 2000.

Singh H.P., Sanjay-Singh, Saxena R.P., Singh R.K. & Singh S. *In vitro* bud break in axillary nodal segments of mature trees of *Acacia nilotica*. Indian J Pl Physiol 1993; 36:21-24.

Skolmen R.G. "Acacia (*Acacia koa* Gray)". In *Biotechnology in Agriculture and Forestry*. Vol. 1: Trees, pp. 375-384. Y.P.S.Bajaj, ed. Berlin: Springer-Verlag, 1986.

Skolmen R.G. & Mapes M.O. Aftercare procedure required for field survival of tissue culture propagated *Acacia koa*. In "Intern Plant Propagator's Society Combined Proceedings". 1978; 28:156-164.

Tassin P. Mimosacées invasives dans l'Océan Indien Occidental. Le Flamboyant 1999; 51:22-24.

Toda T., Tajima M. & Brini P.B. Tissue culture of *Acacia mangium, A. auriculiformis* and their hybrid. Bull Nation For Tree Breed Center 1995; 13: 157-165.

Trigiano R.N., Geneve R.L., Merkle S.A. & Preece J.E. Tissue and Cell Cultures of Woody Legumes. Hortic Rev 1992; 14:265-331.

Tsai L.M. Studies on *Acacia mangium* in Kemasul Forest, Malaysia: I. Biomass and productivity. J Trop Ecol 1988; 4:293-302.

Turnbull J.W., Midgley S.J. & Cossalter C. Tropical Acacias planted in Asia: an overview. In: Turnbull, J.W., Crompton, H.R. and Pinyopusarerk, K. (Editors), Recent Developments in Acacia Planting, Proceedings of the International Workshop. ACIAR-Proc. No. 82 1997. Australian Center for International Agricultural Research, p. 14, Canberra.

Vaidyaratnam P.S. 1994. Indian Medicinal Plants: a Compendium of 500 Species, Vol.1. Orient Longman Ltd., Madras pp.33-35.

Vassal J. "Acacieae". In *Advances in legume systematics*. Part I. R.M. Polhill & P.H. Raven, ed., Kew Royal Botanical Gardens, pp. 169-171, 1981.

Vengadesan G., Ganapathi A., Prem Anand R. & Ramesh Anbazhagan V. *In vitro* organogenesis and plant formation in *Acacia sinuata.* Plant Cell Tiss Organ Cult 2000; 61: 23-28.

Von Arnold S. & Eriksson T. A revised medium for growth of pea mesophyll protoplasts. Physiol Plant 1981; 39:257-260.

White P.R. A handbook of plant tissue culture. Lancaster: Cattell. 1943.

Williams R.R., Taji A.M. & Bolto J.A. Specificity and interaction among auxins, light and pH in rooting of Australian woody species *in vitro*. Hort Sci 1985; 20:1052-1053.

Xie D. & Hong Y. Regeneration of *Acacia mangium* through somatic embryogenesis. Plant Cell Rep. 2001a; 20:34-40.

Xie D. & Hong Y. *In vitro* regeneration of *Acacia mangium* via organogenesis. Plant Cell Tiss Organ Cult 2001b; 66:167-.173.

Yanxiu Z., Dunyi Y., Harris P.J.C. In vitro regeneration of plantlets from explants and callus of *Acacia salicina.* Nitrogen Fixing Tree Res Rept 1990; 8:113-115.

Yap D.K. Introduction of *Acacia* species to Peninsular Malaysia. In: Australian acacia in developing countries. J.W. Turnbull, ed. Proceedings of international workshop. ACIAR-Proc-Ser. Australian Center for International Agricultural Research, 1987, pp. 151-153, Canberra.

10. *IN VITRO* PROPAGATION OF *AEGLE MARMELOS* (L.) CORR., A MEDICINAL TREE

S. Arumugam*, A.S. Rao **and M.V. Rao***
**Department of Biotechnology, J.J.College, Pudukottai. ** Department of Biotechnology, ***Department of Plant Science, School of Life Sciences, Bharathidasan University, Tiruchirapalli - 620 024, Tamilnadu, India. ***Email: mvrao@bdu.ernet.in*

CONTENTS

S.M. Jain and K. Ishii (eds.), Micropropagation of Woody Trees and Fruits, 269–315.

Key words: *Aegle marmelos*, micropropagation, direct organogenesis, indirect organogenesis.

1. INTRODUCTION

Plants are the resources of foods, pharmaceuticals and pesticides. Plant based medicines have served the human race over the ages for various ailments and number of medicinal plants have become important raw material for pharmaceutical industries. The great popularity of herbal drugs has to do with their effectiveness, minimal side effects in clinical experience and moderate prices. The use of medicinal plants in the health care management programme has been recognized internationally. World Health Organization (WHO) has estimated that the global market for medicinal herbs and herbal products is about 62 billion dollars and by the year 2050 will reach 5 trillion dollars. Therefore, there is a need to make an intensive study on medicinal and aromatic plants for their genetic improvement, conservation and cultivation methods.

1.1. Botany and Genetics

Aegle marmelos (L.) Corr., is an important ayurvedic medicinal tree, belongs to the family Rutaceae. It is known as bael tree and holy fruit tree in English. It is a medium sized armed deciduous tree up to 8.0 m tall with straight, sharp, axillary thorns and yellowish brown shallowly furrowed corky bark; leaves trifoliate, aromatic, alternate, leaflets ovate or ovate – lanceolate, crenate, pellucid – punctate, the lateral subsessile and the terminal long petioled; flowers greenish white, sweet scented, in axillary panicles; fruit globose, woody berry with yellowish rind; seeds numerous, oblong, compressed, embedded in orange brown sweet gummy pulp. The tree flowers during May – July. The fruits require a year for ripening. It yields on an average 300 – 400 fruits (200 – 250 kg) per tree. Twelve varieties of bael tree have been identified so far in India. Some have very strong nauseating odor whereas some are mild and sweet scented (Chadha, 1985).

The bael tree is a diploid species with a somatic chromosome number 2n = 18 (Banerji and Pal, 1957; Raghavan, 1957; Nanda, 1962; Sanjappa, 1979) and 36 (Mehra and Khosla, 1973).

1.2. Economic Importance

The bael tree is commonly planted near Hindu temples as a sacred tree. The root is an ingredient of 'dasamula' (ten roots), a medicine commonly used by the ayurvedic practitioners. The roots are varied in their taste and flavour viz., sweet, astringent, bitter and febrifuge. They are useful in diarrhoea, dysentery, dyspepsia, stomachalgia, cardiopalmus, vitiated conditions of *vata*, seminal weakness, uropathy, vomiting, intermittent fever, swelling and gastric irritability in infants. The leaves are astringent, laxative, febrifuge and expectorant, and are useful in opthalmia, deafness, inflammations, catarrh, diabetes and asthmatic complaints. The unripe fruits are bitter, acrid, sour, astringent, digestive and stomachic, and are useful in diarrhoea, dysentery, sweet, aromatic, cooling, febrifuge, laxative and tonic, and are good for heart and brain in dyspepsia (Warrier *et al.*, 1996). The alkaloid 'aegeline' present in the leaf is a potent anti-asthmatic agent (Haravey, 1968). It was also reported that leaf extract of *Aegle marmelos* has been used as antispermatogenic (Sur *et al.*, 1999), to cure jaundice (Gupta *et al.*, 1999) and to enhance wound-healing activity (Jaswanth *et al.*, 2001). Gummy substance around the seeds serves as an adhesive, as a varnish for pictures and adds brilliancy to water-color paints and wood is used as timber (Ambasta, 2000).

1.3. Geographical Distribution

The bael tree is a native of Indo-Malayan region (Hooker, 1975). It is being cultivated in South East Asia, especially in India, Pakistan, Bangladesh, Srilanka, Burma and Thailand (Islam *et al.*, 1996c). It was also introduced into tropical countries of Western Africa and South America for afforestation purposes (Arya, 1986; Zaman, 1988). It can withstand various types of soils, climatic conditions and a pH range of 5-10 (Chadha, 1985).

1.4. Diseases and Pests

There are few reports on diseases and pests of bael tree. Diseases such as shoot hole and fruit canker was caused by *Xanthomonas bilvae*; leaf spot disease by *Cercospora aegle – marmeli*; green rot by *Aspergillus nidulans* and soft fruit rot by *Fusarium solani*. The young leaves and shoots are eaten by caterpillars of *Papilio demoleus* and *P. erithonius*. The beetles of *Amblirrhymus poricollis, Aspidiotus orientalis, Lecanium colemanii, L.viride and Myllocerus discolor* feed on the plant sap and defoliate the trees. The larvae of *Argyroploce illepida, Chaetodacus zonatus* and *Euzophera niveicostella* bore the fruit and feed on the fruit pulp (Chadha, 1985).

1.5. Conventional Practices for Propagation

Aegle marmelos is an out breeder, and is routinely propagated by seeds for cultivation (Singh *et al.*, 1976). Seeds have short viability and are prone to insect attack. Seedlings from seeds show slow growth and are liable to diseases and pests in the initial stages. Vegetative propagation through root suckers is slow, difficult and cumbersome (Ray and Chattarjee, 1996).

1.6. Non-Conventional Approaches (*In Vitro* Propagation)

Conventional methods of propagation either through seed or by root cuttings and layers are associated with problems like seasonal seed production (takes one year), pest problems and low percentage of seed germination. These problems can overcome by non-conventional approaches like *in vitro* propagation methods. These are basically three types – micropropagation, organogenesis and somatic embryogenesis. Micropropagation involves direct proliferation of axillary bud and/or apical bud into plantlet without any formation of callus. Whereas organogenesis involves either direct formation of organs without callogenesis (direct organogenesis) or callus induction is primary aspect followed by morphogenesis (indirect organogenesis). The third method, somatic embryogenesis, involves the development of embryos from somatic cells and results in the production of complete somatic seedling with the potential to grow into a whole plant. The latter two methods i.e. organogenesis and somatic embryogenesis carry some risk of genetic variability, whereas in micropropagation true to type clones will be produced without any genetic changes.

Singh *et al.* (1976) made first attempt on *in vitro* propagation of bael tree, however, failed to regenerate plants from nucellar tissue. Later, Arya *et al.* (1981) regenerated plants from cotyledon and hypocotyl explants through organogenesis, but the regeneration frequency was very low. Further, several researchers have made attempts to produce *in vitro*-propagated plants through micropropagation, organogensis and somatic embryogenesis (Table 1). Somatic embryogenesis in *Aegle marmelos* was reviewed by Arumugam and Rao (2000).

2. INITIATION OF *IN VITRO* PROPAGATION

2.1. Explant Sterilization

Ripened fruits were collected from approximately 40-year-old mature trees (Plate1.a and b), and fruits were immediately scarified and washed thoroughly in running tap water for 45minutes to remove outer mucilaginous sheaths. Later they were surface sterilized with ethanol (70% v/v) for one minute and disinfected with mercuric chloride (0.1% w/v) for 5 minutes and washed 5 times with sterile distilled water. The sterilized seeds were germinated in Erlenmeyer conical flasks (250 ml) containing sterile moist cotton. Initial incubation was done in dark for 5 days, subsequently shifted to cool white fluorescent light at 80 $\mu Em^{-2}s^{-1}$ (16 hrs/day) and 25±2°C temperature condition. After 21 days of seed germination, cotyledonary node (CN),

Table 1. In vitro propagation in Aegle marmelos

Explant	Medium	Result/Response	Author/s
Micropropagation			
Cotyledonary node	MS+BA (3.0 mg/l)	Multiple shoot formation	Arumugam and Rao, 1996a
Shoot	MS+IBA (4.0 mg/l)	Rooting	
Shoot bud	MS+BA (0.5 mg/l)	Shoot elongation	Hazarika *et al.*, 1996a
Shoot	½ MS+IBA (0.5 mg/l)	Rooting	
Axillary bud	MS+BA (2.5 mg/l)+IAA (1.0 mg/l)	Shoot formation	Ajith Kumar and Seeni, 1998
Shoot	1/2MS+IAA (0.5 mg/l) or IBA (10.0 mg/l)	Rooting	
Cotyledonary node, axillary bud and shoot tip	MS+BA (3.0 mg/l)+KN (0.5 mg/l)+ NAA (0.01 mg/l)	Shoot bud initiation and multiple shoot formation	Arumugam, 1998
Shoot	MS+BA (0.1 mg/l)+GA_3 (1.0 mg/l)	Shoot elongation	
Shoot	1/2MS+IBA (1.0 mg/l)+KN (0.1 mg/l)	Rooting	
Direct Organogenesis			
Root segments	MS+BA (0.5 mg/l)+KN (0.5 mg/l)	Shoot regeneration	Rekha Bhati *et al.*, 1992
Shoots	1/4MS+NAA (0.5 mg/l) or IAA (0.5 mg/l)	Rooting	
Nucellus tissue	MS+BA (1.0 mg/l)+NAA (0.5 mg/l)	Adventious bud formation	Hossain *et al.*, 1993
Adventious buds	MS+BA (0. 1 mg/l)	Shoot elongation	
Elongated shoot	1/2MS+IBA (1.0 mg/l)	Rooting	
Leaf	MS+BA (1.5 mg/l)+IAA (0.5 mg/l)	Shoot bud induction	Islam *et al.*, 1993
Shoots	MS+BA (0.5 mg/l)	Shoot elongation	
Elongated shoot	1/2MS+IBA (0.1 mg/l)	Rooting	
Cotyledon	MS+BA (2.0 mg/l)	Shoot bud initiation	Hossain *et al.*, 1994a
Shoot bud	MS+BA (2.0 mg/l)+IAA (0.2 mg/l)	Shoot regeneration	
Shoot	MS+KN (0.5 mg/l)+GA_3 (0.1 mg/l)	Shoot elongation	
Elongated shoot	1/2MS+IBA (0.5 mg/l)	Rooting	

Contd..

Radicle tissue	MS+BA (1.0 mg/l)+NAA (0.2 mg/l)	Shoot bud induction	Islam *et al.*, 1996a
Shoot	MS+BA (0.1 mg/l)	Shoot elongation	
Shoot	MS+IBA (25.0 mg/l) for 1 week and transferred to MS basal medium	Rooting	
Mature seed	MS+BA (1-2 mg/l)	Shoot bud induction	Islam *et al.*, 1996b
Shoot bud	MS+IAA (0.5 mg/l)	Shoot proliferation	
Shoot	MS+IBA	Rooting	
Epicotyl	MS+BA (1.0 mg/l)+KN (0.5 mg/l)+NAA 0.01 mg/l)+ IAA (0.1 mg/l)+GA_3 (3.0 mg/l)+ADS (10.0 mg/l)	Multiple shoot formation	Arumugam and Rao, 1996b
Shoot	1/2MS+IBA (1.0 mg/l)+KN (0.01 mg/l)	Rooting	
Hypocotyl, cotyledon, epicotyl, mature leaflet, stem and root	MS+BA (2.0 mg/l)+KN (1.0 mg/l)+NAA (0.1 mg/l)	Shoot formation	Arumugam, 1998
Shoot	MS+BA (0.1 mg/l)+GA_3 (0.1 mg/l)	Shoot elongation	
Shoot	1/2MS+IBA (1.0 mg/l)+ KN (0.1 mg/l)	Rooting	
Indirect Organogenesis			
Hypocotyl and stem Explants	MS+KN (0.5 mg/l)+NAA (10.0 mg/l)	Callus induction	Arya *et al.*, 1981
Callus	MS+KN (5.0 mg/l)+NAA (0.5 mg/l)+ L-tyrosine (50-200 mg/l)	Shoot regeneration	
Shoots	MS+KN (5.0 mg/l)+IBA (0.5 mg/l)	Rooting	
Axillary buds and hypocotyls	MS+2,4-D+NAA (5.0 mg/l)	Callus induction	Arya and Shekawat, 1986
Callus	MS+KN (1.0 mg/l)+BA (0.5 mg/l)	Multiple shoot formation	
Multiple shoots	White's medium+IBA (0.5 mg/l)	Rooting	
Stem	MS+2,4-D+NAA	Callus induction	Varghese *et al.*, 1993
Callus	MS+KN (1.0 mg/l)+NAA (5.0 mg/l)	Meristemoid formation	
Meristemoids	MS+BA+NAA	Shoot formation	
Shoot	MS+BA+KN+NAA	Multiple shoot formation	
Shoot	MS+IAA	Rooting	

Contd...

Nucellus tissue	MS+NAA (5.0 mg/l)+KN (1.0 mg/l)	Callus induction	Hossain *et al.*, 1994b
Callus	MS+BA (2.0 mg/l)+NAA (0.1 mg/l)+ GA_3 (1.0 mg/l)	Multiple shoot formation	
Shoot	1/2MS+IBA (0.5 mg/l)+NAA (0.5 mg/l)	Rooting	
Cotyledon	MS+BA+IAA or GA_3	Callus induction	Islam *et al.*, 1994
Callus	MS+BA+IAA or GA_3	Shoot bud induction	
Shoot	MS+KN (1.0 mg/l)+IAA (0.1 mg/l)	Shoot elongation	
Shoot	1/2MS+IBA (25.0 mg/l)	Rooting	
Immature embryonic tissue	MS+BA (0.12 mg/l)+2,4-D (2.22 mg/l)	Callus induction	Islam *et al.*, 1995
Callus	MS+KN (1.1mg/l)+IAA (0.1 mg/l)	Shoot formation	
Shoot	MS+IBA (1.0 mg/l)	Rooting	
Cotyledon, hypocotyl	MS salts+B5 vitamins+2,4-D (0.5 mg/l)+ NAA (0.1 mg/l)+KN (0.1 mg/l)	Callus induction	Arumugam *et al.*, 1997
Callus	MS salts+B5 vitamins+BA (2.0 mg/l)+ NAA (0.01 mg/l)+GA_3 (1.0 mg/l)	Shoot induction	
Shoot	1/2MS+BA (1.0 mg/l)+GA_3 (0.1 mg/l)	Shoot elongation	
Shoot	1/2MS+IBA (2.0 mg/l)+KN (0.01 mg/l)	Rooting	
Cotyledon, hypocotyl and immature leaflet	MS+2,4-D (0.5 mg/l)	Callus induction	Arumugam, 1998
Callus	MS+BA (1.0 mg/l)+KN (0.1 mg/l)+ NAA (0.01 mg/l)	Shoot regeneration	
Shoot	1/2MS+IBA (1.0 mg/l)+KN (0.1 mg/l)	Rooting	

Contd..

Somatic embryogenesis-direct			
Embryo axis cotyledon	MS+2,4-D (0.25 mg/l)+BA (0.25 mg/l)+ CW (20%)	Somatic embryo induction (Globular)	Isalm *et al.*, 1996c
Globular embryos	1/2MS+Sucrose (2%)+BA (0.25 mg/l)	Somatic embryo maturation and germination	
Cotyledon and Hypocotyl	MS+NAA (0.2 mg/l)+BA (0.1 mg/l)	Somatic embryo induction (early stages)	Arumugam, 1998
Somatic embryos	MS+BA (0.8 mg/l)+ ascorbic acid (1.0 mg/l)	Somatic embryo maturation and germination	
Somatic embryogenesis – indirect			
Solid medium- Cotyledon , immature leaflet and mature leaf callus	MS+2,4-D (0.2 mg/l)+BA (0.2 mg/l) + glutamine (10.0 mg/l)	Somatic embryo induction (early stages)	Arumugam, 1998
Somatic embryos	MS+BA (0.1 mg/l)+abscisic acid (0.1 mg/l)	Somatic embryo maturation and germination	
Liquid medium hypocotyl, cotyledon, immature and mature leaf let callus	MS+2.4-D (0.1 mg/l)+ BA (0.1 mg/l)+ glutamine (10.0 mg/l)	Somatic embryo induction (early stages)	
Somatic embryos	MS+BA (0.1 mg/l)+absisic acid (0.5 mg/l)	Somatic embryo maturation	
Somatic embryos	MS+BA (0.4 mg/l)+ ascorbic acid (0.01 mg/l) + GA_3 (2.0 mg/l)	Somatic embryo germination	

Plate 1. a) Mature tree of Aegle marmelos (40-years-old), b) A twig with flowers and fruits.

seedling axillary bud (SAB), seedling terminal bud (STB), cotyledon (C), hypocotyl (H), epicotyl (E), immature leaf (IML), mature leaf (ML), stem (S) and root (R) segments were excised from seedlings and were washed with liquid soap solution for 3 minutes and then soaked in 70% (v/v) ethanol for 45 seconds and finally disinfected with 0.5% (w/v) mercuric chloride for 5 minutes and rinsed in sterile distilled water for 5 times. Both axillary and terminal bud explants (1cm length) from young shoots of mature tree were excised, thoroughly washed in running tap water for 5 minutes, followed by 3 minutes wash with liquid soap solution and then soaking in 70% (v/v) ethanol for 45 seconds, disinfecting with mercuric chloride (0.5% w/v) for 5 minutes and finally rinsed in sterile distilled water 5 times. Mature tree axillary buds (MAB) and mature tree terminal buds (MTB) were collected for one year at monthly interval of various seasons to test their response.

2.2. Growth Medium

2.2.1.Growth Medium for Micropropagation

Sterilized explants (CN, SAB, STB, MAB and MTB) 1cm size were inoculated on Murashige and Skoog (1962) (MS) medium, B5 medium (Gamborg *et al.*, 1968), woody plant medium (WPM) (Lloyd and McCown, 1981) and modified MS medium (macro and micronutrients of MS medium + organic supplements of B5 medium), to find out the suitability for micropropagation.

The explants (MAB and MTB) from mature trees were initially cultured on MS medium supplemented with growth regulators and activated charcoal (0.5-3.0%) to check phenolic exudation. Further, subcultures were made on the medium without activated charcoal. The cultures were inoculated on shoot induction and proliferation medium i.e. MS medium fortified with 6-Benzyl adenine (BA) (0.1-4.0 mg/l), Kinetin (KN) (0.001-2.0 mg/l) and α -naphthalene acetic acid (NAA) (0.001-1.0 mg/l). The medium was adjusted to pH 5.8. Agar (0.8% w/v) and sucrose (3% w/v) were added and autoclaved at 120°C (15 lbs pressure) for 15 min. The cultures were incubated at 25 ± 2°C under white fluorescent light at 80 $\mu Em^{-2}s^{-1}$ with 16 hrs photoperiod and 60-65% relative humidity unless otherwise stated. Explants were sub cultured at weekly interval on to the fresh medium of same composition.

Shoots, derived from seedling explants after 25 days, and from mature tree explants after 35 days, were transferred to shoot elongation medium consisting of GA_3 (0.001-3.0 mg/l). The shoots (<3 cm length) after 21 days were transferred to rooting medium consisting of half-strength MS medium supplemented with indole acetic acid (IAA) (0.01-3.0 mg/l), indole butyric acid (IBA) (0.01-3.0 mg/l) and KN (0.01-2.0 mg/l). Initially cultures were kept in dark for 4-6 days, later shifted to 16 hrs photoperiod with 80 $\mu Em^{-2}s^{-1}$ and 25±2°C temperature.

Rooted 15-day-old plantlets were transferred to small plastic containers containing a mixture of sand, garden soil and vermiculite (1:1:2), after removing the adhered culture medium by washing roots in the running tap water. These plantlets were irrigated with half-strength macro and micronutrients of MS medium twice a week for 20 days in controlled conditions (25±2°C) and covered with polythene bags. The polythene bags were removed after 20 days; plants were maintained in the same conditions for a week and later shifted to earthen pots for 3 months under shade conditions of ambient temperature (35±2°C). Subsequently they were transferred to the field.

2.2.2. Growth Medium for Organogenesis

Sterilized explants, cotyledon, hypocotyl, epicotyl, immature leaflet, mature leaf, stem and root explants of 15 mm long from 21-day-old seedlings were cultured on callus induction medium consisting of mMS medium fortified with 2,4-dichlorophenoxy acetic acid (2,4-D) (0.01 – 2.0 mg/l), NAA (0.01 – 1.0 mg/l), and IAA (0.01 – 1.0 mg/l). The cultures were subcultured at 7-day interval for three times on the same fresh culture medium. The cultures were grown under 16 hrs photoperiod with 80 $\mu Em^{-2}s^{-1}$ light intensity and 25±2°C temperature.

After 21 days, yellowish green nodular callus was transferred to shoot regeneration and multiplication medium i.e. MS medium supplemented with BA (0.1 – 3.0 mg/l), KN (0.01 – 1.0 mg/l) and NAA (0.001 – 0.5 mg/l). The cultures were grown under 16 hrs photoperiod with 80 $\mu E\ m^{-2}s^{-1}$ light intensity and 25±2°C temperature with a weekly interval subculture of three times. To find out the effect of photoperiod on callus induction, explants were kept under different dark periods (1, 2, 3, 4, 5, 6 and 7 d).

Various explants (cotyledon, hypocotyl, mature leaf, stem and root) 15 mm in size from 21-day-old seedlings were cultured on direct shoot regeneration (without callus phase) medium combined of MS medium supplemented with BA (0.5 – 3.0 mg/l), KN (0.1 – 2.0 mg/l) and NAA (0.001 -0.5 mg/l). The explants, including cotyledon, hypocotyl, epicotyl, mature leaf, stem and root, were collected sequentially from seedlings of different ages (7, 14, 21, 28, 35 and 42 d), to determine the suitable age of seedling for explant collection. To find out the effect of photoperiod on shoot regeneration response of different explants, cultures were incubated under different photoperiods i.e. 4/20, 8/16,12/12, and 16/8 and 20/4 hrs light/dark per day with 80 $\mu Em^{-2}s^{1}$ light intensity and 25±2°C temperature.

Regenerated shoots (>3.0 cm long) of 35-day-old were transferred to shoot elongation medium containing GA_3 (0.001-3.0 mg/l).

After 15 days, from all treatments, elongated shoots (>3.0 cm) were transferred to rooting medium [half-strength MS + IBA (0.01 – 3.0 mg/l) andKN (0.01 – 0.5 mg/l)]. Shoots were initially incubated under dark for 5 days, later cultured with 16 hrs photoperiod with 80 $\mu Em^{-2}s^{-1}$ light intensity and 25±2°C temperature. Rooted plantlets (21-day-old) were hardened and transferred to the field.

Seven replicates were taken for each treatment. All treatments were repeated thrice. All the data were subjected to Mean±SD and mean separation was by using Duncan's Multiple Range Test (DMRT) (Gomez and Gomez, 1976).

3. MICROPROPAGATION

Micropropagation is an alternative to the conventional methods of vegetative propagation with the objective of enhancing the rate of multiplication. It will reduce time constraint and allow continuous production of propagules. The production of plants from axillary shoots is most generally applicable and reliable method of *in vitro* propagation. Shoot tips from trees, or other woody perennials can be difficult to decontaminate. Thereby, it is preferred to initiate shoot cultures from meristem tips, which could be sterilized more easily. Murashige (1974) described 4 stages in micropropagation, which are-: 1) establishing an aseptic culture (shoot bud initiation), 2) production of suitable propagules (shooting and elongation), 3) preparation for growth and environment (rooting and hardening), and 4) transfer to natural environment.

3.1. Shoot Bud Initiation and Proliferation

3.1.1.Growth Medium

Our results showed that of the four culture media (MS, mMS, B5 and WPM) used for shoot bud induction, MS medium responded well (Data not shown). The use of MS medium for *in vitro* propagation has been well documented (Shekawat *et al.*, 1993; Singh *et al.*, 1993; Pattnaik and Chand, 1997; Sahoo and Chand, 1998; Dhar and Upreti, 1999).

Activated charcoal (1%) was most effective to check the phenolic exudation. Further subcultures (2-3) were made on MS medium without activated charcoal, controlled the browning of the medium. Arya and Shekawat (1986) treated axillary buds and hypocotyl explants of *A. marmelos* with cold antioxidant solution consisting of 100 mg/l and 150 mg/l citric acid for 15 min to reduce the browning of the cut and injured ends of the explants, and increased their rate of survival in culture. Ajithkumar and Seeni (1998) immersed sterilized *Aegle marmelos* stem cuttings for 2 min in a solution containing 50 mg/l ascorbic acid and citrate, respectively, to check phenolic exudation. Reha Bhati *et al.* (1992) incorporated various antioxidants- citric acid (10 mg/l), ascorbic acid (100 mg/l) and polyvinylpyrrolidine (PVP) (50 mg/l) in the medium to check browning in *Aegle marmelos* cultures. Dhar and Upreti (1999) in *Bauhinia vahlii* and Quraishi and Mishra (1998) in *Cleistanthus collinus* made similar reports. The negative effect of activated charcoal and PVP may be due to adsorption of essential components besides the inhibitory factors (Weatherhead *et al.*, 1979); beneficial effect is also reported in *Eucalyptus tereticornis* and *Sesbania sesban* (Das and Mitra, 1990, Shankar and Mohan Ram, 1990). Lloyd and McCown (1981) working with mountain laurel shoots observed that repeated transfer to fresh medium could avoid browning of culture media.

The explants enlarged 2-3 times that of original size after 5 days of inoculation. All the five explants (CN, SAB, STB, MAB and MTB) cultured on growth regulator-free medium did not show any sign of shoot bud development after 15 days of culture. Among the various explants tested, cotyledonary node showed the highest response of shoot bud initiation and proliferation. Variation in the regeneration potential among explants is attributed to the differences in their physiological and genetic make up of cells (Vieitez *et al.*, 1985). In contrast to other explant sources, cotyledonary explants from *in vitro*-grown seedlings of *Pisum sativum* (Mallick and Rashid, 1989; Jackson and Hobbs, 1990), *Phaseolus vulgaris* (Franklin *et al.*, 1991; Mohamed *et al.*, 1992), *Glycine max* (Wright *et al.*, 1986; Kothari *et al.*, 1991), *Cercis canadensis* (Distabanjong and Geneve, 1997) were used. It is also well documented in many other woody plants that only juvenile explants responded well such as *Fagus sylvatica, Quercus robur* (Ahuja, 1986), *Fragrae fragrans* (Lee and Rao, 1986), temperate fruit and nut trees (Hutchinson and Zimmerman, 1987), *Fagus sylvatica* (Vieitez *et al.*, 1993), chestnut (Sanchez *et al.*, 1993), *Cleistanthus collinus* (Quraishi *et al.*, 1996), *Celastrus paniculatus* (Lakshmi and Seeni, 2001).

Our studies demonstrated that shoot proliferation was low on MS medium containing BA alone, whereas BA combined with KN increased the percentage of response. However, addition of NAA with BA and KN gave the highest response with two subcultures on the same medium.

Among the different concentrations of BA (0.1 – 4.0 mg/l) used, 3.0 mg/l induced the highest percentage of shoot proliferation with cotyledonary node (75.7%) followed by axillary bud (72.3%), terminal bud (72.3%); cotyledonary node explant showed the highest response (55.7 shoots/explant) followed by axillary bud (42.0 shoots/explant) and terminal bud explant (39.7 shoots/explant) after 25 days of culture (Table 2 and 3).

The enhanced effect of KN with BA was observed in seedling explants. The highest shoot proliferation response was in cotyledonary node (85.8%), followed by axillary bud (85.0%) and terminal bud (84.6%) on KN (0.5 mg/l) and BA (3.0 mg/l) combination. Cotyledonary node explant produced the highest number of shoots (67.5 shoots/explant) (Table 2 and 3).

A synergistic effect of auxin along with cytokinins (BA and KN) was observed in shoot proliferation and multiplication of seedling explants. The highest response of shoot proliferation and multiplication was observed in cotyledonary node (95.7%) with 78.8 shoots/explant on medium containing BA (3.0 mg/l) + KN (0.5 mg/l) +NAA (0.01 mg/l) followed by SAB and STB (Table 2 and 3) (Plate 2.a, b).

Axillary bud of mature tree was found to be better than terminal bud explant for shoot proliferation response. The shoot proliferation and number of shoots per explant increased with increasing concentration of BA upto 3.0 mg/l. Shoots developed on media (BA>3.0 mg/l) failed to elongate, resulting in rosette shoot clumps. Shoot multiplication rate was found to be high during the first three subcultures.

A synergistic effect of auxin (NAA – 0.01 mg/l) was also observed in increasing shoot proliferation in axillary bud explant to the highest frequency (85.6%) with 55.8 shoots/explant (Table 4 and 5) (Plate 2c, d).

Shoot proliferation and multiplication per explant in all concentrations and combinations was in direct proportion with increasing concentrations of BA, KN and NAA up to an optimal level with 2 subcultures.

Table 2. Shoot bud initiation response of seedling explants of Aegle marmelos grown on MS medium supplemented with BA, KN and NAA for 25 days.

Growth regulators (mg/l)			Explants response (percentage Mean ± SD)		
			CN	SAB	STB
BA					
0.1			35.85 ± 1.29k	31.1 ± 1.25h	38.9 ± 1.44i
0.5			40.1 ± 1.87j	39.5 ± 1.30g	41.7 ± 1.65h
1.0			40.7 ± 1.85j	40.7 ± 1.90g	45.2 ± 1.67g
1.5			52.3 ± 2.10i	47.8 ± 1.97f	51.7 ± 2.08f
2.0			60.2 ± 2.99h	66.4 ± 2.05e	68.7 ± 2.70d
2.5			65.6 ± 3.15g	68.7 ± 2.58e	70.2 ± 2.95d
3.0			75.7 ± 4.24e	72.3 ± 4.02cd	72.3 ± 3.98cd
3.5			62.5 ± 4.75gh	70.9 ± 3.98d	69.6 ± 3.55d
4.0			63.1 ± 3.88g	67.7 ± 3.12e	65.1 ± 3.12e
BA	KN				
3.0	0.001		75.3 ± 3.19e	75.3 ± 3.11c	74.4 ± 3.15cd
3.0	0.01		78.2 ± 3.17d	76.8 ± 3.43c	76.9 ± 3.29e
3.0	0.1		78.7 ± 3.22d	77.5 ± 3.37c	77.6 ± 3.88c
3.0	0.5		85.8 ± 3.77b	85.0 ± 1.57b	84.6 ± 4.95b
3.0	1.0		71.3 ± 3.99ef	72.9 ± 3.90cd	72.9 ± 3.27cd
3.0	2.0		63.2 ± 2.99g	64.8 ± 2.44e	62.9 ± 2.33e
BA	KN	NAA			
3.0	0.5	0.001	81.2 ± 3.85c	77.9 ± 3.09c	79.2 ± 3.05c
3.0	0.5	0.01	95.7 ± 4.67a	94.3 ± 4.50a	92.1 ± 4.21a
3.0	0.5	0.1	77.3 ± 3.23d	73.6 ± 3.17cd	74.3 ± 3.15cd
3.0	0.5	0.5	72.8 ± 3.13ef	70.1 ± 3.08d	70.4 ± 2.90d
3.0	0.5	1.0	66.6 ± 2.97fg	66.4 ± 5.76e	65.3 ± 2.77e

Each value represents the average of 7 replicates mean separation using Duncan's New Multiple Range Test. Means within column with different letters are significant at 5% level. CN – Cotyledonary node; SAB – Seedling axillary bud; STB – Seedling terminal bud.

Growth regulators can bring about *de novo* morphogenesis on media containing sub optimal concentrations and considerably enhance morphogenesis on an otherwise optimal medium (Margara, 1969). The cytokinins usually promote formation of multiple shoots and inhibit root formation (Jouannaeau, 1970, 1975). The effect of cytokinins on tissue or organ culture can vary according to the particular compound used, the type of culture, the variety of plant from which it was derived from juvenile or mature tissues (Messeuger and Mele, 1987; Welsh and Sink, 1981). A low concentration of auxin in conjunction with cytokinin in shoot multiplication is usually required, whereas, in some cases cytokinin alone is sufficient (Skoog and Miller, 1957; Hansen *et al.*, 1985). It is important to choose an auxin at a concentration, which will promote growth without inducing callus formation (George, 1993).

Table 3. Shoot multiplication in seedling explants of Aegle marmelos grown on MS medium supplemented with BA, KN and NAA for 25 days.

Growth regulators (mg/l)			No. of shoots / explant (Mean ± SD)		
			CN	SAB	STB
BA					
0.1			1.7 ± 0.69j	1.0 ± 0.00j	1.7 ± 0.66i
0.5			5.3 ± 1.11i	3.4 ± 0.49i	3.8 ± 0.73h
1.0			21.8 ± 1.72h	17.0 ± 0.70h	16.7 ± 1.08g
1.5			28.2 ± 1.73g	25.6 ± 1.79g	22.7 ± 1.81f
2.0			37.1 ± 1.59f	28.4 ± 1.35f	25.0 ± 1.63ef
2.5			38.8 ± 2.68f	32.4 ± 1.70e	26.2 ± 2.98e
3.0			55.7 ± 2.72c	42.0 ± 1.52d	39.7 ± 1.82d
3.5			39.5 ± 2.87f	32.5 ± 1.29e	30.9 ± 1.82e
4.0			37.5 ± 2.87f	30.6 ± 0.15ef	29.8 ± 1.01e
BA	KN				
3.0	0.001		43.7 ± 0.88e	40.2 ± 1.03d	42.5 ± 1.11c
3.0	0.01		46.4 ± 1.64de	48.8 ± 1.05c	45.7 ± 3.48c
3.0	0.1		49.8 ± 1.16d	46.7 ± 1.05c	42.3 ± 1.56c
3.0	0.5		67.5 ± 1.57b	63.0 ± 2.05b	50.1 ± 1.54b
3.0	1.0		56.6 ± 1.11c	45.5 ± 1.58c	32.7 ± 1.61de
3.0	2.0		43.1 ± 1.44e	32.7 ± 2.84e	21.6 ±0.09f
BA	KN	NAA			
3.0	0.5	0.001	52.6 ± 1.01cd	49.1 ± 0.6c	37.7 ± 1.48d
3.0	0.5	0.01	78.8 ± 3.89a	66.4 ± 3.69a	59.5 ± 2.11a
3.0	0.5	0.1	69.2 ± 1.08b	46.7 ± 2.10c	44.0 ± 0.81c
3.0	0.5	0.5	53.8 ± 4.03c	34.6 ± 2.99d	33.5 ± 0.65d
3.0	0.5	1.0	41.5 ± 3.52e	28.5 ± 3.54f	26.2 ± 0.45e

Each value represents the average of 7 replicates mean separation using Duncan's New Multiple Range Test. Means within column with different letters are significant at 5% level. CN – Cotyledonary node; SAB – Seedling axillary bud; STB – Seedling terminal bud.

Varghese *et al.* (1993) reported that nodal explants of *Aegle marmelos* produced numerous shoot buds in the medium supplemented with KN and NAA. Whereas, Ajithkumar and Seeni (1998) obtained maximum shoot formation from young stem nodal explants of *A. marmelos* mature trees (25-year-old) in BA (1.5 mg/l) and KN (2.5 mg/l). Significantly longer shoots were obtained in BA than KN. Synergistic effect of IAA and KN combination showed maximum number of short shoots. NAA showed inhibitory effects on shoot induction. Our studies showed that the cytokinins (BA and KN) with auxin (NAA) had synergistic effect on shoot bud proliferation and multiple shoot formation. The possibility of genotypic differences between individuals, especially of an out breeding tree species, lead to variation in shoot multiplication rates.

Table 4. Shoot bud initiation response of mature explants of Aegle marmelos grown on MS medium supplemented with BA, KN and NAA for 35 days.

Growth regulators (mg/l)			Explants response (Percentage Mean ± SD)	
			MAB	MTB
BA				
0.1			25.3 ± 0.99j	15.1 ± 0.52k
0.5			31.9 ± 1.29i	22.0 ± 0.89j
1.0			36.8 ± 1.76h	27.2 ± 1.23i
1.5			47.2 ± 1.94g	33.4 ± 1.26h
2.0			54.1 ± 2.68f	46.7 ± 1.76f
2.5			65.6 ± 2.98d	52.5 ± 1.96e
3.0			67.2 ± 3.14d	59.1 ± 2.12d
3.5			64.2 ± 2.87d	52.6 ± 2.21e
4.0			55.2 ± 2.95f	43.7 ± 1.91fg
BA	KN			
3.0	0.001		67.2 ± 2.87d	61.2 ± 2.99c
3.0	0.01		69.1 ± 2.18c	63.9 ± 2.85bc
3.0	0.1		70.3 ± 3.14c	66.1 ± 2.26b
3.0	0.5		76.3 ± 3.19b	69.5 ± 2.96b
3.0	1.0		69.2 ± 2.89c	55.2 ± 2.15de
3.0	2.0		61.1 ± 2.76e	43.5 ± 1.96fg
BA	KN	NAA		
3.0	0.5	0.001	72.1 ± 3.07bc	67.8 ± 2.45b
3.0	0.5	0.01	85.6 ± 3.87a	72.9 ± 3.25a
3.0	0.5	0.1	70.3 ± 2.99c	64.6 ± 2.97bc
3.0	0.5	0.5	69.9 ± 2.86c	57.2 ± 1.85d
3.0	0.5	1.0	61.2 ± 2.45e	41.5 ± 1.27g

Each value represents the average of 7 replicates mean separation using Duncan's New Multiple Range Test. Means within column with different letters are significant at 5% level. MAB – Mature tree axillary bud; MTB – Mature tree terminal bud.

The stimulatory effect of BA on bud break and multiple shoot formation was reported by Kukreja *et al.* (1990) (*Pogostemon cablin*), Joshi *et al.* (1991) (*Anogeissus pendula*), Sen and Sharma (1991) (*Withania somnifera*), Purohit *et al.* (1994) (*Chlorophytum borivilianum*), Bhat *et al.* (1995) (*Piper* sp.), Pattnaik and Chand (1996) (*Ocimum* spp.), Quraishi *et al.* (1996) (*Cleistanthus collinus*), Sahoo and Chand (1998) (*Vitex negundo)* and Sinha *et al.* (2000) (*Albizia chinensis*).

The synergistic effect of BA and KN on shoot bud induction and multiple shoot formation is also well documented in many plant species such as *Kaempferia galanga* (Vincent *et al.*, 1992), *Dalbergia latifolia* (Raghava Swamy *et al.*, 1992), *Commiphora wightii* (Durga and Mehta, 1993), *Feronia limonia* (Hossain *et al.*, 1994c), *Tectona grandis* (Sunitibala *et al.*, 1994) and *Acacia catechu* (Kaur and Kant, 2000).

Table 5. Shoot multiplication in mature explants of Aegle marmelos grown on MS medium supplemented with BA, KN and NAA for 35 days.

Growth regulators (mg/l)			No. of shoots /explant (Mean ± SD)	
			MAB	MTB
BA				
0.1			1.0 ± 0.50i	0.0 ± 0.0
0.5			2.0 ± 0.70i	0.0 ± 0.0
1.0			11.4 ± 0.49h	2.4 ± 0.21i
1.5			20.7 ± 0.85g	8.5 ± 0.97h
2.0			23.7 ± 0.59f	13.2 ± 0.99g
2.5			27.5 ± 2.95e	15.5 ± 1.45g
3.0			35.8 ± 1.36d	19.0 ± 1.56f
3.5			27.7 ± 1.16e	16.1 ± 1.12fg
4.0			26.6 ± 0.01ef	14.5 ± 1.10g
BA	KN			
3.0	0.001		40.7 ± 1.05c	19.2 ± 1.26f
3.0	0.01		42.5 ± 0.96c	22.1 ± 1.85e
3.0	0.1		45.4 ± 1.64cd	23.8 ± 1.87e
3.0	0.5		48.2 ± 1.36b	27.0 ± 1.52d
3.0	1.0		41.6 ± 0.83c	18.6 ± 1.25f
3.0	2.0		36.6 ± 0.03d	12.7 ± 1.21g
BA	KN	NAA		
3.0	0.5	0.001	45.2 ± 1.21cd	35.8 ± 1.27b
3.0	0.5	0.01	55.8 ± 2.55a	44.3 ± 2.17a
3.0	0.5	0.1	42.2 ± 1.31c	39.5 ± 1.87b
3.0	0.5	0.5	30.2 ± 1.79e	32.9 ± 1.45c
3.0	0.5	1.0	24.5 ± 1.99f	21.5 ± 1.29e

Each value represents the average of 7 replicates mean separation using Duncan's New Multiple Range Test. Means within column with different letters are significant at 5% level. MAB – Mature tree axillary bud; MTB – Mature tree terminal bud.

3.1.2. Subculture

Repeated sub cultures of nodes from shoot cultures helped to achieve continuous production of callus- free, healthy shoots through five sub cultures in our study. Thus over 12,376 shoots were obtained from each cotyledonary node explant within five months (Arumugam and Rao, 1996a). Similar results were recorded by Mehra-Palta (1982) in *Eucalyptus* sp, Asokan *et al.* (1984) in *Xanthosoma caracu*, Aitken and Jones (1987) in radiata pine, Sharma and

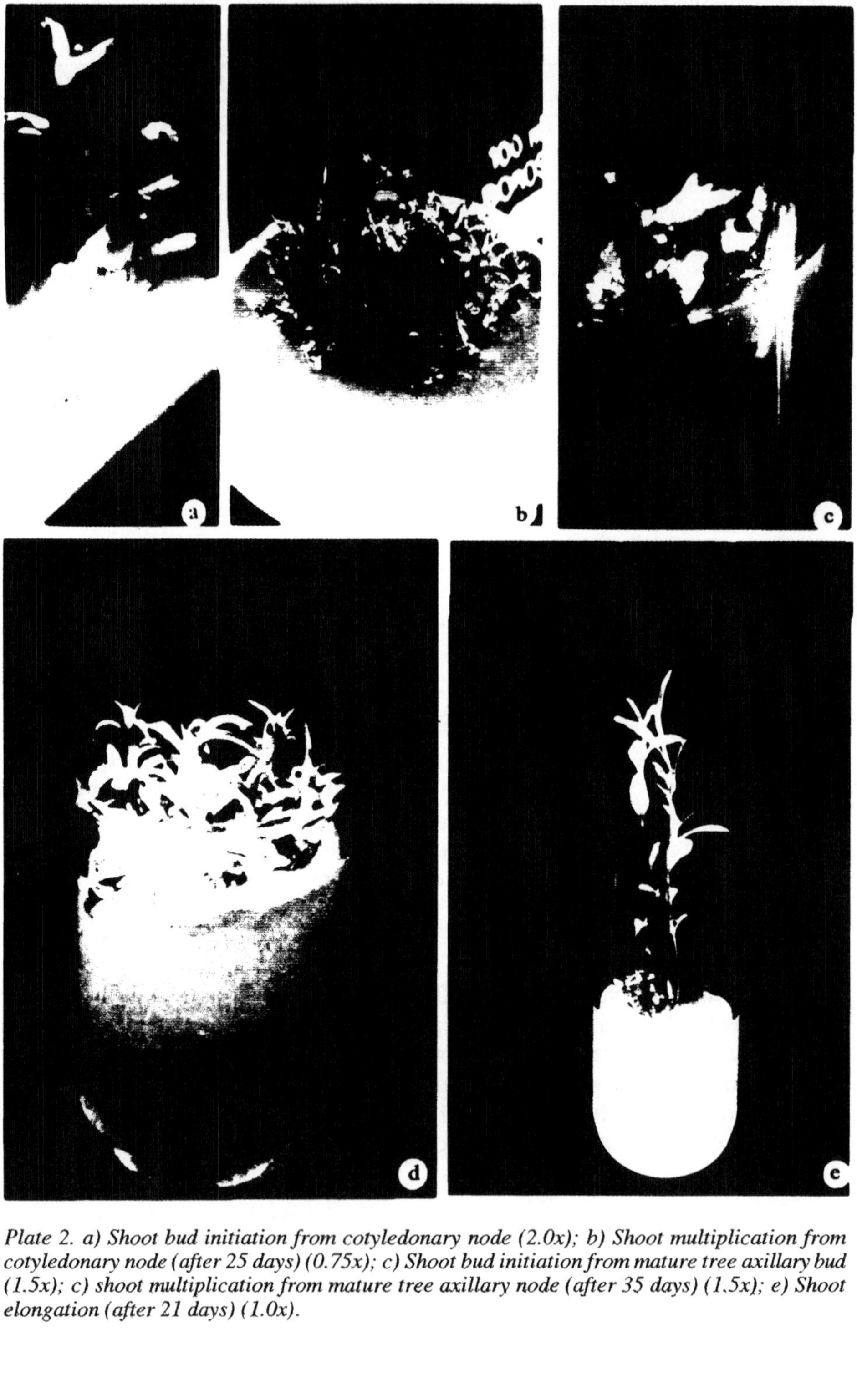

Plate 2. a) Shoot bud initiation from cotyledonary node (2.0x); b) Shoot multiplication from cotyledonary node (after 25 days) (0.75x); c) Shoot bud initiation from mature tree axillary bud (1.5x); c) shoot multiplication from mature tree axillary node (after 35 days) (1.5x); e) Shoot elongation (after 21 days) (1.0x).

Chaturvedi (1988) in *Bougainvillaea* x *buttiana*, Dewan *et al.* (1992) in *Acacia nilotica,* Shekawat *et al.* (1993) in *Prosopis cineraria*, Pattnaik *et al.* (1996) in *Morus australis,* Ajithkumar and Seeni (1998) in *Aegle marmelos* and Lakshmi and Seeni (2001) in *Celastrus paniculatus.* Repeated transferring of woody explants in media containing cytokinins cause rejuvenation of explant tissues taken from adult plants. These treatments release the clone from the developmental effects of past adult operating cycle and also reduce and/ or remove accumulated inhibitory factors (Franclet and Boulay, 1989; Pierik, 1990). Repeated transferring causes activation and conditioning of meristems (Boulay, 1984).

3.1.3. Seasonal Effect

Our results demonstrated that shoot proliferation of mature tree explants (MAB and MTB) were greatly influenced by the time of explant collection. High frequency bud break coupled with highest shoot proliferation were observed in explants collected from May to June, because new shoot formation and flowering started during this period (Data not shown). The seasonal variations influence the level of growth regulators, exudation compounds, amino acids, protein and minerals in the mature tree explants (MAB and MTB). Hence, the variation in bud break and shoot proliferation may be attributed to the above molecular alterations (Ahuja *et al.*, 1982; Hu and Wang, 1983; Tisserat, 1985; Prasad and Chaturvedi, 1988; Das and Mitra, 1990; Shekawat *et al.*, 1993; Durga and Mehta, 1993; Sunitibala *et al.*, 1994; Gars and Calvo, 1996; Quraishi *et al.*, 1996; Pattnaik and Chand, 1996,1997; Sahoo and Chand, 1998; Dhar and Upreti, 1999).

3.2. Shoot Elongation

Species, which have an elongated shoot system *in vivo*, will produce axillary shoots, which can be easily separated as "micro cuttings" and then individually rooted. In rosette habit of growth, it is necessary to elongate shoots before they are rooted. Shoot cluster explants are treated with differential concentrations of GA_3 so as to reduce the axillary bud formation and thereby promote the shoot growth. Gibberellins can influence plant growth and development by increasing stem length, promoting flowering, and inducing fruit set, suppress the synthesis of phosphatase (Zink and Veliky, 1982) and *de novo* synthesis of α-amylase (Thomson *et al.*, 1981). De la Guardia and Benlloch (1980) proposed that growth promotion by GA_3 is also due to the increased transport of K^+ ions. Wickremesinhe (1990) reported that the addition of GA_3 to the medium containing cytokinin and auxin at stage II in *Apios americana* was useful to enhance growth and/or increase the rate of shoot proliferation. GA_3 increase the length of shoots during multiplication

(Boxus, 1974 a, b; Dore, 1975; Boxus *et al*., 1977; Jones *et al*., 1977, 1979; Aldwinckle and Gustafson, 1981; Pua *et al*., 1983; Babic and Neskovic, 1984; Jacoboni and Standardi, 1987; Arora and Bhojwani, 1989; Pattniak and Chand, 1996; Sahoo *et al*., 1997; Sahoo and Chand, 1998).

Our studies revealed that among the various GA_3 concentrations used along with MS and BA (0.1 mg/l), GA_3 (1.0 mg/l) supported highest shoot elongation within 21 days of culture (Data not shown). When increasing the concentrations of GA_3, the shoot elongation was also increased, up to optimum level (1.0 mg/l), afterwards it was decreased with further increase (Plate 2e).

3.3. Rooting

In our study, elongated shoots (>3 cm) were transferred to half-strength MS medium (Pattnaik and Chand, 1997; Roy *et al*., 1998; Dhar and Upreti, 1999) supplemented with IAA (0.01 – 3.0 mg/l), IBA (0.01 – 3.0 mg/l) and KN (0.01 – 2.0 mg/l). The highest percentage of rooting was after 15 days of shoot culture. The dark treatment of 4-6 days stimulated the highest rooting in the presence of IBA and KN. Such dark period requirement has been earlier reported (Hansen *et al*., 1987; Roy *et al*., 1987; Quraishi *et al*., 1996; Brassard *et al*., 1996; Lakshmi and Seeni, 2001).

Indole butyric acid 1.0 mg/l was optimal for rooting in both seedling and mature tree explant derived shoots. Of the different explants tested, CN derived shoots produced high frequency of rooting (87.5%), followed by SAB (81.2%), STB (83.1%), MAB (80.3%) and MTB (71.5%). Of the various concentrations of IAA, 1.0 mg/l was best for rooting response. The highest frequency of rooting noticed was 69.5% in CN derived shoots followed by SAB (67.4%), STB (68.5%), MAB (65.5%) and MTB (58.8%) (Table 6).

The combined effect of KN (0.1 mg/l) and IBA (1.0 mg/l) showed the highest response of rooting in CN derived shoots (94.8%) followed by MAB (94.5%), SAB (93.9%), STB (93.55%) and MTB (83.7%) (Table 6, Plate 3a). These results showed that the level and type of growth regulators highly influence the rooting frequency.

Varghese *et al*. (1993) reported 25% rhizogenesis of *A.marmelos* in the presence of IAA. Hazarika *et al*. (1996 a) rooted the shoots from explants of *in vitro* seedlings of *Aegle marmelos* on half-strength MS medium supplemented with IBA (0.5 mg/l). Pulsing of *in vitro* shoots of *Aegle marmelos* with IBA (10 ppm for 2 min) promoted early rooting in soil rite (soil+vermiculite). Plants were acclimatized successfully without any transplant shock in the soil (Hazarika *et al*., 1996b). Ajithkumar and Seeni (1998) observed 90% root formation in shoots obtained from nodal explants of *A.marmelos* mature trees (25-year-old), on half-strength MS medium supplemented with IBA (10.0 mg/l), while NAA induced callus along with root formation. We have obtained much higher percentage of rooting of cotyledonary node explant as compared to the above reports. This is probably due to the combined effect of KN and IBA.

3.4. Hardening

The rooted plants (15-day-old) were removed from the culture medium and the roots were washed in running tap water, to remove agar. The plantlets were transferred to small plastic containers (5 cm dia) containing garden soil mixed with sand and vermiculite (1:1:2) under controlled conditions (25±2°C, 16 hrs photoperiod with 40 $\mu Em^{-2}s^{-1}$ light intensity and 70-75% relative humidity). These potted plants were covered with polythene bags to ensure high humidity and provided with half-strength MS salt solution devoid of sucrose and myo-inositol twice a week for 20 days. After 20 days, polythene bags were removed from the plastic containers and after a week, the plants were transferred to earthen pots (10 cm dia) containing garden soil and watered at regular intervals. The plants were kept initially under shade for 3 months till the appearance of new buds or shoots and then transplanted to ambient field conditions. The field-transferred plants were irrigated regularly at weekly intervals and maintained up to 10 months (Plate 3b-e).

The survival percentage of plantlets was 90-95%. The *in vitro* raised plantlets initially showed single leaflet later on trifoliate nature was observed only after transferring plants to earthen pots under shade conditions. The regenerated plants did not show any detectable variation in morphology or growth characteristics.

4. ORGANOGENESIS

Organogenesis can be obtained either through direct differentiation of shoot buds from explants (direct organogenesis) or through callus formation in explants and subsequent formation of shoots (indirect organogenesis) and roots (Hicks, 1980).

A wide variety of biological, physical and chemical factors affect the specific pattern of organogenesis and its extent (Thorpe, 1994). Plant growth regulators are centrally involved in control of organogenesis (Skoog and Miller, 1957; Thorpe, 1982; Thorpe and Biondi, 1984; Torrey, 1985; Mantell *et al.*, 1985; Tanimoto and Harada, 1985; Halperin, 1986; Mohnen *et al.*, 1990). The extent of pre-determination or commitment to morphogenesis depends on tissue type and plant genetic background. Street (1977) suggested that although many cells may have the capacity to become "competent", newly formed meristems may preferentially accept essential metabolites from surrounding cells which would then be less likely to follow the same developmental pathway.

Media and growth regulators, which favor rapid cell proliferation and the formation of callus from an explant, are not usually conducive to the initiation of the morphogenetic meristems, which give rise to roots or shoots. However, in cultures of some plants, organogenesis becomes inevitable when callus is maintained on a medium for a prolonged period, but may be prevented if the tissue is sub cultured. In other cases unorganized callus initiated on one medium needs to be transferred to another with different composition of growth regulators (a regeneration medium) for shoot initiation to occur. Cells, from which organs are formed, have arisen from the "reprogramming" of dedifferentiated cells or, more commonly, have retained a commitment to morphogenesis, which existed in the explanted tissues. This may explain why highly organogenic cultures are most frequently obtained from young actively dividing parts of plants (George, 1993).

Table 6. Root induction in elongated shoots of various explants of Aegle marmelos grown on half-strength MS medium supplemented with IBA, IAA and KN for 15 days.

Growth regulators (mg/l)		Rooting response (Percentage Mean ± SD)				
		CN	SAB	STB	MAB	MTB
IBA						
0.01		37.1 ± 2.48h	34.2 ± 1.84j	35.2 ± 2.67j	34.7 ± 1.65i	31.5 ± 1.25g
0.1		45.4 ±2.84fg	39.1 ± 1.98i	45.5 ± 2.87i	45.4 ± 2.24g	34.6 ± 1.40f
0.5		61.4 ± 3.28e	63.8 ± 2.71f	68.5 ± 2.93e	67.5 ± 2.12d	52.0 ± 2.11d
1.0		87.5 ± 3.11b	81.2 ± 2.83b	83.1 ± 3.14b	80.3 ± 2.58b	71.5 ± 3.62b
2.0		47.1 ± 2.48f	55.4 ± 2.18g	52.9 ± 2.99g	53.6 ± 2.27f	51.7 ± 2.74d
3.0		32.3 ± 2.65i	34.2 ± 1.95j	32.5 ± 1.24k	34.2 ± 1.45i	42.2 ± 2.18e
IAA						
0.01		42.8 ± 2.25g	41.9 ± 1.98i	40.3 ± 2.58i	40.5 ± 2.34h	37.3 ± 1.67f
0.1		47.1 ± 2.99f	47.2 ± 1.64h	46.2 ± 2.98h	47.2 ± 2.14g	45.4 ±1.98de
0.5		51.4 ± 3.21f	52.2 ± 3.48g	51.3 ± 3.41g	52.3 ± 3.57f	49.2 ± 1.96d
1.0		69.5 ± 3.44d	67.4 ± 4.87e	68.5 ± 3.96e	65.5 ± 3.99d	58.8 ± 2.25c
2.0		43.1 ± 1.13g	44.4 ± 2.15h	45.3 ± 1.93h	46.4 ± 1.82g	41.6 ± 1.86e
3.0		39.7 ± 1.86h	39.5 ± 1.85i	38.4 ± 1.37ij	38.5 ± 1.34h	29.5 ± 1.90h
IBA	KN					
1.0	0.01	87.51 ±4.55b	86.9 ± 4.76b	86.8 ± 4.34b	84.8 ± 3.83b	72.0 ± 3.43b
1.0	0.1	94.8 ± 4.98a	93.9 ± 4.16a	93.5 ± 4.76a	94.5 ± 4.24a	83.7 ± 3.77a
1.0	0.5	85.3 ± 4.34b	84.3 ± 4.95b	83.8 ± 4.12b	83.6 ± 3.84b	74.9 ± 3.74b
1.0	1.0	74.3 ± 3.87c	73.2 ± 3.98d	73.6 ± 3.76d	74.6 ± 3.27c	68.4 ±3.11bc
1.0	2.0	58.9 ± 2.56e	57.9 ± 2.74g	57.8 ± 2.78f	58.0 ± 1.24e	49.6 ± 2.60d
IAA	KN					
1.0	0.01	42.8 ± 1.65g	42.4 ± 1.45i	43.7 ± 2.38h	42.7 ± 1.43h	41.5 ± 2.21e
1.0	0.1	57.1 ± 2.34e	56.3 ± 1.76g	57.2 ± 2.89f	56.6 ± 2.11e	57.3 ± 2.71c
1.0	0.5	87.5 ± 3.85b	85.2 ± 4.39b	85.6 ± 3.96b	84.3 ± 3.65b	71.0 ± 3.55b
1.0	1.0	74.4 ± 3.62c	78.4 ± 3.43c	78.2 ± 3.87c	75.5 ± 2.89c	55.5 ±2.57cd
1.0	2.0	71.4 ± 2.21c	73.2 ± 3.95d	72.5 ± 3.14d	72.7 ± 3.34c	62.4 ± 2.34c

Each value represents the average of 7 replicates mean separation using Duncan's New Multiple Range Test. Means within column with different letters are significant at 5% level.
CN – Cotyledonary node; SAB – Seedling axillary bud; STB – Seedling terminal bud
MAB – Mature tree axillary bud; MTB – Mature tree terminal Bud.

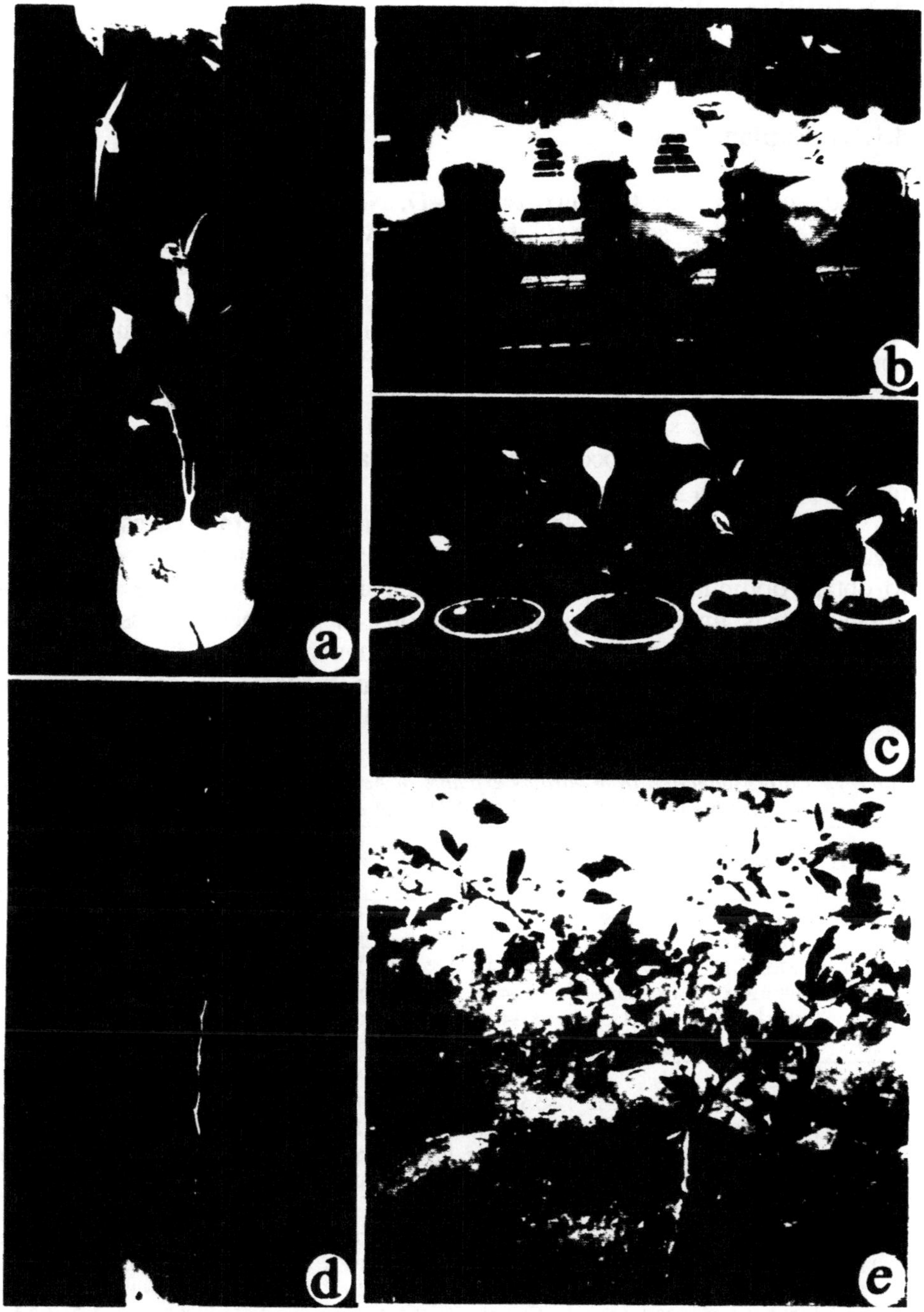

Plate3. a) Rooted plantlets (after 15 days) (0.75x); b) Hardened plantlets covered with polythene bag (25±1°C) (after 20 days) (0.1x); c) Hardened plantlets (25±1°C) (after 1 week) (0.2x); d) Potted plant (3-months-old) (0.1x); e) Field grown plant (10–months-old) (0.08x).

4.1. Indirect Organogenesis

4.1.1. Callus Induction and Proliferation

4.1.1.1. Explant Type

Our studies indicated that auxins (2, 4-D, NAA, and IAA) produced callus from all explants of seedling (C, H, E, IML, ML, S and R) after two weeks of culture on modified MS medium. The callus induction frequency varied with explant type used. The explants began to enlarge in size within 7 days and callus developed from the cut ends of the explant. Among different explants, cotyledon, hypocotyl and immature leaflet showed 100% callus initiation and proliferation followed by epicotyl, mature leaf, stem and root segments (Table 7). Plate 4a, b is shown only for hypocotyl and root explants. Arya *et al.* (1981) reported that hypocotyl and stem explants showed better performance in callus induction and proliferation.

Explants selection plays an important role in successful plant regeneration studies. Explants must be chosen from healthy, vigorous plants to obtain optimum results. Although all parts of the plant are capable of organogenesis under appropriate conditions, it is generally observed that immature tissues and organs are morphogenetically more plastic than the mature differentiated tissues. In some plant groups such as legumes the response is genotype dependent (George, 1993).

Table 7. Callus induction from various explants of Aegle marmelos grown on modified MS medium supplemented with 2,4-D, NAA and IAA for 21 days.

Growth regulators (mg/l)	Explants response (Percentage Mean ± SD)						
	C	H	E	IML	ML	S	R
2,4-D							
0.01	41.4±1.70g	40.4±1.97f	39.4±1.29g	38.4±1.23h	45.7±1.97f	37.1±1.70f	35.7±1.65f
0.1	57.3±2.55f	55.7±2.75e	51.4±2.45e	55.7±2.11e	55.7±2.34e	41.4±1.86e	45.4±1.94e
0.5	100.0±0.00a	100.0±0.00a	95.7±4.65a	100.0±0.00a	95.7±4.84a	85.7±3.65a	85.7±3.65a
1.0	90.4±4.26b	89.0±4.23b	87.1±4.68b	81.0±4.86b	71.4±3.23c	71.4±3.23b	71.4±3.23b
2.0	71.4±3.35d	70.2±3.44c	69.9±2.95c	67.5±2.95d	70.2±2.88c	65.2±2.39c	65.5±2.89c
NAA							
0.01	45.7±1.85g	45.7±1.36f	45.5±1.85f	46.6±1.97g	44.7±1.56f	37.2±1.85f	38.4±1.69f
0.1	58.0±1.95f	57.1±2.34e	56.7±2.31d	57.6±2.47e	57.7±1.87e	56.7±1.67d	55.7±1.65d
0.5	41.3±3.50g	46.1±2.59f	48.7±4.65ef	51.4±3.45f	48.4±3.15f	45.7±3.34e	40.7±3.25e
1.0	44.3±2.11g	45.5±2.21f	45.6±2.87f	42.0±1.00g	44.3±2.45f	40.4±2.45e	41.3±2.76e
IAA							
0.01	66.7±2.45e	65.7±2.12d	58.0±2.45d	54.2±2.27e	53.4±3.23e	53.4±3.23d	51.4±3.23d
0.1	71.4±3.21d	72.7±3.12c	65.7±3.12c	68.2±3.10d	65.7±3.65d	65.7±3.65c	61.3±2.98c
0.5	85.7±3.65c	84.1±2.59b	85.4±3.45b	81.4±3.45b	85.5±4.34b	81.0±4.11a	81.0±4.11a
1.0	75.7±3.40d	85.7±4.12b	85.7±4.12b	75.7±3.65c	71.4±3.56c	75.7±3.56b	71.6±3.40b

Each value represents the average of 7 replicates mean separation using Duncan's New Multiple Range Test. Means within column with different letters are significant at 5% level.
C – Cotyledon; H-Hypocotyl; E-Epicotyl; IML – Immature leaflet; ML – Mature leaf; S- Stem; R – Root.

4.1.1.2. Growth Medium

Our results showed that 2,4-D was most effective for callus induction as well as shoot bud regeneration followed by IAA and NAA. The presence of 2, 4-D (0.5 mg/l) produced prolific callus in all explants. Cotyledonary explant showed the highest response (100%) for callus induction followed by hypocotyl, immature leaflet, epicotyl, mature leaf, stem and root explant (Table 7). The nature of callus was white, friable and shining with moderate proliferation rate.

Arya *et al.* (1981) obtained creamish-yellow to light brown, globular, compact and hard callus from hypocotyl and stem explants grown on MS medium supplemented with KN (0.5 mg/l) and NAA (10 mg/l). Arya and Shekawat (1986) induced callus from axillary bud and hypocotyl explants of *A.marmelos* on MS medium supplemented with 2,4-D (5.0 mg/l) and NAA (5.0 mg/l). Varghese *et al.* (1993) initiated the callus from *Aegle marmelos* stem explants grown on MS medium supplemented with different concentrations of KN, 2, 4 – D and NAA. Hossain *et al.* (1994b) obtained efficient callus production from nucellus tissue of *Aegle marmelos* excised from 90-120-day-old mature fruits. MS medium supplemented with 400 mg/l casein hydrolysate, NAA (5.0 mg/l) and KN (1.0 mg/l) were more efficient for callus production and 3-month-old hard, compact, irregular – shaped callus was transferred to regeneration medium. Islam *et al.* (1995) reported 100% callus formation in embryonic tissues of mature and immature seeds of *Aegle marmelos*, grown on MS medium supplemented with 2, 4-D (2.2 mg/l) and BA (0.12 mg/l). Auxins seem to be capable of erasing the genetically programmed physiology of whole plant tissue, which had previously determined their differentiated state. Cells which respond to auxin revert to a dedifferentiate state and begin to divide. How auxins bring about this reprogramming is not fully understood. Lo Schiavo *et al.* (1989) found that auxins cause DNA to become more methylated than usual and suggested that this might be necessary for the reprogramming of differentiated cells. Tissue-specific programmes specifically associated with differentiation would become eradicated by hypermethylation, with perhaps small fraction of the cells searching an ultimate state of dedifferentiation in which they become capable of morphogenesis, or embryogenesis (Terzi and Lo Schiavo, 1990).

4.1.1.3. Darkness Effect on Callus Induction

The incubation of cultures in dark for various days (1, 2, 3, 4, 5, 6 and 7) promoted abundant callus formation in all explants. Three-day dark treatment showed highest response and cotyledon explant responded 95% followed by other explants (Data not shown). Priyadarshi and Sen (1992) also reported that the complete darkness is an important step to induce callus in Easter lily. Alternative causes of inhibition of growth of callus in the light may be due to reduction in DNA replication by destroying vitamin B_{12}, which inhibits the activity of nucleic acid synthesis (Yeoman and Davidson, 1971), or due to increased production of phenolic compounds which interfere with growth regulator activity (Andersen and Kasperbauer, 1973), or may be due to the destruction of cytochrome oxidase which is involved in respiratory processes (El-Mansy and Salisbury, 1971).

4.1.2. Shoot Regeneration and Multiplication

The callus derived from all explants produced shoots. The callus with tiny shoot buds was sub-cultured onto the medium with similar growth regulators for further shoot proliferation. Thus, shoots were regenerated in each culture tube. In each multiple subculture all dead and discolored tissues or shoots were removed from cultures. The rate of shoot multiplication was high in the initial subcultures (2-3) on the fresh medium, but later, there was a gradual decline. Subculture of regenerated shoots developed further into secondary multiple shoots with little basal callus. The regeneration capacity of various explants varied with various concentrations of cytokinins and auxins. Our results demonstrated that BA suppressed callus formation as well as rooting but induced shoot bud regeneration. Benzyl adenine (1.0 mg/l) produced the highest frequency of shoot regeneration (67.5%) and number of shoots (15.7 shoots/callus) from hypocotyl explant derived callus (Table 8 and 9).

The combination of BA with KN improved shoot regeneration rate and gave number of shoots from hypocotyl explant derived callus (Table 8 and 9). There was a significant increase in shoot regeneration from callus when auxin was added with cytokinins. Of the several combinations tested, BA (1.0 mg/l), KN (0.1 mg/l) and NAA (0.1 mg/l) combination gave the highest percentage of shoot regeneration (97.5%) from hypocotyl explant derived callus with 41.0 shoots/callus (Table 8 and 9). Plate 4c, d is shown only for hypocotyl and root explants. The combined effect of growth regulators stimulated quick response than the individual growth regulator effect. The synergistic effect of cytokinins and auxins also reduced the time taken for shoot regeneration. Arya *et al.* (1981) regenerated shoots (6-7) from the callus (induced from hypocotyl and shoot explants) grown on MS medium supplemented with KN (5.0 mg/l), IBA (0.5 mg/l) and L-tyrosine (50-200 mg/l). Arya and Shekawat (1986) induced shoot bud and multiple shoot formation from the callus (produced from axillary bud and hypocotyl explants) in *A.marmelos* on MS medium supplemented with KN (0.5 mg/l). They further observed that shoot number was doubled at 28-30°C. Varghese *et al.* (1993) observed meristemoids and multiple shoot development from the callus (produced from stem explants) in *Aegle marmelos* when subcultured on medium supplemented with KN and NAA.

Hossain *et al.* (1994b) reported that healthy, green and long multiple shoots were regenerated from callus tissue (obtained from nucellus tissue) of

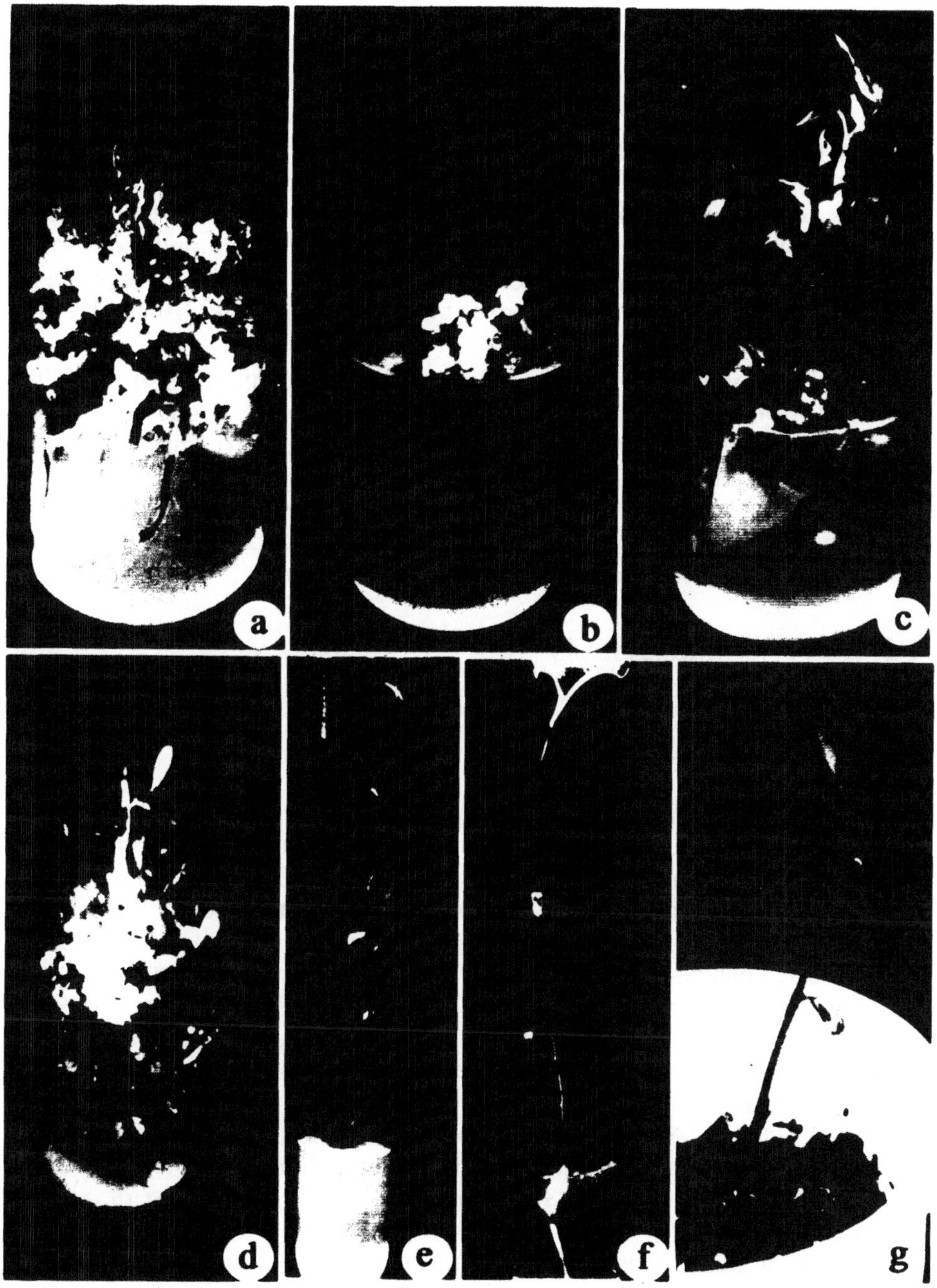

Plate 4. a) Callus induction and shoot bud formation from hypocotyl explant (after 21 days) (1.5x); b) Callus initiation and shoot bud formation from root explant (after 21 days) (1.5x); c) Shoot regeneration and multiplication from hypocotyl callus (after 35days) (1.5x); d) Shoot regeneration and multiplication from root derived callus (after 35 days) (1.5x); e) Shoot elongation (after 15 days) (0.75x); f) Rooted plantlet (after 21 days) (1.5x); g) Acclimatized plantlet (after 25 days) (1.5x).

Aegle marmelos, grown on MS medium supplemented with BA (2.0 mg/l), NAA (0.1 mg/l) and GA_3 (1.0 mg/l). Low concentration of NAA (0.1 mg/l) was capable of sustaining a good growth rate of callus and GA_3 enhanced shoot elongation. They obtained 10,000 shoots in a year from a single nucellus explant with subcultures at 5-6 week interval. Further, elongated shoots (<4 cm long) were excised and cultured on rooting medium. Cotyledon explants of *Aegle marmelos* from 11-15-day-old showed the highest frequency of callus induction, adventious buds and highest number of shoots per explant, on MS medium supplemented with BA, IAA, GA_3. Shoots were elongated on the same basal medium with KN (1.0 mg/l) and IAA (0.1 mg/l) (Islam *et al.*, 1994).

The 6-week-old hard, green and compact irregularly shaped callus (obtained from embryonic tissue) of *Aegle marmelos* produced multiple shoots on MS medium containing KN (1.1 mg/l) and IAA (0.1 mg/l) (Islam *et al.*, 1995). They further reported that in all cases both mature and immature embryo axes-derived calli were capable of producing multiple shoots. However, regeneration frequency and number of shoots per callus were much higher in immature embryo axes than mature embryo axes.

4.2. Direct Organogenesis

4.2.1. Growth Medium

Shoot meristems formed directly on explant tissues are often initiated by the commencement of cell divisions after about 48 hrs in nutrient culture e.g. cotyledons of conifers, meristems in subdermal mesophyll (Sommer *et al.*, 1975; Cheah and Cheng, 1978; Janson and Bornman, 1981; Flinn *et al.*, 1988). There is evidence that a single epidermal cell may give rise to meristematic centers, from one of which up to 22 identical shoots may arise (Broertjes *et al.*, 1976). Broertjes and Van Harten (1978), Broertjes and Keen (1980) suggested that directly regenerated adventious shoot meristems might always form from one or few daughter cells that originated from a single cell. Adventious bud meristems may arise from many cells in sub-epidermal layer (Falavigna *et al.*, 1980; Hussey and Falavigna, 1980). The cytokinins in the culture medium induce adventious bud formation in cotyledons (Villalobos *et al.*, 1983; Abdullah *et al.*, 1987; Noh *et al.*, 1988; Martinez *et al.*, 1990; Lorenzo *et al.*, 1994).

In our studies, the direct shoot regeneration without callus phase was observed from cotyledon, hypocotyl, epicotyl, mature leaf, stem and root explants grown on MS medium supplemented with cytokinins (BA and KN) and auxin (NAA). Benzyl adenine (2.0 mg/l) produced highest frequency of shoot regeneration rate in hypocotyl explant. The shoot formation occurred on the surface of the explant. Kinetin along with BA enhanced shoot regeneration rate. Hypocotyl explant produced the highest frequency of shoot

Table 8. Shoot bud regeneration from different explants derived callus of Aegle marmelos grown on MS medium supplemented with BA, KN and NAA for 35 days.

Growth regulators (mg/l)			Percentage of callus forming shoot buds (Mean±SD)						
			C	H	E	IML	ML	S	R
BA									
0.1			30.0±2.07h	25.0±1.07h	23.1±1.05h	35.4±1.23g	23.1±1.20i	21.2±1.50h	23.1±1.04h
0.5			47.5±3.78g	25.0±1.09h	25.0±1.11h	37.5±1.27g	27.5±1.27h	29.5±1.05g	30.5±1.07g
1.0			66.0±2.20e	67.5±3.1c	60.5±2.71d	65.0±3.11d	66.3±2.35c	65.2±2.09b	62.5±2.12d
2.0			55.0±3.33f	37.5±1.37g	42.8±1.87f	50.0±2.11e	50.0±2.10e	42.5±1.87e	37.7±1.23f
3.0			55.0±2.95f	50.0±2.09e	50.0±2.11e	45.9±2.67f	37.5±1.27g	25.5±1.23h	37.5±1.27f
BA	KN								
1.0	0.01		67.5±2.11e	42.8±1.78f	37.5±1.27g	37.5±1.27g	50.0±2.11e	55.5±2.09c	65.0±2.77d
1.0	0.1		82.5±3.05a	85.5±3.14b	75.5±3.11b	87.5±3.70b	87.5±3.70a	82.3±3.87a	81.2±3.48a
1.0	0.5		70.5±2.85d	50.0±2.11e	42.8±1.78f	65.9±2.55f	65.0±2.71c	62.5±2.16b	72.0±3.14b
1.0	1.0		67.8±2.51e	67.5±1.27c	42.8±1.78f	62.5±2.88d	65.0±2.54c	50.0±2.11d	62.5±2.10d
BA	KN	NAA							
1.0	0.1	0.01	75.0±3.11b	37.5±1.27f	25.0±1.10h	33.5±1.25h	57.3±2.21d	37.5±1.27f	42.5±1.71e
1.0	0.1	0.1	87.5±2.11a	97.5±3.76a	85.5±1.27a	95.5±1.27a	86.2±2.78a	82.0±2.15a	85.5±2.16a
1.0	0.1	0.5	75.0±3.14b	62.5±2.65d	65.2±2.87c	75.0±3.11c	75.0±3.11b	65.7±2.12b	70.3±2.98c
1.0	0.1	1.0	72.5±2.90c	50.0±2.11e	42.8±1.78f	50.0±2.11e	45.2±1.78f	42.5±1.78e	37.5±1.27f

Each value represents the average of 7 replicates mean separation using Duncan's New Multiple Range Test.
Means within column with different letters are significant at 5% level.
C – Cotyledon; H-Hypocotyl; E-Epicotyl; IML – Immature leaflet; ML – Mature leaf; S- Stem; R – Root.

Table 9. Multiple shoot formation from different explants derived callus of Aegle marmelos grown on MS medium supplemented with BA, KN and NAA for 35 days.

Growth regulators (mg/l)			No. of shoots /callus(Mean±SD)						
			C	H	E	IML	ML	S	R
BA									
0.1			2.5±0.43f	2.5±0.83f	2.3±0.47g	2.3±0.47h	2.0±0.89h	1.4±0.48i	1.6±0.48i
0.5			9.0±0.81e	8.2±2.81e	6.6±0.48f	8.1±0.68f	7.1±0.89g	5.0±0.57h	6.1±0.60h
1.0			14.4±0.83d	15.7±1.84d	14.8±0.63de	12.1±1.45e	15.0±1.15d	12.8±0.78f	12.5±0.90e
2.0			11.5±0.52e	12.0±1.19d	13.8±0.74e	11.1±0.68e	12.2±0.74e	11.3±0.74f	10.5±0.90f
3.0			9.6±0.43e	9.7±1.27e	7.8±0.68f	6.2±0.74g	8.0±1.06g	5.4±0.90h	6.8±0.83h
BA	KN								
1.0	0.01		14.8±1.16d	12.1±2.06d	14.0±0.81de	12.6±0.80e	13.2±0.88e	11.0±1.29f	11.0±1.69ef
1.0	0.1		22.4±0.48b	26.0±1.16b	22.8±0.74b	21.5±0.49b	24.0±1.30b	19.0±1.65c	17.5±0.90c
1.0	0.5		18.2±1.27c	17.5±2.06c	16.3±0.94cd	14.3±0.74d	17.5±0.72c	15.8±1.57e	14.5±0.95d
1.0	1.0		19.5±1.17c	19.8±1.34c	15.8±1.34d	17.0±0.75c	15.5±1.15cd	14.3±1.49e	12.7±0.69e
BA	KN	NAA							
1.0	0.1	0.001	21.5±0.50b	20.5±1.29c	23.0±0.75b	22.3±0.74b	23.4±1.17b	22.8±1.12b	24.7±0.69b
1.0	0.1	0.01	39.6±0.48a	41.0±1.64a	35.4±1.39a	36.1±0.68a	37.1±1.64a	33.5±0.76a	32.1±0.89a
1.0	0.1	0.1	18.5±0.95c	18.7±1.38c	17.0±0.81c	17.8±0.33c	16.1±1.84c	17.6±0.94d	15.0±0.57cd
1.0	0.1	0.5	10.8±0.69e	12.0±1.00d	13.6±1.59e	12.8±0.68e	10.0±1.77f	9.4±0.90g	8.6±0.48g

Each value represents the average of 7 replicates mean separation using Duncan's New Multiple Range Test. Means within column with different letters are significant at 5% level.

C – Cotyledon; H-Hypocotyl; E-Epicotyl; IML – Immature leaflet; ML – Mature leaf; S- Stem; R – Root.

regeneration response on MS medium supplemented with BA (2.0 mg/l) and KN (1.0 mg/l). Auxin NAA (0.1 mg/l) showed a combined effect along with BA (2.0 mg/l) and KN (1.0 mg/l) in direct organogenesis (Table 10 and11, Plate 5a).

Rekha Bhati *et al.* (1992) also observed direct regeneration of shoots from root segments of *Aegle marmelos* on MS medium supplemented with BA (0.5 mg/l) and KN (0.5 mg/l), without any callus phase. First the segments of the proximal end of roots differentiated into shoots and subsequent shoot formation occurred all along the surface of the explant. Incorporation of auxin (NAA/IAA) with optimal concentration of cytokinin enhanced the production of shoots per explant as well as their vigour and health.

Cotyledons from *Aegle marmelos* seedlings of 10-day-old showed improved efficiency of shoot bud regeneration on MS medium supplemented with BA (2.0mg/l) and IAA (0.2 mg/l) (Hossain *et al.*, 1994a) or NAA (0.5 mg/l) using nucellus tissue (Hossain *et al.*, 1993). Hossain *et al.* (1995) observed direct organogenesis from hypocotyl explants of *Aegle marmelos* on MS medium supplemented with 0.1 mg/l BA.

High frequency of adventious shoot regeneration from radicle tissues of *Aegle marmelos* was obtained on MS medium supplemented with BA (1.0 mg/l) and NAA (0.2 mg/l) (Islam *et al.*, 1996a). The maximum frequency of shoot organogenesis (61.2%) (without callus formation) and the highest number of shoots/explant (38.4) was obtained from leaf explants of *in vitro* grown seedlings of *Aegle marmelos*, on MS medium supplemented with BA (1.5 mg/l) and IAA (0.5 mg/l). Twenty one –day-old seedlings proved to be the optimum source for explants (Islam *et al.*, 1993).

4.2.2. Effect of Photoperiod

Among the various photoperiods (4/20, 8/16, 12/12, 16/8 and 20/4 hrs light/dark per day) with light intensity 80 $\mu Em^{-2}s^{-1}$ for shoot regeneration, 12/12 hrs for a week showed the highest response of shoot regeneration in cotyledon explant followed by hypocotyl, epicotyl, mature leaf, shoot and root explants (Fig.1). Rekha Bhati *et al.* (1992), Hossain *et al.* (1993) and Ajith kumar and Seeni (1998) were able to obtain *in vitro* propagation response in *A.marmelos* at light intensity 50-70, 60 and 50-60 $\mu Em^{-2}s^{-1}$ respectively.

The photoperiodic effects on plants are energized by relatively low irradiance (Cathey and Campbell, 1980) and are generally associated with the phytochrome system. Length of the day can influence natural growth substance levels within cultured plant tissues (Hillman and Galston, 1961). Murashige (1978) pointed out that plants needing specific photoperiods for vegetative growth and development are likely to manifest this need during *in vitro* culture. However, the photoperiod requisite for flowering may not be the same as that for optimum regeneration of shoots and roots or shoot proliferation. The photoperiod effect on shoot morphogenesis was reported in

Table 10. Direct shoot regeneration response from different explants of Aegle marmelos grown on MS medium supplemented with BA, KN and NAA for 35 days.

Growth regulators (mg/l)			Explants response (Percentage mean ± SD)					
			C	H	E	ML	S	R
BA								
0.5			30.0±1.26f	27.0±1.09e	23.9±1.05f	8.5±1.07f	8.4±0.75f	29.1±1.21e
1.0			37.5±1.27e	35.4±1.28d	30.0±1.25e	10.4±1.14f	11.4±1.05e	37.5±1.31d
2.0			37.5±1.30e	42.5±1.98c	33.7±1.14de	17.5±1.21e	18.9±1.07d	39.6±1.45d
3.0			23.2±1.28g	23.2±1.05e	22.3±1.10f	7.2±1.06f	9.4±1.05f	23.2±1.06f
BA	KN							
2.0	0.1		43.2±2.27d	25.0±2.25e	20.0±1.05f	23.2±2.20d	17.8±0.78d	22.2±2.37f
2.0	0.5		42.5±2.11d	42.5±2.90c	37.5±2.83d	37.5±2.98c	25.3±2.09c	42.3±2.07c
2.0	1.0		55.3±3.12b	60.5±3.06b	50.0±2.28b	47.5±3.27b	38.2±3.15b	54.2±3.14b
2.0	2.0		50.0±1.20c	37.5±2.21d	37.5±3.05d	37.2±1.70c	33.4±2.11bc	42.5±2.12c
BA	KN	NAA						
20	1.0	0.001	25.5±3.21g	23.4±1.50e	22.4±2.25f	25.5±1.25d	18.8±0.95d	22.9±1.09f
2.0	1.0	0.01	42.5±4.21d	45.5±1.80c	42.5±1.17c	45.3±1.20b	35.3±1.34b	37.2±1.33d
2.0	1.0	0.1	65.0±2.87a	67.5±3.85a	62.2±1.54a	67.3±1.57a	65.4±1.17a	65.4±1.79a
2.0	1.0	0.5	37.5±2.92e	37.5±1.44d	42.5±1.64c	35.3±1.88c	27.9±1.24c	35.4±1.24b

Each value represents the average of 7 replicates mean separation using Duncan's New Multiple Range Test.
Means within column with different letters are significant at 5% level.
C – Cotyledon; H-Hypocotyl; E-Epicotyl; IML – Immature leaflet; ML – Mature leaf; S- Stem; R – Root.

Table 11. Multiple shoot formation (direct organogenesis) from different explants of Aegle marmelos grown on MS medium supplemented with BA, KN and NAA for 35 days.

Growth regulators (mg/l)			No. of shoots/explants (Mean ± SD)					
			C	H	E	ML	S	R
BA								
0.5			9.6±0.47d	8.5±0.70d	7.6±0.48e	0.0	0.0	1.6±0.48g
1.0			11.5±0.83d	11.6±0.48c	10.8±0.61d	0.0	0.0	1.6±0.48g
2.0			12.8±0.75c	14.6±0.48c	13.2±0.45c	0.0	0.0	2.1±0.78fg
3.0			12.0±0.78cd	12.3±0.69c	11.1±1.29d	0.0	0.0	1.6±0.69g
BA	KN							
2.0	0.1		4.4±1.15f	4.6±2.11e	4.2±1.12f	1.4±0.94e	1.6±0.66d	3.0±1.15f
2.0	0.5		5.5±0.50ef	5.0±1.05e	4.9±0.95f	2.7±0.66c	1.6±0.99d	6.5±1.28d
2.0	1.0		18.2±0.96b	18.5±0.80b	17.5±0.30b	3.1±37b	2.6±0.24b	14.5±1.82b
2.0	2.0		7.3±1.33e	7.5±1.44d	7.2±1.27e	1.7±0.66d	1.8±0.99c	6.4±1.25d
BA	KN	NAA						
2.0	1.0	0.001	13.1±4.03c	12.5±0.96c	9.8±0.83d	1.3±0.47e	1.2±0.45e	5.4±1.29e
2.0	1.0	0.01	16.0±0.94b	16.8±1.16b	15.3±0.95bc	1.3±0.48e	1.3±0.45e	5.0±1.00e
2.0	1.0	0.1	23.1±4.03a	31.1±2.84a	26.3±1.23a	3.8±0.73a	3.4±0.83a	21.0±1.56a
2.0	1.0	0.5	8.4±0.95de	8.2±2.92d	7.5±1.28e	1.5±0.43e	1.2±043e	10.7±1.98c

Each value represents the average of 7 replicates mean separation using Duncan's New Multiple Range Test. Means within column with different letters are significant at 5% level.

C – Cotyledon; H-Hypocotyl; E-Epicotyl; IML – Immature leaflet; ML – Mature leaf; S- Stem; R – Root.

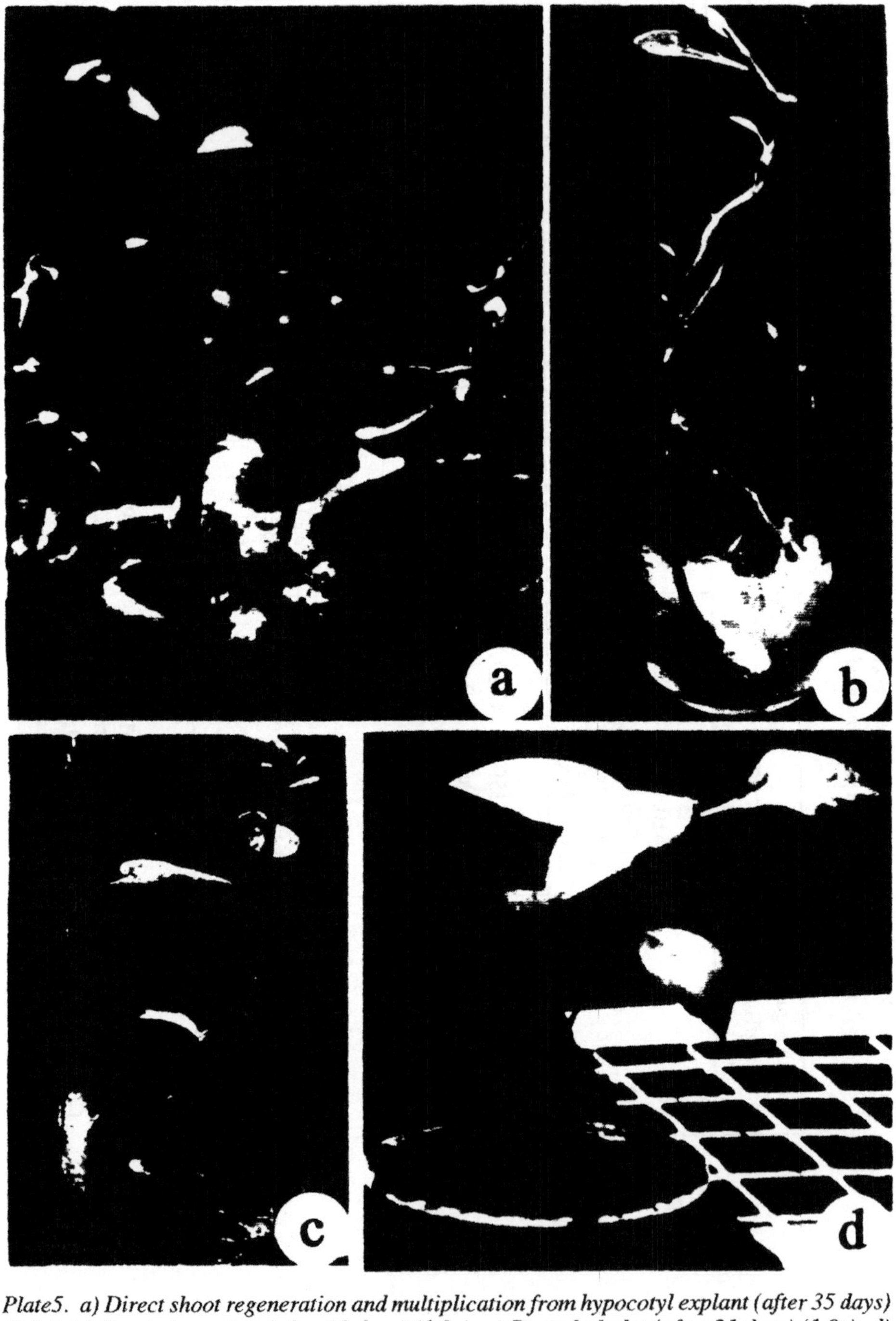

Plate5. a) Direct shoot regeneration and multiplication from hypocotyl explant (after 35 days) (3.0x); b) Shoot elongation (after 15 days) (1.0x); c) Rooted platlet (after 21 days) (1.0x); d) Potted plant (after 25 days) (0.75x).

various plant species by several workers (Heide, 1964; Kukulczanka *et al.*, 1977; Kato, 1978; Stimart and Ascher, 1978; Hussey and Falavigna, 1980; Phillips and Hubstenberger, 1985; Nehra *et al.*, 1988).

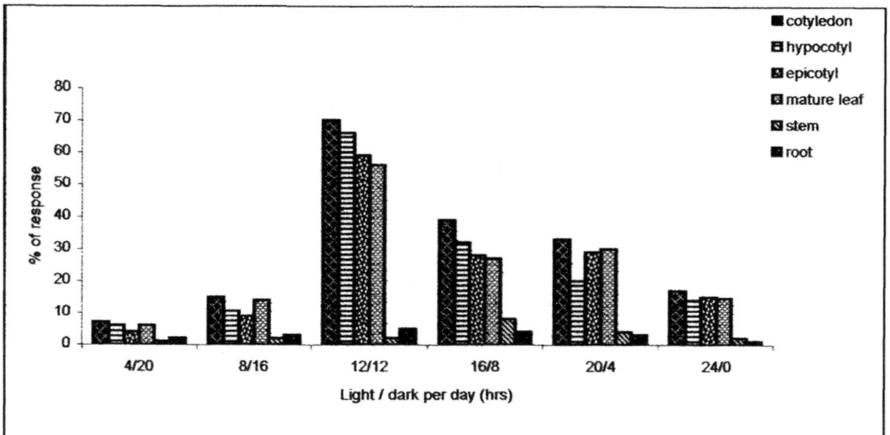

Figure 1 . Effect of photoperiod on direct regeneration

4.2.3. Age of Explants

The age of the seedling explants was important for direct organogenesis. Our studies revealed that there was an increase in shoot regeneration with respect to the age of seedlings, reaching a maximum response at 21 days. Thereafter, the response sharply declined, and shoots differentiated from them did not grow vigorously (Fig.2). So the explants from 21-day-old seedlings were used in all of our studies. Islam *et al.* (1993) also obtained maximum adventious shoot buds from leaf explant of *Aegle marmelos* collected from 21-day-old seedling. Hossain *et al.* (1994a) regenerated *Aegle marmelos*

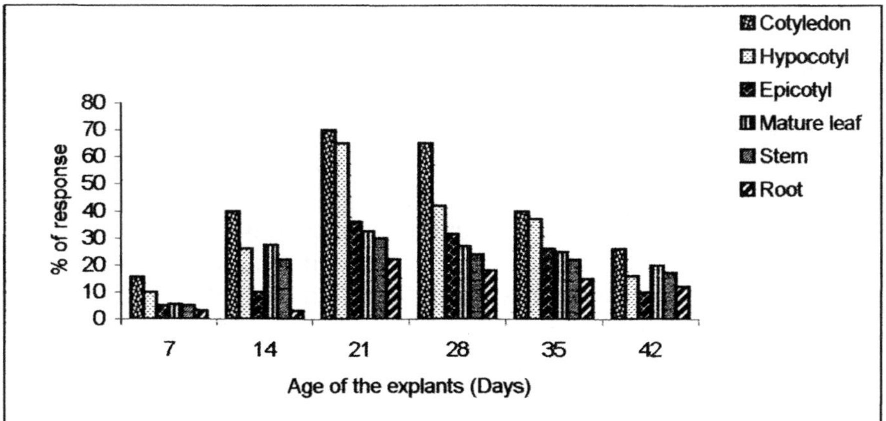

Figure 2 . Effect of age of explants on direct shoot proliferation

plantlets from *in vitro* cultured cotyledons of different ages seedlings. They observed that cotyledon explants derived from 10-day-old seedling performed well. It may be due to higher nutrient reserves.

Explant age and the degree of differentiation of tissues are often inter-related and produce interactive effects *in vitro*. Both the size, and degree of development of certain organs particularly cotyledons, hypocotyls and epicotyls depend on plant age. Toivonen and Kartha (1988) observed that cotyledons excised from 7-8-day-old seedlings of *Pinus glauca* readily form adventious shoots. After 8 days the capacity of the explants declined to produce shoots. Kathal *et al.* (1988) noted that adventive shoots in *Cucumis melo* were more in small leaves (0.3-0.5mm) obtained from 14-day-old seedlings than from older seedlings (21, 28, 35 d) or from larger leaves (0.6-0.9 mm; 1.0-1.2 mm).

4.3. Shoot Elongation

Our studies revealed that 0.1 mg/l GA_3 supported maximum shoot elongation (5.6 cm length/shoot) within 15 days of culture (Data not shown). When increasing the concentrations of GA_3, the shoot elongation was also increased, up to optimum level (0.1 mg/l); afterwards it was decreased with further increase (Plate 4e, 5b).

Multiple shoots regenerated from leaf explant of *Aegle marmelos* seedlings were elongated on MS medium supplemented with BA (0.5 mg/l) (Islam *et al.*, 1993). Whereas, shoots obtained from nucellus tissue were elongated on the medium supplemented with low concentration of BA (0.1 mg/l) (Hossain *et al.*, 1993). Adventitious shoots regenerated from cotyledon explant of 10-day-old *Aegle marmelos* were elongated on MS medium containing 0.5 mg/l KN and 0.1-mg/l GA_3 (Hossain *et al.*, 1994a). Adventitious shoots regenerated from radicle tissue of *Aegle marmelos* were elongated well on MS medium supplemented with low concentration (0.1mg/l) of BA alone (Islam *et al.*, 1996 a).

4.4. Rooting

The elongated shoots (>3.0 cm length) were grown on rooting medium, half-strength MS medium supplemented with IBA (0.01 – 3.0 mg/l) and KN (0.01-0.5 mg/l). It was observed that IBA (1.0 mg/l) induced 66.7% rooting in cotyledon explant derived shoots and thereafter further increase in IBA concentration decreased rooting. The rooting potential was enhanced by synergistic effect of KN with IBA, up to 96.7% in cotyledon explant derived shoots (Table 12, Plate 4f, g, 5c, d).

Rooting of *in vitro* regenerated shoots of *Aegle marmelos* derived from various explants was earlier reported (Arya *et al.*, 1981; Arya and Shekawat, 1986; Varghese *et al.*, 1993; Islam *et al.*, 1993; Hossain *et al.*, 1993; Islam *et al.*, 1994; Hossain *et al.*, 1994a, 1994b; Islam *et al.*, 1995,

Table 12. Root induction of shoots derived from different explants in Aegle marmelos grown on half-strength MS medium supplemented with IBA and KN for 15 days.

Growth regulators (mg/l)		Rooting response (Percentage Mean ± SD)						
		C	H	E	IML	ML	S	R
IBA								
0.01		36.1±2.38j	35.2±1.44g	34.1±2.73f	36.7±1.50g	35.0±1.00g	27.2±1.25h	37.3±1.66f
0.1		47.3±2.74h	37.7±1.60g	48.1±2.42e	43.1±2.94f	45.1±1.87f	44.2±1.95f	45.1±1.97e
0.5		52.7±3.28g	44.6±2.55f	49.5±2.34e	49.4±2.15ef	50.7±2.18e	49.5±1.25e	51.1±1.80d
1.0		66.7±3.48e	59.1±2.65d	61.2±3.40d	66.1±2.22d	64.5±2.67d	64.2±2.12d	65.7±2.23c
2.0		44.1±2.45i	53.5±2.89e	50.4±2.93e	52.4±2.38e	52.4±2.11e	51.3±1.95e	52.2±1.87d
3.0		35.5±2.25j	36.2±1.56g	34.5±1.12f	35.6±1.77g	37.2±1.22g	38.1±1.55g	37.2±1.35f
IBA	KN							
1.0	0.01	88.7±4.94b	87.2±4.30b	84.1±4.63b	86.2±3.37b	80.7±3.98b	85.0±3.45b	87.1±3.90b
1.0	0.1	96.7±4.68a	91.7±4.34a	91.4±4.16a	91.8±4.56a	91.2±4.18a	96.2±4.55a	95.2±4.25a
1.0	0.2	85.3±4.14c	85.2±4.72b	85.7±4.32b	84.4±3.73b	76.8±3.12c	77.2±3.30c	85.2±3.89b
1.0	0.4	73.4±3.43d	75.1±3.48c	75.6±3.73c	70.5±3.37c	68.9±2.98d	67.0±3.08d	68.8±2.97c
1.0	0.5	57.3±2.75f	54.1±2.83e	53.8±2.67e	55.2±1.45e	53.4±2.11e	52.2±2.90e	54.5±2.29d

Each value represents the average of 7 replicates mean separation using Duncan's New Multiple Range Test. Means within column with different letters are significant at 5% level.

C – Cotyledon; H-Hypocotyl; E-Epicotyl; IML – Immature leaflet; ML – Mature leaf; S- Stem; R – Root.

1996a, 1996b). All these have used half-strength MS medium with a suitable combination of IBA, IAA or NAA.

4.5. Hardening

Rooted plantlets (21-day-old) after hardening (as described in micropropagation) were transferred to field.

5. CONCLUSIONS

Aegle marmelos is extensively used in indigenous medicine systems and for afforestation programmes. Its genetic improvement and production can be enhanced effectively by *in vitro* propagation methods. Hence, the present investigation was undertaken. We have been successful in developing efficient protocols for micropropagation using cotyledonary node explant mature tree axillary bud explant (Fig. 3) and organogenesis both direct and indirect via hypocotyl explants (Fig. 4).

Micropropagation studies revealed that MS medium is most suitable growth medium. Cotyledonary node explant was best for efficient shoot bud initiation and proliferation. Shoot bud initiation and proliferation were effective on MS medium amended with BA (3.0 mg/l), KN (0.5 mg/l) and NAA (0.01 mg/l). Shoot elongation was obtained on GA_3 (1.0 mg/l) supplemented medium. Elongated shoots were rooted on half-strength MS medium containing with IBA (1.0 mg/l) and KN (0.1 mg/l).

Hypocotyl explant induced highest direct organogenesis on MS medium supplemented with BA (2.0 mg/l), KN (1.0 mg/l) and NAA (0.1 mg/l). Among photoperiod treatments, 12/12 hrs (L/D) per day for a week period showed the highest frequency of direct shoot regeneration in all explants. Twenty one-day-old seedlings were found to be best for maximum shoot regeneration response. Cotyledon, hypocotyls and immature leaflet explants showed better callus induction response on mMS medium supplemented with 2,4-D (0.5 g/l). Three-day dark pretreatment enhanced callus induction. Shoot bud regeneration and multiplication from the callus was observed on MS medium supplemented with BA (1.0 mg/l), KN (0.1 mg/l) and NAA (0.1 mg/l). Shoot elongation was observed on GA_3 (0.1 mg/l) supplemented medium. The enhanced rhizogenesis was observed on half-strength MS medium supplemented with IBA (1.0 mg/l) and KN (0.1 mg/l).

The following aspects may be considered for future research.

1. To find out the effect of various light intensities on direct organogenesis.
2. To find out somaclonal variations obtained via indirect organogenesis can be explored for plant improvement.
3. To assess the improvement of quantity of active principles (secondary metabolites) *in vitro* raised plants.

4. To quantify the field performance of tissue culture plants in terms of yield performance of different plant parts.
5. More reports are required on *in vitro* propagation of *Aegle marmelos* distributed over geographical regions of the world.

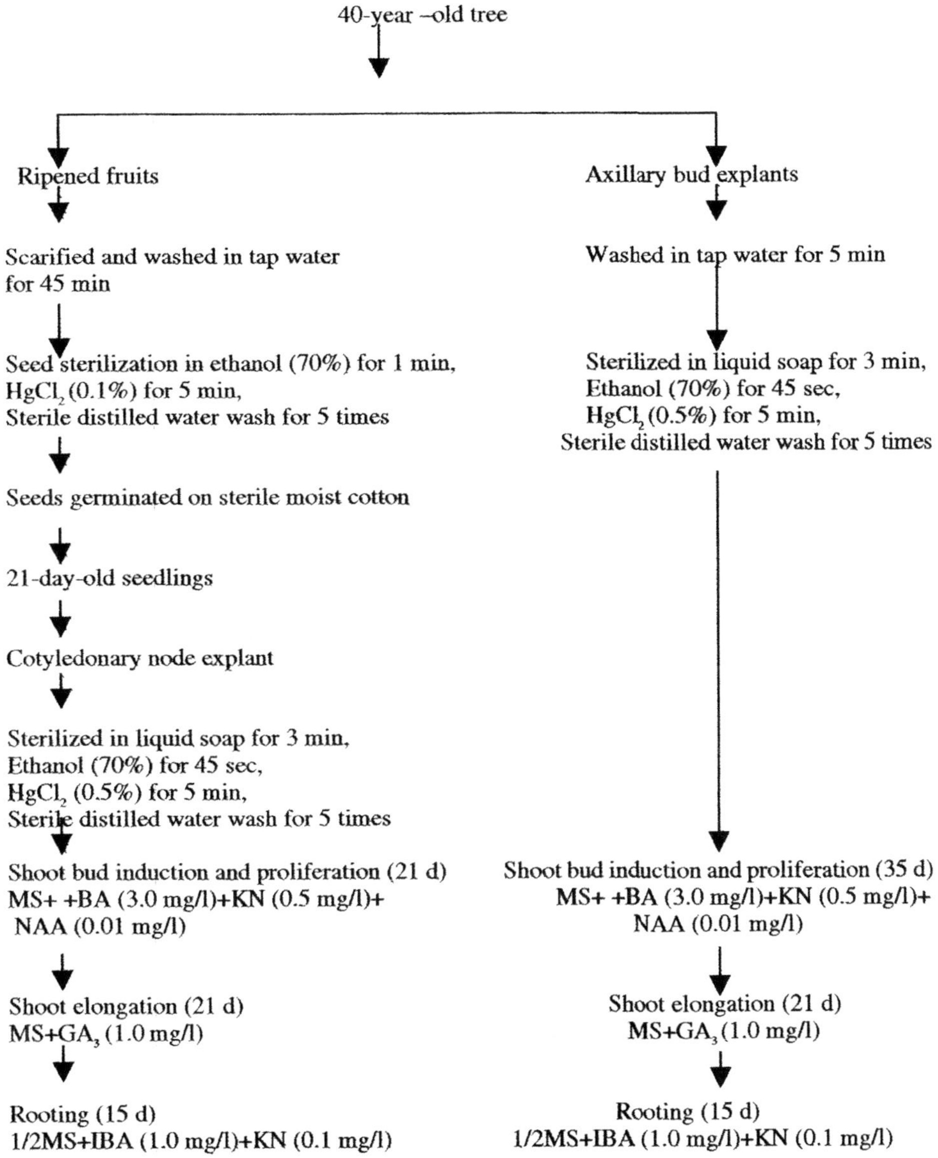

Figure 3. Protocol for micropropagation of Aegle marmelos

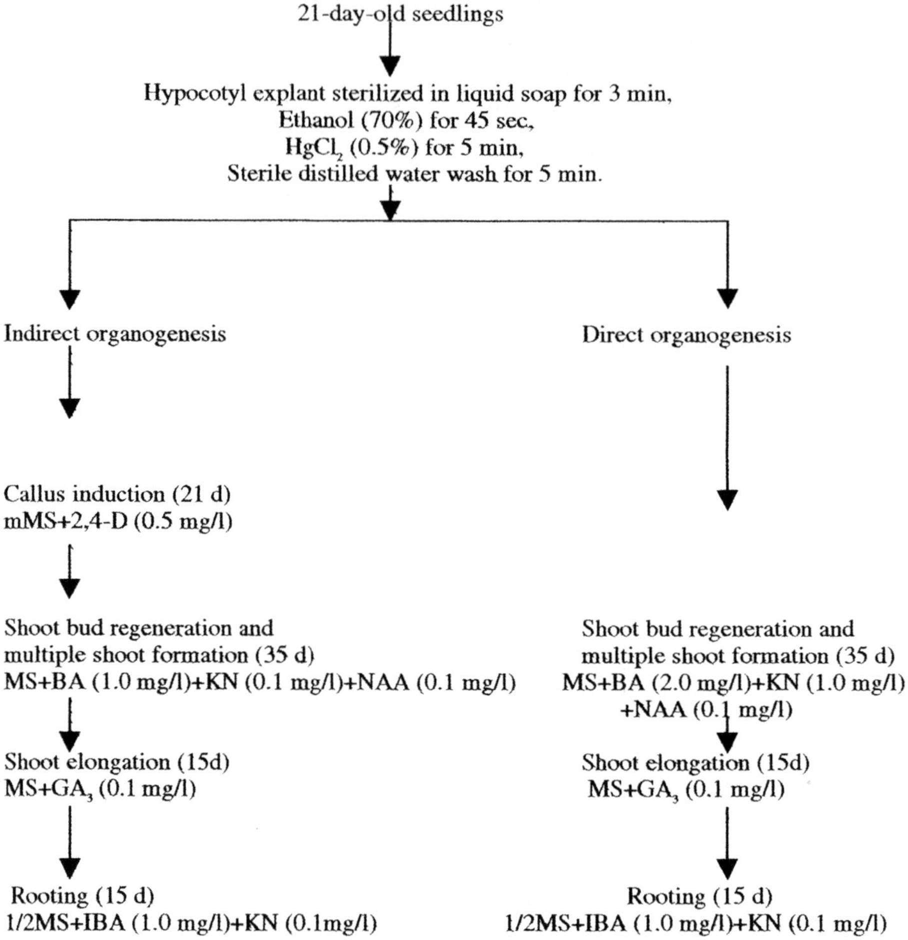

Figure 4. Protocol for Organogenesis in Aegle marmelos

6. ACKNOWLEDGEMENTS

Authors are thankful to Prof.Dr.K.V. Krishnamurthy, Head, Department of Plant Science, for providing the laboratory facilities. S.Arumugam is grateful to the University Grants Commission, New Delhi for providing Junior and Senior Research Fellowships.

7. REFERENCES

Abdullah, A., Yeoman, M. and Grace, J. 1987. Regeneration of Calabrian pine from juvenile needles. Plant Sci. 53:147-155.

Ahuja, A., Verma, M. and Grewal, S.1982. Clonal propagation of *Ocimum* species by tissue culture. Indian J. Exp.Biol.20: 455-458.

Ahuja, M.R.1986. Micropropagation of juvenile and mature beech and oak. In: Abstracts VI Intl. Cong. Plant Tissue and Cell Cult., Minneapolis, Minn. P.31

Aitken-Christie, J. and Jones, C. 1987. Towards automation; radiata pine shoot edges *in vitro*. Plant Cell Tiss. Org. Cult. 8: 195-196.
Ajithkumar, D. and Seeni, S. 1998. Rapid clonal multiplication through *in vitro* axillary shoot proliferation of *Aegle marmelos* (L.) Corr., a medicinal tree. Plant Cell Rep.17: 422-426.
Aldwinckle, H.S. and Gustafson, H.I. 1981. *In vitro* propagation of *Malus prunifolia* cultivar Xantho Caupa. Env. Exp. Bot.21: 440 (Abst.).
Ambasta, S.P. 2000. The useful plants of India. National Institute of Science Communication. CSIR, New Delhi. Pp.16-17.
Anderson, R. and Kasperbauer, M.J. 1973. Chemical composition of tobacco leaves alterd by near –ultra violet and intensity of visible light. Plant Physiol. 51: 723-726.
Arora, R.and Bhojwani, S.S.1989. *In vitro* propagation and low temperature storage of *Saussurea lappa* C.B. Clarke – an endangered medicinal plant. Plant Cell Rep. 8:44-47.
Arumugam, S. and Rao, M.V.1996a. *In vitro* production of plantlets from cotyledonary node cultures of *Aegle marmelos* (L.) Corr. Adv. Plant Sci. 9(2): 181-186.
Arumugam, S. and Rao, M.V. 1996b. *In vitro* regeneration of plantlets directly from epicotyl explants of *Aegle marmelos* (L.) Corr. Int. J. Mendel 13(3&4): 81-82.
Arumugam, S., Alagumanian, S. and Rao, M.V. 1997. Shoot differentiation and development from cotyledonary explants of *Aegle marmelos* (L.) Corr. via - organogenesis. J. Swamy Bot. Cl. 14: 108-112.
Arumugam, S. 1998. *In vitro* studies in *Aegle marmelos* (L.) Corr., a medicinal tree. Ph.D. thesis, Bharathidasan University, Tiruchirapalli, India. P.116.
Arumugam, S. and Rao, M.V. 2000. Somatic embryogenesis in *Aegle marmelos* (L.) Corr., a medicinal tree. In: Somatic embryogenesis in woody plants, Vol.6 (eds) S.M.Jain, P.K.Gupta and R.J.Newton. Kluwer Acad. Publ., Dordrecht, London. Pp.605-656.
Arya, H.C., Ramawat, K.G. and Suthar, K.C. 1981. Culture and differentiation of plants of economic importance. J. Ind. Bot. Soc. 60: 134-137.
Arya, H.C. and Shekawat, N.S. 1986. Clonal multiplication of tree species in the Thar desert through tissue culture. For. Ecol. Manag. 16: 201-208.
Asokan, M.P., O' Hair, S.K. and Litz, R.E.1984. Rapid multiplication of *Xanthosoma caracu* by *in vitro* shoot tip culture. Hort. Sci. 19:885.
Babic, V. and Neskovic, M. 1984. Propagation of three blackberry cultivars from small apical buds *in vitro*. J. Hort. Sci.59: 183-185.
Banerji, I. and Pal, S. 1957. A note on the cytology and pollen of *Aegle marmelos* Corr. Phyton (Bunos Aires) 8: 75-78.
Bhat, S.R., Chandel, K.P.S.and Malik, S.R. 1995. Plant regeneration from various explants of cultivated *Piper* species. Plant Cell Rep. 14:395-402.
Boulay, M.1984. Some practical aspects and applications of the micropropagation of forest trees. In: Proc. International Symposium on *In vitro* propagation of forest tree species, Bolonga, pp.51-81.
Boxus, Ph. 1974a. The production of strawberry plants by *in vitro* micropropagation. J.Hort.Sci. 49:209-210.
Boxus, Ph. 1974b. La micropropagation *in vitro* du fraisier. Proc. Int. Hort. Cong. 19(1a): 65.
Boxus, Ph. Quoirin, M. and Laine, J.M. 1977. Large scale propagation of strawberry plants from tissue culture. In: Plant cell Tissue and Organ Culture (Eds) Reinert, J. and Y.P.S. Bajaj. Springer-Verlag, Berlin. Pp.130-143.
Brassard, N., Brissette, L., Lord, D. and Laiberte, S. 1996. Elongation, rooting and acclimatization of micropropagated shoots from matured material of hybrid larch. Plant Cell Tiss. Org. Cult. 44: 37-44.
Broertjes, C., Roest, S. and Bokelmann, G.S. 1976. Mutation breeding of *Chrysanthemum morifolium* using *in vivo* and *in vitro* adventious bud techniques. Euphytica 25:11-19.
Broertjes, C. and Van Hanten, A.M.1978. Application on mutation breeding in the improvement of vegetatively propagated crops. Elsevier Sci. Publ., Amsterdam, Oxford, New York.

Broertjes, C. and Keen, A. 1980. Adventious shoots-do they develop from one cell? Euphytica 29: 73-87.

Cathey, H.M. and Campbell, L.E. 1980. Light and lightining systems for horticultural plants. Hort. Rev. 2:491-537.

Chadha, Y.R. 1985. Wealth of India: Raw materials. CSIR, New Delhi, Vol. 1- A (rev), pp. 513.

Cheah, K.T. and Cheng, T.Y. 1978. Histological analysis of adventious bud formation in cutured Douglas fir cotyledon. Am.J.Bot.65: 845-849.

Das, T. and Mitra, G.C. 1990. Micropropagation of *Eucalyptus tereticornis* Smith. Plant Cell Tiss. Org. Cult. 22:95-103.

De La Guardia, M.D. and Benlloch, M. 1980. Effects of potassium and gibberellic acid on stem growth of whole sunflower plants. Physiol. Plant. 49: 443-448.

Dewan, A., Nanda, K. and Gupta, S.C.1992. *In vitro* micropropagation of *Acacia nilotica* subsp. *indica* Bernan VI cotyledonary nodes. Plant Cell Rep. 12: 18-212.

Dhar, U. and Upreti. J. 1999. *In vitro* regeneration of a mature leguminous liana (*Bauhinia vahlii* Wight & Arnott) Plant Cell Rep.18: 664-669.

Distabanjong, K. and Geneve, R.L. 1997. Multiple shoot formation from cotyledonary node segments of Eastern redbud. Plant Cell Tiss. Org. Cult. 47:247-254.

Dore, C. 1975. La multiplication clonale de l" asperge (*Asparagus officinalis*) par culture in vitro: Son utilization et selection. Ann. Amel Plantes 25: 201-224.

Durga, M.B. and Mehta, A.R. 1993. Clonal propagation of mature elite trees of *Commiphora wightii*. Plant Cell Tiss.Org. Cult.35: 237-244.

El-Mansy, H.I. and Salisbury, F.B.1971. Biochemical responses of *Xanthium* leaves to ultra violet light. Rad.Bot. 11: 325-328.

Falavigna, A., Hussey, G. and Hilton, J.1980. The effect of colchicine on proliferating adventious shoots in *Allium cepa*. Ann.Rep.John Innes. Inst.1979, p.58.

Flinn, B.S., Webb, D.T. and Necomb, W. 1988. The role of cell clusters and promeristemoids in determination and competence for caulogenesis by *Pinus strobus* cotyledons *in vitro*. Can.J. Bot.66: 1556-1565.

Franklet, A. and Boulay, M. 1989. Rejuvenation and clonal silviculture for *Eucalyptus* and forest species harvested through short rotation. In: eds. Biomass production by fast growing trees (Eds) J.S. Pereira and J.J. Landesberg Pp. 267-274. Dortrecht. The Netherlands: Kluwer Acad. Publ.

Franklin, C.I., Trieu, T.N., Gonzales, R.A. and Dixon.1991. Plant regeneration from seedling explants of green bean (*Phaseolus vulgaris* L.) *via* organogenesis. Plant Cell Tiss. Org. Cult.24: 199-206.

Gamborg, O.L., Miller, R.A. and Ojima, K. 1968. Nutrient requirement of suspension cultures of soybean root cultures. Expt.Cell Res. 50: 151-158.

Gars, M.C.S. and Calvo, M.C. 1996. Micropropagation of *Lavendula latifolia* through nodal bud culture of mature plants. Plant Tiss. Org. Cult. 45:259-261.

George, E.P. 1993. Plant propagation by tissue culture. Part 1. The technology. Exegetics Ltd., England. P. 574.

Gomez, K.Z. and Gomez, K.Z. 1976. Statistical procedure for agricultural research with emphasis of rice. International Rice Research Institute, Los Bans, Phillippines.

Gupta, P.L. and Sharma, R.D. 1999. Role of Bilwa-patra in neonatal jaundice. South East Asian seminar on herbs and herbal medicines, Patna.p.113.

Halperin, W. 1986. Attainment and retention of morphogenic capacity *in vitro*. In: Cell culture and somatic cell genetics of plants. Vol.3. Plant generation and genetic variability. (Ed.) I.K.Vasil. Acad. Press, New York. Pp.3-47.

Hansen, C.F., Meins, F. Jr. and Milani, A. 1985. Clonal and physiological variation in the cytokinin content of tobacco cell lines differing in cytokinin requirement and capacity of neoplasmic growth. Differentiation 29: 1- 6.

Hansen, C.F., Meins, F. Jr. and Milani, A.1987. Hormonal regulations of zeatin-riboside accumulation by cultured tobacco cells. Planta 172: 520-525.

Haravey, S.K. 1968. A preliminary communication of the action of *Aegle marmelos* (bael) on heart. Indian J. Med. Res. 56:327.

Hazarika, B.N., Nagaraju, V. and Parthasarathy, V.A. 1996a. Morphogenetic response of bael (*Aegle marmelos* L.) microshoots to benzyl amino purine. Ann. Pl. Phys. 10 (1): 40-44.

Hazarika, B.N., Nagaraju, V. and Parthasarathy, V.A. 1996b. *Ex vitro* acclimatization of micro shoots of *Aegle marmelos* L. Int. J. Trop. Agri. 14(1-4): 251-253.

Heide, O.M.1964. Effects of light and temperature on regenerative ability of Begonia leaf cuttings. Physiol. Plant. 17: 789-804.

Hicks, G.S.1980. Patterns of organ development in plant tissue culture and the problem of organ determination. Bot.Rev.46: 1-23.

Hillman, W.S. and Galston, A.W.1961. The effect of external factors on auxin content. Encyclopedia Plant Physiol.14: 683-702.

Hooker, J.D. 1975. The flora of British India, Vol.I. Reeve, United Kingdom, pp. 516-517.

Hossain, M., Karim, M.R., Islam, R. and Joarder, O.I. 1993. Plant regeneration from nucellar tissues of *Aegle marmelos* through organogenesis. Plant Cell, Tiss. Org. Cult. 34:199-203.

Hossain, M., Islam, R., Karim, M.R., Joarder, O.I. and Biswas, B.K. 1994a. Regeneration of plantlets from *in vitro* cultured cotyledons of *Aegle marmelos* Corr. (Rutaceae). Sci.Hort.57: 315-321.

Hossain, M., Islam, R., Karim, M.R., Rahman, S.M. and Joarder, O. 1994b. Production of plantlets from *Aegle marmelos* nucellar callus. Plant Cell Rep. 13:570-573.

Hossain, M., Biswas B.K., Karim, M.R., Rahman, S.M., Islam, R. and Joarder, O.1994c. *In vitro* organogenesis of elephant apple (*Feronia limonia*). Plant Cell. Tiss. Org. Cult.39: 265-268.

Hossain, M., Islam, R., Islam, A. and Joarder, O.I. 1995. Direct organogenesis in cultured hypocotyls explants of *Aegle marmelos* Corr. Plant Tiss. Cult. 5(1): 21-25.

Hu, C.Y. and Wang, P.J. 1983. Meristem, shoot tip and bud culture. In: Handbook of Plant Cell Culture, vol.1 (Eds) Evans D.A, Sharp, W.R., Ammirato, P.V. and Yamada, Y. Macmillan Publ. Co., Newyork, pp 177-227.

Hussey, G. and Falavigna, A.1980. Origin and production of *in vitro* adventious shoots in the onion *Allium cepa* L. J.Exp. Bot.31: 1675-1686.

Hutchinson, J.F. and Zimmerman, R.H. 1987. Tissue culture of temperate fruit and nut trees. Hort. Rev. 9:273-349.

Islam, R., Hossain, M., Joarder, O.I. and Karim, M.R. 1993. Adventious shoot formation on excised leaf explants of *in vitro* grown seedlings of *Aegle marmelos* Corr. J. Hort. Sci.68: 495-498.

Islam, R., Karim, M.R., Rahman, S.M., Hossain, M. and Joarder, O.I. 1994. Plant regeneration from excised cotyledon of *Aegle marmelos* Corr. Pak.J.Bot.26 (2): 393-396.

Islam, R., Hossain, M., Karim, M.R. and Joarder, O.I. 1995. Regeneration of *Aegle marmelos* (L.) Corr., plants *in vitro* from callus cultures of embryonic tissues. Curr. Sci. 69:494-495.

Islam, R., Ahad, R., Rahman, M.H., Hossain, M. and Joarder, O.I. 1996a. High frequency adventious plant regeneration from radicle explants of *Aegle marmelos* Corr. Pak. J. Bot. 28:203-206.

Islam, R., Hossain, M., Reza, M.A., Mamun, A.N.K. and Joarder, O.I. 1996b. Adventious shoot regeneration from root tips of intact seedlings of *Aegle marmelos* J. Hort. Sci. 71:995-1000.

Islam, R., Hossain, M., Hoque, A. and Joarder, O.I. 1996c. Somatic embryogenesis and plant regeneration in *Aegle marmelos* – a multipurpose social tree. Curr. Sci.71: 259-260.

Jackson, J.A. and Hobbs, S.L.A. 1990. Rapid multiple shoot production from cotyledonary node explants of pea (*Pisum sativum* L.). *In Vitro* Cell. Dev.Biol. 26:835-838.

Jacoboni, A. and Standardi, A. 1987. Tissue culture of Jojoba (*Simmon chinensis* Link). Acta Hort. 212:557-560.

Janson, E. and Bornman, C.H.1981. *In vitro* initiation of adventious structures in relation to the abscission zone in needle explants of *Picea abies* anatomical considerations. Physiol.Plant.53: 191-197.

Jaswanth, A., Akilandeswari, L.V., Manimaran, S. and Ruckmani.2001. Wound healing activity of *Aegle marmelos*. Indian J. Pharm. Sci. 63(1): 41-44.
Jones, O.P., Hopgood, M.E. and O'Farmell, D.1977. Propagation *in vitro* of M 26 apple rootstocks. J. Hort. Sci. 52: 235-238.
Jones, O.P., Pontikis, C. A. and Hopgood, M.E.1979. Propagation *in vitro* of five apple scion cultivars. J. Hort. Sci. 54:155-158.
Joshi, R., Shekhawat N.S. and Rathore, T.S. 1991. Micropropagation of *Anogeissus pendula* Edgew. - an arid forest tree. Indian J. Exp. Biol. 29:615-618.
Jouannaeau, J.P. 1970. Renouvellement des proteins et effect specifique de la kinetine sur des cultures de cellules de Tabac. Physiol. Plant. 23: 232-244.
Jouannaeau, J.P. 1975. Protein synthesis requirement for the cytokinin effect upon tobacco cell division. Exp.Cell Res.91: 184-190.
Kathal, R., Bhatnagar, S.P. and Bhojwani, S.S. 1988. Regeneration of plants from leaf explants of *Cucumis melo* cv. Pusa Sharibati. Plant Cell Rep.7: 449-451.
Kato, Y.1978. Induction of adventiuos buds on undetached leaves, excised leaves and leaf fragments of *Heloniopsis orientalis* (Liliaceae). Plant Cell Physiol.19:791-799.
Kaur, K. and Kant, U. 2000. Clonal propagation of *Acacia catechu* Willd. by shoot tip culture. Plant Growth Reg. 31: 143-145.
Kothari, S.L., Chandra, N. and Sharma, V.K. 1991. Multiple shoot formation from cotyledonary node and shoot apex cultures of *Glycine max*. Soyabean Genet. Newsl.18: 148-149.
Kukulczanka, K., Klimaszewska, K. and Pluta, H.1977. Regeneration of entire plants of *Peperomia scandens* from different parts of leaves *in vitro*. Acta. Hort.78: 365-371.
Kukreja, A.K., Mathur, A.K. and Zain, M.1990. Mass production of virus free patchouli plants [*Pogostemon cablin* (Blanco) Benth] by *in vitro* culture. Trop. Agric. 67:101-104.
Lakshmi, G. N. and Seeni. S. 2001. Rapid *in vitro* multiplication of *Celastrus paniculatus* Willd. sub sp. *paniculatus* (Celastraceae), a medicinal woody climber. Indian J. Exp. Bot. 39: 697-704.
Lee, S.K. and Rao, A.N. 1986. *In vitro* regeneration of plantlets in *Fragraea fragans* Roxb. – a tropical tree. Plant Cell Tiss. Org. Cult. 7: 43 – 51.
Lloyd, G. and McCown, B. 1981. Commercially feasible micropropagation of Mountain Laurel, *Kalmia latifolia*, by use of shoot tip culture. Int. Plant Prop. Soc. Proc. 30:421-427.
Lo Schiavo, F., Pitto, L., Giuliano, G., Torti, G., Nuti-Ronchi, V., Marazziti, D., Vergara, R., Orselli, S. and Terzi, M. 1989. DNA methylation embryogenenic carrot cell cultures and its variation as caused by mutation, differentiation, hormones and hypomethylating drugs. Theor. Appl. Genet. 77: 325-331.
Lorenzo, G.F., Serrano, L. and Pardos, J.A. 1994. *In vitro* shoot organogenesis from excised immature cotyledons and microcuttings production in Stone Pine. Plant Cell Tiss. Org. Cult. 36:135-140.
Mallick, M.A. and Rashid, A. 1989. Induction of multiple-shoots from cotyledonary node of grain legumes, pea and lentil. Biol. Planta. 31:230-232.
Mantell, S.H., Matthews, J.A. and Mckee, R.A. 1985. Principles of plant biotechnology. Blackwell Sci. Publ., Oxford.
Margara, J. 1969. Etude prilimina des facteurs de la neoformations de bourgeones chez choufleur (*Brassica oleracea* L. Var *botrytis*). Compt. Rend Acad. Sci. Paris. 268D: 686 – 688.
Martinez, C., Harry, I.S. and Thorpe, T.A. (1990). *In vitro* regeneration of plantlets of Canary Island pine (*Pinus canariensis*). Can. J. For. Res.20: 1200-1211.
Mehra, P.N. and Khosla, P.K. 1973. Cytological studies of Himalayan Rutaceae. Silvae Genet. 22: 182-188.
Mehra-Palta, A.1982. Clonal propagation of *Eucalyptus* by tissue culture. Plant Sci Lett.26: 1.
Messeuger, J. and Mele, E.1987. *In vitro* propagation of adult material and seedlings of *Corylus avellane*. Acta Hort. 212: 499 –503.

Mohamed, M.F., Read, P.E. and Coyne, D.P.1992. Dark preconditioning, CPPU, and thidiazuron promote shoot organogenesis on seedling node explants of common and faba beans. J. Amer.Soc.Hort.Sci.117: 668-672.

Mohnen, D., Eberhand, S. and Marfa, V.1990. The control of root, vegetative shoot and flower morphogenesis in tobacco thin cell–layer explants (TCLS). Development 108:191-201.

Murashige, T. and Skoog, F. 1962. A revised medium for rapid bioassay with tobacco culture. Physiol. Plant. 15: 473-493.

Murashige, T. 1974. Plant propagation through tissue cultures. Ann.Rev.Plant Physiol.25: 135-166.

Murashige, T. 1978. Principles of rapid propagation. Pp.14-24. In: Propagation of higher plants through tissue culture (Eds) Hughes, K.W., Henke, R. and Constantan, M. Technical information center, US Dept.Energy, Spring –field, Va.CONF-7804111.

Nanda, P.C. 1962. Chromosome numbers of some trees and shrubs. J. Indian Bot. Sci. 41: 271-277.

Nehra, N.S., Stushnoff, C. and Kartha, K.K.1988. Direct shoot regeneration from leaf disc in straw berry. Hort.Sci.23: 808.

Noh, E.W., Minocha, S.E. and Riemenschneider, D.E. 1988. Adventitious shoot formation from embryogenic explants of red pine (*Pinus resinosa*). Physiol. Plant. 74:119-124.

Pattnaik, S.K. and Chand, P.K. 1996. *In vitro* propagation of the medicinal herbs *Ocimum americanum* L. Syn. *O. canum* Sims (hoary basil) and *Ocimum sanctum* L. (holy basil). Plant Cell Rep.15: 846-850.

Pattnaik, S.K., Sahoo, Y. and Chand, P.K. 1996. Micropropagation of a fruit tree, *Morus australis* Poir. Swin. *M. acidosa* Griff. Plant Cell Rep. 15: 841- 845.

Pattnaik, S.K. and Chand, P.K. 1997. Rapid clonal propagation of three mulberries, *Morus cathayana* Hemsl., *M. ihou* Koiz. and *M.serrata* Roxb., through *in vitro* culture of apical culture of apical shoot buds and nodal explants from mature trees. Plant Cell Rep. 16:503-508.

Phillips, G.C. and Hubstenberger, J.F.1985. Organogenesis on pepper tissue cultures. Plant Cell Tiss. Org. Cult.4: 261-269.

Pierik, R.L.M. 1990. Rejuvenation and micropropagation. In: Progress in plant cellular and molecular biology (Eds) H.J.J. Nijkamp, L.H.W. Plas Vander and V.J. Aartrijk. Kluwer Acad. Publ., Dordrecht, The Netherlands, Pp.91-101.

Prasad, R.N. and Chaturvedi, H.C. 1988. Effect of season of collection of explants on micropropagation of *Chrysanthemum morifolium* Biol.Plant.30: 20-24.

Priyadarshi, S. and Sen, S. 1992. A revised scheme for mass propagation of Ester Lily. Plant Cell Tiss.Org. Cult. 30: 193-197.

Pua, E.C., Chong, C. and Roussefe, G.I. 1983. *In vitro* propagation of Ottawa 3 apple stock. Can.J. Plant Sci. 63:183-188.

Purohit, S.D., Dave, A. and Kukda, G.1994. Micropropagation of safed musli (*Chlorophytum borivilianum*), a rare medicinal herb. Plant Cell Tiss. Org. Cult.39: 93-96.

Quraishi, A., Vijaya, K. and Mishra, S.K. 1996. *In vitro* micropropagation from nodal segments of *Cleistanthus collinus*. Plant Cell. Tiss. Org. Cult. 45: 87-91.

Quraishi, A. and Mishra, S.K. 1998. Micropropagation of nodal explants from adult trees of *Cleistanthus collinus*. Plant Cell Rep.17: 432-433.

Raghavan, R.S. 1957. Chromosome numbers in Indian medicinal plants. Proc. Indian Acad. Sci., Sect. B. 45: 294-298.

Raghava Swamy, B.V., Himabindu, K. and Lakshmi Sita, G. 1992. *In vitro* micropropagation of elite rosewood (*Dalbergia latifolia* Roxb.). Plant Cell Rep. 11: 126.

Ray, D.P. and Chattarjee, B.K. 1996. Effect of different concentrations of growth regulators, etiolation and vigoration treatments on the rooting of stem cutting of bael (*Aegle marmelos* Corr.) Orissa J.Hort. 24(1-2): 36-41.

Rekha Bhati, Shekhawat, N.S. and Arya, H.C. 1992. *In vitro* regeneration of plantlets from root segments of *Aegle marmelos*. Indian J.Exp. Biol. 30:844-845.

Roy, S.K., Pal, P.K. and Das, A.K. 1987. Propagation of a timber tree, *Terminalia bellerica* Roxb. by tissue culture. Bangladesh J. Bot. 16:125.

Roy, S.K., Isalm, M.S. and Hadiuzzaman. 1998. Micropropagatioin of *Elaeocarpus robustus* Roxb. Plant Cell Rep. 17: 810-813.

Sahoo, Y., Pattnaik, S.K. and Chand, P.K. 1997. *In vitro* clonal propagation of a medicinal plant *Ocimum basilicum* L. (sweet basil) by axillary shoot proliferation. *In Vitro* Cell Dev. Biol. Plant. 33:293-296.

Sahoo, Y. and Chand, P.K. 1998. Micropropagation of *Vitex negundo* L., a woody aromatic medicinal shrub, through high-frequency axillary shoot proliferation. Plant Cell Rep.18: 301-307.

Sanchez, M.C., San-jose, M C., Ferro, E., Ballester, A. and Vieitez, A.M. 1993. Improving micropropagation conditions for adult-phase shoots of chestnut, J. Hort. Sci. 72: 433.

Sanjappa, M. 1979. In IOPB chromosome number reports LXIII. Taxon 28: 265-279.

Sen, J.and Sharma, A.K. 1991. Micropropagation of *Withania somnifera* from germinating seeds and shoot tips. Plant Cell Tiss. Org. Cult.26: 71-73.

Shankar, S. and Mohan Ram, H.Y. 1990. *In vitro* plantlet regeneration in the leguminous tree *Sesbania sesban*. Phytomorphology 40: 43-52.

Sharma, A.K. and Chaturvedi, H.C. 1988. Micropropagation of *Bougainvillaea x buttiana* 'Scarlet Queen Variegated' by shoot tip culture. Indian J. Exp. Biol. 26:285-288.

Shekawat, N.S., Rathore, T.S., Singh, R.P., Deora, N.S. and Rao, S.R. 1993. Factors affecting *in vitro* clonal propagation of *Prosopis cineraria*. Plant Growth Reg. 12:273-280.

Singh, U.R., Panday, I.C. and Prasad, R.S. 1976. Propagation of bael (*Aegle marmelos* Corr.). Punjab Hort. J.16: 57-59.

Singh, H.P., Singh, S., Saxena, R.P. and Singh, R.K.1993. *In vitro* bud break in axillary nodal segments of mature trees of *Acacia nilotica*. Indian J. Plant Physiol. 34:21-24.

Sinha, R.K., Majumdar, K. and Sinha, S. 2000. *In vitro* differentiation and plant regeneration of *Albizia chinensis* (Osb.) Merr. *In Vitro* Cell. Dev. Biol.- Plant. 36:370-373.

Skoog, F. and Miller, C.O. 1957. Chemical regulation of growth and organ formation in plant tissues cultured *in vitro*. Symp. Soc. Exp. Bot. 11: 118- 131.

Sommer, A.L., Brown, C.L. and Kormani, K.P.P. 1975. Differentiation of plantlets in longleaf pine (*Pinus palustris*) tissue cultured *in vitro*. Bot.Gaz.136: 196-200.

Stimart, D.P. and Ascher, P.D. 1978. Foliar emergence and growth of *in vitro* generated bulblets from bulb scale explants of *Lilium longiflorum*. Hort. Sci. 13:56(Abstr.140).

Street, H.E. 1977. Cell (suspension) cultures-techniques. In. Plant Tissue and Cell Culture. Bot. Monographs Vol. II (Ed) Street. H.E. Blackwell Scientific Publ., Oxford, London. Pp. 61-102.

Sunitibala, Y.D., Mukherjee, B.B. and Gupta, S. 1994. Rapid cloning of elite teak (*Tectona grandis* Linn.) by *in vitro* multiple shoot production. Indian J. Exp. Biol. 32: 668-671.

Sur, P.J.K. and Pramanik, T. 1999. Antispermatogenic activity of leaves of *Aegle marmelos* Corr. in albino rats: A preliminary report. Biomedicine 19 (3): 199-202.

Tanimoto, S. and Harada, H. 1985. Hormonal regulation of flowering. In: Hormonal regulation of plant growth and development. (Ed) Purohit, S. Martinus Nijhoff, Bordrecht. Pp 41-93.

Terzi, M. and Lo Schiavo, F.1990. Developmental mutants in carrot. In: Progress in plant cellular and molecular biology (Eds) Nijikamp,H.J.J. and Van Der Plas, L.H.W., and Proc. VIIth Int.Cong. Plant Tissue Culture, Amsterdam, The Netherlands. Kluwer Acad. Publ. Dortrecht, The Netherlands, pp. 391-393.

Thomson, B.I., Sink, K.C. and Dennis, F.C.Jr. 1981. A non-destructive assay for gibberellin production *in vitro* and its application to callus cultures of *Phaseolus*. Hort. Sci. 16: 432 (Abst.247).

Thorpe, T.A.1982. Callus organization and *de novo* formation of shoots, roots and embryos *in vitro*. In: Application of plant cell and tissue culture to agriculture and industry. Guelph. University of Guelph: (Eds) Tomes, B.T., Ellis, B.E. and Harney, P.M. Pp 115-138.

Thorpe, T.A. and Biondi, S. 1984. Conifers. In: Handbook of plant cell culture, Vol.2. Macmillan Publ. Co., pp 434-470.

Thorpe, T.A.1994. Morphogenesis and regeneration. In: Plant cell and tissue culture. (Eds) Vasil, I.K. and Thorpe, T.A. Kluwer Acad. Publ., Dorde recht.

Tisserat, B. 1985. Embryogenesis, organogenesis and plant regeneration. In: Plant Cell Culture – a practical approach (Ed) Dixon, R.A. IRL Press, Washington, D.C. Pp 79-105.

Toivonen, P.M.A. and Kartha, K.K. 1988. Regeneration of plantlets from in vitro cultured cotyledons of white spruce(Picea glauca Moench Voss). Plant Cell Rep. 7:318-321.

Torrey, J.G. 1985. The development of plant biotechnology. Ann. Sci.73: 354-363.

Varghese, S.K., Karan, J.A., Kalia, K., Subramanian, R.B. and Nataraj, M. 1993. Micropropagation of *Aegle marmelos*. Phytomorphology 43: 87-92.

Vieitez, A.M., Carmen, S.M and Vieitez, E. 1985. *In vitro* plantlet regeneration from juvenile and mature *Quercus robur*, L. J. Hort. Sci. 60:99-106.

Vieitez, A.M., Ferro, E. and Ballester, A. 1993. Micropropagation of *Fagus sylvatica* L. *In Vitro* Cell Dev. Biol. 29: 183.

Villalobos, V.M., Oliver, M.J., Yeung, E.C. and Thrope, T.A. 1983. Cytokinin-induced switch in development in excised cotyledons of radiata pine cultured *in vitro*. Physiol. Plant. 61: 483-489.

Vincent, K.A., Mathew, K.M. and Hariharan, M. 1992. Micropropagation of *Kaempferia galanga* L., a medicinal plant. Plant Cell Tiss. Org. Cult. 28: 229-230.

Warrier, P.K., Nambiar, V.P.K. and Ramankutty, C. 1996. Indian medicinal plants. Vol.1. Orient Longman, Madras. Pp.62-66.

Weatherhead, M.A., Burdan T.and Henshaw, G.G. 1979. Effects of activated charcoal as an additive plant tissue culture media. Z.Pflanzen Physiol. 94: 399-405.

Welsh, K.J. and Sink, K.C.1981. Morphogenetic responses of *Browelis* leaf sections. Aus. Ann. Bot. 48:583-590.

Wickremesinhe, E.R.M., Blackman, W.J. and Reynolds, B.D. 1990. Adventious shoot regeneration and plant production from explants of *Apios americana.* Hort. Sci. 25: 1436-1439.

Wright, M.S., Koehler, S.M., Hinchee, M.A. and Carnes, M.G.1986. Plant regeneration by organogenesis in *Glycine max.* Plant Cell Rep.5: 150-154.

Yeoman, M.M. and Davidson, A.W. 1971. Effect of light on cell division in developing callus cultures. Ann.Bot. 35: 1085-1100.

Zaman, M.F. 1988. Bangladeshes phaller chash, la Academy Dhaka, Bangaladesh, pp. 189-193.

Zink, M.W. and Veliky, I.A. 1982. Separation and some properties of two acid phosphatases from suspension cultures of *Ipomea* sp. Plant Cell. Tiss. Org. Cult. 1: 265 – 273.

GPSR Compliance

The European Union's (EU) General Product Safety Regulation (GPSR) is a set of rules that requires consumer products to be safe and our obligations to ensure this.

If you have any concerns about our products, you can contact us on ProductSafety@springernature.com

In case Publisher is established outside the EU, the EU authorized representative is:

Springer Nature Customer Service Center GmbH
Europaplatz 3
69115 Heidelberg, Germany

Batch number: 10755284

Printed by Printforce, the Netherlands

MICROPROPAGATION OF WOODY TREES AND FRUITS

FORESTRY SCIENCES

Volume 75

The titles published in this series are listed at the end of this volume.

Micropropagation of Woody Trees and Fruits

edited by

S. Mohan Jain

International Atomic Agency,
FAO/IAEA Joint Division,
Vienna, Austria

and

Katsuaki Ishii

Forestry and Forest Products Research Institute,
Tsukuba, Japan

SPRINGER-SCIENCE+BUSINESS MEDIA, B.V.

A C.I.P. Catalogue record for this book is available from the Library of Congress.

ISBN 978-1-4020-1135-1 ISBN 978-94-010-0125-0 (eBook)
DOI 10.1007/978-94-010-0125-0

Printed on acid-free paper

CONTENTS

Section C

Section D

Preface

Global warming, environmental changes, water shortage, and sustainable development are the most up-to-date issues, which have challenged mankind. Researchers worldwide are engaged in addressing some of these problems, including reduction in carbon dioxide accumulation, and enrichment of perennial woody species on the terrestrial ecosystem. About 12 million hectares of the world's forests disappear every year. By 2025, the world population will reach 7.5 billion, and the forest area will be reduced to well below 50 % of the current area. Reforestation is an important to prevent the loss of forest resources including timber, biodiversity and water resources. Therefore, subsequent volume of reforestation over the deforested land should be followed to safeguard the forests and maintain its size, which will require a continuous supply of planting material. Similarly fruit trees, including tropical and subtropical fruit trees, are consumed both as fresh and in the processed form including juices, beverages, and dried fruits. They are an important source of nutrition *e.g.* rich in vitamins, sugars, aromas and flavour compounds, and raw material for food processing industries. The production, cultivation and maintenance of tree species provide highly sustainable production systems that conserve soils, microenvironment and biodiversity. Fruit trees have long juvenile periods and large tree size. In many fruit trees *e.g.* avocado and others controlled crosses are difficult to make due to massive fruit drop. Moreover, fruit trees are faced with agronomic and horticultural problems in terms of propagation, yield, appearance, quality, diseases and pests control, abiotic stresses and poor shelf life. The available genetic information in fruit crops is very limited and their genetic improvement has relied heavily upon vegetative propagation techniques and classical breeding. The key technology for biomass production of trees is propagation via micropropagation. However, there are many recalcitrant species among useful forest and fruit trees to propagate in large numbers. Recent developments of *in vitro* culture techniques have made it possible to commercially propagate useful trees both forest and fruit trees.

In this book, comprehensive information is provided on micropropagation of economically important forest and fruit trees, which is usually available in scattered literatures. Topics cover a wide range, from tropical forest and fruit trees for paper or food supply to *Prunus* species for local craft bark production. This book is divided into four sections. Section A contains five chapters on different parameters influencing micropropagation, which are quality of light, arbuscular mycorrhizal (AM) fungi, pathogen elimination and contamination management, and molecular markers for genetic fidelity assessment.

Section B has five chapters on micropropagation of tropical woody species, *Fagus* spp., *Euycalyptus*, *Acacia* species, and *Aegle marmelos* medicinal tree. Section C includes 14 chapters, which are dealing mainly with micropropagation of fruit trees such as grapevine, bananas, arid zone fruit trees of India, apples, small fruits, Prunus, avocado, papaya, pistachio, olive, kiwifruit, tea, and litchi. Section D has four chapters mainly on bioreactor, photoautotrophic, thin cell layer and *in vitro* germplasm conservation. Even though few papers are published on litchi micropropagation, we have deliberately included this chapter just to remind the readers that plenty of work is still needed in most of the commercially important tropical fruits, including litchi.

The invited authors were selected world wide from east to west. The editors are grateful to all the contributing authors for timely submission of their manuscripts, which made it possible to realise this book project. We appreciate it very much that our colleagues took some time off to review these manuscripts. We sincerely hope that our hard work will benefit all engaged in research on forest and fruit trees.

S. Mohan Jain

K. Ishii

SECTION C

11. MICROPROPAGATION OF THE GRAPEVINE (*Vitis* spp.)

Alain Bouquet[1] and Laurent Torregrosa[2]

[1]*UMR Diversity and Genome of Cultivated Plants, INRA, Grape Breeding Experiment Station " Le Chapitre ", 34751 Villeneuve lès Maguelone Cedex, France*

[2]*UMR Biology of Development of Cultivated Perennial Plants, ENSAM-INRA, 2 place Viala, 34060 Montpellier Cedex 01, France*

1. INTRODUCTION

During the past decades, rapid progress has been made in the application of plant biotechnology to agriculture and horticulture, mainly with the assistance from botanists, plant physiologists and molecular geneticists who lack expertise in crop plants, crop husbandry or crop breeding. The literature is pervaded by a feeling that only a few trivial technical problems remain and we are at the threshold of a latter-day agrarian revolution. This is wishful thinking. Temperate perennial fruit crops, and grapevines in particular, are subject to special circumstances that can affect or limit the use of tissue culture methods for both plant propagation and plant improvement. Substantial benefits will accrue the grapegrowing industry from further development of tissue culture but advocacy for *in vitro* techniques must be tempered by an appreciation of the historical, biological, technical and economic realities of viticultural production.

S.M. Jain and K. Ishii (eds.), Micropropagation of Woody Trees and Fruits, 319–352.

2. THE GRAPEVINE AND ITS CULTURE

2.1. Taxonomical status of the grapevine

Grapevine and its relatives are dicotyledonous angiosperms that belong to the family *Vitaceae.* They are mostly woody or herbaceous tree-climbing plants or shrubs with liana-like stems. According to the classification of Planchon (1887) revised by Galet (1988), the family comprises 16 genera, but only one, *Vitis*, gives edible berries. Other *Vitaceae* present ornamental interest such as garden climbing plants like *Parthenocissus* (Virginia creeper) and *Ampelopsis* or indoor plants like *Cissus* and *Rhoicissus*.

According to the classification of Galet (1988), the genus *Vitis* (2n=38 chromosomes) comprises 57 species. Unlike the majority of *Vitaceae*, which are primarily intertropical in their distribution, they are found mainly in the temperate zones of the northern hemisphere and distributed almost equally between America and Asia. Only one species, *Vitis vinifera* L., originated in Eurasia. It has been spread throughout the world by man for several millenariums and the cultivated acreage was about 8 millions hectares in 1999.

Another species, *Vitis* (*Muscadinia*) *rotundifolia*, is cultivated on a few thousand hectares in the southwest of the USA (Olien, 1990). According to the classification of Planchon (1887), it belongs to a distinct section of the genus *Vitis*. Small (1903) proposed that the sections *Euvitis* (bunch grapes) and *Muscadinia* (muscadine grapes) be elevated to generic level. This proposal, although well supported by morphological, anatomical, and karyological (2n=40 chromosomes) characteristics, has not yet achieved general acceptance (Mullins *et al.,* 1992). Due to its high resistance to pests and diseases, and despite problems of cross-sterility, muscadine grape presents considerable interest for breeding (Olmo, 1986).

2.2. History of viticulture

It is thought that cultivation of the grapevine began 6000 years ago during the neolithic era along the eastern shores of the Black Sea, but archaeological findings of grape seeds indicate that *Vitis vinifera*, or its wild progenitor *Vitis sylvestris*, was distributed throughout much of Europe (Nuñez and Walker, 1989). It is likely that prehistoric people ate the berries of the vines that climbed on the forest trees, but the domestication of the plant was the result of the discovery of wine (Marinval, 1997). Grape growing and winemaking extended from Transcaucasia to Asia Minor, the Balkan peninsula and the Nile delta. During the first millenium BC, the Phoenicians and Greeks introduced viticulture into North Africa, Southern Italy, Spain and France. But grape cultivation became widespread throughout Europe with the establishment of the Roman domination from the shores of the Atlantic to the valleys of Rhine and Danube. With the fall of the Roman empire, the wine trade was disrupted and grape cultivation declined. The spread of Christianity into Europe and the need for sacramental purposes, led to the establishment of a new international trade of wine, that soon outgrew its religious association. By the end of the Middle Ages, wine drinking was a firmly established social custom over most of Europe (Flandrin, 1997). By 600 AD the consumption of wine had been prohibited under Islamic law which promoted the cultivation of table grapes in the Middle East. During the 17th and 18th centuries, viticulture extended progressively into the New World (South Africa, South and North America, Australia).

2.3. Economic importance of viticulture

Grape berries are eaten fresh, dried (raisins), canned, or as juice or jam, but the main use is winemaking. Total world grape production in 1999 was estimated to be 607 millions quintals (Mqts), from which 76% was used for wine production, 22% consumed fresh, and 2% consumed as raisins. The eight highest wine producing countries with their estimated production and grape acreage are listed in Table 1.

Table 1. Wine production and grape acreage (statistics 1999) of the eight highest wine producing countries (OIV, 2000)

	Wine production (thousands hl)	Grape acreage* (thousands ha)
France	62.235	914
Italy	58.073	909
Spain	32.679	1.180
USA	20.691	374
Argentina	15.888	208
Germany	12.286	106
Australia	8.511	123
South Africa	7.968	115

(*) including acreage for table grape production

Asia has the highest grape production (66 Mqts), with Turkey, Iran and China the leading growers, ahead of Europa (31 Mqts, produced mainly in Italy), North and South America (19 Mqts, produced mainly in California and Chile), and Africa (16 Mqts produced mainly in Egypt). For raisins, the world's largest producers remain Turkey with a production of nearly 3.5 Mqts, ahead of California (3.1 Mqts), Greece (0.9 Mqts), Iran (0.6 Mqts) and South Africa (0.4 Mqts).

2.4. The cultivars : origin, propagation and selection

2.4.1. Winegrape cultivars : the load of tradition

The primary difference between broad-acre agriculture and viticulture is that many crops were improved with plant breeding. New cultivars have been continuously introduced in response to changes in biological or economical factors of production such as epidemics and new technology. In viticulture, most of the world's vineyards are planted with cultivars that have been grown for centuries (Bouquet, 1982). Innovation has been mainly at the level of crop husbandry.

The persistence of the traditional European winegrape cultivars (*Vitis vinifera*), both in their countries of origin and in the New World, is due to a combination of plant and human factors. The breeding of woody perennial fruit plants such as grapevines presents great difficulties. Prior to domestication, vine plants were dioecious and reached a high level of heterozygosity by outbreeding. As a side effect, deleterious recessive genes accumulated in the genome (Olmo, 1976). During domestication of grapes, the selection of successful self-fertile genotypes was directed against homozygosity. The ease with which *vinifera* grapes are propagated by cuttings is likely to have facilitated the perpetuation of superior heterozygous selections (Mullins and Meredith, 1989).

The traditional cultivars represent highly subtle gene combinations that cannot be reproduced by seeds and consequently need to be preserved by vegetative propagation. Their wines have unique and typical characteristics that enjoy a high level of consumer acceptance and are firmly placed in the market. Moreover, the prevalence of international models in wine-trading is often strengthened by customs, rules and laws, especially in Europe. There are few reports of *Vitis vinifera* winegrape cultivars obtained by controlled crosses during the 19th century, but the place of cultivars recently developed in viticulture remains limited, despite considerable breeding work carried out during the second half of the 20th century (Alleweldt and Possingham, 1988; Bouquet *et al.*, 2001).

2.4.2. Tablegrape cultivars : the need for innovation

Tablegrape cultivars are generally of recent origin (19th - 20th century) and many of them have been obtained from controlled crosses. Most of the table grapes grown worldwide are still seeded, but a lot of new seedless cultivars have been obtained during the last 50 years (Ledbetter and Ramming, 1989). In many countries, considerable breeding work is now being carried out using biotechnology, including marker-assisted selection (This *et al.*, 2000), *in vitro* embryo rescue (Bouquet and Davis, 1989), and genetic transformation (Perl *et al.*, 2000).

2.4.3. Rootstock cultivars

Until the end of 19th century, grapevine was propagated by layering or hardwood cuttings. Since the introduction of the root-louse Phylloxera (*Daktulosphaira vitifoliae*) in Europe during 1868, *Vitis vinifera* varieties, killed by the pest, are grafted on resistant rootstocks issued from North American *Vitis* species, mainly *V. riparia*, *V. rupestris* and *V. Berlandieri*, and their hybrids (Pouget, 1990). The selection of Phylloxera-resistant rootstocks adapted to different edaphic and climatic conditions, such as limy soils or drought, is the foundation of viticulture. In limited areas free from Phylloxera, vines are generally grafted as a precaution but also to protect against other pests such as root-knot nematodes (*Meloidogyne* spp.). Most of the rootstock cultivars were selected at the end of 19th century, but breeding work is still continuing in France, Germany, California, Brazil and Australia. One of the main objectives of breeding is to develop resistance against the soil-inhabiting nematode *Xiphinema index*, which is a carrier of the grape fanleaf virus (Bouquet *et al.*, 2000a). Genetic engineering is also largely used to search for virus resistant rootstocks (Mauro *et al.*, 2000).

2.4.4. Disease resistant hybrid wine cultivars

Before the end of the 19th century, European viticulture was endangered by two fungal diseases introduced from North America, powdery mildew (*Uncinula necator*) and downy mildew (*Plasmopara viticola*). Though they could be controlled by fungicides such as sulphur and copper, these diseases led to the selection of many hybrids between *Vitis vinifera* and American *Vitis* species, resistant to the fungi. In fact, the first hybrids were obtained during the 19th century in North America by natural or controlled crosses between introduced European cultivars and native American species, mainly *Vitis labrusca*. Some of them, such as the variety Concord, are still grown on a large acreage in Northeastern American vineyards. Others, called "American varieties" and partially resistant to Phylloxera, were introduced in France that started the breeding work during the first half of the 20th century and led to hundreds of hybrid cultivars. They are currently called "French hybrids" and show complex parentages (Galet, 1988). Though they were grown in France in 1958 on 400,000 hectares, their plantation was progressively discouraged due to the bad wine quality. Hybrids were also forbidden in other European countries, such as Italy and Spain. Nevertheless, the breeding of disease resistant cultivars was carried out in Germany and in Central European countries, using *Vitis* germplasm. In France, it was initiated again 25 years ago, using *Muscadinia* germplasm (Bouquet, 1980 ; Bouquet *et al.*, 2000b).

2.4.5. Clonal selection

As a consequence of the cultivation of predominantly traditional cultivars, the exploitation of genetic variation within these cultivars, called clonal selection, has become the most widely used procedure for improvement of wine grapes. Most of the leading cultivars of wine grapes are of ancient origin and are released from natural crosses within *Vitis vinifera* (Bowers and Meredith, 1997; Bowers *et al.*, 1999). These cultivars might be a mixture of more than one genotype, which is due to mixed plantings and propagation of closely related and phenotypically similar plants. This concept is known as the polyclonal origin of grapes (Rives, 1961). However, recent molecular analysis (Ye *et al.*, 1998) did not detect genetic polymorphism among the different clones of the cultivar Pinot noir, despite this cultivar is being grown since the early Middle-Ages, and the most phenotypically heterogeneous among the *Vitis vinifera* cultivars. Whatever their origin, grapevine cultivars are likely to have accumulated by now a substantial amount of somatic mutations. Though the selection of bud-sports has been an important element in genetic improvement of pome and stone fruits, the occurrence of such bud-sports is less well documented in grapes. There is much debate on the relative importance of genetic and non-genetic factors in clonal variation. One school of thought is that genetic differences within cultivars are minimal and that clonal variation is primarily attributed to differences in the occurrence of pathogens.

2.4.6. Sanitary selection

Grapevines are common with other woody perennial fruit plants in that they are infected by many viruses and virus-like organisms (Walter and Martelli, 1997). These organisms are all transmissible by grafting. To this date, none are known to be transmitted by seed or by pollen. The fanleaf degeneration is the most destructive grapevine virus disease in the world. It is caused by a nepovirus transmitted by the nematode *Xiphinema index* (Hewitt *et al.*, 1958). But several other soil-inhabiting nematodes, belonging to the genera *Xiphinema* or *Longidorus*, are also vectors of less important virus diseases (Taylor, 1971). Widespread virus diseases, such as corky-bark, stem pitting and leaf-roll, that belong to the families of vitivirus (trichovirus) and closterovirus are transmitted by mealybugs (Tanne *et al.*, 1989; La Notte *et al.*, 1997; Cabaleiro and Segura, 1997). Several yellow diseases such as Flavescence dorée and Bois noir are transmitted by leafhoppers (Caudwell, 1990) or planthoppers (Sforza *et al.*, 1998). Pierce's disease caused by the bacterium *Xylella fastidiosa*, is transmitted by many sharpshooter leafhoppers (Purcell, 1979).

The use of pathogen-free planting material is of great importance in viticulture. Grapevines with recognizable visible symptoms of diseases are not used as sources of cuttings or budwood. But, seemingly healthy rootstock mother plants can be symptomless carriers of pathogens, such as the grapevine leafroll associated viruses or the Flavescence dorée phytoplasma that have devastating effects on vineyard productivity, quality and perenniality,

Elaborate and time-consuming virus indexing procedures, using sensitive indicator plants, have been necessary to identify clean propagating material (Garau *et al.*, 1997). New diagnostic tools for virus or virus-like diseases are now developed and include immunological methods and nucleic acid technology (Martin *et al.*, 2000). There has been much research since the 1960s on heat therapy of grapevine viruses and the use of *in vitro* culture for production of healthy propagating material (Gifford and Hewitt, 1961). In most grape growing countries, production of clean stock programs have been established in order to multiply certified true-to-type and pathogen-free propagation material for nursery trade. So, it is reasonable to question whether these programs are amenable to *in vitro* technology.

3. LITERATURE REVIEW

The primary question is why grow grapevine *in vitro*? The first successful attempt to establish grapevine tissue culture was achieved in 1941 by Morel (1948), who selected a method to allow dual culture of vine and downy mildew. This work on *in vitro* dual culture of grapevine and its pathogens or pests was not carried out until the 1980s and calli were replaced by *in vitro* cultured plantlets (Aldwinckle, 1980 ; Lee and Wicks, 1982 ; Klempka *et al.*, 1984 ; Palys and Meredith, 1984 ; Barlass *et al.*, 1986 ; Loubser and Meyer, 1990 ; Bavaresco and Walker, 1994 ; Forneck *et al.*, 1996). Studies on callus induction and proliferation were continued after the Morel's pioneer work but conditions of growth for long term tissue culture of grapevine were not established until the work of Hawker *et al.* (1973). Grapevine cell suspension cultures enabled studying host-parasite interactions, producing secondary metabolites, and monitoring gene expression (for a review see Torregrosa *et al.*, 2001). Nowadays, grapevine *in vitro* culture deals mainly with virus elimination, large-scale propagation, germplasm conservation, and genetic engineering.

3.1. Virus elimination

Gifford and Hewitt (1961) and Galzy (1961) demonstrated the importance of *in vitro* culture to eliminate grape fanleaf virus by heat therapy. To improve sanitation techniques, investigations were undertaken on the culture of apical explants (Galzy, 1972; Harris and Stevenson, 1979; Goussard, 1981). The difficulty to recover plants from entire apices established on solidified medium led to the use of fragmented apices cultured in liquid medium (Barlass and Skene, 1978; Barlass *et al.*, 1982). The technique was improved by micrografting shoot meristems onto the hypocotyls of *in vitro* grown seedlings (Bass *et al.*, 1978) or the internodes of healthy *in vitro* propagated plantlets (Ayuso *et al.*, 1978). The effectiveness of somatic embryogenesis in eliminating viruses from diseased plants has been demonstrated (Goussard *et al.*, 1991).

3.2. Micropropagation

Following the pioneer work of Morel (1963) on propagation of orchids by meristem culture, there has been an increasing interest in the application of tissue culture techniques as an alternative means of asexual multiplication of economically important plants (Murashige, 1974). The propagule size in culture is very small and this *in vitro* asexual propagation technique has been referred to as micropropagation.

Since the early 1960s, grapevine has been the subject of research for developing an ideal procedure for *in vitro* propagation, in order to hasten production of superior selected plants, especially released from emerging clonal and sanitary selection or breeding programs. Depending on the morphogenetic process, *in vitro* propagation techniques either used pre-existing meristems or tissues without pre-existing meristematic structures.

There are two possible routes available for *in vitro* propagule multiplication. The first method, called nodal bud culture, is based on the formation of a single rooted plantlet from a node explant (Figure 1a). To increase the efficiency of micropropagation, Jona and Webb (1978) proposed a second method by enhancing release and proliferation of axillary buds with high levels of cytokinins in the growth medium. These cultures no longer have a defined plant form but appear as dense proliferative clusters of small root-free shoots (Figure 1b).

Two other ways of plant regeneration have been developed : somatic embryogenesis, which provides embryos or embryo-like formations; and adventitious organogenesis, which provides bud-like and caulinary structures. Though these regenerative procedures are not currently used in grapevine for micropropagation *sensu stricto*, they will be succinctly studied in the next sections.

Each method of propagation has its merits and shortcomings. The merit of using nodal bud culture or axillary bud proliferation is that the incipient shoot has already differentiated *in vivo*. Thus, to establish a complete plant, only shoot elongation and root differentiation are required. This method is applicable in many cases. With nodal bud culture, under optimal growing conditions, the multiplication rate can theoretically reach three million plants from a single explant in a single year (Nozeran and Bancilhon, 1972). For axillary bud proliferation, the production of about 75 million plants in a single year can be expected (Chee *et al.*, 1984). Special elaborate machines have been developed for *in vitro* propagation in liquid culture (Harris and Mason, 1983).

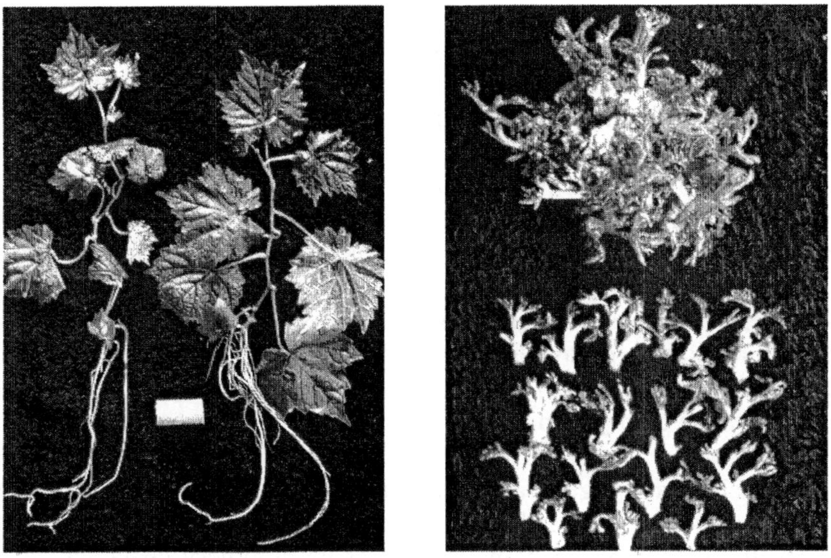

Figure 1. Micropropagation of *Vitis* x *Muscadinia* hybrids. **a** (left): Nodal bud culture.- **b** (right): Axillary bud proliferation (from Torregrosa, 1994).

3.3. Somatic embryogenesis

The regeneration of complete plants via somatic embryogenesis was first obtained by Mullins and Srinivasan (1976), using unfertilized ovules of *Vitis vinifera* cv Cabernet-Sauvignon and Hirabayashi *et al.* (1976) using anther culture of *Vitis thumbergii*. Later, Krul and Worley (1977) obtained plantlets of the complex interspecific hybrid Seyval, using callus from different types of explants. Rajasekaran and Mullins (1979) obtained plantlets of the interspecific hybrid Gloryvine (*Vitis vinifera* x *V. rupestris*) using anther culture. The same technique was used on *Vitis vinifera* (Zou and Li, 1981) and different rootstock cultivars (Bouquet *et al.*, 1982).

First attempts of anther culture in grapevine were made to obtain haploid or dihaploid plants from pollen grains (Gresshoff and Doy, 1974). The gametic origin of embryos and plants from anther culture is controversial (Cao, 1990). In a few cases, haploid chromosome sets (n=19) and mixoploid populations of cells were reported in anther-derived plants (Zou and Li, 1981). But, the somatic origin of the embryos and plantlets was clearly demonstrated (Rajasekaran and Mullins, 1983a).

Anther culture is currently used in many species and cultivars of bunch grapes to induce somatic embryogenesis, however leaf blades also proved to be usable in many cases (Hirabayashi, 1985). Additionally, other explants, such as immature embryos (Stamp and Meredith, 1988) or tendrils (Salunkhe *et al.*, 1997) were used. Most recently, protoplasts that have underwent division were used to produce somatic embryos and regenerated plants. This was possible only when protoplasts were isolated from leaf-derived embryogenic calli or embryos (Reustle *et al.*, 1995; Zhu *et al.*, 1997). Somatic embryogenesis was not limited to bunch grapes. In muscadine grapes, it was induced on immature embryos (Gray, 1992) and leaves (Robacker, 1993). Leaf explants from micropropagated plants were also used to induce somatic embryogenesis on *Vitis* x *Muscadinia* hybrids (Torregrosa *et al.*, 1995). In 25 years, considerable work has been done on somatic embryogenesis on grape, indicated by the publication of several papers, including protocols for induction of embryogenic callus, differentiation (Figure 2a) and development of embryos (Figure 2b), preservation of embryogenic competence, and conversion of embryos in plants.

In somatic embryogenesis, an intermediary callus is involved in the formation of plants. This callus is able to generate somatic embryoids from single cells, both deep in the callus and at its periphery (Altamura *et al.*, 1992). The two types of somatic embryogenesis, "deep genesis" and "superficial genesis", both resembled zygotic embryogeny, but differed from each other. The embryoid's initial cells show meristematic features compared to the confining callus cells. Newton and Goussard (1990) observed pro-embryogenic mass (PEM) clusters composed of at least 5 small cells densely cytoplasmic and lightly stained, but could not prove that the PEMs arose out of single embryogenic cells. As embryoid formation is a widely dispersed event in culture, the probability of genetic changes does not guarantee keeping "true to type" propagation of planting material, especially in classified wine-growing areas. Thus, somatic embryogenesis is not recommended for large-scale propagation of grapevine, despite the fact that the rate of embryo and plantlet multiplication can be astonishing in some genotypes.

Nevertheless, somatic embryogenesis has been widely applied in breeding perspectives. Using *Agrobacterium*-mediated transformation of somatic embryos, Mullins *et al.* (1990) recovered non-chimeric transgenic buds and plants expressing the *gus* reporter gene. A few years later, Le Gall *et al.* (1994) reported the transformation of somatic embryos and regeneration of transgenic plants expressing a gene of agronomical interest. Today, somatic embryogenesis is unquestionably the most suitable tool for genetic transformation in the genus *Vitis* (for reviews see : Perl and Eshdat, 1998; Thomas *et al.*, 2000; Vivier and Pretorius, 2000; Colova-Tsolova *et al.*, 2001; Kikkert *et al.*, 2001; Martinelli and Mandolino, 2001).

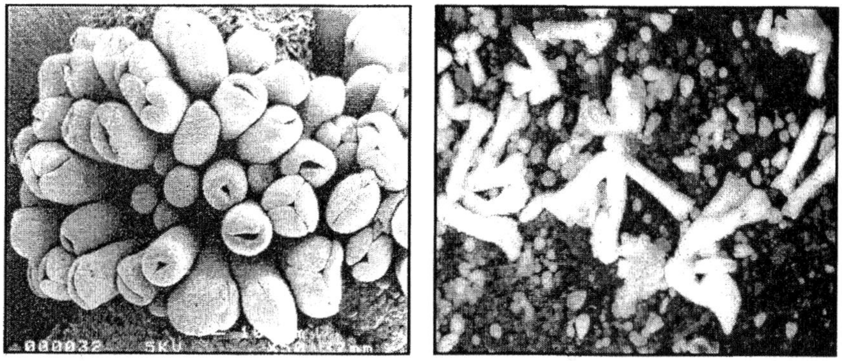

Figure 2. Somatic embryogenesis in grapevine. **a** (left): Cluster of embryoids differentiating on a callus, observed by Scanning Electron Microscopy. - **b** (right): Different developmental stages of somatic embryos in liquid culture (from Torregrosa, 1994).

In plants, genetic variability generated by the use of a tissue culture cycle was called somaclonal variation by Larkin and Scowcroft (1981). The nature and causes of this variation, whether encompassing changes at the cytological, molecular, cytoplasmic, or epigenetic levels (Karp, 1994 ; Jain, 2001), are not yet clear, but have far wider significance than the originally expected contribution to plant breeding.

As individual somatic embryos are converted to plants that can be multiplied by classical micropropagation, somatic embryogenesis was proposed as a way to induce somaclonal variation in grapevine (Kuksova *et al.*, 1997), and specifically to amplify or restore clonal variability in *Vitis vinifera* (Mullins, 1985; Bouquet, 1989a). Embryogenic calli or somatic embryos have been used in screening for resistance or tolerance to abiotic stresses (Lebrun *et al.*, 1985; Bouquet *et al.*, 1985; Bouquet *et al.*, 1990; Netzer *et al.*, 1991). A high degree of tolerance to the bacterial necrosis has been reported on microcuttings of an anther-derived plant of *Vitis vinifera* cv Grenache (Bouquet, 1989b). Similarly, lower susceptibility to the culture filtrate of the fungus *Eutypa lata* was reported on protoplasts of anther-derived plants of *Vitis vinifera* cv Cabernet-Sauvignon (Mauro, 1986). This led to a selection strategy of somaclones tolerant to the eutypa dieback of the grapevine by using fungal toxin (Fallot *et al.*, 1990). In addition, other applications of somatic embryogenesis have been reported or proposed, including virus elimination (Goussard *et al.*, 1991) and germplasm conservation (Dussert *et al.*, 1992; Gray and Compton, 1993).

Somatic embryogenesis and its applications have been included in a number of publications on grape *in vitro* culture and biotechnology (Krul and Mowbray, 1984; Monette, 1988; Gray and Meredith, 1992; Grenan, 1992; Torregrosa, 1995; Torregrosa *et al.*, 2001) and reviews (Gray, 1995; Martinelli and Gribaudo, 2001). In the latest review, the authors focused particularly on crucial aspects of the technique, such as embryo teratology, dormancy and developmental markers, that contribute largely to the understanding of this technique.

3.4. Adventitious organogenesis

The first report on the production of bud-neoformations on leaves is generally attributed to Favre (1977). However, the caulinary structures typically bipolar were more like somatic embryos. Rajasekaran and Mullins (1981) induced organogenesis from internode callus of several hybrid cultivars of bunch grapes but failed with *Vitis vinifera* and muscadine grapes. In grapevine, bud neoformation can be obtained from various types of organs (for reviews see Gray and Meredith, 1992; Torregrosa, 1995) : including leaf primordia from fragmented apices (Barlass and Skene, 1980a), leaves and petiole stubs (Cheng and Reisch, 1989; Clog *et al.*, 1990; Tang and Mullins, 1990; Stamp *et al.*, 1990a), and hypocotyls or cotyledons from somatic embryos (Vilaplana and Mullins, 1989). The organogenesis pathway uses the ability of tissues to form adventitious bud-like structures surfaces (Figure 3a) directly on the explant (Barlass *et al.*, 1981; Colby *et al.*, 1991) or indirectly from callus that develops on the cut. Though direct organogenesis from petioles and leaves of *in vitro* grown plantlets is obviously influenced by the genotype, a reasonable rate of success has been reported for several important *vinifera* and rootstock cultivars (Stamp *et al.*, 1990a; Clog *et al.*, 1990; Tang and Mullins, 1990) and *Vitis* x *Muscadinia* hybrids (Torregrosa and Bouquet, 1996)

Induction media used for direct bud organogenesis contain usually high cytokinin and low auxin levels. However, bud organogenesis obtained through callus formation can be induced by low cytokinin and high auxin levels (Rajasekaran and Mullins, 1981). Petioles are more regenerable than are leaves (Figure 3b). Regeneration from leaves is always from the cut leaf midrib at the site of petiole attachment, suggesting that the regenerable tissue is the same in both organs (Cheng and Reisch, 1989). Tang and Mullins (1990) observed no consistent effects of the growing conditions of the mother plant on adventitious bud formation. Regenerative explants were derived equally from growth chamber-propagated mother plants and from mother plants grown *in vitro*. Stamp *et al.* (1990b) observed that leaf explants taken from newly established shoot-tip cultures produced buds at a consistently higher frequency than did leaf explants from nodal cultures.

Adventitious bud-neoformation has limited applications in grapes. It is not usable in large-scale micropropagation and, contrary to somatic embryogenesis, its efficiency in DNA transfer is restricted by cellular differences in *Agrobacterium* susceptibility and regenerative capacity (Colby *et al.*, 1991).

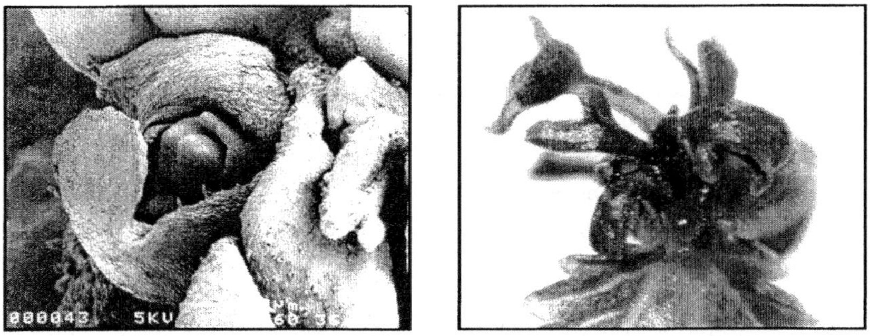

Figure 3. Adventitious organogenesis in *Vitis* x *Muscadinia* hybrids. **a** (left): Adventitious bud-like structure observed by SEM. - **b** (right): Development of shoots on a petiole stub (from Torregrosa, 1994).

3.5. Germplasm preservation

The grape germplasm is currently maintained in field collections where each genotype (species, hybrid, variety or clone) is replicated with a limited number of plants. Management of this germplasm is very expensive and subject to environmental hazards. In particular, the risk of contamination by pathogens transmitted through vegetative propagation is difficult to avoid. To deal with these problems, many laboratories have sought to define conditions allowing *in vitro* conservation of genetic resources.

For maintaining a long-term genetic conformity, research has focused on systems using propagation through direct morphogenesis from organized meristems (Barlass and Skene, 1983; Blaich, 1985; Galzy, 1985). Slowing the growth rate allows increased time between subcultures, reducing the upkeep cost of thousands of genotypes, and the risk of mistake at the time of subculture. Proposals have been made to modify physical factors like gas exchanges (Fournioux and Bessis, 1986), carbon nutrition (Galzy and Compan, 1988), and temperature (Galzy *et al.*, 1990) to reduce growth rate. Comparing plantlet responses to different temperatures and light intensity, Galzy (1985) demonstrated that grapevine cannot be successfully cultured below 8°C. Temperature of 12°C with a light intensity reduced to 12 $\mu e\ s^{-1}\ m^{-2}$ was optimal in slowing plantlet growth without affecting morphogenesis (Galzy *et al.*, 1990). At 9°C, Skene *et al.* (1988) could maintain meristem cultures for 12 months without changing ploidy level. The use of compounds like chlorocholine chloride (CCC) at high levels combined with lower temperatures appeared to reduce plantlet growth by a factor of ten and increase resistance to cold (Alleweldt and Harst-Lagenbucher, 1987).

When growth is slowed by lowering temperature, dry matter content increases greatly in response to stress. To compensate for this effect, Galzy *et al.* (1990) suggested reducing the carbohydrate content of the medium. But the rationing of other organic compounds like vitamins is not advisable since it is detrimental to plantlet quality (Galzy, 1969). Galzy showed that the rationing of minerals, especially nitrogen and potassium, can modulate plantlet growth. These results were later confirmed by Moriguchi and Yamaki (1989). The nitrogen intake form, i.e. the NH_4:NO_3 balance, influenced growth, but there are interactions between nitrogen supply and the uptake of other cations, particularly K and Mg (Vellegas *et al.*, 1992).

Cryopreservation by freezing in liquid nitrogen is a technique successfully used for numerous woody plants. With grapevine, the first studies were carried out on latent buds taken from *in situ* canes (Ezawa *et al.*, 1989; Esensee *et al.*, 1990). The technique was extended to axillary buds from *in vitro*-grown plantlets (Plessis *et al.*, 1991; Matsumoto *et al.*, 1998). Dussert *et al.* (1992) focused on cryoconservation of embryogenic cell suspension cultures but this technique is unsuitable for the preservation of genetic conformity.

4. PROTOCOLS FOR MICROPROPAGATION

4.1. Source material

The proper handling of source material is of the upmost importance before *in vitro* culture initiation in successful establishment and reducing further contamination, especially in hardwood species. The best method is based on the collection of dormant canes from selected vines, their storage at 3°C, and the enhancement of bud-burst and shoot elongation under controlled indoor conditions. This offers the advantage of availability of young tissues with reduced contamination at any time of the year. Additionally, the subjection of canes to hot-water treatments proved efficient for decontamination from most external pathogens (Goussard, 1977), phytoplasma (Caudwell *et al.*, 1997), and bacteria (Burr *et al.*, 1989).

4.2. Choice of explants

When virus elimination is not part of the objective, shoot tips and buds are generally preferred over meristems to initiate *in vitro* culture. Indeed, use of larger explants is desirable as they are easier to dissect and have much higher survival and growth rate than smaller explants. Generally, in most cultivated plants, apical tips are commonly considered the best explants for general micropropagation purposes (Hu and Wang, 1983). However, in grapevine, Yu and Meredith (1986) reported that survival rate and shoot production were greater with axillary buds than shoot tips. Moreover, they obtained better results in micropropagation with explants from shoots grown in shade than from shoots developed under full sun. These results were consistent with those of Fanizza *et al.* (1984). The vigour of the mother plant is important too. Explants are usually obtained from rapidly expanding shoots (Barlass and Skene, 1978; Chee and Pool, 1982) and their survival can be affected if the plants are subjected to hydric stresses (Harris and Stevenson, 1982).

4.3. Sterilization of explants

Explants are generally soaked with sodium or calcium hypochlorite (1 or 5% w/v, respectively) with a wetting agent (Tween 20 at 0.01 to 0.1% v/v) for 10 to 20 minutes and then rinsed three times for 5 minutes with sterile distilled water (Krul and Mowbray, 1984). The disinfectants tend to penetrate into the organs through the cut surfaces, causing tissue bleaching and the injured tissues become brown in a few days producing large amounts of phenolic compounds that diffuse in the medium and hinder further development of the explants. Therefore 5 mm of the exposed ends are removed.

4.4. Prevention of the browning of explants

The browning and necrosis of explants during the first phase of their establishment *in vitro*, even when the injured tissues have been removed, are usually attributed to the production of phenolic compounds. Like other woody species, grapevine tissues exhibit high levels of tannins. Yu and Meredith (1986) observed a strong negative correlation between *in vitro* survival and the pre-existing phenolic content of explants. In addition, lipid peroxidation products lead to oxidative stresses and may contribute to the failure of the explant development (Benson and Roubelakis-Angelakis, 1992). Such problems can be solved using media containing antioxidants. A simple procedure to prevent browning of explants had been proposed by Goussard (1981) and consists in the addition of a thin layer of sterile distilled water to the surface of the solidified medium.

4.5. Culture media, environmental conditions and hormone requirements

Functions of the culture medium are (i) to supply the basic nutritional requirements for continued growth of the isolated explants, and (ii) to manipulate development through the control of growth regulator balance. The components, inorganic salts (macro and micronutrients) and organic compounds (sugar, vitamins and amino-acids) used to establish grapevine explants in culture and sustain further growth of propagules consist generally of Murashige and Skoog (1962) medium, with macronutrients often at half-strength. Galzy *et al.* (1990) proposed macronutrient formulation specifically adapted for nodal bud culture.

Because low lignt intensity and CO_2 availability prevent normal photosynthesis, grapevines cultivated *in vitro* are not able to grow photoautotrophically for long periods. So, a significant part of the carbon and energy supply must therefore be provided through the addition of carbohydrates in the culture medium. This mode of carbon nutrition, simultaneously autotrophic and heterotrophic is called mixotrophic (Galzy and Compan, 1992). The relative importance of autotrophic versus heterotrophic nutrition depends on the genotype, the developmental stage of the explant, and the culture conditions (Galzy, 1990). The evaluation of the contribution of photosynthesis to growth requires the consideration of interactions of carbon coming from exogenous carbohydrates with inorganic carbon incorporated by photosynthesis (Falque *et al.*, 1991; Lima da Silva *et al.*, 1996).

The reduced space imposes new environmental conditions. Düring and Harst (1996) studied the effects of light intensity along with CO_2 concentration on stomata response and their results indicate CO_2 fixation and photorespiration of two *Vitis* hybrids. Stomata displayed an ability to adjust aperture to variations in the light intensity but appeared incapable of complete closing under darkness. Nevertheless, photosynthesis output was positively influenced by light intensity and CO_2 level. Photorespiration intensity was shown to be inversely correlated to CO_2 level. These results clearly indicate that a weak photosynthetic rate is the result of a deficiency of light and/or CO_2. In the same way, Iacono and Martinelli (1998) evaluated the CO_2 assimilation and transpiration balance and estimated the stomatal and cuticular transpiration *in vitro*. Fournioux and Bessis (1986) showed that low light intensity and CO_2 enrichment decrease the plantlet growth due to the inhibition of photorespiration and the reduction of energy supply. Characteristics of O_2 and CO_2 exchanges can have a marked effect on plantlet growth and development. When control over gas composition and light intensity is not possible, the best is air renewal by increasing gas leakage from the closed/airtight vessel (Thomas *et al.*, 1999).

4.5.1. Axillary bud proliferation

The selection of appropriate growth regulators is of critical importance to promote axillary branching for maximum production of uniform plants, especially when several cycles of multiplication are repeated. The main usable cytokinins included are : 6-benzylaminopurine or benzyl adenine (BA), kinetin (Kin), 6-(2-isopentenyl) adenine (2iP), zeatin or zeatine riboside (ZR) and 3-(1,2,3-thidiazol-5-yl)-1-phenylurea or thidiazuron (TDZ). BA proved highly effective for many genotypes (see Table 2) and is being used routinely. Among other cytokinins, ZR resulted in low number of proliferating shoots, but morphological characteristics resembled shoots growing under normal condition and its use can be highly recommended (Goussard, 1982 and 1987).

When cultures are grown on a high cytokinin medium to favour shoot proliferation, root initiation is strongly inhibited and must then be induced *in vitro* before transplantation. Rooting is influenced by several factors, of which the stage of propagule growth and growth regulator requirements are of major importance. Individuals are transferred to a rooting medium lacking cytokinins but containin auxins, either singly in tubes or as batches in larger culture vessels. The most important synthetic auxins used include : 1-naphthaleneacetic acid (NAA), indole-3-butyric acid (IBA), 4-chlorophenoxyacetic acid (CPA), 2,4-dichlorophenoxyacetic acid (2,4-D), 2-naphthoxyacetic acid (NOA), and indole-3-acetic acid (IAA). Barlass and Skene (1978 and 1980b) achieved successful rooting of shoot tips on a medium without auxin, however NAA was still needed (Chee and Pool 1982 and 1988; Lee and Wetzstein, 1990).

IBA and IAA were effective in rooting of propagules whereas 2,4-D was ineffective (Grenan, 1979; Novak and Juvova, 1983). When applied to several genotypes, IBA proved superior to IAA, NAA, and 2,4-D (Roubelakis-Angelakis and Zivanovitc, 1991). Auxin stimulated root initiation but may inhibit subsequent root growth (Galzy, 1969), so the appropriate auxin concentration is of critical importance. Zlenko *et al.* (1995) and Novak and Juvova (1983) observed that the effect of auxins on rooting depends on the mineral composition of the nutrient medium. Root initiation is not influenced by salt concentration, however, root growth is enhanced by reducing salt content of the medium (Harris and Stevenson, 1979). A new culture medium with an enhanced mineral balance has been developed by Roubelakis-Angelakis and Zivanovitc (1991) to promote rhizogenesis of grapevine cuttings, in the absence or in the presence of IBA.

Table 2. Studies on axillary bud proliferation in grapevine

Species	Genotype	Studied factor	Reference
V. vinifera	Sylvaner	BA Kin NAA	Jona and Webb (1978)
V. vinifera	10 cvs	BA GA_3	Silvestroni (1981)
V. vinifera	Chenin Blanc	BA ZR	Goussard (1981, 1982)
Vitis hybrids	18 genotypes	BA IAA IBA NAA	Harris and
V. vinifera	3 cvs	MS strength PO_4 agar Organic compounds	Stevenson (1982)
Vitis hybrid	Craciunel	BA Kin 2iP IBA	Novak and Juvova
V. vinifera	7 cvs	MS strength	(1983)
Vitis hybrid	Rougeon	BA NAA Photoperiod	Chee and Pool (1982)
V. vinifera	Limberger	Culture vessel size	Monette (1983)
Vitis hybrids	Marechal Foch Cascade	BA NAA IBA MS strength	Li and Eaton (1984)
Vitis hybrid	Remaily seedless	Vitamins Amino acids	Chee and Pool (1985)
Vitis hybrids	15 genotypes	BA Kin Picloram	Reisch (1986)
V. labrusca	Alba	Adenine	
V. labruscana	3 cvs	MS strength	
V. vinifera	3 cvs		
Vitis hybrid	Remaily seedless	Light spectrum Mn and KI	Chee (1986)
Vitis hybrid	Remaily seedless	Salt formulation	Chee and Pool (1987)
V. vinifera	Frappato Mortilla	Salt formulation BA Kin 2iP ZR	Lê (1987)
V. rotundifolia	Summit	BA IBA	Lee and Wetzstein (1990)
Vitis hybrid	41B	BA IBA GA_3	Celik and Batur (1990)
V. vinifera	Kalecik Karasi	MS strength Vitamins	
V. vinifera	Barbera	TDZ	Gribaudo and Fronda (1991)
V. rotundifolia	4 cvs	BA TDZ Kin Initial nodal position	Sudarsono and Goldy (1991)
V. rotundifolia	3 cvs	BA TDZ Kin NAA Explant length	Gray and Benton (1991)
Vitis hybrid	5BB	MS strength	Zlenko *et al.* (1995)
V. vinifera	3 cvs	Vitamins Mg and Ca	
Vitis × *Muscadinia* hybrids	5 genotypes	BA Salt formulation	Torregrosa and Bouquet (1995)
Vitis hybrid	Fercal		
V. vinifera	Listan Negro	BA 2iP NAA Darkness	Molina *et al.* (1998)

From Torregrosa *et al.* (2001)

4.5.2. Nodal bud culture

Nodal bud culture is mainly characterized by the use of medium lacking growth substances. Thus shoot and root growth proceed simultaneously. Galzy (1969) demonstrated that the mineral requirements vary with the morphogenetic process. High potassium and nitrogen concentrations proved favourable to shoot development but impeded root growth. It was shown that the need for vitamins increases with temperature and growth rate.

Bernard and Mur (1979) demonstrated that axillary buds of grapevine plantlets are composed of two types of lateral meristematic structures : a "prompt bud" and a "latent bud". On the adult vine, the "prompt bud" grows out in the season of its formation to produce a short shoot known as a "summer lateral". The "latent bud" will lie dormant until spring. It was shown that *in vitro* shoots emerging from microcuttings generally derive from "prompt buds" and are not subjected to a physiological dormancy (Fournioux, 1995). However, few shoots do not develop (Martin *et al.*, 1987), due to a translocation problem of carbohydrates and water within the explant (Guerrier-Julien *et al.*, 1996).

Generally, carbohydrates are added to the medium in the form of sucrose, but other forms could be convenient. Galzy *et al.* (1990) showed that most of the sucrose is hydrolysed into fructose and glucose during culture. Despite a low consumption of sucrose by plantlets, initial sucrose concentration positively influenced total dry matter and root growth. Soulié *et al.* (1994) studied the effect of ethylene on microcutting development. Results revealed a significant negative correlation between the level of ethylene inside culture tubes and the explant development. When inhibitors of ethylene synthesis were added to the medium, shoot and root growth were restored confirming its effect. Similarly, Moncousin *et al.* (1989) showed early changes in auxin and ethylene production in vine cuttings before adventitious rooting.

4.6. Acclimatization and transplant to soil

There is insufficient information available on the biological behaviour of plantlets during acclimatization when slowly growing plants are extremely sensitive to environmental stress conditions, especially to water stress. Studies on photosynthesis and water and gas exchanges during this stage are of particular interest (Lakso *et al.*, 1986; Fila *et al.*, 1998; Slavtechva and Dimitrova, 2000). Numerous recommendations have been given, including *in vitro* hardening of plantlets during the last micropropagation phase, addition of polyethylene glycol (Dami and Hughes, 1997), or paclobutrazol in the culture medium with lowering the relative humidity in specifically adapted culture vessels (Smith *et al.*, 1992; Gribaudo *et al.*, 2001). Specific techniques have been developed for rapid acclimatization, with minimal labour and inexpensive equipment (Goussard and Wiid, 1989), however, there is a lack of well-defined guidelines.

5. PROBLEMS IN GRAPE MICROPROPAGATION

5.1. Variability of response

When *in vitro* grapevine propagation is based on regenerative processes, the tissues must undergo developmental changes that usually involve the formation of undifferentiated callus with subsequent reorganization in plantlets. This has not been easy to achieve with all the species and cultivars of grapevine. The competence of explant regeneration is generally considered to be greatly linked to the genotype. Moreover, the ability of some species, such as *Vitis rupestris*, to produce somatic embryos from cultured anthers is likely heritable (Rajasekaran and Mullins, 1983b).

The potential of micropropagation of grapevine cultivars by nodal bud culture or axillary bud proliferation is also under genetic control, though there are very few reports (Chee and Pool, 1983). Peros *et al.* (1998) observed a large variability among 32 *Vitis vinifera* cultivars micropropagated on two different media. It is suggested that differences in *in vitro* response between genotypes may be related to differences in endogenous level of growth regulators. However, organogenic competence, measured by adventitious bud regeneration from leaves was not correlated with micropropagation ability. The hypothesis of a relationship between endogenous hormone content and the *in vitro* behaviour of cultivars is supported by the differential response of *Vitis* genotypes observed when IBA is present in the micropropagation medium (Roubelakis-Angelakis and Zivanovitc, 1991).

5.2. Juvenility

In vitro culture induces anatomical and morphological adaptations, of which the most obvious is the strong reduction in the size of organs. Histological organization of tissues is greatly altered and Dami and Hughes (1995) showed that *in vitro* leaves contain larger mesophyll cells and lack normal palisade layer formation. A major feature of rejuvenation is a shift to spiral phyllotaxy (2/5) and a gradual reduction in the production of tendrils (Mullins *et al.*, 1979; Thomas, 1999). *In vitro* production of flowers and berries is extremely rare and observed only in a few cultivars (Favre and Grenan, 1979). However, the "adult" morphology of grapevine (distichous phyllotaxy of 1/2 and presence of tendrils) can be restored *in vitro* by CO_2 enrichment (Fournioux and Bessis, 1993).

Rejuvenation can be a serious limitation for many potential uses of *in vitro* culture of grapevine, especially for large-scale propagation or genetic improvement of winegrape cultivars. Modifications induced by *in vitro* culture may affect leaf morphology (Nozeran *et al.*, 1983) and remain stable under vineyard conditions (Grenan and Caraux, 1984). Moreover, on certain cultivars, fertility of the buds can be dramatically reduced (Grenan, 1982). Given the strict obligation to ensure trueness to type, especially in classified wine-growing areas, vine micropropagation could not be extensively used in practice until these problems are solved.

After planting micropropagated plants in the field, reversion from juvenile to mature state is not uniform and is dependent on the pruning system. Grenan (1984) reported that morphological alterations due to juvenility can be eliminated in producing plants with the uppermost nodes of acclimatized plantlets growing under greenhouse in conditions (hydroponics) that are highly favourable to vigour.

It has been proposed that rejuvenation could be caused by the miniaturisation of stem meristems (Nozeran and Bancilhon, 1972) or by a reduction in the distance between the apical bud and roots (Mullins *et al.*, 1979). According to these authors, as the shoot apex becomes more remote from the roots, as a consequence of extension growth, there are changes in its hormonal environment. Specifically, levels of endogenous gibbereelins in apices appear to decline with increasing distance from roots. During the *in vitro* subculture procedure, actively-growing shoot apices were kept permanently in close proximity to the root system. With this treatment, the adult morphology of the initial plant was gradually replaced by the juvenile form. Endogenous polyamines could also be involved in the juvenility process (Martin-Tanguy and Carre, 1993; Heloir *et al.*, 1998). Whatever the origin of the phenomenon, it is important to limit the number of *in vitro* propagation cycles and transfer plants periodically to the greenhouse to check their phenotypic conformity.

5.3. Genetic instability

In grapevine, changes have commonly been observed among plants derived from *in vitro* culture. Rejuvenation involves recurring effects that can be reversible, while somaclonal variation (Larkin and Scowcroft, 1981; Karp, 1994; Jain, 2001) corresponds to modifications that are unpredictable and generally stable (For a review of the changes observed in somatic seedlings of grapevine, see Torregrosa *et al.*, 2001). Somaclones occur more frequently among plants obtained from regenerative processes, including bud-neoformation, that need high concentrations of growth regulators in the media, than among those obtained from organized meristems.

However, Chee and Pool (1985) pointed out that within high-density shoot clusters produced by axillary branching on media supplemented with cytokinins, the occurrence of adventitious organogenesis cannot be ruled out. Moreover, when the fragmented shoot apex technique is used to initiate cycles of micropropagation, the regenerative capacity of leaf primordia leads to form adventitious buds (Barlass and Skene, 1980a; Barlass *et al.*, 1981) and separate periclinal chimera (Skene and Barlass, 1983). Thus, to avoid risks of genetic instability, especially when *in vitro* culture is used for germplasm conservation, nodal bud culture technique is recommended on media without growth substances.

The molecular basis of age-related changes due to *in vitro* culture is poorly understood. Until now, the only one example of somaclonal variation detected by molecular markers (RAPD-PCR) was an altered phenotype observed *in vitro* on a grapevine regenerant from protoplasts (Schneider *et al.*, 1996). Although no molecular marker can provide absolute evidence that somaclonal variation has not occurred in culture, there is a great need for methods to detect genetic instability during medium or long term storage of *Vitis* germplasm and during the accelerated cycles of micropropagation of a new improved genotype.

It was demonstated that significant changes in DNA methylation can occur during initiation and maintenance of grapevine shoot cultures (Harding *et al.*, 1996). These results suggest that DNA methylation plays a role in modification of gene expression during *in vitro* culture and may be implicated in rejuvenation. In oil palm (*Elaeis guinensis* Jack), DNA methylation was involved in somaclonal variation and possibly linked to modifications in cytokinin metabolism (Jaligot *et al.*, 2000). It is likely that somaclonal variation currently observed in the flower-type (cytokinin-dependent) of grapevine rootstocks or hybrids regenerated from somatic embryos are due to similar causes (Mullins, 1987; Torregrosa *et al.*, 2001). Further work is needed to identify appropriate markers of this phenomenon, using particularly RNA differential display, for better understanding of the factors responsible for somaclonal variation in the grapevine cultured *in vitro*.

5.4. Practical use in viticulture

The worldwide exchange of micropropagated *Vitis* germplasm (including improved genotypes) needed for breeders, nurserymen, researchers or others, can be the safest option. *In vitro* cultures that are free from exogenous fungal and bacterial contaminants could simplify the problems of quarantine, and may be virus-free if produced from meristem culture. But, it should not be assumed that *in vitro* cultures, when derived from meristem culture, are necessarily disease-free and indexing of tissue should precede distribution.

Though *in vitro* culture offers great possibilities for grapevine germplasm conservation, virus sanitation, and genetic improvement, its application in large-scale commercial propagation is limited by the fact that *Vitis vinifera* rooted plants produced by micropropagation cannot be used in most worldwide vineyards that are infested with Phylloxera. Grapevine plantings require grafting on resistant rootstocks. *In vitro* micrograftings are unsuitable for practical use in the vineyards. Furthermore, due to the small size of their basal internodes, that leads to the presence of numerous buds and the emission of suckers difficult to suppress in the field, micropropagated and acclimatized rootstock plants are difficult to use directly even for establishment of mother plantations. That explains why there are very few reports on field experimentation of *in vitro* propagated grapevine plants. There is only concern over plants derived from somatic embryos (Mowbray *et al.*, 1985; Martinez *et al.*, 1997).

The green-grafting-rooting technique could be a powerful tool as a complementary method to the grapevine micropropagation (Boubals, 1987; Martin *et al.*, 1987). But, to follow viticultural practices, the rootstock cuttings have to be at least 20 cm long with one or two internodes. To produce herbaceous cuttings, a separate step is undertaken in the greenhouse after acclimatization of the *in vitro* plantlets. This extra step and the practice of green-grafting with a special machine, require technical skills that increases the cost of propagation considerably. Consequently, this technique is not currently used now for commercial plantings but rather to propagate rapidly rare or improved genotypes. However, in the future, we cannot exclude the possibility that grafting will be replaced by genetically-modified *Vitis vinifera* cultivars that will be resistant to Phylloxera and express genes encoding plant proteins that are effective against homopteran insects (Viss and Driver, 1996).

6. FUTURE PROSPECTS

There has been extensive research on micropropagation of grapevine for 40 years. But, unlike many horticultural crops, the applications can be considered rather disappointing. The main practical shortcoming of this technique has been in the production of pathogen-free stocks or the multiplication of seedlings issued from *in vitro* embryo rescue, before their acclimatization in greenhouse and growing in the field. Actually, there is little propsect for the development of grapevine micropropagation for commercial purposes. In fact, the future of this technique depends on the development and success of DNA transfer in grapevine. It could be used to propagate and release rapidly cultivars with a high value, particularly if the necessity of grafting is discarded by the use of Phylloxera-resistant, genetically modified winegrape cultivars. Such a possibility would be another cultural revolution leading to a renewal of interest for grapevine micropropagation.

Although the risk of genetic instability may be minimized through the use of highly organized tissue systems, careful monitoring of cultures will still be required. This may prove difficult while plantlets remain in culture vessels and most of the mutations may only be recognized at the later stages of plant development. Moreover, the risk of mixings cannot be excluded during the sub-cultures. Some regular assessment of plants must be made in field conditions as well as during *in vitro* culture. There is currently a great deal of interest in molecular genetic markers for such monitoring. PCR-related technologies may be highly suitable for this purpose. The simplest of these PCR techniques, RAPDs, may not be very reproducible. AFLPs may be more appropriate for reliable markers, as they give broad coverage of the genome, using specific probes and are more sensitive than microsatellites, particularly to identify somatic mutants (Scott *et al.*, 2000).

7. CONCLUSION

In the area of cell and tissue physiology, a lot of references have involved the grapevine. As with all woody perennials, *in vitro* technologies are relatively delicate. Nevertheless, remarkable progress in research has produced a situation where practical applications of results can be expected, even if some of them are hampered by technical constraints. Reciprocal understanding has yet to be established between scientists developing micropropagation with strict conformity to varietal type and those using *in vitro* technology for variety improvement. Although the potential of the last is continuously increasing, it is not sure that in the future the second objective will dominate the first. Instead of being exclusive, they must be complementary and sustained investment in research is needed to develop biotechnology programs in viticulture.

REFERENCES

Aldwinckle, H.S. (1980) *In vitro* techniques for studying obligate pathogens of *Vitis*. *In*: Proceed. 3rd Intern. Symp. Grape Breeding, Davis (California), June 1980, pp. 87-91.

Alleweldt, G. and M. Harst-Lagenbucher (1987) The effect of growth inhibitors on long-term storage of *in vitro* cultures of grapevines. Vitis 26: 57-64.

Alleweldt, G. and J.V. Possingham (1988) Progress in grapevine breeding. Theor. Appl. Genet. 75: 669-673.

Ayuso, P. and A. Peña-Iglesias (1978) Shoot apex (meristem) grafting : a novel and promising technique for regeneration of virus infected grapevines. *In* : Proceed. 6th ICVG Conference, Cordoba (Spain) 1976, Monografias INIA 18: 319-323.

Barlass, M. and K.G.M. Skene (1978) *In vitro* propagation of grapevine (*Vitis vinifera* L.) from fragmented shoot apices. Vitis 17: 335-340.

Barlass, M. and K.G.M. Skene (1980a) Studies on the fragmented shoot apex of grapevine. I. The regenerative capacity of leaf primordial fragment *in vitro*. J. Exp. Bot. 31: 483-488.

Barlass, M. and K.G.M. Skene (1980b) Studies on the fragmented shoot apex of grapevine. II. Factors affecting growth and differentiation *in vitro*. J. Exp. Bot. 31: 489-495.

Barlass, M. and K.G.M. Skene (1983) Long term storage of grape *in vitro*. IBPGR Plant Genet. Ress. Newslett. 53: 19-21.

Barlass, M., Skene, K.G.M., and P.R. Clingeleffer (1981) Studies on the fragmented shoot apex of grapevine. III. A scanning electron microscope study of adventitious bud formation *in vitro*. J. Exp. Bot. 32: 1079-1083.

Barlass, M., Skene, K.G.M., Woodham, R.C., and L.R. Krake (1982) Regeneration of virus-free grapevines using *in vitro* apical culture. Ann. Appl. Biol. 101: 291-295.

Barlass, M., Miller, R.P., and A.J. Antcliff (1986) Development of methods for screening grapevines for resistance to infection by downy mildew. I. Dual culture *in vitro*. 37: 61-66.

Bass, P., Vuittenez, A., and R. Legin (1978) Amélioration de la thermothérapie des vignes virosées au moyen de la culture d'apex sur milieux nutritifs ou par greffage sur plantules de semis cultivées aseptiquement *in vitro*. *In* : Proceed. 6th ICVG Conference, Cordoba (Spain) 1976, Monografias INIA 18: 325-332.

Bavaresco, L. and M.A. Walker (1994) Techniques for successfully establishing *Xiphinema index* in dual culture with grape. Am. J. Enol. Vitic. 45: 273-277.

Benson, E.E. and K.A. Roubelakis-Angelakis (1992) Fluorescent lipid peroxidation products and antioxidant enzymes in tissue cultures of *Vitis vinifera* L. Plant Sci. 84: 83-90.

Bernard, A.C. and G. Mur (1979) Observations sur l'organogenèse des bourgeons de plants de *Vitis vinifera* cultivés *in vitro*. Ann. Amélior. Plantes 29: 311-323.

Blaich, R. (1985) Recherche sur les cultures de méristèmes et d'organes de vigne *in vitro* en vue de la sélection et de la conservation des génotypes. Bull. OIV 650-651: 391-395.

Boubals, D. (1987). La culture *in vitro* et la production de plants greffés-soudés. Progr. Agric. Vitic. 104: 311-314.

Bouquet, A. (1980) *Vitis* x *Muscadinia* : A new way in grape breeding for disease resistance in France. *In* : Proceed. 3rd Intern. Symp. Grape Breeding, Davis (California), June 1980, pp. 42-61

Bouquet, A. (1982) Origine et évolution de l'encépagement français à travers les siècles. Progr. Agric. Vitic. 99: 110-121.

Bouquet, A. (1989a) Intérêt des techniques de culture *in vitro* pour l'amélioration génétique de la vigne, Bull. OIV 697-698: 179-192.

Bouquet, A. (1989b) Application de la culture *in vitro* à l'amélioration génétique de la vigne: Culture d'embryons *in ovulo*, embryogénèse somatique et variation somaclonale. Quad. Vitic. Enol. Univ. Torino 13: 51-64.

Bouquet, A. and H.P. Davis (1989) Culture *in vitro* d'ovules et d'embryons de vigne (*Vitis vinifera* L.) appliquée à la sélection de variétés de raisins de table sans pépins. Agronomie 9: 565-574.

Bouquet, A., Piganeau, B., and A.M. Lamaison (1982) Influence du génotype sur la production de cals, d'embryoïdes et de plantes entières par culture d'anthères *in vitro* dans le genre *Vitis*. C.R. Acad. Sci. Paris 295: 569-574.

Bouquet, A., Danglot, Y., and A.M. Lamaison (1985) Amélioration génétique des porte-greffes de vigne par culture *in vitro*. *In* : Amélioration de la vigne et culture *in vitro*, Colloque Moet et Hennessy, Paris, April 1985, pp. 145-159.

Bouquet, A., Chabbert, A.M., and Y. Danglot (1990) *In vitro* selection for tolerance to magnesium deficiency in grapevine rootstocks. *In*: Proceed. 5th Intern. Symp. Grape Breeding, St-Martin/Pfalz (Germany) Sept. 1989, *Vitis* (Special Issue) p. 445.

Bouquet, A. , Danglot, Y., Torregrosa, L., Bongiovani, M., Castagnone-Sereno, P., Esmenjaud, D., and A. Dalmasso (2000a) Breeding rootstocks resistant to grape fanleaf virus spread using *Vitis* x *Muscadinia* hybridization. *In* : Proceed. 7th Intern. Symp. Grapevine Genetics and Breeding, A. Bouquet and J.M. Boursiquot (Eds), Montpellier (France) July 1998. Acta Hort. 528: 517-526.

Bouquet, A., Pauquet, J., Adam-Blondon, A.F., Torregrosa, L., Merdinoglu, D., and S. Wiedemann-Merdinoglu (2000b) Vers l'obtention de variétés de vigne résistantes à l'oïdium et au mildiou par les méthodes conventionnelles et biotechnologiques. Bull. OIV 833-834: 445-441.

Bouquet, A., Adam-Blondon, A.F., and C. Schneider (2001) Métis ou hybrides : Des variétés nouvelles pour une viticulture de qualité. J. Intern. Sci. Vigne Vin (hors série), pp. 109-115.

Bowers, J.E. and C.P. Meredith (1997) The parentage of a classic wine grape, Cabernet-Sauvignon. Nature Genet. 16: 84.

Bowers, J.E., Boursiquot, J.-M. , This, P., Chu, K., Johansson, H., and C.P. Meredith (1999) Historical Genetics : The parentage of Chardonnay, Gamay and other wine grapes of Northeastern France. Science 285: 1562-1565.

Burr, T.J., Ophell, K., Katz, B.H., and A. Kerr (1989) Effect of hot water treatment on systemic *Agrobacterium tumefaciens* biovar 3 in dormant grape cutttings. Plant Disease 73: 242-245.

Cabaleiro, C. and A. Segura (1997) Field transmission of grapevine leafroll associated virus 3 (GLRaV-3) by the mealybug *Planococcus citri*. Plant Disease 81: 283-287.

Cao, Z. (1990) Grape : anther culture. *In* : Handbook of Plant Cell Culture, Vol.VI, Perennial crops, Zenghua Cheng *et al.* (Eds), MacGraw-Hill, New-York, pp. 300-311.

Caudwell, A. (1990) Epidemiology and characterization of Flavescence dorée (FD) and other grapevine yellows. Agronomie 10: 655-663.

Caudwell, A., Larrue, J., Boudon-Padieu, E., and G.D. McLean (1997) Flavescence dorée elimination from dormant wood of grapevines by hot-water treatment. Aust. J. Grape Wine Res. 3: 21-25.

Celik, H. and M. Batur (1990) Meristem culture for clonal micropropagation of grapevines. *In*: Proceed. 5th Intern. Symp. Grape Breeding, St-Martin/Pflaz (Germany), Vitis (Special Issue) pp. 455-462.

Chee, R. (1986) *In vitro* culture of *Vitis*: The effects of light spectrum, manganese sulfate and potassium iodide on morphogenesis. Plant Cell Tiss. Org. Cult. 7: 121-134.

Chee, R. and R.M. Pool (1982) The effects of growth substances and photoperiod on the development of shoot apices of *Vitis* cultures *in vitro*. Scientia Hort. 16: 17-27.

Chee, R. and R.M. Pool (1983) *In vitro* propagation of *Vitis*: application of previously defined culture conditions to a selection of genotypes. Vitis 22: 363-374.

Chee, R. and R.M. Pool (1985) *In vitro* propagation of *Vitis*: The effects of organic substances in shoot multiplication. Vitis 24: 106-118.

Chee, R. and R.M. Pool (1987) Improved inorganic media constituents for *in vitro* multiplication of *Vitis*. Scientia Hort. 32: 85-95.

Chee, R. and R.M. Pool (1988) Sucrose and NAA influence. Growth of subcultured shoots and *in vitro* production of roots in *Vitis*. HortScience 23: 776.

Chee, R., Pool, R.M., and D. Bucher (1984) A method for large scale *in vitro* propagation of *Vitis*. N.Y. Food Life Sci. Bull. 109: 1-9.

Cheng, Z.M. and B.I. Reisch (1989) Shoot regeneration from petioles and leaves of *Vitis* × *labruscana* "Catawba". Plant Cell Rep. 8: 403-406.

Clog, E., Bass, P., and B. Walter (1990) Plant regeneration by organogenesis in *Vitis* rootstock species. Plant Cell Rep. 8: 726-728.

Colby, S.M., Juncosa, A.M., and C.P. Meredith (1991) Cellular differences in *Agrobacterium* susceptibility and regenerative capacity restrict the development of transgenic grapevines. J. Amer. Soc. Hort. Sci. 116: 356-361.

Colova-Tsolova, V., Perl, A., Krastanova, S., Tsvetkov, I., and A. Atanassov (2001) Genetically engineered grape for disease and stress tolerance. *In* : Molecular Biology and Biotechnology of Grapevine, K.A. Roubelakis-Angelakis (Ed.), Kluwer Publish. Co., Amsterdam, pp. 411-431.

Dami, I. and H. Hughes (1995) Leaf anatomy and water loss of *in vitro* PEG-treated 'Valiant' grape. Plant Cell Tiss. Org. Cult. 42: 179-184.

Dami, I. and H.G. Hughes (1997) Effects of PEG-induced water stress on *in vitro* hardening of 'Valiant' grape. Plant Cell Tiss. Org. Cult. 47: 97-101.

Düring, H. and M. Harst (1996) Stomatal behaviour, photosynthesis and photorespiration of *in vitro*-grown grapevines : effects of light and CO_2. Vitis 35: 163-165.

Dussert, S., Mauro, M.-C., and F. Engelman (1992) Cryopreservation of grape embryogenic cell suspensions. Cryo-letters 13: 15-22.

Esensee, V., Stushnoff, C., and P.L. Forsline (1990) Cryopreservation of dormant grape (*Vitis* sp.) buds. HortScience 25: 90.

Ezawa, T., Harada T., and T. Yakuwa (1989) Studies on freeze-preservation of fruit tree germplasm : III- Freeze preparation of grape shoot tips. J. Fac. Agr. Hokkaido Univ. 64: 51-55.

Fallot, J., Tey-Rulh, P., Coutouly, P., Petitprez, M., Roustan, J.-P., Philippe, I., and R. Tabacchi (1990) Cultures *in vitro*, étude de l'eutypiose et stratégies de création de somaclones de vigne tolérants. Les Colloques de l'INRA, Paris, 51: 151-159.

Falque, M., Compan, D., and R. Galzy (1991) A method for the study of CO_2 exchanges in *in vitro* cultured *Vitis rupestris* plantlets. Plant Cell Tissue Organ Cult. 27: 175-181.

Fanizza, G., Tanzarella, O.A., Carrozzo, G., and B. Greco (1984) Influence of *Vitis* source on *in vitro* shoot apex culture. Ann. Appl. Biol. 104: 577-578.

Favre, J.M. (1977) Premiers résultats concernant l'obtention *in vitro* de néoformations caulinaires chez la Vigne. Ann. Amélior. Plantes 27: 151-169.

Favre, J.M. and S. Grenan (1979) Sur la production de vrilles, de fleurs et de baies chez la vigne cultivée *in vitro*. Ann. Amélior. Plantes 29: 247-252.

Fila, G., Ghashghaie, J., Hoarau, J., and G. Cornic (1998) Photosynthesis, leaf conductance and water relations of *in vitro* cultured grapevine rootstock in relation to acclimatization. Physiol. Plant. 102: 411-418.

Flandrin , J.-L. (1997) Usages et fonctions du vin en Occident du Moyen-Age à nos jours. *In*: L'histoire du Vin. OIV ed., pp. 73-82

Forneck, A., Walker, M.A., and N. Merkt (1996) Aseptic dual culture of grape (*Vitis* spp.) and grape phylloxera (*Daktulosphaira vitifoliae* Fitch). Vitis 35: 95-97.

Fournioux, J.-C. (1995) Facteurs de l'édification de la tige de *Vitis vinifera* L. dans différentes conditions de culture, Thèse Université de Bourgogne, France, 554 p.

Fournioux, J.-C. and R. Bessis (1986) Influence de la teneur en gaz carbonique sur la morphogenèse de la vigne en culture *in vitro*. Can. J. Bot. 64: 2608-2616.

Fournioux, J.-C. and R. Bessis (1993) Use of carbon dioxide enrichment to obtain adult morphology of grapevine *in vitro*. Plant Cell Tiss. Org. Cult. 33: 51-57.

Galet, P. (1988) Les Vignes Américaines, *In* : Cépages et vignobles de France, Tome II, Déhan ed. (Montpellier) 533 p.

Galzy, R. (1961) Confirmation de la nature virale du court-noué de la vigne par des essais de thermothérapie sur des cultures *in vitro*. C.R. Acad. Sci. Paris 253: 706-708.

Galzy, R. (1969) Remarques sur la croissance de *Vitis rupestris* cultivée *in vitro* sur différents milieux nutritifs. Vitis 8: 191-205.

Galzy, R. (1972) La culture *in vitro* des apex de *Vitis rupestris*. C.R. Acad. Sci. Paris 274: 210-213.

Galzy, R. (1985) Les possibilités de conservation *in vitro* d'une collection de clones de vignes. Bull. OIV 650-651: 376-390.

Galzy, R. and D. Compan (1988) Growth and nutrition of grapevine during *in vitro* long-terme storage. Plant Cell Tissue Organ Cult. 13: 229-237.

Galzy, R. and D. Compan (1992) Remarks on mixotrophic or autotrophic carbon nutrition of grapevine *in vitro*. Plant Cell Tiss. Org. Cult. 31: 229-244.

Galzy, R., Haffner, V., and D. Compan (1990) Influence of three factors on the growth and nutrition of grapevine microcuttings. J. Exp. Bot. 41: 295-301.

Garau, R., Padilla, V., Rumbos, I., Walter, B., and V. Savino (1997) Indexing for the identification of virus and virus-like diseases of the grapevine. *In* : Sanitary selection of the grapevine. Protocols for detection of viruses and virus-like diseases, B. Walter (Ed.), Les Colloques de l'INRA 86: 97-117.

Gifford, E.M. and W.B. Hewitt (1961) The use of heat therapy and *in vitro* shoot tip culture to eliminate fanleaf virus from the grapevine. Am. J. Enol. Vitic. 12: 129-135.

Goussard, P.G. (1977) Effect of hot-water treatments on vine cuttings and one-year-old grafts. Vitis 16: 272-278.

Goussard, P.G. (1981) Effects of cytokinins on elongation, proliferation and total mass of shoots derived from shoot apices of grapevine cultured *in vitro*. Vitis 20: 238-234.

Goussard, P.G. (1982) Morphological responses of shoot apices of grapevine cultured *in vitro*: Effects of cytokinins in routine subculturing. Vitis 21: 293-298.

Goussard, P.G. (1987) Anatomical responses to cytokinins of abscised grapevine shoot apices cultured *in vitro*. S. Afr. J. Enol. Vitic. 8: 53-59.

Goussard, P.G. and J. Wiid (1989) A revised approach to the acclimatization of grapevine plantlets cultured *in vitro*. Dec. Fruit Grower 39: 29-31.

Goussard, P.G., Wiid, J., and G.G.F. Kasdorf (1991) The effectiveness of *in vitro* somatic embryogenesis in eliminating fanleaf virus and leafroll associated viruses from grapevine. S. Afr. J. Enol. Vitic. 12: 77-81.

Gray, D.J. (1992) Somatic embryogenesis and plant regeneration from immature zygotic embryos of muscadine grapevine (*Vitis rotundifolia*) cultivars. Amer. J. Bot. 79: 542-546.

Gray, D.J. (1995) Somatic embryogenesis in grape. *In* : Somatic embryogenesis in woody plants, S. Jain, P. Gupta, and R. Newton (Eds), Kluwer Acad. Publish, Dordrecht, Vol. 2, pp. 191-217.

Gray, D.J. and C.M. Benton (1991) *In vitro* micropropagation and plant establishment of muscadine grape cultivars (*Vitis rotundifolia*). Plant Cell Tiss. Org. Cult. 27: 7-14.

Gray, D.J. and M.E. Compton (1993) Grape somatic embryo dormancy and quiescence: potential of dehydrated seeds for germplasm conservation. *In*: Synseeds, applications of synthetic seeds in crop improvement, K. Redenbaugh (Ed.), CRC Press Inc., pp. 367-379.

Gray, D.J. and C.P. Meredith (1992) Grape. *In*: Biotechnology of perennial fruit crops, F.A. Hammerschlag and R.E. Litz (Eds), CAB Int., Wallingford, pp. 229-261.

Grenan, S. (1979) Rhizogenèse de bourgeons apicaux de boutures de vignes cultivées *in vitro*. Conn. Vigne Vin 2: 125-136.

Grenan, S. (1982) Quelques réflexions à propos de modifications morphogénétiques consécutives à la culture *in vitro* chez la vigne (*Vitis vinifera* L.). Ann. Sci. Nat. Bot., Paris 13: 135-146.

Grenan, S. (1984) Polymorphisme foliaire consécutif à la culture *in vitro* de *Vitis vinifera* L. Vitis 23: 159-174.

Grenan, S. (1992) Micropropagation of grapevine (*Vitis vinifera* L.). *In*: Biotechnology in agriculture and forestry: High tech and micropropagation, Y.P.S. Bajaj (Ed.), Springer-Verlag, Berlin, pp. 371-398.

Grenan, S. and G. Caraux (1984) Influence de la thermothérapie *in vitro* sur le polymorphisme foliaire de variétés de *Vitis vinifera* L. Bull. Soc. Bot. Fr. 131: 277-300.

Gresshoff, P.M. and C.H. Doy (1974) Derivation of a haploid cell line from *Vitis vinifera* and the importance of the stage of meiotic development of anthers for haploid culture of this and other genera. Z. Pflanzenphysiol. 73: 132-141.

Gribaudo, I. and A. Fronda (1991) Effects of thidiazuron on grapevine axillary buds cultivated *in vitro*. HortScience 26: 1083.

Gribaudo, I., Novello, V., and M. Restagno (2001) Improved control of water loss from micropropagated grapevines (*Vitis vinifera* cv. Nebbiolo). Vitis 40:137-140.

Guerrier-Julien, J., Deloire, A., and A. Coudret (1996) Water relations in grapevine microcuttings grown *in vitro*. Biol. Plant. 38: 149-152.

Harding, K., Benson, E.E., and K.A. Roubelakis-Angelakis (1996) Methylated DNA changes associated with the initiation and maintenance of *Vitis vinifera in vitro* shoot and callus cultures : A possible mechanism for age-related changes. Vitis 35: 79-86.

Harris, R.E. and J.H. Stevenson (1979) Virus elimination and rapid propagation of grapes *in vitro*. Proceed. Int. Plant Prop. Soc. 29: 85-108.

Harris, R.E. and J.H. Stevenson (1982) *In vitro* propagation of *Vitis*. Vitis 21: 22-32.

Harris, R.E. and E.B.B. Mason (1983) Two machines for *in vitro* propagation of plants in liquid media. Can. J. Plant Sci. 63: 311-316.

Hawker, J.S., Downton, W.J.S., Wiskich, D., and M.G. Mullins (1973) Callus and cell culture from grape berries. HortScience. 8: 398-399.

Heloir, M.C., Fournioux, J.-C., Barbier, M., Jeandet, P., and R. Bessis (1998) Endogenous polyamine concentration in juvenile, adult and micropropagated grapevine (*Vitis vinifera* L. cv Pinot noir). Vitis 37: 61-62.

Hewitt, W.B., Raski, D.J., and A.C. Goheen (1958) Nematode vector of soil-borne fanleaf virus of grapevines. Phytopathology 48: 586-595.

Hirabayashi, T. (1985) Somatic embryogenesis from leaf tissues of grape. *In* : Amélioration de la vigne et culture *in vitro*, Colloque Moet et Hennessy, Paris, April 1985, pp. 75-81.

Hirabayashi, T., Kozaki, I., and T. Akihama (1976) *In vitro* differentiation of shoots from anther callus in *Vitis*. HortScience 11: 511-512.

Hu, C.Y. and P.J. Wang (1983) Meristem, shoot tip and bud culture. *In* : Handbook of Plant Cell Cultures, W.R. Sharp, D.A. Evans, P.V. Ammirato, and Y. Yamada (Eds), Vol. 1, Macmillan Publishing Co, New York., pp. 177-227.

Iacono, F. and L. Martinelli (1998) CO_2 assimilation and transpiration balance in species of genus *Vitis* cultivated *in vivo* and *in vitro*. Estimation of stomatal and cuticular transpiration *in vitro*. J. Int. Sci. Vigne Vin 32: 91-97.

Jaligot, E., Rival, A., Beulé, T., Dussert, S. and J.L. Verdeil (2000) Somaclonal variation in oil palm (*Elaeis guineensis* Jacq.) : the DNA methylation hypothesis. Plant Cell Rep. 19 : 684-690.

Jona, R. and K.J. Webb (1978) Callus and axillary-bud culture of *Vitis vinifera* « Sylvaner Riesling ». Scientia Hort. 9: 55-60.

Karp, A. (1994) Origin, causes and uses of variation in plant tissue cultures. *In* : Plant Cell and Tissue Culture, K. Vasil and T.A. Thorpe (Eds), Kluwer Acad. Publish., Dordrecht, pp. 139-151.

Kikkert, J.R., Thomas, M.R., and B.I. Reisch (2001) Grapevine genetic engineering. *In* : Molecular Biology and Biotechnology of Grapevine, K.A. Roubelakis-Angelakis (Ed.), Kluwer Publish. Co., Amsterdam, pp. 327-351.

Klempka, K.C., Meredith, C.P., and M.A. Sall (1984) Dual culture of grape powdery mildew (*Uncinula necator* Burr.) on its host (*Vitis vinifera* L.). Am. J. Enol. Vitic. 35: 170-174.

Krul, W.R. and J.F. Worley (1977) Formation of adventitious embryos in callus cultures of « Seyval », a French hybrid grape. J. Am. Soc. Hort. Sci. 102: 360-363.

Krul, W.R. and G.H. Mowbray (1984) Grapes. *In* : Handbook of Plant Cell Cultures, W.R. Sharp, D.A. Evans, P.V. Ammirato, and Y. Yamada (Eds), Vol 2, Crop species, Macmillan Publishing Co., New York., pp. 369-395.

Kuksova, V.B., Piven, N.M., and Y.Y. Gleba (1997) Somaclonal variation and *in vitro* induced mutagenesis in grapevine. Plant Cell Tiss. Organ. Cult. 49: 17-27.

Lakso, A.N., Reisch, B.I., Mortensen, J., and M.H. Roberts (1986) Carbon dioxide enrichment for stimulation of growth of *in vitro* propagated grapevines after transfer from culture. J. Amer. Soc. Hort. Sci. 111: 634-638.

La Notte, P., Buzkan, N., Choueiri, E., Minafra, A., and G.P. Martelli (1997) Acquisition and transmission of grapevine virus A by the mealybug *Pseudococcus longispinus*. J. Plant Pathol. 79: 79-85.

Larkin, P.J. and W.R. Scowcroft (1981) Somaclonal variation : a novel source of variability from cell culture for plant improvement. Theor. Appl. Genet. 60: 197-214.

Lê, C.L. (1987) Multiplication végétative *in vitro* de la vigne (*Vitis vinifera* L.). Schweiz. Landw. Fo. 26:507-517.

Lebrun, L., Rajasekaran, C., and M.G. Mullins (1985) Selection *in vitro* for NaCl tolerance in *V. rupestris* Scheele. Ann. Bot. 56: 733-739.

Ledbetter, C.A. and D.W. Ramming (1989) Seedlessness in grapes. Hort. Rev. 11: 159-184.

Lee, N. and Y. Wetzstein (1990) *In vitro* propagation of muscadine grape by axillary shoot proliferation. J. Amer. Sci. Hort. Sci. 115: 324-329.

Lee, T.C. and T. Wicks (1982) Dual culture of *Plasmopara viticola* and grapevine and its application to systemic fungicide evaluation. Plant Disease 66: 308-310.

Le Gall, O., Torregrosa, L., Danglot, Y., Candresse, T., and A. Bouquet (1994) *Agrobacterium*-mediated genetic transformation of grapevine somatic embryos and regeneration of transgenic plants expressing the coat protein of grapevine chrome mosaic nepovirus (GCMV). Plant Sci. 102: 161-170.

Li, J.R. and G.W. Eaton (1984) Growth and rooting of grape shoot apices *in vitro*. HortScience 19: 64-66.

Lima da Silva, A., Hariscain, P., Ollat, N. and J.-P. Doazan (1996) Evaluation of the photoautotrophic capability of *Vitis* plantlets (rootstock var. Gravesac): Set-up of a special open flow gas exchange system. Vitis 35: 73-78.

Loubser, J.T. and A.J. Meyer (1990) Dual cultures of *Meloidogyne javanica* and grapevine rootstocks on artificial media. S. Afr. J. Enol. Vitic. 11: 42-45.

Marinval, P. (1997) Vigne sauvage et vigne cultivée dans le bassin méditerranéen : Emergence de la viticulture. Contribution archéo-botanique. *In* : Histoire du Vin. OIV Ed., pp. 137-172.

Martin, C., Vernoy, R., Carre, M., Vesselle, G., Collas, A., and C. Bougerey (1987) Vignes et techniques de culture *in vitro*, quelques résultats d'une collaboration entre recherche publique et entreprise privée. Bull. OIV 675-676: 447-458.

Martin, R.R., James, D., and C.A. Levesque (2000) Impacts of molecular diagnostic technologies on plant disease management. Ann. Rev. Plant Pathol. 38: 207-239.

Martin-Tanguy, J. and M. Carre (1993) Polyamines in grapevine microcuttings cultivated *in vitro*. Effects of amines and inhibitors of polyamine biosynthesis on polyamine levels and microcutting growth and development. Plant Growth Regul. 13: 269-280.

Martinelli, L. and I. Gribaudo (2001) Somatic embryogenesis in grapevine. *In* : Molecular Biology and Biotechnology of Grapevine, K.A. Roubelakis-Angelakis (Ed.), Kluwer Publish. Co., Amsterdam, pp. 327-351.

Martinelli, L. and G. Mandolino (2001) Transgenic grapes (*Vitis* species). *In* : Biotechnology in Agriculture and Forestry, Vol. 47, Transgenic crops II, Y.P.S. Bajaj (Ed.), Springer-Verlag Berlin Heidelberg, pp. 325-338.

Martinez, M.C., Boursiquot J.-M., Grenan, S. and R. Boidron (1997) Etude ampélométrique de feuilles adultes de somaclones du cv Grenache noir (*Vitis vinifera* L.). Can. J. Bot. 75: 333-345.

Matsumoto, T., Sakai, A., and Y. Nako (1998) Cryopreservation of *in vitro* cultured axillary shoot tips of grape (*Vitis vinifera*) by vitrification. J. Japanese Soc. Hort. Sci. (Suppl.) 67: 78.

Mauro, M.-C. (1986) Embryogenèse somatique chez *Vitis vinifera* cv Cabernet-Sauvignon. Application à la sélection de clones résistant à *Eutypa lata*. Thèse Institut National Polytechnique de Toulouse, France, 176 p.

Mauro, M.-C., Walter, B., Pink, L., Valat, L., Barbier, P., Boulay, M., and P. Coutos-Thevenot (2000) Analysis of 41 B grapevine rootstocks for grapevine fanleaf resistance.

In : Proceed. 7th Intern. Symp. Grapevine Genetics and Breeding, A. Bouquet and J.-M. Boursiquot (Eds), Montpellier (France) July 1998. Acta Hort. 528: 313-319.

Molina, S.M., Trujillo Jacinto del Castillo, M.I., and F. Gutierrez-Jerez (1998) *In vitro* propagation of *Vitis vinifera* cv Listan negro. Plant Tissue Cult. 8: 165-172.

Moncousin, C., Favre, J.M., and T. Gaspar (1989) Early changes in auxin and ethylene production in vine cuttings before adventitious rooting. Plant Cell Tiss. Org. Cult. 19: 235-242.

Monette, P.L. (1983) Influence of size of culture vessel on *in vitro* proliferation of grape in a liquid medium. Plant Cell Tiss. Org. Cult. 2: 327-332.

Monette, P.L. (1988) Grapevine (*Vitis vinifera* L.), *In* : Biotechnology in Agriculture and Forestry, Y.P.S. Bajaj (Ed.), Vol.VI, Crops, Springer-Verlag, Berlin, pp. 3-37.

Morel, G. (1941) Sur les possibilités de réaliser la culture indéfinie de tissus de vigne. C.R. Acad. Sci. Paris 219: 36-37.

Morel, G. (1948) Recherches sur la culture associée de parasites obligatoires et de tissus végétaux. Ann. Epiphyties 14: 123-234.

Morel, G. (1963) La culture *in vitro* du méristème apical de certaines orchidées. C.R. Acad. Sci. Paris 256: 4955-4957.

Moriguchi, T. and S. Yamaki (1989) Prolonged storage of grape nodal culture using low concentration of ammonium. HortScience 24: 372-373.

Mowbray, G.H., Sherrer, C.E., and M.H. Desmonds (1985) Agronomic characteristics of vitroplants from somatic embryos. *In* : Amélioration de la vigne et culture *in vitro*, Colloque Moet et Hennessy, Paris, April 1985, pp. 121-126.

Mullins, M.G. (1985) Amplification of clonal variation in the grapevine. Progress and prospects. *In* : Amélioration de la vigne et culture *in vitro*, Colloque Moet et Hennessy, Paris, April 1985, pp. 63-71.

Mullins, M.G. (1987) Propagation and genetic improvement of temperate fruits : the role of tissue culture. *In* : Plant tissue and cell culture, C.E. Green, D.A. Somers, Hackett W.P. and D.D. Biesboer (Eds), Plant Biology, Vol. 3, Alan R. Liss, Inc. New York, pp. 395-406.

Mullins, M.G. and C.P. Meredith (1989) The nature of clonal variation in winegrapes : a review. *In* : Proceed. 7th Austr. Wine Ind. Tech. Conference, Adelaïde (Australia), pp. 14-16.

Mullins, M.G. and C. Srinivasan (1976) Somatic embryos and plantlets from an ancient clone of the grapevine (cv Cabernet Sauvignon) by apomixis *in vitro*. J. Exp. Bot. 27: 1022-1030.

Mullins, M.G., Nair, Y., and P. Sampet (1979) Rejuvenation *in vitro* : Induction of juvenile characters in an adult clone of *Vitis vinifera* L. Ann. Bot. 44: 623-627.

Mullins, M.G., Tang, F.C.A., and D. Facciotti (1990) *Agrobacterium*-mediated genetic transformation of grapevines: transgenic plants of *Vitis rupestris* Scheele and buds of *Vitis vinifera* L. Bio/Technology 8: 1041-1045.

Mullins, M.G., Bouquet, A., and L.E. Williams (1992) Biology of the grapevine. Cambridge University Press, 239 pp.

Murashige, T. (1974) Plant propagation through tissue cultures. Annu. Rev. Plant. Physiol. 25: 135-166.

Murashige, T. and F. Skoog (1962) A revised medium for rapid growth and bioassays with tobacco tissue cultures. Physiol. Plant. 15: 473-497.

Netzer, M.-H., Guellec, V., and M. Branchard (1991) Etude du comportement, au cours de l'embryogenèse somatique, de divers porte-greffes de vigne cultivés *in vitro* en conditions chlorosantes. Agronomie 11: 125-131.

Novak, F.J. and Z. Juvova (1983) Clonal propagation of grapevine through *in vitro* axillary bud culture. Scientia Hort. 18: 231-240.

Nozeran, R. and L. Bancilhon (1972) Les cultures *in vitro* en tant que technique pour l'approche des problèmes posés par l'amélioration des plantes. Ann. Amélior. Plantes 22: 167-185.

Nozeran, R., Grenan, S., Truel, P., and J.-M. Favre (1983) Morphogenèse à partir du stade juvénile de *Vitis vinifera* L. issu de graine ou de culture *in vitro*. Agronomie 3: 681-684.

Nuñez, D.R. and M.J. Walker (1989) A review of paleobotanical findings of early *Vitis* in the Mediterranean and the origins of cultivated grapevines, with special reference to prehistoric exploitations in the western Mediterranean. Rev. Paleobot. Palynol. 61: 205-237.

OIV (2000) The state of vitiviniculture in the world and the statistical information in 1999. Bull.OIV 73 (Suppl.) 94 p.

Olien, W.C. (1990) The muscadine grape : Botany, viticulture, history and current industry. HortScience 25: 732-739.

Olmo, H.P. (1976) Grapes. *In* : Evolution of Crop Plants, N.W. Simmonds (Ed.), Longman, London, pp. 294-298.

Olmo, H.P. (1986) The potential role of *vinifera* x *rotundifolia* hybrids in grape variety improvement. Experientia 42: 921-926.

Palys, J.M. and C.P. Meredith (1984) Dual cultures of grape (*Vitis* sp.) and the lesion nematode (*Pratylenchus vulnus*) for studies of nematode tolerance. Plant Disease 68: 1058-1059.

Perl, A. and Y. Eshdat (1998) DNA transfer and gene expression in transgenic grapes. *In* : Biotechnology and genetic engineering reviews, M.P. Tombs (Ed.), Vol. 15, Intercept, Andover, UK, pp. 365-386.

Perl, A., Sahar, N., Spiegel-Roy P., Gavish, S., Elyasi, R., Orr, E., and H. Bazak (2000) Conventional and biotechnological approaches in breeding seedless table grapes. *In* : Proceed. 7th Intern. Symp. Grapevine Genetics and Breeding, A. Bouquet and J.-M. Boursiquot (Eds) Montpellier (France) July 1998. Acta Hort. 528: 607-612.

Peros, J.-P., Torregrosa, L., and G. Berger (1998) Variability among *Vitis vinifera* cultivars in micropropagation, organogenesis and antibiotic sensitivity. J. Exp. Bot. 49: 171-179.

Planchon, J.E. (1887) Monographie des *Ampelidae* vraies. *In* : Monographia Phanerogamerum, 5: 305-364.

Plessis, P., Leddet, C., and J. Dereuddre (1991) Résistance à la déshydratation et à la congélation dans l'azote liquide d'apex enrobés de vigne (*Vitis vinifera* L. cv Chardonnay). C.R. Acad. Sci. Paris 313: 373-380.

Pouget, R. (1990) Histoire de la lutte contre le Phylloxéra de la vigne en France. INRA Ed. (Paris) 157 p.

Purcell, A.H. (1979) Leafhopper vectors of xylem-borne plant pathogens. *In* : Leafhoppers and Plant Disease Agents, K. Maramorosch and K.F. Harris (Eds), Academic Press, New-York, pp. 603-625.

Rajasekaran, K. and M.G. Mullins (1979) Embryos and plantlets from cultured anthers of hybrid grapevines. J. Exp. Bot. 30: 399-407.

Rajasekaran, K. and M.G. Mullins (1981) Organogenesis in internode explants of grapevines. Vitis 20: 218-227.

Rajasekaran, K. and M.G. Mullins (1983a) The origin of embryos and plantlets from cultured anthers of hybrid grapevines. Am. J. Enol. Vitic. 34: 108-113.

Rajasekaran, K. and M.G. Mullins (1983b) Influence of genotype and sex-expression on formation of plantlets by cultured anthers of grapevines. Agronomie 3: 233-238.

Reisch, B.I. (1986) Influence of genotype and cytokinins on *in vitro* shoot proliferation of grapes. J. Amer. Soc. Hort. Sci. 111: 138-141.

Reustle, G., Harst, M., and G. Alleweldt (1995) Plant regeneration of grapevine (*Vitis* sp.) protoplasts isolated from embryogenic tissues. Plant Cell Rep. 15: 238-241.

Rives, M. (1961) Bases génétiques de la sélection clonale chez la vigne. Ann. Amélior. Plantes 11: 337-348.

Robacker, C. (1993) Somatic embryogenesis and plant regeneration from muscadine grape leaf explants. HortScience 28: 53-55.

Roubelakis-Angelakis, K.A. and S.B. Zivanovitc (1991) A new culture medium for *in vitro* rhizogenesis of grapevine (*Vitis* spp.) genotypes. HortScience 26: 1551-1553.

Salunkhe, C.K., Rao, P.S., and M. Mhatre (1997) Induction of somatic embryogenesis and plantlets in tendrils of *V. vinifera* L. Plant Cell Rep. 17: 65-67.

Schneider, S., Reustle, G., and E. Zyprian (1996) Detection of somaclonal variation in grapevine regenerants from protoplasts by RAPD-PCR. Vitis 35: 99-100.

Scott, K.D., Ablett, E.M., Lee, L.S. and R.J. Henry (2000) AFLP markers distinguishing an early mutant of Flame seedless grape. Euphytica 113: 245-249.

Sforza, R., Clair, D., Daire, X., Larrue, J., and E. Boudon-Padieu (1998) The role of *Hyalesthes obsoletus* (Hemiptera : cixiidae) in the occurrence of Bois noir of grapevine in France. J. Phytopathol. 146: 549-546.

Silvestroni, O. (1981) Prime esperienze sulla micropropagazione della vite europea. VigneVini 8: 31-37.

Skene, K.G.M. and M. Barlass (1983) Studies on the fragmented shoot apex of grapevine. IV. Separation of phenotypes in a periclinal chimera *in vitro*. J. Exp. Bot. 34: 1271-1280.

Skene, K.G.M., Goodwing, D.R., and M. Barlass (1988) Ploidy stability in grapevine following long term storage *in vitro*. Vitis 27: 41-46.

Slavtcheva, T. and V. Dimitrova (2000) Photosynthesis, dark respiration and transpiration of *in vitro* cultured vine plants during their acclimatization. *In*: Proceed. 7th Intern. Symp. Grapevine Genetics and Breeding, A. Bouquet and J.-M. Boursiquot (Eds), Montpellier (France) July 1998, *Acta Hort.* 528: 427-434.

Small, J.K. (1903) *Vitis* and *Muscadinia*. In Flora of the southeastern United States, New York. pp.752-757.

Smith, E. F., Gribaudo, I., Roberts, A.V., and J. Mottley (1992) Paclobutrazol and reduced humidity improve resistance to wilting of micropropagated grapevine. HortScience 27: 111-113.

Soulié, O., Dulenc, V., Roustan, J.-P. and J. Fallot (1994) Effet de l'éthylène sur le développement *in vitro* de *Vitis vinifera* cv Ugni Blanc. C.R. Acad. Sci. Paris 317: 253-256.

Stamp, J.A. and C.P. Meredith (1988) Proliferative somatic embryogenesis from zygotic embryos of grapevine. J. Amer. Soc. Hort. Sci. 113: 941-945.

Stamp, J.A., Colby, S.M., and C.P. Meredith (1990a) Direct shoot organogenesis and plant regeneration from leaves of grape (*Vitis* spp.). Plant Cell Tiss. Org. Cult. 22: 127-133.

Stamp, J.A., Colby, S.M., and C.P. Meredith (1990b) Improved shoot organogenesis from leaves of grape. J. Amer. Soc. Hort. Sci. 115: 1038-1042.

Sudarsono, X. and R.G. Goldy (1991) Growth regulator and axillary bud position effects on *in vitro* establishment of *Vitis rotundifolia*. HortScience 26: 304-307.

Tang, F.A. and M.G. Mullins (1990) Adventitious bud formation in leaf explants of some grapevine rootstocks and scion cultivars. Vitis 29: 151-158.

Tanne, E., Ben-Dov, Y., and B. Raccah (1989) Transmission of the corky-bark disease by the mealybug *Planococcus ficus*. Phytoparasitica 17: 55.

Taylor, C.E. (1971) Nematodes as vectors of plant viruses. *In* : Plant Parasitic Nematodes, B.M. Zuckerman, W.F. Mai and R.A. Rhode (Eds), Academic Press, New York, pp. 185-211.

This, P., Lahogue, F., Adam-Blondon, A.-F., Doligez, A., and A. Bouquet (2000) Towards marker-assisted selection for seedlessness in grapevine. *In* : Proceed. 7th Intern. Symp. Grapevine Genetics and Breeding, A. Bouquet and J.-M. Boursiquot (Eds), Montpellier (France) July 1998. Acta Hort. 528: 221-229.

Thomas, P. (1999) Relationship between tissue growth, CO_2 level and tendril formation during *in vitro* culture of grape (*Vitis vinifera* L.). Vitis 38: 25-29.

Thomas, M.R., Franks, T., and P. Iocco (2000) Transgenic grapevines : status and future. *In* : Proceed. 7th Intern. Symp. Grapevine Genetics and Breeding, A. Bouquet and J.-M. Boursiquot (Eds), Montpellier (France) July 1998. Acta Hort. 528: 279-288.

Torregrosa, L. (1994) Transformation génétique de la vigne : Mise au point de protocoles de régénération par organogenèse et embryogenèse chez des hybrides *Vitis* x *Muscadinia*. Obtentions chez deux porte-greffes de racines et de plantes transgéniques produisant la protéine capsidiale du virus de la mosaïque chromée de la vigne (GCMV). Thèse Ecole Nationale Supérieure Agronomique de Montpellier (France) 162 p.

Torregrosa, L. (1995) Biotechnologies de la vigne : les techniques de régénération *in vitro*. Progr. Agric. Vitic. 112: 479-489.

Torregrosa, L. and A. Bouquet (1995) *In vitro* propagation of *Vitis* × *Muscadinia* hybrids by microcuttings or axillary budding. Vitis 34: 237-238.

Torregrosa, L. and A. Bouquet (1996) Adventitious bud formation and shoot development from *in vitro* leaves of *Vitis* × *Muscadinia* hybrids. Plant Cell Tiss. Org. Cult. 45: 245-252.

Torregrosa, L., Torres-Viñals, M., and A. Bouquet (1995) Somatic embryogenesis from leaves of *Vitis* x *Muscadinia* hybrids. Vitis 34: 239-240.

Torregrosa, L., Bouquet, A., and P.J. Goussard (2001) *In vitro* culture and propagation of grapevine. *In* : Molecular Biology and Biotechnology of Grapevine, K.A. Roubelakis-Angelakis (Ed.), Kluwer Publish. Co., Amsterdam, pp. 195-240.

Vellegas, A., Mazuelos, C. and A. Troncoso (1992) Influence of $N\text{-}NO_3$ and $N\text{-}NH_4$ on the mineral composition of grapevine rootstocks cultured *in vitro*. *In* : Proceed. 23rd Intern. Hort. Congress, Florence (Italy), Acta Hort. (Special Issue), pp. 119-123.

Vilaplana, M. and M.G. Mullins (1989) Regeneration of grapevines (*Vitis* spp.) *in vitro*: formation of adventitious buds on hypocotyls and cotyledons of somatic embryos. J. Plant Physiol. 134: 413-419.

Viss, W.J. and J.A. Driver (1996) Key grapevine pests and diseases in North America and resistance through genetic engineering. Brighton Crop Protection Conference Pests and Diseases 3B2: 125-130.

Vivier, M. and I.S. Pretorius (2000) Genetic improvement of grapevine : Tailoring grape varieties for the third millenium. S. Afr. J. Enol. Vitic. 21: 5-26.

Walter, B. and G.P. Martelli (1997) Clonal and sanitary selection of grapevine. *In* : Sanitary selection of the grapevine. Protocols for detection of viruses and virus-like diseases, B. Walter (Ed.), Les Colloques de l'INRA 86: 43-95.

Ye, G.N., Söylemezoglu, G., Weeden, N.F., Lamboy, W., Pool, R.M., and B.I. Reisch (1998) Analysis of the relationship between grapevine cultivars, sports and clones via DNA fingerprinting. Vitis 37: 33-38.

Yu, D.H. and C.P. Meredith (1986) The influence of explant origin of tissue browning and shoot production in shoot tip cultures of grapevine. J. Amer. Soc. Hort. Sci. 111: 972-975.

Zhu, Y.M., Hoshino, Y., Nakano, M., Takahashi, E., and M. Mii (1997) Highly efficient system of plant regeneration from protoplasts of grapevine (*V. vinifera* L.) through somatic embryogenesis by using embryogenic callus culture and activated charcoal. Plant Sci. 123: 151-157.

Zlenko, V.A., Troshin, L.P., and I.V. Kotikov (1995) An optimized medium for clonal micropropagation of grapevine. Vitis 34: 125-126.

Zou, C.J. and P.F. Li (1981) Induction of pollen plants of grape (*V. vinifera* L.). Acta Bot. Sinica 23: 79-81.

12. MICROPROPAGATION OF BANANAS

Kazumitsu Matsumoto
Embrapa - Genetic Resources and Biotechnology, Pq-EB, Final, Av W3 Norte, 70770-900 Brasilia/DF, Brazil

Sebastião Pedro da Silva Neto
Cia de Promoção Agrícola – CAMPO, Department of Biotechnology, SEPN 516, Bl A, 49, 70770-515 Brasilia/DF, Brazil

Contents

1. INTRODUCTION

Bananas (*Musa* spp), including plantains and cooking bananas, are among the most important horticultural crops in tropical and subtropical regions of the world. They are cultivated in areas of about 9 millions hectares spread throughout more than 100 countries, with an annual production of around 89 million metric tons (FAO, 2000). Approximately 90% of the bananas are directed to local consumption. Only 10 % of the total production are for export, but these accounts for revenue of approximately 5 billion US dollars annually, benefiting many developing countries (Sharrock & Frison, 1999). Besides providing a staple food for millions of people, bananas have an excellent nutritional value, being one of the most easily

S.M. Jain and K. Ishii (eds.), Micropropagation of Woody Trees and Fruits, 353–380.

digested foods. Banana also provides nutrition to infants and useful to people suffering from various intestinal disorders (Sharrock & Lusty, 2000).

Global *Musa* production has almost doubled in the last 30 years. The increase in production has mainly resulted from an expansion in the plantation areas, cultivating approximately 5 million hectares in 1968 and 9 million hectares in 1998 (INIBAP, 1999). Most commercial cultivars of edible bananas are triploids (AAA, AAB and ABB genome type), with some few diploid landraces (AA and AB genome type) and tetraploid cultivars (AAAA and AAAB genome type, mostly artificial hybrids). These genotypes are virtually or completely seed-sterile. Consequently, they can only be propagated by vegetative means. Materials which can be used for propagation include 'button buds', young shoots called 'sword suckers', and portions of the corm, so called 'bits' (Stover and Simmonds, 1987). They are main cause of dissemination of pests and diseases to new plantation areas. Banana production has been threatened by a series of diseases and pests transmitted by contaminated planting materials, such as *Fusarium* wilt (caused by *Fusarium oxysporum,* f.sp. *cubense*), black Sigatoka (*Mycosphaerella fijiensis*), nematodes (*Radopholus similis*, *Helicotylenchus multicinctus* and others), bacterial wilt (*Ralstonia solanacearum*) and several viruses. Once such problems are disseminated by conventional planting materials, they can be avoided or partially controlled throughout the use of clean planting materials. Therefore, there is a high demand for rapid propagation systems for pathogen-free plantlets and safe germplasm exchange. *Musa* micropropagation system has been developed responding to this demand, and also became a very important tool for *Musa* breeding.

In this article, a recent *Musa* micropropagation system and its application are reviewed.

2. SHOOT TIP CULTURE

Shoot tip (meristem plus a few attached leaf primordia) culture of *Musa* may be considered simple, easy, and applicable to a wide range of *Musa* genotypes (Cronauer and Krikorian, 1984; Vuylsteke, 1998). *In vitro* propagation has many advantages, such as higher propagation rates for multiplying clean (pest and pathogen-free) planting material, small space requirements regardless of season, and short time requirements. Multiple shoot or bud formation is easily achieved by culturing shoot tips on culture media of several compositions (Table 2). Multiplication rates range from 2 to 10 or more shoot or bud propagules per month, resulting in potential propagation rates of several thousands or millions of plants per year (Crouch et al., 1998; Silva Neto, 2000). Such rates are several orders of magnitude greater than those achievable through conventional propagation and are disease free. Several shoot tip culture protocols for *Musa* are reported in the literature and many others have been used in commercial laboratories with small adaptations according to convenience and market needs. Basically, all laboratory tissue culture procedures may be divided into phases such as: selection of mother plants, initiation, multiplication, rooting and acclimatization.

2.1. Selection of Stock Plants and Preparation

2.1.1. Selection of Stock Plants

Selection of stock plants is needed to avoid multiplication of undesirable genotypes as well as to take advantage of the useful variability that may exist in banana plant populations. It is strongly suggested that the selection of stock plants for micropropagation should be done by experienced people working under environmental conditions similar to where the plantlets to be produced will be grown. Desirable genotypes with superior fruit characteristics such as a cylindrical bunch, long and thick fingers and better shipment quality must be found. Moreover, higher performance clones, vis-a-vis soil, climate and diseases, can be selected. If possible, it is advisable to select plants based on records of historic performance for at least three cycles of production. Stock plants should also be free of disease and growing vigorously. Explant material should only be collected from flowering plants, to ascertain that they are true-to-type.

2.1.2. Virus Indexing

After the selection of the desirable stock plants they must be certified for virus-free status (virus indexing). Herbaceous indicator plants, electron microscopy, serological tests and nucleic acid analysis are now employed for this purpose which vary somewhat in their applicability, according to the virus being tested (George, 1993). Virus indexing in bananas is indispensable because bananas worldwide are affected by a number of viral diseases (Table 1) that may be spread even through micropropagated materials. Some of those viruses do not show clear symptoms and may be carried by propagating material produced by micropropagation. The virus indexation by herbaceous indicator plants are the most simple, but time consuming in giving result. Electron microscopy is not always available, particularly for laboratories in developing country. Serological techniques are based upon the specific recognition of viral antigens by antibodies and are frequently preferred because of their speed, specificity and simplicity. But, in many cases, nucleic acid based assays such as PCR (Polymerase Chain Reaction) based techniques are necessary (Rowhani and Falk, 1995). The advantages of serological techniques such as ELISA (Enzyme-Linked Immunosorbent Assay) in simplicity for handling large number of samples and providing results in a short time, are well known. However, it is not enough to detect some virus strains and virus particles presented at low titer, neither to detect viroids. In those cases, PCR based techniques are needed, though being more complicated, time-consuming and requirement of more lab facilities. The sensitivity of the dot blot hybridization, ELISA and RT-PCR (Reverse Transcription Polymerase Chain Reaction) assays were compared for detection of CMV in banana (Hu *et al.*,1995). They found that RT-PCR was the most sensitive and might be used to confirm samples that were inconclusive in ELISA and dot blot tests. Singh *et al.* (1995) also suggest that RT-PCR should be used where the detection of very low levels of virus infection is required. The PCR has proved to be about 1000 times more sensitive than ELISA or dot blots for detection of BBTV (Xie and Hu, 1995). IC-PCR (Immuno-Capture Polymerase Chain Reaction) technique was developed to detect encapsidated episomal forms of the BSV - banana streak virus (Frison and Sharrok, 1998).

Table 1. Viruses affecting banana and tests for indexing

Viruses		Diagnostic test
BBTV	Banana Bunch Top Virus	ELISA
BBrMV	Banana Brack Mosaic Virus	ELISA, PCR
CMV	Cucumber Mosaic Virus	ELISA, RT-PCR
BSV	Banana Streak Virus	ELISA, ISEM, IC-PCR
BanMMV	Banana Mild Mosaic Virus	EM, RT-PCR
BDBV	Banana Die-Back Virus	ELISA
AbaMV	Abaca Mosaic Virus	ELISA, PCR

ELISA (Enzyme-linked immunosorbent assay); EM (electron microscopy); IC (Immuno-capture); ISEM (Immunosorbent electron microscopy); PCR (Polymerase chain reaction); RT (Reverse transcription).
Source: Jones (2000).

2.2. Establishment of Aseptic Culture

After the selection and virus indexing of stock plant, shoot tips can be obtained from all plant parts that contain a shoot meristem (Vuylsteke, 1998). Growth response and explant survival in culture do not differ among shoot apices obtained from the parental pseudostem, suckers, lateral buds or even very small eyes (Jarret *et al.*, 1985; Vuylsteke and De Langhe, 1985). However, buds and small sword suckers are the preferred source materials due to the easiness of handling (Jarret *et al.*, 1985) and because there is less damage to the parent stool during their removal.

The size of the primary explant depends on the purpose of culture establishment. As virus elimination is not the goal, excision and culture of only apical meristem dome should not be attempted because it increases the mortality rate and grows very slowly. Conversely, a larger shoot tip (4-5 mm) has proven to be the most appropriate size for culture initiation (Sandoval and Muller, 1987). However, explants more than 10 mm are not recommended because they cause more browning and microbial contamination. Several sterilization methods have been employed for disinfecting of initial explant tissues (Hamill *et al.*, 1993). The tissues are usually treated with 0.5 to 1.0% sodium hypochloride solution during 5 to 45 minutes, depending on the size. The shoot tips are then excised and inoculated to the medium. When the shoot tip is disinfected after excision, a lower sodium hypochloride concentration (0.0525%) and a shorter treatment time (5 min.) must be used (Krikorian and Cronauer, 1984). However, surface disinfection before excising the shoot tip is preferable, so that tissues damaged by the sodium hypochloride solution can be cut off from the explant (Vuylsteke, 1989).

2.3. *In Vitro* Multiplication

The aseptic shoot tips are cultured on media with high levels of cytokinins (Table 2) which reduce the apical dominance and stimulate the arising of adventitious and/or axillary buds directly from the explants (Vuylsteke, 1998). Some researchers fragmentize shoot tips as a form of stimulating multiple shoot cultures (Ma and Shii, 1972; Jarret *et al.*, 1985). However, the practical results show that this approach is not essential. Conversely it may increase the oxidation. After 12-18 days in culture, clusters of small shoots or clump of tiny corms covered by several small shoot primordia are formed. Then, leaves grow and new buds appear. The multiplication of propagules is accomplished by subdividing the shoot or bud cluster with a few scalpel incisions and re-culturing the resultant explants on fresh medium. The multiplication rate depends on genotype, concentration of cytokinin, photoperiod, explant size and time in culture (George, 1993; Vuylsteke, 1998). By repeatedly subculturing, the multiplicated shoots or buds increase exponentially from cycle to cycle (Silva Neto, 2000). However, the number of proliferation cycles must be limited to 5-7, depending on the genotype, in order to limit the multiplication of stable variants and to reduce the incidence of "off-types" (Côte, 1994; Sandoval *et al.*, 1996; Mendes *et al.*, 1999).

2.4. Plants Regeneration

After the multiplication phase, shoots must be elongated, rooted and hardened to adapt to the external environment. Roots can be initiated on the same medium employed for multiplication or on media with lower ionic concentration. Shoots are elongated and rooted spontaneously after 20 days in culture on the MS medium containing 2.5 mgL^{-1} of kinetin (Wong, 1986). Shoots are often placed on a medium without growth regulators, or with a low concentration of auxin (Table 2). White cord roots appear between 4 and 14 days after transfer to rooting medium when leafy shoots have had at least 1 cm long (Vuylsteke, 1998). When the explants are bud-like structures, shoot elongation is required to achieve rooting. The elongation may be obtained by transferring cultures to medium containing 20 times less 6-benzylaminopurine than the multiplication medium (Vuylsteke, 1998) or even completely without the cytokinin (Jarret *et al.* 1985), although Gupta (1986) and Wong (1986) observed that the use of the cytokinin during elongation promoted rooting.

2.5. Acclimatization

Before transferring plantlets to field, they must be adapted to varying light levels, changes in temperature, reduced humidity and low availability of nutrients, in a transition phase called acclimatization. After the elongation and rooting, plantlets should be taken up from flasks, washed up in tap water to remove the agar medium and separated into individuals. The plantlets can be taken from flasks at various sizes: 30-50 mm (Jarret *et al.*, 1985); 80-100 mm (Vuylsteke, 1998). However, the best results are usually achieved by keeping plants *in vitro* to grow to 50-70 mm with an active root system (Daniells and Smith, 1991).

Table 2. Media for banana shoot tip culture

Medium	Sucrose ($g\ l^{-1}$)	Gelling agent ($g\ l^{-1}$)	Plant growth regulator (μM)	Reference
A – Initiation				
SM	30	Agar (8)	IAA (11.4) + Kin (9.3)	Ma & Shii (1972)
KC	41	Agar (7)	BA (22.2) + 15% CM	Krikorian & Cronauer (1984)
LS	30	Agar (7)	IAA (5.7) + BA (13.3)	Jarret et al. (1985)
W	20	Agar (7)	IBA (0.5) + BA (22.2)	Wong (1986)
MS	30	Phytagel(2)	BA (4.44)	Acuña (1993)
MS	41	Liquid	IBA (1.0) + BA (22.2)	Singh & Singh (1995)
MS	30	Phytagel(2)	BA (2.7~17.7)	Silva Neto (2000)
B – Multiplication				
SM	30	Agar (8)	IAA (11.4) + Kin (9.3)	Ma & Shii (1972)
HM	40	Agar (7)	BA (22.2)	Cronauer & Krikorian (1984)
LS	30	Agar (7)	BA (22.2)	Jarret *et al.* (1985)
MS	30	Agar (4.5)	IAA (1.0) + BA (10.2)	Banerjee *et al.* (1986)
W	20	Agar (7)	BA (22.2)	Wong (1986)
MS	30	Liquid	BA (22.2)	Ganapathi *et al.* (1992)
MS	30	Phytagel(2)	BA (8.9~13.3)	Acuña (1993)
MS	30	Liquid	BA (8.9)	Alvard *et al.* (1993)
MS	30	Phytagel(2)	BA (2.7~17.7)	Silva Neto (2000)
C – Elongation and rooting				
MT	15	Agar (6.5)	NAA (0.5) + IBA (9.8)	Mante & Tepper (1983)
HM	40	Agar (7)	IAA (5.7)	Cronauer & Krikorian (1984)
KC	41	Agar (7)	IBA (0.1)	Krikorian & Cronauer (1984)
LS	30	Agar (7)	None	Jarret *et al.* (1985)
MS	30	Agar (4.5)	IBA (1.0)	Banerjee *et al.* (1986)
MS	30	Phytagel(2)	None	Acuña (1993)
MS	30	Agar (6)	NAA (2.7)	Silva Neto (2000)

SM (Smith & Murashige, 1970); KC (Krikorian & Cronauer, 1984); LS (Linsmaier & Skoog, 1965); W (Wong, 1986); MS (Murashige & Skoog, 1962); HM (Heinz & Mee, 1969); MT (Mante & Tepper, 1983).

Acclimatization is achieved by several means. Cronauer and Krikorian (1984) potted the plantlets in plastic pots with a commercial substrate and vermiculite (1:1) mixture and placed them on a misting bed of a 15-min cycle for 7 to 10 days. Afterwards they were moved to greenhouse conditions (23 °C day / 18 °C night). Jarret *et al.* (1985) transferred plantlets directly in plastic bags filled with sterilized soil, or firstly planted them into a bench-type box containing substrate mixture or sterilized soil until they reach 10-15 cm height, and then transferred to the plastic bags. All plants were maintained under 50% shade. After two months plants of 30-50 cm tall, ready to plant in the field, were obtained. Krikorian *et al.* (1993) used plastic trays filled with a substrate mixture for the first stage of acclimatization. Plantlets were grown under intermittent mist and received fertilization weekly. After four weeks, plantlets were transplanted to black polyethylene bags filled with 1500-2000 cm^3 of sterilized topsoil. They were grown under 50% shade, watered by a sprinkler system and received fertilizers for 4 additional weeks. Thereafter, the plants received one application of a granular fertilizer (N: P_2O_5: K_2O: MgO = 10:5:20:3) of 15 g per plant. The plants were ready for transplanting to the field 10-12 weeks after the beginning of acclimatization. Most of commercial laboratories have used the above described acclimatization methods, modifying slightly.

2.6. Field Performance of Micropropagated Banana Plants

The practical experience in several countries has led to the conclusion that material derived from micropropagation performs equally well or superior to conventional material (Smith and Drew, 1990; Vuylsteke, 1998). Micropropagated plants grow faster and more vigorously, have a shorter and more synchronized production cycle, and have higher yield than conventional propagules (Drew and Smith, 1990; Robinson *et al.*, 1993). A maximum yield gain of the micropropagated plants comes up 20% in bananas and 70% in plantains (Crouch *et al.*, 1998). Generally, the superior performance of micropropagated bananas is attributable on the propagules that already possess an active shoot and root system at planting time and are free of most diseases and pests (Drew and Smith, 1990).

Results obtained in Cuba and Brazil, comparing large populations of conventional and micropropagated plantlets, showed that the yield of marketable fruits was 20 to 25% higher in plantations established with the micropropagated material (Orellana *et al.*, 1991; Silva Neto, 2000). Moreover, advantages of the micropropagated plantlets are observed in survival rate in the field, cost of labor for planting and disease control, quality of exportable fruits and economy in chemicals for disease control. Considering such advantages accumulate during a whole production life span, it is concluded that micropropagated plantlets is highly cost effective not so economically but also environmentally.

2.7. Protocols

The following is the protocol used in our laboratory.

A – Selection of stock plants based on historic performance (Figure 1-A).
B – Virus indexing.
C - Establishment of aseptic culture.
(1) Small sword suckers are collected from selected and indexed stock plants and taken to the laboratory
(2) Suckers are washed in tap water and tissue blocks of 30 x 30 x 30 mm^3 with shoot tip are extracted
(3) Surface-sterilization treating by commercial ethanol for 15-30 seconds, followed by 3.5% sodium hypochloride (NaOCl) with a few drops of Tween 20 for 15 minutes.
(4) External tissues are removed leaving blocks (10 x 10 x 20 mm^3) with intact apex, leaf primordia and corm. The blocks are re-treated by the NaOCl solution for 5 minutes in a laminar flow cabinet (Figure 1-B).
(5) External tissues are again removed leaving blocks of 5 x 5 x 8 mm^3.
(6) They are transferred into glass vessels containing MS basal medium supplied with 30 gL^{-1} sucrose, 4.4 µM BA, 2 gL^{-1} Phytagel and, kept in a culture room at 25 °C and 16-hour photoperiod supplied by white fluorescent lamps (27 µmol $m^{-2}s^{-1}$)
(7) After 2 weeks, the cultures are checked to detect and discard microbial contamination
B – *In vitro* multiplication
(8) After additional 2 weeks, the explants are cut longitudinally in two parts and transferred to the multiplication medium (same as the establishment medium but 17.7 µM BA) and cultured in the culture room for 4 weeks
(9) Induced multiple shoots (Figure 1-C) are divided into one or two shoots clusters, transferred to and cultured in vessels containing fresh medium of the same composition (this procedure is repeated for 5-7 cycles).
C – Elongation and rooting
(10) After the 5-7 cycles of the multiplication, the resulting multiple shoots are cut and divided into each ones, and transferred to half strength of MS basal medium supplied with 30 gL^{-1} sucrose, 2.7 µM NAA and 6 gL^{-1} agar. The cultures are maintained during 2 weeks in the culture room. Depending on the natural climate, the culture room of this stage can be substituted by a greenhouse. For example, the condition provided by 50% shade in greenhouse (17-30 °C) is effective to promote satisfactory results in Brazilian savanna region (Figure 1-D).

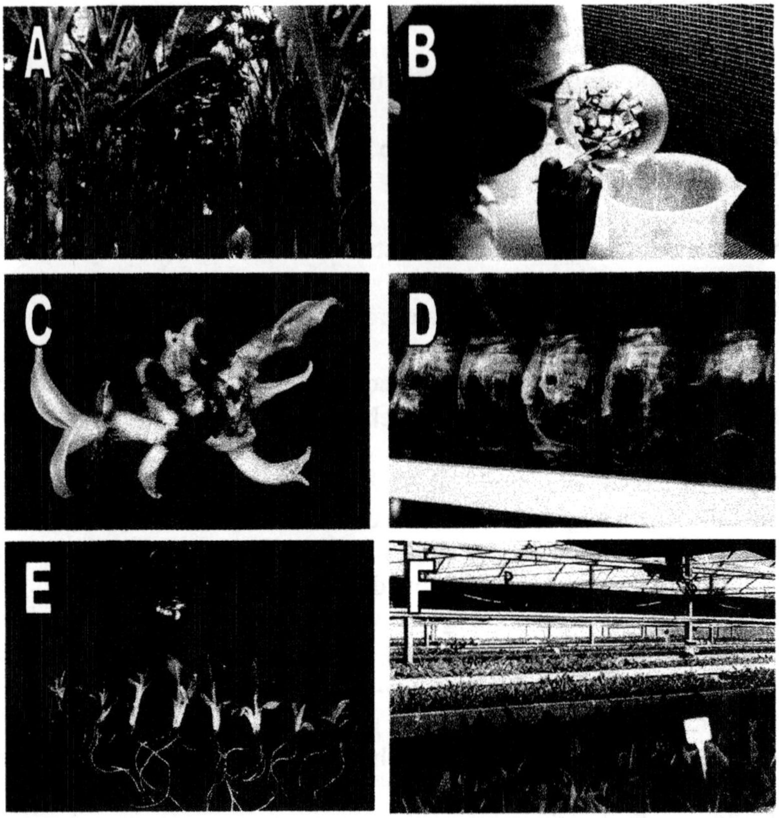

Figure 1. Shoot tip culture for mass production of pathogen-free plantlets. A: Selection of stock plants; B: Sterilization of initial explants; C: Induced multiple shoots; D: shoot elongation and rooting *in vitro*; E: Plantlets ready for acclimatization; F: Large-scale acclimatization.

D – Acclimatization

(11) Plantlets obtained are taken up from the culture vessels, and agar medium of the roots is washed out by tap water. The plantlets are then sorted according to the size (Figure 1-E) and transplanted onto a bench-type box containing substrate mixture (substrate: vermiculite = 1:1) in a 50% shaded greenhouse. Initially, the plantlets should be covered with a plastic sheet for 7 to 10 days to preserve moisture. After this period the plastic cover is removed and the plantlets are irrigated by a mist-type sprinkler system until they are 20 cm high (Figure 1-F).

(12) The plantlets of 20-cm high are then transferred to black polyethylene bags filled with about 1500-2000 cm^3 of sterilized top soil, and allowed to complete acclimatization in a shade house (50% of sun light), watered by a sprinkler system and fertilized weekly.

(13) When plants reach 50-60 cm high, off-types are eliminated and true-to-type plants are considered ready to be transplanted into the field.

3. CELL SUSPENSION CULTURE

Cell suspension culture consists in maintenance and propagation of cells in a form of suspension in liquid culture medium. Since a plant cell has totipotency (potential capacity to regenerate a whole plant), the cell suspension culture is a very useful tool for micropropagation, plant breeding and genetic resources conservation. Cell suspension culture for micropropagation has been studied in many species due to its high propagation efficiency, easy adaptation to automation, including synthetic-seed production, and generally low propagation cost (Kitto and Janick, 1985; Redenbaug et al., 1986). In banana, plant regeneration from cell suspension via somatic embryogenesis was reported (Novak et al., 1989; Dhed'a et al., 1991; Krikorian and Scott, 1995; Côte et al., 1996). Somaclonal variation, which is the most serious problem in micropropagation through somatic embryogenesis, is quite low in some banana varieties (Schoofs, 1997, Côte et al., 2000a). The micropropagation by cell suspension culture and somatic embryogenesis is, therefore, one of the most promising techniques for obtaining large amounts of qualified plantlets with low production costs.

3.1. Establishment of Aseptic Culture

Cell suspension cultures are mainly established from immature zygotic embryos (Cronauer-Mitra and Krikorian, 1988; Escalant and Teisson, 1989), *in vitro* multiple shoots or buds (Novak et al., 1989; Dhed'a et al., 1991) and flower buds (Escalant et al., 1994; Côte et al., 1996; Grapin et al., 1996 and 1998). Zygotic embryos, however, are not normally used as initial explants for clonal propagation due to the possibility of genetic segregation. In this review, cell suspension cultures of the latter two explant types are described.

When *in vitro* multiple shoots or buds are used as initial explants, no sterilization process is needed and any laboratory where there is a controlled-environment room or phytotron may be used to obtain them. To induce cells or calli, thin layer tissue (1 - 2 mm) of about 5 x 5 mm^2 size are extracted from meristematic rhizomes of multiple shoots or buds (Novak et al., 1989). The proliferating meristematic tissue explants are also called "scalps" (Dhed'a, et al., 1991). To obtain high quality scalps, shoots or buds can be preliminary cultured in a culture medium with high concentration (100 μM) of BA (6-benzylaminopurine) (Schoofs et al., 1998). On the other hand, explants of flower buds may be used only by those who live in or near a banana-growing region. Advantages of using this explant type are that there is not a need for *in vitro* maintenance and a possibility of the plant evaluation before collection. A young male inflorescence is collected from a field growing plant and surface disinfected with 70% alcohol. The immature male flowers from positions 1 to 15 (position 0 was the floral apex) are isolated under a dissecting microscope in a laminar flow cabinet and used as explants (Escalant et al., 1994; Côte et al., 1996; Grapin et al., 1996). The immature male buds are also used together, without separation by the positions. In this case, tissues (about 3 x 3 x 3 mm^3) with floral apex and adjacent male buds are extracted from a male inflorescence and cut longitudinally into two parts to be used (Matsumoto et al., 2002).

For embryogenic callus or cell induction, most of the studies use culture media consisted of an intact or half strength of MS salts, vitamins of MS or its modification, sucrose (20 – 30 gL^{-1}) and auxins, particularly 2,4-D (2,4-dichlorophenoxyacetic acid), in dark or low intense illumination. The 2,4-D, or similar synthetic auxins such as picloram and dicamba, is generally used at a concentration range of 5 to 30 μM (Novak et al., 1989; Dhed'a, et al., 1991; Panis et al., 1993; Côte et al., 1996). Exceptionally, a very high concentration (414 μM) is used when the medium is supplemented with activated charcoal (0.2%) to reduce the problem of tissue browning (Conceição et al., 1998). Many kinds of calli or cells are induced. It is important to select embryogenic calli or cells. The embryogenic calli are a white, globular and friable (Figure 2-A). They are induced directly from the explants or indirectly from yellow compact callus formed. By means of microscopic analysis, globular cells with dense cytoplasm and small vacuoles should be observed (Schoofs et al., 1998; Georget et al., 2000) (Figure 2-B).

3.2 *In Vitro* Multiplication and Plant Regeneration

For multiplication and maintenance of the suspension cells, auxin concentration in a culture medium generally should be reduced to about 5 μM with or without cytokinin. The cells or microcalli are passed through a sieve to obtain a homogenous cell suspension. The cultures are maintained on a shaker at 100 – 120 rpm (Figure 2-C). Embryogenesis from the suspension cells following plant regeneration is observed when they are transferred into a medium without growth regulator or with only low concentration of cytokinin (Dhed'a, et al., 1991; Panis et al., 1993), though combinations of different kinds of growth regulators, carbohydrates, amino acids and complex organic supplements may increase the efficiency (Côte et al., 1996; Grapin et al., 1996). Embryogenesis occurs in a liquid medium on a shaker at about 70 rpm or on a semisolid medium (Figure 2-D and E).

The plantlet regeneration rate depends on cultivated varieties and culture process, but is quite high. From 1 ml of packed embryogenic cells, 10,000 to 70,000 plantlets were obtained after 5 months of regeneration and germination processes (Côte et al., 1996; Grapin et al., 1996).

3.3. Acclimatization and Synthetic Seeds

The acclimatization process is the same as that of micropropagation via shoot tip culture. Details of this process was already reported elsewhere (Daniells and Smith, 1991; Vuylsteke, 1998). Aiming to facilitate or eliminate this costly process, synthetic seeds have been studied. A synthetic seed is a somatic embryo or shoot tip encapsulated in a coating material to be protected from mechanical damage and/or microbial infection, for the purpose of use as a sexual seed. Initially, banana shoot tips were encapsulated in calcium alginate beads (Ganapathi et al., 1992; Matsumoto et al., 1995). However, using cell suspension culture and somatic embryos, synthetic seed formation should be much easier than shoot tip culture (Redenbaugh et al., 1986). Although synthetic seeds are still not practically applicable, they can be used in acclimatization, making automation of the micropropagation process possible.

3.4. Protocols

The following is the protocol used in our laboratory.

A- Cell suspension establishment from a male inflorescence.

(1) A banana "heart" (male inflorescence) is collected just after the last fruit set and surface-disinfected.

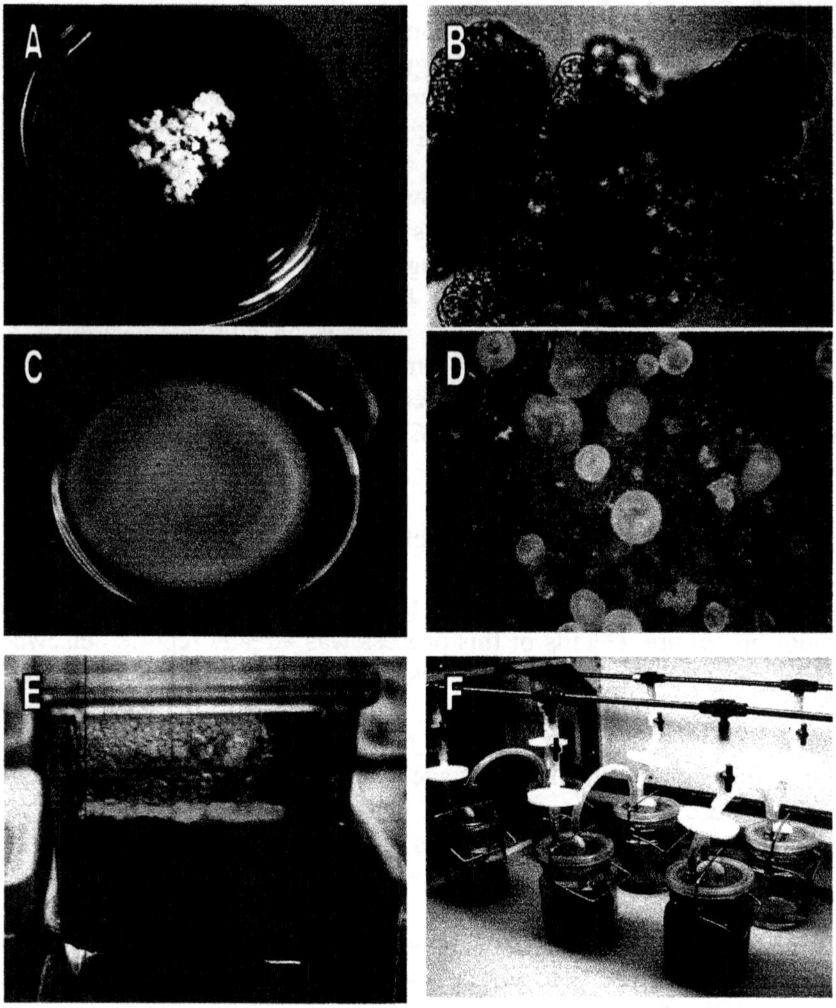

Figure 2. Suspension cell culture and culture system. A: Banana embryogenic calli. B: Embryogenic suspension cells. C: Embryogenic suspension cells in an Erlenmeyer flask. D: Somatic embryos induced from suspension cells. E: Shoot primordia regenerated from suspension cells via embryogenesis. F: Temporary Immersion System.

(2) A tissue of about 3 x 3 x 3 mm^3 floral apex with male bud primordia is extracted.

(3) The tissue is cut longitudinally into two parts and put on a cell induction medium (Table 3-A), contacting the cut surface with the medium.

Table 3. Media for cell suspension culture.

Component	Medium					
	A	B	C	D	E	F
Major nutrient and Fe-EDTA	MS	½ MS	½ MS	½ MS	½ MS	MS
Minor nutrient and vitamins	MS	MS	MS	MS	MS	MS
Ascorbic acid	-	10 mgL^{-1}	10 mgL^{-1}	10 mgL^{-1}	10 mgL^{-1}	-
Auxin	414 µM Picloram	5 µM 2,4-D	-	-	-	2 µM IAA
Cytokinin	492 µM 2iP	1 µM Zeatin	-	-	10 µM Zeatin	2 µM BA
Activated charcoal	0.2%	-	0.05%	-	-	-
L-proline	-	-	10 mM	10 mM	10 mM	-
Sucrose	30 gL^{-1}	30 gL^{-1}	30 gL^{-1}	30 gL^{-1}	30 gL^{-1}	20 gL^{-1}
Phytagel	2 gL^{-1}	2 gL^{-1}	2 gL^{-1}	-	-	2 gL^{-1}
Other components	300 µM L-arginin; 15.3 mM MES	-	-	-	-	-
pH	5.8	5.8	5.8	5.8	5.8	5.8

MS: Murashige & Skoog (1962)

(4) After 2 to 3 months of culture, when yellow globular calli are formed, they are transferred to a semisolid cell multiplication medium (Table 3-B).

(5) After 2 to 3 further subcultures of one month each, white globular friable calli will be obtained. They are transferred into a liquid cell multiplication medium (without Phytagel) on a shaker at 100 – 120 rpm, in dark.

Obs: To maintain embryogenic capacity of calli or cells, they are cultured on a cell maintenance medium (Table 3-C) under low illumination.

(6) The liquid medium is renewed weekly. The calli are occasionally passed through a 0.6-mm sieve to select only microcalli.

(7) After about 3 – 5 months, embryogenic cell suspension should be obtained. The period varies depending on banana varieties.

B- Plant regeneration.

(8) 0.5 ml PVC (Packed cell volume) of embryogenic cells is transferred into 20 ml of an embryo induction medium (Table 3-D) in a 250 ml erlenmeyer flask on a shaker (50 – 60 rpm) in low intensity light.

(9) After 2 weeks of culture, the cells are transferred into an embryo maturation medium (Table 3-E).

(10) After one more week, when pro-embryos are observed, they are transferred to a liquid or semisolid regeneration medium (Table 3-F).

(11) After 2 months, plantlets above 3 cm in height are obtained and may be transferred to a greenhouse for acclimatization.

4. TEMPORARY IMMERSION SYSTEM

Although *in vitro* propagation techniques in banana are widely used, a lower cost propagation system is still needed. Increasing the micropropagation capacity is one way to indirectly reduce the production cost. Several factors influenced this. In addition to culture medium composition, its contact area with explants and oxygenation also influence micropropagation efficiency. Early works showed banana multiple shoots could grow very well in liquid culture media with low intense shaking (Novak et al., 1986) or on inert absorbent materials (Matsumoto and Yamaguchi, 1989). However, a temporary immersion system developed recently revolutionized the liquid medium use, due to the extraordinary increase in propagation ratio (Alvard et al. 1993; Escalant et al., 1994). The system is now applied not only to banana but also to many other plant species such as potato (Jimenez et al., 1999), coffee (Etienne-Barry et al., 1999), pineapple (Escalona et al., 1999), sugarcane (Lorenzo et al., 1998), Citrus (Cabasson et al., 1997), tea (Akula et al., 2000), and others. The system consists in a transferring, at periodic intervals, the liquid culture medium from one compartment (250-500 ml) to another that contains explants. The explants are then maintained immersed in the medium for one period of previously determined time and, subsequently the medium is drained back to the original compartment. Scaled-up systems were developed by Cuban researchers, using two culture vessels interconnected (Pérez-Ponce, et al., 1998; Escalona et al., 1999; Daquinta et al., 2000) (Figure 2-F). This system permits the use of vessels of up to 20 liters in volume, so that it becomes a real bioreactor for plantlet production.

One of the advantages of the temporary immersion system is the high culture capacity of explants in one recipient, reducing the laboratory space occupied by micropropagation. However, if microbial contamination occurs, this advantage is null. Instead of field-grown explants, *in vitro* multiple shoots or buds were, then, generally used as initial explants aiming to reduce the risks of the contamination. For multiple shoot propagation, an explant (8 x 8 x 8 mm^3) containing 4-5 shoots is inoculated in 250 ml medium. When the temporary immersion system is used for somatic embryo germination, a quantity of 250 - 500 mg embryos is initially inoculated in 250 ml medium.

The transfer cycle of the liquid medium must be adjusted depending on culture medium, variety (species), vessel volume and objective. Alvard et al. (1993) immersed the shoot explants in the medium for 20 min every 2 h, while Daquinta et al.(2000) did for 3 min every 3 h. Generally immersion cycles of 10-min intervals per 4 h of drain functioned well for various species (Lemos et al., 2000). Aiming at somatic embryo multiplication, embryogenic cultures were immersed in a liquid medium for 1 min every 6 h (Escalant et al., 1994)

Since many explants must be cultured in one culture vessel using the temporary immersion system, elongated plantlets make manipulation difficult. Paclobutrazol, a growth inhibitor, is used to reduce the elongation of explants and to make handling easier (Escalona et al., 1999; Daquinta et al., 2000).

Other culture conditions, such as medium composition, temperature and luminosity, are the same as in traditional micropropagation systems, but some growth regulators could be reduced. The acclimatization process is also the same as that of traditional micropropagation systems.

5. SOMACLONAL VARIATION AND GENETIC FIDELITY

Somaclonal variation is a genetic variation that appears during clonal propagation of plants. It is frequently observed in banana and plantain micropropagation and can be useful to plant breeding by increasing genetic variability. However, it causes serious problems in clonal plantlet production and germplasm conservation. Although many biochemical variations must appear, mostly morphological changes (off-types) are studied in order to identify and eliminate them in the acclimatization phase of the propagation. The rates of somaclonal variation of *Musa* plants produced by shoot tip culture vary mostly from 0 to 25%, but could reach to 70% in some plantain cultivars or 100% in extreme case (reviewed by Côte et al., 1993). Various off-types are reported such as dwarfism, gigantism, mosaic-like, deformed fruit, short fruit, stem color abnormality, inflorescence abnormality, rosette structure, etc (Israeli et al., 1991; Côte et al., 1998). However, about 80% of the somaclonal variation are manifested as dwarfism (Reuveni & Israeli, 1990).

The origin of somaclonal variation may be explained by the following 3 mechanisms: (1) preexisting genetic changes in the explant tissue, (2) mutagenic actions of tissue culture media and (3) results of stresses induced by the conditions of the tissue culture environment (Novak,

1992). The frequency of somaclonal variation varies according to cultivated varieties, clonal lines (families) and sources of propagules (Reuveni & Israeli, 1990). More recently, a molecular-based analysis (PCR/RAPD) of micropropagated populations revealed that substantial amounts of the variation exists within the original culture meristems, which are hypothetically sectorial chimeras (Crouch *et al.,* 1998). It has been suggested that some somaclonal variation must be preexistent in explanted tissues. High concentration of cytokinin (mainly 6-benzylaminopurine) and prolonged subculture induce mitotic instability (Shepherd et al., 1996) and morphological changes (Reuveni & Israeli, 1990; Rodrigues et al., 1998; Vuylsteke, 2001).

Early detection of the somaclonal variation is quite important. In case of dwarf and giant off-types, *in vivo* or *in vitro* applications of gibberellic acid can facilitate their identification (Damasco et al., 1996a; Sandoval et al., 1999). However, other off-types are more difficult to identify and can be observed only in the field stage. Molecular marker technologies have then become a useful tool for early detection. Isozyme profile polymorphism is useful for this purpose (Carvalho et al., 1998). Restriction fragment length polymorphism (RFLP) is also a useful technology (Gawel & Jarret, 1991). However, it is not widely used for somaclonal variant detection due to a relatively high cost, timewasting and laborious work. Molecular markers based on the polymerase chain reaction (PCR) have therefore been more frequently used. The development of PCR technology provided a rapid, sensitive and specific alternative to hybridization based makers. Exploiting PCR, molecular marker methods, such as Random Amplified Polymorphic DNA (RAPD), Amplification Fragment Length Polymorphism (AFLP), Simple Sequence Repeats (SSR) and Sequence Tagged Site (STS), have been developed. These techniques differ in their reliability, ease of reproduction in different laboratories, the level of polymorphism revealed and genetic dominance (Henry, 1998). They have been used to classify *Musa* cultivars or wild species (Kaemmer et al., 1992; 1997; Howell et al., 1994) and, are also used for somaclonal variant detection (Engelborghs et al., 1998; Grajal-Martin et al., 1998; Damasco et al., 1996b; 1998; Silva Neto, 2000).

In chromosome level variations, flow cytometric ploidy analysis can also be applied. Although it can not detect a point mutation, the method can verify ploidy and aneuploidy variations caused by micropropagation (Dolezel et al., 1997; Lysak et al., 1999). The great advantage of this method over the PCR-based methods is its simplicity and speed. Only a few minutes per sample are needed for evaluation.

Somaclonal variation can be used positively for plant breeding. With or without mutagenic treatments, agriculturally useful traits are selected from somaclonal variants, such as dwarfism, early flowering, aluminum-tolerance, disease resistance, etc. Details are discussed in the next section.

6. USES OF MICROPROPAGATED PLANTS

6.1. Commercialization Experiences

Tissue culture laboratories have supplied the regular replanting needs of plantlets for the large export-banana plantations in Central America, Asia and Australia (Smith and Drew, 1990). Despite the clear advantages of the use of tissue cultured planting materials, a wide-scale adoption of this method has been hampered by high levels of somatic mutations derived from tissue culture and by extra care required to the young tissue cultured plants in the field (Daniells and Smith, 1991). For most farmers that cultivate bananas for the domestic market, the cost of micropropagated plantlets is still relatively high, compared to sucker and other conventional planting material. Cost reductions in the micropropagation of bananas, as well as of other species, are difficult for commercial laboratories because micropropagation is extremely labor-intensive and the cost of labor is directly associated with the transfer and maintenance of cultures. One solution adopted by some laboratories is to move to lower-cost labor sites, but problems such as distance to the market and other difficulties may arise.

In India, since 1988, the micropropagation industries have been increasing to supply pathogen-free banana plantlets to the growers in domestic market and for export. As a result, the yield of bananas in that country has increased from 8 to 14 ton/ha/year, i.e., almost doubled in the last decade (FAO, 2000). Banana micropropagation in India was performed at 14 of the 17 commercial tissue culture units working in 1996 (Govil and Gupta, 1997). In Taiwan, the needs of pathogen-free plantlets to overcome *Fusarium* wilt disease and to increase the production of bananas to meet the fruit demand of Japanese market made the industry produce more than 2 million banana plantlets a year (Taiwan Banana Research Institute, 1992). In Brazil, laboratories have produced around 5 million banana plantlets a year, and a half of million have been imported, mainly from Israel. Since the 1980's, Israel laboratories have experienced the adaptation of the *in vitro* method to commercial use. Through a governmental incubator program for incentive of laboratory based enterprises, Israel has become a worldwide supplier of *in vitro* banana plants. One company currently exports approximately 6 million plantlets to countries all over the world. Usually banana plantlets produced in Israel are shipped *in vitro* and acclimatized in the imported countries until be ready to plant in the field.

6.2. Germplasm Conservation and International Exchange

One of the advantages of micropropagation is to obtain pathogen-free plantlets, except virus pathogens. Germplasm conservation and international exchange have become much easier using micropropagation techniques (Banerjee & Langhe, 1985). Since 1994, International Network for the Improvement of Banana and Plantain (INIBAP) has conserved and exchanged *Musa* germplasm using a slow growing technique for micropropagation (INIBAP, 1995; 2001). Cryopreservation techniques of meristem tissue and embryogenic cells were also developed for the germplasm conservation (Panis, 1995; Panis et al., 1996; Côte et al., 2000b). At present, a major problem is mainly virus contamination, which has not been eradicated by conventional micropropagation techniques. Through micropropagation, the frequency of the virus infected plants could be reduced (Drew et al., 1989; 1992; Krikorian et al., 1999) but the viruses might be disseminated with symptomless infected plants. A virus indexing process is, then, indispensable for germplasm conservation and safe movement.

6.3. Plant Breeding

Plant breeding in *Musa* has derived great benefit from micropropagation techniques, particularly from embryo culture or embryo rescue technique. The technique was first reported by Cox et al. (1960) and has been used by almost all of the *Musa* cross-breeding programs in the world (Stover & Simmonds, 1987; Shepherd, 1987). Since banana hybrid seeds hardly germinated in natural condition, it constitutes the main technique for plantlet acquisition.

Somaclonal variation can increase genetic variability and can be used for *Musa* plant breeding. With or without mutagenic treatments, agriculturally useful traits have been selected from somaclonal variants, such as early-flowering (Novak et al., 1990), aluminum-tolerance (Matsumoto &

Yamaguchi, 1990), *Fusarium* wilt resistance (Hwang & Ko, 1988; Bhagwat & Duncan, 1998; Matsumoto et al., 1999), etc.

In banana, genetic transformation techniques have already been reported (Sagi et al., 1995; May et al., 1995; Becker et al., 2000). The application of these techniques to banana breeding depends on the molecular-level knowledge of useful genes, particularly genes related to mechanisms of resistance to the major diseases. Without this knowledge, somatic hybridization by protoplast fusion has also been tried (Matsumoto et al., 1992).

Sexual and somatic hybridization can help to identify useful genes. Point mutants derived from somaclonal variation or induced mutation are also very important for gene identification (Damasco et al., 1996; Ferreira et al., 1998; Tulmann Neto et al., 1998). On the other hand, transgenic plants with a marker gene, such as antibiotic resistance, make hybrid identification easier (Gerdemann-Knorck et al., 1995; Ramulu et al., 1995; Samoylov & Sink, 1996). Therefore these techniques should interact with each other and be applied in *Musa* breeding as complementary tools.

7. CONCLUSIONS AND PROSPECTS

Banana micropropagation has already been used for rapid multiplication of elite clones, and for conservation and international exchange of *Musa* germplasm. However, for the purpose of more common and extensive use, the problems of high production cost and somaclonal variation in micropropagated plants should be overcome. The micropropagation system using embryogenic cell suspension culture and the bioreactor with a temporary immersion system will reduce the production cost. Molecular marker techniques applied for somaclonal variant detection will make quality control of the micropropagated plants easier. A banana micropropagation system, in future, may be composed of the following processes: (1) simple cryopreservation technique of embryogenic cells or multiple shoots to maintain initial explants; (2) cell or shoot multiplication and plant regeneration in a bioreactor using a temporary immersion system; (3) synthetic seed formation with or without acclimatization; and (4) quality control with molecular markers. All of these systems are expected to be available for our use in the near future.

REFERENCES

Acuña P.I., 1993. Micropropagación de banano a partir de ápices vegetativos. CORBANA 17(39): 9—12.

Akula A., Becker D., Bateson M., 2000. High-yielding repetitive somatic embryogenesis and plant recovery in selected tea clone, 'TRI-2025', by temporary immersion. Plant Cell Rep. 19(12): 1140—1145.

Alvard D., Cote F., Teisson C., 1993. Comparison of methods of liquid medium culture for banana micropropagation, Effects of temporary immersion of explants. Plant Cell Tissue Organ Cult. 32: 55—60.

Banerjee N., Langhe E., 1985. A tissue culture technique for rapid clonal propagation and storage under minimal growth conditions of *Musa* (banana and plantain). Plant Cell Rep. 4: 351—354.

Banerjee N., Vuylsteke D., De Langue E., 1986. Meristem tip culture of *Musa*: histomorphological studies of shoot bud proliferation. In: Winthers L.A. and Alderson P.G. (eds.) Plant tissue culture and its agricultural applications. Butherworths, London, pp139—147.

Becker, D.K., B. Dugdale, M.K. Smith, R.M. Harding & J.L. Dale, 2000. Genetic transformation of Cavendish banana (*Musa* spp. AAA group) cv 'Grand Nain' via microprojectile bombardment. Plant Cell Rep 19(3): 229—234.

Bhagwat B., Duncan E.J., 1998. Mutation breeding of banana cv. Highgate (*Musa* spp., AAA group) for tolerance to *Fusarium oxysporum* f.sp. *cubense* using chemical mutagens. Scientia Horticulturae 73: 11—22.

Cabasson C., Alvard D., Dambier D., Ollitrault P., Teisson C., 1997. Improvement of *Citrus* somatic embryo development by temporary immersion. Plant Cell Tissue Organ Cult. 50(1): 33—37.

Carvalho M.T.V., Derbyshire E., Martins-Corder M.P., Mendes B.M.J., Tulmann Neto A., 1998. Isozyme and dissociated protein profiles of normal plants and somaclonal variants of banana obtained by *in vitro* culture. Acta Hort. 490: 437—444.

Conceição A.S., Matsumoto K., Bakry F., Bernd-Souza R.B., 1998. Plant regeneration from long-term callus culture of AAA-group dessert banana. Pesq. Agropec. Bras., Brasília, 33(8)1291—1296.

Côte F.X., 1994. Mathematical approach of somaclonal variations. *In vitro* culture of tropical plants. Repères. CIRAD-GERDAT (Biotrop). 43—46.

Côte F.X., Folliot M., Domergue R., Dubois C., 2000a. Field performance of embryogenic cell suspension-derived banana plants (*Musa* AAA, cv. Grande naine). Euphytica 112: 245—251.

Côte F.X., Domergue R., Monmarson S., Schwendiman J., Teisson C., Escalant J.V., 1996. Embryogenic cell suspensions from the male flower of *Musa* AAA cv. Grand nain. Physiol. Plant. 97, 285—290.

Côte F.X., Goue O., Domergue R., Panis B., Jenny C., 2000b. In-field behavior of banana plants (*Musa* AA sp.) obtained after regeneration of cryopreserved embryogenic cell suspensions. CryoLetters 21: 19—24.

Côte F.X., Sandoval J.A., Marie P., Auboiron E., 1993.Variations chez les bananiers et les plantains multipliés *in vitro*: analyse des données de la littérature. Fruits 48(1): 15—23.

Côte F.X., Sandoval J.A., Marie P., Auboiron E., 1998. Phenotypic variation in micropropagated bananas and plantains. Corbana 23(50): 177—198.

Cox E.A., Stotzky G., Goods R.D. 1960. *In vitro* culture of *Musa balbisiana* Colla embryos. Nature 185(4710): 403—404.

Cronauer S.S. and Krikorian A.D., 1984. Multiplication potential of *Musa* from excised stem tips. Annals of Botany 53: 321—328.

Cronauer-Mitra S.S., Krikorian A.D., 1988. Plant regeneration via somatic embryogenesis in the seeded diploid banana *Musa ornata* Roxb. Plant Cell Rep. 7: 23—25.

Crouch J.H., Vuylsteke D., Ortiz R., 1998. Perspectives on the application of biotechnology to assist the genetic enhancement of plantain and banana (*Musa* spp.). Electronic Journal of Biotechnology , Universidad Católica de Valparaíso, Chile, 1(1): issue of April 15.

Damasco O.P., Adkins S.W., Godwin I.D., Smith M.K., 1998. Use of SCAR-based marker for the early detection of dwarf off-types in micropropagated Cavendish bananas. Acta Hort. 461: 157—164.

Damasco O.P., Godwin I.D., Smith M.K., Adkins S.W., 1996a. Gibberellic acid detection of dwarf off-types in micropropagated Cavendish bananas. Australian J. Exper. Agri. 36(2): 237—241.

Damasco O.P., Graham G.C., Henry R.J., Adkins S.W., Smith M.K., Godwin I.D., 1996b. Random amplified polymorphic DNA (RAPD) detection of dwarf off-types in micropropagated Cavendish (*Musa* spp. AAA) bananas. Plant Cell Rep. 16: 118—123.

Daniells J., Smith M., 1991. Post-flask management of tissue-cultured bananas. ACIAR Technical Reports 18, 8p.

Daquinta M., Lezcano Y., Escalona M., Santos R., 2000. *In vitro* multiplication of the banana FHIA-18 with paclobutrazol and thidiazuron, using different ways of cultivation. Rev. Bras. Frutic. Jaboticabal-SP 22(1): 86—88.

Dhed'a, D., Dumortier, F., Panis, B., Vuylsteke, D., De Langhe, E., 1991. Plant regeneration in cell suspension cultures of the cooking banana cv. 'Bluggoe'(*Musa* spp. ABB group). Fruits 46(2): 125—135.

Dolezel J., Lysák M.A., Van den Houwe I., Dolezelová M., Roux N., 1997. Use of flow cytometry for rapid ploidy determination in *Musa* species. Infomusa 6(1): 6—9.

Drew R.A., Moisander J.A., Smith M.K., 1989. The transmission of banana bunchy-top virus in micropropagated bananas. Plant Cell Tissue Organ Cult. 16: 187—193.

Drew, R.A. and Smith, M.K., 1990. Field evaluation of tissue-cultured bananas in south-eastern Queensland. Austral. J. Expt. Agr. 30: 569—574.

Drew R.A., Smith M.K., Anderson D.W., 1992. Field evaluation of micropropagated bananas derived from plants containing bunchy-top virus. Plant Cell Tissue Organ Culture 28: 203—205.

Engelborghs I., Swennen R., Van Campenhout S., 1998. Capacidad del AFLP para detectar diferencias genéticas y variantes somaclonales en *Musa* spp. Infomusa 7(2): 3—6.

Escalant J.V., Teisson C., 1989. Somatic embryogenesis and plants from immature zygotic embryos of the species *Musa acuminata* and *Musa balbisiana*. Plant Cell Rep. 7: 665—668.

Escalant J.V., Teisson C., Cote F., 1994. Amplified somatic embryogenesis from male flowers of triploid banana and plantain cultivars (*Musa* spp.). In Vitro Cell. Dev. Biol. 30P: 181—186.

Escalona M., Lorenzo J.C., Gonzalez B., Daquinta M., Gonzalez J.L., Desjardins Y., Borroto C.G., 1999. Pineapple (*Ananas comosus* L. Merr) micropropagation in temporary immersion systems. Plant Cell Rep. 18(9): 743—748.

Etienne-Barry D., Bertrand B., Vasquez N., Etienne H., 1999. Direct sowing of *Coffea arabica* somatic embryos mass-produced in a bioreactor and regeneration of plants. Plant Cell Rep. 19(2): 111—117.

FAO, 2000. FAOSTAT Database, http://apps1.fao.org/servlet.

Ferreira M.E., Caldas L.S., Pereira E.A., 1998. Aplicações da cultura de tecidos no melhoramento genético de plantas. In: A.C. Torres, L.S. Caldas & J.A. Buso (Eds).

Cultura de Tecidos e Transformação Genética de Plants. Embrapa-SPI / Embrapa-CNPH, Brasília, pp. 21—43.

Frison E.A., Sharrock S.L., 1998. Banana streak virus: a unique virus-Musa interaction? Proceedings of a workshop of the Promusa virology working. Montpellier, France.

Ganapathi T.R., Suprasanna P., Bapat V.A., Rao P.S., 1992. Propagation of banana through encapsulated shoot tips. Plant Cell Rep. 11: 571—575.

Gawel N., Jarret R.L., 1991. Cytoplasmic genetic diversity in bananas and plantains. Euphytica 52: 19—23.

George E.F., 1993. Plant propagation by tissue culture. London, Exegetics. 1361 p.

Georget F., Domergue R., Ferrière N., Côte F.X., 2000. Morphohistological study of the different constituents of a banana (*Musa* AAA, cv. Grande naine) embryogenic cell suspension. Plant Cell Rep. 19(8): 748—754.

Gerdemann-Knorck M., Nielen S., Tzscheetzsch C., Iglisch J., Schieder O., 1995. Transfer of disease resistance within the genus *Brassica* through asymmetric somatic hybridization. Euphytica 85: 247—253.

Govil S., Gupta S.C., 1997. Commercialization of plant tissue culture in India. Plant Cell Tissue Organ Cult. 51: 65—73.

Grajal-Martin M.J., Siverio-Grillo G., Marrero-Domínguez A., 1998. The use of randomly amplified polymorphic DNA (RAPD) for the study of genetic diversity and somaclonal variation in *Musa*. Acta Hort. 490: 445—454.

Grapin A., Ortiz J.L., Domergue R., Babeau J., Monmarson S., Escalant J.V., Teisson C., Côte F.X., 1998. Establishment of embryogenic callus initiation and regeneration of embryogenic cell suspensions from female and male immature flowers of *Musa*. Infomusa 7: 13—15.

Grapin A., Schwendiman J., Teisson C., 1996. Somatic embryogenesis in plantain banana. In Vitro Cell. Dev. Biol. - Plant 32: 66—71.

Gupta P.P., 1986. Eradication of mosaic disease and rapid clonal multiplication of bananas and plantains through meristem tip culture. Plant Cell Tissue Organ Cult. 6: 33-39

Hamill S.D.. Sharrock S.L., Smith M.K., 1993. Comparison of decontamination methods used in initiation of banana tissue cultures from field-collected suckers. Plant Cell, Tissue Organ Cult. 33: 343—346.

Henry R.J., 1998. Practical applications of molecular markers to tropical and subtropical species. Acta Hort. 461: 107—111.

Heinz D.J., Mee G.W.P., 1969. Plant differentiation from callus tissue of Saccharum species. Crop Sci. 9: 346—348.

Ho J.S., Li H.P., Barry K., Wang M. and Jordan, R. 1995. Comparison of dot blot, ELISA, and RT-PCR assays for detection of two cucumber mosaic virus isolates infecting banana in Hawaii. Plant Dis. 79: 902-906.

Howell E.C., Newbury H.J., Swennen R.L., Withers L.A., Ford-Lloyd B.V., 1994. The use of RAPD for identifying and classifying *Musa* germplasm. Genome 37: 328—332.

Hwang S.C., Ko W.H., 1988. *In vitro* somaclonal variation in banana and its application for screening for resistance to fusarial wilt. Technical Bulletin No. 107, Food & Fertilizer Technology Center, Taipei, Taiwan, 8p.

INIBAP, 1995. Networking Banana and Plantain: INIBAP Annual Report 1994. International Network for the Improvement of Banana and Plantain, Montpellier, France, 56p.

INIBAP, 1999. Networking Banana and Plantain: INIBAP Annual Report 1998. International Network for the Improvement of Banana and Plantain, Montpellier, France, 64p.

INIBAP, 2001. Networking Banana and Plantain: INIBAP Annual Report 2000. International Network for the Improvement of Banana and Plantain, Montpellier, France, 68p.

Israeli Y., Reuveni O., Lahav E., 1991. Qualitative aspects of somaclonal variations in banana propagated by in vitro techniques. Scientia Horticulturae 48:71—88.

Jarret R.L., Rodrigues W., Fernandez R., 1985. Evaluation, tissue culture propagation and dissemination of 'Saba' and 'Pelipita' plantains in Costa Rica. Scientia Hort. 25: 137—147.

Jimenez E., Perez N., Feria M. de, Barbon R., Capote A., Chavez M., Quiala E., Perez J.C., 1999. Improved production of potato microtubers using a temporary immersion system. Plant Cell Tissue Organ Cult. 59(1): 19—23.

Jones D.R. (ed.), 2000. Diseases of banana, abacá, and enset. CABI Publishing, Wallingford, UK, 544p.

Kaemmer D., Afza R., Weising K., Kahl G., Novak F.J., 1992. Oligonucleotide and amplification fingerprinting of wild species and cultivars of banana (*Musa* spp.). Bio/Technology 10: 1030—1035.

Kaemmer D., Fischer D., Jarret R.L., Baurens F.C., Grapin A., Dambier D., Noyer J.L., Lanaud C., Kahl G., Lagoda P.J.L., 1997. Molecular breeding in the genus *Musa*: a strong case for STMS marker technology. Euphytica 96: 49—63.

Kitto, S.L., Janick, J., 1985. Production of synthetic seeds by encapsulating asexual embryos of carrot. J. Amer. Soc. Hort. Sci. 110(2), 277—282.

Krikorian A.D., Cronauer S.S., 1984. Aseptic culture techniques for banana and plantain improvement. Econ. Bot. 38: 328—331.

Krikorian A.D., Irizarry H., Cronauer-Mitra S., Rivera E., 1993. Clonal fidelity and variation in plantain (*Musa* AAB) regenerated from vegetative stem and floral axis tips *in vitro*. Annals of Botany 71: 519—535.

Krikorian A.D., Irizarry H., Goenaga R., Scott M.E., Lockhart B.E.L., 1999. Stability in plant and bunch traits of a 'French-type' dwarf plantain micropropagated from the floral axis tip and five lateral corm tips of a single mother plant: good news on the tissue culture and bad news on banana streak virus. Scientia Horticulturae 81(2): 159—177.

Krikorian A.D., Scott M.E., 1995. Somatic embryogenesis in bananas and Plantains (*Musa* clones and species). In: Y.P.S. Bajaj (Ed.), Biotechnology in Agriculture and Forestry, Vol.31, Somatic Embryogenesis and Synthetic Seed II. Springer-Verlag, Berlin Heidelberg, pp. 183—195.

Lemos E.E.P., Ferreira M.S., Magalhães V.S., Alencar L.M.C., Oliveira G.L., 2000. Uso de bioreatores de imersão temporária para incrementar a micropropagação de banana. ABCTP Notícias 36: 2—5.

Linsmaier E.M., Skoog F., 1965. Organic growth factor requirements of tobacco tissue culture. Physiol. Plant. 18: 100—127.

Lorenzo J.C., Gonzalez B.L., Escalona M., Teisson C., Espinosa P. Borroto C., 1998. Sugarcane shoot formation in an improved temporary immersion system. Plant Cell Tissue Organ Cult. 54(3): 197—200.

Lysák M.A., Dolezelová M., Horry J.P., Swennen R., Dolezel J., 1999. Flow cytometric analysis of nuclear DNA content in *Musa*. Theor. Appl. Genet. 98: 1344—1350.

Ma S.S., Shii C.T., 1972. *In vitro* formation of adventitious buds in banana shoot apex following decapitation. J. Chinese Soc. Hort. Sci. 18: 135—142.

Mante S., Tepper H.B., 1983. Propagation of *Musa textilis* Née plants from apical meristem slices *in vitro*. Plant Cell, Tissue Organ Cult. 2: 151—159.

Matsumoto K., Barbosa M.L., Souza L.A.C., Teixeira J.B., 1999. In vitro selection for *Fusarium* wilt resistance in banana. II. Resistance to culture filtrate of race 1 *Fusarium oxysporum* f.sp. *cubense*. Fruits 54:151—157.

Matsumoto K., Hirao C., Teixeira J.B., 1995. *In vitro* growth of encapsulated shoot tips in banana (*Musa* sp.). Acta Hort. 370: 13—19.

Matsumoto K., Kobayashi S., Barbosa, M.L., Teixeira J.B., 1992. Isolamento e fusão dos protoplastos de banana. Rev Bras Frutic – Cruz das Almas 14(1): 27—33.

Matsumoto K., Morais L.S., Vianna G.R., Aragão F.J.L., Rech E.L., 2002. Genetic transformation of banana embryogenic cells through particle bombardment, using a herbicide resistance gene as selectable marker. Acta Hort., in press.

Matsumoto K., Yamaguchi Y., 1989. Nonwoven materials as a supporting agent for *in vitro* culture of banana protocorm-like bodies. Trop. Agric. (Trinidad) 66(1): 8—10.

Matsumoto K., Yamaguchi Y., 1990. Selection of aluminium-tolerant variants from irradiated protocorm-like bodies in banana. Trop. Agric. (Trinidad) 67(3): 229—232.

May G.D., Afza R., Mason H.S., Wiecko A., Novak, F.J., Arntzen C.J., 1995. Generation of transgenic banana (*Musa acuminata*) plants via *Agrobacterium*-mediated transformation. Bio/Technology 13: 486—492.

Mendes B.M.J., Filippi S.B., Demétrio C.G.B., Rodriguez A.P.M., 1999. A statistical approach to study the dynamics of micropropagation rates, using banana (*Musa* spp.) as an example. Plant Cell Rep. 18: 967—971.

Murashige T., Skoog F., 1962. A revised medium for rapid growth and bio-assays with tobacco tissue cultures. Physiol. Plant. 15: 473—497.

Novak F.J., 1992. *Musa* (bananas and plantains). In: Hammerschlag F.A., Litz R.E. (eds.), Biotechnology of Perennial Fruit Crops, C.A.B. International, Wallingford, UK, pp.449—488.

Novak F.J., Afza R., Phadvibulya V., Hermelin T., Brunner H., Donini B., 1986. Micropropagation and radiation sensitivity in shoot-tip cultures of banana and plantain. In: IAEA, Nuclear Techniques and In Vitro Culture for Plant Improvement, Vienna, IAEA, pp167—174.

Novak F.J., Afza R., Van Duren M., Omar M.S., 1990. Mutation induction by gamma irradiation of *in vitro* cultured shoot-tips of banana and plantain (*Musa* cvs.). Trop. Agric. (Trinidad) 67(1): 21—28.

Novak F.J., Afza R., Van Duren M., Perea-Dallos M., Conger B.V., Tang Xiaolang, 1989. Somatic embryogenesis and plant regeneration in suspension cultures of dessert (AA and AAA) and cooking (ABB) bananas (*Musa* spp.). Bio/Technology 7, 154—159.

Orellana P., Perez Ponce J., Agramonte D., Gomez , Jimenez E., Martinez S., Almaguer E., Gomez P., 1991. La micropropagacion del platano a escala comercial en Cuba. Associason Colombiana de Estudios Vegetales In Vitro, Boletin Cientifico 3 (3): 29—38.

Panis B., 1995. Cryopreservation of banana (*Musa* spp.) germplasm. Ph.D. thesis, K. U. Leuven, Belgium. Dissertationes de Agricultura 272, 201p.

Panis B., Totte N., Van Nimmen K., Withers L.A., Swennen R., 1996. Cryopreservation of banana (*Musa* spp.) meristem cultures after preculture on sucrose. Plant Science 121: 95—106.

Panis B., Wauwe A.V., Swennen R., 1993. Plant regeneration through direct somatic embryogenesis from protoplasts of banana (*Musa* spp.). Plant Cell Rep. 12: 403—407.

Pérez Ponce J.N., Jiménez González E.A., Agramonte Peñalver D., 1998. Capítulo 10 Aumento de la eficiencia en la micropropagación. In: Pérez Ponce J.N. (ed.), Propagación y mejora genética de plantas por biotecnología, Instituto de Biotecnología de las Plantas, Santa Clara, Cuba, pp179—191.

Ramulu K.S., Dijkhuis P., Rutgers E., Blaas J., Verbeek W.H.J., Verhoeven H.A., Colijn-Hooymans C.M., 1995. Microprotoplast fusion technique: a new tool for gene transfer between sexually-incongruent plant species. Euphytica 85: 255—268.

Redenbaugh K., Paasch B.D., Nichol J.W., Kossler M.E., Viss P.R., Walker K.A., 1986. Somatic seeds: encapsulation of asexual plant embryos. Bio/Technology 4, 797—801.

Reuveni O., Israeli Y., 1990. Measures to reduce somaclonal variation in *in vitro* propagated bananas. Acta Hort. 275: 307—313.

Robinson J.C., Fraser C., Eck K., 1993. Field comparison of conventional tissue culture banana planting three crop cycles. J. Hort. Sci. 68

Rodrigues P.H.V., Tulmann Neto A., Cassieri Neto P., Mendes B.M.J., 1998. Influence of the number of subcultures on the somaclonal variation of banana plantlets cv. nanicão in Vale do Ribeira – SP. Rev. bras. Frutic., Cruz das Almas, 20(1): 74—79.

Rowhani A., Falk B.W., 1995. Enzyme-linked immunosorbent assay (ELISA) methods to certify pathogen (virus)-free plants. *In:* Gamborg O. L., Phillips G.C. (eds.) Plant cell, tissue and organ culture – fundamental methods. Springer Verlag, Berlin Heidelberg, New York. pp 267—280.

Sagi L., Panis B., Remy S., Schoofs H., De Smet K., Swennen R., Cammue B.P.A., 1995. Genetic transformation of banana and plantain (*Musa* spp.) via particle bombardment. Bio/Technology 13: 481—485.

Samoylov V.M., Sink K.C., 1996. The role of irradiation dose and DNA content of somatic hybrid calli in producing asymmetric plants between an interspecific tomato hybrid and eggplant. Theor. Appl. Genet. 92: 850—857.

Sandoval J.A., Côte F., Doumas P., 1999. *In vitro* recognition of high stature somaclonal variants in banana (cv. 'Grand Naine', *Musa* AAA). Response to a GA_3 treatment. Corbana 24(51): 11—20.

Sandoval J.A., Côte F.X., Escoute J., 1996. Chromosome number variations in micropropagated true-to-type and off-type banana plants (*Musa* AAA Grande Naine cv.). *In vitro* Cell. Dev. Biol.-Plant 32: 14—17.

Schoofs H., 1997. The origin of embryogenic cells in *Musa*. Ph.D. thesis, K. U. Leuven, Belgium. Dissertationes de Agricultura 330, 258p.

Schoofs H., Panis B., Swennen R., 1998. Competence of scalps for somatic embryogenesis in *Musa*. Acta Hort. 490: 475—483.

Sharrock S., Frison E., 1999. *Musa* production around the world – trends, varieties and regional importance. Networking Banana and Plantain: INIBAP Annual Report 1998. International Network for the Improvement of Banana and Plantain, Montpellier, France, pp 42-47.

Sharrock S., Lusty S., 2000. Nutritive value of banana. Networking Banana and Plantain: INIBAP Annual Report 1999. International Network for the Improvement of Banana and Plantain, Montpellier, France, pp 28-31.

Shepherd K., 1987. Banana breeding – past and present. Acta Hort. 196: 37—43.

Shepherd K., Souza F.V.D., Silva K.M., 1996. Mitotic instability in banana varieties. IV. BAP concentration and effects of number of subcultures. Fruits 51(4): 211—216.

Silva Neto S.P., 2000. Studies on mass production of pathogen-free bananas (*Musa* spp.) through shoot cultures in Brazil. PhD. thesis, Tokyo University of Agriculture, Tokyo, 127p.

Singh B., Singh N.S., 1995. Role of liquid versus agar-gelled media in mass propagation and *ex vitro* survival of bananas. Plant Cell Tissue Organ Cult. 41: 71—73.

Singh Z., Jones R.A.C., and Jones, M.G.K. 1995. Identification of cucumber mosaic virus Subgroup I isolates from banana plants affected by infectious chlorosis disease using RT-PCR. Plant Dis. 79: 713-716.

Smith M.K., Drew R.A., 1990. Current applications of tissue culture in plant propagation and improvement. Austral. J. Plant Physiol. 17: 267—289.

Smith M.K., Murashige T., 1970. *In vitro* development of the isolated shoot apical meristem of angiosperms. Amer. J. Bot. 57: 562—568.

Stover R.H., Simmonds N.W. 1987. Bananas. 3rd ed. – Tropical Agriculture Series, Longman Scientific & Technical, England, 468p.

Taiwan Banana Research Institute, 1992. Procedure and application of banana micropropagation technique in Taiwan. Extension Bulletin No. 81-2, 10p.

Tulmann Neto A., Mendes B.M.J., Ando A., 1998. Progressos na indução e uso de mutações *in vitro*. In: A.C. Torres, L.S. Caldas & J.A. Buso (Eds). Cultura de Tecidos e Transformação Genética de Plants. Embrapa-SPI / Embrapa-CNPH, Brasília, pp. 21—43.

Vuylsteke D.R., 1998. Shoot-tip culture for the propagation, conservation, and distribution of *Musa* germplasm. International Institute of Tropical Agriculture, Ibadan, Nigeria. 82p.

Vuylsteke D.R., 2001. Strategies for utilization of genetic variation in plantain improvement. Ph.D. thesis, K. U. Leuven, Belgium. Dissertationes de Agricultura, 213p.

Vuylsteke D., De Langhe E. A., 1985. Feasibility of *in vitro* propagation of bananas and plantains. Trop. Agric. (Trinidad) 62 (4): 323—328.

Xie W.S. and Hu J.S., 1995. Molecular cloning, sequence analysis, and detection of banana bunchy top virus in Hawaii. Phytopathology 85: 339-347.

Wong W.C., 1986. *In vitro* propagation of banana (*Musa spp.*): Initiation, proliferation and development of shoot tip cultures on defined media. Plant Cell Tissue Organ Cult. 6: 159—166.

13. MICROPROPAGATION OF ARID ZONE FRUIT TREES OF INDIA

R. RAJ BHANSALI AND MANJIT SINGH
Division of Plant Sciences and Biotechnology
Central Arid Zone Research Institute
Jodhpur-342 003, Rajasthan, India.
E-Mail: **bhansali@cazri.raj.nic.in**

Contents

I. INTRODUCTION

About 12 per cent of India's total geographical area is hot arid zone and the extent of arid area is about 31.7 million hectares. Significantly more than 61 % of the area falls in Rajasthan and extends to Haryana and Punjab (9 %), Gujarat (20 %) and some pockets in Andhra Pradesh, Karnataka and Maharashtra (10 %). Over 50 % hot arid zone area is adversely affected by either moderate or severe desertification. The arid areas are characterised by low and erratic rainfall (100-420 mm/year), frequent droughts, high summer temperatures (45-50° C), high wind velocity (30-40 km/h), and high evapo-transpiration (1500-2000 mm/year). The sandy soils have poor fertility and low water retention (Krishnan, 1977). The increasing populations of both human and livestock perpetually put stress on natural resources available in these areas. This has resulted in over exploitation of resources causing rapid and wide spread land, water and biological degradation that lead to decline in productivity of arid zone (Faroda *et al.*, 1997).

S.M. Jain and K. Ishii (eds.), Micropropagation of Woody Trees and Fruits, 381–432.

Biodiversity

In arid zone, only 1 % of the area is covered with trees as compared to nearly 10 % average in the country. Nevertheless, these regions are well abounding in valuable germplasm of some of the hardiest flora having high nutritive and medicinal value. The great biodiversity present in arid ecosystem can gainfully be utilized for various economical uses (Saxena, 1977). The endemic biota, including horticultural plants, has perfect harmony with the harsh arid environment. They may serve as basic gene pool for the improvement of indigenous germplasm. Arid horticulture encompasses gene pools of fruit, vegetables, and ornamental and medicinal plants favourably grown in arid zone. A rich gene pool consisting of several varieties of ber (603), pomegranate (475), date palm (128), custard apple (51), aonla (151), fig (35) and bael (74) has been collected and maintained in field gene banks at different locations in the arid zone of the country (Pareek, 1997).

Many indigenous species that have been neglected such as *Capparis decidua* (ker), *Citrullus lanatus*, (matera), *Cucumis callosus* (kachri), *Cucumis melo* (snap melon), *Salvadora oleoides* (mitha jal or pilu), *Prosopis cineraria* (khejri) and *Grewia subinaequalis* (phalsa), *Cordia myxa* (gonda or lasoda), *Cordia gharaf* (gundi) could be improved and used as future potential plants for horticultural and industrial use in the future (Paroda, 1979; Goodin and Northington, 1985). Some of the introduced species such as *Tamarindus indica, Malphighia glabra, Borassus flabellifer* and *Morus* species have acclimatized like native plants. A number of exotic plants has performed well in arid climate. Elite plant types in this gene pool from existing variability need to be identified for extensive cultivation in arid areas (Raj Bhansali and Jindal, 1997).

Issues

The conventional breeding techniques in improving arid zone fruit crops are slow and time consuming because of the long juvenile phase and out crossing nature of these species. There has been a little success in improvement of tropical arid fruits, despite the significance of perennial tropical plants for food and export income. Most superior cultivars have resulted from individual selection within natural seedling populations derived from controlled crosses or have been derived from useful somatic mutants within the existing clones. As a result of most important fruit, plant characteristics are conferred by gene complexes, the genetic integrity of many tropical crop cultivars has been maintained by means of vegetative propagation (Litz *et al*., 1985). The fragile arid zone ecosystem faces several biophysical constraints (poor soil conditions, extreme high temperature, hot winds, high solar radiation and very low precipitation). These lead to high demand of atmospheric moisture in this region. The arid zone ecosystem is prone to serious imbalance even with the slightest disturbance due to mismanagement of resources or influx of external factors (Faroda *et al.*, 1997). Therefore, agro climatically suitable areas are needed for conducting experiment for the improvement of arid zone fruit crops. Thus, biotechnology (tissue culture and greenhouse technologies) can facilitate experiments to develop new variants (Raj Bhansali, 1989).

II. TISSUE CULTURE

The most popular and widely commercialized application of plant biotechnology has been in the sphere of plant tissue culture. It is a highly useful facet of biotechnology, which has been used in commercial agri-horti business, especially in the form of micropropagation of tropical and temperate fruit trees (Litz *et al.*, 1985; Zimmerman, 1987; Raj Bhansali, 1994). *In-vitro* micropropagation has been demonstrated as an excellent method of large-scale production of true-to-type and disease-free planting material, unhindered by seasonal constraints (Murashige, 1974; Jain *et al.*, 1997 and 2000; Raj Bhansali, 2001). Arid zone fruits such as date palm, ber, pomegranate, bael, custard apple, ker and jamun could be micropropagated in same way (Raj Bhansali, 1998a). The three important factors that warrant use of tissue culture in arid zone horticultural industry are: I). a million fold increase in multiplication rate over the conventional method of propagation. It also remarkably reduces the time and space required for the introduction of desired varieties in fruit industry, II). Planting material can be available round the year; and III). Quick multiplication or development of new insect pest and disease resistant and pathogen-free variants (Litz *et al.*, 1985; Zimmerman, 1987; Raj Bhansali, 1995). Moreover, biotechnological research can be easily conducted in controlled conditions in spite of harsh arid environment. Arid zone fruits are generally of woody perennial trees having long juvenile phase periods. Therefore, breeding programmes for such plants can involve the professional life times of several generations of scientists.

Recent advances in tissue culture technology offer an excellent opportunity for the improvement of many staple crops in India, including arid zone plants (Chopra and Sharma, 1991; Chopra, 1997; Raj Bhansali, 1997b; Jain *et al.*, 1995; Raj Bhansali and Singh, 2000). The use of *in-vitro* systems as an alternative propagation technique is likely to have the greatest immediate impact on the productivity of arid zone agroforestry plants (Raj Bhansali, 1993a; 1998b). Many arid zone fruit trees can be propagated asexually but the production is limited to some extent. The micropropagation of selected, high yielding and disease resistant superior trees can increase the efficiency of propagation techniques as well as the quality of products. Moreover, the long-term conservation of the rich biodiversity would be greatly facilitated by the development of tissue culture for the exchange of micropropagated plants. The importance of *in-vitro* propagation of arid zone fruit trees is obvious, particularly since disease, pest and environmental stresses are constant threats to the production of elite germplasm for domestic use and for international exchange programme.

Currently, date palm, ber and pomegranate cultivation are threatened due to the outbreak of various diseases such as black scorch, false smut, powdery mildew, leaf and fruit spot in arid regions (Raj Bhansali, 1997a). Epidemics of powdery mildew of ber and fruit spot diseases of pomegranate are largely responsible for the limited yield and quality of fruit in various parts of arid and semi arid regions in India (Singh *et al.*, 1998). The *in-vitro* selection of fungal disease resistant and virus and phytoplasma-free plants will help in producing pathogen resistant and free plants (Ingram and Helgeson, 1980; Hall, 1987; Raj Bhansali, 1993b; Raj Bhansali and Singh 1991). The selection of pathotoxin resistant cell lines and mutant cells for the introduction of somaclonal variants into crop improvement schemes could lower the genetic vulnerability and increase the effectiveness of breeding programmes (Earle, 1982; Raj Bhansali, 1995). Clonal trueness of the desirable crop would remain intact with the exception of desirable mutants if efficient *in-vitro* selection system could be developed. The arid zone fruit species discussed in this chapter are listed in Table-1.

III. CURRENT STATUS OF TECHNOLOGY

The most significant successes in tissue culture of arid zone trees have been with species having natural polyembryony. Possibly, the greatest obstacle in developing *in-vitro* systems for woody plant regeneration is the absence of juvenile phase in mature tree (Bonga, 1982; Jain *et al.*, 1988). The woody trees grown in arid environment are extremely slow growing and xerophytic in nature. Their juvenile phase lasts for a very limited period when *in-vitro* systems could be developed. The germinating seeds and young seedling parts retain the highest regeneration capabilities in many plants, including the arid zone woody tree species. These tissues in ber, date palm, pomegranate and jamun even possess potentiality for somatic embryogenesis (Sharma *et al.*, 1986; Raj Bhansali and Kaul, 1991; Raj Bhansali, 1995; Dodeman *et al.*, 1997).

The nucellus parts particularly in citrus, have high morphogenetic potential (Raj Bhansali, 1997b). Somatic embryos generated *in-vitro* from cultured citrus ovules were first reported in 1958. Many subsequent advances in development of *in-vitro* systems for citrus and other arid zone fruit species have been the result of this early observation. This helps to overcome the difficulties of working with mature phase of woody trees *in-vitro*. The techniques developed then could be applied to mature clones. Another approach that has been taken into consideration is to rejuvenate mature clones by severe prunning of the stock plants or by sequential grafting. This technique is successful with many *Ziziphus* species and *Punica granatum*.

Table 1. Economically important fruits of arid and semi-arid regions in India

Plant family	Genus	Species	Common name	Local name
Annonaceae	*Annona*	*squamosa*	Custard apple	Sitaphal, Sharifa
Capparidaceae	*Capparis*	*decidua*	Caper-bush	Ker
Ehretiaceae	*Cordia*	*myxa*	Cordia	Lasoda
Euphorbiaceae	*Emblica*	*officinalis*	Indian gooseberry	Aonla, Amla
Leguminosae	*Prosopis*	*cineraria*	Mesquite	Khejri
Myrtaceae	*Syzygium*	*cumini*	Jambolana	Jamun
Punicaceae	*Punica*	*granatum*	Pomegranate	Anar
Rhamnaceae	*Ziziphus*	*mauritiana*	Chinese date, Jujube	Ber
		rotundifolia		Bore
		nummularia	Wild jujube	Bordi, Jharberi
Rutaceae	*Citrus*	*sinensis*	Sweet lime	Mosumbi
		aurantifolia	Acid lime	Neembu
	Aegle	*marmelos*	Bengal quince, Bael	Bel, Bilva
Salvadoraceae	*Salvadora*	*persica*	Mustard tree	Khara Jal
		oleoides		Mitha Jal, Pilu
Tiliaceae	*Grewia*	*subnaequalis*		Phalsa

Plants rejuvenated reversion to juvenile leafy forms in culture can be easily rooted. The resulting plants are quickly readapted to greenhouse or field conditions by this way and the mature form of shoots will develop again. These plants exhibit more vigorous vegetative growth in the fields than conventionally propagated plants at the initial phase of growth. They may even flower and fruit at an early age with increased productivity (Zimmerman, 1987; Jain *et al.*, 1999).

Adaptation in culture has been the most difficult task in many woody fruit tree species of arid zone such as *P. granatum, Z. mauritiana, C. decidua, A. squamosa* and *Phoenix dactylifera*. The length of the establishment stage may extend over several months to years before they become active in cultures. From mature phase of woody plants, cultures will grow after successful establishment and then proliferate rapidly on repeated subculturing in tissue culture system. The mature explants obviously undergo various physiological and morphological changes during *in-vitro* culture environment. These changes can be noticed in leaf morphology and development of adventitious shoots and roots. The explants that have an erratic growth habit are usually more difficult to establish in tissue culture than those that respond continuously. Nevertheless, once cultures of such woody species are established in tissue culture, then growth and proliferation proceed as continuous processes (Lloyd and Mc Cown, 1981; Raj Bhansali, 1995).

Micropropagation methods developed for a particular genus, species or cultivars of woody trees will not respond in the same way to all cultural conditions and media composition. This has been well documented in somatic embryogenesis in date palm and pomegranate (Raj Bhansali *et al.*, 1988; Raj Bhansali, 1990). Differences among cultivars have been found in response to induction of organogenesis, somatic embryogenesis, proliferation of shoots and induction of roots. Nutritional requirements (nitrogen and hormones) and cultural conditions (temperature and photoperiods) also vary largely among species and even among different parts of the same plant. Various *Ziziphus* species and cultivars of date palm had different optimal auxin and cytokinin concentrations for maximum shoot proliferation and in induction of somatic embryos. Many fold differences occur between prolific and non-prolific species or cultivars. Several citrus cultivars vary greatly in this respect; some are readily established on the first attempt and others are unsuccessful even after repeated attempts with varied media and cultural conditions (Raj Bhansali and Arya, 1978c).

IV. MICROPRPAGATION

A. <u>Major fruit trees:</u>

Arid zone fruit plants that have been most studied with respect to micropropagation include: *Phoenix dactylifera, Punica granatum, Ziziphus mauritiana, Z. rotundifolia, Z. nummularia,* various *Citrus* species, *Aegle marmelos, Emblica officinalis* and *Prosopis cineraria*.

1. Date palm

Date palm (*Phoenix dactylifera* L.) is the oldest among the cultivated fruit trees. The date palm is mostly adapted to the areas with long, very hot summers with little rain, and low humidity, but requires abundant water. As per an Arabic saying, "date palm should be grown with its feet in running water and its head in fire of sky". These conditions are found in oases and river valley in arid sub-tropical deserts of the Middle East. The date palm is most probably originated from the area of northeastern Africa (the Nile delta), northern Arabia, Iraq and western Iran (Kruger, 1998). This is the area known as "Fertile Crescent" (ancient Mesopotamia), where agriculture in Old World is thought to have arisen. Indeed the date palm has been cultivated in this area from ancient times, possibly being on of the first crops domesticated.

Date seeds at least 5000 years old have been found in the storage godowns at Mohenjo Daro, the ancient city along the Indus River in Sind, and the date palm was used in the construction of the Temple of the Moon God in Ur (Iraq) some 4000-5000 years ago. By the several millennia BP, the date palm culture spread to other parts of world including western India. In India, it is believed that the date palm was introduced by soldiers of Alexander, in the 4th century, in the Indus valley. Date palm cultivation by planting regular orchards in Rajasthan was first started by the then ruler of erstwhile Bikaner State, Maharaja Ganga Singh. He planted the suckers of cultivars e.g. Halawy, Khadrawy and Zahidi at Sriganganagar.

Table 2. Date palm explants and their morphogenesis

Explant source	**Availability**	**Morphogenetic potential**	**Remark**
Apical tip	One per tree/ offshoot	Very high	Sacrifice whole plant or offshoot
Inflorescence	Abundant	Immature-High Mature-Low	Dependent on flowering time
Lateral bud	Several	Variable Young-High Old-Low	Sacrifice the plant to obtain
Zygotic embryo	Abundant	Variable Immature-High Mature-Low	Non-clonal, specific genotype

It requires prolonged hot dry summers and rain-free area, particularly at the time of fruit ripening, for attaining Pind stage (July-August period)

(Chandra *et al.*, 1994). Western Rajasthan comprising Jaiselmer, Barmer, Bikaner and Jodhpur districts is ideally suited for date palm cultivation. Date palm varieties including Halawy, Shamran, Khadrawy, Medjool, Barhee, Zagloul, Hayani, Zahidi, Khalas and Sewi are important in reference to requirement of heat summation unit to reach colour-turning stage (early Doka and Pind stage). These have been considered suitable in the states of Rajasthan, Gujarat, Haryana and Punjab (Manohar and Chandra, 1995).

Date palm is commercially propagated from suckers (offshoots). It can also be propagated from seeds but being dioceous in nature, it produces male and female flowers on separate palms. Therefore, propagation through suckers is preferred over seeds. An offshoot is a lateral branch, which develops from the base of the trunk of a tree. Basal offshoots usually have a well-developed root system as compared to aerial suckers. Date palm produces 10 to 25 suckers during the lifetime (Chandra *et al.*, 1994). According to Pareek and Nath (1996), about 8-20 offshoots of 8-15 kg size can be obtained from each mother palm during its 4th-10th year and none thereafter.

1.1 *In-vitro* propagation:

1.1.1 Organogenesis:

Date palm can be propagated successfully by tissue culture methods. Serious efforts have been made in date palm tissue culture as a mean to mass produce high-yielding disease resistant clones in large number for plantation, beginning in the 1970's in Israel (Oppenheimer and Reuveni, 1972; Reuveni, *et al.*, 1972; Reuveni and Lilyen Kipnis, 1974; Raj Bhansali, 2001). Several explant sources and types of media were used to obtain morphogenetic responses *in-vitro* (Table 2). Initially, several problems occurred in starting date palm tissue culture. These included browning of media and explants, resulting in premature death of tissues. The growing tip produced roots, whereas the callus was short-lived and could not survive in subculture.

Refinement in cultural conditions, type of media, addition of charcoal and anti oxidizing chemicals, type of growth hormones, and media compositions have improved the methods of mass propagating date palm through tissue culture. Induction of vegetative bud development from shoot-tip or axillary bud explants in culture has been reported (Tisserat, 1984; Bouguedoura *et al.*, 1990: Raj Bhansali and Kaul, 1991; Bekheet and Saker, 1998) (Fig. 3). However, the number of buds produced per explant tissues was limited. In date palm, clonal propagation through suckers is the normal way of propagation and a more efficient method of cloning would be highly desirable.

1.1.2 Somatic embryogenesis:

Asexual embryogenesis has been obtained from excised zygotic embryos (Ammar and Benbadis, 1977, Reynold and Murashige, 1979) and also from somatic tissues (Tisserat *et al.*, 1979; Mater, 1986; Sharma *et al.*, 1984,1986; Raj Bhansali, *et al.*, 1988, Daguin and Letouze, 1988; Dass *et al.*, 1989; Raj Bhansali and Kaul, 1991; Bhaskaran and Smith, 1992; Sudhersan *et al.*, 1993; Letouze et al., 1998). During the embryogenic callus growth and development *in-vitro*, the date palm explants release excessive browning substances, which became a great problem (Zaid, 1987; Raj Bhansali *et al.*, 1988). These substances (phenols) have profound physiological effects on the establishment and growth of embryogenic callus. Browning of explant tissues and culture medium is assumed to be due to oxidation of polyphenols and formation of quinones. These are highly reactive and toxic to the tissues (Maier and Metzlier, 1965; Zaid, 1987; Raj Bhansali and Singh, 1982). The inhibitory effects may result from the bonding of phenols with proteins and their subsequent oxidation into quinones.

Murashige (1974) suggested the presoaking of explants in ascorbic acid and citric acid solutions and adding them to the culture medium for curtailing the oxidation of phenols. Incorporation of PVP, cysteine-HCl and ascorbic acid also helped in minimizing browning problems in several plant species including sugarcane and date palm (Raj Bhansali and Singh 1982; Dass *et al.*, 1989). Zaid and Tisserat (1983) soaked date palm explants in an antioxidant solution (150 mg/l citric acid and 100 mg/l ascorbic acid) prior to surface sterilization treatment. Raj Bhansali and Kaul (1991) have also used these antioxidant solutions for 30 minutes in cold storage (0° C). Furthermore nutritionally balanced media containing activated charcoal (3g/l) have significantly checked the browning problems in date palm explants. Raj Bhansali *et al* (1988) found that shoot tips and lateral bud cultures grow successfully on frequent transfers (after a period of incubation of 7 to 15 days) them onto the fresh media.

Somatic embryogenesis was induced in date palm by several researchers (Rhiss *et al.*, 1979; Drira, 1983; Sharma *et al.*, 1986; Raj Bhansali, *et al.*, 1988; Sudhersan *et al.*, 1993) (Fig. 1&2). Free-living plants have been established in the pots and field following tissue culture at Haryana Agriculture University, Hissar and at this Institute in Jodhpur (Fig. 4A&B). The repetitive somatic embryogenesis (RSE) process developed by the author is highly efficient in raising date palm tissue culture of certain varieties (Raj Bhansali, 1998a,b) (Fig. 5). Shabana *et al* (1998) compared the quality of fruits of Hilali cultivar grown from tissue culture and from the offshoots. They did not found significant differences in the chemical and physical characteristics of fruits obtained from tissue culture and offshoots raised plants. Thus the results obtained clearly indicate that tissue culture technique is a viable method of date palm propagation.

1.1.3 Suspension culture:

Various workers have also tried to establish suspension culture of date palm friable callus for rapid embryogenesis (Sharma *et al.*, 1986; Bhaskaran and Smith, 1992). The proembryo masses developed into embryos after passing through several sequential and distinct embryo development phases. Hundreds of embryos could be developed from suspension culture within 3 weeks. These embryos had to be subcultured after one month to promote further growth. Approximately 1000 embryos could be obtained from 200 mg embryogenic friable callus cultured per vessel (Bhaskaran and Smith, 1992). These embryos (up to 40 %) germinated into normal plantlets on plating on solid medium. These could be transferred to the soil. The somatic embryogenesis in date palm will lend this method suitable for studies on developmental embryogenesis or for encapsulation of embryos for long-term storage and shipment for export.

2. Molecular characterization:

Improvement of date palm is very difficult due to its long life cycle, strongly heterozygous nature and impossibility to determine sex at an early stage of development (Mourey and Saker, 1996). Most of earlier studies on genetic characterization, detection of genetic variation and gene mutation have been conducted on the basis of variation in chromosome number, isoenzyme polymorphism and biochemical diversity. But, chromosomes are numerous and small and mitotic examination of tissue culture derived palm plants is unreliable (Corley et al., 1976). Recently, various workers have demonstrated the utility of RAPD markers for the analysis of genetic diversity among cultivars and within plant population (Sarker and Moursy, 1998; Sedra et al., 1998). Similarly molecular techniques including mitochondrial plasmid-like DNAs markers have used for molecular characterization of Tunisian date palm varieties and to develop preventive procedure for their protection against Bayoud disease (*Fusarium oxysporum albedinis*) causing vascular fusariosis (Mokhtar et al., 1998).

3. <u>PROFILE OF TECHNIQUE</u>

Stage I. Processing of plant material prior to culturing

- Small pieces of date palm shoot tip from healthy offshoots are cut and cleaned in 5-10 % liquid soap solution and washed in running tap water.

- Soaked in an antioxidant solution of 150 mg /l citric acid and 200 mg/l ascorbic acid for 30-45 min.
- Cleaned and treated plant pieces are surface sterilized with mercuric chloride ($HgCl_2$) 0.1 % for 5-7 minutes and subjected to a final 3-4 times repeated rinsing in sterilized distilled water.

Stage II. Nutrient medium for initiating cultures

- Culture medium for date palm comprises of inorganic salts (macro elements: NH_4NO_3, $CaCl_2$ $2H_2O$, $MgSO_4$ $7H_2O$, KH_2PO_4; and micro elements: $MnSO_4$ $4H_2O$, $ZnSO_4$ $7H_2O$, H_3BO_3, KI, Na_2MoO_4 $2H_2O$, $CuSO_4$ $5H_2O$, $CoCl_2$ $6H_2O$, Na_2EDTA, $FeSO_4$ $7H_2O$, vitamins, growth regulators (auxin and cytokinin), sucrose and charcoal. Agar (0.64 %) is used for solidifying the medium adjusted to pH 5.7. The medium is then sterilized by autoclaving at 15 lb pressure for 20 minutes; 30 ml of medium is dispensed into glass culture tubes or flasks (125 ml).
- A contaminant-free culture is initiated by inoculating sterile plant tissues in to autoclaved nutrient medium under clean air-flow hood (aseptic worktable). Air-flow hood keeps away the dust and contaminants (most air-borne microbes) by forcing air through sub-micron filters that provide a sterile atmosphere during inoculation, cutting and surface sterilization of plant tissues.
- The glass vessels containing plant tissues and medium are incubated in clean growth chamber provided with artificial light (4000 lux) at a temperature of 25°-28° C and at relative humidity 60-70 % initially for 60 days. Subculturing is done onto fresh medium every 25-35 days.

Stage III. Repetitive somatic embryogenesis (RSE)

- Small nodular embryogenic calli (ascertained by staining developed) obtained from explant tissues are transferred to medium with low auxin and high cytokinin contents. The embryogenic cultures are multiplied by regular sub-culturing onto fresh medium devoid of auxins.
- Embryogenic callus produces a large number of somatic embryos repetitively at given 16 hr light, while organogenetic cultures produce few shoots from callus or shoot-tip tissues. Although such shoots may develop leaves, but they have no root primordia.
- Embryos developed in loose clumps are easily separable and elongate further on sub-culturing on fresh medium.
- Under constant light (4000 lux), well developed, elongated, white, mature embryos germinate by producing first root then shoot, akin to the germination of true seeds in nature.

Stage IV. In-vitro hardening of the test-tube plants

1. Germinated plantlets are delicate and usually die if transplanted directly. They are very susceptible to even non-pathogenic diseases. Pre-planting hardening by keeping them under high illumination and low nutrient level (¾ and ½ base medium) becomes compulsory. The hardened plantlets can tolerate moisture stress and diseases. They also become more self-reliant and manufacture their food during hardening process.

Stage V. Pre-planting conditioning

2. The hardened tissue-culture-derived plants are ready to be taken out of the laboratory for their transfer into sterilized soil. These plants have to be kept under high humidity or low temperature (not more than 30° C in mist chambers). After 8 weeks, they can be kept under net-house benches for further hardening process. Such areas should be provided with shaded light and extremely high humidity (fog-like atmosphere). After spending a few weeks, they are prepared for a final planting in the open field.

2. Ber

Ber or jujube (*Ziziphus mauritiana*), of Rhamnaceae family is the hardiest cultivated fruit tree in the northern Indian plains. The fruits have been in use since ancient times. It is distributed all over the warm arid and semi-arid regions due to the drought and heat tolerance abilities. It is now becoming more popular because of the low price and high nutritional value for its high content of vitamin C, A and B complex (Chandra *et al*., 1994). Ber leaves are rich in proteins and minerals and are nutritious fodder for the animals. The major grafted ber growing states are Haryana, Punjab, Uttar Pradesh, Rajasthan, Gujarat, Madhya Pradesh, Bihar, Maharashtra, Andhra Pradesh and Tamil Nadu. Seedlings are found growing extensively in arid and semi-arid areas (Pareek, 1997).

There are two prominent species of ber viz., Chinese ber *(Z. jujuba)* and Indian ber (*Z. mauritiana)*. Multiplication of true-to-type superior cultivars with good fruit quality and high yield is important factors for profitable cultivation. Popular methods of propagation are by I-or T-shield, ring and patch budding. The most common commercial method is by the T-shield budding because it is easier and yields 87 to 97 % success. Rootstocks of several species (*Z. rotundifolia, Z. nummularia, Z. oenoplia, Z. rugosa* and *Z. xylocarpa*) can be used but *Z. rotundifolia* is most vigorous and result in the maximum plant height, spread, collar diameter and canopy area.

Suitable cultivars in ber are: Gola for very dry condition, Kaithali for slightly higher rainfall; and Umran for moderate climatic conditions. Cultivars

"Dharakhi-1", "Dharakhi-2", "Villaiti", and seedless showed resistance to powdery mildew at Rahuri whereas Illachi, Jhajjar, Katha, Phal, Safed Rothak, and Sanaur-5 at Hissar (Pareek, 1997).

2.1 *In-vitro* propagation:

2.1.1 Organogenesis:

Protocols for cloning of *Z. nummularia* and *Z. mauritiana* (Goyal and Arya 1985; Raj Bhansali 1998; Rathore *et al.*, 1992, Kabir *et al.*, 1994; Mathur *et al.*, 1995), and *Z. xylopyra* (David and Vasikar, 1980) through tissue culture have been developed. Various types of explants such as nodal segments, shoot tip, seedling and root suckers can be used for clonal propagation of *Ziziphus* species. Goyal and Arya (1985) described methods for cloning mature *Z. mauritiana* (cultivars 'Apple' and 'Gola') plants. A large number of shoots per nodal explant tissues was developed on shoot proliferating medium (Murashige and Skoog (MS) medium containing 2.5-5.0 mg/l BAP and 0.03 mg IAA through axillary bud sprouting method) (Fig. 6&7). More than 60 % of field-transferred plants of *Z. nummularia* and 40 % of *Z. mauritiana* survived. Similarly, shoots of other *Ziziphus* species could be multiplied. The isolated shoots produced roots on White liquid medium with a filter paper bridge containing 25 mg/l IBA for 48 h; these shoots were then transferred onto hormone-free medium. Rathore *et al* (1992) reported successful field transplantation of regenerated ber plantlets. Mathur and Vyas (1995) multiplied mycorrhizal fungus *Glomus deserticola* on *Z. nummularia* cultures.

2.1.2 Somatic embryogenesis:

Asexual embryogenesis in *Z. mauritiana* was induced in seedling explant on 2.5 mg/l 2,4-D, 0.5 mg/l BA and 0.25 mg/l KN on incubation in dark condition for initially 2-3 subcultures at 26° C (Raj Bhansali, 1989 and Islam *et al.*, 1994). The cellular embryos were in various stages of development but recovery of plantlets was very rare on medium containing 2,4-D. However, Sung and Song (1992) have reported induction of somatic embryogenesis in Chinese jujube (*Ziziphus jujuba* Miller) from immature cotyledons and seedling parts on MS containing 2.5 mg/l 2,4-D, BA and 50% sucrose and complete regeneration on 1 mg/l zeatin and 0.01 mg/l 2,4-D. Embryos developed into rooted plantlets on medium supplemented with 2-5 mg/l BA without auxin under 16 hr photoperiod. Cell lines resistant to *Fusarium oxysporum* and *Colletotrichum* were also developed in *Z. rotundifolia* rootstock (Raj Bhansali, 2001). Mitrofanova *et al* (1997) also induced somatic embryogenesis and regenerated complete plants of *Z. jujuba* (Chinese date) on half strength MS with 1 mg/l 2,4-D through direct somatic embryogenesis. The plantlets were regenerated on modified Pierik medium without hormone and acclimatized in *in-vitro* conditions.

3. Pomegranate:

Pomegranate (*Punica granatum*), of Punicaceae family is a favorite table fruit in India. It is popularly known as Anar or Dalima or Dadima in India. It was brought to India from the Middle East via Persia or Afghanistan, presumably in the first century. The tree is large shrub with cylindrical branches and spines. It acclimatized very well in the tropical and sub-tropical regions. The fruit is very nutritious and refreshing. The edible portion is the seed aril, which contains mineral matter, particularly phosphorus and iron, besides thiamine, riboflavin and vitamin C. It is also highly valued for secondary metabolites of pharmacological interest.

Pomegranate cultivation became commercially important in Maharashtra, Gujarat and Rajasthan, Andhra Pradesh, Uttar Pradesh, Karnataka and Tamil Nadu. The prominent cultivars are Ganesh and Musket in Maharashtra, Basin seedless in Karnataka, Dholka in Gujarat and Jalore seedless in Rajasthan. Jodhpur-Red has occupied considerable area in Rajasthan but its seeds are hard. Plants can be easily multiplied by cuttings or by air layering. Hard cuttings taken from fully matured wood of one-year-old plants are most suitable for vegetative propagation. Dipping of stem cutting in 200 ppm IAA has been found helpful to induce profuse rooting.

3.1 *In-vitro* propagation:

3.1.1 Organogenesis:

Micropropagation of a dwarf variety of pomegranate has been reported using stem portion (Omura *et al*., 1987a), leaf explant (Omura *et al*., 1987b), cotyledons, hypocotyls and embryos (Jaidka and Mehra, 1986; Raj Bhansali, 1991). Additional tissue cultures of anther (Moriguchi *et al*., 1987) and petals (Nataraja and Neelambica, 1996) for haploid and somatic embryogenesis have been reported. Browning of explants and media is the biggest problem to initiate pomegranate culture from mature plants. The growth of cultures could be maintained successfully by using charcoal containing media followed by quick subculturing on fresh media.

The apical and nodal segments of pomegranate can be used for clonal multiplication of Jalore seedless and Ganesh by using 0.5-1.0 mg/l BA and 0.1 NAA or without auxin. Higher concentration of BA (2-3 mg/l) induced multiple shoot formation from nodal bud as well as from apical tips. Elongated shoots initiated roots from the basal cut end when soaked in 200-500 mg/l IBA and then cultured on medium without auxin. In this way, a large number of plants could be regenerated, hardened and successfully grown in the fields.

3.1.2 Somatic embryogenesis:

It has been reported in seed explant and petal tissue (Raj Bhansali 1990; Nataraja and Neelambica, 1996) that embryogenic cell lines can be selected by using staining techniques developed by Raj Bhansali *et al.* 1990 for peach somatic embryogenesis (Fig. 8).

Embryos produced roots after 8 weeks on MS medium containing 5 mg/l IAA, IBA or NAA. However, shoots were developed when embryogenic cultures were placed on medium having 5 mg/l BA or KN (Fig. 9). Embryos rapidly proliferated into secondary embryos on medium containing 5 mg/l BA and 0.1 mg/l IAA (Raj Bhansali, 1990). Complete plantlets were obtained on half strength MS medium containing 1 mg/l IAA or IBA. Large number of fully developed plantlets could be produced by using nodal bud segments or via somatic embryogenesis in several pomegranate species, including the ornamental types.

4. Bael

Bael (*Aegle marmelos* Linn. Corr.). is a hardy fruit tree that can be grown in arid climate. It belongs to Rutaceae family. The tree is deciduous and medium sized, with characteristic trifoliate leaves. It has both mythological (commonly grown in temple or religious compounds) and medicinal significance (good for stomach ailments, diarrhoea and dysentery). The ripe fruit is a tonic, laxative and good for heart and brain. The fruit has high nutritive value, having riboflavin, carotene, niacin and vitamin C.

Bael tree is easily grown even in acidic, alkaline and stony soils with pH ranging from 5 to 10. There is no standard name for bael cultivars. They are known by the name of the city, such as Mirzapuri, Rampuri, Kaghji Gonda, Kaghji Etawah, Deoria large and Basti No.1. The common method of propagation of bael is through seeds, which are sown in June and seedlings become ready for transplantation a year later. The seedling trees are not true-to-type. Plants can also be propagated by means of budding, grafting, and air layering. Budding (T-budding, patch budding, chip budding) is most popular for raising saplings for plantation in new orchards.

4.1 *In-vitro* propagation:

4.1.1 Organogenesis:

Micropropagation of selected trees enables rapid propagation (and hastens the availability) of new cultivars. Regeneration from seedling or seed explants tissues such as hypocotyl (Arya *et al*., 1981, Hossain *et al*., 1993a), nucellus (Hossain *et al*., 1993b), cotyledons (Hossain *et al*., 1994), leaf (Islam *et al*., 1992) zygotic embryos (Islam *et al*., 1994, 1996a) have been reported. High frequency of plant regeneration from radicle tissues has been achieved on 1 mg/l BA and 0.2 mg/l NAA for shoot proliferation and elongation on low BAP (0.1 mg/l). Rooting up to 80% in regenerated shoots has been obtained on medium containing 25 mg/l IBA for a week period. The number of shoots increased to 75.2 shoots/cotyledonary explant tissue on medium containing 3 mg/l BA (Arumugam and Rao, 1996). Shoot production could further be increased using nodal buds of *in vitro* regenerated shoots. Production of up to 12376 shoots has been reported from single explant. Hazarika *et al* (1996) have reported multiplication of shoots by using microshoots. Similarly, stem explants produced meristemoids and multiple shoots on MS medium containing BA, KN and NAA.

Biochemical changes, particularly in the composition of membrane lipids, in relation to differentiation in callus cultures have been reported (Bhardwaj *et al*., 1995). Decrease in phosphatidylglycerol content and presence of 3-trans-hexadecenoic acids in phosphatidylglycerol was found related to greening and shoot bud differentiation. Several workers have reported successful regeneration of complete plantlets and transferred to soil (Hazarika *et al*., 1996: Arumugham and Rao, 1996; Islam *et al*., 1996a).

4.1.2 Somatic embryogenesis:

Asexual embryogenesis and regeneration of plantlets have been reported using seeds obtained from unripe fruits of 20-year-old *A. marmelos* tree (Islam *et al*., 1996b). In this case, embryonic callus has been established from zygotic embryos on a medium supplemented with 2,4-D and BA (1μm). Concentration of BA above 1 μm had inhibitory effect on the induction of somatic embryos.

5. Khejri

Khejri (*Prosopis cineraria*) is one of the most important leguminous trees of the Indian desert. This is highly esteemed and valued tree of the dry forest ecosystem as a renewable source of energy and biomass (Leaky and Last, 1980). It is a multipurpose tree, mostly used as food (vegetable and dry fruit), fodder, and fuel (Paroda, 1979 and Raj Bhansali and Jindal, 2000). It enriches the soil with nitrogen and provides canopy covers to the grain cereal and legume crops of the region. Khejri is now available only in protected land areas. It is also a source of gum and other products. Above all, it combats the drought and counters the salinity and alkalinity. It tolerates both high and low temperatures. *Prosopis* species are among the trees, which have very high rate of CO_2 fixation. It is no exaggeration to say that it is the backbone of rural economy of arid zone of India (Arya *et al.*, 1995).

5.1 *In-vitro* propagation:

Plant tissue culture provides an excellent system to study the factors involved in vegetative multiplication and the possibilities of obtaining improved plants by genetic engineering. Vegetative multiplication through tissue culture has emerged as a major applied aspect of plant tissue culture research (Murashige, 1974; Bajaj, 1986; Ramawat and Nandwani, 1991; Raj Bhansali, 2001a). Now, a large number of research groups are working on plant biotechnology for the mass regeneration *Prosopis* species through plant tissue culture (Table 3). This could be an alternative mean for afforestation of arid lands. *Prosopis* species exhibit a wide range of characters due to their heterozygous nature. It is essential to select fast growing individual trees with desirable additional characters viz., high leaf protein, absence of protease inhibitors, thorniness/thornless, fast growing foliage and better in timber quality. The tree shows moderate to high salinity resistance. But, these plants are frequently attacked by variety of insects and root rotting (*Ganoderma lucidum*) and wood rotting (*Fomes* sp. and *Phillinus* sp.) fungi. After careful germplasm selection, improvement programme for pest and disease resistance can be initiated by conventional breeding methods or in combination with plant tissue, cell and protoplast culture techniques.

A. Seedling explants

Various parts of young seedlings have been used for tissue culture. Goyal and Arya (1981) have reported culture from hypocotyl segments (0.8 cm to 1.0 cm) from seven to ten days old *in-vitro* grown seedlings on MS medium containing kinetin and NAA/IAA. Shoot bud differentiation was recorded in these explants. These shoots developed roots on basal White's medium containing IBA (3.0 mg/l). However, this technique is not applicable for multiplication of mature trees. Further, Goyal and Arya (1984) employed lateral buds leafy shoots of a seven-year-old *P. cineraria* strain (CAZRI elite selection) for initiation of tissue cultures. Multiple shoot differentiation was achieved from such explants on MS medium containing 3.0 mg/l IAA+0.05

mg/1 kinetin (Fig. 10). The developed shoots produced roots on White's basal medium containing IBA (3.0 mg/1) and kinetin (0.05 mg/1). Cotyledon, epicotyl, hypocotyl and apical bud explants excised from seedlings were suitable for multiple axillary shoot regeneration in *P. cineraria* (Puri *et al.*, 1992; Nandwani, 1990; Nandwani and Ramawat, 1993; Raj Bhansali, 2001). A maximum of 11 shoots was obtained per explant obtained from 2-week-old seedlings on MS medium supplemented with 0.5 mg/l IAA and 2-5 mg/l BAP/kinetin (Raj Bhansali, 2001). A greater number of healthy shoots was produced from brown, spiny and woody explants than from green and young stem explants. Regeneration has not been reported from callus. Explants of different genotypes varied in their morphogenetic response *in-vitro* and produced varied number of shoots on the same medium. Regenerated shoots were rooted on MS medium containing 5 mg/l IBA. Plantlets transferred to pots survived well under field conditions (Nandwani and Ramawat, 1993).

B. Mature adult tree

1. Nodal segment

In-vitro multiplication of shoots from mature trees is highly desirable for mass propagation of selected tree. Several workers have reported methods for rapid propagation of *P. cineraria* through axillary branching. Single node segments from actively growing branches of elite trees (4.5-years-old) have been cultured on MS basal medium containing 3 mg/l each of 2-naphthoxyacetic acid (NOA) and NAA (Shekhawat *et al.*, 1993; Raj Bhansali, 2001). They produced several axillary shoot buds within 7-10 days when 2 cm shoots were transferred to modified MS medium containing 3 mg NOA/l (Kackar *et al.*, 1991 and 1992). Eighty per cent of shoots showed rooting to a length of about 8 cm within 30 days. Such shoots were then cut into 5-7 segments and planted, individually, on MS + NOA supplemented medium. Each segment produced a plantlet. By following this procedure, 5-7 fold multiplications of plants could be achieved within a month (Fig. 11) thirty per cent of plants survived after transplantation.

2. Root segment

Root segments from 3-4-month-old seedlings were also used for production of multiple shoots on MS medium containing 3 mg/l NOA (Kackar *et al.*, 1992). Shekhawat *et al* (1993) have also used root segments for development of multiple shoots on MS medium containing 0.5 mg/l IBA + 2.5 mg/l BA. However, all these workers describe methods for propagation of young trees no older than 7- year- old.

The main problem is that these methods cannot be applied to multiply field-tested old tree, which were selected for various purposes. Moreover, selection in *P. cineraria* is not possible for any character at ages less than 12-15 years because of its slow expression of characters in nature (Shekhawat *et al.*, 1993; Raj Bhansali, 2001). Several intrinsic and extrinsic factors influence the behaviour of explant in culture. The age of the tree, lopping season, harvesting period and sterilization procedure have more influence than the medium. Various factors affecting *in-vitro* clonal propagation of *P. cineraria* have been studied in detail (Shekhawat *et al.*, 1993). Genotype, age of tree, nature and size (length and diameter) of explant, season of collection, explant position on medium, plant growth regulators and certain additives (ascorbic and citric acids, adenine sulphate, L-arginine, glutamine and ammonium citrate), incubation conditions, and frequency of subculturing period greatly influenced *in-vitro* clonal propagation of *P. cineraria* (Shekhawat *et al.*, 1993). A maximum number of 10-12 shoots were induced from the nodal shoot segments from pruned thorny adult trees on MS medium containing 0.1 mg/l IAA + 2.5 mg/l BA + additives. Differentiated shoots multiplied best on MS medium containing 0.1 mg/l NAA + 1.0 mg/l BA + additives.

Root development

Rooting in developed shoots is achieved by pulsing treatment with 100 mg/l IBA for 4 h and then culturing on hormone-free half strength MS medium. Initial dark incubation for 5 days at high temperature ($33 \pm 2^\circ$ C) was found essential for root induction and up to 63 % rooting was obtained within two weeks. The rooted plantlets contained a consistent number of chromosomes (2n=28). It is suggested that the protocol developed could be useful for cloning mature and trees of *P. cineraria* (Shekhawat, *et al.*, 1993). Large number of shoots (8-12 per explant) were harvested from 10-year-old spineless *P. cineraria* on the MS medium containing IAA, BA and other additives (Nandwani and Ramawat, 1991). Some of these shoots could be rooted but success is highly unpredictable.

Thus, more experiments are needed in order to define protocols, which can give successful and reproducible results on tissue cultures of *Prosopis* sp.

Table 3. Regeneration studies with various species of *Prosopis*

***Prosopis* species**	**Explant**	**Medium**
P. cineraria	Hypocotyl, Stem, Bud explant, Root	MS medium+ IAA (0.25mg/l)/NOA (3mg/l)/NAA(0.1mg/l)+ Kinetin (4.5mg/l)/BA (1mg/l)+additives
P. tamarugo	Hypocotyl, Nodal section, Shoot tip,	MS medium+ NAA (0.25mg/l)+ BA (2-5mg/l)
P. alba	Shoot tip, Lateral bud explant	MS medium+ IAA (0.25mg/l) BA (2-5mg/l)
P. chilensis	Shoot bud, Nodal segment, Shoot tip, callus	MS/B5 medium+ 2,4-D (2mg/l)+NAA (0.25mg/l)+ BA (2-5mg/l)
P. juliflora	Stem, Meristem, Shoot tip, Nodal segment	MS medium+ IAA (0.1mg/l)+ BA (2-5mg/l)

Other *Prosopis* species such as *P. tamarugo, P. chilensis* and *P. alba* have also been cultured on MS medium with 5-10 mg NAA/l and produced rooted plantlets (Jordan, 1985; Tabone *et al*., 1986; Green *et al*., 1990; Nandwani and Ramawat, 1992a &b) (Table 3). It has been observed that rooting can be induced in *P. tamarugo* and *P. juliflora* with relative ease as compared to *P. cineraria* (Nandwani 1990; Goyal and Arya 1984). The system requires more refinement for increasing frequency and subsequent viability of rooted plants in field conditions. However, in the case of *P. cineraria,* the culture medium, genotype and age of shoots markedly affect the rooting behaviour of *in-vitro* regenerated shoots. Therefore, in order to develop a reproducible rooting system, all factors must be taken into considerations. Once shoots are rooted, establishment of these plants does not pose difficulty except that initial low watering is required.

C. Callus culture

Callus was induced in three *Prosopis* species (*P. cineraria, P. juliflora* and *P. tamarugo*) on medium with several components including growth regulators and using various types of explants (Nandwani and

Ramawat 1989 & 1991). It is possible to induce and maintain callus culture of *P. cineraria* and *P. tamarugo* (Nandwani and Ramawat, 1989) and cell suspension cultures of *P. chilensis* and *P. tamarugo* (Jordan *et al.* 1987; Jordan 1988). However, all attempts to regenerate whole plantlet from the isolated and subcultured callus failed. It is interesting to note that Jordan *et al.* (1987) observed a single embryo/shoot apex like structure in cultures grown on B5 medium supplemented with 2,4-D. *P. chilensis* also formed differentiating calli. Production of embryo like structures has been reported in cell suspensions in a few cultures, but need to be confirmed. However, confirmation of somatic embryogenesis/organogenesis has not been reported from callus tissues, so far. Further efforts are urgently needed to overcome this problem and to open new avenue for biotechnological research in *Prosopis* improvement programmes. Tissue culture methods should also be attempted to overcome the problem of self-incompatibility and to generate variability in these species (Jordan, 1985). Efforts were also made to isolate protoplasts from *P. cineraria* (Shekhawat and Kackar, 1987).

Callus morphogenesis

It was reported that in *P. cineraria* once callus isolated from explants and maintained separately, it did not regenerate through somatic embryogenesis/organogenesis. However, from juvenile explants of *P. tamarugo*, in a few cases, highly regenerative tissues were obtained spontaneously. It was suspected that such a high regeneration could not be obtained without interface of callus morphogenesis. Histological observations of the regenerative mass showed shoot bud development from callus tissues in *P. tamarugo*. This is the first evidence of callus regeneration in any *Prosopis* species. Increased nitrogen content in the medium enhanced shoot bud proliferation while added inhibitors of auxin synthesis and auxin transport enhanced callus formation in the regenerative tissues (Nandwani 1990).

D. Bud sprouting

Bud break from nodal explants in different species of *Prosopis* can be obtained on diverse media (Goyal and Arya 1984; Jordan *et al.* 1987; Wainright and England 1987; Yao *et al.* 1989; Nandwani 1990).Murashige and Skoog (MS) medium containing high concentration of cytokinin (Nandwani 1990), auxin (Goyal and Arya 1984) or a combination of cytokinin and auxin (Batchelor *et al.* 1989) support shoot bud proliferation from stem explants of *P. cineraria*. These findings suggest that response is not precisely controlled by concentration and quality of exogenous growth regulators. Perhaps the requirement is quite different from that provided exogenously, but subsequent changes in metabolic level and endogenous level of growth regulators promote shoot formation. This conclusion is supported by the fact that, in most cases, only single shoot formation has been achieved. The complete bud stimulation in explant was lacking, thereby production of adventitious shoots remained suppressed (Nandwani and Ramawat, 1991).

P. cineraria has produced many evidences that explants, from either different or the same genotypes (as well as explants of same genotype) differ in their regenerative potential on same medium. These explants showed considerable variation in phenotypic expression in relation to delayed regeneration, size, length and vigour of regenerated shoots and leaf expansion (Nandwani, 1990). Thus, results obtained by various workers are mostly due to different material (genotype) used, making the reproducibility of results very difficult. Therefore, these studies require a detailed account of explant source used in the investigation. Some of these species are very recalcitrant to regenerate. In the case of *P. juliflora,* even several hundred permutations and combinations of medium factors could not produce subculturable callus (Nandwani and Ramawat 1989 & 1991). Similarly, shoot number could not be enhanced to more than seven in direct adventitious bud formation. Low number of shoot formation was also obtained with *P. chilensis, P. cineraria* and *P. juliflora* nodal explants by Yao *et al* (1989) and Nandwani and Ramawat (1989 & 1991).

E. Multiple shoot formation

In *P. tamarugo,* when various growth hormones are incorporated into MS medium, single shoot from hypocotyl explants can be induced. In some cases, hypocotyls grown on high cytokinin medium, when subjected to growth hormone without MS medium, produced multiple shoot buds at the base of developing shoot. Nandwani and Ramawat (1989 & 1991) have designed a sequence of experiments to evaluate the precise role of growth hormones, particularly cytokinin, in adventitious shoot formation. This has resulted in the formation of multiple shoot buds from hypocotyls in successive passages through hormone free and cytokinin supplemented media. Embryonic explants isolated from zygotic embryos produced multiple shoots when grown on hormone free or cytokinin-supplemented media. It was observed that proliferation of shoot buds increased in the second passage. Reduction in cytokinin concentration in the medium resulted in elongation of shoots from the clump of organogenetic mass. Elongated shoots produced roots on rooting medium containing auxin.

It has also been reported that if both cotyledons are left attached to the embryonic axis, multiple shoot formation in the explants is enhanced (Nandwani and Ramawat 1991). In absence of cotyledons, the number of shoots remained low. The growth of explant is delayed but the number of shoots increased by formation of adventitious buds. It appears that these buds developed adventitiously at a later stage and formed newly proliferating callus cells. In certain cases, apical meristems in axillary position proliferate to produce a phyllody like structure.

F. Cultures of insect gall tissues

Plant tissue culture methods have also been used for *in-vitro* studies of insect induced plant galls. Stem and rachis galls of *Prosopis cineraria* are caused by a *Chalcid* and *Lobopteromyia prosopidis [Contarinia prosopidis]*. The galled tissues were used for induction of callus and differentiation (Kant and Ramani, 1990). Laboratory studies were carried out to induce gall formation from larvae on the leaf rachis of germinated seedlings and callus tissues of *P. cineraria*. Gall induction took place on the leaf rachis of seedlings germinated *in-vitro*. Galls formed in tissue cultures were soft and lacked the differential structures, which are normally associated with the gall. Normal and gall callus inoculated with insect larvae showed nodule-like structures with high meristematic activity. This is the first report on *in-vitro* gall formation by a Cecidomyiid larva (Ramani and Kant, 1993).

B. Minor Fruit Trees

1. Lasoda

The genus *Cordia* has been (found to be of) difficult (nature in) to initiate of cell, callus and organ culture for mass multiplication (Raj Bhansali, 2001c). Recently, in Central American and Caribbean countries, the pioneer work in biotechnology of forest trees has been focused on the development of *in-vitro* micropropagation protocols for *Cordia* and many other woody plant species. Callus can be raised on MMS (modified Murashige and Skoog) medium containing 2-5 mg/l 2,4-D and 0.25 mg/l kinetin plus additives. Organogenesis was observed in very few callus tissues obtained from 7-day-old seedling explant on 5 mg/l BAP and kinetin after 3-4 subcultures. These activities have been linked to conservation of biological diversity of endangered and/or economically important species (Kanninen *et al.*, 1998). The CATIE (Centro Agronomico Tropical de Investigacion y Ensenanza), Costa Rica, as a regional research institution, is willing to promote the establishment and functioning of the proposed network of plant biotechnology. The activities of this proposed network is linked to international organizations such as IUFRO, FAO and IPGRI. Tissue culture work could be coordinated with other Latin American networks and South-East Asian countries where *Cordia* and other woody tree species are frequently grown.

Successful propagation of *C. alliodora* has also been made for the vegetative propagation of selected clones of woody tree species grown in Costa Rica. The material used was a juvenile cutting with leaves, about 5 cm long, derived from shoots of hedges (Mesen *et al.*, 1992). In India, a large genetic variability exists in different species of *Cordia,* in varied ecological zones. This could be exploited for improvement and development of superior varieties by using biotechnological tools. Protocols need standardization in order to develop quick multiplication methods through proliferation of axillary or apical buds because various problems have been encountered during the tissue culture process of *Cordia* species. Organogenesis and somatic embryogenesis processes can be attempted for the propagation of high yielding genetically superior clones of *C. myxa*. Its leaves are often infected with fungal diseases and weevil on fruits causes holes in stones. These are the problems most frequently found in *Cordia* plantation. Somaclonal variation could be exploited to generate variation with regard to pest and disease resistances. The application of tissue culture technology will not only help in mass multiplication but also in conservation of selected germplasm of *C. myxa* and *C. gharaf* for breeding purposes.

2. Ker

Ker or caper (*Capparis decidua*) is an important arid zone fruit plant, which offers an ample scope for conserving large gene pool available in the arid regions for the improvement. The application of tissue culture technology will not only help in *in-vitro* conservation but also in rapid multiplication of selected superior germplasm. The major concern is on how to develop high yielding varieties, which can produce green, soft seeded/seedless small fruits through conventional and biotechnological approaches (Raj Bhansali, 2001b; Raj Bhansali and Singh, 2000). There is considerable scope for the use of molecular markers to assess the genetic variability and characterize the genes of soft seeded germplasm. Recently, wide variability was recorded in desert areas of western part of Rajasthan. Plants with yellow flowers and large fruits have been marked for *in-vitro* conservation (Fig. 12). Efforts are being in progress to develop methods of somatic embryogenesis from single cell to transfer desirable genes in caper plants.

The role of tissue culture in clonal propagation of capers is limited to three major areas: (a) rapid multiplication of selected elite material; (b) *in-vitro* conservation of genetic variability; and (c) micro grafting of scion to rootstock cultivars. Ever since the first report of Ancora and Cuozzo (1985), *in-vitro* propagation of *Capparis spinosa* L. has generated extensive interest on developing method for mass multiplication. Cuttings and fruits taken from field-grown plants, buds sprouted in a growth chamber and *in-vitro* grown seedlings (nodal explants, cotyledonary nodes, cotyledons and hypocotyls) have been used (Ancora and Cuozzo, 1985, Rodriguez *et al.*, 1990, Raj Bhansali, 1994 and Tyagi *et al.*, 1997).

Raj Bhansali (2001) has described the shoot multiplication and organogenesis procedures for fruit and forest trees of arid zone. Development of new composition of media, rapid methods for callus induction, shoot multiplication and rooting of micropropagated shoots of *C. decidua* have been reported (Raj Bhansali, 1994 and 2001, Deora and Shekhawat, 1995 and Tyagi *et al.*, 1997) (Fig. 13). A 20-fold multiplication in every 20 days was achieved in *C. spinosa* L. by culturing nodal shoot segments Murashige and Skoog (MS) medium containing BA, IAA and GA_3 (Rodriguez *et al.*, 1990). Various concentrations of BA(1, 2.5 and 5 mg/l) in combination of KN 1 mg/l promoted the development of multiple shoots and shoot buds in the primary cultures of *C. decidua* within 30 days (Fig. 14). Initially, the medium containing BA (5 mg/l) was best for shoot bud proliferation from the nodal as well as from seedling explants. The shoot multiplication process was quicker in cotyledonary nodes. With nodal explants, shoot multiplication was best on medium supplemented with 5 mg /l BAP and, after a second and subsequent subculturing, further multiplication of shoot buds was highest on the medium containing 3 mg /l BA (Raj Bhansali, 1994 and Tyagi *et al.*, 1997).

Rapid plantlet regeneration can be induced on shoots taken from proliferating clusters subcultured for 20 days on a medium with reduced BAP (2 mg/l) without auxin or gibberellin. The resulting shoots were subcultured *in-vitro* on BAP supplemented MS medium for secondary shoot proliferation. Primary and secondary axillary shoots developed on medium containing BAP + GA_3, (and less IAA) can be used for root initiation on rooting medium. An increase in the content of $CaCl_2$ over that in the original medium had a beneficial effect on leaf expansion. Proliferation was mainly due to axillary shoot bud development as revealed by histological studies. The extensive meristematic activities observed indicated the enormous morphological potential of this species.

Higher rooting responses (70 %) were obtained after a 20-day incubation period in darkness on semi-solid half-strength modified MS (MMS mineral salts with 0.5 mg/l m-inositol and 1 mg/l thiamine) medium plus IAA (1 mg/l). Rooted shoots were further subcultured for 20-30 days on half-strength MMS basal medium to increase root development. Shoots separated from primary/ secondary cultures were used for rooting. Rooting was best achieved in medium containing 1 mg/l IBA. Rooted plantlets were transferred to pots with garden soil and peat moss mixture (Tyagi *et al*., 1997).

3. Custard apple

Custard apple (*Annona squamosa)* is a medium sized tree, mostly cultivated in arid regions of Tamil Nadu, Karnataka, Andhra Pradesh and south Rajasthan. It can thrive in dry condition and also withstand frost to some extent, but fruit ripening is inhibited during winter. Several reports have described *Annona* tissue culture by using nodal, leaf, endosperm, anther (Nair *et al*., 1983; Nair *et al*., 1984a,b; Nair *et al*., 1986; Bejoy and Hariharan, 1992; Lemos and Black, 1996a). Haploid plants were induced from anther-derived callus on Nitsch medium supplemented with BAP or NAA. Initially, tissue culture medium with high sucrose content incubated in the dark, followed by low sucrose in light are required to induce haploid plants (Nair *et al*., 1984b). Endosperms excised from 2-4-day-old seedlings produced calli on White medium. Shoot and root developed after subculturing on Nitsch medium. The resulting plants showed triploidy (Nair *et al*., 1986). Micropropagation with complete protocols have been reported in *A. sequamosa* using hypocotyl of seedling and nodal cuttings from 3-year-old plants on woody plant medium with BAP (Lemos and Black, 1996b). Silver thiosulphate (0.5 mg/l) was added to control abscission. Shoot produced roots when cultured first on medium with 43 μM NAA or 39 μM IBA and then placed on activated charcoal (10 g/l). Galactose instead of sucrose improved rooting and the plantlets acclimatized easily in soil. Similarly, *A. muricata* can be micropropagated from either juvenile or mature explant tissues (Lemos and Black, 1996a).

1. Jamun

Jamun (*Syzygium cumini* Skeel) belongs to Myrtaceae family and is generally grown along roadsides or adjacent to orchards. Production of multiple shoots from nodal and shoot-tip segments of young seedlings has been reported in culture on MS with BAP alone or in combination of auxins (Yadav *et al*., 1990; Raj Bhansali, 1994, 1998b) (Fig. 15A&B). Shoots developed roots on medium containing NAA and or IBA and then successfully transferred into soil. Roy *et al* (1993) has reported high frequency shoot formation and plant regeneration from mature embryos.

IV. REQUIREMENT OF LABORATORY FACILITY

1. Plant tissue culture room: Temperature (22-25° C) and photoperiod are regulated. Lighting must be bright (1-5 W/m^2), discontinuous (16hr of light/dark or complete dark condition) and provided by fluorescent tubes (40 watt).

2. Explant from actively growing tips of plants, seeds, callus or cell suspension.
3. Culture medium: The mineral solution of Murashige and Skoog/Gamborg/Woody plant/Raj Bhansali media can be used for micropropagation and initiation of embryogenic cultures (Table 4).
4. Medium should be fixed at pH 5.5-5.7 before autoclaving. In addition, it should contain NH_4^- and K^+ ions and auxins (2,4-D/ IAA/NAA) and cytokinins (2ip/BAP/kinetin/zeatin/TDZ), ABA, complex addenda (CW, ME and CH) and activated charcoal for induction, development and maturation of somatic embryogenesis. For maturation of embryos, half-strength basal medium containing sucrose at 5g/l is helpful.
5. Sieve (1-mm grid) made of stainless steel screen to fraction suspension culture. Transfer only small developing cellular embryos into the new flask having fresh medium. Sieve can also be made from nylon mesh filtration cloth.
6. Subsequent plantlet regeneration requires full, half, or one forth strength media with filter paper bridge/semi-solid medium with and without hormone and then conditioning, hardening and habituating in small pot containing soil mixture.

VI. GENERAL PROTOCOLS FOR MICROPROPAGATION

1. Processing of explant material:

1. Collection of explant material.
   a. Collect explant material from seedlings raised from seeds.
   b. Select and cut young twigs from mature elite tree having desired characters.
   c. Collect young axillary buds along with nodal portions from long, well developed twigs.

Table 4. Selected plant culture media for micropropagation of arid zone fruit plants

Compound	Murashige & Skoog	Woody plant Medium	Gambor g B 5	Raj Bhansali
Concentration (mg /litre)				
NH_4NO_3	1650	400	-	275.38
KNO_3	1900	-	2500	2337
$Ca(NO3)_2.\ 4H_2O$	-	556	-	-
$(NH_4)_2SO_4$	-	-	134	-
$MgSO_4.\ 7H_2O$	370	370	250	185
$CaCl_2.\ 2H_2O$	440	96	150	220
KH_2PO_4	170	170	-	85
$NaH_2PO_4.\ H_2O$	-	-	150	-
$K_2\ SO_4$	-	990	-	-
$FeSO_4.\ 7H_2O$	27.8	27.8	27.8	13.92
Na_2 EDTA	37.3	37.3	37.3	18.62
$MnSO_4.\ H_2O$	-	22.3	10	-
$MnSO_4.\ 4H_2O$	22.3	-	-	8.4
$ZnSO_4.\ 7H_2O$	8.6	8.6	2.0	4.3
H_3BO_3	6.2	6.2	3.0	3.1
KI	0.83	-	0.75	0.415
$Na_2MoO_4.\ 2H_20$	0.25	0.25	0.25	0.125
$CuSO_4.\ 5H_2O$	0.025	0.25	0.025	0.0125
$CoCl_2.\ 6H_2O$	0.025	-	0.025	0.0125
myo-Inositol	100	100	100	500
Nicotinic acid	0.5	0.5	1.0	0.5
Pyridoxine HCl	0.5	-	1.0	0.5
Thiamine HCl	0.1	1.6	10.0	1.0
Glycine	2.0	-	-	2.0
L-Glutamine	-	-	-	500
Casein hydrolysate	-	-	-	500
Sucrose	30,000	20,000	20,000	30,000

- Commercial preparations of M&S and B 5 salt (minus sucrose glutamine and hormones) are available in packets that make up 1 L of media.
- pH adjusted using 0.1M HCl and 0.1M KOH before agar added and prior to autoclaving.
- The amount of agar needed varies depending on the brand used.
- Varying the hormones allows the above media to be used for micropropagation of wide range of arid zone species.

2. Remove the leaves from the twigs containing shoot buds, cut them into segments of about 2.5 - 5.0 cm.
3. Transfer the buds to a sterile 250 ml conical flask and surface sterilize the explants. The surface sterilized shoot and axillary buds are kept in sterile water till they are used for culture.
4. Treat the explant material with antioxidant solution, if browning problem exists.
5. Sterilize Petri dishes, scalpel and blades, forceps, filter paper and medium.
6. Transfer plant materials into sterile Petri dishes with the help of sterile forceps and cut the explants into small pieces each containing axillary buds/apical buds in laminar air-flow system.
7. Inoculate 2 pieces to each tube or flask on sterilized medium.
8. Incubate the cultures in an environmentally controlled growth chamber at $25 \pm 2°$ C at 500 lux light intensity for 3 to 5 days.
9. Keep the cultures in growth chamber at $25 \pm 2°$ C and 15000 lux light intensity and 16 h photoperiod.
10. After 3-4 weeks, the axillary buds start sprouting.
11. 10-15mm long shoots are transferred into fresh medium and incubated at $25 \pm 2°$C.

    Or

    Sprouted buds are cultured in liquid medium in 100 ml flask for multiple bud initiation. Incubate the flask on shaker at 120rpm and 500 lux light intensity.
12. Multiple shoots are formed and subcultured after 3-4 weeks of incubation.

2. Multiplication of shoots:

1. Take out the multiple shoots from the flask to sterile Petri dish; separate the shoots under sterile conditions in laminar air-flow cabinet and cut them into pieces containing 1 or 2 nodes with axillary buds.
2. Inoculate 1 or 2 pieces on fresh medium.
3. Incubate the culture at 1500 lux light intensity with 16 h photoperiod at 25°C.
4. 6-10 shoots/explant tissue can be observed after 3-4 weeks in culture.
5. Separate the shoots aseptically and culture them on rooting medium.
6. Incubate in dark at 25° C for 2-5 days, depending upon explant of particular species.
7. Again, incubate at 25° C under 1500 lux light intensity with 16 h photoperiod.
8. Formation of roots observed within 3-4 weeks.
9. Remove rooted plants after 15 days and transfer them to polythene tubes/pots containing sterilized sand and soil mixture (1:1).

10. Cover the plastic polythene tubes/pots with empty polythene tube or glass beaker on plastic pots to increase humidity and prevent desiccation.
11. Incubate the transferred plants in polythene tubes/pots in a cabinet at 25° C with 15000 lux light intensity 16 h photoperiod.
12. A number of rooted plantlets surviving in pots/polythene tube and also in the field soil can be obtained.

3. Somatic embryogenesis:

Generally, three steps are required to obtain somatic embryogenesis in plant tissue cultures of arid zone plants (Table 5).

1. Induction of somatic embryogenic cultures from explants/callus tissues on agar medium with high auxin content.

2. Repetitive embryogenesis on medium with low auxin in agar semi-solid medium or liquid medium for cell suspension.

3. Transfer of developing somatic embryos into a medium without hormone for development of embryos and plantlet regeneration.

I. Induction:

Growth initiation for embryo induction is better on solidified medium containing (2,4-D, 1-5 mg/l). The embryogenesis occurs directly on the explants or the callus induced from them within 3-4 subcultures by initially incubating in dark phase.

IIA. Establishment:

1. Embryogenic cultures should be removed from flask or tube with sterile forceps and transferred to Petri dish containing filter paper. Discard the agar medium and browning parts of callus and inoculate a fragment of approx. 0.25-2g in 100-250 ml Erlenmeyer flask containing 50 ml liquid or 30 ml semi-solid (agar) medium. Smaller Erlenmeyer flask can be used, but the volume of liquid medium in relation to the size of flask must be kept for adequate aeration. About 20 % of the volume of the flask can be used for medium for suspension culture. These flasks are placed on gyratory or orbital incubating shakers and agitated at 100-150 rpm for aeration and growth of suspension culture.

2. When plant material is first placed on medium, there is an initial lag period prior to cell division. After 2 or 3 sub-cultures, an exponential rise

Table 5. General protocol for somatic embryogenesis and plant regeneration

Stage/process	Factors
1.Selection of material	Genetic base, zygotic tissues, leaf, petal, shoot tip and stage of development
2. Conditions	Light, dark, low temperature treatment
3. Media	Conditioning, inductive, sequence, frequency of transfer, charcoal, complex addenda, growth regulators
4. Somatic embryogenesis	Direct, adventitious, repetitive, indirect, suspension of cells, protoplasts, screening
5. Somatic embryos	Maturation, selection, dormancy, cold treatment, abnormalities, growth requirement
6. Germination	Primary root and shoot, cotyledon, expansion and greening, epicotyl growth, leaves
7. Propagation in pot/soil	Transfer to non-sterile environment, dilute nutrient medium, day length, ambient humidity, fungicides
8. Acclimatization	Gradual leaf development in ambient temperature and humidity, hardiness to atmospheric fluctuations
9. Evaluation	Test for clonal or variant characters, selection and screening of plants

in growth of cultures follows. Finally, the cells in suspension culture enter in a stationary phase. In order to maintain viability of the cultures, the cell should be sub-cultured at the beginning of the stationary phase. Usually, it is obtained within 2-3 weeks. The suspension has to be transferred to fresh medium at regular intervals within this period, keeping a minimum density to maintain embryogenic potential.

3. To subculture embryogenic suspension, wait several minutes until cells settle at the bottom of the flask and decant almost all the medium. Re-suspend the culture by gently rotating the flask and transfer 1/4 of the population into fresh medium by using micropipet.
4. A population of cells/callus tissues shows a wide range of pro-embryogenic stages visible under the microscope. It is a mixture of single cells, pro-embryogenic clusters and new embryogenic centers and old embryos (Fig. 16). To obtain a certain degree of uniformity, it is necessary to sieve the inoculum before transferring to fresh medium.

The pro-embryo suspension is passed through a 200 μm and 100 μm sieves. In the second sieve, differentiated embryos from late globular stage are retained and may be microscopically examined, while pro-embryonic cells pass through. The cell suspension mass which has passed through the sieves is settled. Decant most of the medium to obtain a suspension with high density of embryogenic cells.

IIB. Maturation:

The embryo inoculum at the globular stage is transferred into a medium containing the same composition but with either reduced level of auxin and L-glutamine or completely lacking both. Cells quickly exhibit their embryogenic potential. Different developing stages of somatic embryos appear if initial inoculum had non-uniform embrygenic cells. They attain mature stage within 2 wk. These embryos can be dispersed in a Petri dish on agar medium without auxin, but with cytokinin (0.1-1.0 mg/l of BA or 2ip or zeatin).

III. Germination:

Mature embryos are placed in tubes either on filter paper bridge or on semi-solid medium with low concentration of sucrose (5 mg/l). To stimulate the formation of leafy shoot apices, the tube must be well illuminated. The germinated seedlings with 5-7 leaves and well developed roots are placed into pots with a mixture of peat, soil, and vermiculite for subsequent development and making hardy to regenerated plantlets. After their transfer, plantlets have to be protected from desiccation by covering the pots with plastic film for 2-3 wk.

VII. FUTURE PROSPECTS

Improvement in productivity, profitability, stability, and sustainability of arid zone farming system is a major mandate to be achieved by using modern methods of agricultural sciences. Arid and semi-arid regions of India have wide biodiversity and diversity of agroecological situations, and offer some distinctive features of great potential for growth and development of horticultural crops. A rich gene pool consisting of many economically important popular varieties of ber, pomegranate, date palm, custard apple, anola, fig, bael and ker is available in various regions of the country as field gene bank. Arid zone horticulture is one area where biotechnological breakthroughs can yield the fruits more efficiently, particularly in mass cloning of superior fruit trees, free from pest and diseases. By applying advanced technologies, including post-harvest management, and fruit processing, packaging industries can further help in boosting returns. Value addition to products would help in increasing economy of rural areas by contributing in the form of increased earning in foreign exchange by export, which is now an urgent need of the nation. The agri-horti business will not only improve the income opportunities but also generate employment for rural and urban masses living in arid and semi-arid areas. This makes it imperative that a long-term strategy for biotechnological research be formulated to identify specific priorities to achieve goals, proper approaches to attain them quickly and assess the total requirement of infrastructure and financial inputs.

Until recently, only a very small area was under fruit cultivation in the arid regions of India. Various technologies such as development of promising cultivars, selection of rootstocks, raising of nursery plants, and vegetative propagation methods including micropropagation and plant protection measures developed by Central Arid Zone Research Institute (CAZRI) have increased more and more agri-horti business in this region. Many of these technologies are commercially viable and large number of nurseries has been established in arid region. They are now supplying the plants throughout the India. But, to meet out the growing demand, the present production level has to be increased 3-5 folds over the next 25 years. In spite of climatic constraints, arid regions offer excellent potential to gainfully exploit indigenous as well as introduced horticultural plants. Conservation of the available gene pool by using *in-vitro* conservation methods will be required to protect them from extinction and for breeding of new desired varieties.

The perspective plan for horticultural development in arid regions would be through incorporation of modern scientific technology including biotechnology for the improvement in productivity of ber, pomegranate, date palm, citrus, custard apple, ker and bael for growing under prevailing abiotic and biotic stresses, and optimization of the use of water and solar radiation for high productivity. Most of arid zone fruit plants have become national and internationally important owing to their rapid spread in arid and semi-arid areas, by providing income sustainability even in poor or degraded lands. These fruit crops are drought resistant, have low cost of production and are available in the market when other fruits are scarce.

Until now, woody plants have been micropropagated almost exclusively by proliferation of axillary shoots from nodal buds or apical-tip explant tissues. Though these *in-vitro* techniques are likely to continue to dominate in the production of arid zone woody fruit trees, future application of tissue culture technology will be broadly based on developing simple, rapid and economically viable methods. Rooting of micropropagated shoots directly under mist or fog can be used in the future to reduce space required in growth room, there by curtailing the number of steps in which strictly aseptic condition are needed. It has an additional advantage of enabling the shoots to acclimatize at the same time that they are rooting. This could be a highly efficient micropropagation technology.

Research efforts are needed to simplify or automate harvesting of shoots in the future. They are needed to reduce the number of transfers and synchronize proliferating cultures for mass cloning of desired plant material in a short period. Somatic embryogenesis has been used by A.V. Thomas Group (1998) for commercial production of date palm. This group is professionally managed and has a commercial plant tissue culture unit for fruit and plantation crops in the country, based at Cochin Export Processing Zone. A few other commercial (R& D labs) biotech industries are rapidly being established as joint ventures with multinational companies in the country. Though asexual embryogenesis will continue to be a minor but rapid method until the requisite field-testing of the fruit plants are completed, which may requires many years. However, the technology developed for less economically important plants or needed for reforestation in arid and semi-arid or wastelands, then somatic embryogenesis could become a major micropropagation technique.

Eradication of bacteria from the explants from mature woody fruit plants of date palm, ber and pomegranate is difficult even by treating with mixtures of antibiotics. Most antibiotics tested have caused tissue damage at concentration effective to control the bacteria. A mixture of antibiotics (streptocycline and penicillin) has been found effective to eliminate bacteria from date palm explants from offshoots (Raj Bhansali and Singh, 1982; Dass *et al.*, 1989). Most bacteria isolated from *in-vitro* grown arid zone plants are non-pathogenic. But, it is not desirable to maintain contaminated cultures, regardless of the organism.

The cost of chemicals, labour and temperature controlled culture room are major factors contributing to the cost of micropropagation via tissue culture technology. Reducing cost of production by improving efficiency of the technique will result in added improvement in many facets of micropropagation. Year-round utilization of tissue culture facilities for other cash crops is essential to maximize return from capital investment required for establishment and maintenance of tissue culture laboratory in arid climatic conditions. High valued crops having export potential will generate more income and commercially viable venture.

VIII. CONCLUSION

The development of tissue culture systems for the micropropagation and improvement of arid zone fruit plants has been restricted due to non-availability of efficient and economically feasible technology. The feasibility of plant regeneration via somatic embryogenesis from tissue cultures of the highly valued date palm species has been demonstrated by several workers in India and abroad (Sharma *et al.*, 1986; Raj Bhansali and Kaul, 1991), but field performance has to be undertaken extensively before starting their commercial cultivation. The tissue culture techniques are highly advantageous for in- vitro conservation and international exchange of sterile disease-indexed material. The embryo-rescue or *in-vitro* fertilization methods can complement conventional plant breeding approaches. The exploitation of *in-vitro* propagation techniques may lead to the upgrading not only the newly introduced and disease-indexed horticultural plants, but also the recalcitrant or intractable woody tree species to production in commercial scale. *In-vitro* propagation of elite selections may be one solution for economic exploitation of fruit crops of arid lands.

A significant potential benefit of *in-vitro* approaches to improvement of tropical trees also exists in the use of mutant cell selection techniques and somaclonal variations for inducing disease resistance and as a means for lowering genetic vulnerability of clonally propagated fruit trees, without altering unique characters of elite selections (Jain *et al.*, 1997). This would greatly enhance the effectiveness of tropical arid zone fruit tree improvement schemes, particularly in the face of constant disease and pest, and arid environmental stresses. As the human and livestock populations continue to increase rapidly in arid zones, greater attention must be focused on research for quick improvement of staple food, fodder and fruit crops of this region. Undoubtedly, more studies are required to realize the potential of biotechnology, particularly in the field of genetic engineering. In India, work on transgenic plants is at a rudimentary stage. It will be a highly efficient technology in coming years, sometime before assessment of such projected potential becomes available and viable. One reason for this is the inevitable time it takes to produce and identify transformants that adequately express the transgenics to produce functionally relevant quantities of gene products. The other and more important point is to make available cloned genes, which are important for arid zone plants in India.

REFERENCES

A. V. Thomas and Co. Ltd. 1998. Tissue culture Date palm from AVT. In: Group meet on the "Progress of Date palm tissue culture" April 1998 at Rajasthan Agricultural University, Beechwal, Bikaner.

Ammer, S. and Benbadis, A. 197. Multiplication vegetative du palmier dattier *(Phoenix dactylifera* L.) par la culture de tissue de jeunes palnts Issues de Semis. C. R. Acad. Sci. Series D., 284: 1789-1794.

Ancora, G. and Cuozzo, L. 1985 *In-vitro* propagation of caper (*Capparis spinosa* L.). Genetica Agraria, 39: 504-305.

Arumugam, S. and Rao, M. V. 1996 *In-vitro* production of plantlets from cotyledonary node cultures of *Aegle marmelos* (L.) Corr. Advances in Plant Sciences, 9: 181-186.

Arya, H. C., Ramawat, K. G. and Suthar, K. C. 1981 Culture and differentiation of plants of economic importance II. *Aegle marmelos*. J. Ind. Bot. Soc., 60: 134-137.

Arya, S., Kumar, N. and Toky, O. P. 1995 Khejri (*Prosopis cineraria* L. Druce): Its value, research and extension. HDRA-ODA Project, India.

Bajaj, Y. P. S. 1986 Biotechnology of tree improvement for rapid propagation and biomass energy production. In: Biotechnology in Agriculture and Forestry Trees: Vol. 1. (Bajaj, Y. P. S. ed.), Springer-Verlag, Heidelberg, Berlin, pp. 1-23.

Batchelor, C. A., Yao, D., Koechler, M. J. and Harris, P. J. C. 1989 *In-vitro* propagation of *Prosopis* species (*P. chilensis, P. cineraria* and *P. juliflora*). Annals des Sciences Forests, 46: 110-112.

Bekheet, S. K. and Saker, M. M. 1998 *In vitro* propagation of Egyptian date palm: II Direct and in direct shoot proliferation from shoot-tip explants of *Phoenix dactylifera* L. cv, Zaghlool, Proceedings The First International Conference on Date palms, Al-Ain Hilton Hotel, Al-Ain, United Arab Emirates, march 8-10, 1998, pp.150-157.

Bejoy, M. and Hariharan, N. 1992 *In-vitro* plantlet differentiation in *Annona mauricata*. Plant Cell, Tissue and Organ Culture, 31: 245-247.

Bhandari, M. M. 1990 Flora of Indian Desert. Scientific Publishers, Jodhpur, pp. 215-216.

Bhardwaj, L., Merrillan, J. M. and Ramawat, K. G. 1995 Changes in the composition of membrane lipids in relation to differentiation in *Aegle marmelos* callus cultures. Plant cell, Tissue and Organ culture, 3: 33-37.

Bhaskaran, S. and Smith, R. H. 1992 Somatic embryogenesis from shoot-tip and immature inflorescence of *Phoenix dactylifera* cv. Barhee. Plant Cell Reports, 12: 22-25.

Bonga, J. M. 1982 Vegetative propagation in relation to juvenility, maturity and rejuvenation. In: Tissue culture in Forestry, pp. 387-412. (J.M. Bonga and Durzan, D.J. eds.), Martinus Nijhoff, Dr W. Junk Publishers, The Hague.

Bouguedoura, N., Imechaux-ferriere, N. and Bompar, J. L. 1990 Comportment *in-vitro* de bourgeons axillaries de type in determine du palmier dattier (*Phoenix dactylifera* L). Can. J. Bot., 68: 2004-2009.

Chandra, A., Chandra, A. and Gupta, I. C. 1994 Arid Fruit Research. Scientific Publishers, Jodhpur.

Chopra, V. L. 1997 Development in crop biotechnology. 50-years of ICAR Research, ICAR, pp. 28-36.

Chopra, V. L. and Sharma, R. P. 1991 Biotechnology in plant improvement. Curr. Sci., 60: 543-547.

Corley, V., Barrett, H. and Jones, H. 1976 Vegetative propagation of palm via tissue culture, In: Malaysian International Agriculture Oil Palm Coference Incorporated Society of Plants, Kulala Lumpur, pp. 1-7.

Daguin, F. and Letouze, R. 1088 Regeneration of date palm (*Phoenix dactylifera* L.) by somatic embryogenesis improved by dipping in stirred liquid medium, Fruits, 43: 191-194.

Dass, H. C., Kaul, R. K., Joshi, S. P. and Raj Bhansali, R. 1989 *In-vitro* regeneration of Date palm plantlets. Curr. Sci., 58: 22-24.

David, S. B. and Vasikar, S. M. 1980 Rapid multiplication of economically important plants by tissue culture and organ culture. In: Plant Tissue Culture, Genetic Manipulation of Plant Cell, (P. S. Rao., M. R. Heble and M. S. Chadda eds.) BARC, Bombay, pp. 285-289.

Deora, N. S. and Shekhawat, N. S. 1995 Micropropagation of *Capparis decidua* (Forsk) Edgew, A tree of arid horticulture. Plant Cell Reports, 15: 278-281.

Dodeman, V. L., Ducreux, G. and Kreis, M. 1997 Zygotic embryogenesis versus somatic embryogenesis. Journal of Experimental Botany, 48: 1493-1509.

Drira, N. 1983 Multiplication vegetative du palmier (*Phoenix dactylifera* L.) par la culture *in-vitro* de bourgons axillaires et de feuilles qui en dervant. C. R. Acad. Sci. Paris, 296: 1077-1082

Earle, E. D. 1982 Application of plant tissue culture in the study of host-pathogen interaction. In: Application of plant, cell and tissue culture to Agriculture and Industry (D. T. Tomes, B.E. Ellis, P.M. Harney, K.J. Kasha, R.L, Peterson eds). The University of Guelph, Guelph, Canada.

Faroda, A. S., Singh, H. P., Singh, M., Lodha, S., Kumar, S., Bhati, T. K. and Harsh, L. N. 1997 Vision-2020. CAZRI-Perspective Plan. ICAR.

Goodin, J. R. and Northington, D. K. 1985 Plant resource of arid and semi-arid lands - A global perspective. Academic Press Inc., New York.

Goyal, Y. and Arya, H. C. 1984 Tissue culture of desert trees: I. Clonal multiplication of *Prosopis cineraria* by bud culture. Journal of Plant Physiology, 115:183-189.

Goyal, Y. and Arya, H. C. 1981 Differentiation in cultures of *Prosopis cineraria* Linn. Current Science, 50:468-469.

Goyal, Y. and Arya, H. C. 1985 Tissue culture of desert trees II. Clonal multiplication of *Ziziphus in-vitro*. J. Plant Physiol., 119: 399-404.

Green, B., Tabone, T. and Felker, P. 1990 A comparison of amide and ureide nitrogen sources in tissue culture of tree legume *Prosopis alba* clone B2V50. Plant Cell, Tissue and Organ Culture, 21: 83-86.

Hall, R. 1987 Biotechnology for plant disease control: Application of biotechnology to plant pathology. Canadian Journal of Plant Pathology, 9: 152-155.

Hazarika, B. N., Nagaraju, V. and Parthasarthy, V. A. 1996 Organogenetic responses of Bael (*Aegle marmelos*) micro shoots to benzyl aminopurine. Ann. Plant Physiol., 10: 40-44.

Hossain, M., Islam, R., Karim, M. R., Rahman, S. M. and Jorder, O. I. 1994 Production of plantlets from *Aegle marmelos* nucellar callus. Plant Cell Reports, 13: 570-573.

Hossain, M., Karim, M. R., Isalam, R. and Jorder, O.I. 1993a Plant regeneration from nucellar tissues of *Aegle marmelos* through organogenesis. Plant Cell, Tissue and Organ Culture, 34: 199-203.

Hossain, M., Karim, M. R., Islam, R. and Joarder, O. I. 1993b Adventitious shoot formation on nucellus, embryonic tissues and seedlings of *Aegle marmelos* Corr., Intern. Plant Tissue Culture Conf., Deptt of Bot., Dhaka Univ., Dhaka, December, 19-21, 1993.

Ingram, D.S. and Helgeson, J.P. 1980 Tissue culture methods for plant pathologists. Blackwell Scientific Publications, Oxford.

Islam, R., Hossain, M., Reza, M. A., Mamun, A. N. K. and Joader, O. I. 1996a Shoot regeneration from root-tips of intact seedlings of *Aegle marmelos*. J. of Hort. Sci., 71: 995-1000.

Islam, R., Hussain, M., Hoque, A. and Joarder, O. I. 1996b Somatic embryogenesis and plant regeneration in *Aegle marmelos*-a multipurpose social tree. Curr. Sci., 71: 259-260..

Islam, R., Karim, M. R., Rahman, S. M., Hussain, M. and Joarder, O. I. 1994 Plant regeneration from excised cotyledons of *Aegle marmelos* Corr. Pakistan J. Bot., 26: 393-396.

Islam, R., Karim, M. R. Hossain, M. and Joarder, O. I. 1992 Plant regeneration from leaf derived callus in *Aegle marmelos* Corr. Plant Tissue Culture, 2: 55-59.

Jain, S. M., Newton, R. J. and Soltes, B. J. 1988 Enhancement of somatic embryogenesis in Norway spruce (*Picea abies*). Theor. Appl. Genet. 76: 501-506.

Jain, S. M., Brar, D. C. and Ahloowalia, B. S. 1997 Somaclonal variation and induced mutation in crop improvement. Kluwer Academic Publishers, The Netherlands.

Jain, S. M., Gupta, P. K. and Newton, R. J. 1999 Somatic embryogenesis in woody plants. Vol. IV-V, Kluwer Academic Publishers, The Netherlands.

Jain, S. M., Gupta, P. K. and Newton, R. J. 2000 Somatic embryogenesis in woody plants. Vol. VI, Kluwer Academic Publishers, The Netherlands.

Jaidka, K. and Mehra, P. N. 1986 Morphogenesis in *Punica granatum* (Pomegranate). Can. J. Bot., 64: 1644-1653.

Jordan, M. 1985 Propagation of *Prosopis* by tissue culture, Biotecnologia en las Americas. II. Aplicaciones en agricultura tropical. 1985, Session 1, paper No. 4, 3pp. San Jose, Costa Rica, CONICIT.

Jordan, M. 1988 *In-vitro* culture of *Prosopis* species. In: Cell and Tissue Culture in Forestry. Vol. 3., (Bonga, J. M. and Durzan, D. J., eds.), Martinus Nijhoff Pub. Boston, pp. 370-384.

Jordan, M., Cortes, I. And Goreux, A. 1987 Potentialities of cell and callus tissue culture to regenerate two mesquite species (*Prosopis tamarugo* and *P. chilensis*). Gartenbauwi ssenschaft, 52: 166-169.

Kabir, A., Islam, R., Hoque, A., Khalekuzzaman, M. and Joarder, O.A. 1994 Micropropagation of *Ziziphus mauritiana* from nodal segments of mature trees. Bull.Seri. Cult. Res., 5.

Kackar, N. L., Solanki, K. R., Singh, M. and Vyas, S. C 1991 Micropropagation of *Prosopis cineraria.* Indian Journal of Experimental Biology, 29: 65-67.

Kackar, N. L., Vyas, S. C., Singh, M. and Solanki, K. R. 1992 *In-vitro* regeneration of *Prosopis cineraria* (L.) Druce using root as explant. Indian Journal of Experimental Biology, 30: 429-430.

Kanninen, M., Elena-Aguilar, M., Navarro, C., Cote, F., Bruns, S. 1998 Recent advances in biotechnology for tree conservation and management. Proceedings of an IFS Workshop, Florianopolis, Brazil, 15-19 September 1997, (Mantell, S., Tragardh, C and Viana, A. M. eds.), International Foundation for Science (IFS), Stockholm, Sweden), 319-327.

Kruger, R. 1998 Date palm germplasm: Overview and utilization in the USA. Proceedings The First International Conference on Date palms, Al-Ain Hilton Hotel, Al-Ain, United Arab Emirates, march 8-10, 1998, pp.2-37.

Kant, U. and Ramani, V. 1990 Insect induced plant galls in tissue culture. Proceedings Indian Academy of Sciences-Animal Sciences, 99: 257-265.

Krishnan, A. 1977 A climatic analysis of the arid zone of North-West India. In: Desertification and its control. ICAR Publication, pp. 42-57.

Leaky, R. R. B. and Last, R. T. 1980 Biology and potential of *Prosopis* species in arid environments, with particular reference to *P. cineraria*. Journal of Arid Environment, 3: 9-24.

Letouze, R., Daguin, F., Satour, P., Hamama, L. and Marrionnet, F. 1998 Somatic embryogenesis and mass micropropagation of dare palm (*Phoenix dactylifera* L.), characterization and genetic stability pg regenerated plantlets by RAPD markers, Proceedings The First International Conference on Date palms, Al-Ain Hilton Hotel, Al-Ain, United Arab Emirates, march 8-10, 1998, pp.158-167.

Lemos, E. E. P., and Black, J. 1996a Micropropagation of juvenile and mature *Annona muricata* L. J. Hort. Sci., 91: 395-403.

Lemos, E.E.P. and Black, J. 1996b Micropropagation of juvenile and adult *Annona squamosa*. Plant Cell, Tissue and Organ Culture, 46: 77-79.

Litz, R.E., Moore, G.A. and Chinnathambi, S. 1985 *In-vitro* system for propagation and improvement of tropical fruits and palm. In: Horticultural Reviews, vol.7 (ed. J. Janick) API Publication Co., 137-200.

Lloyd, G. and B. McCown 1981 Commercially feasible micropropagation of mountain laurel *(Kalmia latifolia)* by use of shoot tip culture. Proc. Int. Plant Propagation Society, 30: 421-427.

Maier, V. and Metzlier, D. M. 1965 Quantitative changes in Date polyphenols and their relation to browning. J. Food Sci., 30: 80-84.

Manohar, M. S. and Chandra, A. 1995 Date palm culture in Rajasthan. Directorate of Research, Rajasthan Agricultural University, Bikaner.

Mater, A. A. 1986 *In-vitro* propagation of *Phoenix dactylifera* L. Date Palm J., 4: 137-151.

Mathur, N. and Vyas, A. 1995 *In-vitro* production of *Glomus deserticola* in association with *Ziziphus nummularia*. Plant Cell Reports, 4: 735-737.

Mathur, N., Ramawat, K. G. and Nandwani, D. 1995 Rapid *in-vitro* multiplication of Jujube through mature stem explant. Plant Cell, Tissue and Organ Culture, 43: 75-77.

Mesen, F., Leakey, R. R. B. and Newton, A. C. 1992 Towards the development of clonal silviculture techniques for small farms. Chasqui (Spanish), 28: 6-18.

Mitrofanova, I. V., Mitrofanova, O. V. and Panderi, D. K. 1997 Somatic embryogenesis and plant regeneration in *Ziziphus jujuba* Mill. *In-vitro*. Russian J. Plant Physiol., 44: 84-94.

Mokhtar, T., Benslimane, A. A., Rhouma, A. M., Rode, A. and Marrakchi, M. 1998 Molecular characterization of Tunisian date palm varieties, The First International Conference on Date palms, Al-Ain Hilton Hotel, Al-Ain, United Arab Emirates, march 8-10, 1998, pp.183-193.

Moriguchi, T., Omura, M., Matsura, N. and Kozaki, I. 1987 *In-vitro* adventitious shoot formation from anthers of Pomegranate. Hort Science 22 : 947-948.

Moursy, H. A. and Saker, M. M. 1996 Date palm problems and need for biotechnology, Fifth International Conference on Desert Development, Texas Tech University, August 12-17, 1996, USA.

Murashige, T. 1974 Plant propagation through tissue cultures. Ann. Rev. Plant Physiol., 25: 135-166.

Nair, S., Gupta, P. K. and Shirgurkar, M.V. and Mascarenhas, A.F 1984b. *In-vitro* organogenesis from leaf explant of *Annona squamosa* Linn. Plant Cell, Tissue and Organ Culture, 3: 29-40.

Nair, S., Gupta, P. K., Mascarenhas, A. F. 1983 Haploid plants from *in-vitro* anther culture of *Annona squamosa* L. Plant cell Reports, 2: 198-200.

Nair, S., Gupta, P. K., Mascarenhas, A. F. 1984a *In-vitro* propagation of *Annona* hybrid (*Annona squamosa* L. x *Annona cherimola* L.). Indian Journal of Horticulture, 41: 1670-163.

Nair, S., Shirgurkar, M. V. and Mascasrenhas, A. F. 1986 Studies on endosperm culture of *Annona squamosa* Linn. Plant Cell Reports, 5: 132-135.

Nandwani, D. 1990 Tissue culture of forest plants of Rajasthan I. *Prosopis* spp, Ph. D. Thesis, University of Jodhpur, Jodhpur, pp. 101.

Nandwani, D. and Ramawat, K. G. 1989 *In-vitro* growth of three species of *Prosopis*. In Proc. Tissue Culture Conference VII., (Tandon, P., ed.), NEHU, Shillong.

Nandwani, D. and Ramawat, K. G. 1991 Callus cultures and plantlets formation in *Prosopis juliflora* (Swartz.) D.C. Indian Journal of Experimental Biology, 29: 253-257.

Nandwani, D. and Ramawat, K. G. 1992a *In-vitro* regenerative potential of explants and callus morphogenesis in *Prosopis tamarugo.* Gartenbauwissenschaft, 57: 106-111.

Nandwani, D. and Ramawat, K. G. 1992b High frequency plantlets regeneration from seedling explants of *Prosopis tamarugo.* Plant Cell Tissue and Organ Culture, 29:173-178.

Nandwani, D. and Ramawat, K. G. 1993 *In-vitro* plantlets formation through juvenile and mature explants in *Prosopis cineraria,* Indian Journal of Experimental Biology, 31:156-160.

Nataraja, K. and Neelambica, G. K. 1996 Somatic embryogenesis and plantlet from petal cultures of Pomegranate, *Punica granatum* L. Indian Journal of Experimental Biology, 34: 719-721.

Omura, M., Matsuta, N., Moriguchi, T., Kozaki, I. and Sanada 1987a Establishment of tissue culture methods of dwarf Pomegranate (*Punica granatum* L var. Nana) and application for induction of varieties. Bull. Fruit Tree Res. Stn. Japan A. 14: 17-44.

Omura, M., Matsuta, N., Morikguchi, T. and Kozaki, I. 1987b Adventitious shoots and plantlet formation from cultured Pomegranate leaf explants. Hort. Science, 22: 133-134.

Oppenheimer, C. and Reuveni, O. 1972 Development of a method for quick propagation of new and superior Date varieties. The Valconi Centre, Bet-Dagan, Israel, pp. 1-92.

Pareek, O. P. 1998 National Research Centre for Arid Horticulture. In: 50 Years of Horticulture Research, (Ghosh, S. P. and Bhatanagar, P. S., eds.), Division of Horticulture, ICAR, New Delhi, Pp. 95-99.

Pareek, O. P. and Nath, V. 1996 Coordinated fruit research in Indian arid zone. National Research Centre for Arid Fruits, Bikaner

Pareek, O.P. 1997 Arid horticulture: its potential. Indian Horticulture, 42: 42-46.

Paroda, R. S. 1979 Plant resource of Indian arid zones for industrial uses. In: Arid Land Plant Resources, (Goodin J. R. and Northington, K.D., eds.), Inter. C.A.S.L.A.S., Texas pp. 261-281.

Puri, S., Jain, M. and Sharma, P. 1992 *In-vitro* regeneration of *Prosopis cineraria.* Nitrogen Fixing Tree Research Reports, 10: 189.

Raj Bhansali, R. 1989 Micropropagation of temperate and tropical fruit trees through repetitive somatic embryogenesis. XII Plant Tissue Culture Conf., North-Eastern Hill University, Shillong, p. 16.

Raj Bhansali, R. 1990 Somatic embryogenesis and regeneration of plantlets in Pomegranate. Annals of Botany, 66: 249-l253.

Raj Bhansali, R. 1993a Bud culture of shoot multiplication and plantlet formation of *Tecomella undulata* (Rohida) woody tree of arid zone. Tropical Science, 33: 1-8.

Raj Bhansali, R. 1993b *In-vitro* selection of disease resistant embroygenic cell lines of Date palm by using pathotoxins. International Conference on Biotechnology in Agriculture and Forestry. Kanpur, India.

Raj Bhansali, R. 1994 Micropropagation of temperate and tropical fruit trees through somatic embryogenesis. In: Advances in Plant Tissue Culture in India (P. Tandon, ed.), Pragati Prakashan, Meerut.

Raj Bhansali, R. 1995 A Manual of Plant-Tissue Culture (Prepared for the short-term training course). Department of Biotechnology, New Delhi.

Raj Bhansali, R. 1997a Recent advances in management of diseases of fruit and forest trees of arid zone-Principles and Strategy. Summer Inst. on Sustainable Dry land Agric., CAZRI, Jodhpur, pp 294-299.

Raj Bhansali, R. 1997b Role of somatic embryogenesis in micropropagation and artificial seed production. IUFRO Symposium on Innovations in Forest Tree Seed Sciences and Nursery Technology, pp. 69-70.

Raj Bhansali, R. 1998a Repetitive somatic embryogenesis and enhanced axillary proliferation methods for multiplication of arid zone fruit plants. Intern. Seminar on Micropropagation and Spread of Superior Genetic Material of Forest Trees, WWF, New Delhi, p. 17.

Raj Bhansali, R. 1998b Micropropagation of Date palm through repetitive somatic embryogenesis (RSE) process. In: Group meet on the "Progress of Date palm tissue culture" April, 1998 at Rajasthan Agricultural University, Beechwal, Bikaner.

Raj Bhansali, R. 2001 Biotechnology in arid agriculture and forestry. In: Priorities of Research for Development of Arid Regions, (Verma, S. K., Nahar, N. M., Raj Bhansali, R., Jindal, S. K. and Satya Vir, eds.), Scientific Publishers (India), Jodhpur, pp. 113-127.

Raj Bhansali, R. 2001 Date palm. In: Biotechnology of Horticultural Crops Vol. I (Parthasarathy, V. A., Bose, T. K., Deka, P. C., Das, P., Mitra, S. K. ans Mohandas, S. eds.), Naya Prokash, Calcutta, India, pp. 344-350.

Raj Bhansali, R. 2001a Prosopis. In: Horticultural Crops Vol. II (Bose, T. K. ed.), Naya Prokash, Calcutta, India (In press).

Raj Bhansali, R. 2001b Capparis. In: Horticultural Crops Vol. II (Bose, T. K. ed.), Naya Prokash, Calcutta, India (In press).

Raj Bhansali, R. 2001c Cordia. In: Horticultural Crops Vol. II (Bose, T. K. ed.), Naya Prokash, Calcutta, India (In press).

Raj Bhansali, R. and Kaul, R.K. 1991 Into future - Date through tissue culture. Indian Horticulture, 36: 6-10.

Raj Bhansali, R. and Singh, K. 1982 Callus and shoot formation from leaf of sugar cane in tissue cultures. Phytomorphology, 12: 167-170.

Raj Bhansali, R. and Singh, K. 1991 Grassy shoot mycoplasma-free sugarcane plants from apical meristem of lateral buds. Indian Phytopathology, 44: 291-294.

Raj Bhansali, R. and Singh, M. 2000 Somatic embryogenesis in fruit and forest trees of arid zone. In: Somatic Embryogenesis in Woody Plants, Vol. 6, (Jain, S. M., Gupta, P. K. and Newton, R. J. eds.), Kluwer Academic Publishers, Dordrecht / Boston / London, pp. 141-168.

Raj Bhansali, R. and Jindal, S. K. 2000 Role of farmers in promotion of eco-friendly multipurpose trees of arid zone. In: Environmental Protection, (Chaudhary, V., Singh, K. Kakralya, B. L., eds.), forwarded by Sunder Lal Bhuguna, Pointer Publication, Jaipur, pp. 92-101.

Raj Bhansali, R., Driver, J. A. and Durzan, D. J. 1990 Rapid multiplication of adventitious somatic embryos in peach and nectarine by secondary embryogenesis. Plant Cell Reports, 9: 280-284.

Raj Bhansali, R., Kaul, R.K. and Dass, H.C. 1988 Mass cloning of Date palm plantlets through repetitive somatic embryogenesis process. J. Plant Anat. and Morphology 5: 73-79.

Ramani, V. and Kant, U. 1993 *In-vitro* "gall" formation on the leaf rachis of *Prosopis cineraria* (Linn.) Druce (Mimosaceae) by *Lobopteromyia prosopidis mani* (Diptera: Cecidomyiidae). Phytophaga, 5: 7-11.

Ramawat, K. G. and Nandwani, D. 1991 Propagation of *Prosopis* species: Problems perseverance and perspectives. Annals of Arid zone, 30:247-258.

Rathore, T. S., Singh, R. P., Deora, N. S. and Sekhawat, N. S. 1992 Clonal propagation of *Ziziphus* species through tissue culture. ScientiaHort, 51: 165-168.

Reuveni, O. and Lilyen-Kipnis, H. 1974 Studies of the *in-vitro* culture of Date palm (*Phoenix dactylifera* L.) tissue and organ. The Valconi Agric. Res. Pamphlet no 145, Bet-Dagan, Israel, pp. 1-4.

Reuveni, O., Adato, Y. and Lilyen-Kipnis, H. 1972 A study of new and rapid methods for the vegetative propagation of Date palms. Date Grower's Inst. Report, 49: 17-24.

Reynolnd, J. F. and Murashige, T. 1979 Asexual embryogenesis in callus cultures of palm. *In-vitro* Cell Dev. Biol., 15: 383-387.

Rhiss, A., Poulin, C. and Beauchesne, G. 1979 La culture *in-vitro* applique a la multiplication vegetative du palmier dattier (*Phoenix dactylifera* L.). Fruits, 34: 551-554.

Rodriguez, R., Rey, M., Cuozzo,L. and Ancora,G. 1990 *In-vitro* propagation of caper (*Capparis spinosa* L.).*In-vitro* Cellular and Developmental Biology, 26: 531-536.

Roy, S. K., Islam, M. S. and Sen, J. 1993 High frequency shoot formation and plant regeneration from mature embryos of *Syzygium cumuni.* Intern. Plant Tissue Culture Conf., Dhaka Univ., Deptt. of Bot., Dhaka, December, 19-21.

Saker, M. M. and Moursy, H. A. 1998 Molecular characterization of Egyptian date palm, The First International Conference on Date palms, Al-Ain Hilton Hotel, Al-Ain, United Arab Emirates, march 8-10, 1998, pp.173-182.

Saxena, S. K. 1977 Vegetation and its succession in the Indian desert. In: Desertification and its control. ICAR Publications, pp. 176-192.

Sedra, M.H., Lashermas, P. and Hamon, s. 1998 Identification and genetic diversity analysis of date palm (*Phoenix dactylifera* L.) varieties from Morocco using RAPD markers, The First International Conference on Date palms, Al-Ain Hilton Hotel, Al-Ain, United Arab Emirates, march 8-10, 1998, pp.208-224.

Shabana, R. H., Shiraqui, M.K. R., Mansoor, M. L. and Safadi, M.A.W. 1998 Comparision between date palm fruits produced from tissue culture and offshoot plants (Hilali cultivar), The First International Conference on Date palms, Al-Ain Hilton Hotel, Al-Ain, United Arab Emirates, march 8-10, 1998, pp.168-172.

Sharma, D. R., Dawara, S. and Chowdhary, J. B. 1984 Somatic embryogenesis and plant regeneration in Date palm (*Phoenix dactylifera* Lin.) cv. 'Khadravi' through tissue culture. Indian Journal of Experimental Biology, 22: 596-598.

Sharma, D. R., Deepak, S. and Chowdhary, J. B. 1986 Regeneration of plants from somatic tissues of Date palm (*Phoenix dactylifera* Linn.). Indian Journal of Experimental Biology, 24: 763-766.

Shekhawat, N. S. and Kackar, A. 1987 Protoplasts of *Prosopis cineraria*: Isolation and culture. Nitrogen fixing tree Research Reports, 5: 51-53.

Shekhawat, N. S., Rathore, T. S., Singh, R. P., Deora, N. S. and Rao, S.R.1993 Factors affecting *in-vitro* clonal propagation of *Prosopis cineraria.* Plant Growth Regulation, 12: 273-280.

Singh, M.P., Raj Bhansali, R., Kaul, R.K. and Tripathi, R.S. 1998 Integrated pest management in perennial plants. In: Fifty years of Arid Zone Research in India (A.S. Faroda and M. Singh eds), CAZRI, Jodhpur pp. 375-392.

Sudhersan, C., Abo El Nil, M. M. and Al Baiz, A. 1993 Occurrence of direct somatic embryogenesis on sword leaf *in-vitro* plantlets of *Phoenix dactylifera* L. cv. Barhee. Curr. Sci., 65: 687-689.

Sung, D. O. and Song, W. I. 1992 Regeneration of chinese jujube (*Ziziphus jujuba* Miller) from immature cotyledons and segments of seedling. HortScience, 27: 210.

Tabone, T. J., Felker, P., Bingham, R. L., Reyes, I. and Loughrey, S. 1986 Techniques in the shoot multiplication [tissue culture] of the leguminous tree *Prosopis alba* clone B2V50. Forest Ecology and Management, 16: 191-200.

Tisserat, B. 1979 Propagation of Date palm (*Phoenix dactylifera* L.) *in-vitro.* J. Exp. Bot., 30: 1275-1283.

Tisserat, B. 1984 Clonal propagation. In: Cell Culture and Somatic Cell Genetics of Plants, (I. K. Vasil, ed.), Academic Press, New York, pp. 74-80.

Tisserat, B., Foster, G. and De Mason, D. 1979. Plantlet production *in-vitro Phoenix dactylifera* L. Dates Growers Reports, 54: 19-23.

Tyagi, P., Chandra, N. Pareek, L. K. and Kothari, S. L. 1994 Micropropagation of *Capparis decidua.* National Symposium on Recent Advances in Plant Tissue Culture and Biotechnology, Department of Botany, University of Jaipur, p. 57.

Tyagi, P., Kothari, S. L. and Tyagi, P. 1997 Micropropagation of *Capparis decidua* through *in-vitro* shoot proliferation on nodal explants of mature tree and seedling explants. J. Plant Biochemistry and Biotechnology, 6: 19-23.

Wainright, H. and England, N. 1987 The micropropagation of *Prosopis juliflora* (Swartz) D.C.: Establishment *in-vitro.* Acta Horticulture, 212: 49-53

Yadav, U., Lal, M. and Jaiswal, V. S. 1990 *In-vitro* micropropagation of the tropical fruits tree *Syzygium cumuni* L. Plant Cell, Tissue and Organ Culture, 21: 87-92.

Yao, D., Batchelor, C. A., Koechler, J. J. and Harris, P. J. C. 1989 *In-vitro* regeneration of *Prosopis* species (*P. cineraria* and *P. juliflora*) from nodal explants. Chinese Journal of Botany, 1: 89-97.

Zaid, A. 1987 *In-vitro* browning of tissue and media with special reference to Date palm cultures- A review. Acta Hort., 212: 561-566.

Zaid, A. and Tisserat, B. 1983 Morphogenetic responses obtained from a variety of somatic explant tissue of Date palm. Bot. Mag. (Tokyo), 96: 67-73.

Zimmerman, R. H. 1987 Regeneration of woody ornamentals and fruit trees. In: Cell Culture and Somatic Cell Genetics. Vol.3. Academic Press Inc.

1

1. **Induction of somatic embryos from apical tips of date palm off shoots**

2. **Regeneration of complete plant let from embryogenic callus cultures of date palm**

3

3. Development of leafy shoots from axillary bud of date palm

4. (A) Free-living date palm plant transferred in polythene tubes and (B) pots regenerated from RSE method

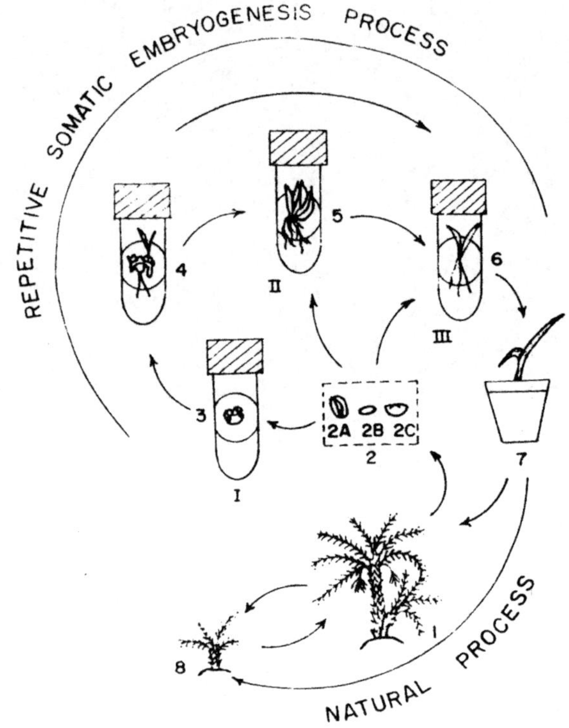

5 A BIOTECHNIQUE FOR DATE PALM MICROPROPAGATION.

5. Process of date palm micropropagation through repetitive somatic embryogenesis (RSE) process

6. **Proliferation of large number of shoots from nodal bud of *Z. mauritiana* on MS medium containing 2.5 mg/l BAP**

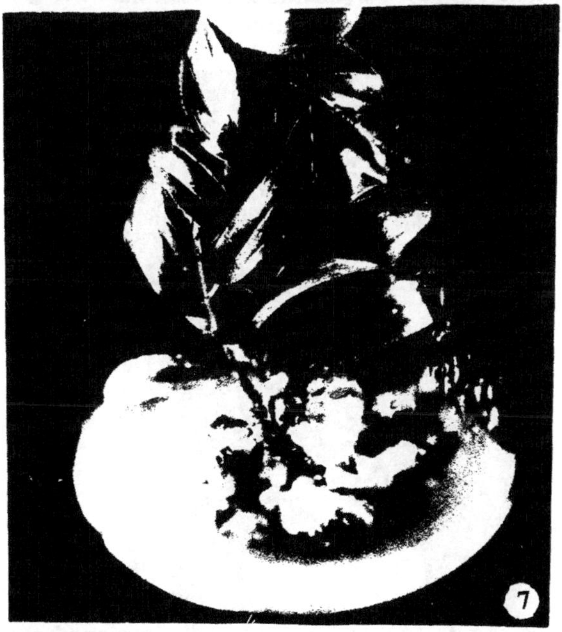

7. **Regeneration of complete plantlet from somatic embryos developed from cotyledonary explant tissues of *Z. mauritiana***

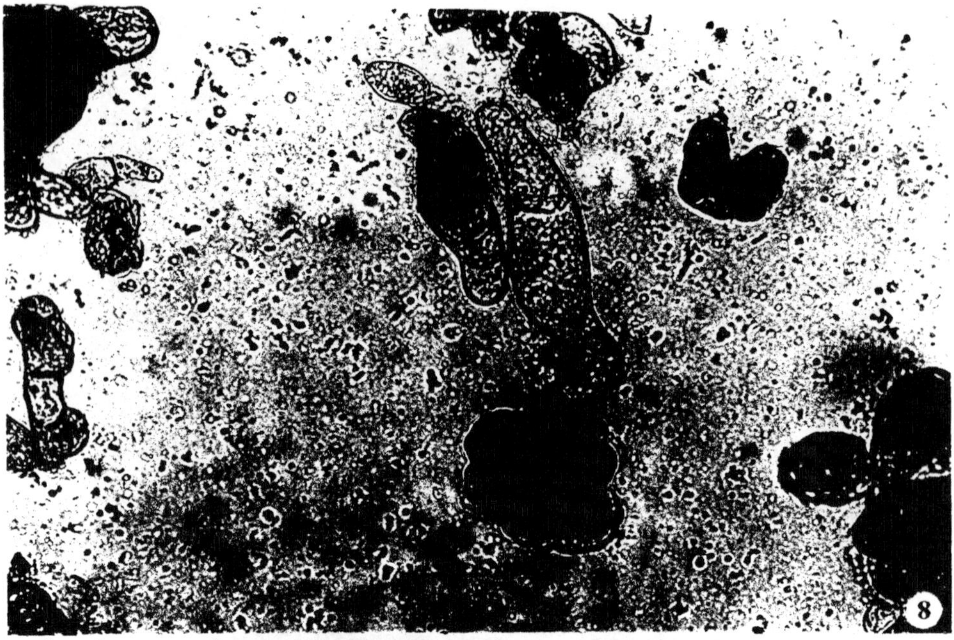

8. **Acetocarmine staining technique indicating red stained cellular somatic embryos of pomegranate**

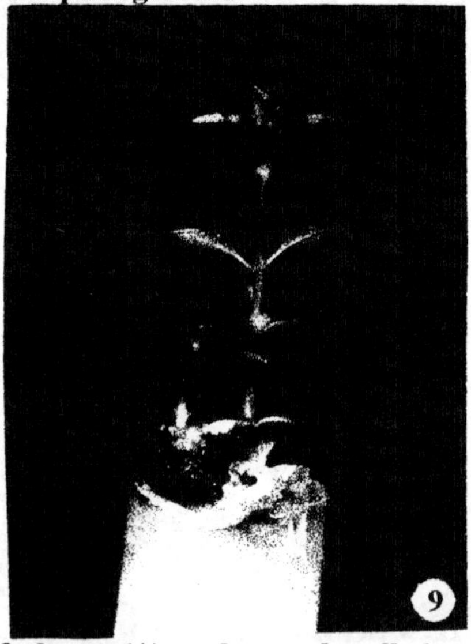

9. **Development of shoots (A) and rooted well developed shoot (B) from embryogenic cultures of pomegranate placed on medium having 5 mg/l BAP**

10. Multiple shoot differentiation from seed explant of *Prosopis cineraria* on MS medium containing 3.0 mg/1 IAA+0.05 mg/1 KN

11. Initiation of large number of shoots from stem segment obtained from in-vitro proliferated shoots

12. Rare *Capparis decidua* plants having yellow flowers producing large sized fruits have been used for *in-vitro* conservation

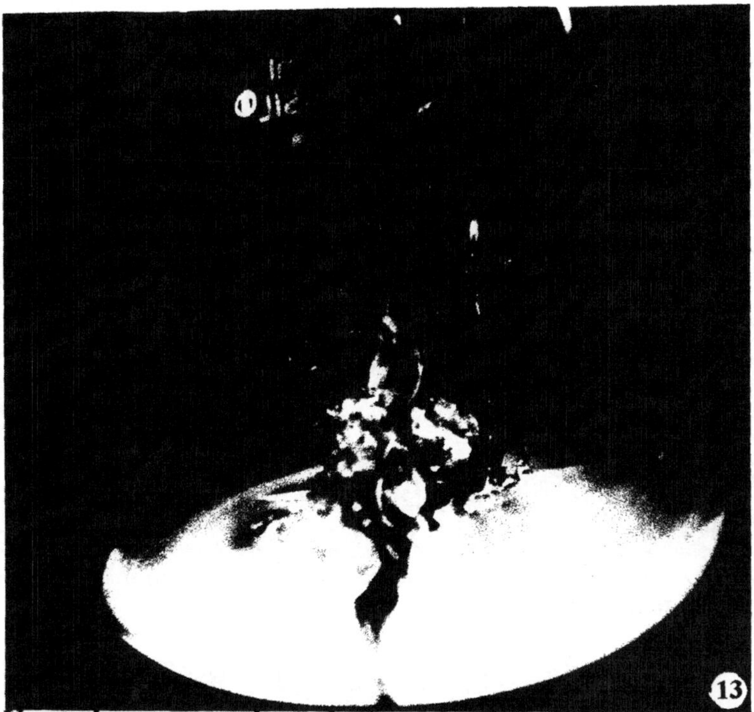

13. Vegetative propagation of *Capparis decidua* plants through sprouting of lateral bud from seedling explant tissues

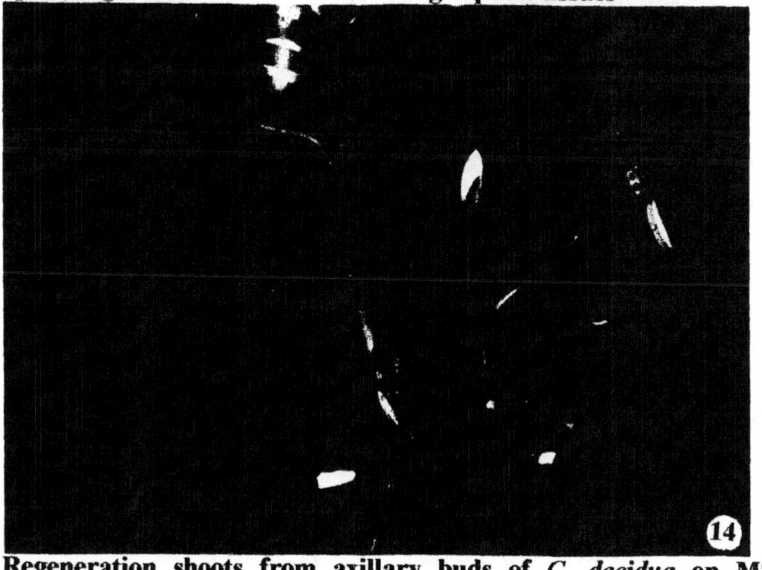

14. Regeneration shoots from axillary buds of *C. decidua* on MS medium containing 2.5 mg/l BAP and 1 mg/l KN

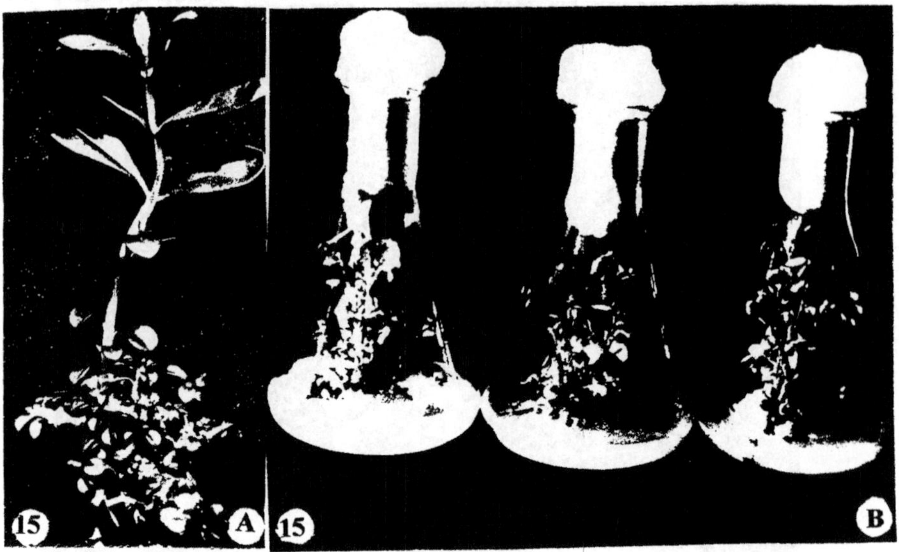

15. (A) Multiple shoots formation from nodal-bud/ (B) shoot-tip segments of young seedlings of *S. cumuni*

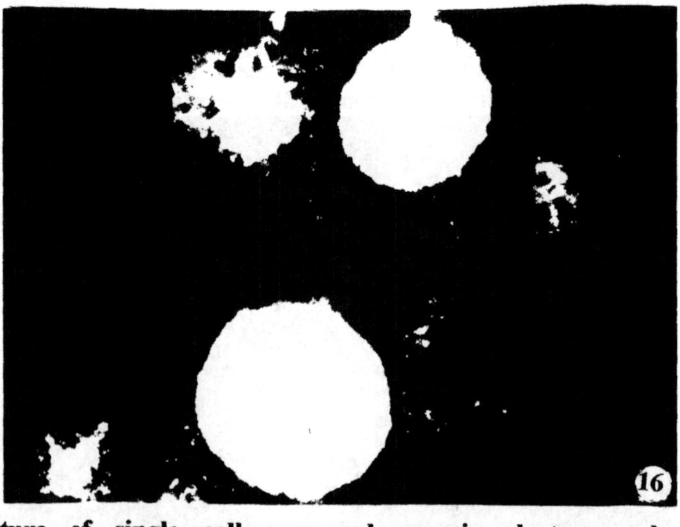

16. Mixture of single cells, pro-embryogenic clusters and new embryogenic centers

14. Micropropagtion of apples (*Malus sp.*)

Ph DRUART
CRA-Department Biotechnology, 234 chaussée de Charleroi, 5030 Gembloux, Belgium.

Chapter contents

S.M. Jain and K. Ishii (eds.), Micropropagation of Woody Trees and Fruits, 433–463.

1. INTRODUCTION

Apple (*Malus x domestica* Bork) is probably the most widely grown fruit crop in the world, being cultivated throughout the northern and the southern temperate zones. Selections from the wild have been made for several thousands years, since the period when apples were cultivated by the Greeks and Romans.

Recent cultivars are the result of inter-crossing among several *Malus* species (Oraguzie *et al.*, 2000). Rootstocks are basically selected on: ease of vegetative propagation by the conventional techniques, characteristics of regulating the tree growth, tolerance against soil diversity and diseases, and cropping performances.

Micropropagation is done for the following reasons:

- production of pathogen free plants when meristem-tip is the initial explant
- rapid increase of select genotype for use in breeding programs
- large-scale production of rootstocks and new cultivars for more rapid introduction into commercial trade
- propagation of plants for which conventional vegetative methods are ineffective or unprofitable.

Tissue culture also facilitates the international exchanges of plant material, since the certification free of any insects and pathogens is easier than for the scionwood (Lane, 1992).

Apple is one of the first woody plants successfully propagated by *in vitro* technique. The scientists have considered single *Malus* rootstock or cultivar as a model plant for the study of physiological disorders related to *in vitro* growth conditions like hyperhydricity ('vitrification'), rooting failure, variability about the propagation performances amongst laboratories or technologies in the frame of a scaling-up application. Research and progress with *in vitro* propagation has probably been more extensive with apple than any other type of tree.

The greatest use of tissue culture has been the production of clonal rootstocks. However, the high cost of these plants was the limiting factor for an extended application of the technology. Studies have now covered the all-long plant growth cycle till the field production for years (Zimmerman and Steffens, 1996).

A number of review articles provide comprehensive overviews of progress with apple propagation *in vitro* (see Zimmerman, 1986, Hammerschlag, 1986, Skirvin *et al.*, 1986, Lane, 1992, Jones 1993) and genetic manipulation (Hanke *et al.*, 2000, Dolgov *et al.*, 2000). The present paper will concern general principles with emphasis on significant developments during the last years and on the optimisation of the propagation technology developed in our laboratory.

2. ABOUT PROTOCOLS

Several different protocols have been published for each stage of apple micropropagation. In our laboratory, the original protocol was developed from meristem-tip culture (Quoirin, 1974) to apply to most cultivars and rootstocks of *Malus* and other fruit trees of the *Rosaceae* family including *Prunus, Cydonia* and *Pyrus* species as well (Druart 1980, Druart *et al.*, 1982, Druart, 1984, Druart, 1985, Boxus et Druart, 1986, Duron *et al.* 1989, Druart, 1992)

3. *IN VITRO* ESTABLISHMENT

Apple trees can be established *in vitro* from meristems, shoot tips or nodal buds, the initial explant length ranging from 0.1mm to several cm.

3.1. Meristem culture

Meristems are optimally excised from the vegetative buds of annual scionwood at the post-dormancy stage. The branches are thus collected prior to bud flushing, at the end of wintertime and they are stored in plastic bags at +2 to +4°C as 3 nodes size fragments pwdered with fungicide (Captan, TMTD). The differentiation culture can then start immediately. Cold storage of post-dormant buds allows the extension of the picking period for at least 4 months, while it stimulates the meristem organogenesis when the plant material is cut from the trees 2 to 3 months earlier.

The fragments of branches are sterilised by successive immersion in ethanol (95°) for 5 min and in Ca hypochlorite (9%) for 20 min, then rinsed in sterile distilled water and kept there until meristem picking. The efficiency of the sterilization method depends on the origin of the branches. The contamination rate is less than 5% when they originated from the tree crown whereas it frequently rises to 90% when they come from stool beds. In that case, the end parts of the cleanest layers should be chosen with extreme care.

Meristem-tip dissection begins by removing the bark away all around the vegetative bud. The sequential removal of the bud scales and then the leaf primordia finally expose the meristem-tip. The sterile scalpels and forceps used for that dissection are changed several times during the procedure, with the aim to reduce the risk of contaminating the meristem-tip. The meristem dome (0.1mm size) alone or together with one or two leaf primordia, is finally excised with a razor blade and immediately cultured in solidified nutrient medium. The vessel is a screw cap vial put on an inclined support exposing the explant to whole lighting suitably provided by Sylvania gro lux fluorescent tubes. The whole manipulation is achieved in a quiet corner of the room where a laminar flow is continuously working. Meristem-tips grow initially into small buds and further into well-developed rosettes of leaves within 4 to 8 weeks. Then, the explant is subcultured in the multiplication medium in order to induce axillary branching for monthly subcultures.

The principal factors influencing the survival rate and the development of the rosettes are: period of sampling the branches and later on the meristems; origin and size of the meristematic explant; genotype; composition of the culture medium. The presence of the leaf primordia is favorable for bud formation, however, decreases the efficiency of the virus elimination (Boxus and Druart, 1986).

The culture medium for meristem contains half-strength Murashige and Skoog's macroelements (1962), Nitsch's microelements (1969) with 10^{-1} mM Fe ($FeNa_2$ EDTA), vitamins of Jacquiot (Gautheret, 1959), 1 $mg.l^{-1}$ benzyladenine (BA), 0.1 $mg.l^{-1}$gibberellic acid (GA3), 10^{-3} $mg.l^{-1}$ 2,4-Dichlorophenoxyacetic acid (2,4 D), 2% sucrose, 7 to 5 $g.l^{-1}$ Merck agar (art. 1613), pH 5.8. This medium is adapted to the meristem establishment of most fruit species.

M 9 is one of the most recalcitrant apple genotype to establish from meristems. A comparison between different clones of such rootstock has been carried out using meristem dome as explant (Table 1). If the number of meristems developing rosettes remained relatively low, sorbitol added together with sucrose in the culture medium improved organogenesis and establishment of clones.

Table 1 . Comparison of M9 clones about the rosette formation from meristem domes cultivated in presence of sucrose or sucrose + sorbitol.

Clone	Sucrose (2%)		Sucrose(1%) + Sorbitol (1%)	
8	2/59	3.39 %	6/56	10.71 %
9	6/56	10.71 %	6/59	10.16 %
19	9/59	15,25 %	12/60	20.00 %
25	2/56	3,57 %	3/59	5.09 %
28	3/59	5,09 %	8/59	13.56 %
29	10/38	26,32 %	24/56.	42.86 %
J 12	0/20		0/39	
J 15	1/48	2.08 %	2/51	39.22 %
J 31	2/47	4,26 %	10/50	20.00 %
J 36	0/48		4/51	7.84 %
J 442	0/40		8/40	20,00 %
J 453	0/55		2/60	3,33 %
EM 513	11/47	23,40 %	4/50	8,00 %
Total				
Est. clones	9/13	69,2 %	12/13	92,3 %
Est. meristems	46/632	7,28 %	89/690	12,9 %

Number of rosettes/ Number of cultivated meristems

3.2. Shoot-tip culture

The use of shoot apices from the main plant axis or from flushing lateral buds as primary explants has been quite successful. Shoot tips (3 to 5cm long) are collected from shoots actively growing in the greenhouse, in the field or forced in growth chamber.

Dormant shoots can readily be exchanged internationally. Scions can also be grafted onto rootstocks growing in the greenhouse, providing a long-term supply of explants. Buds at the basal end of dormant shoots are usually smaller and are slower to initiate growth. Starch reserves within the scionwood are generally sufficient to support growth for a week or two (Lane, 1992).

Leafy shoots should be prevented from wilting before shoot tips are excised in the laboratory. The leaves are snapped off and the stems are put than in distilled water. After disinfection with Ca-hypochlorite solution containing 0.01% Tween 20, stirred for 20 min, stems are rinsed briefly in sterile distilled water followed by a second 5-min rinse with stirring. The shoots are kept in fresh sterile distilled water until final trimming before placing them in culture medium. Shoots are cut below a node, to 1-2 cm long and placed in 15ml liquid culture medium contained in Erlenmeyer flasks and rotated for 2-4 days in a culture room at 26°C. This is to detect bacterial and fungal contaminations, to rinse phenolic compounds leached from explant and to provide a maximum contact between the explant and the culture medium. The explants are then transferred to a solid culture medium of the same composition. Alternative procedures have been developed (Jones, 1993, Yepes and Aldwinckel, 1994) where shoot tips are taken shortly after bud-break.

Nodal segments from leafy shoots are collected either in the field or from potted plants cultivated in the glasshouse. An efficient and very simple sterilisation method deals with treatment of the young leafless shoots to formaldehyde (37%) vapour. Leaves are removed from three nodes stem pieces by cutting the petiole at the blade insertion level. The sterilization takes 10 to 20 minutes according to the explant tolerance. Basal and apical sliced parts of the stem on one hand, and distal part of the petioles on the other hand, are then removed before culture in the gelled medium. Generally, the cultivated explant is limited to the middle node. Any direct or indirect contact with the formaldehyde causes the necrosis of the plant tissue. Therefore, the original plant material has to be dry enough to prevent detrimental injuries of the dissolving vapours during the treatment. Jehangir *et al.*(1998) used 0.2 to 0.5% (W/v) $HgCl_2$ to sterilise MM106 rootstock axillary buds. Antibiotics have bacteriostatic and not bactericidal effect.

The shoot tips and buds are usually placed directly on the multiplication medium. Spring is the ideal season for establishing *in vitro* cultures with minimum contamination (Hammerschlag, 1986) and browning under control (Wang *et al.*, 1994). From axillary buds, the reduction of phenol exudation by adding antioxidants and absorbents improved the survival rate of the apple cv. Tydeman's Early Worcester (Mogdil *et al.*, 1999). Ascorbic acid, citric acid, 8-hydroxy-quinolinol-sulfate and activated charcoal are effective during initial days of *in vitro* culture (Laimer *et al.*, 1991, Yepes and Adwinckel, 1994).

When established *in vitro*, buds grow directly into foliated shoots or remain at the leaf rosette stage as a function of both the harvesting period of the twig and their original position on it. From M9 clones growing in greenhouses, shoots elongation was obtained from sub-apical buds in June and from basal ones in September. Such different growth response could be relevant to the dominance correlations regulating buds from the same branch according to the growth season. The addition of silver thiosulfate (STS) to the culture medium increased shoot length of Jork 9 rootstock after budbreak (Van Telgen *et al.*, 1992).

4. MULTIPLICATION BY AXILLARY BRANCHING

In 1993, Jones mentioned that about 70 apple rootstocks and cultivars had successfully been propagated by shoot culture methods. On an average, multiplication rates were 4 to 6, by regular subculture at one-month interval. Since then, the technique has been obviously tried on new genotypes but without any further drastic improvement.

Recently, preconditioning of shoots with high concentration of cytokinin followed by stem slicing has been reported to improve the propagation of Gala cultivar (Bommineni *et al.*, 2001). Multiple shoots were regenerated with thidiazuron in the nutrient medium and 2 weeks culture in darkness. This technique is also successful in genetic transformation experiments. The regenerants require a strict control of the genetic conformity (Zimmerman, 1997, Virscek *et al.*, 1999) before applying on a large scale. Molecular analysis would allow efficient screening (Watillon *et al.*, 1991, Tignon *et al.*, 2000, Benson *et al.*, 2001).

Actually, the most common and trust technique remains the axillary branching. Shoots of axillary origin develop successively two distinct growth events:

- first " budburst" when at the beginning of the culture, the axillary buds start to form the rosettes of leaves due to cytokinin effect,
- second "shoot development" when the newly formed internodes elongate, shoots are spontaneously formed.

4.1. Plant material

In general, regulation of axillary branching is genotypic dependent, as well the number as the growth and development of newly formed shoots.

Axillary branching can be significantly stimulated after:

- temporary shaking of newly established cultures in the liquid medium prior to grow the shoots on a solid medium
- a short-term cold storage (2 to 6 months at +4°C) of the shoot clusters maintained on the culture medium or of the shoot explants isolated on a fresh medium free of growth regulator.
- the partial overlays of the shoot clusters with a liquid phase (water, basal medium, propagation medium) added on upper

surface of the solidified culture medium (Ali *et al.*, 1996). This approach was initiated with pear (Viseur, 1987).

Slow multiplication rate and lack of vigour can be reverted by discarding the basal part of the shoots either by removing the leaves or by inverting the shoots in agar. When cultivated horizontally on the culture medium, decapitated leafy shoots of Golden Delicious and Jonagold cultivars, of M111 and M9 rootstocks, formed the highest number of axillaries and consequently the highest number of elongated shoots. The improvement reached minimum 10% and maximum 70%. Leaf removal was unfavourable for shoot growth.

Shoot elongation starts during the second half of the monthly propagation cycle. It is generally limited to few shoots within clusters as a consequence of the progressive declining effect of the cytokinin as a dominancy correlation establishing between growing shoots. Both « budburst » and « shoot development » steps are term relative and consequently determine a variable amount of handlable explants from subculture to subculture. A certain homogeneity of the axillary shoot population would apparently be restored (at least partly) by keeping the apical part of the elongated shoot as a new explant sized on an average length of the shorter axillaries.

4.2. Environmental factors

Proliferating shoot cultures are being grown at a light intensity of 50-70 $\mu E\ m^{-2}s^{-1}$ (400-700 nm) provided by Philips daylight or by Sylvania gro lux fluorescent lamps, 16h photoperiod and constant temperature of 23 ± 1°C. The level of irradiance and spectral composition influenced the multiplication rate according to the clone in Golden Delicious variety (Noe *et al.*, 1997).

Container design interferes with the shoot development mainly due to the atmosphere (volume, gaseous exchange, volatile compounds). With M26 rootstock, high atmospheric volume of the vessel allows easier control of the vitrification symptoms. Moreover, shoot size deviation increased significantly from 460 ml to 710 ml.
With MM111 and M9 Lambardi *et al.* (1997) found that high exogenous concentration of ethylene in the culture flasks had negative effects on shoot proliferation. The first one rootstock appeared to be more sensitive to ethylene than the second one

4.3. Composition of the culture medium

A very efficient medium to multiply *Malus* (and other fruit tree species belonging to *Prunus, Cydonia* and *Pyrus* genera) has already been published (Druart 1987, 1988). It generally allows a high monthly multiplication rate while the vitrification remains at a low level (Druart, 1988). The composition is based on Quoirin and Lepoivre macroelements (0.4g.l^{-1} NH_4NO_3, 1.8 g.l^{-1} KNO_3, 0.36 g.l^{-1} $MgSO_4.7H_2O$, 0.27 g.l^{-1} KH_2PO_4, 1.2 g.l^{-1} $Ca(NO_3)_2$) and modified microelements (8.6 mg.l^{-1} $ZnSO_4$. $7H_2O$, 12 mg.l^{-1} H_3BO_3, 1 mg.l^{-1} $MnSO_4.4H_2O$, 0.025 mg.l^{-1} $CuSO_4.5H_2O$, 0.080 mg.l^{-1} KI, 0.25 $Na_2MoO_4.2H_2O$, 0.025 mg.l^{-1} $CoCl_2.6H_2O$), Walkey vitamins (0.4 mg.l^{-1} $thiamine_{HCl}$, 100 mg.l^{-1} meso-inositol), 100 mg/1 L-methionine and L-tyrosine amino acids, 1 mg.l^{-1} BA, 0.1 mg.1^{-1} IBA_{KOH}, a carbohydrate solution resulting from a previous autoclaving of the sucrose (3%) in the presence of activated charcoal (1%), pH 5.5, Pastagar (5 g.1^{-1}).

There is a correlation between the internode length and the carbohydrate content in dwarf apple trees (Wang and Faust, 1987). Exogenous C sources may disturb the internal C balance of feeding vitro-shoots and consequently could affect their growth regulation. Significant differences in shoot development were seen in cultivars McIntosh, Make and Jaspi whereas the cv. Gala was little affected by carbohydrates (Karhu, 1995). In our experiment (Druart, 1995), fructose stimulated the budburst but simultaneously extended hyperhydrical disorders. M9 clone 29 and M26 formed their tallest shoots, respectively, on glucose (3%) and on fructose + glucose (both 1.25%) without or with sucrose (0.5%). Hydrolysed sucrose is a very efficient C source for proliferation (Seingre *et al.*, 1991b, Druart and Dewulf, 1993)

Ribaux *et al.* (1995) measured the sugar absorption during *in vitro* proliferation of the apple rootstock M9 from the rate of carbohydrate disappearance from culture media. Fructose was absorbed most rapidly and sorbitol was the slowest, whereas sucrose was hydrolysed during the whole culture period into glucose and fructose. Sorbitol is the major translocation form of carbon in *Malus* and may be more effective than sucrose with some apple cultivars (see Jones, 1993, Seingre *et al.*, 1991b). Axillary branching enhanced in McIntosh and Jaspi when sorbitol was provided as the carbon source (Karhu, 1995 and 1997).

Several studies demonstrated that the gelling agent could influence the anatomy and the physiology of cultured shoots. The modified response of explants is independent of the media components. The gel rigidity influences the hormone and water availability (Seingre *et al.*, 1991a, Orlikowska and Olszewski, 1993, Zimmerman *et al.*, 1995). Amongst the C-sources, fructose was related with a drastic reduction in the gel firmness of the culture medium.

Thidiazuron has also been shown to stimulate apple proliferation at concentration considerably lower than benzyladenine. However, shoot proliferation was shown to continue after the explants were transferred to cytokinin-free medium.

Several subcultures are frequently required to normalize the explant reaction to the cytokinin. Experiments carried out over 25 monthly subcultures with already well-established lines of M26 and Golden Delicious showed that the multiplication tended to their lowest rate after 5 to 10 cultures (according to the genotype) under uniform conditions. Temporary improvement can be restored within 3 to 5 subcultures when the medium composition is modified (macroelements concentration, vitamin D2, amino acids or any other supply) (Druart, 1988). Due to putative hyperhydricity consequences, the optimal number of stimulative subcultures is critical to determine.

5. ROOTING

The success of the rooting is subordinated to several factors which interfere together. Those include the conditions during preceding subcultures, handling of the shoots at the time of their isolation, the components of the culture medium and the environmental conditions of the rooting phase. They make high rates and quality of the rooting sometimes difficult to realize when an optimal reproducibility is required.

5.1. Plant material

The rooting response is cultivar dependent. Apple genotypes could be classified according to their requirements for rooting. The easy to root genotypes are able to form a high quality rooting system after a single auxin treatment. A second group requires a complementary rhizogenous stimulus, e.g. a darkness treatment of the shoots in the beginning of the rooting stage. The difficult to root group needs additional treatment, e.g. vitamin D2 or L-proline (Druart, 1987 and 1997, Lisek, 1996), L-arginine (Orlikowska, 1992a) or potassium humate (Baraldi *et al.*, 1991), etc.,in the culture media to increase the auxin effect in the darkness without declining the root quality.

Vigorously growing shoots from proliferating cultures are most often used to make cuttings about 2 cm long. However, according to the cultivar, an elongation stage of the microcuttings is suitably considered as a preparative phase to rooting. It results from the transfer of the clusters to a specific medium containing gibberellic acid or better, from the addition of a liquid medium onto the multiplication solidified medium. Partial immersion stimulates both branching and shoot growth , and increases then drastically the rootable shoots from 36% to 60% (Table 2). With M26 rootstock, the addition of ethrel (0.1%) together with GA_3 (1mg.l^{-1}) tripled the number of shoots to root while the liquid multiplication medium gave similar results however, less favourable for further root formation.

Table 2. Influence of the elongation (EL) culture condition on the axillary branching and shoot elongation performances of M26 rootstock.

Parameters	Gelified (EL) medium	Liquid medium added onto the gelified multiplication medium (MM)		
		EL	EL + 1% Ethrel	MM medium
Multipl. rate	6.77 ± 1.24 *100*	9.16 ± 1.52 *135.3*	12.13 + 0.55 *179.2*	12.11 ± 1.70 *178.9*
Elong. rate	2.44 ± 0.67 *100*	5.40 ± 1.00 *221.3*	7.65 ± 0.98 *313.5*	7.20 ± 2.59 *295.1*
Ratio (%)	36.0	59.0	63.1	59.5

EL medium is the basal medium supplied with 1mg.$l^{-1}$$GA_3$.

With the cv. compact Spartan, the first 3 subapical leaves (mainly the second and the third ones) are essential although removal of apex only delayed root emergence (Druart, 1994).

5.2. Composition of the culture medium

In our laboratory, "NK" rooting medium is the most efficient to root woody fruit trees in general. It consists of 1/2 Quoirin et Lepoivre macronutrients and full Quoirin and Lepoivre micronutrients, 0.4 mg.l^{-1}thiamine HCl, 100 mg.l^{-1} meso-inositol, 2 mg.l^{-1} indolebutyric acid (IBA_{KOH}), 2 % sucrose, pH adjusted to 5.6, as previously described (Druart *et al.*, 1982), completed by a vitamin and amino acid complex called "NK complex" (Druart, 1992). This complex contains riboflavin (1 mg.l^{-1}), pyridoxine (0.1 mg.l^{-1}), nicotinic acid (25 mg.l^{-1}), Ca panthothenate (1 mg.l^{-1}), ascorbic acid (1 mg.l^{-1}), and D2 vit. watersoluble (5 ml.l^{-1}), K5 vitamin (5 mg.l^{-1}) and L-prolin (100 mg.l^{-1}).

The auxin treatment covers root induction and initiation and initiates in complete darkness. Lowering the minerals to 1/2, 1/3 even 1/4 of the initial concentration generally leads to improved *in vitro* rooting. With some cultivars, further root emergence and elongation take place on watered (100 ml distilled water/250 ml vermiculite) vermiculite (grade 3) or on the solidified basal medium. NH_4NO_3 was then an inhibiting factor for cultivars like compact Spartan, which is highly recalcitrant (Druart, 1997). On the contrary, Ca $(NO_3)_2$ stimulated root initiation, emergence and growth, and improved root viability. The level of inhibition due to NH_4NO_3 depended on the agar brand. The physical structure of the culture medium plays a very important role in high quality rooting (Navatel and Bourrain, 1994). Sucrose was essential to root compact Spartan on agar medium but unnecessary on watered vermiculite. The superiority of sucrose over fructose, glucose, sorbitol and xylitol for rooting has been proven (Karhu and Ulvinen (1995).

Ethylene does not promote root initiation. ACC (1-amino cyclopropane carboxylic acid) strongly inhibits the induction phase (Harbage and Stimart, 1996a). Ma *et al.* (1997) reported that ACC delays root emergence and increases callus formation at the base of the shoots while $AgNO_3$ stimulates root emergence and enhances root growth. Like $AgN0_3$, AVG (aminoethoxyvinyl glycine) increases the number of roots/shoot. $CoCl_2$ slightly increases root number and rooting efficiency. This may result from a reduction in ethylene concentration or inhibition of ethylene action. Silver thiosulfate at certain concentration relative to the auxin content of the culture medium, would stimulate shoot rooting but not the rooting of stem segments or slices (de Klerk *et al.*, 1999).

The pH of unbuffered rooting medium increases from 5.5 to 7 within 2 days culture (Harbage and Stimart, 1996b, Harbage *et al.*, 1998). As pH decreases, lower concentrations of IBA are required to induce high number of roots. Root count differences between easy-to-root and difficult-to-root cultivars are not consistent and could not be explained solely by IBA loss from root induction medium.

5.3 Controversial effect of phloroglucinol

Phloroglucinol has stimulated rooting of some apple rootstocks and cultivars while having no effect on others (see Hammerschlag, 1986, Caboni *et al.*, 1994, Lisek, 1996). Phloroglucinol (162 $mg.l^{-1}$) increased rooting rate of MM106 microshoots (Aklan *et al.*, 1997) and at 100 $mg.l^{-1}$ during the whole rooting process proved beneficial for early and better root development of apple cv. Tydeman's Early Worcester (Mogdil *et al.*, 1999). With M26, phoroglucinol had no significant effect (Shawky *et al.*, 1993). Its beneficial effect depends on the duration of subcultures; that would mean a certain degree of tissue rejuvenation (Jones, 1993).

Zimmerman pointed out that the different rooting responses could be due to different environmental conditions at the time when the explant is taken and its physiological state. However, we cannot exclude the indirect effect of the phloroglucinol on contaminant putatively present in the initial bud explant as responsible of the rooting improvement (Savela and Uosukainen, 1994).

Amongst the phenolic compounds, dithiothreitol stimulates the rooting of difficult-to-root Cepiland clone in the presence of IAA (indol-3-acetic acid). Roots were longer and more robust than in presence of IAA alone (Auderset *et al.*,1996 and 1997).

5.4. Environmental factors

Several studies have shown the positive effect of dark treatment on rooting of apple (see Hammerschlag, 1986). Optimal period of darkness at the beginning of the rooting stage is from 3 to 9 days according to the ambient temperature and specific root abilities (Druart *et al.*, 1982, Druart, 1987).

5.5. Control of the rooting stimulus

The total auxin activity resulting from cumulative physiological, physical and chemical factors have to be controlled with the aim to ensure optimal root induction and initiation. Long auxin treatment generally acts like excessive auxin concentration that means callus formation and root inhibition and even hindrance in shoot apical growth. However, some recalcitrant cultivars like compact Spartan looked different:

- our most efficient rooting conditions which combined IBA_{KOH} (2 $mg.l^{-1}$), 5 days of darkness and the addition of the specific organic NK complex did not succeed in rooting the shoots,
- the unrooted shoots didn't form any friable callus which commonly characterizes root inhibition,
- the root emerged after transfer on watered vermiculite as expected in case of root inhibition.

Total absence of friable callus at the stem section theoretically excluded any excessive auxin activity as a potential cause for root growth inhibition. On the contrary, such observation would entail a stronger auxin treatment. Only the complete *in vitro* rooting medium where the hormonal compounds, NH_4NO_3 and the gel were distinctly inhibitors, prevented root emergence and growth.

The auxin treatment could be appropriately interrupted by transplanting the induced shoots on hormone free medium possibly combined with the addition of activated charcoal although it could reduce the number of roots/shoot (Shawki *et al.*, 1993), or some antioxidants (Standardi and Romani, 1990) or by increasing the photoxydation of the auxin in the presence of riboflavin.

5.6. Phases of adventitious root formation

Several combined histological analysis, biochemical characterisation and/or molecular tools (mainly as part of the european COST Action 87) were done to investigate the rooting process on apple rootstock (Auderset *et al.*, 1994, Seifert *et al.*, 1994, De Klerk and Ter Brugge, 1992, Welander and Pawlicki, 1993, De Klerk *et al.*, 1995, 1997 and 1999, Jasik and De Klerk, 1997, Kamboj *et al.*, 1997, Butler and Gallagher, 1998, Vegvari, 1999, Kromer and Gamian, 2000). Successive organogenic events have been then described very precisely on shoots, stem segments (1cm) and on stem discs (1mm) models. The successive phases called dedifferentiation, induction, initiation (or re-differentiation) and elongation need different hormonal balances. High level of auxin and low level of cytokinin are required only during the induction phase.

Histological observation showed after 24 h, that cells with swollen nuclei and dense cytoplasm appeared in the regions of the stem where the roots formed. The first cell divisions were observed after 48h (De Klerk *et al.*, 1995). Root primordium is identifiable about 100 hours after auxin treatment and originates from the interfascicular parenchyma neighbouring the vascular bundles. The morphogenic area appeared close to the activated cambium in a discontinued manner. It is localised in the 1.5-2 mm basal segment of the stem (Auderset *et al.*, 1994). The initiating signal may be the wounding of the stem at the shoot transfer; the cut effect may be amplified by exogenous supply of auxin. IAA uptake by 1mm thick slices of Jork 9 stem is dependent on the thickness and on the medium layer (Guan *et al.*, 2000).

The influence on root and callus formation (M9 Jork stem discs) of IBA concentration (15-65μM), culture duration (45 min to 11h), explant position and orientation, temperature and light intensity has been investigated. Formation of smooth callus was increased by light intensities below 0.85 W/m^2. Rooting efficiency decreased and filamentous callus increased at 19°C and 25°C. Explants close to the shoot apex displayed reduced rooting efficiency and profuse filamentous callus (Seifert *et al.*, 1995). Sorbitol reduced callus formation while combination of sucrose/mannitol gave 100% rooted discs, the highest number of roots per disc and the lowest callus formation (Pawlicki and Welander, 1995)

In practice, all well-grown shoots are able to root after a three to five-day-dark treatment at low auxin concentration. This period varies from one species to another and even from one shoot to another. All the shoots are physiologically different at the onset of rooting. With the apple rootstock M9, only 12 to 48 h is the required period of shoots contact with indolebutyric acid. Quick dip in high concentrated auxin solution reveal sometimes satisfactory according to the cultivar (Collet *et al.*, 1994). Induced shoots are able to keep their rooting ability for several months (at least 6) at low temperature (+ 4°C). The latter aspect makes easier the plant sales and acclimatisation.

6. ACCLIMATISATION

Plantlets are removed from the culture vessel after 3-4 weeks on the rooting medium. Elongated roots provide sufficient cytokinin for plant growth (Lane, 1992). The agar is washed off the roots and the plantlets are placed in plastic boxes and suitably stored at 4°C before transplantation in a well-drained soil mixture (1vol. vermiculite/2vol. compost).

Usually misting or fogging prevent leaves desiccation which occurs after short time exposure to the ambient atmosphere. Easiest way is to keep the plants growing under a plastic film at the ambient temperature of the glasshouse. The height of the plastic cover allows warm air to rise from the plantlets maintaining their temperature at the plantlet level at 16-20°C (Lane, 1992). When acclimatisation takes place in growth chamber, the temperature is regulated around 21-23°C.

Care must be taken to avoid over-watering the soil mixture and a spray with fungicide prevents fungal contamination. Once the shoot-tip grows actively, the plastic film can be progressively removed. Several pre-treatment have been proposed to quicken the shoot growth: pre-exposure to high light intensity at high humidity; removing the caps 3-5 days before removing the plants from the tubes; transfer of plants to a lower concentrated medium (see Zimmerman, 1986). Storage at 4°C for 7 to 12 days hardens the plantlets against the transplantation stress, synchronises shoot apex regrowth, hasten root functions and soil solution exchanges.

Stomata are observed in great number on the lower epidermis cells of J-TE-F rootstock (Gyorgy, 1996) and close in response to reduced relative humidity, demonstrating they are functioning. Stomata density was highest in leaves of the M 9 rootstock exposed to continuous light and lowest in those exposed to continuous dark (Zacchini *et al.*, 1997). It decreases from the 1st to the 3rd to the 5th leaf counting down from the apex. While Diaz-Perrez *et al.* (1995) observed that roots formed *in vitro* are associated with a higher plant water status, suggesting that these roots are functional in water uptake. Vegvari and Vertesy (1999) reported that roots of MM106 formed *in vitro* generally lose their hairs during acclimatisation, but new roots of full value develop out of the stem. On our side, the root hairs appeared during the cold storage of the plantlets. Minerals of the rooting medium are favourable for leaf development and the choice of the carbohydrate has to be taken into consideration for further growth into soil (Karhu and Ulvinen, 1995). Inoculation with arbuscular mycorrhizae both at the rooting and the acclimatization stages, enhances plant weaning, mineral uptake and further plant growth (Uosukainen, 1992, Uosukainen and Vestberg,

1994, Schubert and Lubraco,2000). With MM106 apple, growth enhancement and percentage of actively growing apices Glomus–inoculated plantlets were comparable to those obtained in phosphate fertilized plantlets (Fortuna *et al.*, 1996).

After 1 week of root induction in the presence of IBA and darkness followed by incubation under light for root elongation, Bolar *et al.* (1998) transferred plantlets to Jiffy-7 supplemented with biological plant protectants and fertilizers in plastic humidity tray, and later in pots covered by plastic bags. Changes in the chlorophyll fluorescence values and the differences in shoot growth could not be explained only by the number of roots: shoot ratio.

The direct transfer to the dry culture conditions at the end of spring/beginning of summer time stresses leafy and just acclimatised plants. Consequently, a great number of them stop their growth for unpredictable term. Once again, a chilling period at 4°C for 3weeks prevent such a rest and achieves further uniform growth. Gibberellic acid (0.1-0.3g.l^{-1}) applied at 10-days-interval may be beneficial (Lane, 1992) but sometimes causes a bushy development. Micorrhizal inoculation can overcome blocked apical growth reducing chemical inputs (Fortuna *et al.*, 1996).

7. FIELD BEHAVIOUR

7.1. Improved conventional propagation

The propagation of rootstocks took place mainly to improve conventional propagation of cuttings or stool beds. *In vitro* origin of the mother-plants gives the advantage of a higher number of rooted plants and a higher quality of rooting system (Quamme and Hogue, 1994). Production of P22 was about twice that from traditional propagated mother plants and gave less thorny rootstocks (Czynczyk *et al.,* 1994). Much improved rooting has been observed with both winter and summer cuttings from field hedges of the apple rootstock M9 (see Jones, 1993, Sharma and Webster, 1993, Isutsa *et al.*, 1998). This improved propagation has been sustained for years. Drastic pruning allows some physiological adjustment but nevertheless, renewal of the mother-plants is justified at a regular frequency.

7.2. Own-rooted cultivars

Self-rooted scions need several years of field evaluation to determine tree vigour, flowering precocity and intensity, fruit quality and yield, plant conformity, before the producers can accept the technology. Self-rooting could be considered for spur types but is not convenient for vigorous varieties. Own-rooted cultivars hold great promise for the areas where tree top is damaged due to cold temperature leading to death, and would only require regeneration from the established root system.

Micropropagated cultivars of Gala and Triple Red Delicious make more total growth than trees on clonal rootstocks (Hirst and Ferree, 1995). In Beltsville (U.S.A), higher vigour has been measured at the crown and trunk developments, and greater lateral branching. The differences have persisted for several years. In East Malling (England), small tissue culture tree, allowed to go dormant while still in pots then planted in the nursery, did not perform well in the first and sometimes during the second year after planting out. In Bologna (Italy), spur-type strain of Golden Delicious was self-rooted. Other micropropagated trees of Redspur Delicious apple lacked uniformity in tree size, appearance and flowering (Zimmerman, 1997). No differences in fruit quality were apparent.

Vigorous tissue culture trees are susceptible to leaning and require a better balance between top and roots. Increased trees densities and plant growth regulators sprays (like paclobutrazole, uniconazole and daminozide + etephon) are used to control the tree size and flowering potential (Buban *et al.*, 1993, Zimmerman and Steffens, 1995).

In Gembloux (Belgium), Cox's Orange variety grafted on M26 plants issued from *in vitro* did not show significant differences about the fruit properties. Self rooted trees of Golden Delicious confirmed higher tree vigour (Druart and Boxus, 1987). Other varieties like Jonagold, Mutsu, Jonathan showed similar behaviour. Most of the cultivars started to flower one year later than the grafted trees, during third or fourth growing season. Fruit yield vary with cultivar and increase as tree age increases. Low fruit yield is due to lower number of fruits instead of fruit weight. Excessive leafy twigs decrease the fruit colouration and commercial quality mainly due to a lack of exposure to the sunlight during maturation. Meristem lines of a red mutant of Jonagold and a columnar mutant of McIntosh were true-to-type when propagated by axillary branching.

8. SHORT AND LONG-TERM PRESERVATION

8.1. Shoot preservation at low temperature

Storage of the cultures at + 4°C allows interruption of propagation at the multiplication stage for 3 to 6 months.

The preservation of apple microcuttings for 3 to 4 years takes place at + 2°C in darkness. Defoliated and tapped shoots are half-immersed in a nutritive basal culture medium without growth regulator but enriched with 100 $mg.l^{-1}$ L-prolin. Wounding the stem and petioles sometimes causes phytotoxic phenol exudation. The adaptation of previous culture conditions for proper preservation of cultivars, amended with the addition of an antioxidant to the medium. Alternatively, transfer the propagules to a new fresh medium in order to save them.

8.2. Node encapsulation

Uni-nodal microcuttings of cv. Starkspur Red can survive up to 2 months in microvessels without culture medium (Negri *et al.*, 1995). Ninety percent single nodes from M26 rootstock produced plantlets with well-formed shoots and a root system after 30 days of culture *in vitro* and a 24h pre-treatment (24.6 μM IBA + 15 g sucrose /litre) in darkness.
Culture of single nodes for 3-9 days in darkness before encapsulation together with the addition of growth regulators to the artificial endosperm is required to reach 58 to 66% of plantlets production. The presence of activated charcoal could be detrimental to the survival of the encapsulated explants when it is added in presence of the other nutrients.

Synthetic seeds from apical buds of M26 reached conversion rates of 85% when compared with 25% obtained with encapsulated axillary buds. Direct encapsulation after auxin treatment had an inhibitory effect on rooting. Postponed encapsulation for a few days allowed the formation of root primordia and the subsequent successful conversion of the synseeds (Piccioni, 1997). Root induction treatment (5 $mg.l^{-1}$ IBA), followed by root initiation culture for 3-6 days prior to encapsulation with a nutrient-enriched artificial endosperm, was the best protocol for inducing rooting of the encapsulated explants. Cold treatment for 60 days reduced the formation of undesired callus that is negatively correlated with rooting (Capuano *et al.*, 1998).

8.3. Cryopreservation

Cryopreservation was reported very efficient method for long-term preservation. It has been carried out recently with several kinds of apple micropropagules (Niino *et al.*, 1992, Piccioni and Standardi, 1995, Standardi and Piccioni, 1998, Paul *et al.*, 2000, Siocurani *et al.*, 2001). The physiological state of the plant material directly affects the results and the procedure. Cold - hardening of mother plants for 3 weeks at 5°C improved regrowth rate whatever the cryopreservation technique employed (two-step freezing, vitrification, and encapsulation-dehydration) (Wu *et al.*, 1999).
Encapsulation-dehydration method was the best performable. After encapsulation with 3 % Na-alginate with 100mM $CaCl_2$, embedded shoot-tips are dehydrated for 4 to 6 hours (21% residual water) by exposure to a sterile air flow before freezing in liquid nitrogen (Wu *et*

al., 1999, Paul *et al*., 2000). A pre-culture of encapsulated apices in 0. 75 M sucrose MS medium led to the highest shoot regrowth. Sucrose and sorbitol presented the best cryoprotective effect. However, increasing the time during which mother-plants were maintained without subculture before sampling of apices decreases the water content of shoot-tips and the duration of the sucrose pregrowth treatment of encapsulated apices. After 6 months, sucrose (progressive increase concentration: 0.1, 0.3, 0.7 M) preculture of explant became unnecessary to achieve regrowth after immersion in liquid nitrogen and the dehydration period was shortened (Zhao *et al*., 1999a).
70 to 92 % of the frozen shoot-tips regrow directly without callus formation after using a droplet freezing method (Zhao *et al*., 1999b). After 3 weeks of cold-hardening at 5°C, the shoot tips are pre-cultured in liquid medium with 1 % DMSO (dimethyl sulfoxide) for 24 hours at 5° C, treated for 75 min. at room temperature with 0.3M sucrose and 15 % DMSO, placed in 5 µl droplets of cryoprotectant medium containing 0.3 M sucrose and 15% DMSO on aluminium foil and cooled to – 40°C at 0.2°C/min before immersion in liquid nitrogen.

Dormant vegetative buds from cold-tender apple germplasm require stabilization by encapsulation in 5% alginate and step-wise imbibition treatments of 0.5 to1.0M sucrose and 0.2 M raffinose solution for protecting against the injury during forced air desiccation at 0°C and cryopreservation at – 30°C. This procedure uses non-toxic cryoprotectants. Sucrose levels did not increase following stabilization treatment, but levels of glucose and fructose and of an unknown disaccharide increased (Seufferheld *et al*., 1999).

9. PROBLEMS AND INTERESTS

9.1. Bacterial contamination

Sudden appearance of bacterial contaminations after several months of culture has been reported. These are external contaminants from soil and water, surviving in alcohol. Usual surface-sterilization do not apparently eliminate these bacteria. To detect the presence of bacteria and to discard contaminated explants, the freshly cut shoots are screened on a specific bacteriological agar medium. Clean apices (about 1.0 cm size) are then excised with care, avoiding any contact of the tools with the shoot or culture medium areas contaminated by the bacteria. In case of failure, a shoot pre-etiolating in the dark at the room temperature is very often successful.

9.2. Variability of response

Cultivars do not respond in the same way and temporarily require appropriate treatments during establishment, proliferation or rooting in order to restore optimal growth performances. Furthermore, different explants of the same cultivar respond specifically to similar culture conditions applied in the same laboratory. Proliferation rate vary with cultivar in a range of 3 to 6 in 4 week cycle subculture. Successive nutritive stress would be recommended to keep the propagation at an optimal and more stable rate according to the hyperhydrical sensitivity.

9.3. Hyperhydricity

Erroneously called "vitrification", hyperhydricity is a well-known physiological disorder affecting vitro-cultures. However, it remains very complex phenomenon, cannot be kept totally under control and requires efficient means allowing the recovery of plant material.

Hyperhydricity affects the explants from their *in vitro* establishment and appears spontaneously during the multiplication cycles. Vitrified shoots are characterised by malformed, strap-like and curled leaves, and inhibition of further stem elongation. Cuticle was thickened, chapped and often separated from the epidermis of the J-TE-F rootstock (Gyorgy, 1996).

Vegvary (1999) analysed the dynamics of the nutrient uptake by the shoots of normal aspect. Cu, P and Zn are utilised totally, but only 50% of K and 20-40% of Ca and Mg are taken. Ca would be utilised at the beginning of the subculture, B in the later phases. P and K are taken at variable rates during the culture period. In case of hyperhydrical disorder, our spectrometry analysis made on the M27 rootstock, showed that the total mineral content of the shoots, increased significantly except for Mn that remained unchanged. Total nitrogen also increased from 15 to 20 % and aspartic acid accumulated in the leaves of another M26 rootstock. There is a similarity with hyperhydrical plant material of *P. cerasus* cv. Rhexii that accumulated Ca and Mg macroelements and Zn microelement when cultivated in the same conditions (Druart, 1987).

Hydrolysed agar protects against hyperhydricity (Marga *et al.*, 1995) and inhibits growth at the same time. Marga *et al.* (1997) demonstrated that oligosaccharides isolated from Difco agar reduced the occurrence of hyperhydricity. The occurrence of changes in pectic content, percentage of uronic acid and associated ions took place in the M26 shoots cultures growing on media with or without hydrolysed agar during a span of several days. The most anti-hyperhydric fraction had molecular weight slightly less than 1900 daltons and contained methylated and sulphated galactose derivatives. We demonstrated the significant effect of galactose on the control of the hyperhydricity and on shoots recovery, when supplied at 0.5 to 1% to the multiplication medium (Druart, 1988). Using labelled compounds, we observed that galactose was translocated from the culture medium directly to the apex of M26 microcuttings where the leaflets of the vitrified shoots recovered normal aspect after about 10 days culture. Galactose was detected in the vacuole as in the cell wall but didn't affect significantly benzyladenine or sucrose uptakes. The anti-hyperhydric effect was confirmed in other culture circumstances with several cultivars like M26 and M9 clones cultivated in the presence of 1% fructose + 1% glucose (Table 3). However, the enrichment of the culture medium with galactose would be recommended only when hyperhydricity becomes out of control.

Table 3. Comparison of M 26 and M 9 clone 29 rootstocks about the propagation and the rate of hyperhydricity* in function of the addition of galactose to a culture medium containing fructose (1%) + glucose (1%) as basic C-source.

Galactose (%)	Rootstock	Number of axillary shoots	Vitrification (%)		
			Rate	Min.	Max.
0	M 26	5.55 ± 1.08	47,2 ± 11.8	19.0	83.5
	M 9 clone 29	5.39 ± 0.86	32.2 ± 9.0	5.2	53.0
0.5	M 26	7,61 ± 0.36	16.8 ± 6.8	7.0	43.4
	M 9 clone 29	4,60 ± 0.29	14.1 ± 5.7	1.3	32.1
1	M 26	7.16 ± 0.76	8.3 ± 3.7	1.3	22.2
	M 9 clone 29	4.54 ± 0.38	4.2 ± 2.8	0	14.5

*: Average of 5 subcultures and 42 axillary shoot clusters each ± standard error.

Zimmerman (1986) reported that tissue hyperhydricity is a greater problem with the WPM (Woody Plant Medium) medium. Replacing the salt formulation, BA from the propagation medium with 2ip (Δ^2-isopentenyladenine), the gelling agent and lowering temperature, contribute to control hyperhydricity.

9.4. Phenotypic un*stability*

After micropropagation, a few Redspur Delicious trees lost their typical spur-type growth habit (Zimmerman, 1997). However, further micropropagation of the trees that have maintained clonal fidelity gave no more variation.

McMeans *et al.* (1998) observed no difference between Royal Gala and Gala fruits from trees of axillary or adventitious origin. On the contrary, several variants of Greensleaves (Chevreau *et al.*, 1998), Jonagold and of aneuploïd genotypes (Druart, 2000) have been regenerated from leaves. Any variability would be advantageously detected before the plants reach the field.

Varieties could be identified by various types of biochemical and molecular markers (Weeden, N.F. and Lamb, R.C., 1985, Guilford, P. *et al.*, 1997, Hokanson, S.C. *et al.*, 1997.Grandfranceschi *et al.*,1998). The isoenzymatic systems were the first used (Bournival and Korban, 1987). Their use however is limited by the paucity of informative isozymes, the potential sensitivity of those to the environmental variations and the stage of development of the plant, and do not allow to reach variability potentially present in the genome. Among the molecular markers, the techniques of RFLP, AFLP and of amplification of the microsatellites sequences constitute methods of references for the identification of polymorphisms. RFLP technique was applied successfully to the identification of apple clones (Watillon *et al.,* 1991) and of seedlings (Nybom and Schaal, 1990). The variable levels of polymorphism are closely related to the choice of the probe used. Only one probe (pW-1) in combination with three enzymes of restriction thus made it possible to identify 10 varieties of apple trees. Various combinations of ADNc probes and restriction enzymes used by Tignon (2001) resulted in revealing to ten polymorphic fragments by combination. The profiles were of a clear and reliable legibility since they consisted of a limited number of tapes, contrary to other more polymorphic probes. The RFLP needs however more time and much labour, it requires large amount of DNA. Selective AFLP primers give highly polymorphic fragments. Automatizable, requiring the extraction of little DNA amount and allowing the treatment of a great number of samples, AFLP is very reproducible and more powerful for the identification of cultivars of narrowly related apple trees (Tignon, 2001). The sequences of microsatellites constitute very robust and highly informative tools for identification. It however requires a preliminary study of the loci presenting microsatellites. The unit of plant genetic resources at Geneva (New York,U.S.) developed eight SSRs markers that constitute today, a method of reference for the evaluation of the genetic diversity of *Malus* species (Hokanson *et al.,* 1998).

The identification by molecular tools requires the pre-definition of standard profiles on certified cultivars but does not make it possible to identify mutants for example.

9.5. Juvenility

In this phase, trees grow very actively without initiation of flowering. Its duration vary in years with the species and the growth conditions. Changes in the leaf serration, size and shape of the leaves are typical seedling traits mentioned in the literature concerning the reversion to the juvenile phase. The juvenile tissue culture trees are characterised by the appearance of more abundant suckers and burr-knots than conventionally produced counterparts and this may lead to increased susceptibility to pathogens and more expensive tree management, respectively.

9.6. Cost of plant production

The major problem of *in vitro* micropropagation for the production of scion cultivars is the high labor cost. The work of *in vitro* transplantation, sorting and transfer of the plantlets in acclimatization represented nearly 95% of a production cost limited to the manipulation of the plant material during multiplication and rooting phases; the remainder corresponded to the culture media expenses. Rooting and transfer to acclimatization together reached 65 to 71 % of this labour cost for the rootstock M26 and 55 to 63 % for the variety Golden Delicious respectively. The influence of the rooting success is more important than the multiplication phases. For a ± 22 % variation of the average rate of multiplication, the final cost of the plantlet varied only from ± 10% with M26 and ± 16% with Golden Delicious. With Golden Delicious that the awaited rooting rate could pass from 80 % to 33 %, the final cost of the plantlet can be multiplied by 2.4 coefficient. This phase of *in vitro* rooting thus requires a meticulous preparation of the plant material and qualified workers for handling the micropropagules. A 15 to 20 % difference of the rooting efficiency has been registered between peoples who cut the shoots of M 27. At this stage and during further acclimatization, contamination occurrence is also dramatically prejudicial.

Replacement of rooting *in vitro* by direct method, nutrient replenishment through adding fresh liquid culture medium, automation or robotic reduce the cost of *in vitro* plants.

Rooting could be initiated in liquid medium. Jones (1993) reported that Zimmerman induced roots in darkness in an auxin-sucrose liquid medium and root formation in preformed peat plugs, a method applied to several cultivars (Zimmerman and Fordham, 1985). With

M26, dipping shoots into IBA powder and rooting directly into potting compost was more successful than *in vitro* rooting (Simmonds, 1993) and improve *ex vitro* establishment. Liquid medium with cuttings supported by vermiculite, perlite or sand are also used with or without auxin for some cultivars.

Apple trees have been produced by grafting scion plants on small potted rootstock approximately 50 cm height and issued from *in vitro*. They can be produced at any time of the year and develop large enough tree for field planting within a six months growth period. Other studies have been carried out with only the scion or the rootstock component of the grafted trees originating from *in vitro* (see Jones, 1993). Grafting microcuttings on pre-acclimatized rootstocks speed up the procedure of virus elimination after meristem culture and the production of virus-free trees.

11. CONCLUSIONS

Success is achieved with most apple cultivars and mainly subjects such as scion cultivars difficult or impossible to propagate by conventional methods. A progressive increase in the propagation potential is cultivar dependent, and generally reported with increasing subcultures.

Cost reduction, control of physiological disorders occurring during the *in vitro* culture and reversion of the juvenile characteristics induced by excessive cytokinin supply would continue to enlarge the application of this method of propagation.

Acknowledgement

The author is very grateful to Etienne, N. and Richet, S. for their technical assistance.

REFERENCES

Aklan, K., Cetiner, S.,Aka Kacar, Y.and Yalcin-Mendi, Y. 1997. *In vitro* multiplication of clonal apple rootstocks M 9, M 26 and MM 106 by meristem culture. *Acta Hort*. 441: 325-327.

Ali, Ahmad, M., Hughes, H.G., Safadi, F., and Dami, I. 1996. Liquid medium overlays enhance growth and multiplication of *in vitro* grapes and apples. *Fruit Varieties Journal* 50 (2): 86-87.

Auderset, G., Gavillet, S., Micheli, J., O'Rourke, J., Ribaux, M., Moncousin, Ch. 1994. Histological analysis and the evolution of biochemical markers during the *in vitro* rooting of *Malus domestica* Borkh. 'Jork 9'. *Adv. in Hort. Sc.* 1: 5-11.

Auderset, G., Moncousin, Ch., O'Rourke, J. and More. D.J. 1996. Stimulation of root formation by thiol compounds. *HortScience* 31 (2): 240-243.

Auderset, G., Moncousin, Ch., O'Rourke, J and More., D.J. 1997. Stimulation of root formation in difficult-to-root woody cuttings by dithiothreitol. *Int. J. Plant Sci.* 158 (2): 132-135.

Baraldi, R., Malavasi, F., Predieri, S., Castagneto, M. 1991. Effect of potassium humate on apple cv. "Golden Delicious" cultured *in vitro*. *Plant Cell, Tissue and Organ Cult.* 24: 187-191.

Benson, L.L., Lamboy, W.F. and Zimmerman, R.H. 2001. Molecular identification of *Malus hupehensis* (Tea Crabapple) accessions using simple sequence repeats. *HortScience* 36 (5): 961-966.

Bolar, J.P., Norelli, J.L. and Aldwinckel, H.S. 1998. An efficient method for rooting and acclimatization of micropropagated apple cultivars. *HortScience* 33 (7): 1251-1252.

Boxus, Ph. and Druart, Ph. 1986. Virus-free trees through tissue culture. In "*Biotechnology in Agriculture and Forestry* " vol. 1. *Trees* I. ed. Y.P.S. Bajaj, Springer-Verlag Berlin : 24-30.

Bournival, B.L. and Korban, S.S., 1987. Electrophoretic analysis of genetic variability in the apple. *Sci. Hortic.*, 31: 233-243.

Bommineni, V.R., Mathews, H., Samuel, S.B., Kramer, M. and Wagner, D.R. 2001. A new method for rapid *in vitro* propagation of apple and pear. *HortScience* 36 (6) : 1102-1106.

Buban, T., Csubak, E., Balla, I. and Tokes, G. 1993. Some doubts about using the micropropagated trees in apple production. *Acta Hort.* 349: 277-281.

Butler, E. and Gallagher, T.F. 1998. Gene expression during adventitious root formation in apple. In : *"Control of plant development : genes and signals"* eds. Greenland, A.J., Meyerowitz, E.M., Steer, M. . Symp. of the Soc. for Exp. Bio. 51: 79-84.

Caboni, E., Damiano, C. and Tonnarini, S. 1994. Effect of phenols and peroxidase activity and *in vitro* rooting of 'M9 Jork'. *Adv. in Hort. Sci.* 8 (1): 49-51.

Caboni, E., Boumis, G. and Damiano, C. 1992. Effects of phenols, gibberellic acid and carbohydrates on the rooting of the apple rootstock M9 Jork. *Agronomie* 12: 789-794.

Capuano, G., Piccioni, E. and Standardi, A. 1998. Effect of different treatments on the conversion of M26 apple rootstock synthetic seeds obtained from encapsulated apical and axillary micropropagated buds. *J. Hort. Sci. and Biotec.* 73 (3): 299-305.

Chevreau, E., Brisset, M.N., Paulin, J.P. and James, D.J. 1998. Fire-blight resistance and genetic trueness-to-type of four somaclonal variants from the apple cultivar Greensleaves. *Euphytica* 104 (3): 199-205.

Collet, G.F., Nowbuth, L. and Le, C.L. 1994. Comparison of the easy-to-root Jork 9 and Cepiland and the difficult-to-root EMLA 9 Lancep *Malus* M9 rootstocks *in vitro*. *Adv. in Hort. Sci.* 8 (1) : 45-48.

Czynczyk, A., Hodun, G. and Bielicki, P. 1994. Influence of micropropagation on the performance and quality of P 22 rootstock in mother plantation. *J. Fruit and Orn. Plant research* 2 (3) : 91-100.

De Klerk, G.J. and Ter Brugge, J. 1992. Factors affecting adventitious root formation in microcuttings of *Malus*. *Agronomie* 12 : 747-755.

De Klerk, G.J., Keppel, M., Ter Brugge, J. and Meekes, H. 1995. Timing of the phases in adventitious root formation in apple microcuttings. *J. Exper. Bot.* 46 (289) : 965-972.

De Klerk, G.J., Ter Brugge, J. and Marinova, S. 1997. Effectiveness of indolacetic acid, indolebutyric acid and nahthaleneacetic acid during adventitious root formation *in vitro* in *Malus* 'Jork 9'. *Plant Cell, Tissue and Organ Cult.* 49: 39-44.

De Klerk, G.J., Paffen, A., Jasik, J. and Haralampieva, V. 1999. Dual effect of ethylene during rooting of apple microcuttings. In *"Current Plant Science and Biotechnology in Agriculture"* Proceedings of the IXth International Congress of the IAPTC. eds. Altman, A., Ziv, M., Izhar, S., Jerusalem, Israël 36: 41-44.

Diaz-Perez, J.C., Shackel, K.A. and Sutter, E.G. 1995. Effects of *in vitro*-formed roots and acclimatization on water status and gas exchange of tissue-cultured apple shoots. *J. Amer. Soc. Hort. Sci.* 120 (3): 435-440.

Dolgov, S.V., Miroschnichenko, D.N. and Schestibratov, K.A. 2000. Agrobacterial transformation of apple cultivar and rootstock. *Acta Hort.* 538: 619-624.

Druart,Ph. 1980. La micropropagation des nouveaux sujets porte-greffe nanifiants chez le cerisier. In "*La culture du cerisier.* " Int. Symp. Gembloux , Belgium : 13-24.

Druart, Ph. 1984. *In vitro* germplasm preservation technique for fruit trees. Proc. EEC Seminar *'In vitro techniques: propagation and long term storage,'* ed. A. Schafer-Menuhr, Braunshweig, Germany: 167-171.

Druart, Ph. 1985. Multiplication conforme de sujets porte-greffe et de variétés de cerisiers par la culture *in vitro*. Int. Cherry Symp. *Acta Hort.* 169: 319-328.

Druart, Ph. 1987. Contribution à l'élaboration de techniques de production en masse *in vitro* d'espèces ligneuses utilisables en culture fruitière. *PHD Thesis*. Gembloux, Faculté des Sciences agronomiques: 174 p.

Druart, Ph. 1988. Regulation of axillary branching in micropropagation of woody fruit species. Proc. Int. Symp. "*Vegetative propagation of woody species*". Acta Hort. 227: 369-380.

Druart, Ph., 1992. *In Vitro* culture and micropropation of Plum (*Prunus spp.*). In "*Biotechnology in Agriculture and Forestry*.Vol.18 *High-Tech and Micropropagation* II" ed. Y.P.S. Bajaj. Springer-Verlag Berlin Heidelberg: 279-3O3.

Druart, Ph. 1994. Role of the leaves and apex in the *in vitro* rooting of *Malus domestica* Borkh. cv. Compact Spartan. *Adv. in Hort. Sci.* 1: 43-44.

Druart, Ph. 1995. C-source and growth response of *Prunus glandulosa 'Sinensis'* Thund. and *Malus pumila* Mill. M26 and M9 clone 29 during in vitro propagation. *Bull. Rech. Agron. Gembloux* 30 (1-2): 29-37.

Druart, Ph. 1997. Optimization of culture media for *in vitro* rooting of *Malus domestica* Borkh. cv. Compact Spartan. *Biol. Plant.* 39 (1): 67-77.

Druart, Ph. 2000. Aneuploïds and variants of apple (*Malus domestica* Borkh.) through *in vitro* culture techniques. *Acta Hort.* 520: 301-307.

Druart, Ph. and Boxus, Ph., 1987. Comportement au champ de *Malus domestica* Borkh. cv. 'Golden Delicious' issu de micropropagation. In : "*Plant micropropagation in horticultural industries*". Syposium Florizel 87, Arlon, Belgium: 185-189.

Druart, Ph. and De Wulf, O. 1993. Activated charcoal catalyses sucrose hydrolysis during autoclaving. *Plant Cell, Tissue and Organ Cult.* 32: 97-99.

Druart, Ph., Kevers, C., Boxus, Ph. and Gaspar, Th. 1982. *In vitro* promotion of root formation by apple shoots through darkness effect on endogenous phenols and peroxidases. *Z. Pflanzenphysiol. Bd.* 1085: 429-436.

Duron, M., Decourtye, L. and Druart, Ph. 1989. Quince (*Cydonia oblonga* Mill.) In "*Biotechnology in Agriculture and Forestry*". vol. 5. *Trees* II. ed. Y.P.S. Bajaj, Springer-Verlag Berlin: 42-58.

Fortuna, P., Citernesi,A.S., Morini, S., Vitagliano, C. and Giovanetti, M. 1996. Inluence of arbuscular mycorrhizae and phosphate fertilization on shoot apical growth of micropropagated apple and plum rootstocks. *Tree Physiol.* 16 (9): 757-763.

Gautheret, R.J. (ed) 1959. La culture des tissus végétaux. Masson, Paris. 863 p

Grandfranceschi,L., Seglias, N., Tarchini, R., Kimjanc, M. and Gessler, C., 1998. Simple sequence repeats for the genetic analysis of apple. *Theor. Appl. Genet.*, 96:1069-1076.

Guan, H.Y. and De Klerk, G.J. 2000. Stem segments of apple microcuttings take up auxin predominantly via the cut surface and not via the epidermal surface. *Sci. Hortic.* 86 (1): 23-32.

Guilford, P., Prakash, S., Zhu, J.M., Rikkerin, E., Gardiner, S., Basset, H. and Forster, R., 1997. Microsatellites markers in *Malus x domestica* (apple): abundance, polymorphism and cultivar identification. *Theor. Appl. Genet.*, 94: 249-254.

Gyorgy,V. 1996. Studies of stomata of *in vitro* propagated apple and the morphological changes of vitrificated plants. *HortScience* 28 (3-4): 21-24.

Hammerschlag, F.A. 1986. Temperate fruits and nuts. In "*Tisssue culture as a plant production system for Horticultural* crops" eds: Zimmerman, R.H., Griesbach,R; J., Hammerschlag, F.A., and Lawson,R.H.. Martinus Nijhoff Publishers, Dordrecht, The Netherlands: 221-236.

Hanke, V., Norelli, J.L., Aldwinckel, H.S., Hiller, I., Klotzsche, G., Winkler, K., Egerer, J. and Richter, K. 2000. Transformation in apple for increased disease resistance. *Acta Hort.* 538: 611-616.

Harbage, J.F., Stimart, D.P. and Auer, C. 1998. ph affects 1H-indol-3-butyric acid uptake but not metabolism during the initiation phase of adventitious root induction in apple microcuttings. *J. Amer. Soc. Hort. Sci.* 123 (1): 6-10.

Harbage, J.F. and Stimart, D.P. 1996a. Ethylene does not promote adventitious initiation on apple microcuttings. *J. Amer. Soc. Hort. Sci.* 121 (5): 880-885.

Harbage, J.F. and Stimart, D.P. 1996b. Effect of ph and 1H-indol-3-butyric acid (IBA) on rooting of apple microcuttings. *J. Amer. Soc. Hort. Sci.* 121 (6): 1049-1053.

Hirst, P.M. and Ferree, D.C. 1995. Effect of rootstock and cultivar on the growth and precocity of young apple trees. *Fruit varieties Journal* 49: 34-41.

Hokanson,S.C., Szewc-McFadden, A.K., Lamboy, W.F. and McFerson, J.R., 1997. Simple sequence repeat (ssr) variation in a collection of *Malus* species and hybrids. *HortScience*, 32: 440.

Hokanson, S.C., Szewc-McFadden, A.K., Lamboy, W.F. and McFerson, J.R., 1998. Microsatellite (SSR) markers reveal genetic identities, genetic diversity and relationships in a *Malus x domestica* Borkh. Core subset collection. *Theor. Appl. Genet.*, 97: 671-683.

Isutsa, D.K., Pritts, M.P. and Mudge, K.W. 1998. A protocol for rooting and growing apple rootstock microshoots. *Fruit varieties Journal* 52 (2): 107-116.

Jasik, J. and De Klerk, G.J., 1997. Anatomical and ultrastructural examination of adventitious root formation in stem slices of apple. *Biol. Plant.* 39 (1): 79-90.

Jehangir, K., Ihsanul, H., Khattak, M.S., Mukhtar, A. 1998. *In vitro* bud culture of MM 106 apple rootstock. *Sarhad J. of Agric.* 14 (3): 207-209.

Jones, O.P. 1993. Propagation of apple *in vitro*. In *"Micropropagation of woody plants"* ed M.RM Ahuja, Kluwer Academic publishers. Dordrecht,,The Netherlands) : 169-186.

Kamboj, J.S., Browning, G., Quinlan, J.D., Blake, P.S. and Baker, D.A. 1997. Polar transport of [^{3}H]-IAA in apical shoot segments of different apple rootstocks. *J. Hort. Sci.* 72 (5): 773-780.

Karhu, S.T. 1995. The quality of applied carbohydrates affects the axillary branching of apple microshoots. *Bull. Rech. Agron. Gembloux* 30 (1-2): 21-27.

Karhu, S.T. 1997. Sugar use in relation to the shoot induction by sorbitol and cytokinin in apple. *J. Amer. Soc. Hort. Sci.* 122 (4): 476-480.

Karhu, S.T. and Ulvinen, S.K. 1995. The effect of different carbohydrates on the rooting of micropropagated apple shoots and their adaptation after transplantation. *Bull. Rech. Agron. Gembloux* 30 (1-2): 87-101.

Kromer, K. and Gamian, A. 2000. Analysis of soluble carbohydrates, proteins and lipids in shoots of M7 apple rootstock cultured *in vitro* during regeneration of adventitious roots. *J. Plant Physiol.* 16 (5-6): 775-782.

Laimer da Camara Machado, M., da Camara Machado, A., Hanzer, V., Dalthoff, B., Weib, H., Mattanovitch, R. and Katinger, H. 1991. A new, efficient method using 8-hydroxyquinolinol-sulphate for the initiation and the establishment of tissue cultures of apple from adult material. *Plant Cell, Tissue and Organ Cult.* 27: 155-160.

Lambardi, M., Beneeli, C. and Fabbri, A. 1997. *In vitro* axillary proliferation of apple rootstocks under different ethylene conditions. *In vitro Cell. and Develop. Biol. Plant* 33 (1): 70-74.

Lane, W.D. 1992. Micropropagation of Apple (Malus domestica Barkh.). In *"Biotechnology in Agriculture and Forestry, Vol 18. High-Tech and Micropropagation II* ed. Y.P.S. Bajaj. Spinger-Verlag, Berlin Heidelberg: 229-243.

Lisek, A. 1996. Rooting *in vitro* of Polish dwarf apple rootstock P59. *J. Fruit Orn. Plant Res.* 4 (1): 1-9.

Ma, J.H., Yao, J.L., Cohen, D. and Morris, B. 1997. Ethylene inhibitors enhance *in vitro* root formation from apple shoot cultures. *Plant Cell Rep.* 17 (3): 211-214.

Marga, F., Morvan, C. and Morvan, H. 1995. Pectins in normal and vitreous apple microplants cultured in liquid medium. *Plant Physiol. and Bioch.* 33: 81-86.

Marga, F., Vebret, L. and Morvan, H. 1997. Agar fractions could protect apple shoots culture in liquid media against hyperhydricity. *Plant Cell, Tissue and Organ Cult.* 49: 1-5.

MCMeans, O., Skirvin, R.M., Otterbacher, A. and Mitiku, G. 1998. Assessment of tissue culture-derived 'Gala' and 'Royal Gala' apples (*'Malus* X *domestica* Borkh.). *Euphytica* 103 (2): 251-257.

Mi-W.H., Tetsuro, S. and Zhu, D.W. 1995. Experiment on the cryopreservation of apple shoots *in vitro*. *Acta Hort.* 403: 59-62.

Mogdil, M.,Sharma, D.R. and Bhardwaj, S.V. 1999. Micropropagation of apple cv. Tydeman's Early Worcester. *Sci. Hort.* 81 (2): 179-188.

Murashige, T. and Skoog, F. 1962. A revised medium for rapid growth and bioassays with tobacco tissue culture. *Physiol Plant.* 15: 473-497.

Navatel, J.C. and Bourrain, L. 1994. Influence of the physical structure of the medium on *in vitro* rooting. *Adv. in Hort. Sci.* 1: 57-59.

Negri, V., Hammad, H.A. and Standardi, A. 1995. A technique to save storage place for *in vitro* cultures of apple. *J. Gen. and Breed.* 49 (2): 127-132.

Niino, T., Sakai, A., Yakuwa, H. and Nojiri, K. 1992. Cryopreservation of *in vitro*-grown shoot tips of apple and pear by vitrification. *Plant Cell, Tissue and Organ Cult.* 28: 261-266.

Nitsch,J.P. 1969. Experimental androgenesis in *Nicotiana*. *Phytomorphlogy* 19: 389-404.

Noe, N., Eccher, T. and Bonini, L. 1997. Effect of irradiance and spectral composition of radiation on *in vitro* shoot proliferation in *Malus domestica* Borkh. . *Biol. Plant.* 39 (4): 491-497.

Nybom, H. and Schaal, B.A. 1990. DNA "fingerprintings" applied to paternity analysis in apples (*Malus x domestica*). *Theor. Appl. Genet.* 79: 763-768.

Oraguzie, N.C., Hofstee, M.E., Alspach, P.A., Brewer, L.R. and Howard, C. 2000. Quantitative genetic studies in a recurrent selection program in apple. *Acta Hort.* 538: 191-195.

Orlikowska, T. 1992. Influence of arginine on *in vitro* rooting of dwarf apple rootstock. *Plant Cell, Tissue and Organ Cult.* 31: 9-14.

Orlikowska, T. and Olszewski, T. 1993. Influence of agar brands on the multiplication of apple rootstock and on chemical and physical attributes of the media. *J. Fruit Orn. Plant Res.* 1 (4): 115-125.

Paul, H., Daigny, G. and Sangwan-Norrel, B.S. 2000. Cryopreservation of apple (*Malus* X *domestica* Borkh.) shoot tips following encapsulation-dehydration or encapsulation-vitrification. *Plant Cell Rep.* 19 (8): 768-774.

Pawlicki, N. and Welander, M. 1995. Influence of carbohydrate source, auxin concentration and time of exposure on adventitious rooting of the apple rootstock Jork 9. *Plant Sc. Limerik* 106 (2): 167-176.

Piccioni, E. and Standardi, A. 1995; Encapsulation of micropropagated buds of six woody species. *Plant Cell, Tissue and Organ Cult.* 42:221-226.

Piccioni, E. 1997. Plantlets from encapsulated micropropagated buds of M26 apple rootstock. *Plant Cell, Tissue and Organ Cult.* 47: 255-260.

Quamme, H.A. and Hogue, E.J. 1994. Improved rooting of 'Ottawa 3' apple rootstock by softwood cuttings using micropropagated plants as a cutting source. *Fruit Varieties Journal* (3): 170-173.

Quoirin, M. 1974. Premiers résultats obtenus dans la culture *in vitro* du méristème apical de sujets porte-greffe de pommier. *Bull. Rech. Agron. Gembloux*, 9 (2): 189-192.

Ribaux, M., O'Rourke, J., Gavillet, S. and Moncousin, C. 1995. Carbohydrate absorption during *in vitro* proliferation of *Malus* EM IX. *Bull. Rech. Agron. Gembloux*. 30 (1-2): 103-111.

Savela, M.L. and Uosukainen, M., 1994. Characterization of bacteria contaminating tissue cultures of apple rootstock 'YP'. *J. Applied Bacteriology* 76 (4): 368-376.

Schubert, A. and Lubraco, G. 2000. Mycorrhizal inoculation enhances growth and nutrient uptake of micropropagated apple rootstocks during weaning in commercial substrates of high nutrient availability. *Applied Soil Ecology* 15 (2): 113-118.

Seifert, G.J., Kanzler, P., da Camara Machado, A., Katinger, H. and Laimer da Camara Machado, M. 1994. Hormonal optimization and morphological characterization of the M9 'Jork ' stem and disc system. Towards its biochemical. *Adv. in Hort. Sci.* 1:61-66.

Seifert, G.J., Kanzler, P., da Camara Machado, A., Katinger, H. and Laimer da Camara Machado, M. 1995. Rhizogenesis in stem discs of *Malus pumila* M9 'Jork ' : I. Hormonal and environmental effects on root induction and callus formation. *Plant Cell Rep.* 14: 679-683.

Seingre, D., O'Rourke, J., Gavillet, S. and Moncousin, Ch. 1991a. Influence of gelling agent and carbon source on the *in vitro* proliferation of apple rootstock EM IX. *Acta Hort.* 289: 151-156.

Seingre, D., O'Rourke, J., Gavillet, S. and Moncousin, Ch. 1991b. Influence of carbon source and type of vessel on the *in vitro* proliferation of apple rootstock EM IX. *Acta Hort.* 289: 157-160.

Seufferheld, M.JM, Stushnoff, C., Forsline, P.L., Gonzales, G.H.T. 1999. Cryopreservation of cold-tender apple germplasm. *J. Amer. Soc. Hort. Sci.* 124 (6): 612-618.

Sharma,S. and Webster,A.D. 1993. Influence of shoot blanching treatment on adventitious root development in four clones of M9 rootstock. *Gartenbauwissenschaft* 58 (5): 214-216.

Shawky, I., Damiano, C., El-Hennawy, H.M., Chearioti, A. and El Wakeel, H.F. 1993. Studies on the behaviour of proliferated shoots of M26 apple rootstock *in vitro. Annals of Agric. Sci. Cairo* 38 (2): 691-697.

Simmonds, J. 1984. Direct rooting in micropropagted M26 apple rootstock. *Sci. Hortic.* 21: 233-241.

Siocurani, M., Piccioni, E. and Standardi, A. 2001. Micropropagation and preparation of synthetic seed in M26 apple rootstock I: Attempts towards saving labor in the production of adventitious shoot tips suitable for encapsulation. *Plant Cell, Tissue and Organ Cult.* 66: 207-216.

Skirvin, R.M., Kouidir, M., Joung,H. and Korban, S.S. 1986. The tissue culture of apple (*Malus x domestica* Borkh). In *"Biotechnology in agriculture and forestry".* V.1. (*Trees 1*). ed. Y.P.S. Bajaj Springer-Verlag, Berlin, Heidelberg: 183-189.

Standardi, A. and Romani, F. 1990. Effects of some antioxydants on *in vitro* rooting of apple shoots. *HortScience* 25: 1435-1436.

Standardi, A. and Piccioni, E. 1998. Recent perspectives on synthetic seed technology using nonembryogenic in vitro-derived explants. *Int. J. Plant Sci.* 159 (6): 968-978.

Tignon, M., Kettmann,R. and Watillon, B. 2000. AFLP: Use for the identification of apple cultivars and mutants. *Acta Hort.* 521: 219-226.

Tignon, M. 2001. Utilisation de marqueurs moléculaires pour l'identification de variétés et de sujets porte-greffe de pommiers (*Malus domestica* L. Borkh.). *PHD thesis,*, Faculté des Sciences agronomiques, Gembloux. : 176 p.

Uosukainen, M. 1992. Rooting and weaning of apple rootstock YP. *Agronomie* 12: 803-806.

Uosukainen, M. and Vestberg, M. 1994. Effect of inoculation with arbuscular mycorrhizas on rooting, weaning and subsequent growth of micropropagated *Malus* (L.) Moench. *Agricultural Science in Finland* 3 (3): 269-279.

Van Telgen, H.J., Elagoz, V., Van Mil, A., Paffen, A., De Klerk, G.J. 1992. Role of plant hormone in lateral bud growth of rose and apple *in vitro. Acta Hort.* 319: 137-142.

Vegvari, G. 1999. Dynamics of the uptake of nutrient aliments from the medium of *in vitro* cultured apple rootstocks. *Intern. J. Hort. Sci.* 5 (3-4): 50-53.

Vegvari, G. and Vertesy, J. 1999. Further information to the acclimatization of *in vitro* plants. *J. Hort. Sci.* 5 (3-4): 54-58.

Virscek, M.M., Javornok, B., Stampar, F., Bohanec, B. and Tobutt, K.R. 1999. Assessment of genetic variation among regenerants from *in vitro* apple leaves using molecular markers. *Acta Hort.* 484: 299-303.

Viseur, J. 1987. Micropropagation of pear, *Pyrus communis* L., in a double-phase culture medium. *Acta Hort.* 212: 117-124.

Wang, S.Y. and Faust, M. 1987. The relationship of internode length to carbohydrate content in genetic dwarf apple trees. *Sci. Hortic.* 33: 197-203.

Wang, Q.C., Tang, H.R., Quan, Q. and Zhou, G.R. 1994 Phenol induced browning and establishment of shoot-tip explants of 'Fuji' apple and 'Jinhua' pear cultured *in vitro. J. Hort. Sc.* 69 (5): 833-839.

Watillon, B., Druart, Ph., Du Jardin, P., Kettemann, R., Boxus, Ph. and Burny, A., 1991. Use of random cDNA probes to detect restriction fragment length polymorphism among apple clones. *Sci. Hort* 46: 235-243.

Weeden N.F. and Lamb, R.C., 1985. Identification of apple cultivars by isozyme phenotypes. *J. Amer. Soc. Hort. Sci.* 110 (4): 509-515.

Welander, M. and Pawlicki, N. 1993. A model system for studying root regeneration in woody species. *Acta Hort.* 336:225-230.

Wu, Y.J. 1997. Cryopreservation of apple in vitro cultured shoot tips. *IPGRI newsletter for Asia, the Pacific and Oceania* 23: 9.

Wu , Y.J., Engelmann, F., Zhao, Y.H., Zhou, M.D., Chen, S.Y. 1999. Cryopreservation of apple shoot tips: importance of cryopreservation technique and of conditioning of donor plants. *Cryo-letters* 20 (2): 121-130.

Yepes, L.M. and Aldwinckel, H.S. 1994. Micropropagation of thirteen *Malus* cultivars and rootstocks, and effect of antibiotics on proliferation. *Plant Growth Reg.* 15: 55-67.

Zacchini, M., Morini, S. and Vitagliano, C. 1997. Effect of photoperiod on some stomata characteristics of *in vitro* cultured fruit tree shoots. *Plant Cell, Tissue and Organ Cult.* 49 (3): 195-200.

Zhao,Y.H., Wu, Y.J., Engelmann, F., Zhou, M.D., Zhang, D.M., Chen, S.Y. 1999a. Cryopreservation of apple shoot tips by encapsulation-dehydration: effect of preculture, dehydration and freezing procedure on shoot regeneration. *Cryo-letters* 20 (2): 103-108.

Zhao,Y.H., Wu, Y.J., Engelmann, F., Zhou, M.D., Zhang, D.M., Chen, S.Y. 1999b. Cryopreservation of apple *in vitro* shoot tips by the droplet freezing method. *Cryo-letters* 20: 2, 109-112.

Zimmerman, R.H. 1986. Propagation of fruit, nut and vegetables crops- overview. In " *Tisssue culture as a plant production system for Horticultural crops.* eds: ZimmermanR.H., Griesbach,R; J., Hammerschlag, F.A., and Lawson,R.H.. Martinus Nijhoff Publishers, Dordrecht, The Netherlands: 183-200.

Zimmerman, R.H. 1997. Orchard variation in micropropagated trees of 'Redspur Delicious' apple. *HortScience* 32 (5): 935-936.

Zimmerman, R.H. and Fordham, I.M. 1985. Simplified method for rooting apple cultivars *in vitro*. *J. Amer. Soc. Hort. Sci.* 110:34-38.

Zimmerman, R.H., Bhardwaj, S.V. and Fordham, I.M. 1995. Use of starch-gelled medium for tissue culture of some fruit crops. *Plant Cell, Tissue and Organ Cult.* 43 (3): 207-213.

Zimmerman, R.H. and Steffens, G.L. 1995. Cultivar, planting density and plant growth regulator effects on growth and fruiting of tissue-cultured apple trees. *J. Amer. Soc. Hort. Sci.* 120: 183-193.

Zimmerman, R.H. and Steffens, G.L. 1996. Long-term evaluation of micropropagated apple trees: vegetative growth, cropping, and photosynthesis. *Sc. Hortic.* 66 (1-2): 69-76.

Shoot growth of Golden Delicious during multiplication stage

Recovery of hyperhydrical M26 from the apex (left) and from axillary buds (right) in presence of 1% galactose

Rooting of compact Spartan on Agar, Vermiculite and Agar + 2% Sucrose

Rooting of compact Spartan on Agar, Agar and 2% Sucrose + Mg SO4.7H2O or + Ca(NO3)2

Plantlets of compact Spartan on Agar Difco (left) or Merck 1613 (right) + 2% Sucrose

Growth of own-rooted columnar McIntosh'Wijcik'

Figure 1. Some features relevant to the *in vitro* propagation of apple.

15. MICROPROPAGATION OF SMALL FRUITS

Samir C. Debnath
Atlantic Cool Climate Crop Research Centre
Agriculture and Agri-Food Canada
P.O. Box 39088, St. John's, NF A1E 5Y7 CANADA

CONTENTS

1. INTRODUCTION

The small fruit plants are predominantly woody perennial dicot angiosperms, bear small to moderate-sized fruits on herbs, vines, or shrubs; and are usually vegetatively propagated to maintain true-to-type. The importance of small fruits in horticulture lies in their dual role as in the landscape and of food. The fruits themselves are highly prized for their varying shapes, textures, flavors, and colors. Small fruits are highly nutritious and are used as snack and dessert foods and in beverages. The principal plant genera cultivated for their berry-like fruit are *Actinidia* (kiwifruit or Chinese gooseberry), *Fragaria* (strawberries), *Ribes* (currants and gooseberries), *Rubus* (raspberries, blackberries, loganberries, etc.), *Vaccinium* (blueberries, bilberries, cranberries, lingonberries, whortleberries, etc.), and *Vitis* (grapes). Other small fruit species cultivated regionally include *Amelanchier* (June- or service berries or saskatoons), *Sambucus* (elderberries), and *Viburnum* (highbush cranberry or American cranberry bush).

S.M. Jain and K. Ishii (eds.), Micropropagation of Woody Trees and Fruits, 465–506.

Being genetically heterozygous, small fruit species do not reproduce individuals from seed that are similar to the seed parent (i.e., they do not "come true" from seed). Traditionally, most small fruit species are vegetatively propagated, which ensure that desired genetic characteristics are preserved and a fruit bearing condition is rapidly achieved. Considerable interest has been expressed in utilizing tissue culture method of propagation in various small fruit crops. The commercial use of this technology is for clonal mass propagation of specific clone and of parental stocks for hybrid seed production, maintenance of pathogen-free (indexed) germplasm, use as the initial step in a nuclear stock crop production system, and year-round production of plants. Anderson's (1975, 1978) early developmental work on micropropagation of ericaceous plants was the foundation for commercial micropropagation of these crops. However, protocols are continued to develop that has broaden the range of plants for commercial micropropagation economically. This chapter will attempt to assess the current status of micropropagation for the principal genera of small fruits.

2. GENERAL PROBLEMS IN MICROPROPAGATION OF SMALL FRUITS

The general problems associated with the tissue culture of small fruit plants are in:

- establishing explants,
- obtaining contamination-free explants from field-grown plants,
- obtaining growth and shoot proliferation from mature explants,
- selecting type and concentration of growth regulators,
- selecting appropriate basal media.

Frequently, explants taken from field-grown plants are difficult to sterilize to establish *in vitro* cultures due to high degree of contamination. Explant size is also critical to avoid contamination; as explant size increases, so does the chances of contamination. Reduction in explant size, however, must be tempered by the fact that larger explants generally display a greater survival rate *in vitro*. Also, the explants often turn brown and die, partly due to the strong sterilization conditions, that must be used to avoid contamination. It is usually recommended to take explants from plants grown under controlled conditions such as growth room or greenhouse, or from buds which flush from dormant shoots stored indoors. As with other woody perennials, shoot tips or nodal explants of small fruit plants are liable to release undesirable phenolic substances when placed onto a growth medium. Buds taken from mature parts of the plant could be recalcitrant to grow *in vitro*. Juvenile explants are readily induced into growth and proliferative shoot cultures.

3. *ACTINIDIA*

The genus *Actinidia* (family: Actinidiaceae) has over 60 species. The kiwifruit, *A. deliciosa* (A. Chev.) C.F. Liang et A.R. Ferguson, is cultivated extensively and large quantities of fruit of another, *A. chinensis* Planch., are collected in China. *Actinidia* species are found from the equator to about 50° north, from Indonesia in the south, through Malaysia, India, Thailand and Indo-China and China to Siberia, Korea and Japan in the north. The kiwifruit is available worldwide today, and is produced in New Zealand, the United States, Italy, Japan, France, Greece, Spain, Australia, and Chile. All *Actinidia* species are perennial climbing or scrambling plants. Most species are deciduous, but a few that occur in more subtropical regions are evergreen. The kiwifruit is unique and differ from other fruits. It is rich in folic acid, high in fiber, and an excellent source of minerals such as potassium, magnesium, calcium, iron, and phosphorus. Its most valuable attribute is its content of vitamin C (ascorbic acid) of up to 300 milligrams per 100 grams of fresh fruit (Schroeder and Fletcher, 1967; Sale, 1983). Although *Actinidia* species can be propagated through the seed, since all members of the genus appear to be functionally dioecious (male and female flowers on different plants) approximately one half of the progeny will be males, and as they are heterozygous there will be lot of variation within the progeny and even some of the desirable characters may be lost. Traditionally kiwifruit propagation is carried out by cuttings or grafting (Sale, 1983). The use of micropropagation techniques eliminates the seasonal limitations encountered with these methods (Standardi, 1983) and is favored when a large number of plants are required in a short time (Wessels et al., 1984).

3.1. Micropropagation

Actinidia species are largely propagated by shoot culture, although plant regeneration is also possible from petioles (González et al., 1995), stem segments (Uematsu et al., 1991), hypocotyls (Uematsu et al., 1991), leaf explants (Rugini et al., 1991; Janssen and Gardner, 1993), stamens (Brossard-Chriqui and Tripathi, 1984), endosperm (Gui et al., 1982), mature embryos (Gui et al., 1982), roots (Harada, 1975), and from protoplasts (Zhang et al., 1998). Cultures are initiated readily from nodal segments or shoot tips (Harada, 1975; Velayandom et al., 1985; Piagnani et al., 1986; Wiyaporn et al., 1990; Pedroso et al., 1992; González et al., 1995; Moncaleán et al., 2001) and dissected meristem tips, both apical and lateral (Standardi and Catalano, 1985; Monette, 1986; 1987).

3.1.1. Culture Establishment and Shoot Proliferation

Explants are excised from greenhouse-grown plants, stored for 6-8 weeks at 4 °C. The rapid growth of forced buds from these plants, resulting in the formation of new leaf buds, offers relatively contamination-free sources of nodal sections as explants. Explants used for initiating shoot culture can also be obtained from hardwood canes with dormant buds taken from mother stocks in the field in winter. They are thoroughly washed in running tap water, dipped shortly in a fungicide solution (a 0.4% Captan solution and 8% ferrous sulphate, w/v, in water; Moncaleán et al., 2001) and then stored in plastic bags in a refrigerator at 2 to 4 °C until use. Cut ends were protected with parafilm and the branches were stored at 4°C for 1 month to fulfil the chilling requirement by Moncaleán et al. (2001). Shoot-tip material that had been cold-stored for a year remained viable (Monette, 1987). Shoots are induced in the greenhouse. Prechilled bud material is superior to shoot-tip material (Lionakis and Zirari, 1991). Young roots (Xiao et al., 1991) and root segments (Harada, 1975) have also been used for micropropagation. Lionakis and Zirari (1991) reported that storage of buds at 6 °C for 900 hours was needed to produce cultures with the highest number of axillary shoots and high growth rate. They found lateral buds to produce more axillary shoots in culture than shoot tip explants.

3.1.1.1. Explant Sterilization

The effectiveness of explant sterilization depends on the degree of penetration of the disinfection agent into the explant. Explants obtained from field- or greenhouse-grown plant material are generally disinfected by treating with dilute calcium hypochlorite solution: 5%, 40 min for petiole segments (Pias et al., 1987), 6%, 20 min, for shoot sections (Hirsch, 1970; Hirsch and Bligny-Fortune, 1979) and 4%, 20 min for nodal segment, previously immersed in ethanol, 80%, (Standardi, 1981, 1983). Sodium hypochlorite (0.6%) accompanied by few drops of a wetting agent, Tween 20, has been used by Monette (1986) for shoot tip disinfection. Other sterilizing agents include: Domestos (7.5%, v/v, 15 min) for petiole segments (Revilla and Power, 1988), and solutions of $HgCl_2$ (0.4%, 5 min) with the addition of 2-3 drops of Tween 20, rinsed with $CaCl_2$ solution (0.25%), previously immersed in stéricline (0.001%), and ethanol (70%, 1 min) for shoot nodal segments (Revilla et al., 1992).

3.1.1.2. Culture Initiation

For culture initiation, explants are planted onto agar-solidified MS (Murashige and Skoog, 1962) (Wiyaporn et al., 1990) or 3/4 strength MS medium supplemented with 128 mg l^{-1} $NaH_2PO_4.H_2O$ (Monette, 1986; 1987; Lionakis and Zirari, 1991), and cultured at 22-25°C under 16-h photoperiod provided by cool-white fluorescent lamps at a photon flux density of 26-33 μmol $m^{-2}s^{-1}$ (Pedroso et al., 1992; Moncaleán et al., 2001). Moncaleán et al (2001) cultured the primary explants in KH medium (Cheng, 1977) without plant growth regulators for 7 days followed by transferring on to shoot initiation medium. Standardi and Catalano (1985) used an unique media for culture initiation and shoot proliferation. An auxin, either α-naphthalene acetic acid (NAA) (0.1 μM) or 3-indolebutyric acid (IBA) (0.1-1 μM) and one or more cytokinins: benzyladenine (BA) (2.2-8.8 μM) along with 2.9 μM gibberellic acid (GA_3); or BA (8.8 μM) and adenine sulfate (0.33mM) are added to initiate shoot initiation (Standardi and Catalano, 1985; Monette, 1986; 1987; Lionakis and Zirari, 1991). González et al. (1990) obtained axillary shoot-bud development in KH medium with 3-indolyl-acetic acid (IAA) (0.14 μM) and zeatin (4.56 μM) from nodal explants.

3.1.1.3. Shoot Proliferation

Shoot proliferation can be obtained on agar-solidified MS (Chiarotti et al., 1991; Pedroso et al., 1992; Moncaleán et al., 2001) and in LS (Linsmaier and Skoog, 1965) medium supplemented with 170 mg l^{-1} $NaH_2PO_4.H_2O$ (Lionakis and Zirari, 1991). Liquid cultures produced more shoots which were slightly vitrified when compared with agar solidified medium (Monette, 1986; 1987). Liquid culture medium using cellulose plugs as a support of explants have been used by Moncaleán et al. (1999) where MS medium was too rich for kiwifruit cultures except for the phosphate, which should be increased more. They proposed to use automatizated culture systems to allow renovation of the culture medium, as well as addition of growth regulators when necessary. Shoot proliferation is promoted by either 4.4 μM BA and 0.6-2.9 μM GA_3 (Standardi and Catalano, 1985; Moncaleán et al., 2001) or BA (8.8 μM), adenine sulfate (0.43mM) and IBA (0.15-0.3 μM) (Monette, 1986; 1987; Lionakis and Ziratri, 1991). From nutritional point of view, the MS medium is able to meet the needs of *Actinidia* species for one month, and resuming regularly subculturing every month to reach the highest rate of proliferation (Mezzetti et al., 1991). Addition of coconut milk and cytokinins to media enhanced shoot proliferation in *Actinidia* (Boase et al., 1993).

Tanaka et al. (1997) used fruit galls to micropropagate *A. polygama*. The fruit galls of *A. polygama* have been used for analgesic and tonic purposes as a folk medicine in Japan. Callus was induced on the B5 medium (Gamborg et al., 1968) supplemented with NAA (5.4 μM), BA (4.4 μM) and kinetin (4.6 μM). Shoots developed when shoot primordia appeared on the surface of the callus, and were transferred to MS medium containing 2.3 μM 2,4-dichlorophenoxy acetic acid (2,4-D) and 2.2 μM BA.

3.1.2. Rooting and Acclimatization

Both *in vitro* and *ex vitro* methods have successfully been used to root and acclimatize micropropagated *Actinidia* shoots. Although *in vitro* rooting offers several advantages, including reduced exposure to disease and environmental stress during rooting process and the production of rooted plantlets for *in vitro* experiments, plantlets can be produced more rapidly with less cost by eliminating *in vitro* rooting (Monette, 1986; Pedroso et al., 1992).

For *in vitro* rooting, shoots are dipped briefly into an IBA solution, or cultured them for a short while (e.g. 10 days) on a medium containing 4.9 μM IBA, and then placed them onto an auxin-free medium (Standardi and Catalano, 1985). *In vitro*-regenerated shoots can also be rooted on growth regulator free MS medium or supplemented with NAA (5.4 μM) (Tanaka et al., 1997). Moncaleán et al. (2001) immersed shoots briefly (15 s) in IBA (4.9 mM) solution and then cultured in KH liquid medium without growth regulators for rooting. Zhang et al. (1998) rooted shoots obtained from leaf protoplasts by immersion in 20 ppm IBA solution before culturing on half-strength MS medium lacking growth regulators.

For *ex vitro* rooting, shoots are either dipped into a 2-2.5 mM IBA solution for a few seconds (Monette, 1986; 1987) or placed in a more dilute solution for a longer period (e.g. 98 μM IBA 24 h; Pedroso et al., 1992) and then planted in a compost, such as 2 peat : 1 vermiculite : 1 perlite (v/v/v) (Monette, 1987).

Microcuttings, moved from *in vitro* culture conditions, must be acclimatized gradually to ambient conditions to avoid mortality that might otherwise occur under an abrupt change in relative humidity, temperature, or irradiance. For *in vitro*-rooted plantlets, standard procedure is to wash the plantlets and transfer to pots containing a 1 peat : 1 perlite (v/v) (Moncaleán et al., 2001) or 1 sand : 1 soil mixture (v/v) (Pedrosa et al., 1992), and maintained in a chamber with 80% relative humidity with 16-h photoperiod under reduced light (8.6 μmol $m^{-2}s^{-1}$) (Pedroso et al., 1992).

3.1.3. Field Performance of Micropropagated Plants

After hardening in the greenhouse, the plants can be transferred to the field for evaluation of morphologic and genetic stability. In a field trial, micropropagated vines were less susceptible to frost than vines raised from cuttings, and when the plants were pruned in the spring, the new leaves on the micropropagated stock contained more chlorophyll and higher assimilation rates than the controls (Massai et al., 1991). Piccotino et al. (1991) reported that root system of micropropagated vines were 50% larger than those produced by vines from cuttings. The cumulated yield at the end of the seventh year was higher in the micropropagated vines than those from cuttings (Monastra and Testoni, 1991).

3.2. Concluding Remarks and Prospects for the Future

Clonal micropropagation of *Actinidia* is necessary to preserve genetic uniformity of material. Shoots multiplied clonally may be used as a source of explants for induction of somaclonal variants through callus culture (Revilla and Power, 1988), protoplast fusions, or gene transfer (Rugini et al., 1991; Uematsu et al., 1991). Clonal multiplication of novel material regenerated from breeding experiments is also desirable to produce clonal plants rapidly for field evaluation. Plants can be mass propagated in a liquid medium using flask or bioreactor (Levin et al., 1988; Ziv, 1991; 1992). Liquid cultures and automate manipulation eliminate some of the disadvantages associated with agar-solidified medium, such as its difficulty to follow physical and chemical changes occurring along the culture (Preil, 1991), the alteration of the elemental components of the basal medium caused by agar composition (Singha et al., 1987), and nutrient changes caused by the gel matrix (Debergh, 1983). However, optimal medium composition for shoot cultures in liquid medium should be established to increase growth rates *in vitro* and to direct morphogenesis more efficiently.

4. *FRAGARIA*

Members of the genus *Fragaria* L. (family: Rosaceae) are dicotyledonous perennial herbs, widely distributed throughout Europe, northern Asia, North America, and northern Africa (Reed, 1966). The cultivated or garden strawberry (*Fragaria* × *ananassa* Duch.), a hybrid between the Scarlet or Virginia strawberry (*F. virginiana* Duch.) and the pistillate South American *F. chiloensis* Duch., is the most widely grown species, although the unique aromas and flavors of the Wood (*F. vesca*) and Musky (*F. moschata*) strawberries account for their continued but limited culture today. Strawberries are predominantly used as fresh fruit; their use in processed forms such as cooked and sweetened preserves, jams or jellies, and frozen whole berries or sweetened juice extracts or flavorings, and their use in making a variety of other processed products made them one of the most popular berry crops, more widely distributed than any other fruit, including grape (Childers, 1980). The berry is valued for its low-calorie carbohydrate along with high vitamin C and fiber contents.

Conventionally, strawberries are vegetatively propagated by runners arising from axillary leaf buds on the plant crown. Plant propagation through runner produces a limited number of propagules. Although production of propagules through runner has been reported to contribute 90% of total Dutch strawberry production, the product in 'Elsanta' cultivar was found to be susceptible to several fungal diseases (Dijkstra, 1993). If virus-free plants are used to initiate the cultures, it offers the advantage of producing disease-free plants. Micropropagated strawberry plants can be stored under refrigeration (Mullin and Schlegel, 1976), making it a valuable technique for storage of germplasm. Micropropagation is used in commercial propagation of strawberries and in breeding programs to produce many plants rapidly since the mid 1970s. Successful proliferation have been obtained from single meristems (Boxus, 1974; Boxus et al., 1977), meristem callus (Nishi and Oosawa, 1973), and from node culture (Bhatt and Dhar, 2000).

4.1. Meristem Culture

Most commonly used explant for strawberry micropropagation is the meristem from the tip of runners (Sowik et al., 2001). Meristem-tip culture, sometimes combined with heat treatment (Yoshino and Hashimoto, 1975) is widely used to obtain virus and fungus-free strawberry plants (Molot et al., 1972).

Cultures can be initiated and maintained on Boxus (Boxus, 1974) medium containing Knop's (Knop, 1965) macronutrients and MS micronutrients and organic components, or MS medium supplemented with 2.2-4.4 μM BA, 0.5-2.5 μM IBA and 0.3 μM GA_3 (Cerović and Ružić, 1989) at 23-25 °C during the light period, and 17 °C in dark; the quantum irradiance is 46 μmol $m^{-2}s^{-1}$ for a 16 h photoperiod (Sowik et al., 2001). Runners can be initiated from *in vitro* culture on media containing $AgNO_3$ (10-20 mg l^{-1}). GA_3 (28.9-57.7 μM) increased the efficiency of $AgNO_3$ significantly (Zatykó et al., 1989). Borkowska (2001) initiated strawberry cultures according to Boxus (1974, 1992), but modified medium by lowering the concentration of cytokinin (BA, 2.2 μM) and auxin (IBA, 0.5 μM). Cultures were maintained at 23 °C under a photosynthetic photon flux density of 30 μmol $m^{-2}s^{-1}$ from 'warm-white' fluorescent lamps and 16 h photoperiod. Proliferated shoots can be rooted *in vitro* on Boxus (Borkowska, 2001; Sowik et al., 2001) or half-strength MS medium (Yue et al., 1993) without growth regulators, or on half-strength MS with activated charcoal (0.6 g l^{-1}) and IAA (5.7 μM) (Moore et al., 1991).

In conventional strawberry micropropagation, the rooting phase is carried out *in vitro*. The *in vitro*-formed roots are thick, posses no hairy roots, grow horizontally, and are fragile and easily damaged. *In vitro*-grown plantlets have low photosynthetic activity, poor water balance and their anatomy and morphology are far from being optimal (Grout and Millam, 1995; Van Huylenbroeck and Debergh, 1996; Borkowska, 2001). Kozai (1991) and Hayashi et al. (1997) developed photoautotrophic micropropagation system for plant multiplication by simultaneously increasing the CO_2 concentration and light quality.

Rooting can be induced *ex vitro* with complete success (Hayashi et al., 1997; Borkowska, 1999; 2001; Borkowska et al., 1999) by transferring microshoots directly in rockwool (Borkowska et al., 1999). After one month, the plantlets rooted *ex vitro* are planted in 1 peat : 1 rockwool (v/v) and grown in the greenhouse for acclimatization (Borkowska, 2001).

4.2. Node Culture

Bhatt and Dhar (2000) reported an efficient method of node culture of wild strawberry (*F. indica* Andr.). Nodal segments (2.5-3 cm), treated with Savlon (an antiseptic containing 3% antimicrobial agent centrimide + detergent) for15 min, were immersed in 80% ethanol for 30 sec prior to surface sterilization in $HgCl_2$ (0.05%) containing few drops of Tween 20 for 5 min and then cultured on MS medium supplemented with 4 μM BA and 0.1 μM NAA. The same medium was used for shoot multiplication. Excised shoots rooted on half-strength agar-gelled medium with 1 μM NAA. Rooted shoots with fully expanded leaves when transferred to a mixture of soilrite and sand (1:1 ratio), about 70% of the plantlets survived *ex vitro*.

4.3. Field Performance of Micropropagated Plants

There are concerns about genetic changes resulting from strawberry micropropagation. Discrete morphological variants have been observed in micropropagated strawberry plants, e.g., leaf variegation consisting of a narrow white streak in the leaf blade (Sansavini and Gheradi, 1979; Scott and Zanzi, 1980; Swartz et al,. 1981), chlorosis of the leaves (Scott and Zanzi, 1980; Swartz et al., 1981), and growth changes including dwarfs, compact trusses, lack of runner production, and female sterility (Swartz et al., 1981). Moore et al. (1991) observed variability among micropropagated subclones of 'Olympus' which were most likely transient responses to the micropropagation environment, not genetic. Generally, micropropagated plants have greater vigor, runner production, and yields than runner-propagated plants (Swartz et al., 1981). However, not all cultivars exhibit a yield increase (Cameron et al., 1985).

4.4. Concluding Remarks

Strawberry micropropagation finds its main practical application in mass propagation of virus-free plants. The highest genetic, physiological and morphological value of micropropagated plants is guaranteed by the method elaborated and recommended by Boxus (1974, 1992). However, micropropagation can profoundly alter the physiology of micropropagated strawberry plants. Many of the observed changes are related to rejuvenation; the causes of others is not yet understood. Somaclonal variation (Larkin and Scowcroft, 1981) can arise from callus formed at the base of shoot clumps. The possibility of manipulating plants towards greater vigor, runner production and yield in strawberry micropropagules is of particular interest.

5. *RIBES*

The genus *Ribes* L. (currants and gooseberries) belongs to the Saxifragaceae family and consists of approximately 150 species, distributed mainly in temperate regions of the northern hemisphere, and extends into the southern hemisphere in the Andes of South America. Being very hardy, they can be grown in areas having very cold winter temperatures. The cultivated species can withstand −30 to −35 °C during the period of dormancy and −5 to −7 °C at bud opening. In warm climates at 30 °C or higher, leaves are frequently injured; and often the plants are defoliated. The genus comprises two distinct groups: the currants and the gooseberries. While currants mostly lack nodal spines, gooseberries possess spines and are usually prickly (Bailey, 1976). The major crops of economic importance are the European black currant (*R. nigrum* L.), found in northern Europe and central and northern Asia up to the Himalayas, and gooseberry (*R. Grossularia* L.) (Lanham et al., 2000). Some other species grown for their edible fruit include: *R. grossularia* L. (= *R. uva-crispa* L.) (European gooseberry), *R. hirtellum* (American gooseberry), *R. odoratum* (flowering currant), and *R. sativum* (red and white currants). Currants and gooseberries grow as bushes, and produce small berries rich in vitamins A, B, and C, in pectins, mineral elements, citric acid, fructose, and in anthocyanins. Most of the crop is used for juice production and other processes including the production of jams and jellies, liqueurs, and for the conversion of white wines to rose and to impart color to yoghourt and other dairy products.

The traditional vegetative propagation of currants and gooseberries is from cuttings, either hardwood in the autumn, softwood in the spring, or single-bud cuttings (Thomas and Wilkinson, 1962), although rooting is only 50% or below for gooseberry hardwood cuttings. The use of micropropagation is becoming increasingly popular. Work on *in vitro* cultures has shown that they respond well to sterile culture, both for germplasm storage and for propagation purposes. Tissue culture techniques in *Ribes* have been reviewed by Brennan et al. (1989; 1990) and Orlikowska et al. (1991).

5.1. Meristem Culture

Meristem culture in *Ribes* has been reported by many researchers for the elimination of virus diseases and for mass and fast propagation of disease-free plants (Jones and Vine, 1968; Harvey and Grasham, 1977; Welander, 1982; Tyulenev et al., 1987). Media commonly used are derived from MS supplemented with cytokinin except those used for rooting. Auxin is a component of most proliferating and all rooting media. Plant regeneration from apical meristem pass through callus phase and thus not recommended for true-to-type propagation (Blenda and Kirilenko, 1982), although Kochetkova et al. (1981) obtained shoot regeneration from gooseberry meristems bypassing any callus stage.

5.2. Shoot Culture

Micropropagation of cultivated species of *Ribes* is done by culturing shoot tips or nodal segments on MS or modified MS medium containing half-strength MS macrosalts and full strength microsalts. (Arena and Martínez Pastur, 1995). Cultures are grown at 22±2°C with a 16 h photoperiod (57 μmol $m^{-2}s^{-2}$). While BA at 2.2 μM is an optimal concentration for shoot multiplication, GA_3 is unnecessary in the shoot proliferation medium (Arena and Martínez Pastur, 1995).

5.3. Root Induction

Ribes shoots from meristem or shoot cultures can be rooted in modified MS medium containing half-strength MS macrosalts and full strength microsalts (Arena and Martínez Pastur, 1995). Percentage of rooting >95% were attained by Arena and Martínez Pastur (1995).

5.4. Concluding Remarks

Research on micropropagation in currants and gooseberries has important potential applications for mass propagation and germplasm storage. Long-term storage and survival of micropropagated *Ribes* shoots using cryogenic techniques have been successfully achieved (Sakai and Nishiyama, 1978a,b).

6. *RUBUS*

The genus *Rubus* (Tourn.) L. (family: Rosaceae), the members are also called as brambles, contains approximately 740 species (Ying et al., 1990) that are particularly common in cold and temperate regions of the northern hemisphere. Plants are mostly deciduous perennial shrubs varying in habit from erect to trailing with usually biennial canes, but few species produce perennial or annual canes. Fruits, each of which proceeds from a single flower, consists of several adhering drupelets. Important bramble crops are raspberries and blackberries.

6.1. Raspberry

The present cultivars of the red raspberry (*R. idaeus* L.) are hybrids, developed by hybridization between *R. idaeus* L. var *vulgatus* Arrhen. (the original European red raspberry) and *R. idaeus* L. var. *strigosus* Michx. (the native American red raspberry) (Qurecky, 1975). Most red raspberries require relatively cool summer and moderate winter temperatures for best production. Black raspberries are vulnerable to cold temperatures, and diseases, and as a result production is poor, and have small market potential. Purple raspberries are usually F_1 hybrid selections from crosses between red and black raspberries.

Red raspberry plants are conventionally propagated by planting shoots (suckers) with attached roots that grow from adventitious roots of established plants, or from root cuttings. The traditional method of propagation for black and purple raspberries is by tip layering. *In vitro* clonal propagation avoids some of the limitations of conventional methods of raspberry propagation, such as seasonal growth patterns and the available numbers of canes or adventitious buds on root cuttings.

6.1.1. Meristem Cultures

Meristem tips (0.3-0.8 mm in size), isolated from apical and axillary buds from heat-treated plants, are cultured on MS or modified MS medium containing BA (0.88-8.8 μM) and IBA (0.25 μM) with or without GA_3 (0.29 μM) (Pyott and Converse, 1981; Welander, 1987; Sobczykiewicz, 1987). Reed (1990) used basal MS medium supplemented with 4.4 μM BA and 0.49 μM IBA for the establishment of meristem tips of 256 different *Rubus* genotypes. Once virus-free cultures have been obtained, plant propagation is done by shoot culture.

Proliferation is accomplished by inhibition of apical dominance and promotion of axillary bud development using growth regulators. Although shoot proliferation has been achieved successfully either on LS (James et al., 1980), MS (Sobczykiewicz, 1987; Hoepfner, 1989) or modified MS (Palonen and Buszard, 1998) medium, some investigators suggested that better results could be obtained with preparations of lower ionic composition such as Anderson's (1980) medium in which the concentrations of nitrate, ammonium, and potassium ions were one quarter that in MS medium, but the levels of PO_4^{3-} and iron were doubled (Anderson, 1980; Pyott and Converse, 1981; Sobczykiewicz,1992). Vertesy (1979) used the medium of Morel with micronutrients of Heller. The presence of BA and IBA is very stimulative for shoot proliferation. Reed (1990) observed that while majority (62%) of a diverse *Rubus* germplasm induced shoot proliferation on the MS basal medium supplemented with BA (4.4 μM), IBA (0.49 μM) and GA_3 (0.29 μM), another 18% responded on Anderson's medium with same growth regulator levels except GA_3. The remaining 20% required higher concentration of BA (8.8 μM) for shoot proliferation. Ascorbic acid (50 mg l^{-1}) can be added to the medium for preventing from browning around the tissue and to stimulate shoot growth (Sobczykiewicz, 1992).

6.1.2. Node Culture

Gonzalez et al. (2000) reported a protocol for micropropagation of adult field-grown raspberry plants from nodal segments. Nodal explants from softwood cuttings, collected in spring, were surface sterilized with 0.5% (v/v) sodium hypochlorite (20 min) after a pre-sterilization treatment with 70% ethanol (30 s). Explants were cultured in MS medium for 15 days and then transferred to the same medium, but with 4 μM BA and 0.25 μM IBA. After 30 days of culture, shoot-buds were established in a modified MS medium (Anderson's macronutrients except calcium, with Sequestrene as the iron source) with the same growth regulators. For multiplication, shoot-buds were monthly subcultured on the same medium containing 4-8 μM BA plus 0.25 μM IBA. On average, 1.6±0.1 shoots per nodal explant were obtained in *R. idaeus* cv. Gradina after 30 days of culture (Gonzalez et al., 2000).

6.1.3. Rooting

The proliferated shoots are transferred onto MS medium with one half or one quarter of major salts enriched with either IBA (9.8-14.8 μM) or BA (4.9 μM) for rooting (Sobczykiewicz, 1992). Microshoots can also be directly rooted *ex vitro* in a mixture of peat and perlite (1:1, v/v) in a mist chamber (Gonzalez et al., 2000). The *in vitro* rooting percentage (73-85%, Sobczykiewicz, 1987) was better than *ex vitro*(67%, Gonzalez et al., 2000), and 100% of rooted shoots survived (Gonzalez et al., 2000).

6.1.4. Field Performance

The *ex vitro* plantlets, once established and acclimatized, can be planted directly in the field after danger of spring frost has passed (Bartel et al., 1988). The success of transplant and survival of raspberry plants under field conditions depend on the quality of roots. The actively growing plants produce berries after three years of transfer in the field (Gonzalez et al., 2000).

6.1.5. Concluding Remarks

Micropropagation has become an important technique for the commercial propagation of raspberry (Donnelly and Doubeny, 1986). It is now widely used for the rapid multiplication of advanced selections and new cultivars, and for germplasm preservation. Raspberry growers had traditionally used cold-treated bare root stock for planting until plants from tissue culture became an option during the mid-1980s. Today, tissue-cultured plants provide a greater source of disease-free plant material, grow more rapidly and uniformly than conventionally propagated transplant, and generally produce higher yields early in the life of the planting. Meristem culture combined with heat treatment is used to obtain virus-free plants of older cultivars.

6.2. Blackberry

Blackberries are usually distinguished from raspberries by fruit that adheres completely or partially to the receptacle, species range from evergreen subtropical types to deciduous Arctic type. Cultivated blackberries are derived from a wide group of basic *Rubus* species (Daubeny, 1996).

6.2.1. Micropropagation

Micropropagation procedures for blackberry, in many respects, are similar to those used for raspberries. Successful *in vitro* propagation has been achieved with blackberry cultivars (Broome and Zimmerman, 1978; Skirvin et al., 1981; Fernandez and Clark, 1991; Bobrowski et al., 1996; Gonzalez et al., 2000). Fernandez and Clark (1991) observed that when nodal segments were cultured on the initial medium (one-half strength MS) for 3 to 5 days; 3 weeks in the proliferation medium (full strength MS plus 0.3 g l^{-1} acid-washed activated charcoal, or full strength MS, or one-half strength MS); and finally, 2 weeks on the rooting medium (full strength MS plus 0.3 g l^{-1} acid-washed activated charcoal); the entire propagation procedure required 10 to 11 weeks for thornless 'Navaho' blackberry. Blackberry cultivars can also be proliferated *in vitro* on NCGRRUB medium (Reed, 1990) which includes MS salts and vitamins with double MS iron, 0.29 μM GA_3, 0.49 μM IBA, and 4.45 μM BA. Growth room conditions may be 25°C with a 16-h photoperiod (25 μmol $m^{-2}s^{-1}$) (Tsao and Reed, 2002). Gonzalez et al. (2000) achieved shoot-bud establishment by culturing nodal explants in MS medium for 15 days, and then in the same medium with 4 μM BA and 0.25 μM IBA. This medium was also best for shoot multiplication. Microshoots rooted directly in a mixture of peat and perlite (1:1, v/v) in a mist chamber, and a 100% of rooted shoots were acclimatized under field conditions. MS medium supplemented with only BA (4.4-8.8 μM) was suitable for shoot multiplication, and one-third strength MS plus IBA (1.5-3.9 μM) was ideal for rooting of blackberry genotypes by Bobrowski et al. (1996).

6.2.2. Concluding Remarks

Blackberries are often established through direct planting of root cuttings or from containerized plants grown from root cuttings (Caldwell, 1984). Shoot emergence from root cuttings is poor. The use of tissue-cultured plants would eliminate the need for root pieces and allow uniform spacing of plants in the field.

7. *VACCINIUM*

The genus *Vaccinium* L. belongs to the family Ericaceae and consists of about 450 species, which are interspersed through out the world except Australia, but concentrated in the montane tropics (Vander Kloet, 1988; Luby et al., 1991; Dierking and Dierking, 1993). Plants may be terrestrial or epiphytic and are generally found on acidic, sandy, peaty, or organic soils. The berries produced by the genus are edible; their flavour varies from insipid to tart to sweet. Blueberry, cranberry and lingonberry are three *Vaccinium* fruit crops of economic importance which have been domesticated in the twentieth century.

Although plant regeneration *in vitro* may be accomplished by employing callus, organ, cell and protoplast culture, shoot culture with the proliferation of axillary shoots is not only the generally accepted method of *Vaccinium* propagation to maintain the genetic integrity of a clone, but also the most commercially viable means of propagating selected cultivars.

7.1. Blueberries

Blueberries (*Vaccinium* section Cyanococcus) are a diverse group of perennial, woody shrubs, mostly deciduous, which bear fruit in clusters. Currently, five major groups of blueberries are commercially grown: 1) lowbush (*V. angustifolium* Ait., *V. myrtilloides* Michx., *V. boreale* Hall and Aald.), 2) highbush (*V. corymbosum* L.), 3) half-high, which are hybrid or backcross derivatives of highbush-lowbush hybridizations; 4) southern highbush, which were developed from hybridization of *V. corymbosum* with one or more species (mainly *V. darrowi* Camp and *V. ashei* Reade) (Lyrene, 1990); and 5) rabbiteye (*V. ashei*). Traditionally, softwood, semi-hardwood and hardwood cuttings have been used to propagate blueberries, although cloning by micropropagation is more demanding and potentially more effective method for improving blueberry fields, comparable in its requirements with growing and setting out seedlings (Morrison et al., 2000). The juvenile characteristics of tissue-cultured lowbush blueberry plantlets may facilitate rhizome production for quick spread into bare areas of field. Morrison et al. (2000) observed that micropropagated lowbush blueberry plants from shoots that passed through several subcultures produced ten-fold more rhizomes than for stem cuttings.

In vitro propagation of *Vaccinium* sp. was initiated by Zimmerman and Broome (1980) to culture the highbush blueberry on modified Anderson's (1975) *Rhododendron* medium. In fact, Barker and Collins (1963) grew lowbush blueberry rhizome pieces on White's (1943) medium and the first callus formation in *in vitro* culture of lowbush blueberry was described by Nickerson and Hall (1976). Subsequently, tissue culture techniques have been developed for lowbush (Nickerson, 1978; Frett and Smagula, 1983; Brissette et al., 1990), highbush (Cohen, 1980; Zimmerman and Broome, 1980; Wolfe et al., 1983; Grout and Reed, 1986; Eccher and Noè, 1989; Noè and Eccher, 1994; Gonzalez et al., 2000) and rabbiteye (Lyrene, 1980) blueberry.

Shoot cultures can be initiated from nodal segments or from shoot tips, juvenile explants are most suitable for shoot proliferation (Lyrene, 1980). Hardwood cuttings from adult blueberry, collected in winter, are surface sterilized and stored at low temperature. Cuttings are stimulated to budbreak by placing them in water at 26°±2°C (20°±2°C night temperature) with a 16 h photoperiod (60μmol m^{-2}s^{-1} for 15 days) (Gonzales et al., 2000). Lateral shoots are then used as the explant source. Softwood cuttings, collected in spring from field-grown mother plants (Gonzales et al., 2000), shoots from a new flash of growth of greenhouse-grown plants (Morrison et al., 2000), seedling explants (Lyrene, 1980), and rhizome pieces (Barker and Collins, 1963) have also been used for *in vitro* propagation of blueberries.

Nodal explants or shoot tips of lengths ranging from about 0.3 to 2 cm are surface sterilized, usually by washing in solutions of sodium (0.5%; Reed and Abdelnour-Esquivel, 1991; Morrison et al., 2000) or calcium hypochlorite (6%; Brissette et al., 1990) for 10 to 20 min followed by rinsing with sterile water. Gonzales et al. (2000) pre-sterilized highbush blueberry nodal explants in 70% ethanol and 1% Tween-20 for 30 s, and then immersed the explants in 0.1% (w/v) benomyl for 15 min, surface sterilized in 0.5% (v/v) sodium hypochlorite for 20 min, and finally washed four times with sterile distilled water.

Several basal media supplemented with growth regulators such as N^6-[2-isopentenyl]adenine (2iP), zeatin, zeatin riboside or thidiazuron (TDZ), and possibly some auxin (Table 1) have been used for shoot culture of *Vaccinium* species. Media with low ionic concentrations were suitable for *Vaccinium* culture. Growth regulators responded according to the blueberry cultivars, 2iP responded well at 49.2-73.8 μM concentration in the establishment phase, and at 24.6-34.5 μM in the proliferation phase (Eccher and Noè, 1989; Isutsa and Pritts, 1994; Noè and Eccher, 1994). Some researchers routinely use cold treatment at 4°C in the darkness with 2iP for initiation of new growth for explants (Orlikowska, 1986). However, 2iP shows some phytotoxicity at high doses, especially in the establishment phase, when many explants turn brown and eventually die. Mixtures of zeatin and 2iP are less phytotoxic to new explants than 2iP alone and produces higher initiation rates (Eccher and Noè, 1989). Zeatin is more effective for shoot initiation in *Vaccinium* species (Reed and Abdelnour-Esquivel, 1991; Gonzales et al., 2000) and for shoot proliferation of highbush blueberry (Chandler and Draper, 1986; Eccher and Noe, 1989), although Gonzales et al. (2000) observed best shoot multiplication of highbush blueberry with 25 μM 2iP in the culture medium. A low concentration of an auxin is beneficial and added to the induction medium, e.g., 5.7 μM IAA (Morrison et al., 2000). However, Frett and Smagula (1983) suggested to use 2iP alone in the transfer medium before rooting. Shoot cultures are maintained on culture medium in illuminated growth rooms at 24-26°C under 16-h photoperiod provided by cool-white fluorescent lamps at a photon flux density of 50-60 μmol $m^{-2}s^{-1}$, where they grow rapidly and their axillary buds grow prematurely to produce new shoots. They are excised at 4-6 weeks intervals and are then transferred to fresh medium where further axillary shoot proliferation occurs; shoot cultures may be multiplied indefinitely by sequential subculture. The production of plantlets is completed by excising shoots and then inducing them to produce adventitious roots either *in vitro* or, most frequently, in *ex vitro* condition on an acidic substrate such as 1 peat : 1 perlite (v/v) (Gonzalez et al., 2000), 4 peat : 2 vermiculite : 1 perlite (v/v/v) (Morison et al., 2000) without an auxin-pretreatment. An auxin-pretreatment was unnecessary for *ex vitro* rooting of blueberries (Gonzalez et al., 2000). For *ex vitro* rooting, the microcuttings are generally maintained in a mist chamber with very high relative humidity and then transferred to a greenhouse for acclimatization. Rooting blueberry microcuttings *ex vitro* can reduce costs (Zimmerman, 1987), although the process is often slower than *in vitro* (Wolfe et al., 1983).

7.2. Cranberries

The cranberry (*V. macrocarpon* Ait.) is a slender, creeping, woody, evergreen perennial vine; native to North America (Vander Kloet, 1983).

Cranberries contain high levels of vitamin C, cellulose, and pectin, and also possess organic acids that are beneficial to the digestive and urinogenital tracts (Eck, 1990). Cranberries are traditionally propagated by stem cuttings into freshly sanded bogs where they readily form adventitious roots and later spread by producing horizontal shoots.

Few reports are available on propagation of *V. macrocarpon* by tissue culture. Scorza et al. (1984) reported that the herbicide glyphosate, alone or in combination with 2iP, increased axillary shoot formation in cultured cranberry nodes. However, they observed fairly low proliferation rates (5 to 7 times). Marcotrigiano and McGlew (1991) and Smagula and Harker (1997) used various commercial salt formulations and recommend using high 2iP concentrations (98.4 μM to 0.39 mM) along with an auxin (IAA or IBA) in the culture media to increase shoot proliferation.

Debnath and McRae (2001a) developed an efficient protocol for *in vitro* clonal propagation and maintenance of cranberry genotypes (Table 1). From nodal and shoot-tip explants obtained under aseptic conditions, cultures are established and maintained in the medium of Debnath and McRae (2001a) containing 12.3-24.6 μM 2iP. Nodal explants are more productive than shoot tips. Shoots are rooted *in vitro* in the same media used for shoot proliferation, but without any growth regulators. Rooted shoots are removed from tissue culture, rinsed to remove culture medium, and planted to potting medium containing 2 peat:1 perlite (v/v). Plantlets are placed in a humidity chamber with a vaporizer, and acclimatized by gradually lowering the humidity over 2 to 3 weeks (temperature 24±2 °C, humidity 95%, PPF = 55 μmol m^{-2} s^{-1}, 16 h photoperiod). Hardened-off plants can be maintained in the greenhouse (temperature 20±2 °C, humidity 85%, maximum PPF = 90 μmol m^{-2} s^{-1}, 16 h photoperiod) (Debnath and McRae, 2001a). However, rooting can also be done under *ex vitro* conditions (Marcotrigiano and McGlew, 1991). *In vitro* rooting is quicker and can also be better controlled in less space, and 85-95% of the proliferated shoots can be acclimatized under greenhouse conditions (Debnath and McRae, 2001a).

Shoot-proliferating cultures of cranberry can easily be stored in the refrigerator. We have stored shoot-proliferating cultures of cranberry cultivars and clones in the dark at 4±2 °C continuously for one year without any loss of viability under normal culture conditions (Debnath unpubl.). Stored cultures continue to grow slowly under dark, refrigerated conditions, and resume normal growth when retrieved from storage. Normally, cultures should routinely be transferred to fresh medium every 6-8 months during storage. However, transfer or subculture period can go up to 12 months or even longer without serious losses in viability.

7.3. Lingonberries

The lingonberry (*Vaccinium vitis-idaea* L.) is a circumboreal woody, dwarf to low-growing, rhizomatous, evergreen shrub (Vander Kloet, 1988; Luby et al., 1991). It is commercially important as a fruit crop, a medicinal plant, and a landscape ornamental ground cover (Dierking and Dierking, 1993). Berries and leaves are used as bladder and kidney disinfectants, to lower cholesterol levels, and to treat rheumatic diseases. When eaten raw, it stimulates the production of gastric juices (Dierking and Dierking, 1993) and an extract of arbutin taken from lingonberry leaves is used for stomach disorders (Racz et al., 1962). Lingonberry has also high anthocyanin content (Stark et al., 1978) and is a good source of antioxidants (Wang et al., 1997), which are believed to reduce the risk of various human degenerative diseases (Prior and Cao, 2000).

Lingonberry can be vegetatively propagated by cuttings or by division of the rhizome. Propagation by rooted stem cuttings is nonproductive (Fernqvist, 1977) due to their inability to produce rhizomes (Holloway, 1985). Plants derived from tissue culture were superior to those obtained by stem cuttings for fruit yield, rhizome production, and vigor in the lingonberry cultivar, 'Sanna' (Gustavsson and Stanys, 2000).

Since initial reports of *in vitro* propagation of lingonberry (Hosier et al., 1985), various culture conditions, basal media, and growth regulators have been investigated (Gebhardt and Friedrich, 1986; Reichers and Bunemann, 1989; Reed and Abdelnour-Esquivel, 1991; Serres et al., 1994; Kutas, 1998; Debnath and McRae, 2001b; Jaakola et al., 2001). Examples of basal media and plant growth regulators used for lingonberry micropropagation are presented in Table 1.

As in cranberries, lingonberries can be propagated (Figure 1) from nodal or shoot-tip explants in a number of media (Table 1). Debnath and McRae (2001b) compared four different media for shoot proliferation of three lingonberry cultivars ('Regal' 'Splendor', and 'Erntedank') and found that modified MS medium was more effective than the Woody Plant Medium (WPM, Lloyd and McCown, 1980) for shoot multiplication. In terms of overall effects on growth as measured by shoot height and leaf number per shoot, modified MS was only moderate in performance. WPM was much less effective for the shoot proliferation of 'Regal' and 'Erntedank'. 'Splendor' was not much affected by the media composition for shoot proliferation. A reasonable balance in terms of shoot multiplication rate and desirable growth characteristics was attained in a new medium (BM-C) formulated in the author's laboratory (Debnath and McRae, 2001b). In the presence of a cytokinin (2iP or zeatin), explants proliferate to produce multiple shoots arising principally from axillary buds (Table 1). Although very high concentration of 2iP (79-98 μM) have been recommended by Hosier et al. (1985) and Norton and Norton (1985), Serres et al. (1994) found WPM containing 10 μM 2iP was an effective *in vitro* shoot proliferation medium for lingonberry cultivars. Gebhardt and Friedrich (1986) recommended a 2iP concentration of 4.9 μM for 'Erntesegen', 'Erntedank' and 'Koralle'. Reichers and Bunemann (1989) suggested a combination of 5.7 μM

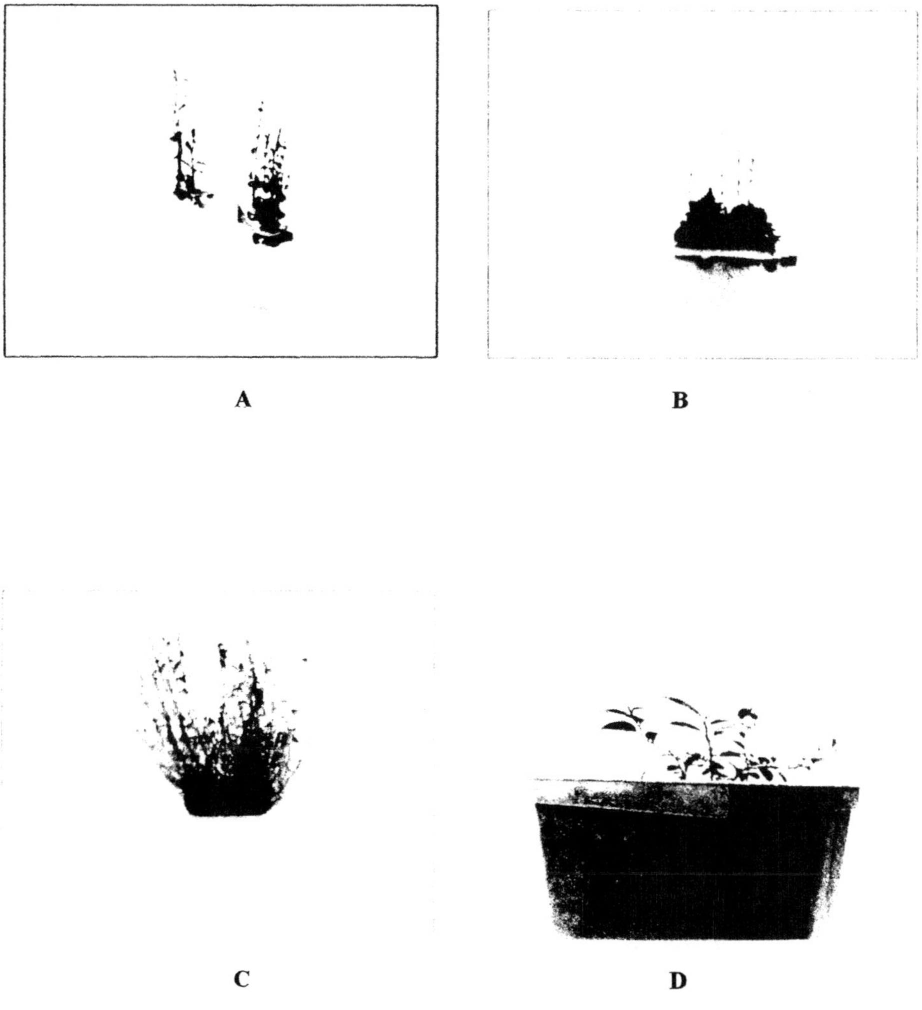

Figure 1. Micropropagation of lingonberry cultivar 'Regal'. A. *In vitro* shoot proliferation after 8 weeks on BM-A (Debnath and McRae, 2001b) containing 12.3 μM 2iP (left) or 5.7 μM zeatin (right), B. Adventitious shoots from dense callus growing at the base of the shoots after 14 weeks on BM-A containing 73.8 μM 2iP. C. *In vitro* shoot proliferation of basal nodal segment subcultured without removal of basal callus on BM-C (Debnath and McRae, 2001b) containing 11.4 μM Zeatin, D. Greenhouse-grown micropropagated plant after 12 weeks under *in vitro* condition.

Table 1. Examples of basal media[a] and plant growth regulators[b] (PGR) used for *in vitro* shoot culture of *Vaccinium* species.

Species	Basal medium	Shoot initiation	Shoot multiplication	Rooting	Reference
V. angustifolium	Z-2	73.8 μM 2iP + 5.7 μM IAA	73.8 μM 2iP + 5.7 μM IAA	*Ex vitro*	Morrison (2000)
V. angustifolium × *corymbosum*	WPM	--	24.6 μM 2iP	*Ex vitro*	Isutsa and Pritts (1994)
V. ashei	Modified Knop (1965)	73.8 μM 2iP	73.8 μM 2iP	148 mM IBA dip, *ex vitro*	Lyrene (1980)
V. corymbosum	PMN	36.9 μM 2iP	36.9 μM 2iP	4.9 mM IBA dip, *ex vitro*	Noè and Eccher (1994)
V. corymbosum	WPM salts + MS vitamins	18 μM zeatin	25 μM 2iP	*Ex vitro*	Gonzalez et al. (2000)
V. macrocarpon	Anderson's (1975) macrosalts + MS microsalts and organics	150 μM 2iP + 1 μM IBA	150 μM 2iP + 1 μM IBA	*Ex vitro*	Marcotrigiano and McGlew (1991)
V. macrocarpon	Z-2 salts	98.4 μM 2iP + 5.7 μM IAA	98.4 μM 2iP + 5.7 μM IAA	98.4 μM 2iP + 5.7 μM IAA	Smagula and Harker (1997)
V. macrocarpon	Unique macrosalts and vitamins + MS microsalts	12.3-24.6 μM 2iP	12.3-24.6 μM 2iP	No PGR	Debnath and McRae (2001a)
V. vitis-idaea	Modified MS	2.2-4.4 μM BA or 4.9 μM 2iP	4.9 μM 2iP	2.2 μM BA or dip in 1mM IAA	Gebhardt and Friedrich (1986)
V. vitis-idaea	WPM + 1.3 g l^{-1} calcium gluconate	5 μM 2iP	10 μM 2iP	39.4 mM IBA dip, *ex vitro*	Serres et al. (1994)

Table 1 (Continued)

Species	Basal medium	Shoot initiation	Shoot multiplication	Rooting	Reference
V. vitis-idaea	Modified MS (establishment), BM-C (proliferation)	5.7 μM Zeatin	11.4 μM Zeatin or 24.6 μM 2iP	39.4 mM IBA dip, *ex vitro*	Debnath and McRae (2001b)
V. vitis-idaea	Modified MS	24.6 μM 2iP	24.6 μM 2iP	2.07 mM IBA dip, *ex vitro*	Jaakola et al. (2001)
Vaccinium spp.	Modified WPM	18.2 μM Zeatin	24.6 μM 2iP	Not reported	Reed and Abdelnour-Esquivel (1991)

[a]**Media:** BM-C = Debnath and McRae (2001b), MS = Murashige and Skoog (1962), WPM = Lloyd and McCown (1980), Z-2 = Zimmerman and Broome (1980), PMN = Eccher et al. (1986).
[b]**PGR:** IAA = 3-indolyl-acetic acid; IBA = 3-indolebutyric acid, 2iP = N^6-[2-isopentenyl]adenine.

IAA and 24.6 μM 2iP in WPM for micropropagation of 'Red Pearl'. Zeatin induces proliferation of 2 to 3 times as many shoots as 2iP, although shoot vigor is better with 2iP than with zeatin (Figure 1A). Shoots growing for more than 12 weeks on media that contained more than 24.6 μM 2iP or 11.4 μM zeatin occasionally produced adventitious shoot masses, which appeared to arise from dense calluses growing at the base of the shoots (Figure 1B) (Debnath and McRae, 2001b).When basal segments are subcultured without the basal callus removed, shoots exhibit very vigorous growth (Figure 1C). Debnath and McRae (2001b) proposed a protocol for lingonberry micropropagation (Figure 2) by using a zeatin-supplemented culture medium for shoot initiation and proliferation, followed by culturing on 2iP-supplemented medium before rooting, more than one million plants could be produced from a single stem cutting in a year.

Like blueberries, rooting of micropropagated shoots are easily induced under *ex vitro* conditions within 4-6 weeks of transfer of shoots on potting medium, and rooted plants can be acclimatized readily in the greenhouse with 80 to 100% survival rate (Serres et al., 1994; Debnath and McRae, 2001b). Incubation of microshoots for 5 min in 2.07 mM IBA solution before planting on peat increases rooting efficiency (Jaakola et al., 2001).

Explants: Stem segments from greenhouse-grown plants

⇓

Washing in a mild liquid detergent (2% v/v) (2 min)

⇓

Surface sterilization in a solution of 0.75-1% sodium hypochlorite and 0.1% Tween 20 (20-25 min)

⇓

A quick rinse in 70% ethanol

⇓

Washing three times in sterilized deionized water

⇓

Culture initiation on BM-A (half MS macro-salts, MS micro-salts and vitamins, 25 gl^{-1} sucrose) containing 5.7 μM zeatin

⇓

Shoot proliferation on BM-A containing 11.4 μM zeatin (6 weeks)

⇓

Transfer of proliferated shoots onto BM-C (Debnath and McRae, 2001b) containing 24.6 μM 2iP

⇓	⇓
Rooting of shoot tips in 2 peat: 1 perlite mix after dipping in 39.4 mM IBA	**Shoot multiplication of nodes with basal mass on BM-A with 11.4 μM zeatin followed by shoot transfer onto BM-C containing 24.6 μM 2iP**
⇓	⇓
Acclimatization	**Continuation of rooting of shoot tips and shoot multi-plication of nodes with basal mass**

Figure 2. Schematic representation of explant procedure and micropropagation protocol of lingonberry

7.4. Concluding Remarks

Propagation of the *Vaccinium* species by tissue culture is becoming increasingly common commercially, since this is a reliable and efficient method, especially for rapid introduction of new cultivars. However, the initiation of new growth on *Vaccinium* explants is often the limiting factor for *in vitro* culture, especially when field- or natural-grown plant materials are used. The main reason for this is the high degree of contamination. Also, the explants often turn brown and die, partly due to the strong sterilization conditions, that must be used to avoid contamination. Because of the great variability within this genus, some species and cultivars would require further study to optimize micropropagation protocol.

8. *VITIS*

There are 65 species in the genus *Vitis* L. (family: Vitaceae, vine family), all of which are deciduous woody shrubs or climbers. *V. vinifera* L., the common grape vine, is the most important species of the genus that is of commercial interest; and is cultivated as dessert fruit; for the preparation of dried currants, raisins and sultanas; and for the making of wine. *V. rotundifolia* Michx., the muscadine grape, is important in wine production in the South Eastern United States (Sudarsono and Goldy, 1991). Grape plants are confined primarily to the temperate zones of the world, although they are also grown in subtropical and tropical climates. Conventionally, grape cultivars are propagated by dormant hardwood cuttings (36 to 46 cm long) collected during the winter from well-developed current season's canes. They are planted in the nursery during the spring, and one season's growth gives plants to transplant to the vineyard (Hartmann et al., 1997). Micropropagation has offered great potential for rapid propagation of promising new cultivars, and to provide stocks of virus-free plants (Monette, 1988; Cao, 1990; Hartmann et al., 1997).

Two main approaches have been attempted in grapevine micropropagation: (1) the enhancement of axillary shoot development (enhanced apical and axillary meristem growth and branching) and (2) the adventitious *de novo* induction of shoots, roots, and somatic embryos directly.

8.1. Micropropagation via Axillary Branching

8.1.1. Culture Establishment and Shoot Proliferation

In vitro culture of *Vitis* sp. started with the work of Morel (1944 a, b). Numerous publications on micropropagation of grapevines have appeared since the first report of obtaining entire plantlets from microcuttings (Galzy, 1961). Grapevine can be rapidly propagated *in vitro* from shoot tips or nodal segments, isolated from actively growing vines of greenhouse- (Mhatre et al., 2000), or field-grown material (Lewandowski, 1991). Figure 3 presents a general micropropagation protocol of *Vitis*. Node cuttings collected from vineyard can be placed in pots filled with vermiculite and then grown in a controlled environment chamber. When the shoots develop 8-10 leaves, axillary buds are surface disinfected with 5-7% calcium hypochlorite containing a few drops of Tween-20 (15 min) and used as explants for culture initiation (Heloir et al., 1997; Péros et al., 1998). Mhatre et al. (2000) used $HgCl_2$ (0.1% w/v, 10 min) for sterilizing nodal explants isolated from 3-month-old greenhouse-grown vines.

Explants: Shoot tips or nodal segments from field- or greenhouse-grown plants

⇓

Washing thoroughly in a mild liquid detergent

⇓

Surface sterilization in a solution of 5-7% $Ca(OCl)_2$ containing a few drops of Tween-20 (15 min) or in 0.1% $HgCl_2$ (10 min)

⇓

Washing three times in sterilized distilled water

⇓

Culture initiation on MS (Murashige and Skoog, 1962)/ modified MS/M64 (Péros et al., 1998) containing 4.4-17.7 μM BA and 0.2-0.4 mM adenine sulphate

⇓

Shoot proliferation on the same medium used for shoot initiation containing same or half-strength BA plus 1 μM IAA and 0.1 μM GA_3

⇓

Rooting of microshoots on full-, half- or quarter-strength MS containing no growth regulators or with 0.5-2 μM NAA/2.5 μM IBA

⇓

Acclimatization in a mixture of 1 peat : 1 perlite (v/v)

Figure 3. Schematic representation of the general procedures for explant preparation and micropropagation of *Vitis*

A medium containing MS salts has been used most frequently (Roubelakis-Angelakis and Zivanovitc, 1991; Gribaudo et al., 2001), although some researchers have used modified MS (Lewandowski, 1991), M64 (Péros et al., 1998), modified Nitsch and Nitsch (1969) (Mhatre et al., 2000) or G90 (Galzy et al., 1990; Péros et al., 1998) medium for culture initiation. Cultures are usually established on agar solidified media supplemented with 4.4-17.7 μM BA and 0.2-0.4 mM adenine sulphate (Lewandowski, 1991;Mhatre et al., 2000). Sometimes low concentration auxin (IBA or NAA) (Mhatre et al., 2000) or GA_3 (Lewandowski, 1991) has also been added to the culture medium. Sudarsono and Goldy (1991) reported that thidiazuron (2.3-4.5 μM) alone or in combination with BA (1-5 μM) or kinetin (1-5 μM) was effective for establishing axillary buds in muscadine grape. The culture can be subcultured at 1-3 month interval on the same fresh culture medium using shoot tip or leaf-bearing single node microcuttings (Thomas, 2000; Thomas and Schiefelbein, 2001). However, Sudarsono and Goldy (1991) suggested to transfer cultures every 20 days onto fresh medium to reduce browning and senescence. Heloir et al. (1997) used MS medium supplemented with 8.9 μM BA for *in vitro* culture establishment from axillary-bud microcuttings, and the concentration of BA was reduced to 4.4 μM to prevent hyperhydricity in subcultures for *V. vinifera* cv 'Pinot noir'. During subculturing, IAA (Mhatre et al., 2000; Thomas and Schiefelbein, 2001) or GA_3 (Lewandowski, 1991) at 1 μM or both IAA (1 μM) and GA_3 (0.1 μM) (Thomas, 2000) have also been added in the culture media. Explant factors such as age, size and position on the stock shoot influence shoot growth and multiplication rate *in vitro* in grapevine (Botti et al., 1993).

8.1.2. Rooting and Acclimatization

In grapes, rooting *in vitro* is generally practiced (Lewandowski, 1991; Amâncio et al., 1999; Mhatre et al., 2000; Thomas, 2000; Gribaudo et al., 2001) although there are some reports on *ex vitro* rooting (Gray and Klein, 1989; Gray and Benton, 1991). The *in vitro* rooting percentage is better than *ex vitro* (Gray and Benton, 1991). For rooting, 1-2 cm long microcuttings can be trimmed of basal leaves and transferred to culture vessels containing rooting media.

There have been reports for using full-, half- or quarter-strength MS or Nitsch and Nitsch or White medium for rooting of shoots of *Vitis* genotypes. Rooting media have been supplemented with 0.5-2 µM NAA (Barlass and Skene, 1980; Amâncio et al., 1999), 0.1 µM IBA (Novak and Juvova, 1983), 5 µM 2,4-D (Rajasekaran and Mullins, 1981) or 0.57 µM IAA (Mhatre et al., 2000). Among the two auxins (NAA and IBA) tested for *in vitro*-rooting, 2.5 µM IBA induced an excellent rooting-system in the *V. vinifera* cv 'Pinot noir' shoots (Heloir et al., 1997). Additive root-stimulating effects from the auxin combination of NAA (0.0054 µM) and IBA (0.0246 µM) in half-strength MS resulted in >95% rooting of *V. labrusca* L. 'Delaware' microcuttings in 10 days (Lewandowski, 1991). Roubelakis-Angelakis and Zivanovitc (1991) and Péros et al. (1998) rooted *Vitis* genotypes on growth regulator-free culture media. Leafy microcuttings promote early and higher rooting percent, production of vigorous roots and healthier shoots, and higher output of plantlets suitable for hardening (Thomas, 1997; 1998; 2000). Grapevine plantlets can be grown *in vitro* after addition of 0.5-1 mg l^{-1} paclobutrazol to the liquid culture medium and by reducing relative humidity from 100% to 94% inside the culture vessel during the last micropropagation phase to allow an easier acclimatization (Smith et al., 1992; Gribaudo et al., 2001).

Micropropagation in grape is accomplished through induction of both root and shoot growth in microcuttings on a culture medium supplemented with an auxin (Cao, 1990; Thomas, 1997; 1998). Thomas and Schiefelbein (2001) has recently developed a rapid propagation method of grapevine employing a combination of *in vitro* and *in vivo* approaches. The *in vitro* part includes induction of root and shoot growth in shoot tip and nodal microcuttings on MS medium containing 1 μM IAA. The sequential pruning of shoots is done at the intervals of 1, 1.5, and 2 months, leaving the basal one to two nodes, resulting in fresh sprouts on the stump, and use the remaining stumps for *in vivo* establishment. The *in vivo* part includes acclimatization of *in vitro* rooted plantlets and stumps; use of single node cuttings from 1.5- to 2-month-old *in vivo* shoots for the subsequent propagation; and utilizing the fresh sprout growth from these cuttings and *in vivo* stumps for further propagation.

Ex vitro rooting is done by transfer of *in vitro*-grown shoot directly into 'compost'. Troncoso et al. (1988) obtained higher proportion of rooted shoots than *in vitro* rooting, and Gray and Benton (1991) reported that acclimatized plants can be produced more readily through *ex vitro* rooting.

For acclimatization, plantlets rooted *in vitro* can be planted in a mixture of 1 peat : 1 perlite (v/v) and the acclimatization phase can take place at 40 or 90 μmol $m^{-2}s^{-1}$ photon flux density for 12 or 16 h per day, respectively (25±2°C during the light and 22±2°C during the dark period) at plant level (Amâncio et al. 1999).

8.2. Adventitious Bud Regeneration Using Shoot Apices

Adventitious bud regeneration from shoot apices is used for rapid shoot multiplication (Barlass and Skene, 1978). The procedure comprises two major steps: firstly, the surface sterilized shoot apices (ca. 1 mm in length) containing 2-3 leaf primordia when cultured in liquid MS medium supplemented with BA (10 μM), the apical fragments grow into individual leaf-like structures within 1 month of culture; and secondly, adventitious shoots proliferate at the basal end of these leaves when transferred onto the same medium solidified with agar. Repeated shoot proliferation occurred when explants are transferred to this medium (Skene and Barlass, 1981). The differentiating cell clumps can also be transferred to solid medium after excising a part of the apices. The leaf-like structures on solid medium increase in size and shoots proliferate from the basal swelling of leaf-like structures. Excision of this area from the leaf-like structures, followed by division and subculturing to the same medium, results in prolific formation of adventitious buds. Despite originating adventitiously, plants originating from divided apices were true-to-type, and the technique can be used for the multiplication of disease-free clones (Barlass, 1991).

8.3. Somatic Embryogenesis

Somatic embryogenesis, is rapid and well suited for clonal multiplication of plants (Murashige, 1977; 1978; Bajaj, 1986); has been reported for various *Vitis* species and several important cultivars of *V. vinifera* L. and their hybrids (see reviews by Gray and Meredith, 1992; Gray, 1995; Martinelli and Gribaudo, 2001). The large-scale production of plants (Denchev and Atanassov, 1995) using bioreactors (Nishimura et al., 1993; Bajaj, 1995) is more efficient when somatic embryos are used as propagules because they lend themselves to automation and to sorting techniques such as digital image analysis (Smith, 1995). The rapid multiplication of plants via synthetic seed (Redenbaugh, 1993; Gray et al., 1995), encapsulated somatic embryos for direct seeding of crops, is a potential method of clonal propagation in grapevine.

In grape, somatic embryogenesis was first reported by Mullins and Srinivasan (1976) and Hirabayashi et al. (1976), and thereafter with a variety of *Vitis* explants such as anthers (Rajasekaran and Mullins, 1979; Popescu, 1996; Salunkhe et al., 1999; Martinelli et al., 2001b), unfertilized ovules (Mullins and Srinivasan, 1976; Srinivasan, 1980), ovaries (Martinelli et al., 2001b), stem (Krul and Worley, 1977), leaves (Martinelli et al., 1993; Robacker, 1993; Das et al., 2002), petioles (Martinelli et al., 1993; 2001a), and tendrils (Salunkhe et al., 1997). Examples of protocols of somatic embryo formation in *Vitis* are shown in Table 2.

Table 2: Examples of explants, basal media[a], and growth regulators[b] used for the induction of embryogenesis and the formation of somatic embryo in *Vitis*.

Species	Explant	Embryogenic callus induction/ formation	Callus proliferation	Embryo development	Plant regeneration	Reference
V. latifolia	Anthers	NN + 20 μM 2,4-D + 9 μM BA	NN + 10 μM NAA	NN + 10 μM NAA + 9 μM BA	NN	Salunkhe et al. (1999)
V. rotundifolia	Immature leaves & petioles	NN + 9 μM 2,4-D + 4.4 μM BA	NN +10.7 μM NAA + 0.9 μM BA	NN	NN	Robacker (1993)

Table 2 continued

Species	Explant	Embryogenic callus induction/ formation	Callus proliferation	Embryo development	Plant regeneration	Reference
V. vinifera	Leaves	MS + 0.5 μM 2,4-D + 4.4 μM BA	--	NN + 0.5 μM IBA + 4.4 μM BA	NN	Das et al. (2002)
V. vinifera	Tendril	NN salts + LS vitamins + 10 mM BA + 0.4 μM NAA + 2.8 μM GA_3	--	ER + 1 μM BA	1/2MS + 1 μM IAA	Salunkhe et al. (1997)

[a]The media above are: B5 = Gamborg et al. (1968); ER = Emershad and Ramming (1994) ; LS = Linsmaier and Skoog (1965); NN = Nitsch and Nitsch (1969); and MS = Murashige and Skoog (1962).

[b]Growth regulators are: BA = benzyladenine; GA_3 = gibberellic acid; 2,4-D = 2,4-dichlorophenoxy acetic acid; IAA = 3-indolyl-acetic acid; IBA = 3-indolebutyric acid; and NAA = α-naphthaleneacetic acid.

9. CONCLUSIONS

Micropropagation is being used increasingly in small fruits. Shoot culture with the proliferation of axillary shoots is reliable and simple method of *in vitro* propagation to maintain the genetic integrity of clones. Meristem culture, sometimes combined with heat treatment is widely used to obtain virus- and disease-free plants. Recently, methods of regenerating adventitious shoots or embryos from explants, callus or protoplasts have met with increasing success with small fruits and may have potential for rapid propagation, provided genetic integrity of the micropropagated plants with respect to the mother plants is retained. Where the objective is to maintain genetic uniformity, micropropagation by enhancement of axillary shoot growth is preferred over adventitious shoot formation and somatic embryogenesis. The former involves continued growth of organized tissue whereas adventitious organogenesis and somatic embryogenesis involve plant regeneration from one or few cells. Somatic embryogenesis and enhanced axillary branching have been claimed to give rise to genetically uniform and true-to-type plants (Rani and Raina, 2000). Enhanced axillary branching involves the abolition of apical dominance resulting in the de-repression and multiplication of axillary buds, and has become a very important method due to its simplicity of approach and faster propagation rate (Debergh and Reed, 1991; Wang and Charles, 1991). Although somatic embryogenesis is amenable to mechanization, making possible the use of bioreactors for large-scale production of somatic embryos, and their delivery through encapsulation into artificial seeds; it often exhibits abnormalities with respect to morphology (Halperin, 1966) as well as physiology. Jayasankar et al. (2002) observed a unique somatic embryo morphotype, characterized by having a single cotyledon and an enlarge hypocotyl, among the *in vitro* cultures of grapevine somatic embryos.

The tissue-culture-associated variation, termed somaclonal variation (Larkin and Scowcroft, 1981), has frequently been observed in long-term cultures of small fruits. Genetic stability during micropropagation is controlled by numerous factors, such as the internal plant factors (genotypes, presence of chimeral tissue, explant origin, etc.) and external factors (tissue culture protocol, media sequences, and growth substance types and concentrations). Micropropagation systems using organized meristems are not absolutely free from genetic changes that occur in *in vitro* conditions. Regardless of the micropropagation method, it is important to investigate the morphogenic and genetic stability of the plants produced via organogenesis or somatic embryogenesis. Somaclonal variations can be distinguished by their morphological, biochemical, physiological, and genetic characteristics. Molecular markers, such as restriction fragment length polymorphism (RFLP), random-amplified polymorphic DNA (RAPD), arbitrary primed PCR (AP-PCR), DNA amplified fingerprinting (DAF), simple (short) sequence repeat (SSR), short tandem repeat (STR), sequence characterized amplified region (SCAR), sequence-tagged sites (STSs), and amplified fragment length polymorphism (AFLP) (Henry, 1997), generate special classes of markers sensitive for genetic analysis of tissue culture-raised plants. They are powerful tools in genetic identification of somaclonal variation with greater precision and less effort than phenotypic and karyologic analysis (Cloutier and Landry, 1994), and have been used for genetically analyzing micropropagated plantlets (Goto et al., 1998; Rani and Raina, 1998; Watanabe et al., 1998; Rani et al., 2000; Tang, 2001).

Genetic variants or somaclones are undesirable for propagation but may be exploited for crop improvement if they have desirable agronomic characters (Brar and Jain, 1998; Jain, 2001). Excessive concentrations of auxins and/or cytokinins in media, however, can result in regenerated plants deviating morphologically from the normal (George, 1996). The appearance of somaclonal variation in the *in vitro*-regenerated plants can be greatly reduced by using lower levels of growth regulators in the culture media used for micropropagation of small fruits.

10. REFERENCES

Amâncio S, Rebordão JP and Chaves MM (1999) Improvement of acclimatization of micropropagated grapevine: Photosynthetic competence and carbon allocation. Plant Cell Tiss. Org. Cult. 58:31-37

Anderson WC (1975) Propagation of rhododendrons by tissue culture. Part I: Development of a culture medium for multiplication of shoots. Comb. Proc. Int. Plant Prop. Soc. 25:129-135

Anderson WC (1978) Rooting of tissue cultured rhododendrons. Comb. Proc. Int. Plant Prop. Soc. 28:135-139

Anderson WC (1980) Tissue culture propagation of red and black raspberries, *Rubus idaeus* and *R. occidentalis*. Acta Hort. 112:13-20

Arena ME and Martínez Pastur GJ (1995) *In vitro* propagation of *Ribes magellanicum* Poiret. Sci. Hort. 62:139-144

Bailey LH (1976) Hortus third. Macmillan, New York

Bajaj YPS (1986) Biotechnology of tree improvement for rapid propagation and biomass energy production. In: Bajaj YPS (ed.) Biotechnology in agriculture and forestry 1: Trees I (pp. 1-23). Springer-Verlag, Berlin, Heidelberg, Germany

Bajaj YPS (1995) Somatic embryogenesis and its application for crop improvement. In: Bajaj YPS (ed.) Biotechnology in agriculture and forestry, vol 30. Somatic embryogenesis and synthetic seed, I (pp 105-125). Springer, Berlin, Heidelberg, New York

Barker WG and Collins WB (1963) The blueberry rhizome: *in vitro* culture. Can. J. Bot. 41:1325-1329

Barlass M (1991) Commercial production in Australia and New Zealand. In: Debergh PC and Zimmerman RH (eds.) Micropropagation: Technology and application (pp. 191-198). Kluwer Acad. Publ., Dordrecht, Boston, London

Barlass M and Skene KGM (1978) *In vitro* propagation of grapevine (*Vitis vinifera*) from fragmented shoot apices. Vitis 17:335-340

Barlass M and Skene KGM (1980) Studies on the fragmented shoot apex of grapevine. II. Factors affecting growth and differentiation *in vitro*. J. Exp. Bot. 31:489-495

Bartel S, Bartholomew H, Ellis MA, Funt RC, Nameth ST, Overmyer RL, Schneider H, Twarogowski WJ and Williams RN (1988) Brambles: production, management and marketing. Ohio Coop. Ext. Serv. Bul. 783

Bhatt ID and Dhar U (2000) Micropropagation of Indian wild strawberry. Plant Cell Tiss. Org. Cult. 60:83-88

Blenda VF and Kirilenko ED (1982) Regeneration of black currant from apical meristem (brief report) (in Russian). Fiziolog. Biokhim. Kul'turnykh Rastenii 14:244-247

Boase M, Wright R and McLeay L (1993) Coconut milk enhancement of axillary shoot growth *in vitro* of kiwifruit (*Actinidia deliciosa* var *deliciosa*). New Zealand J. Crop Hort. Sci. 21:171-176

Bobrowski VL, Mello-Farias PC and Peters JA (1996) Micropropagation of blackberries (*Rubus* sp.) cultivars. Rev. Bras. de AGROCIÊNCIA 2:17-20

Borkowska B (1999) Developmental and physiological status of micropropagated strawberry plants rooted *ex vitro* and planted to different substrates. In: Proc. ISHS Conf. on methods and markers for quality assurance in micropropagation (pp. 163-164), Cork, Ireland

Borkowska B (2001) Morphological and physiological characteristics of micropropagated strawberry plants rooted *in vitro* or *ex vitro*. Sci. Hort. 89:195-206

Borkowska B, Szczygiel A and Pierzga K (1999) Direct *ex vitro* rooting technique of micropropagated strawberry shoots. J. Fruit Ornam. Plant Res. 7:1-10

Botti C, Garay L and Reginato G (1993) The influence of culture dates, genotype and size and type of shoot apices on *in vitro* shoot proliferation of *Vitis vinifera* cvs. Thompson Seedless, Ribier and Black Seedless. Vitis 32:125-126

Boxus P (1974) The production of strawberry plants by *in vitro* micropropagation. J. Hort. Sci. 49:209-210

Boxus P (1992) Mass production of strawberry and new alternatives for some horticultural crops. In: Kurata K and Kozai T (eds.) Transplant production systems (pp. 151-162). Kluwer Acad. Publ., The Netherlands

Boxus P, Quoirin M and Laine JM (1977) Large scale propagation of strawberry plants from tissue culture. In: Reinert J and Bajaj YPS (eds.) Plant cell, tissue and organ culture (pp. 130-143). Springer-Verlag, New York

Brar DS and Jain SM (1998) Somaclonal variation: mechanism and applications in crop improvement. In: Jain SM, Brar DS and Ahloowalia BS (eds.) Somaclonal variation and induced mutations in crop improvement (pp. 15-38). Kluwer Acad. Publ., The Netherlands

Brennan RM, Davidson D, Wilshin A and Millam S (1989) An assessment of the *in vitro* multiplication rates of fourteen black current cultivars. J. Hort. Sci. 64:679-681

Brennan RM, Millam S, Davidson D and Wilshin A (1990) Establishment of an in vitro *Ribes* germplasm collection and preliminary investigations into long-term low temperature germplasm storage. Acta Hort. 280:109-112

Brissette L, Tremblay L and Lord D (1990) Micropropagation of lowbush blueberry from mature field-grown plants. HortSci. 25:349-351

Broome OC and Zimmerman RH (1978) *In vitro* propagation of blackberry. HortSci. 13:151-153

Brossard-Chriqui D and Tripathi BL (1984) Comparaison des aptitudes morpho génétiques des étamines fertiles ou stériles d'*Actinidia chinensis* cultivées *in vitro*. Can. J. Bot. 62:1940-1946

Caldwell JD (1984) Blackberry propagation. HortSci. 19:193-195

Cameron JS, Hancock JF and Nourse TM (1985) The field performance of strawberry nursery stock produced originally from runners or micropropagation. Adv. Strawberry Prod. 4:56-58

Cao Z. (1990) Grape: Micropropagation. In: Chen Z, Evans DA, Sharp WR, Ammirato RV and Sondhal MR (eds.) Handbook of Plant Cell Culture Vol 6, Perennial Crops (pp. 321-328). Academic, New York.

Cerović R and Ružić D (1989) Micropropagation of strawberry cvs Čačanska Rana and Senga Sengana: pomological-biochemical characteristics of micropropagated plants. Acta Hort. 265:353-356

Chandler CK and Draper AD (1986) Effect of zeatin and 2iP on shoot proliferation of three highbush blueberry clones *in vitro*. HortSci. 21:1065-1066

Cheng TY (1977) Factors affecting adventitious bud formation of cotyledon culture of Douglas fir. Plant Sci. Lett. 9:179-187

Chiarotti A, Caboni E and Frattarelli A (1991) Shoot regeneration from *in vitro* roots of kiwi. Acta Hort. 289:97-99

Childers NF (1980) Foreward. In: Childers NF (ed.) The strawberry: Cultivars to marketing (p ix). Hort. Publ., Gainesville, FL.

Cloutier S and Landry BS (1994) Molecular markers applied to plant tissue culture. In Vitro Cell. Dev. Biol. Plant 30:32-39

Cohen D (1980) Application of micropropagation methods for blueberries and tamarillos. Comb.

Proc. Int. Plant Prop. Soc. 30:144-146

Das DK, Reddy MK, Upadhyaya KC and Sopory SK (2002) An efficient leaf-disc culture method for the regeneration via somatic embryogenesis and transformation of grape (*Vitis vinifera* L.). Plant Cell Rep. 20:999-1005

Daubeny HA (1996) Brambles. In: Janick J and Moore JN (eds.) Fruit breeding, vol. II: Vine and small fruits (pp. 109-190). John Wiley and Sons, Inc., New York

Debergh, P (1983) Effect of agar brand and concentration on the tissue culture medium. Physiol. Plant. 59:270-276

Debergh PC and Reed PE (1991) Micropropagation. In: Debergh PC and Zimmerman RH (eds.) Micropropagation: Technology and application (pp. 1-13). Kluwer Acad. Publ., Dordrecht, Boston, London

Debnath SC and McRae KB (2001a) An efficient *in vitro* shoot propagation of cranberry (*Vaccinium macrocarpon* Ait.) by axillary bud proliferation. In Vitro Cell. Dev. Biol. Plant 37 :243-249

Debnath SC and McRae KB (2001b) *In vitro* culture of lingonberry (*Vaccinium vitis-idaea* L.): The influence of cytokinins and media types on propagation. Small Fr. Rev. 1:3-19

Denchev PD and Atanassov AI (1995) Micropropagation through somatic embryos. In: Bajaj YPS (ed.) Biotechnology in agriculture and forestry, vol 30. Somatic embryogenesis and synthetic seed, I (pp. 193-206). Springer, Berlin, Heidelberg, New York

Dierking WJr. and Dierking S (1993) European *Vaccinium* species. Acta Hort. 241:299-304

Dijkstra J (1993) Research on strawberries focuses on healthy plant material. Expensive cultural method requires excellent material. Fruitteelt-Den-Haag 83:14-15

Donnelly DJ and Doubeny HA (1986) Tissue culture of *Rubus* species (a review). Acta Hort. 183:305-314

Eccher T and Noè N (1989) Comparison between 2iP and zeatin in the micropropagation of highbush blueberry (*Vaccinium corymbosum).* Acta Hort. 441:185-190

Eccher T, Noè N, Piagnani C and Castellis S (1986) Effects of increasing concentrations of BAP and 2iP on *in vitro* culture of *Vaccinium corymbosum*. Acta Hort. 179II:879-881

Eck P (1990) The American Cranberry. New Brunswick, NJ, Rutgers Univ. Press

Emershad RL and Ramming DW (1994) Somatic embryogenesis and plant development from immature zygotic embryos of seedless grapes (*Vitis vinifera* L.). Plant Cell Rep. 14:6-12

Fernandez GE and Clark JR (1991) *In vitro* propagation of the erect thornless "Navaho' blackberry. HortSci. 26:1219

Fernqvist I (1977) Results of experiments with cowberries and blueberries in Sweden. Acta Hort. 61:295-300

Frett JJ and Smagula JM (1983) *In vitro* shoot production of lowbush blueberry. Can. J. Plant Sci. 63:467-472

Galzy R (1961) Confirmation de la nature virale du court-noué de la vigne par des essais de thermothérapie sur des cultures *in vitro*. C. R. Acad. Soc. Biol. Paris 253:706-708

Galzy R, Haffner V and Compan D (1990) Influence of three factors on the growth and nutrition of grapevine microcuttings. J. Expt. Bot. 41:295-301

Gamborg OL, Miller RA and Ojima K (1968) Nutrient requirements of suspension cultures of soybean root cells. Exp. Cell Res. 50:151-158

Gebhardt K and Friedrich M (1986) *In vitro* shoot regeneration of lingonberry clones. Gartenbauwissenschaft 51:170-175

George EF (1996) Plant Propagation by Tissue Culture, Part 2: In Practice. Exegetics Ltd., Edington

González MV, Rey M and Rodríguez R (1995) Plant regeneration from petioles of kiwifruit microshoot. HortSci. 30:1302-1303

Gonzalez MV, Lopez M, Valdes AE and Ordas RJ (2000) Micropropagation of three berry fruit species using nodal segments from field-grown plants. Ann. Appl. Biol. 137:73-78

González MV, Rey MA, Revilla MA and Rodríguez R (1990) Morphogenetic patterns in kiwi tissue culture and after *Agrobacterium* co-culture. Acta Hort. 282:359-365

Goto S, Thakur RC and Ishii K (1998) Determination of genetic stability in long-term micropropagated shoots of *Pinus thunbergii* Perl. using RAPD markers. Plant Cell Rep. 18:193-197

Gray DJ (1995) Somatic embryogenesis in grape. In: Gupta PK, Jain SM and Newton RJ (eds.) Somatic embryogenesis in woody perennials, vol 2 (pp 191-217). Kluwer Acad. Publ., Dordrecht

Gray DJ and Benton CM (1991) *In vitro* micropropagation and plant establishment of muscadine grape cultivars (*Vitis rotundifolia*). Plant Cell Tiss. Org. Cult. 27:7-14

Gray DJ and Klein LM (1989) *In vitro* micropropagation and plant establishment of 'Blanc du Bois' grapes. Proc. Fla. State Hort. Soc. 102:221-223

Gray DJ and Meredith CP (1992) Grape. In: Hammerschlag FA and Litz RE (eds.) Biotechnology of perennial fruit crops (pp. 229-262). CAB Int., Wallingford, UK

Gray DJ, Compton ME, Harrell RC and Cantliffe DJ (1995) Somatic embryogenesis and the technology of synthetic seed. In: Bajaj YPS (ed.) Biotechnology in agriculture and forestry, vol 30. Somatic embryogenesis and synthetic seed, I (pp 126-151). Springer, Berlin, Heidelberg, New York

Gribaudo I, Novello V and Restagno M (2001) Improved control of water loss from micropropagated grapevines (*Vitis vinifera* cv. Nebbiolo). Vitis 40:137-140

Grout BWW and Millam S (1995) Photosynthetic development of micropropagated strawberry plantlets following transplanting. Ann. Bot. 55:129-131

Grout JM and Reed PE (1986) Influence of stock plant propagation method on tissue culture and leaf-bud propagation of 'Northblue' blueberry. J. Am. Soc. Hort. Sci. 111:368-371

Gui YL, Mu XJ and Xu TY (1982) Studies on morphological differentiation of endosperm plantlets of Chinese gooseberry in vitro. Acta Bot. Sin. 24:216-221

Gustavsson BA and Stanys V (2000) Field performance of 'Sanna' lingonberry derived by micropropagation vs. stem cuttings. HortSci. 35:742-744

Halperin W (1966) Alternative morphogenetic events in cell suspensions. Am. J. Bot. 53:443-453

Harada H (1975) *In vitro* organ culture of *Actinidia chinensis* as a technique for vegetative multiplication. J. Hort. Sci. 50:81-83

Hartmann HT, Kester DE, Davies FT and Geneve RL (1997) Plant propagation – Principles and Practices. Sixth Edn. Prentice Hall, NJ

Harvey AE and Grasham JL (1977) Callus proliferation and differentiation of *Ribes* spp. *in vitro*. USDA For. Serv. Note INT-222

Hayashi M, Kogami N, Takatuji M and Honjo T (1997) Growth of strawberry plantlets rooted by the direct *ex vitro* rooting method in environmental control unit. J. Soc. High Tech. Agric. 9:67-71

Heloir M-C, Fournioux J-C, Oziol L and Bessis R (1997) An improved procedure for the propagation *in vitro* of grapevine (*Vitis vinifera* cv. Pinot Noir) using axillary bud microcuttings. Plant Cell Tiss. Org. Cult. 49:223-225

Henry, RJ (1997) Practical applications of plant molecular biology. Chapman & Hall, London

Hirabayashi T, Kozaki I and Akihama T (1976) *In vitro* differentiation of shoots from anther callus in *Vitis* grape research. Hort. Sci. 11:511-512

Hirsch AM (1970) Identification et dosage des acides aminés libres de fragments de tiges d'*Actinidia chinensis* Planch. Mise en culture *in vitro* de ces fragments. C R Acad. Sci. Paris Ser. D 270:1462-1464

Hirsch AM and Bligny-Fortune YD (1979) Organogenèse dans les cultures de tissus de deux plantes appartenant au genre *Actinidia*. Relations entre organogenèse et péroxydases.C R Acad. Sci. Paris Ser. D 288:1159-1162

Hoepfner AS (1989) In vitro propagation of red raspberry. Acta Hort. 262:285-288.

Holloway PS (1985) Rooting of lingonberry, *Vaccinium vitis-idaea*, stem cuttings. Plant Prop. 31:7-9

Hosier MA, Flatebo G and Read PE (1985) *In vitro* propagation of lingonberry. HortSci. 20:384-385

Isutsa DK and Pritts MP (1994) Rapid propagation of blueberry plants using *ex vitro* rooting and controlled acclimatization of micropropagules. HortSci. 29:1124-1126

Jaakola L, Tolvanen A, Laine K and Hohtola A (2001) Effect of N^6-isopentenyladenine concentration on growth initiation *in vitro* and rooting of bilberry and lingonberry microshoots. Plant Cell Tiss. Org. Cult. 66:73-77

Jain SM (2001) Tissue culture-derived variation in crop improvement Euphytica 118: 153-166

James DJ, Knight VH and Thurbon IJ (1980) Micropropagation of red raspberry and the influence of phloroglucinol. Sci. Hort. 12:313-319

Janssen B-J and Gardner RC (1993) The use of transient GUS expression to develop an *Agrobacterium*-mediated gene transfer system for kiwifruit. Plant Cell Rep. 13:28-31

Jayasankar S, Bondada BR, Li Z and Gray DJ (2002) A unique morphotype of grapevine somatic embryos exhibits accelerated germination and early plant development. Plant Cell Rep. 20:907-911

Jones OP and Vine SJ (1968) The culture of gooseberry shoot tips for eliminating virus. J. Hort. Sci. 43:289-292

Knop W (1965) Quantitative Untersuchungen über die Ernahrungsprozesse der Pflanzen. Landwirtsch. Vers. Stn. 7:93-107

Kochetkova NI, Aleshkevich LV and Kochetov YV (1981) Aspects of regenerating gooseberry plants *in vitro* (in Russian). Vest. Sel'skok. Nauk. 2:80-82

Kozai T (1991) Micropropagation under photoautotrophic conditions. In: Debergh PC and Zimmerman RH (eds.) Micropropagation technology and application(pp. 447-469). Acad. Publ., Dordrecht

Krul WR and Worley JF (1977) Formation of adventitious embryos in callus cultures of Seyval, a French hybrid grape. J. Am. Soc. Hort. Sci. 102:360-363

Kutas E (1998) Morphogenesis of introduced varieties of *Vaccinium vitis-idaea* in aseptic culture, In: Proc. Int. Conf. Wild Berry Culture: An Exchange of Western and Eastern Experiences (pp. 87-93). 10-13 Aug., 1998, Tartu, Estonia

Lanham PG, Korycinska A and Brennan RM (2000) Genetic diversity within a secondary gene pool of *Ribes nigram* L. revealed by RAPD and ISSR markers. J. Hort. Sci. Biotech. 75:371-375

Larkin PJ and Scowcroft WR (1981) Somaclonal variation – a novel source of variability from cell cultures for plant improvement. Theor. Appl. Genet. 60:197-214

Levin R, Gaba V, Tal B, Hirsch S and DeNola D (1988) Automated plant tissue culture for mass propagation. Bio-technol. 6:1035-1040

Lewandowski VT (1991) Rooting and acclimatization of micropropagated *Vitis labrusca* 'Delaware'. HortSci. 26:586-589

Linsmaier EM and Skoog F (1965) Organic growth factor requirements of tobacco tissue cultures. Physiol. Plant. 18:120-127

Lionakis SM and Zirari A (1991) The effect of kind of explant and chilling pretreatment of plant material on shoot proliferation of kiwi fruit *in vitro*. Acta Hort. 297:205-210

Lloyd G and McCown B (1980) Commercially feasible micropropagation of mountain laurel, *Kalmia latifolia*, by use of shoot-tip culture. Comb. Proc. Int. Plant Prop. Soc. 30:421-

427
Luby JJ, Ballington JR, Draper AD, Pliszka K and Austin ME (1991) Blueberries and cranberries (*Vaccinium*). Acta Hort. 290:391-456
Lyrene PM (1980) Micropropagation of rabbiteye blueberries. HortSci. 15:80-81
Lyrene PM (1990) Low-chill highbush blueberries. Fruit Var. J. 44:82-86
Marcotrigiano M and McGlew SP (1991) A two-stage micropropagation system for cranberries. J. Amer. Soc. Hort. Sci. 116:911-916
Martinelli L and Gribaudo I (2001) Somatic embryogenesis in grapevine (*Vitis* spp.). In: Roubelakis-Angelakis (ed.) Molecular biology and biotechnology of grapevine (pp. 327-352). Kluwer Acad. Publ. Co., Amsterdam
Martinelli L, Bragagna P, Poletti V and Scienza A (1993) Somatic embryogenesis from leaf- and petiole-derived callus of *Vitis rupestris* Plant Cell Rep. 12:207-210
Martinelli L, Candioli E, Costa D and Poletti V (2001a) Morphogenic competence of *Vitis rupestris* S. secondary somatic embryos with a long culture history. Cell Biol. Morphogen. 20:279-284
Martinelli L, Gribaudo I, Bertoldi D, Candioli E and Poletti V (2001b) High efficiency somatic embryogenesis and plant germination in grapevine cultivars Chardonnay and Brachetto a grappolo lungo. Vitis 40:111-115
Massai R, Piccotino D, Baroni G and Xiloyannis C (1991) Response to frost in kiwifruit propagated by different btexchniques. Acta Hort. 297:237-246
Mezzetti B, Rosati P and Casalicchio G (1991) *Actinidia deliciosa* C.F. Liang *in vitro*. I. Growth and mineral uptake by explants. Plant Cell Tiss. Org. Cult. 25:91-98
Mhatre M, Salunkhe CK and Rao PS (2000) Micropropagation of *Vitis vinifera* L: towards an improved protocol. Sci. Hort. 84:357-363
Molot PM, Leroux JP and Nourrisseau JG (1972) Régéneration par culture d'apex de clones de fraisiers infectés de façon chronique par le *Phytophthora cactorum*. pp 415-419 in Actas do III Congresso da Uniao Fitopatologica Mediterranea Oeiras, Portugal, 1972
Monastra F and Testoni A (1991) Horticultural performance and quality characteristics of fruit from kiwifruit plants (cv. Hayward) obtained by *in-vitro* propagation, by cuttings and by grafting. Acta Hort. 297:197-203
Moncaleán P, Cañal MJ, Feito I, Rodríguez A and Fernández B (1999) Cytokinins and mineral nutrition in *Actinidia deliciosa* (Kiwi) shoots cultured in vitro. J. Plant Physiol. 155:606-612
Moncaleán P, Rodríguez A and Fernández B (2001) *In vitro* response of *Actinidia deliciosa* explant to different BA incubation periods. Plant Cell Tiss. Org. Cult. 67:257-266
Monette PL (1986) Micropropagation of kiwifruit using non-axenic shoot tips. Plant Cell Tiss. Org. Cult. 6:73-82
Monette PL (1987) Organogenesis and plantlet regeneration following *in vitro* cold storage of kiwifruit shoot tip cultures. Sci. Hort. 31:101-106
Monette PL (1988) Grapevine (Vitis vinifera L.) In: Bajaj YPS (ed.) Biotechnology in Agriculture and Forestry, Vol 6, Crops II (pp. 3-37.) Springer-Verlag, Berlin
Moore PP, Robbins JA and Sjulin TM (1991) Field performance of 'Olympus' strawberry subclones. HortSci. 26:192-194
Morel G (1944a) Le développement du mildiou sur des tissus de vigne cultivés *in vitro*. C. R. Acad. Sci. Paris 218:50-52
Morel G (1944b) Sur le développement de tissus de vigne cultivés *in vitro*. C. R. Acad. Soc. Biol. Paris 138:62
Morrison S, Smagula JM and Litten W (2000) Morphology, growth, and rhizome development of *Vaccinium angustifolium* Ait. seedlings, rooted softwood cuttings, and micropropagated plantlets. HortSci. 35:738-741

Mullin RH and Schlegel DE (1976) Cold storage maintenance of strawberry meristem plantlets. HortSci. 11:100-101

Mullins MG and Srinivasan C (1976) Somatic embryos and plantlets from an ancient clone of the grapevine (cv 'Cabernet-Sauvignon') by apomixis *in vitro*. J. Exp. Bot. 27:1022-1030

Murashige T (1977) Clonal crops through tissue culture. In: Barz W, Reinhard E and Zenk MH (eds.) Plant tissue culture and its bio-technological applications (pp. 393-403). GSF, München

Murashige T (1978) The impact of plant tissue culture on agriculture. In: Thorpe TA (ed.) Frontiers of plant tissue culture (pp. 15-26). Univ Calgary, Canada

Murashige T and Skoog F (1962) A revised medium for rapid growth and bioassays with tobacco tissue cultures. Physiol. Plant. 15:473-497

Nickerson NL (1978) *In vitro* shoot formation in lowbush blueberry seedling explants. HortSci. 13:698

Nickerson NL and Hall IV (1976) Callus formation in stem internode sections of lowbush blueberries cultured on a medium containing plant growth regulators. Hort. Res. 16:29-35

Nishi S and Oosawa K(1973) Mass production method of virus-free strawberry plants through meristem callus. Jap. Agr. Res. Quart. 7:189-194

Nishimura S, Terashima T, Higashi K and Kamada H (1993) Bioreactor culture of somatic embryos for mass propagation of plants. In: Redenbaugh K (ed.) SYNSEEDS: Application of synthetic seeds to crop improvement (pp.175-181). CRC Press, Boca Raton

Nitsch JP and Nitsch C (1969) Haploid plants for pollen grains. Science 163:85-87

Noè N and Eccher T (1994) Influence of irradiance on *in vitro* growth and proliferation of *Vaccinium corymbosum* (highbush blueberry) and subsequent rooting *in vivo*. Physiol. Plant. 91:273-275

Norton ME and CR Norton (1985) *In vitro* propagation of Ericaceae: a comparison of the activity of the cytokinins N6-benzyladenine and N6-isopentenyladenine in shoot proliferation. Sci. Hort. 27:335-340

Novak FJ and Juvova Z (1983) Clonal propagation of grapevine through *in vitro* axillary bud culture. Sci. Hort. 18:231-240

Orlikowska T (1986) Micropropagation of highbush blueberry. Fruit Sci. Rep. 13:105-115

Orlikowska T, Lisek L and Serwik M (1991) Micropropagation of 9 genotypes of black current. Fruit Sci. Rep. 18:51-62

Palonen P and Buszard D (1998) *In vitro* screening for cold hardiness of raspberry cultivars. Plant Cell Tiss. Org. Cult. 53:213-216

Pedroso MC, Oliveira MM and Pais MSS (1992) Micropropagation and simultaneous rooting of *Actinidia deliciosa* var. *deliciosa* 'Hayward'. HortSci. 27:443-445

Péros J-P, Torregrosa L and Berger G (1998) Variability among *Vitis vinifera* cultivars in micropropagation, organogenesis and antibiotic sensitivity. J. Expt. Bot. 49:171-179

Piagnani C, Eccher T and Castelli S (1986) Micropropagation of *Actinidia chinensis*: Effect of growth regulators on proliferation rate. Acta Hort. 179:887-890

Pias MS, Oliveira MM and Barroso J (1987) Use of petiole segments of *Actinidia chinensis* (Kiwi) for plant differentiation and production of friable calli for protoplast isolation. Acta Hort. 212:687-690

Piccotino D, Massai R, Baroni G and Bovo M (1991) Root system conformation and growth of kiwifruit as affected by propagation technique. Acta Hort. 297:391-401

Popescu CF (1996) Somatic embryogenesis and plant development from anther culture of *Vitis vinifera* (L). Plant Growth Reg. 20:75-78

Preil G (1991) Application of bioreactor in plant propagation. In: Debergh PC and Zimmerman RH (eds.) Micropropagation (pp 426-445). Kluwer Acad. Publ., Dordrecht, Boston, London

Prior RL. and Cao G (2000) Antioxidant phytochemicals in fruits and vegetables: diet and health implications. HortSci. 35:588-592

Pyott JL and Converse RH (1981) *In vitro* propagation of heat treated red raspberry clones. HortSci. 3:308-309

Qurecky DK (1975) Brambles. In: Janick J and Moore JN (eds) Advances in fruit breeding (pp.98-129). Purdue Univ. Press, W. Lafayette, IN

Racz G, Fuzi I and Fulop L (1962) A method for determination of the arbutin content of cowberry leaves (*Folium vitis-idaea*). Rumanian Med. Rev. 6:88-90

Rajasekaran K and Mullins MG (1979) Embryos and plantlets from cultured anthers of hybrid grape vines. J. Exp. Bot. 30:399-407

Rajasekaran K and Mullins MG (1981) Organogenesis in internode explants of grapevines. Vitis 20:218-227

Rani V and Raina SN (1998) Genetic analysis of enhanced-axillary-branching-derived *Eucalyptus tereticornis* Smith and *E. camaldulensis* Dehn. plants. Plant Cell Rep. 17:236-242

Rani V and Raina SN (2000) Genetic fidelity of organized meristem-derived micropropagated plants: A critical reappraisal. In Vitro Cell. Dev. Biol. Plant 36:319-330

Rani V, Singh KP, Shiran B, Nandy S, Goel S, Devarumath RM, Sreenath HL and Raina SN (2000) Evidence for new nuclear and mitochondrial genome organizations among high-frequency somatic embryogenesis-derived plants of allotetraploid *Coffea arabica* L. (Rubiaceae). Plant Cell Rep. 19:1013-1020

Redenbaugh K (1993) Synseeds: Applications of synthetic seeds to crop improvement. CRC Press, Boca Raton, FL

Reed BM (1990) Multiplication of *Rubus* germplasm *in vitro*: a screen of 256 accessions. Fruit Var. J. 44:141-148

Reed BM and Abdelnour-Esquivel A (1991) The use of zeatin to initiate *in vitro* cultures of *Vaccinium* species and cultivars. HortSci. 26:1320-1322

Reed CF (1966) Wild strawberry species of the world. In: Darrow GM (ed.) The strawberry: History, breeding and physiology (pp. 108-121). Holt, Rinehart and Winston, New York

Reichers U and Bunemann G (1989) Micropropagation of lingonberry (*Vaccinium vitis-idaea*). Erwerbsobstbau 31:129-132.

Revilla MA and Power JB (1988) Morphogenic potential of long-term callus cultures of *Actinidia*. J. Hort. Sci. 63:541-545

Revilla MA, Rey MA, Gonzalez-Rio F, Gonzalez MV, Diaz-Sala C and Rodriguez R (1992) Micropropagation of kiwi (*Actinidia* spp.). In: Bajaj YPS (ed.) Biotechnology in agriculture and forestry. Vol. 18. High-tech and micropropagation II (pp. 399-423). Springer-Verlag, Berlin

Robacker C (1993) Somatic embryogenesis and plant regeneration from muscadine grape leaf explants. HortSci. 28:53-55

Roubelakis-Angelakis KA and Zivanovitc SB (1991) A new culture medium for *in vitro* rhizogenesis of grapevine (*Vitis* spp.) genotypes. HortSci. 26:1551-153

Rugini E, Pellegrineschi A, Mencuccini M and Mariotti D (1991) Increase of rooting ability in the woody species kiwi (*Actinidia deliciosa* A. Chev.) by transformation with *Agrobacterium rhizogenes rol* genes. Plant Cell Rep. 10:291-295

Sakai A and Nishiyama Y (1978a) Cryopreservation of winter vegetative buds of hardy fruit trees in liquid nitrogen. HortSci. 13:225-227

Sakai A and Nishiyama Y (1978b) Storage in liquid nitrogen of winter buds of deciduous fruit trees (in Japanese). J. Jpn. Soc. Hort. Sci. 47:27-30

Sale PR (1983) Kiwifruit culture. Williams DA (ed.) Government Printer, Wellington, NZ

Salunkhe CK, Rao PS and Mhatre M (1997) Induction of somatic embryogenesis and plantlets in tendrils of *Vitis vinifera* L. Plant Cell Rep. 17:65-67

Salunkhe CK, Rao PS and Mhatre M (1999) Plantlet regeneration via somatic embryogenesis in anther callus of *Vitis latifolia* L. Plant Cell Rep. 18:670-673

Sansavini S and Gheradi G (1979) Selezionc clonale e stabilita genetica di fragola micropropagate. S.O.I. Incontro Nazionale Sulla Coltura Dell Fragola. Ferrara, pp. 153-168

Schroeder CA and Fletcher WA (1967) The Chinese gooseberry (*Actinidia chinensis*) in New Zealand. Econ. Bot. 21:81-92

Scorza R, Welker WV and Dunn LJ (1984) The effect of glyphosate, auxin, and cytokinin combinations on *in vitro* development of cranberry node explants. HortSci. 19:66-68

Scott DH and Zanzi C (1980) Rapid propagation of strawberries from meristems. In: Chiders NF (ed.) The strawberry, cultivars to marketing (pp. 213-221). Hort. Publ., Gainesville, FL

Serres RA, Pan S, McCown BM and Stang EJ (1994) Micropropagation of several lingonberry cultivars. Fruit Var. J. 48:7-14

Singha S, Oberly GH and Towsend EC (1987) Changes in nutrient composition and pH of the culture medium during *in vitro* shoot proliferation of crabapple and pear. Plant Cell Tiss. Org. Cult. 11:209-220

Skene KGM and Barlass M (1981) Micropropagation of grapevine. Comb. Proc. Int. Plant Prop. Soc. 1980, 30:564-570

Skirvin RM, Chu MC and Gomez E (1981) *In vitro* propagation of thornless trailing blackberries. HortSci. 16:310-312

Smagula JM and Harker J (1997) Cranberry micropropagation using a lowbush blueberry medium. Acta Hort. 446:343-347

Smith EF, Gribaudo I, Roberts AV and Mottley J (1992) Paclobutrazol and reduced humidity improve resistance to wilting of micropropagated grapevine. HortSci. 27:111-113

Smith MAL (1995) Machine vision analysis of plant cells and somatic embryos. In: Bajaj YPS (ed.) Biotechnology in agriculture and forestry, vol 30. Somatic embryogenesis and synthetic seed, I (pp. 87-101). Springer, Berlin, Heidelberg, New York

Sobczykiewicz D (1987) Mass propagation of raspberry plants by meristem culture. Acta Hort. 212:607-609

Sobczykiewicz D (1992) Micropropagation of raspberry (*Rubus idaeus* L.). In: Bajaj YPS (ed) Biotechnology in agriculture and forestry, vol. 18, High-tech and micropropagation II (pp. 339-353). Springer-Verlag, Berlin, Heidelberg

Sowik I, Bielenin A and Michalczuk L (2001) *In vitro* testing of strawberry resistance to *Verticillium dahliae* and *Phytophthora cactorum*. Sci. Hort. 88:31-40

Srinivasan C (1980) High frequency somatic embryo production from unfertilized ovules of grape. Sci. Hort. 13:245-252

Standardi A (1981) Micropropagazione dell' *Actinidia chinensis* Pl. mediante coltura *in vitro* di apici meristematici. Frutticoltura 43:23-27

Standardi A (1983) La 'micropropagzione' nella moltiplicazione dell'actinidia. Frutticoltura 45:17-22

Standardi A and Catalano F (1985) Tissue culture propagation of kiwifruit. Int. Plant Prop. Soc. Comb. Proc. 34:236-243

Stark R, Hall IV and Hendrickson PA (1978) The partridgeberry of Newfoundland. Canadex (Hort. Crops) 230

Sudarsono and Goldy RG (1991) Growth regulator and axillary bud position effect on *in vitro* establishment of *Vitis rotundifolia*. HortSci. 26:304-307

Swartz HJ, Galletta GJ and Zimmerman RH (1981) Field performance and phenotypic stability of tissue culture-propagated strawberries. J. Am. Soc. Hort. Sci. 106:667-673

Tanaka H, Shoyama Y, Sasaki Y and Sashida Y (1997) Micropropagation of *Actinidia polygama* from fruit galls. Plant Cell Tiss. Org. Cult. 48:135-137

Tang W (2001) *In vitro* regeneration of loblolly pine and random amplified polymorphic DNA analysis of regenerated plantlets. Plant Cell Rep. 20:163-168

Thomas P (1997) Increase in clonal propagation of 'Arka Neelamani' grape (Vitis vinifera L.) through induction of axillaries in *in vitro* layering technique. Indian J. Agr. Sci. 67:594-596

Thomas P (1998) Contribution of leaf lamina of grape nodal microcuttings to rooting, root vigour and plantlet growth *in vitro*. Plant Physiol. 153:727-732

Thomas P (2000) Microcutting leaf area, weight, and position on the stock shoot influence root vigor, shoot growth and incidence of shoot tip necrosis in grape plantlets *in vitro*. Plant Cell Tiss. Org. Cult. 61:189-198

Thomas P and Schiefelbein JW (2001) Combined *in vitro* and *in vivo* propagation for rapid multiplication of grapevine cv. Arka Neelamani. HortSci. 36:1107-1110

Thomas GG and Wilkinson EH (1962) Propagation of black currents from single-bud cuttings. J. Hort. Sci. 37:115-123

Troncoso A, Cantos M, Linan J, Prieto J and Sarmiento R (1988) The use of *in vitro* culture and tubular container system to propagate selected grapevine plants for sherry wine production. Acta Hort. 227:358-362

Tsao CWV and Reed BR (2002) Gelling agents, silver nitrate, and sequestrene iron influence adventitious shoot and callus formation from *Rubus* leaves. In Vitro Cell. Dev. Biol. Plant 38:29-32

Tyulenev VM, Zhukova NV and Naftaliev PM (1987) Clonal micropropagation of disease-free plants of black currant (in Russian). Byul. Nauch. Tsentr. Ord. Trud. Krasn. Znam. Genet. Lab. I. V. Michurina 45:41-43

Uematsu C, Murase M, Ichikawa H and Imamura J (1991) *Agrobacterium*-mediated transformation and regeneration of kiwifruit. Plant Cell Rep. 10:286-290

Vander Kloet SP (1983) Taxonomy of Vaccinium section Oxycoccus. Rhodora 85:1-43

Vander Kloet SP (1988) The Genus *Vaccinium* in North America. Res. Branch Agric. Can. Publ. 1828

Van Huylenbroeck JM and Debergh PC (1996) Physiological aspects in acclimatization of micropropagated plantlets. Plant Tiss. Cult. Biotech. 2:136-141

Velayandom L, Hirsch AM and Fortune D (1985) Tissue culture of nodal stem segments of *Actinidia chinensis* (L) Planchon, as a method of micropropagation. C. R. Acad. Sci. Ser III Vie 301:597-600

Vertesy J (1979) Experiments on the production of virus-free raspberry propagation material by meristem culture. Acta Hort. 95:77-79

Wang H, Cao G and Prior RL (1997) Oxygen radical absorbing capacity of anthocyanins. J. Agric. Food Chem. 45:304-309

Wang PJ and Charles A (1991) Micropropagation through meristem culture. In: Bajaj YPS (ed.) Biotechnology in agriculture and forestry, vol. 17, High-tech and micropropagation I (pp. 32-52). Springer-Verlag, Berlin, Heidelberg

Watanabe A, Araki S, Kobari S, Sudo H, Tsuchida T, Uno T, Kosaka N, Shimomura K, Yamazaki M and Saito K (1998) *In vitro* propagation, restriction fragment length polymorphism, and random amplified polymorphic DNA analysis of *Angelica* plants. Plant Cell Rep. 18:187-192

Welander M (1982) Micropropagation of raspberry and current (in Swedish) In: Rep. Balsgård Fruit Breeding Inst (pp. 102-115), 1980-81

Welander M (1987) *In vitro* culture of raspberry (*Rubus idaeus*) for mass propagation and virus elimination. Acta Hort. 212:610

Wessels E, Nel DD and von Staden DFA (1984) *In vitro* propagation of *Actinidia chinensis* Pl. cultivar Hayward. Decid. Fruit Grow. 34:453-457

White PR (1943) A handbook of plant tissue culture. The Jacques Cattel Press, Lancaster, Penn.

Wiyaporn S, Subhadrabandhu S and Sahavacharin O (1990) *In-vitro* vegetative multiplication of kiwi plant. Acta Hort. 279:447-459

Wolfe DE, Eck P and Chin C (1983) Evaluation of seven media for micropropagation of highbush blueberry. HortSci. 18:703-705

Xiao XG, Hirscha AM and Fortune D (1991) Régénération du kiwi (*Actinidia deliciosa* cultivar Hayward) à partir de jeunes feuilles. Freuits 46:57-66

Ying G, Zhao CM and Jun W (1990) On *Rubus* resources in Hunan and Fujian provinces. XXIII Int. Hort. Cong. Abst. 4014

Yoshino M and Hashimoto K (1975) Occurrence of aphid-transmitted strawberry viruses in Saitama Prefecture and production of virus-free clones by means of meristem culture. Bull. Saitama Hort. Exp. Sta. 5:46-61

Yue D, Gosselin A and Desjardins Y (1993) Re-examination of the photosynthetic capacity of *in*

vitro-cultured strawberry plantlets. J. Am. Soc. Hort. Sci. 118:419-424

Zatykó JM, Kiss G, Radics Z and Simon I (1989) Initiation of strawberry runner formation in vitro. Acta Hort. 265:349-352

Zhang Y-J, Qian Y-Q, Mu X-J, Cai Q-G, Zhou Y-L and Wei X-P (1998) Plant regeneration from *in vitro*-cultured seedling leaf protoplasts of *Actinidia eriantha* Benth. Plant Cell Rep. 10:819-821

Zimmerman RH (1987) Micropropagation of woody plants: Post tissue culture aspects. Acta Hort. 227:489-499

Zimmerman RH and Broome OC (1980) Blueberry micropropagation. In: Proc. Conf. on Nursery Prod. of Fruit Plants Through Tiss. Cult.- Applications and Feasibility (pp. 44-47). U. S. Dep. Agric. Sci. and Ed. Adm. ARR-NE-11

Ziv, M (1991) Morphogenic patterns of plants micropropagated in shaken flasks or large scale bioreactor cultures. Isr. J. Bot. 40:145-153

Ziv, M (1992) Morphogenic control of plants micropropagated in bioreactor cultures and its possible impact on acclimatization. Acta Hort. 319:119-124

16. Micropropagation of *Prunus sargentii* REHDER and *P. verecunda* (KOIDZ) Koehe.

Yoh Sasaki
Akita Prefecture Forestry Technical Center, Idojiridai 47-2, Toshima, Kawabe, Akita 019-2611, Japan

Contents

1. Introduction

Kabazaiku is a traditional Samurai handicraft from Akita Prefecture, dating back to the Edo period. Cherry tree bark is used to laminate trays, tea caddies, cigarette cases and other items. Today, kabazaiku souvenirs earn about one billion yen a year. Two wild species of cherry tree, *P. sargentii* and *P. verecunda*, are used for kabazaiku. Other species, such as cherry blossom or fruit cherry species, are not used. *P. sargentii* grows naturally in the mountains of Japan (except Kyusyu and Okinawa), the Korean peninsula, the southern valley of Usury, and southern Sakhalin (Okuda et al. 1993). *P.*

S.M. Jain and K. Ishii (eds.), *Micropropagation of Woody Trees and Fruits*, 507–518.

verecunda grows naturally in the mountains of Japan (except Kyusyu and Okinawa), the Korean peninsula and in China (Okuda et al. 1993). Generally, *P. sargentii* grows at higher altitudes than *P. verecunda*.

Recent increased demand has resulted in a shortage of the bark used for kabazaiku. Since 1973, 10,000 *P. sargentii* seedlings have been planted annually in the mountains in an attempt to protect supplies. However, the bark of these trees has proved to be of poor quality and therefore unsuitable for kabazaiku. One explanation for the poor quality of the bark is the genetic characteristics of the seedlings (Sasaki et al. 2000). Therefore, it was thought that genetic improvement of the kabazaiku cherry tree nursery was necessary (Sasaki et al. 2000). Generally, tree-breeding programs take over 10 years. However, if highly desirable trees with superior bark characteristics were to be cloned, it is expected that improvements in both tree production and bark quality could be accomplished more rapidly (Ritchie, 1984, Timmis, et al., 1987).

Presently, only *P. sargentii* seedlings can be purchased from nurseries in Hokkaido, which is a major production district for cherry blossoms in Japan. Production in Hokkaido in 1998 was 250,000 trees (Satoh & Nishikoori, 2000), but the bark quality of those seedlings was poor. The alternative *P. verecunda* seedlings cannot be purchased, because there is no nursery production of this species in Japan. Thus, it is necessary to develop a clone nursery production system for kabazaiku cherry trees. Producing clones by conventional methods is difficult because the quantities and quality of branches for use are insufficient. Competition with many other trees prevents the lateral branch development of cherry trees grown in mountain areas. Moreover, it is difficult to control production levels using conventional methods in *P. verecunda*, and it is impossible in *P. sargentii*.

Therefore, micropropagation is thought to be a necessary tool for kabazaiku nursery production. Micropropagation can produce plants from small amounts of material, allowing easy control of production quantities. The process of *in vitro* plant production consists of primary culture, shoot proliferation and elongation, rooting, plantlet hardening, and nursery growing. However, there is no practical system for this *in vitro* process in Japan, especially for plantlet hardening and nursery growing. To solve this problem, *in vitro* production of cherry trees as plug plantlets, and improvements in plug plantlet culture, were developed in *P. sargentii* and *P. verecunda*. In addition, lowering the cost of *in vitro* cherry tree production for commercialization was studied.

This chapter describes *in vitro* cloned cherry tree nursery production for two wild species of Japanese *Prunus*.

2. *In vitro* propagation and field nursery

2.1. Plant material

Mother cherry trees were selected considering genetic diversity and geographical location. The bark quality and estimated age of mother plants are summarized in Table 1. Nodal explants containing an axial bud from a current year branch (a sprouting branch is preferable) were used as starting material. Branches must be collected after several days of clear skies, before the rainy season. Branches are harvested in Akita Pref. from late May to early June. A branch from the sunny side of the tree is preferable. Sterilization is performed by rinsing the branch with 300 ml distilled water containing 4-5 drops of non-ionic detergent for 10 minutes using gentle stirring, followed by sterilizing with 70% EtOH for 60-90 seconds, rinsing for 5 minutes 3 times with sterilized distilled water and stirring gently.

Table 1. Mother plants of cherry trees selected for kabazaiku.

Species	Individual No.	Estimated age (year)	Quality of the bark
Prunus verecunda	PV0	30	not superior
	PV1	30	superior
	PV2	40	superior
	PV3	30	superior
	PV4	30	superior
Prunus sargentii	PS1	100	extra superior
	PV2	50	extra superior
	PS3	70	extra superior

2.2. Medium and culture conditions

Solidified MS (Murashige & Skoog, 1962) agar was used as culture medium. For primary culture and shoot proliferation and elongation, the culture medium was supplemented with 4.71 μM diphenyl urea, and 1-5.62 μM BA. The same medium was used for shoot proliferation and elongation. For *in vitro* rooting of the proliferated shoot, half-strength (in terms of inorganic substances) MS, with 15 g L^{-1} sucrose, with or without 1 μ M BA was used as rooting medium. Culture conditions were as follows. Primary culture, shoot proliferation and rooting: 16 hr photoperiod under cool white fluorescent light at 26°C and 70 μ E s^{-1} m^{-2}; plugging and hardening: 16 hr photoperiod at 24°C and 36 μ E s^{-1} m^{-2}.

2.3. *Shoot proliferation and elongation*

Sterilized nodal explants containing an axial bud were cultured. Shoots that developed from the axial bud of the explants were used for shoot proliferation using adventitious bud formation. The culture period was 2 months for shoot proliferation and 1 month for elongation. This study defined a proliferated shoot as one more than 1 cm long. For mass shoot proliferation, 2 shoots were cultured in a flask (max. volume of 130 ml) containing 20-30 ml of the culture medium (Fig. 1A). For *P. verecunda*, the maximum shoot proliferation ratio was obtained when a flask contained 2 shoots (Sasaki, 1995c). A similar result has been reported for sugarcane (Walker, 1995).

The shoot proliferation was low compared to the plug seedling production cost (see Section 3.1 in this chapter). The low shoot proliferation ratio was improved in the following way. Using 25 × 150-mm culture tubes, the shoot proliferation ratio in the primary proliferation test of *P. sargentii* (PS1) was low, only 1.7. When the culture medium contained 10 μM BA, many short shoots proliferated. Since epigenetic off-types are dependent on an excess of exogenous cytokinin in the culture medium (Werbrouck & Debergh, 1994), the shoot proliferation ratio was examined with less BA in the culture medium. The shoot proliferation ratio increased to 2.6 when the inoculated shoot had a small amount of callus, and increased to 5.1 when 2 shoots were cultured in a beaker. This change in the proliferation ratio was sufficient to lower the cost of plug plantlet production.

In more than a third of *P. verecunda* subcultures, chlorosis (loss of green pigmentation) was observed. Generally, chlorosis in cherry tissue culture is due to iron deficiency or high concentrations of exogenous cytokinin in the culture medium (Skirvin, 1984). This problem was solved by adding 4.71 μM diphenyl urea to the culture medium. As a result, shoot proliferation improved without using high BA in the culture medium. The optimal BA concentration in the culture medium was 5.62 μM for *P. verecunda* PV0-3 and 3.16 μM for *P. sargentii* PS1. However, insufficient shoot proliferation was obtained with *P. verecunda* PV4 and *P. sargentii* PS2 and PS3, even using various BA concentrations in the culture medium.

2.4. *Plugging and hardening of* in vitro *plants*

Hardening, between *in vitro* growth and growth in the field, is an important process that increases survival. The survival of plantlets strongly affects the cost of *in vitro* plant production. Thus, hardening techniques must be

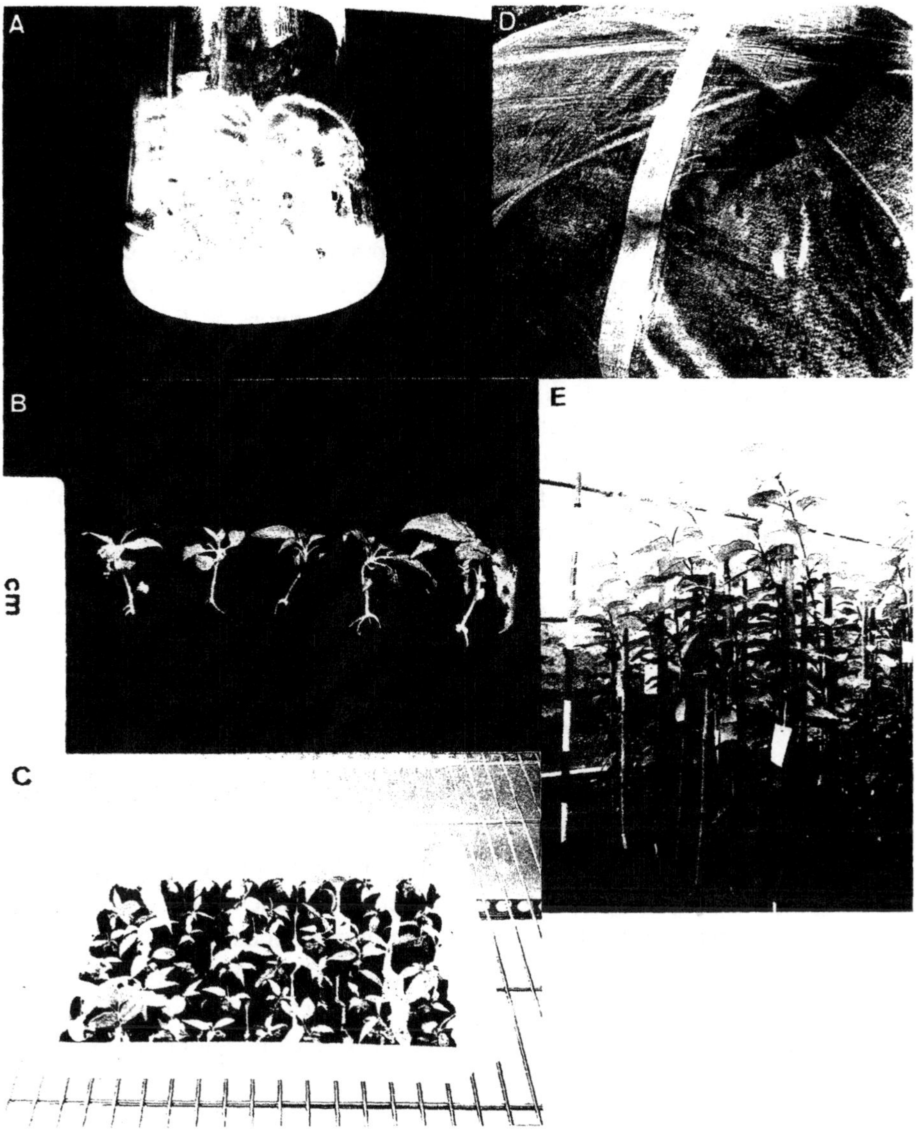

Figure 1. In vitro plant production of *P. verecunda*. (A) Shoot proliferation and elongation in a flask. (B) Shoots used for plugging after short-term *in vitro* rooting. (C) Plug plantlets (2-3 cm tall on average) after hardening in a polystyrene foam container. (D) Plug plantlets planted in the field. (E) Nursery of plug plantlets in the field.

considered to increase the survival of *in vitro* plantlets that are transplanted into the field. Attention must also be paid to the size of the *in vitro* plantlets. Usually, the micropropagated shoot is small;

however, the size and volume of *in vitro* plantlets are dependent on the manufacturing process. Both high survival rates and small plant size during hardening are important for lowering the cost of *in vitro* cherry tree production.

Full rooting in agar medium produced *in vitro* plants that frequently did not grow after hardening (Preece & Sutter, 1991). The survival rate of *P. verecunda* was less than 50% when *in vitro* rooting lasted for 1 month. In an attempt to improve the low survival rate, a combination of short-term *in vitro* rooting and plugging, which combines the advantages of *in vitro* and *ex vitro* rooting, was examined (Welander, 1983; Zimmerman & Fordham, 1985; Bennet & Davies Jr., 1986). After shoot proliferation and elongation, 5-10 shoots were cultured in a flask containing 20-30 ml of rooting medium. The initial root primordial was observed after 10 days of *in vitro* rooting in culture (Fig. 1B). Just after rooting (root length less than 5 mm), the shoot was soaked in Jiffy 9 pellets made of peat moss (ϕ3.5-5 cm). The rooted shoot was then cultured in a non-sterile polystyrene foam case covered with vinyl sheets for 30 days in a culture room. During the 7-10 days of hardening, the vinyl sheets were gradually removed (Sasaki 1995b, 1995c). After hardening, plug plantlet survival exceeded 85% (Fig. 1C).

2.5. Field performance from plug plantlet to nursery

Maintaining and growing cherry plug plantlets in the field of a nursery is costly, and is the most important step in *in vitro* plant production. Observations of plug plantlets that developed during plugging revealed that leaves were not completely adapted to the field. This phenomenon is similar to that observed when *in vitro* leaves are grown in greenhouses (Preece & Sutter, 1991). When abnormal leaf morphology developed *in vitro*, plants did not recover (Preece & Sutter, 1991). However, new leaves that developed from plantlets could adapt to *ex vitro* environments (Preece & Sutter, 1991). Thus, for better field performance it is most important to develop new shoots from the small plug plantlets quickly.

Generally, 10°C is the minimum temperature at which to plant in the field in Akita Pref. After hardening, the plug plantlets (2-3 cm tall on average) were planted directly in the field in early June. The effect of using a covering sheet on plug plantlet survival and growth was examined (Sasaki 1996a). Cloudy and calm days were preferable to sunny days for field planting of the plug plantlets. Plantlets were covered with a vinyl sheet to retain warmth, and a sunshade tent was used to protect plug plantlets from intense sunshine (Fig. 1D). The outer vinyl sheet was removed in early July, and the inner sunshade tent was removed in mid July. The survival and growth of the plug plantlets in the field was improved with these vinyl sheets and tent (Sasaki 1996a). More than 85% of transplanted *P. verecunda* (PV0) plugs survived in the field. The plug plantlets (2-3 cm tall on average) grew to a height of 130 cm (satisfactory for planting out in the mountains) after one growing season (Fig. 1E).

3. Economics of *in vitro* plant production

Generally, the production costs of growing *in vitro* plants are high compared to seedling production (Kozai, 1991). Thus, studying how to lower the costs of *in vitro* plant production is very important for commercialization, especially the costs involved in *in vitro* plant production (Standaert-de-Metsenaere, 1991). This will help to determine whether certain improvements are really cost effective. Production costs are markedly affected by legislation of individual nations. For example, the costs of labor (ca. 700 yen/hr) and facilities (twenty thousand yen / 3.3 m^2) in Japan are high compared to those in many foreign countries. Therefore, the cost of *in vitro* plant production must be evaluated with a simulation model based on the system used that considers various factors of the domestic economy. Our model sums the direct costs of *in vitro* plug plantlet production, including the costs of facilities, equipment, instruments, electricity, water, labor, and supplies (Sasaki, 1996b, 1997, 1999). The square of the cost of laboratory construction was calculated back successively across the number of flasks, the number of cultivation containers and the shelf area for them using the values for shoot proliferation ratio, plugging efficiency (survival after hardening), and annual plug plantlet production.

Plug plantlet production cost was estimated for an annual production of 10,000 plug plantlets, which is the number of cherry trees planted in the mountains of Akita Pref. annually (Table 2). It does not seem to be practical to use a culture room for hardening because of the cost. Instead, it is preferable to use greenhouses for the production of the 10,000 plug plantlets. The acceptable price of *in vitro* cherry tree production is 500 yen per plant per year. Thus, the upper limit of plug plantlet production cost becomes roughly 400 yen, if the cost of cultivating plug plantlets in the field is 100 yen.

3.1. Effect of shoot proliferation ratio

A low shoot proliferation ratio is not cost effective for *in vitro* plug plantlet production. A simulation of *in vitro* plant production cost revealed that it rose sharply for shoot proliferation ratios less than 3 when plugging efficiency was 0.9 (Fig. 2). However, the production cost barely changed when the shoot proliferation ratio exceeded 5 (Fig. 2). These results suggest that a higher shoot proliferation ratio is not always necessary for lowering the cost of *in vitro* plant production. It is believed that 2 is the minimum shoot proliferation ratio to keep the production cost of cherry plug plantlets below 400 yen. Therefore, the production of PS1 and PV0-3 plug plantlets was cost effective (shoot proliferation ratio 4-6), while the production of PS 2, PS3, and PV4 was not cost effective because of the low shoot proliferation ratio (1.5). These results suggest that clone selection is important in shoot proliferation and in lowering the cost of *in vitro* cherry tree production.

Table 2. Tissue culture parameters of simulation of the production of 10,000 plug plantlet.

Stage	Process	Condition	Place	Period	Productivity	Transfer rate (/ hr/person)
2	Shoot proliferation	*in vitro*	Culture room	2 months	4~6	30 shoots
3a	Shoot elongation	↓	↓	1 months		50 clusters of shoots
3b	Rooting	↓	↓	10 days	0.85	50 shoots
4	Plugging	*ex vitro*	↓	30 days		50 plugs
4	Hardening	↓	↓	10 days		250 plugs
not defined	Nursery	↓	Field	4-5 months	0.85	40 plugs

Stage number is from Debergh & Read (1991).

Productivity of shoot elongation incorporates the shoot proliferation ratio and the survival rate.

Figure 2. Effect of the shoot proliferation ratio on *in vitro* cherry tree production cost.

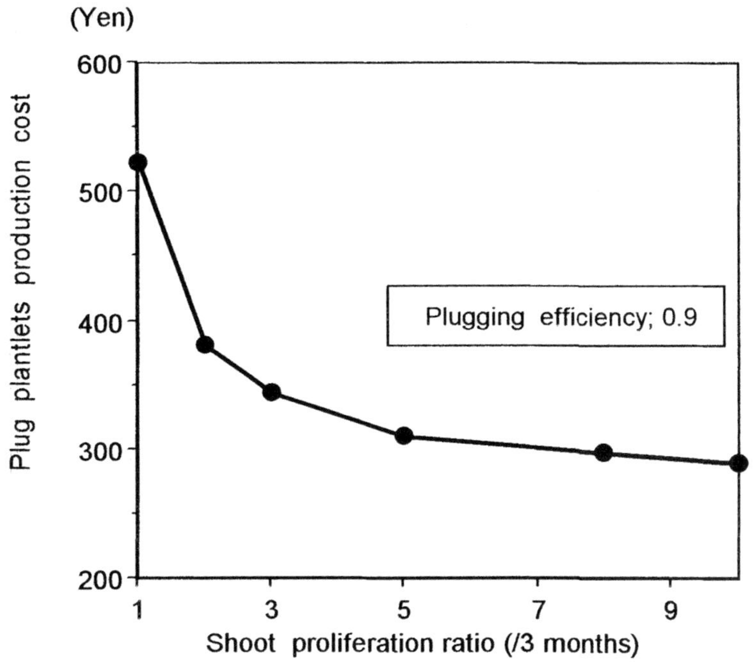

Incidentally, the culture period required for shoot proliferation and elongation had little influence on the plug plantlet production cost. Even when the shoot proliferation and elongation period was increased from 3 to 6 months in the initial simulation, the production cost was only 4.5%. Therefore, it was not necessary to consider shortening the period of shoot proliferation and elongation in order to lower the cost of *in vitro* plant production.

3.2. Effect of in vitro *rooting*

It is important to know what type of rooting procedures are effective for lowering the costs of *in vitro* plant production. Generally, *ex vitro* rooting is preferable to *in vitro* rooting for commercialized *in vitro* plant production (Debergh & Read, 1991). *Ex vitro* rooting with *in vitro* plant production omits the process of *in vitro* rooting, which is a labor-intensive operation. Certainly, the cost of rooting and hardening when using *in vitro* rooting with *in vitro* plantlet production is a significant portion of the total cost (35-70%)(Preece & Sutter, 1991). Thus, it appears uneconomical to use *in vitro* rooting with *in vitro* cherry tree production. When using microcutting (Sasaki, 1995a), however, the cost of *in vitro* plug plantlet production (370 yen) was no lower than the cost using *in vitro* rooting (313 yen). This is because plugging efficiency (survival after hardening) with microcutting was low (0.64) compared with the *in vitro* rooting efficiency (0.85). Using the simulation, it appeared that plugging efficiency was strongly affected by the production cost of plug plantlets. Similar results were obtained with *P. sargentii*. This suggested that it is necessary to use the simulation to determine rooting procedures to judge whether *in vitro* plant production is cost-effective for commercialization (Sasaki, 1997).

The simulation suggested that plugging and hardening in the field are the most effective way to lower the cost of *in vitro* plant production (at 188 yen, total cost was reduced to ca. 40%) (Sasaki, 1996b; 1997). Moreover, scaling up the production quantity of *in vitro* plants or using disposable culture containers to reduce the washing costs should effectively lower the cost of *in vitro* plant production.

4. Conclusions and further prospects

A method of *in vitro* cherry tree production from plug plantlets, and the cultivation of plug plantlets in a nursery and the field was developed for *P. sargentii* and *P. verecunda*. Furthermore, ways of lowering the cost of *in vitro* cherry tree production were studied, and a practical clone nursery production system for bark production for kabazaiku was designed. A simulation was used to examine ways to lower the cost of *in vitro* cherry tree production, and it was determined that *in vitro* plant production of PS1 and PV0-3 was cost effective, given a production cost limit of 400 yen. Currently, the commercialization of the *in vitro* cherry tree technology developed in this study is under way at a nursery in Japan.

The cost of the 120-cm seedlings that are currently planted out in the mountains is 250 yen. In comparison, the estimated cost of planting an *in vitro* cherry tree of that size is 400-500 yen (the 300-400 yen cost of the plug plantlet, plus 100 yen field costs). It therefore seems uneconomical to plant *in vitro* cherry trees. However, the *P. sargentii* seedlings that have been planted in the mountains over the last twenty years have not produced good quality bark. Considering the labor costs of planting cherry trees, clone nursery plantations are an important way to provide good quality cherry bark for kabazaiku. To date, no morphological abnormalities have been observed in bark, leaves, or flowers of *in vitro* plants, after 3 years in the field. The higher price of the *in vitro* plants compared to that of seedlings should prove acceptable over time.

It is very difficult to find superior clones for kabazaiku cherry tree breeding in nature. This is because the more lucrative Japanese cedar has already been planted where wild cherry trees grew previously. Therefore, a seed orchard for genetic improvement was created using micropropagation with 7-superior cherries (Table 1). Seed production has already been observed from several clones after growth in the field for three years. New, superior clones will be generated from seeds from this orchard in the future.

5. Acknowledgement

The author wish to thank Dr. K. Ishii (Forestry and Forest Product Research Institute), Dr M. Nishikoori, Dr. T. Satoh (Hokkaido Forest Research Institute), Dr. M. Kyo (Kagawa Univ.) and Mr. T. Richard (Textcheck.com) for critical reading of the manuscript. Financial support from the Ministry of Agriculture, Forestry and Fisheries of Japan is gratefully appreciated.

6. References

Bennet, L.K. & Davies Jr., F.T., 1986. *In vitro* propagation of *Quercus shumardii* seedlings. Hort Sci. 21: 1045-1047.

Debergh P.C. & Read, P.E., 1991. Micropropagation. In: Debergh, P.C. & Zimmerman, R.H. (Eds.) Micropropagation Technology and Application, pp. 71-93. Kluwer Academic Publishers, Dordrecht. Kawasaki, T., Okuda, M. & Kihara, H. 1993. Flowering cherries of Japan (in Japanese), pp. 197-199. YAMA-KEI Publishers, Tokyo.

Kozai, T., 1991. Micropropagation under photoautotrophic conditions. In: Debergh, P.C. (Eds.) Micropropagation Technology and Application, pp. 447-469. Kluwer Academic Publishers, Dordrecht.

Preece, J.E. & Sutter, E.G., 1991. Acclimatization of Micropropagated Plants to the Greenhouse and Field. In: Debergh, P.C. & Zimmerman, R.H. (Eds.) Micropropagation Technology and Application, pp. 71-93. Kluwer Academic Publishers, Dordrecht.

Ritchie, G.A., 1994. Commercial Application of Adventitious Rooting in Forestry. In: Davis, T.D. & Haissig, B.E. (Eds.) Biology of adventitious root formation, pp. 37-52. Plenum Publishing Corporations, New York.

Sasaki, Y., 1993. The shoot propagation and root induction of *Prunus vercunda* on MS medium (in Japanese). Transactions.Tohoku Branch.Jpn.Forest.Soc. 45: 181-182.

Sasaki, Y., 1995a. Low production cost of clone propagation using tissue culture (I) Direct rooting of *Prunus vercunda* (in Japanese with English summary). Sakura Sci. 5: 1-12.

Sasaki, Y., 1995b. Low production cost of clone propagation using tissue culture (II) Direct rooting of *Prunus vercunda* (in Japanese with English summary). Sakura Sci. 5: 13-20.

Sasaki, Y., 1995c. In vitro propagation of *Prunus vercunda* shoot and its transplant production system (in Japanese). Transactions.Jpn.Forest.Soc. 106: 425-426.

Sasaki, Y., 1996a. Field performance of micropropagules of *Prunus vercunda* Kohne using plug and pot (in Japanese with English summary). J.Jpn.For.Soc. 78: 10-14.

Sasaki, Y., 1996b. Cost estimation of micropropagated plug production using *Prunus vercunda* (in Japanese with English summary). J.SHITA 8: 244-252.

Sasaki, Y., 1997. Reduction in micropropagated plug cost by simulation study on *Prunus vercunda* (in Japanese with English summary). J.SHITA 9: 20-28.

Sasaki, Y., 1999. Effects of clonal differences of phytohormonal responce on the estimated cost of wild cherry (*Prunus sargentii* Rehder) Micropropagation. J.SHITA 11: 237-244.

Sasaki, Y., 2000. Tissue culture of the Japanese wild cherry. (in Japanese). Ringyo gijyutsu 701: 24-26.

Sasaki, Y., Tomiki, K. & Sasaki,.S., 2000. Character and evaluation of the Japanese wild cherry (*Prunus sargentii* Rehder) bark for kabazaiku (in Japanese with English summary). J.SHITA 12: 117-125.

Satoh, T. & Nishikoori, M., 2000. Commercialization of tissue culture in Hokkaido (in Japanese). Forest Breed. 196: 11-12.

Skirvin, R.M., 1984. Stone Fruits. In: Ammilator, P.V., Evans, D.A., Sharp, W.R. & Yamada, Y. (Eds.) Handbook of Plant Cell Culture Vol. 3, pp. 402-452. Macmillan Publishing Company, New York.

Standaert-de Metsenaere, R.E.A., 1991. Economic considerations. In: Debergh, P.C. & Zimmerman, R.H. (Eds.) Micropropagation Technology and Application, pp. 123-140. Kluwer Academic Publishers, Dordrecht.

Timmis, R., Abo El-nil, M.M. & Stonecypher R.W., 1987. Potencial genetic gain thorough tissue culture In: Bonga,, J.M., & Durzan, D.J. (Eds.) Cell and Tissue Culture in Forestry vol.1, pp. 198- 215. Matinus Nijhoff Publishers, Dordrecht.

Walker, P.N., 1995. System analysis and enginering. In: Aitken-Christie, J., Kozai,T., & Smith, M.A.L. (Eds.) Automation and Environmental Control in Plant Tissue Culture, pp. 65-85. Kluwer Academic Publishers, Dordrecht.

Werbrouck, S.P.O. & Debergh, P.C. 1994. Applied aspects of plant regeneration. In: Dixon, R.A. & Gonzales, R.A. (Eds.) in Plant Cell Culture, pp. 127-134. IRL Press, Oxford.

Welander, M., 1983. *In vitro* rooting of the apple rootstock M 26 in adult and juvenile growth phases and acclimatization of the plantlets. Physiol.Plant. 58: 231-238.

Zimmerman, R.H. & Fordham, I., 1985. Simplified method for rooting apple cultivars *in vitro*. J.Amer.Soc.Hort.Sci. 110: 34-38.

17. MICROPROPAGATION OF AVOCADO (*Persea americana* Mill.)

A. Barceló-Muñoz[1] and F. Pliego-Alfaro[2]
[1]*C.I.F.A. Cortijo de la Cruz, Churriana, 29140 Málaga, Spain.*
[2]*Dpto. Biología Vegetal, Facultad de Ciencias, Campus de Teatinos s/n, 29071 Málaga, Spain.*

Chapter Contents

1. INTRODUCTION

1.1. General Description, Importance and Distribution

Avocado (*Persea americana* Mill.) is a tropical tree, of the *Lauraceae* family, genus *Persea*. On the basis of the morphology of

S.M. Jain and K. Ishii (eds.), Micropropagation of Woody Trees and Fruits, 519–542.

reproductive structures, this genus was divided into two subgenera, *Eriodaphne* and *Persea* (Koop, 1966); the subgenus *Eriodaphne* includes many species with small fruits and of no commercial importance, whereas the subgenus *Persea* includes only a small number of species characterized by large fruits. The subgenus *Persea* includes the avocado, of which three horticultural races, named for their presumed region of origin, can be distinguished: Mexican, West Indian and Guatemalan. These races could have evolved under different climatic conditions with geographical barriers between them (Storey *et al.*, 1986).

The West Indian and Guatemalan races were initially grouped within a botanical variety (*Persea americana* var. *gratissima*) clearly distinguished from the Mexican race (*Persea americana* var. *drymifolia*) (Popenoe, 1920). Bergh (1975), however, suggested that each race is equally distinct from the other two. Hence, the designation of each as a botanical variety is more logical, i.e., *Persea americana* var. *drymifolia* (Mexican race), var. *americana* (West Indian or Antillean race) and var. *guatemalensis* (Guatemalan race). These three botanical varieties differ in ecotype adaptation, field distribution, and morphological and physiological traits (Bergh and Ellstrand, 1986; Pliego-Alfaro and Bergh, 1992).

The avocado is the most economically important species of the *Lauraceae* family. It is grown commercially in America, Africa, Europe, Asia and Oceania. The estimated world production was over 2.5 million Mt in 2001. The major avocado growing countries are Mexico, USA, Colombia, Indonesia, Chile and the Dominican Republic (FAOSTAT, 2001).

1.2. Diseases

Fungal diseases are of extreme importance in avocado. However, there are a few other diseases which do not cause economic damage to the crop as a whole but which may still cause severe individual plant or fruit damage, e.g., those caused by bacteria: *Pseudomonas syringae*, *Erwinia* spp. and *Xanthomonas campestris* (Myburgh and Kotzé, 1982; Pegg, 1991; Cooksey *et al.*, 1993); viroids: "avocado sunblotch viroid" and the potato spindle tuber viroid (PSTVd) (Dale and Alen, 1979; López Herrera *et al.*, 1987; Querci *et al.*, 1995); algae: *Cephaleuros* sp. (Anónimo, 1991); or nematodes: *Pratylenchus* sp., *Meloidogyne* sp. and *Xiphinema* sp. (Pegg, 1991). Fungal diseases can be divided into post-harvest diseases: *Colletotrichum gloeospoiroides*, *Dothiorella aromatica* and *Thyronectria pseudotricha,* aerial system diseases: *Botryodiplodia theobromae*, *Phytophthora citricola*, *Phytophthora heveae*, *Sphaeropsis tumefaciens*, *Sphaceloma perseae*, *Oidium* spp., *Verticillium dahliae* and *Cercospora purpurea*, and root system diseases: *Phytophthora cinnamomi*, *Rosellinia necatrix* (*Demathophora necatrix*), *Armillaria mellea*, *Rhizoctonia solani*, *Polyporus* spp. and *Formes* spp. (Pérez Jiménez, 1997).

Among the avocado diseases, *Phytophthora* root rot causes extensive damage to avocado trees in countries growing avocado commercially (Zentmyer, 1984). Control measures include the use of tolerant rootstocks, and breeding programs are being carried out in California (USA), the South African Republic, and Israel, to develop rootstocks tolerant to this pathogen (Bergh *et al.*, 1976; Kellam and Coffey, 1985; Köhne, 1992).

Rosellinia necatrix has occasionally been reported as a pathogen of large avocado roots (Zentmyer, 1984), though its impact in Spain is analogous to *P. cinnamomi* (López Herrera, 1998). Raabe and Zentmyer (1955) reported that Mexican and Guatemalan seedlings were susceptible to this fungus, and since chemical control is unavailable, control measures should include the use of tolerant rootstocks. A breeding program with this objective was initiated in Spain in 1998.

1.3. Conventional Methods of Propagation

The use of seedlings as avocado rootstocks has been linked to a lack of uniformity in orchards in terms of production, vigour, and/or tolerance to soil-borne pathogens or adverse soil conditions (Gustafson and Kadman, 1970). In the avocado, vegetative propagation is difficult; there are differences in rooting capacity due to the ontogenetic age of plant material (Eggers and Halma, 1937; Kadman and Ben Ya'acov, 1970) or genotype (Kadman and Ben Ya'acov, 1965; Reuveni and Raviv, 1976), with Antillean cuttings being more difficult to root than cuttings of Mexican race, while those of Guatemalan origin show an intermediate behaviour. Currently, rootstock selections are commercially propagated by the Frolich method, a relatively expensive and time-consuming procedure which requires grafting in a nurse seedling followed by etiolation of the shoot to be rooted (Frolich, 1951; Frolich and Platt, 1972; Brokaw, 1988).

2. MICROPROPAGATION

Avocado micropropagation might be of interest in order to mass propagate selected rootstocks or to revitalize adult material which could later be multiplied by conventional techniques.

Plant multiplication through axillary branching is the most reliable micropropagation method in terms of genetic stability of elite selections (George, 1993). With woody plants such as avocado, the morphogenetic capacity is generally lower in adult than in juvenile material (Favré and Junker, 1987; Pliego-Alfaro and Murashige, 1987; Arrillaga *et al.*, 1991); hence, micropropagation is usually developed using juvenile explants, and the results can be used as a guide to propagating material of adult origin.

2.1. Culture Initiation

2.1.1. Factors Affecting Explant Establishment

During the initiation phase there may be several problems, such as high contamination and low sprouting rates, exudate production and browning (Cooper, 1987; Zirari and Lionakis, 1994). Moreover, the great variability present in this species usually gives rise to very different *in vitro* behaviour, making it difficult to establish an initiation procedure for general application. The incidence of these problems depends greatly on the ontogenetic stage of the mother plant, being less relevant when using juvenile material and/or explants derived from actively growing branches (Pliego-Alfaro *et al.*, 1987). However, according to Shall (1987), good results can also be obtained using dormant buds, or buds showing slow growth if treated with BA (1-2 mg/l).

Pretreatment of mother plants is of key importance for successfully establishing adult material. Etiolation has been successfully used in the cvs. Duke-7 and Duke (Cooper, 1987; Zirari and Lionakis, 1994), enabling sprouting rates in the range of 70-80% versus 8% in non-etiolated material; moreover, the contamination rate was also lowered, from 70% to 0% and no browning could be observed (Cooper, 1987). Severe pruning of mother plants has also been used (Photographs 1-3) to establish cultures of the IV-8 rootstock (Pliego-Alfaro *et al.*, 1987; Barceló-Muñoz *et al.*, 1999) as well as of Fuerte, Hass, Topa-Topa and Duke cultivars (Zirari and Lionakis, 1994).

Single node segments with a lateral bud (Photograph 4) (Pliego-Alfaro *et al.*, 1987; Cooper, 1987; Capote del Sol *et al.*, 2000) seem to be more suitable for culture initiation than shoot tips (Zirari and Lionakis, 1994). Regarding explant sterilization, rinsing with 96% alcohol prior to disinfection with Na-hypochlorite, noticeably decreases contamination rates (70% versus 16%) (Cooper, 1987). Its use, therefore, is highly recommended (Gonzalez Rosas and Salazar, 1984; Cooper, 1987; Schall, 1987), although other researchers consider alcohol causes explant oxidation (Capote del Sol *et al.*, 2000).

2.1.2. Culture Medium

Mineral formulations are of key importance in culture establishment. The MS formulation (Murashige and Skoog, 1962) has been used occasionally (Gonzalez-Rosas and Salazar, 1984; Pliego-Alfaro *et al.*, 1987; Barceló *et al.*, 1990), although it seems to induce the appearance of toxic symptoms. Mineral formulations of lower ionic content, such as WPM (Lloyd and McCown, 1980) (Cooper, 1987), MS with macroelements at half strength (Pliego-Alfaro *et al.*, 1987; Zirari and Lionakis, 1994; Barceló-Muñoz *et al.*, 1999) or DF (Dixon and Fuller, 1976), (Harty, 1985; Capote del Sol *et al.*, 2000) are therefore more suitable. The lack of toxicity observed by Gonzalez-Rosas and Salazar (1984), Pliego-Alfaro *et al.* (1987) and Barceló-Muñoz *et al.* (1990) after using the MS formulation may be due to the fact that the materials used were fairly tolerant to salinity, e.g. seedlings of the Antillean race or the GvarAm-13 rootstock.

Benzyladenine (BA) is the most commonly used cytokinin, either alone, at concentrations in the range 0.1 - 2 mg/l (Barceló-Muñoz and Pliego-Alfaro, 1987; Cooper, 1987; Schall, 1987; Barceló-Muñoz *et al.*, 1990), or combined with gibberellic acid (GA_3), (Capote del Sol *et al.*, 2000). The cytokinin kinetin, combined with indole-3-butyric acid (IBA) has also been used in the initiation phase (González Rosas and Salazar, 1984).

Photograph 1. Adult avocado tree.

Photograph 2. Severely pruned avocado tree.

Photograph 3. Shoot sprouting on a severely pruned avocado tree.

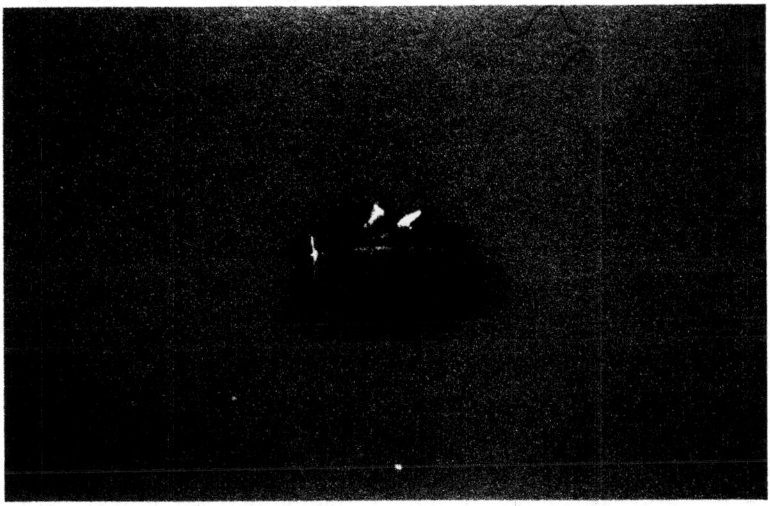

Photograph 4. Single node segment of avocado.

2.2. Shoot Multiplication and Elongation

Ontogenetic age is of key importance in culture during the proliferation phase. Using juvenile material, established procedures allow the maintenance of shoot cultures in active proliferation (Cooper, 1987;

Barceló-Muñoz *et al.*, 1990). However, with adult material, problems such as low growth rates and apical necrosis or lack of response have prevented satisfactory results (Harty, 1985; Cooper, 1987; Schall, 1987). Only in the case of the IV-8 rootstock has it been possible to maintain adult material in active proliferation (Pliego-Alfaro *et al.*, 1987; Barceló-Muñoz and Pliego-Alfaro, 1997; Barceló-Muñoz *et al.*, 1999).

2.2.1. Juvenile Material

Several mineral formulations have been used during the multiplication phase, e.g. $N_{45}K$ (Margara, 1984), a modification of MS macroelements with 45 mM total N and a 4/1 NO_3^-/NH_4^+ ratio, on shoots derived from a GvarAm-13 seedling (Barceló-Muñoz *et al.*, 1990), WPM on material derived from two-year-old seedlings of cvs. Hass and Hopkins (Cooper, 1987) and DF on material obtained from Duke-7 (Harty, 1985) and Duke (Capote del Sol *et al.*, 2000) seedlings.

The cytokinin BA is the most widely used growth regulator, either alone, at concentrations in the range 1-5 mg/l (Schall, 1987; Barceló-Muñoz *et al.*, 1990; Capote del Sol *et al.*, 2000), or combined with IBA (0.1 mg/l) at 0.3 mg/l (Cooper, 1987). However, Harty (1985) used 10 mg/l kinetin, a relatively high concentration.

Use of high BA concentrations in proliferation media and/or maintenance of material throughout several subcultures in the presence of this hormone generally causes a reduction in the proliferation rate or a miniaturization of the shoots, or both. In these cases, an elongation phase could be necessary prior to rooting. Use of lower BA dosages in solid (Schall, 1987) or liquid (Barceló-Muñoz *et al.*, 1990) media or simply subculturing at 6-8 week intervals (Cooper, 1987; Witjaksono et al., 1999) allows shoot elongation. According to Barceló-Muñoz *et al.* (1990) a liquid medium doubles the length of miniaturized shoots after two weeks, but propagules should not be maintained for longer periods to avoid hyperhydricity.

2.2.2. Adult Material

Generally, efforts to maintain adult shoots in active proliferation have had negative results (Harty, 1985; Cooper, 1987; Schall, 1987; Capote del Sol *et al.*, 2000). For example, after the initiation phase, following the separation of the shoot from the nodal section or after several subcultures in multiplication medium, shoots loose vigor, develop symptoms of necrosis and die.

Apical necrosis, an important problem in the proliferation phase, can be avoided by weekly recultures into fresh medium. Adult shoots of the cv. GvarAm-13 have been maintained for over two years with this method in the presence of 1 mg/l BA. The recorded values for average axillary shoot number, axillary shoot length and main shoot length were 3.3, 0.9 and 2.7, respectively (Pliego-Alfaro *et al.*, 1987). Moreover, the use of double phase medium has led to good results in preventing necrosis of adult material of the IV-8 rootstock (Pliego-Alfaro *et al.*, 1987). This medium, with a liquid and a solid phase, includes MS macroelements diluted at half strength and a BA supplement (0.65 mg/l in the solid phase and 0.1 mg/l in the liquid phase) and it has been used by Zirari and Lionakis (1994) to multiply several cultivars (Fuerte, Hass, Topa-Topa and Duke). These authors obtained acceptable results in terms of shoot elongation and axillary shoot formation, but no data on the behaviour of the material throughout successive subcultures have been presented.

Continuous culture under double phase conditions enables acceptable shoot proliferation, but induces formation of shoots with a succulent aspect, thickened bases with the presence of callus, and symptoms of hyperhydricity (Pliego-Alfaro *et al.*, 1987). Use of a liquid medium with agitation in a rotor has been used to reactivate growth of shoots multiplied in double phase medium and, similar to the behaviour observed in juvenile material, shoot length is doubled after a 2-week period (Barceló-Muñoz *et al.*, 1999), while maintenance of shoots in a liquid medium for longer periods increases elongation but induces the appearance of severe symptoms of hyperhydricity. The combined use of liquid medium (2 weeks) and double phase (4 weeks) avoids the appearance of succulent cultures with callus at the base, although 70% of cultures appear hyperhydric. However, increasing the culture interval in double phase conditions from 4 to 6 weeks (Photograph 5) decreases the percentage of hyperhydricity from 70 to 20%. It should be noted that subculture intervals of 8 weeks induce the appearance of apical necrosis (Barceló-Muñoz *et al.*, 1999).

In a comparative study of MS formulation with macroelements at half strength versus Gamborg B5 formulation, using the above mentioned method, Barceló-Muñoz *et al.* (1999) found a general improvement in the morphology of shoots proliferating in the presence of Gamborg's formulation. Moreover, replacement of the cytokinin BA by zeatin drastically reduced the proliferation rate (0.2 axillary shoots/culture in the presence of zeatin versus 3.2 axillary shoots in the presence of BA).

Photograph 5. Adult shoots cultured in liquid and double phase media.

2.3. Hyperhydricity

Hyperhydricity is a serious physiological disorder occurring during the proliferation phase which has strong negative effects on the rooting and acclimatization capacities of the shoots. The avocado is very sensitive to this disorder, generally caused by continuous culture in medium supplemented with BA (Schall, 1987), and/or the use of liquid (Barceló-Muñoz *et al.*, 1990) or double phase (Pliego-Alfaro *et al.*, 1987) media. Although it affects material at different ontogenetic ages, hyperhydricity in juvenile shoots can easily be overcome by increasing subculture intervals or by rooting the shoots, whereas in adult material the problem is much more difficult to overcome.

De la Viña *et al.* (2001) studied the effect of medium texture, solid vs double phase in combination with irradiance level (35-85 μmol $m^{-2}s^{-1}$), on shoot quality during the proliferation phase. As expected, continuous culture in double phase conditions induced hyperhydricity, giving rise to deformed stomata and poor development of epicuticular waxes. The increase in the irradiance level decreased chlorophyll content but did not affect hyperhydricity. Hyperhydric leaves show closed stomata and are thicker than normal leaves, with a high degree of disorders in the different tissues. A higher proliferation capacity of spongy parenchyma tissue, linked to a higher cell size and large intercellular spaces, may account for this anomaly. Cells of spongy parenchyma also show an accumulation of starch in chloroplasts (Barceló-Muñoz, 1995), probably due to inability to export sugars. The size of hyperhydric shoots is also higher than normal, showing a slightly lignified vascular system with vessels distributed over a discontinuous ring.

Two types of hyperhydricity can be distinguished in the avocado. In the first type the shoots present an abnormal aspect with symptoms of succulency, callus at the base, pale green shoots and leaves, and a drastic decrease in the elongation and proliferation capacities, corresponding to what has been defined as succulent (Paques and Boxus, 1987) or pathological (lethal) hyperhydricity (Debergh, 1988). This type of material, which commonly appears under double phase conditions, shows a negligible rooting capacity.

The second type of hyperhydricity, appearing after the use of liquid medium for short periods (2 weeks), gives rise to dark green shoots with a humid aspect, which do not lose their proliferation or rooting capacities; these symptoms coincide with those of the well-known humid, non-succulent hyperhydricity (Paques and Boxus, 1987).

2.4. Rooting

2.4.1. Juvenile Material

For rooting, shoots must have an adequate size and a minimal foliar surface; these traits can be obtained after maintaining shoots over 6 weeks in proliferation medium (Cooper, 1987) or after culturing in liquid medium with agitation in a rotor (Barceló-Muñoz *et al.*, 1990).

As observed in other woody species, dilution of salts in the rooting medium is highly recommended (Pliego-Alfaro, 1981; Cooper, 1987; Pliego-Alfaro, 1988; Barceló-Muñoz *et al.*, 1990). Exogenous auxins do not seem to be a key factor in rooting juvenile avocado shoots (Pliego-Alfaro, 1988), although their presence increases rooting capacity. The rooting process must include two phases: the first, with auxin, for root induction and the second, for root elongation, without growth regulators. Auxin exposure, which in some cases takes place in darkness (de la Viña, 1996), can be carried out for a short period in the presence of a low auxin concentration, 1 mg/l IBA, (Barceló-Muñoz *et al.*, 1990), or by immersion of shoots for a few seconds in a highly concentrated auxin solution, IBA or naphthalene acetic acid (NAA), (1000 or 3000 mg/l), (Cooper, 1987). Rooting percentage obtained with juvenile material is high, 90-100%, if the two-step procedure is followed (Cooper, 1987; Barceló-Muñoz *et al.*, 1990) (Photograph 6), while it decreases drastically, to 40-60%, if shoots are maintained in the presence of auxin, even if it is at a low concentration (Cooper, 1987). As expected, auxin favours root induction but inhibits elongation of preformed root primordia. In this respect, Pliego-Alfaro (1988) recommends the use of activated charcoal in the root elongation medium, while Schall (1987), who recommends the use of 5 mg/l IBA throughout the rooting process, improves rooting percentages from 26% to 87% after inclusion of activated charcoal in the medium.

Changes in endogenous auxin content during the rooting process of juvenile shoots have been studied by García-Gómez *et al.* (1994). In shoots treated with IBA, endogenous IAA levels increase two fold during the first six days of rooting; moreover, the content of IAA-aspartate increases before root differentiation takes place, decreasing afterwards. Treating the shoots with TIBA (2,3,5 triiodobenzoic acid), an inhibitor of polar auxin transport, inhibits the accumulation of IAA and IAA-aspartate as well as the formation of roots. According to García-Gómez *et al.* (1994), an increase in the endogenous level of IAA must occur during the first part of the rooting process (activation of cambial cells and division of cells derived from the cambium), while the formation of the conjugate would be a mechanism to eliminate the excess of free auxin. Peroxidase activity remained constant during the first 3 days of rooting and increased from days 3 to 6, remaining at high levels afterwards

(García-Gómez *et al.*, 1995). Apparently, these enzymes could play a role in the degradation of free auxin as well as the conjugate IAA-aspartate.

Ex vitro rooting has only been accomplished with juvenile material. Rooting percentages up to 100% have been obtained using microcuttings of two-year-old seedlings of Hass and Hopkins cvs., following 1 sec. dipping in 3000 mg/l NAA. Rooted plants showed adequate survival and subsequent growth (Cooper, 1987).

2.4.2. Adult Material

The rooting capacity of adult shoots is very low (Barceló-Muñoz and Pliego-Alfaro, 1987; Cooper, 1987) and generally, after transfer to a rooting medium, microshoots detain their growth, lose their leaves and die (Zirari and Lionakis, 1994).

Rooting capacity does not increase with subculturing (Pliego-Alfaro *et al.*, 1987). However, rooting can be obtained following micrografting of adult buds onto rootstocks derived from *in vitro* germinated seeds (Barceló-Muñoz and Pliego-Alfaro, 1987; Pliego-Alfaro and Murashige, 1987) or after successive pruning of mother plants (Barceló-Muñoz *et al.*, 1999; Capote del Sol *et al.*, 2000).

In the case of rootstock GvarAm-13, with rooting rates in the range 0-5% maintained after 29 subcultures, 90% rooting could be observed following successive micrografts *in vitro*. The restored rooting capacity was stable after 9 subcultures but the proliferation rate was very low (Barceló-Muñoz, 1995).

With the IV-8 rootstock, after severe pruning of the mother plant, a 30% rooting rate could be obtained. However, this percentage increases up to 70% if the root induction step takes place in liquid medium with agitation in a rotor (Photograph 7), and reaches 90% when the shoots proliferate in the presence of Gamborg salt mixture instead of the MS formulation with macroelements at half strength (Barceló-Muñoz *et al.*, 1999). In an attempt to quantify the effect of the physiological state of the mother plant on rooting, Barceló-Muñoz *et al.* (1999) compared the rooting capacity of shoots derived from the IV-8 rootstock subjected to successive severe pruning versus shoots from a copy of this tree obtained by the Frolich technique (Frolich and Platt, 1972) taken to the field and allowed to flower; the observed rooting rates obtained for both types of material, under optimum rooting conditions, were 92% and 50%, respectively.

Photograph 6. Rooted juvenile avocado shots.

Photograph 7. Rooted adult avocado shoots.

2.5. Acclimatization

Several studies have been carried out to ascertain the photosynthetic behaviour of *in vitro* rooted juvenile avocado. De la Viña *et al.* (1999) measured the photosynthetic rate and Rubisco content of plantlets grown in the presence of high/low sucrose concentrations and CO_2 levels, finding that at high sucrose and CO_2 levels the Rubisco content and maximum photosynthetic rate were drastically reduced. Along this line, Witjaksono *et al.* (1999) found a 45% decrease in CO_2 assimilation efficiency when plantlets were grown in a CO_2 enriched environment.

Premkumar *et al.* (2002a, 2002b) studied the effect of varying the sucrose concentration in the culture medium for four weeks, followed by subsequent acclimatization, on several physiological parameters of juvenile avocado microplantlets. Changes in the sucrose levels did not exert any influence on total chlorophylls or carotenoids, either before or after *ex vitro* transfer; however, the transition from heterotrophic to autotrophic conditions strongly influenced the relative proportions of chlorophyll a and b, as well as total carotenoids, reflecting the occurrence of important modifications in the light absorption equipment of the leaves (Premkumar *et al.*, 2002a). In contrast, levels of Rubisco subunits were seriously affected by both sucrose treatments and *ex vitro* transfer; a gradual reduction in both subunits was observed with increasing sucrose in vitro while the pattern was completely reversed after transplanting to *ex vitro* conditions (Premkumar *et al.*, 2002a). With regards to carbohydrates, although the endogenous sucrose levels and starch in leaves during the *in vitro* phase were not affected by varying sucrose concentrations in the medium, during acclimatization, however, endogenous sucrose decreased while the starch content showed a marked increase, suggesting improvements in sucrose utilization efficiency and starch synthesis machinery during *ex vitro* transfer (Premkumar *et al.*, 2002b).

Generally, acclimatization of *in vitro* rooted juvenile shoots is not a problem (Photograph 8) (Cooper, 1987; Schall, 1987; Barceló-Muñoz *et al.*, 1990), although the survival rate can be improved after application of paraffin emulsions on aerial parts (Schall, 1987). Growth and development of *ex vitro* plants can be improved after inoculation with mycorrhiza forming fungi (Photograph 9). Inoculation with *Glomus fasciculatum* and *Glomus deserticola* enhances growth of aerial parts (5.16 / 0.93 g dry weight in plants with and without mycorrhiza, respectively) and development of the root system, increasing the shoot/root ratio as well as N, P, K contents (Barceló-Muñoz *et al.*, 1990; Azcón Aguilar *et al.*, 1992; Vidal *et al.*, 1992; de la Viña *et al.*, 1996).

Adult shoots show low survival and slow growth rates under *ex vitro* conditions. Marín and Gella (1988) have pointed out that desiccation is not the cause of poor survival in plants of *P. cerasus* during acclimatization, since plants start dying 15 days after transfer to soil. A similar behaviour has been found in adult avocado shoots, decreasing survival rates from 100% (2 weeks) to 30-40% (6 weeks). Higher survival rates (70-80%) have been obtained on plantlets of adult IV-8 rootstock (Photograph 10) after brief exposures of rooted plants to low RH during the acclimatization process (Marín and Gella, 1987). Acclimatized plants show good subsequent growth (Photograph 11).

Photograph 8. View of acclimatized juvenile avocado plants.

Photograph 9. Acclimatized juvenile avocado plants inoculated (left) and non-inoculated (right) with *Glomus fasciculatum*.

Photograph 10. Acclimatized adult avocado plantlets.

Photograph 11. View of acclimatized adult avocado plants.

2.6 Micropropagation Procedure for Adult Avocado

2.6.1. Initiation Phase

Nodal sections with lateral buds derived from actively growing basal sprouts should be used for culture establishment. Basal sprouts, in avocado, can be obtained after severe pruning of the tree to ground level. After elimination of the leaves, shoots(20-25 cm in length) are scrubbed with cotton soaked in a 96% ethanol solution, and submerged for 10 min. in a sodium hypochlorite solution (0.5%) with Tween 20 (1drop/100ml). Then, they are washed in running water and divided into 1.5-cm long nodal sections with a lateral bud. Depending on the size of the bud, 2-4 of the leaves covering the bud are eliminated and nodal sections are again deinfested as described above. Afterwards, nodal sections are washed with sterile distilled water, and following elimination of the tissue located at both sides of the nodal section, which has generally been damaged by

the chlorox, the explants are cultured individually in a horizontal position in test tubes with the following medium:
MS mineral formulation with macroelements at half strength and, in mg/L: sucrose (30,000), i-inositol (100), thiamine-HCl (1), pyridoxine-HCl (0.5), nicotinic acid (0.5), glycine (2), benzyladenine (0.3) and Sigma A-1296 agar (6,000).

Nodal sections are incubated with a 16 h photoperiod under 45 $\mu mol/m^2s$ (Sylvania Grolux lamps) at 25ºC. After 6-8 weeks shoots 2-4 cm in length are obtained and divided into 1.5 cm long sections which are used for the multiplication phase. Nodal sections are recultured again to obtain more shoots.

2.6.2. Multiplication Phase

Fifteen mm long shoots are firstly cultured in liquid multiplication medium under agitation in a rollordrum (5 rpm) for 2 weeks. They are then transferred to double-phase medium for an addittional 6-week period. Incubation conditions are those used for the initiation phase. Media composition is as follows:
Liquid medium: Gamborg's mineral formulation and, in mg/L: sucrose (30,000), i-inositol (100), thiamine-HCl (1), pyridoxine-HCl (0.5), nicotinic acid (0.5), glycine (2) and benzyladenine (0.3).
Double-phase medium: Liquid phase: The same composition as the liquid medium indicated above, except for benzyladenine, the concentration of which is lowered to 0.1 mg/L.
Solid phase: The same composition as the liquid medium except for benzyladenine, the concentration of which is increased to 0.65 mg/L and the presence of 6,000 mg/L Sigma A-1296 agar.

2.6.3. Rooting Phase

Fifteen-twenty mm long shoots with 2-3 leaves are adequate for rooting. Shoots are kept for 3-days in liquid medium under agitation in a rollordrum (5 rpm), and subsequently transferred to solid medium. Incubation conditions are the same as indicated for the initiation and shoot multiplication phases. Media composition is as follows:
Liquid medium: MS mineral formulation with macroelements at one third, and in mg/L: sucrose (30,000), i-inositol (100), thiamine-HCl (1), pyridoxine-HCl (0.5), nicotinic acid (0.5), glycine (2) and indole-3-butyric acid (1).

Solid medium: The same compositon as the liquid rooting medium except for the absence of indole-3-butyric acid, which is replaced by Sigma-activated charcoal (1,000 mg/L) and the presence of Sigma A-1296 agar (6,000 mg/L).

2.6.4. Acclimatization Phase

Rooted shoots are transferred to greenhouse conditions under 100% relative humidity. To harden the plantlets, daily exposures of increasing duration to ambient relative humidity are recommended (until leaves show signs of water stress).

3. CONCLUSIONS AND PERSPECTIVES

Avocado micropropagation procedures are well established for material of juvenile origen, although proliferation of adult derived explants presents serious problems. Generally, adult shoots show signs of apical necrosis when cultured in solid medium while severe hyperhydricity symptoms appear when they are continuously kept in liquid medium. Partial success in the multiplication phase can be obtained by culturing shoots in liquid medium followed by transfer to double phase conditions. For root induction, the use of an auxin-supplemented liquid medium is recommended. This procedure is genotype dependent, although our research group have successfully used it to propagate several adult seedlings selected for their tolerance to soil-borne diseases.

Further efforts concerning the development of revitalization techniques prior to establishment of adult explants in culture are needed to develop reliable procedures which would allow *in vitro* mass propagation of rootstocks selected for their tolereance to fungal diseases or to salinity and/or limestone conditions.

Acknowledgements. Part of the results presented in this chapter have been obtained within the framework of research projects AGF95-0588-C02-01 (CICYT), SC98-042 (INIA) and RTA02-023 (INIA).

REFERENCES.

Anónimo (1991) Common names for plant diseases. Plant Disease 3: 225-230

Arrillaga I., Marzo T., Segura J. (1991) Micropropagation of juvenile and adult *Sorbus domestica* L. Plant Cell, Tissue and Organ Culture 27(3): 341-348.

Azcón-Aguilar C., Barceló-Muñoz A., Vidal M.T., de la Viña G. (1992) Further studies on the influence of mycorrhizae on growth and development of micropropagated avocado plants. Agronomie 12: 837-840.

Barceló-Muñoz A. (1995) Micropropagación de aguacate (*Persea americana* Mill.). PhD Tesis. Facultad de Ciencias, Universidad de Málaga. España.

Barceló-Muñoz A., Encina C.L., Simón-Pérez E., Pliego-Alfaro F. (1999) Micropropagation of adult avocado. Plant Cell, Tissue and Organ Culture 58: 11-17.

Barceló-Muñoz A., Pliego-Alfaro F. (1987) Multiplicación *in* vitro de tallos adultos de aguacate (*Persea* americana Mill.). Actas II Congreso Nacional S.E.C.H., 1986. Córdoba, España. Vol. II: 915-918.

Barceló-Muñoz A., Pliego-Alfaro F. (1997) Método para la micropropagación de aguacate (*Persea americana* M.) adulto. Patente nº P9700700/5. España.

Barceló-Muñoz A., Pliego-Alfaro F., Barea J.M. (1990) Micropropagación de aguacate (*Persea americana* Mill.) en fase juvenil. Actas de Horticultura Vol.1: 503-506.

Bergh B.O. (1969) Avocado (*Persea americana* Miller). In: Outlines of Perennial Crop Breeding in the Tropics. Ferwerda F.P., Wit F. (eds.). Agricultural University Wageningen, The Netherlands. pp: 23-51.

Bergh B.O. (1975) Avocados. In: Advances in Fruit Breeding. Janick J., Moore J.N. (eds.). Purdue University Press, West Lafayette. pp: 541-567.

Bergh B.O., Ellstrand N.C. (1986) Taxonomy of the avocado. California Avocado Society Yearbook 70: 135-145.

Bergh B.O., Zentmyer G.A., Whitsell R.M., Boswell S.B., Storey W.B. (1976) Avocado rootstock breeding, specially in relation to *Phytophthora*. Acta Horticulturae 57: 237-240.

Brokaw W.H. (1988) Avocado clonal rootstock propagation. International Plant Propagators' Society Combined Proceedings 37: 97-103.

Capote M., Rodríguez N.N., Blanco M. (2000) *In vitro* propagation of avocado. Tropical Fruits Newsletter 36/37: 3-7.

Cooksey D.A., Ohr D.H., Azad H.R., Menge J.A. and Korsten, L. (1993) *Xanthomonas campestris* associated with avocado canker in California. Plant Disease 77: 95-99.

Cooper P.A. (1987) Advances in the micropropagation of avocado (*Persea americana* Mill.). Acta Horticulturae 212 (II): 571-575.

Dale J.L., Allen R.N. (1979) Avocado affected by sunblotch disease contains low molecular weight ribonucleic acid. Australasian Plant Pathology 8(1): 3-4.

de la Viña G. (1996) Estudio de algunos factores que afectan la capacidad de aclimatación de plantas de aguacate (*Persea americana* Mill.) micropropagado. PhD Thesis, Facultad de Ciencias, Universidad de Málaga, España.

de la Viña G., Azcón Aguilar C., Barea J.M., Barceló A., Pliego-Alfaro F. (1996) Effects of arbuscular mycorrhizas in the acclimatization of micropropagated avocado plants. In: Micorrhyzas in Integrated Systems. (From Genes to Plant Development) Azcón Aguilar, C., Barea, J.M. (eds.). EUR Report 16728-EN. Directorate-General XII Science; Research and Development, Brussels pp: 491-494.

de la Viña G., Barceló-Muñoz A., Pliego-Alfaro F. (2001) Effect of culture media and irradiance level on growth and morphology of *Persea americana* Mill. microcuttings. Plant Cell, Tissue and Organ Culture 65: 229-237.

de la Viña, G., Pliego-Alfaro, F., Driscoll, S.P. Mitchel, V.J., Parry, M.A., Lawlor, D.W. (1999) Effects of CO_2 and sugars on photosynthesis and composition of avocado leaves grown *in vitro*. Plant Physiology and Biochemistry 37: 587-595.

Debergh P. (1988) Micropropagation of woody species –State of the art on *in vitro* aspects. Acta Horticulturae 227: 287-295.

Dixon A., Fuller K.W. (1976) Effect of syntetic auxin levels on *Phaseolus vulgaris* L. Physiological Plant Pathology 11: 287-292.

Eggers E.R., Halma F.F. (1937) Rooting Avocado Cuttings. California Avocado Society Yearbook 22: 121-125.

FAOSTAT (2001) FAOSTAT Database Results 2001. http://apps.fao.org.

Favré J.M., Junker B. (1987) *In vitro* growth of buds taken from seddlings and adult plant material in *Quercus robur* L. Plant Cell, Tissue and Organ Culture 8: 49-60.

Frolich E.F. (1951) Rooting guatemalan avocado cuttings. California Avocado Society Yearbook 36: 136-138.

Frolich E.F., Platt R.G. (1972) Use of the etiolation technique in rooting avocado cuttings. California Avocado Society Yearbook 55: 97-109.

Gamborg O.L. (1966) Aromatic metabolism in plants. II Enzymes of the shikimate pathway in suspension cultures of plant cells. Canadian Journal Biochemistry 44: 791-799.

García-Gómez M.L., Sánchez-Romero C., Barceló-Muñoz A., Heredia A., Pliego-Alfaro F. (1994) Levels of endogenous indole-3-acetic acid and indole-3-acetyl-aspartic acid during adventitious root formation in avocado microcuttings. Journal of Experimental Botany 45(275): 865-870.

García-Gómez M.L., Sánchez-Romero C., Barceló-Muñoz A., Heredia A., Pliego-Alfaro F. (1995) Peroxidase activity during adventitious root formation in avocado microcuttings. Canadian Journal of Botany 73: 1522-1526.

George E.F. (1993) Plant Propagation by Tissue Culture. I. The Technology. Exegetics Ltd., Edington, England.

Gonzalez-Rosas H., Salazar S.G. (1984) Root induction and vegetative development from avocado plantules (*Persea americana* Mill.). California Avocado Society Yearbook 46:167-171.

Gustafson C.D., Kadman A. (1970) Effect of some plant hormones on the rooting capacity of avocado cuttings. California Avocado Society Yearbook 53: 97-100.

Harty P. A. (1985) Propagation of avocado by tissue culture: Development of a culture medium for multiplication of shoots. South African Avocado Growers' Association Yearbook 8: 70-71.

Kadman A., Ben Ya'acov A. (1965) A review of experiments on some factors influencing the rooting of avocado cuttings. California Avocado Society Yearbook 49: 67-72.

Kadman A., Ben Ya'acov A. (1970) Rooting of leaf-bearing cuttings. In: Agricultural Research Organization (10 years report, 1960-69). Bet Dagan, Israel. pp: 47-50.

Kellam M.K., Coffey M.D. (1985) Quantitative comparison of the resistance to *Phytophthora* root rot in three avocado rootstocks. Phytopathology 75 (2): 230-234.

Köhne J.S. (1992) Field evaluation of "Hass" avocado grown on Duke-7, G6 and G755C rootstocks. In: Proceedings of the Second World Avocado Congress., Lovatt, C., Holthe, P.A., Arpaia, M.L. (eds.) Vol. 1 University of California, Riverside, California, pp: 301-303.

Kopp L.E. (1966) A taxonomic revision of the genus *Persea* in the Western Hemisphere (*Persea-Lauraceae*). Memoirs N. Y. Bot. Garden 14 nº1.

Lloyd G.B., McCown B.H. (1980) Commercially feasible micropropagation of mountain laurel, *Kalmia latifolia*, by use of shoot-tip culture. International Plant Propagators' Society Combined Proceedings, 30: 421-427.

López Herrera C.J. (1998) Hongos de suelo en el cultivo del aguacate (*Persea americana* Mill.) del litoral andaluz. V Jornadas de Frutos Tropicales. Colección de Congresos y Jornadas 47/98. Consejería de Agricultura y Pesca. Junta de Andalucía. pp: 139-152.

López Herrera C.J., Flores Pedauyé R., Pliego-Alfaro F., Farré Massip J.M. (1987) El manchado solar del aguacate. Comunicaciones agrarias. Serie Protección Vegetal, nº6. D.G.I.E.A. Junta de Andalucía. 32 pp.

Margara J. (1984) Bases de la Multiplication Végétative. INRA. Versalles. Paris. 262 pp.

Marín J.A., Gella R. (1987) Acclimatization of the micropropagated cherry rootstock "Masto de Montañana" (*Prunus cerasus* L.). Acta Horticulturae 212: 603-606.

Marín J.A., Gella R. (1988) Is desiccation the cause of the poor survival rate in the acclimatization of micropropagated *Prunus cerasus* L.? Acta Horticulturae 230: 105-112.

Murashige T., Skoog F. (1962) A revised medium for rapid growth and bioassays with tobacco tissue cultures. Physiologia Plantarum 15: 473-497.

Myburgh L., Kotzé J.M. (1982) Bacterial disease of avocado. South African Avocado Growers' Association Yearbook 5: 105-106.

Pâques M., Boxus Ph. (1987) A model to learn "vitrification", the rootstock apple M.26. Present results. Acta Horticulturae 212: 193-210.

Pegg K.G. (1991) Avocado pests and disorders. Queensland Department of Primary Industries ed., Information Series Q090013, 74 pp.

Pérez Jiménez R.M. (1997) Podredumbres radiculares del aguacate (*Persea americana* Mill.) en el sur de Andalucía. PhD thesis, Universidad de Málaga. España.

Pliego-Alfaro, F. (1981) A morphogenetic study of the avocado (*Persea americana* Mill.) *in vitro*. I. Development of a rooting bioassay and its application to studying restoration by grafting of rooting competence in adult shoots. II. Somatic embryogenesis in callus. PhD thesis, University of California, Riverside, USA.

Pliego-Alfaro F. (1988) Development of an *in vitro* rooting bioassay using juvenile-phase stem cuttings of *Persea americana* Mill. Journal of Horticultural Science 63 (2): 295-301.

Pliego-Alfaro F., Bergh B.O. (1992) Avocado. In: Biotechnology of Perennial Fruit Crops. Biotechnology in Agriculture nº8. Hammerschlag, F.A., Litz, R.E. (eds.) CAB International. UK. pp: 323-333.

Pliego-Alfaro F., López-Encina C., Barceló-Muñoz A. (1987) Propagation of avocado rootstocks by tissue culture. South African Avocado Growers' Yearbook 10: 36-39.

Pliego-Alfaro F., Murashige T. (1987) Possible rejuvenation of adult avocado by graftage onto juvenile rootstocks *in vitro*. Hort Science 22: 1321-1324.

Popenoe W. (1920) Manual of Tropical and Subtropical Fruits. MacMillan, New York. 474 pp.

Premkumar A., Barceló-Muñoz A., Pliego-Alfaro F., Quesada M.A., Mercado J.A. (2002a) Influences of exogenous sucrose on juvenile avocado during *in vitro* cultivation and subsequent *ex vitro* acclimatization. Tree Physiology (In press).

Premkumar A., Barceló-Muñoz A., Pliego-Alfaro F., Quesada M.A., Mercado J.A. (2002b) Physiological influence of sucrose on juvenile avocado during *in vitro* cultivation and subsequent *ex vitro* acclimatization. Acta Horticulturae (In press).

Querci M., Owens R.A., Vargas C., Salazar L.F. (1995) Detection of potato spindle tuber viroid in avocado growing in Peru. Plant Disease 79: 196-202.

Raave R.D., Zentmyer G.A. (1955) Susceptibility of avocados to *Dematophora* root rot. California Avocado Society Yearbook 39: 172-174.

Reuveni O., Raviv M. (1976) The role of leaves in the rooting of avocado cuttings. Special Publication. Agricultural Research Organization. Israel 65: 33-35.

Schall S. (1987) La multiplication de l'avocatier (*Persea americana* Mill. cv. Fuerte) par microbouturage *in vitro*. Fruits 42 (3): 171-176.

Storey W.B., Bergh B., Zentmyer G.A. (1986) The origin, indigenous range, and dissemination of the avocado. California Avocado Society Yearbook 70:127-133.

Vidal M.T., Azcón-Aguilar C., Barea J.M., Pliego-Alfaro F. (1992) Mycorrhizal inoculation enhances growth and development of micropropagated plants of avocado. HortScience 27(7): 785-787.

Witjaksono, Schaffer, B., Colls, A., Litz, R.E., Moon, P.A. (1999) Avocado shoot culture, plantlet development and CO_2 assimilation in an ambient and enhanced CO_2 environment. *In vitro* Cell Developmental Biology-Plant 35: 238-244.

Zentmyer G.A. (1984) Avocado diseases. Tropical Pest Management 30(4): 388-400.

Zirari A., Lionakis S.M. (1994) Effect of cultivar, explant type, etiolation pretreatment and the age of plant material on the *in vitro* regeneration ability of avocado (*Persea americana*). Acta Horticulturae 365: 69-76.

18. MICROPROPAGATION OF CARICA PAPAYA AND RELATED SPECIES

Roderick Drew
Griffith University, Nathan Q4111, Australia

Chapter contents

1. INTRODUCTION

1.1 Botany, Distribution, Importance

The genus *Carica* is a member of the dicotyledonous family Caricaceae that contains 35 species of small latex-producing trees and shrubs. They all have terminal clusters of leaves and most are dioecious (Purseglove, 1968). Papaya, *Carica papaya* L., also known as papaw and pawpaw, belongs to the genus *Carica* (Badillo, 1971). This genus included 22 species that range in distribution from Argentina and Chile to southern Mexico (Manshardt, 1992), however Badillo (2000) has recently recommended

S.M. Jain and K. Ishii (eds.), Micropropagation of Woody Trees and Fruits, 543–564.

dividing the genus into 2: *Carica* containing *papaya* and *Vasconcella* containing the other 21 species. This classification is supported by recent estimates of genetic distance between papaya and related species using DNA markers that suggest papaya should be in a separate genus (Jobin-Décor et al., 1996; Kim et al., 2002). Available evidence indicates that the centre of origin of *papaya* is the Caribbean coast of Central America (Manshardt, 1992; Purseglove, 1968), however the large-fruited domesticated form is quite different to the small-fruited dioecious types that now grow wild in Central America (Manshardt, 1992).

Papaya is a quick-growing herbaceous tree with a strait cylindrical stem comprised of soft spongy tissue. The stem hollows with age and is characterised by prominent half-moon shaped leaf scars. Papayas are strongly apically dominant and do not branch unless the apical meristem is damaged or removed. Leaves are large, spirally arranged, clustered at the crown, palmately and deeply lobed. The lobes themselves are pinnately lobed. Flowers are borne in the leaf axils, are cream to yellow in colour, can be staminate (bearing 10 stamens), pistillate or hermaphrodite, having large ovaries with a central cavity and numerous ovules (Purseglove, 1968). Despite earlier reports to the contrary (Purseglove, 1968), wind-pollination does not occur in papayas. The pistillate form is stable, however staminate and hermaphrodite trees are subject to seasonal sex reversals, particularly in the subtropics, and produce varying populations of staminate, hermaphrodite and pistillate flowers (Story, 1976). The fruit is a large fleshy berry, varying in weight from a few hundred grams to 9kg (Purseglove, 1968). It is covered with a smooth thin green skin that turns yellow or orange when mature. The flesh is thick, succulent and easily bruised, and varies in colour from yellow to orange to red. Hundreds of black wrinkled seeds are contained within the fruit. They are attached in 5 rows to the wall of the ovary and are enclosed in a gelatinous sarcotesta (Purseglove, 1968).

Cultivated papayas have a narrow genepool and commercial production is limited by a number of serious diseases. By contrast, the wild related *Vasconcella* (formerly *Carica*) species have a wide range of useful characteristics including resistance to most diseases that infect papaya. Potentially useful traits that have been identified in *Vasconcella* species are PRSV-P resistance in *cauliflora, pubescens*, *quercifolia* and *stipulata* (Conover, 1964; Horovitz and Jiménez, 1967), *Phytophthora* resistance in *goudotiana,* papaw dieback resistance in *parviflora* (Drew et al., 1998), black-spot resistance and cold-tolerance in *pubescens*, the pleasant fragrance of *stipulata* and ornamental qualities of *parviflora* (Manshardt and Wenslaff, 1989b). Consequently there have been numerous attempts to hybridise *papaya* with *Vasconcella* species. With the help of embryo rescue and culture it has been possible to produce F1 hybrids, however, because of the genetic distance between *papaya* and its relatives (Jobin-Decor et al., 1996), the F1 plants are generally infertile and of limited value. The only success of note is a fertile backcross between a female *papaya* and a male *papaya* x *quercifolia* hybrid (Drew et al., 2002).

Seed of papaya was taken from Central America to Southeast Asia in the 16th century by the Spaniards and from there it spread to most tropical and subtropical countries by the 18th century (Purseglove, 1968). It is now a major tropical fruit crop worldwide and annual world production in 2001 was 5.47 million metric tones, according to FAO statistics database. Papaya is grown for both its fruit and the milky latex that contains the proteolytic enzyme, papain. Dioecious genotypes are preferred for papain production because female fruit produce higher yields of papain that is also of higher proteolytic activity than hermaphrodite fruit (Madrigal, 1980). Collection of the latex is labour intensive. In the first half of the twentieth century, Sri Lanka was the primary producer of papain, however Central Africa has now assumed that role (Purseglove, 1968).

A number of factors contribute to papaya being a major fruit crop in many countries. Firstly, it is readily distributed by seed and easy to grow. Because it is already popular in backyards in most tropical and subtropical countries, planting material is readily available. Secondly, unlike most tropical fruit crops that have short fruiting seasons, papaya can produce all year round in the tropics. Thirdly, it grows quickly and can set fruit in 4-6 months. Fruit yields are generally high and of good quality. Fourthly, the papaya is adaptable to a wide range of climates. Large crops are produced in tropical lowlands, however good yields are produced near the Equator at elevations of up to 1500 metres (Purseglove, 1968). Good crops are also produced in subtropical regions and in frost-free temperate regions on the east coast of Africa, in Australia and South America. However, in temperate zones there is a definite seasonal trend for fruit production to occur in late summer and autumn, as tree growth is minimal in cold winters and fruit set is poor in early spring because of low pollen viability.

In tropical regions commercial production of papaya is based on hermaphrodite varieties. Hermaphrodites are preferred because they can be inbred, allowing for the development of named varieties that are reasonably true-to-type from seedling populations, e.g. Hawaiian Solo. However, the effect of season on sex expression makes hermaphrodites unsuitable in subtropical and temperate zones, where dioecious cultivars are used.

1.2 Conventional Propagation

In contrast to most horticultural crops that are produced by a wide range of vegetative propagation techniques, the majority of commercial papaya plantations worldwide are based on seedlings. With dioecious lines, this results in a mixture of genotypes that exhibit considerable variability in fruit shape and quality, yield and disease susceptibility (Drew, 1988). Dioecious plants are over-planted and then thinned to the required ratio of male:female plants i.e. 1:10. Even with hermaphrodite cultivars, which are more uniform, variability can still exist between seedlings. Hermaphrodite lines, whose seedlings comprise 1/3 females, are often over-planted and female plants removed to leave 100% hermaphrodite populations. This can be done to achieve consistent shape and quality in production. In Taiwan, where papayas are grown under insect-proof netting, hermaphrodite plants are essential, as no pollination and hence no seed-set, occurs in female plants.

1.3 Need for Micropropagation

Micropropagation would facilitate commercial production and overcome many of the problems associated with seedlings. Firstly, elite genotypes of female or hermaphrodite plants can be cloned and marketed as named cultivars. This is a priority in subtropical and temperate regions where dioecious lines have a reputation for poor flavour and quality, which limits market acceptability. Similarly, micropropagation allows more rapid release of transgenic genotypes resistant to papaya ringspot virus (PRSV-P). Secondly, micropropagation overcomes the problems resulting from overplanting and thinning. Overplanting causes competition between plants and results in weak spindly plants that have weak root systems, long internodes and increased height to the first flower. By contrast, micropropagated plants are characterized by strong root systems that have multiple primary roots. They also exhibited a reduced juvenile stage (Drew, 1988; Drew and Vogler, 1993). In subtropical Queensland this resulted in greatly reduced time to fruiting, depending on time of planting. Seedlings are normally planted in early autumn so that slow growth of the juvenile phase during the cooler months reduces the internode length and thus height to first flower. Seedlings in this climate take 18 months to come into commercial production. Micropropagated papaya planted in spring, were in full production in 10-12 months (Drew and Vogler, 1993). It should be noted that this occurred only when the papaya clones were established from adult material. No consistent differences were observed between seedlings and plants micropropagated from juvenile material (Drew and Vogler, 1993). Thirdly, in vitro storage can facilitate conservation of papaya genetic resources. In vitro slow growth systems (Drew, 1992) are an excellent way to maintain elite genotypes and parents of breeding programs. Recent successes in cryopreservation of papaya meristems and seeds, and subsequent plantlet regeneration (Ashmore et al., 2001), also hold promise for long-term conservation of unique germplasm.

In the author's laboratory, micropropagation is used routinely as an aid to research programs. Over the past 10 years an extensive program of interspecific hybridization between papaya and some of its wild relatives has been conducted. We routinely micropropagate F1 hybrids and backcross plants after embryo rescue and culture, before glasshouse and field evaluation. In this way we can rapidly multiply any genotypes that are required for further research, easily export them to other countries, and conserve unique germplasm.

2. OVERVIEW OF IN VITRO CULTURE OF PAPAYA

Papaya is a quick-growing plant, and if conditions are optimal, it will grow quickly in vitro. However, achieving optimal conditions for papaya has been challenging and a number of problems have hindered development of reliable protocols. Papaya is difficult to establish in culture free of contamination. It is sensitive to plant growth regulators (PGRs), leaving small margin for error in controlling shoot and root growth. Maintaining actively growing cultures is not easy. Adult material behaves differently in culture to juvenile material. Papaya is prone to fungal and bacterial infections, making acclimatization of plantlets from culture far from routine. Papaya is not an easy species to culture.

However, as a result of considerable research attention in recent decades, there are now numerous published reports of successful culture of papaya. Table 1 provides a summary of significant published reports in the history of papaya tissue culture and indicates when various stages of culture were achieved.

In general, three approaches have been used for in vitro culture of papaya: callus culture followed by embryogenesis or organogenesis, proliferation of axillary buds by repeated subculture on multiplication medium, and production of micro-cuttings from axillary buds of apically dominant shoots. Each of these systems has their advantages and limitations.

An overview of the limiting stages in the culture of papaya is presented below, along with protocols developed to address these problems.

2.1 Establishment and Endogenous Contamination

High levels of exogenous contaminants are present in explants from field-grown trees and all plants contain endogenous bacteria that make establishment of "clean" explants in vitro a difficult procedure (Litz and Conover, 1978; Drew, 1988). It is important when establishing papaya explants to minimise contamination and to use actively-growing shoots. In early research, it was reported that explant establishment was influenced by season, sex type and contamination levels (Litz and Conover, 1981). Antibiotics such as rifampicin (Reuveni et al., 1990) and gentamycin (Mondal et al., 1990) have been used as a pretreatment or have been added to establishment media, to control bacterial contaminants. In the author's

TABLE 1. Important reports in the history of papaya tissue culture

Explant	In vitro system	Comments	Reference
Petiole	Embryogenesis	No plants	DeBruijne 1974
Stem sections	Embryogenesis	Plants produced	Yie and Liaw 1977
Axillary buds	Shoot proliferation	Proliferation from shoot explants	Litz and Conover 1977
Apical shoots	Shoot proliferation	Established plants in pots in nursery	Litz and Conover 1978
Seedling roots	Embryogenesis	Plant regeneration from root explants	Chen et al. 1987
Axillary buds from cuttings	Enhanced axillary branching	Plants grown to maturity in field	Drew 1988
Leaf discs	Callus	Transgenic calli *A. tumerfaciens*	Pang and Sanford 1988
Zygotic embryos	Embryogenesis	Transgenic plants via microprojectile bombardment	Fitch and Manshardt 1990
Axillary buds from cuttings	Shoot proliferation	Large field plantings	Reuveni et al. 1990
Axillary buds from cuttings	Micro-cuttings	Slow growth for *in vitro* conservation	Drew 1992
Axillary buds from cuttings	Micro-cuttings	>90% root initiation, good quality roots	Drew et al. 1993
Hypocotyl sections	Embryogenesis	Transgenic plants via microprojectile bombardment	Fitch et al. 1993
Petiole explants	Organogenesis	Plant regeneration	Hossain et al. 1993
Hypocotyls	Embryogenesis	Artificial seeds	Ye et al. 1993
Immature seeds	Embryogenesis	Embryogenesis from maternal tissue	Monmarson et al. 1995
Leaf explants	Embryogenesis	Somatic embryos from transformed roots, *A. rhizogenes*	Cabrera-Ponce et al. 1996
Seedling shoot tips	Shoot proliferation	Improved shoot quality via Fe source	Castillo et al. 1997
Axillary buds	Shoot proliferation	Elucidated effect of ethylene *in vitro*	Lai et al. 1998
Shoot meristems	Micro-cuttings	Cryopreservation	Ashmore et al. 2001
Shoot tips	Shoot proliferation	Efficient rooting *ex vitro*	Lai et al. 2000

laboratory best results have been achieved with rifampicin. It greatly reduced bacterial contamination in the short term, but served only to delay endogenous bacterial growth for 3 to 6 months. It is virtually impossible to eliminate all bacteria from the plant tissue by use of antibiotics. Contaminated cultures eventually produce a cloudy exudate around the base of the explant. Contaminated cultures may exist for several years, however they grow poorly and are difficult to proliferate. This problem can be diagnosed at an early stage by squashing a few leaves or explants, and plating them on a media that promotes growth of bacteria.

Initial establishment of cultures free of microbes is critical, and pre-treatment of mother stock is the best way of achieving this. Small lateral bud explants from side-shoots of actively-growing cuttings are routinely used in the author's laboratory (Drew, 1988). Mother stock should be grown in a glasshouse, where water and fertilizer can be applied carefully to the soil so that the shoots and leaves can be kept dry. Reuveni et al. (1990) emphasized proper growth of stock plants in a greenhouse and used axillary buds from lateral shoots of hedged rooted cuttings.

2.2 Growth of Adult Tissue In Vitro

Micropropagation of elite genotypes for research or commercial purposes requires selection based on field performance. This necessitates culture of explants from mature trees in the field. Establishment and early growth of mature tissue is much more difficult than that of juvenile tissue (Drew and Smith, 1986; Rajeevan and Pandey, 1986). Once papaya have commenced flowering, growth of the apex slows, nodes become progressively closer together with age, lateral vegetative buds are dormant and explants are difficult to disinfest.

Apical dominance can be overcome and actively-growing side shoots produced by three applications of BA (500mg/l) and GA (100mg/l) at one week intervals (Reuveni and Shlesinger, 1990). Once these side shoots are rooted as cuttings and are growing in a glasshouse, similar applications of BA and GA can produce actively-growing side shoots from which apical and axillary buds can be used as explants. If explants are from older trees, they may grow very slowly in vitro. Accelerated growth of adult tissue was obtained by alternate culture of shoot explants in solution containing BA and NAA on a roller drum for 3 to 7 days, and on solid medium free of PGRs for a month (Drew, 1988). Other techniques that facilitate growth of adult tissue in vitro include use of media that limits callus production on explants (Drew and Smith, 1986) and use of GA_3 (1mg/l) in initiation medium (Mondahl et al., 1990).

2.3 Viability at Proliferation Stage and Shoot Quality

A problem that is often encountered with culture of papaya is maintenance of proliferating cultures with time, and inability to subculture indefinitely (Litz and Conover, 1981; Rajeevan and Pandey, 1986; Drew,

1988). Papaya tissue is very sensitive to PGRs and even low concentrations of herbicides containing PGRs, e.g. 2,4-D and 2,4,5-T, can be detrimental to plants in the field. Initially, excessive exposure to cytokinins, e.g. BAP (Drew, 1992) and kinetin (Mondal et al., 1990), was identified as a factor that caused inhibition of growth, shoot damage, low proliferation rates and eventual death of cultures. Another consequence of repeated subculture on cytokinin is complete loss of apical dominance, resulting in short bushy shoots that are difficult to root. Methods to overcome this problem included the use of an elongation stage with reduced cytokinin (Reuveni et al., 1990; Kataoka and Inoue, 1992) and the use of gibberellins to elongate shoots (Mondal et al., 1990; Reuveni et al., 1990) prior to rooting. A multiplication technique based on subculture of nodal sections from apically dominant shoots, and rooting of micro-cuttings derived from these axillary shoots was devised to overcome these problems (Drew, 1992).

More recently, other factors have been shown to affect the quality of shoots grown in vitro, and improvements in quality of proliferating cultures have resulted. A common symptom of papaya shoots in vitro is chlorosis (De Winnaar, 1988). Castillo et al. (1997) related the cause of chlorosis to photoderegulation of iron. Successful explant establishment was achieved when a low concentration of FeEDTA was used at low irradiance levels (5 $\mu mol.m^{-2}s^{-1}$) (Castillo et al., 1997). However, when two iron sources (FeEDTA and FeEDDHA) were added to the proliferation medium, high proliferation and chlorophyll concentrations were obtained at high irradiance levels (30-40 $\mu mol.m^{-2}s^{-1}$) (Castillo et al., 1997).

Growth rates and quality of papaya cultures have been related not only to the composition of the culture media, but also to the composition of the gaseous environment of the culture flask, and the rate of exchange of gases for particular flasks (Lai et al., 1998). Papaya nodal cultures modify the atmosphere in culture vessels by production of ethylene and this has been shown to reduce shoot growth and leaf size, reduce shoot quality and increase rate of leaf senescence (Magdalita et al., 1997). This problem can be overcome to some extent by the use of the ethylene-suppressant AVG, and the ethylene-antagonist STS (Magdalita et al., 1997). Lai et al. (1998, 2000) have separated multiplication into two stages: axillary bud swelling stage in week 1 and shoot growth in subsequent weeks. They further demonstrated that ethylene was beneficial in stage 1, and detrimental in stage 2. A 41% increase in shoot number and a 42% increase in leaf expansion was obtained, as well as increased quality of shoots, by addition of 0.2 or 0.4 ppm ethylene in week 1, followed by aeration of flasks in subsequent weeks to remove ethylene and increase oxygen levels. The beneficial effect of ethylene in week 1 could be obtained by the use of ethylene biosynthesis inhibitors AVG and $CoCl_2$ (Lai et al., 2000).

2.4 Rooting Efficiency

The sensitivity of papaya tissue to PGRs and the detrimental effects of cytokinins in proliferation media have been discussed above. Similarly, prolonged exposure to auxins causes stunting and thickening of roots, low rooting percentages, poor shoot quality and consequent loss of plants at the acclimatization stage (Drew et al., 1993; Teo and Chan, 1994). The most critical factor in producing high percentages (>90%) of good quality roots that are conducive to acclimatization, is optimal duration of exposure to auxin for root initiation. Auxin is essential for root initiation in papaya (Drew et al., 1993), however it has been known since 1937 that auxin is not required for, and can be inhibitory to, root growth (Went and Thimann, 1937; Went, 1939).

The best auxin for root initiation in papaya is IBA (Drew, 1987; Drew et al., 1993) and the optimum duration of exposure to 10μM IBA is 3 days. If shoots are left on IBA for longer periods, callus proliferation and root damage result. When shoots were removed from medium containing IBA and subcultured in PGR-free medium, a carry-over effect of the IBA was still observed on root growth and quality (Drew et al., 1993). This effect was removed by the addition of 10μM riboflavin to the PGR-free medium, as riboflavin rapidly photo-oxidises IBA in light incubation (Drew et al., 1991).

An alternative method to minimise exposure of shoots to IBA is to dip micro-cuttings in 12.3mM IBA (Teo and Chan, 1994) before transfer to a medium containing distilled water and agar; however, this method needs to be optimised as only 59.5% rooting was achieved. Greater than 90% rooting was obtained when micro-cuttings were dipped in 11.1mM IBA then grown ex vitro in vermiculite. Yu et al. (2000) have produced an efficient rooting protocol that combines in vitro and ex vitro treatments. Papaya shoots were cultured on MS (1962) medium containing 2.5μM IBA for 1 week in darkness, followed by two weeks in aerated flasks in vermiculite medium containing ½ strength MS, followed by ex vitro growth in small plastic bags containing vermiculite (Yu et al., 2000).

Optimal IBA treatments are critical, however there are other factors that also influence rooting in vitro. The use of actively-growing shoots, promoted by regular transfer, and removal of 1-2 mm of the stem base to remove suberized tissue were important factors in promoting root initiation (Drew and Miller, 1989). Shoot size and presence of leaves is important for root initiation. Shoots of 5mm or less produced low rooting percentages, (30%) (Miller and Drew, 1990). Nodal sections of apically dominant shoots with leaves intact rooted easily; however, nodal explants with leaves removed were difficult to root (author unpublished). Both the size of micro-cuttings and leaf area affected root initiation ex vitro (Kataoka and Inoue, 1992). Root initiation and growth increased with increasing agar concentration (Kataoka and Inoue, 1992; Teo and Chan, 1994) and was optimum at an incubation temperature of 27°C (Drew and Miller, 1989).

2.5 Acclimatization

Papaya plants are herbaceous and quick-growing, but are susceptible to many diseases in the field. Similarly in ex vitro environments, they are subject to soft rots, they wilt quickly and can be difficult to acclimatize. It is important to follow all recommendations for acclimatization of tissue-cultured plants and to harden them slowly. Because papaya are susceptible to root rots, an open potting mix that allows plenty of aeration is essential. A 1 peat : 1 perlite : 1 polystyrene bead mixture has been used routinely to acclimatize micropropagated papaya plantlets for many years (Drew, 1988). Vermiculite is a good substrate for acclimatization of papaya plants (Kataoka and Inoue, 1992; Yu et al., 2000). Papaya plants will not tolerate acid soils, and similarly, the pH of the potting mix should be 6.5. Addition of lime to increase pH is necessary if peat is added to the mix. It is important in the first week after culture to keep the relative humidity (RH) >90% (Drew, 1988; Yu et al., 2000). Leaves should remain dry, as wet leaves are prone to soft rots. This can be achieved with a humidifier or fogger. Alternatively, RH can be reduced in the last stage of culture by the use of membrane lids (Yu et al., 2000). RH should be reduced after the first week, as plants are difficult to harden if it is maintained >90% for more than that period.

Ideally, the potting mix should be steam pasteurized to remove root rot organisms (unnecessary with vermiculite). Drenching the potting mix with fungicides that control root-rotting fungi can be useful but is no substitute for good hygiene. Liquid nutrients should be applied at weekly intervals, firstly at ¼ strength, increasing to ½ strength in week 2 and full strength by week 4.

3. FIELD PERFORMANCE OF MICROPROPAGATED PLANTS

There are only two published reports of comprehensive studies on field performance of micropropagated papaya plants; however, the results are quite consistent. In India, Pandey and Singh (1988) studied the field performance of tissue-cultured plants and seedlings for 6 months. The most comprehensive study was in Australia by Drew and Vogler (1993) who compared the growth of micropropagated plants established from both adult and juvenile tissue, and compared these with seedlings. The experiment involved 4 planting dates in a year, was of three years duration and included a ratoon crop.

In India, the tissue-cultured plants had greater stem diameters, larger leaf areas, more stomata per unit leaf area, longer petioles, higher chlorophyll content, flowered and fruited earlier (2 months) and produced higher yields (x1.5) when compared to seedlings (Pandey and Singh, 1988). Factors such as leaf number, petiole length and chlorophyll content, which were greater in tissue-cultured plants at 2 months, did not differ significantly from seedlings at 6 months. The photosynthetic rate was higher in seedlings at 2 months, but higher in micropropagated plants after 6 months. Seedlings were taller than micropropagated plants. The authors suggested that the greater vigour and earlier flowering and fruiting might have been due to greater rate of photosynthesis or the stage of juvenility of tissue that was cultured for the experiment (Pandey and Singh, 1988).

In the Australian study, all tissue-cultured plants had strong root systems and consequently established more quickly in the field than seedlings (Drew and Vogler, 1993). Micropropagated plants established from adult tissue (i.e. that had commenced flowering) had a reduced juvenile phase in the field. They were characterized by increased stem circumference, reduced height, lower height to first flower, reduced time to harvest, and more fruit per metre of stem. Planting date affected node number of the first flower for the tissue-cultured plants but not seedlings; and, tissue-cultured plants produced higher yields than seedlings when planted in spring. These differences were not observed between seedlings and tissue-cultured plants established from juvenile tissue (Drew and Vogler, 1993).

Plants micropropagated from adult tissue had 3 advantages when compared with seedlings. Firstly, the shorter, thicker form of tissue-cultured plants that resulted from reduced juvenility, combined with strong root systems to provide plants with greater wind resistance. Secondly, reduced juvenility caused reduced height to first flower and fruit. In subtropical regions seedlings are planted in autumn to reduce juvenility. Micropropagated plants can be planted in spring. The time to fruiting can thereby be greatly reduced in this climate. The higher, earlier yields can reduce production costs and increase returns. This would also be an advantage in tropical regions where papaya plants characteristically have long internodes and the first flower can be high above ground level. Thirdly, micropropagated plants allowed improved cultural practices for commercial papaya production. The practice of over-planting dioecious cultivars in subtropical regions, before thinning to the desired female: male ration for maximum fruit production, can be eliminated. Using micropropagated plants, elite female and male plants can be planted singly at any site as required. This would also be an advantage in tropical regions where hermaphrodite cultivars are grown and hermaphrodites are preferred over females.

An adverse effect of the reduced juvenile phase was observed in the ratoon crop (Drew and Vogler, 1993). The reduced internode lengths in tissue-cultured plants resulted in large numbers of small unmarketable fruit. In summary, however, the advantages of using micropropagated plants greatly outweighed this disadvantage.

4. MICROPROPAGATION OF PAPAYA (recommended protocols)

4.1 Explant Establishment

Juvenile tissue is relatively easy to establish. The apical tips of glasshouse-grown seedlings can be disinfested with sodium hypochlorite solution and grown readily in vitro. However, if micropropagation of elite female or hermaphrodite genotypes is required, it is necessary to culture tissue from adult field-grown trees. It is much more difficult to establish explants of mature trees and a two stage process, using cuttings, is highly recommended (Drew, 1988; Reuveni et al., 1990). As with seedlings, the best way to minimize exogenous contamination is to grow the mother stock in a glasshouse where water and fertilizer can be applied carefully to the potting mix.

Production of cuttings from trees in the field has been thoroughly researched for many years by Allan and excellent protocols have been developed (Allan and MacMillan, 1991). Two 20ml solutions containing 500mg/l BA and 100mg/l GA_3 were injected carefully into the base of stems to avoid the hollow centers. One or two days after treatment the stems were topped to remove the shoot apex. All leaves, fruit and flowers were removed from nodes. Large numbers of small pen-sized side shoots were produced. The advantage of this protocol is that it has been applied to trees between 1 and 15 years of age (Allan and MacMillan, 1991). Cuttings had excess leaves removed but petioles were left intact to prevent bacterial infection. The cuttings were immersed in a solution containing two fungicides – Dithane 2g/l and Benlate 1g/l. After being allowed to dry for 20 mins, the base of cuttings was dipped in IBA rooting powder, Captan and Benlate at the ratio of 9:9:2. They were then transferred to a mist bed in an open potting mix (Allan recommends a coarse perlite mix). Full details of these procedures are available (Allan and MacMillan, 1991). Papaya cuttings are prone to fungal infections and bacterial soft rots, and these procedures have been developed to successfully manage these problems.

Rooted cuttings should be repotted and grown in a dry environment in a glasshouse and again stimulated to produce multiple side shoots using published procedures (Drew, 1988; Reuveni et al., 1990; Allan and MacMillan, 1991). Small explants of apical or axillary buds from the new growth produced in the glasshouse, should be used as explants.

In the author's laboratory, DS (Drew and Smith, 1986) medium has been found to be superior to MS for shoot growth, particularly in the early stages of culture and for adult material. Numerous reports consistently agree that shoot growth of papaya, unlike most species, requires an auxin, preferably NAA, in addition to cytokinin in the media. Apical or axillary bud explants can be established on DS (Drew and Smith, 1986) medium containing 1μM BA, 1μM NAA, and 2% sucrose. Growth of mature tissue can be enhanced by alternatively culturing for 3-7 days in liquid medium (containing 1μM BA and 1μM NAA) and 3-4 weeks on solid PGR-free DS medium. Alternatively, the use of a temporary emersion system (Alvard et al., 1993) has shown promise in the author's laboratory.

Finally, iron source and irradiance level should be optimised for explant establishment (Castillo et al., 1997). Castillo et al. (1997) reported that the highest percentage of successful establishment resulted when 100μM of FEEDTA was used as the iron source and cultures were incubated at a low irradiance level of 5$\mu Mm^{-2}s^{-1}$.

4.2 Multiplication

If large numbers of plants of an elite genotype are required, then it is advisable to use a shoot proliferation system. Regeneration from callus results in genetic instability and somaclonal variation in many species and no exhaustive research has been reported to demonstrate the proportion of aberrants that is likely to occur with papaya.

Two systems are available and are recommended for the multiplication stage. A technique based on production of micro-cuttings from nodes of apically dominant plants has been developed in the author's laboratory (Drew, 1992) and has been extensively field-tested. Fourteen thousands plants of an elite female clone have been grown on a commercial plantation in Southeast Queensland and no off-type plants were detected (Manshardt and Drew, 1998). The second system, based on repeated subculture on proliferation medium, has been developed in Taiwan (Lai et al., 1998). Good quality shoots are produced and the system is used routinely in Taiwan (Yeh, pers comm.). Although there are no published reports of field stability of large numbers of progeny, this system is currently employed on a commercial basis in Taiwan and off-types are not a problem (Yeh, pers comm.).

The micro-cutting system is routinely used in the author's laboratory for cloning of elite genotypes and for plants used in research programs where genetic integrity must be maintained, e.g. transgenic plants, parents in breeding programs, rare interspecific hybrids and valuable backcross plants. Single node sections from apically dominant plants are cultured for 4 weeks on DS medium containing 0.5μM of both BA and NAA. Shoots that develop from axillary buds are rooted as micro-cuttings (using procedures described in section 3.3) and apically dominant plants that develop can be acclimatized after 3 weeks or subcultured after 4 weeks. An advantage of this system is that apically dominant plants can be maintained for 4-6 months without transfer, or 12 months if 1% fructose is substituted for 2% sucrose in the PGR-free DS medium (Drew et al., 1993).

The second multiplication system involves repeated subculture of multishoots on 50 ml MS medium (Murashige and Skoog, 1962) containing

0.88μM BA and 0.1 μM NAA in 250 ml erlinmeyer flasks (Yang and Ye, 1992; Lai et al., 1998). Previous shoot proliferation systems were limited by an inability to subculture indefinitely (Litz and Conover, 1981; Rajeevan and Pandey, 1986) or the need for an elongation phase to overcome the adverse effects of cytokinin before rooting (Mondal et al., 1990; Reuveni et al., 1990). The advantage of the system proposed by Lai et al. (1998) is twofold. Firstly, the concentrations of PGRs has been optimized and, in particular, concentration of auxin in the proliferation medium is reduced. Secondly, in elucidating the positive and negative effects of ethylene in papaya cultures, they have optimizied the gaseous atmosphere as well as the media composition. Growth rates and quality of shoots in vitro have been improved. One week on proliferation medium without aeration is recommended, followed by two-weeks aeration using membrane lids (Lai et al., 1998).

Whichever approach is taken to the shoot multiplication stage, active growth of good quality shoots must be maintained. Close attention should be paid to minimizing cytokinin levels, understanding the effects of ethylene, and optimising the level of irradiance and source of iron (Castillo et al., 1997). Castillo et al. (1997) recorded their highest multiplication rates and best quality shoot growth when 100μM of both $FeNa_2EDTA$ and FeNaEDDHA (sequestrene) were added to the culture medium and cultures were incubated at an irradiance level of $30\mu Mm^{-2}s^{-1}$.

4.3 Rooting and Acclimatization

If shoots are of good quality, actively growing, and are >5mm in length (Drew and Miller, 1989), then they are relatively easy to root. Poor quality and slow-growing shoots result from poor selection of proliferation system, the additive effect of excess cytokinin after repeated subcultures, endogenous bacterial contamination or senescence caused by cultures being left for too long before transfer, usually > 4 weeks. High rooting percentages of good quality roots can be achieved only if duration of exposure to auxin is optimised. Failure to observe this principle, irrespective of optimisation of all other steps, will result in low rooting percentages and poor quality roots that are difficult to acclimatize.

In the author's laboratory, actively-growing shoots are cultured for 3 days on RM (rooting medium: DS medium containing 10μM IBA) before being transferred to PGR-free DS medium containing 10μM riboflavin. To avoid a transfer after 3 days, 1ml of 300μM riboflavin can be injected onto the surface of 10 mls of RM (1:10 :: 300μM riboflavin : volume of RM) after 1 day (Drew et al., 1993). Alternatively, shoots can be incubated for 2 days on RM containing 31μM ($10^{-4.5}$M) riboflavin in darkness before transfer to light incubation (Drew et al., 1993). Daily analysis of tissue culture media by HPLC has shown that IBA concentration decreased steadily in light and that photo-oxidation of IBA was greatly accelerated by the addition of riboflavin to the medium (Drew et al., 1991). Rooted shoots can be acclimatized after 3 weeks in a perspex humidity cabinet using the conditions defined previously (Drew, 1988).

An alternative rooting system that minimizes exposure to IBA has been proposed by Yu et al. (2000). Shoots were cultured on MS medium containing 2.5μM IBA for one week in dark incubation (Yu et al., 2000). The dark conditions apparently stop degradation of the IBA by light and the concentration of IBA is maintained at 2.5μM IBA for the week. Shoots were transferred for root development to vermiculite media containing ½ strength MS medium under aerated conditions using membrane lids. Shoots were then acclimatized for 1 week in vermiculite at 100% RH and then under ambient conditions for two weeks in a temperature controlled glasshouse.

Other factors that favour root initiation and growth in vitro are reducing mineral concentration in the medium (Drew, 1987), incubating cultures at a temperature of 27°C (Drew and Miller, 1989), increasing the agar concentration of the medium (Kataoka and Inoue, 1992; Teo and Chan, 1994) and not removing all leaves from the shoots.

Papaya is one of the more difficult species to acclimatize; however, in the author's laboratory, 100% of rooted plants are regularly acclimatized. Good quality root systems that are not thickened or callused (caused by excess exposure to IBA) are essential, and care should be taken not to allow them to wilt after removal from culture flasks. The use of aerated flasks and vermiculite instead of agar-based culture medium for the last two weeks in vitro (Yu et al., 2000) is recommended as a first step in hardening micropropagated plantlets. Other important factors to be considered when acclimatizating papaya shoots have already been described in Section 2.5 above.

5. REGENERATION FROM CALLUS

There have been numerous reports on callus culture and regeneration of papaya. Most authors report embryogenic systems although there have been occasional reports of regeneration via organogenesis (Arora and Singh, 1978; Litz et al., 1983; Hossain et al., 1993). Organogenesis is most commonly observed from the cut end of stem segments.

Embryogenesis was first reported in the 1970s when embryogenic callus was established from petiole sections (De Bruijne et al., 1974). Later, plants were successfully regenerated via embryogenesis from seedling stem explants (Yie and Liaw, 1977). Subsequently, embryogenesis has been achieved with a range of explants including zygotic embryos (Fitch and Manshardt, 1990), hypocotyls sections (Fitch et al., 1993), integuments of maternal origin from immature seed (Monmarson et al., 1995) and roots from leaf discs that had been transformed with *Agrobacterium rhizogenes* (Cabrera-Ponce et al., 1996).

Good protocols for regeneration of papaya plantlets via embryogenesis have been described by Fitch (1995). Immature zygotic embryos of four hermaphroditic cultivars in Hawaii were rescued from fruits, 90-114 days after anthesis. When placed on ½ strength MS medium and supplemented with 1-10 mg/l 2,4-D, 400 mg/l glutamine and 6% sucrose, up to 20 embryos were produced on apical domes, cotyledonary nodes and radicle meristems between 3 and 6 weeks of culture (Fitch and Manshardt, 1990). After a further 5 months in vitro, each embryo yielded hundreds of somatic embryos. These embryos matured and germinated on MS medium containing 5g/l kinetin (Fitch and Manshardt, 1990).

A system for producing highly embryogenic calli from hypocotyls of 10 day-old seedlings has been described (Fitch, 1995). Hypocotyl sections 2-3 mm in length were cultured on medium as defined for zygotic embryos above. They produced embryogenic calli when cultured in darkness for 6-8 weeks. Cotyledons, leaves, shoot tips and roots have all been used to induce embryogenic callus, however these explants do not consistently produce highly embryogenic calli when compared to zygotic embryos and hypocotyls (Chen et al., 1987; Fitch, 1995).

The proliferation of embryos from the apical domes of zygotic embryos has been used in many laboratories worldwide and is the basis for most papaya transformation protocols. Useful modifications of this system have been reported in the 1990s. Synchronization of rapidly growing secondary somatic embryos was achieved when primary embryos were spread onto filter paper on induction medium for 4 weeks and then again 3 days before particle bombardment (Gonsalves et al., 1998). This procedure improved transformation efficiency. An almost 4-fold increase in somatic embryogenesis from hypocotyls was achieved when embryogenic calli were cultured in a liquid phase, and the addition of 0.5μM abscisic acid in the maturation phase increased mean embryo length from 1.5mm to 2.5mm (Castillo et al., 1998a). Embryos proliferated in this liquid system were encapsulated in 2.5% sodium alginate containing ½ MS salts (Castillo et al., 1998b). A 10 minute treatment in $CaCl_2$ facilitated uniform encapsulation of embryos and a high frequency of germination. Germinated artificial seeds grew into normal plants (Castillo et al., 1998b).

A problem inherent in all of the above protocols is that they depend on tissue derived from zygotic tissue or young seedlings. Whether the protocol is used for micropropagation or transformation, an elite plant that has been field-tested cannot be used as an explant source. Consequently the finding that maternal tissue in the form of integuments of immature seeds can produce embryogenic calli is very important (Momarson et al., 1995). However a procedure is needed to easily separate callus from zygotic and maternal origins if these findings are to be applied for the improvement of elite papaya genotypes.

Of the three reports on regeneration of papaya plants via organogenesis, that of Hossain et al. (1993) is the most significant. Petiole segments, 5-6 mm in length, were excised from in vitro-grown plants and grown on MS medium containing 0.5μM NAA and 2μM BAP. Shoots were regenerated on medium containing 0.1μM NAA and 2μM BAP plus 100mg/l casein hydrolysate before transfer to PGR-free MS medium for shoot elongation.

Regeneration from callus via embryogenesis or organogenesis is not recommended as a procedure for micropropagation because of the possibility of resultant genetic off-types. However, embryogenesis has been essential for transformation systems that have resulted in PRSV-P resistant plants.

6. APPLICATION TO RELATED` SPECIES AND INTERSPECIFIC HYBRIDS

The micropropagation protocol based on micro-cuttings from nodes of apically dominant plants (Drew, 1992; Drew, 1996) has been used successfully on related *Vasconcella* (formerly *Carica*) species, interspecific hybrids and backcross plants resulting from crosses between *papaya* x *quercifolia* and *papaya*. To date we have micropropagated the *Vasconcella* species *cauliflora, goudotiana, quercifolia, parviflora* and *pubescens*; and a species from a related genera: *Jacaratia spinosa.* Numerous hybrids between papaya and the 5 related species listed above have been routinely micropropagated. The protocol described for papaya (Drew, 1992) has been used without modification and actively-growing cultures with good shoot quality and high rooting percentages have resulted.

Regeneration via callus and embryogenesis has been reported for one wild *Vasconcella* species. Callus has been induced from the stem at the basal end of axillary buds of fruit-bearing plants of *candamarcensis (pubescens)* grown in a glasshouse. Somatic embryos developed from this callus. High yields of somatic embryos were produced from cell suspension cultures that were developed from the callus and some plantlets were recovered (Jordan, 1986; Jordan and Velozo, 1996).

Immature zygotic embryos of interspecific hybrids between papaya and other *Vasconcella* species are highly embryogenic and readily produce secondary embryos that can be regenerated into plantlets with high efficiency. This has been reported for interspecific hybrids between papaya and *cauliflora* (Horovitz and Jimenez, 1967; Moore and Litz, 1984; Manshardt and Wenslaff, 1989a; Chen et al., 1991; Magdalita et al., 1996), *pubescens* and *quercifolia* (Manshardt and Wenslaff, 1989b; Drew et al., 1998), *stipulata* (Manshardt and Wenslaff, 1989b), *goudotiana* and *parviflora* (Drew et al., 1998). The highly embryogenic nature of zygotic embryos from interspecific crosses has been demonstrated by Chen and Chen (1992), who isolated protoplasts from rapidly proliferating and highly regenerable suspension cultures derived from *papaya* x *cauliflora* zygotic embryos. Protoplast-derived somatic embryos proliferated rapidly and some developed into plantlets (Chen and Chen, 1992).

From the successes that have been reported for these wild *Vasconcella* species, it is apparent that the wide range of in vitro protocols that have been developed for *papaya* can be either used directly or readily adapted to these related species.

7. MOLECULAR MARKERS

Molecular markers have been used in papaya, however limited information is available at this stage. Their most common use has been to ascertain interspecific hybrids and to estimate genetic differences between papaya and related *Vasconcella* (formerly *Carica*) species. Initially, isozymes were used to confirm hybridity after crossing programs between papaya and other species (Moore and Litz, 1984; Manshardt and Wenslaff, 1989a; Chen et al., 1991), however the availability of RAPD (randomly amplified polymorphic DNA) markers for this task (Magdalita et al., 1997) has rendered isozymes redundant. DNA markers have also been used in the 1990s to estimate genetic distances between papaya and related species (Sharon et al., 1992; Stiles et al., 1992; Jobin-Décor et al., 1996).

Some preliminary work has been done on genetic mapping in papaya. RAPD markers were identified for sex determination, flowering height and fruit carpeloidy (Sondur et al., 1996). More recently, DNA marker protocols have been established that produce a much larger number of polymorphisms than RAPDs. DAF (DNA amplification fingerprinting) reactions have produced 5 times as many markers as RAPDs in an analysis of papaya DNA and a large number of DAF markers were shown to be present in male or hermaphrodite populations but not in female populations (Somsri, 1998). A microsatellite has also been identified in male papaya plants and used to separate male and female seedlings of a dioecious cultivar of papaya (Parasnis et al., 1998).

As mentioned above, a limitation of callus-based systems to multiply elite papaya clones is the probability of genetic off-types. With the ongoing development of more powerful DNA marker techniques, identification of somaclonal variants in the seedling stage is a possibility. More research is warranted in this field.

8. CONCLUSIONS

In the 1970s papaya was considered to be a recalcitrant species in vitro. As a result of considerable research effort in the past 25 years, it has become, in terms of application of biotechnology, a model for other tropical fruit crops (Manshardt and Drew, 1998). It remains a challenging species to establish or grow in culture, however, a wide range of reliable protocols have now been developed by many researchers that address the critical problems peculiar to the culture of papaya and its related species. Papaya can now be micropropagated in large numbers, giving plants of high quality that can be established with high success rates ex vitro.

9. REFERENCES

Allan P., MacMillan C.N. Advances in propagation of Carica papaya L. cv. Honey Gold cuttings. Journal of South African Society for Horticultural Science 1991; 1(2):69-72

Alvard D., Cote F., Teisson C. Comparison of methods of liquid medium culture for banana micropropagation. Plant Cell, Tissue and Organ Culture 1993; 32:55-60

Arora I.K., Singh R.N. Growth hormones and in vitro callus formation of papaya. Scientia Horticulturae 1978; 8:357-361

Ashmore S.E., Azimi M., Drew R.A. Cryopreservation trials in *Carica papaya*. Acta Hortuculturae 2001; 560:117-120

Badillo V.M. Monografia de la Familia Caricaceae. Universidad Central de Venezuela, Maracay, Venezuala, 1971

Badillo V.M. *Carica* L. vs. *Vasconcella* St.-Hil. (Caricaceae) con la rehabilitacion de este ultimo. Ernstia 2000; 10(2):74-79

Cabrera-Ponce J.L., Vegas-Carcia A., Herrera-Estrella L. Regeneration of transgenic papaya plants via somatic embryogenesis induced by *agrobacterium rhizogenes*. In vitro Cellular and Development Biology - Plant 1996; 32(2):86-90

Castillo B., Smith M.A.L., Madhavi D.L. Interactions of irradiance level and iron chelate source during shoot tip culture of *Carica papaya* L. HortScience 1997; 32(6):1120-1123

Castillo B., Smith M.A.L., Yadava U.L. Liquid system scale up of *Carica papaya* L. somatic embryogenesis. Journal of Horticultural Science and Biotechnology 1998a; 73:307-311

Castillo B., Smith M.A.L., Yadava U.L. Plant regeneration from encapsulated somatic embryos of *Carica papaya* L. Plant Cell Reports 1998b; 17:172-176.

Chen M.H., Wang P.J., Maeda, E. Somatic embryogenesis and plant regeneration in *Carica papaya* L. tissue culture derived from root explants. Plant Cell Reports 1987; 6:348-351

Chen M.H., Chen C.C., Wang D.N., Chen F.C. Somatic embryogenesis and plant regeneration from immature embryos of *Carica papaya* x *Carica cauliflora* cultured *in vitro*. Canadian Journal of Botany 1991; 69:1913-1918.

Chen M.H., Chen C.C. Plant regeneration from *Carica* protoplasts. Plant Cell Reports 1992; 11:404-407

Conover R.A. Distortion ringspot, a severe virus disease of papaya in Florida. Proceedings of the Florida State Horticultural Society 164; 77:447-448

De Bruijne E., De Langhe E., Van Rijck R. Actions of hormones and embryoid formation in callus cultures of *Carica papaya*. International Symposium Fytofarmacia en Fytiatrie 1974; 26:637-645

De Winnaar W. Clonal propagation of papaya in vitro. Plant Cell, Tissue and Organ Culture 1988; 12:305-310

Drew R.A. The effects of media composition and cultural conditions on *in vitro* root initiation and growth of papaya (*Carica papaya*, L.). Journal of Horticultural Science 1987; 62(4):551-556

Drew R.A. Rapid clonal propagation of papaya *in vitro* from mature field-grown trees. HortScience 1988; 23(3):609-611

Drew R.A. Improved techniques for in vitro propagation and germplasm storage of papaya. HortScience 1992; 27:1122-1124

Drew R.A. Gene transfer and interspecific hybridisation: two approaches to virus resistance in papaw (Carica). International Plant Propagators' Society Combined Proceedings 1996; 47:59-62

Drew R.A., McComb J.A., Considine J.A. Rhizogenesis and root growth of *Carica papaya* L.in vitro in relation to auxin sensitive phases and use of riboflavin. Plant Cell, Tissue and Organ Culture 1993; 33:1-7

Drew R.A., Miller R.M. Nutritional and cultural factors affecting rooting of papaya (*Carica papaya* L.) *in vitro*. Journal of Horticultural Science 1989; 64:767-773

Drew R.A., Persley D., O'Brien C., Bateson M. Papaya ringspot potyvirus in Australia and production of PRSV-P resistant plants. Proceedings of 2nd International Symposium on Biotechnology of Tropical Fruit Species, Taiwan 2002: in press

Drew R.A., O'Brien C.M., Magdalita P.M. Development of interspecific *Carica* hybrids. Acta Horticulturae 1998; 461: 285-292

Drew R.A., Simpson B.W., Osborne W.J. Degradation of exogenous indole-3-butyric acid and riboflavin and their influence on rooting response of papaya *in vitro*. Plant Cell, Tissue and Organ Culture 1991; 26:29-34

Drew R.A., Smith N.G. Growth of apical and lateral buds of papaw (*Carica papaya* L.) as affected by nutritional and hormonal factors. Journal of Horticultural Science 1986; 61:535-543

Drew R.A., Vogler J.N. Field evaluation of tissue-cultured papaya clones in Queensland. Australian Journal of Experimental Agriculture 1993; 33:475-479

Fitch M.M.M. 1995. "Somatic Embryogenesis in Papaya (*Carica papaya* L.)." In *Biotechnology in Agriculture and Forestry*, Vol. 30, Y.P.S. Bajaj ed. Springer-Verlag, Berlin, 1995

Fitch M., Manshardt R. Somatic embryogenesis and plant regeneration from immature zygotic embryos of papaya (*Carica papaya* L.). Plant Cell Reports 1990; 9:320-324

Fitch M.M.M., Manshardt R.M., Gonsalves D., Slightom J.L. 1993. Transgenic papaya plants from *Agrobacterium*-mediated transformation of somatic embryos. Plant Cell Reports 1993; 12:245-249

Gonsalves C., Cai W., Tennant P., Gonsalves D. Effective development of papaya ringspot virus resistant papaya with untranslatable coat protein gene using a modified microprojectile transformation method. Acta Horticulturae 1998; 461:311-314

Horovitz S., Jimenez H. Cruzamientos interspecificos y intergenericos in *Carica* ceasy sus implicaciones fitotecnicas. Agron. Trop., Maracay, 1967; 17:323-343

Hossain M., Rahman S.M., Islam R., Joarder O.I. High efficiency plant regeneration from petiole explants of *Carica papaya* L. through organogenesis. Plant Cell Reports 1993; 13:99-102

Jobin-Decor M.P., Graham G.C., Henry R.J., Drew R.A. RAPD and isozyme analysis of genetic relationships between *Carica papaya* and wild relatives. Genetic Resources and Crop Evolution 1996; 44:1-7

Jordan M. Somatic embryogenesis from cell suspension cultures in *Carica candamarcensis*. Plant Cell, Tissue and Organ Culture 1986; 7:257-261

Jordan M., Velozo J. Improvement of somatic embryogenesis in highland-papaya cell suspensions. Plant Cell, Tissue and Organ Culture 1996; 44:189-194

Kataoka I., Inoue H. Factors influencing ex vitro rooting of tissue-cultured papaya shoots. Acta Horticulturae 1992; 321:589-597

Kim M.S., Moore P.H., Zee F., Fitch M.M., Steiger D.L., Manshardt R.M., Paull R.E., Drew R.A., Sekioka T., Ming R. Genetic and molecular characterisation of *Carica papaya* L. Genome 2002; 45: 503-512.

Lai C-C., Yu T-A., Yeh S-D., Yang J-S. Enhancement of *in vitro* growth of papaya multishoots by aeration. Plant Cell, Tissue and Organ Culture 1998; 53:221-225

Lai C-C., Yeh S-D., Yang J-S. Enhancement of papaya axillary shoot proliferation in vitro by controlling the available ethylene. Botanical Bulletin of Academia Sinica 2000; 41:203-211

Litz R.E., Conover R.A. Tissue culture propagation of papaya. Proceedings of the Florida State Horticultural Society 1977; 90:245-246

Litz R.E., Conover R.A. *In vitro* propagation of papaya. HortScience 1978; 13:241-242

Litz R.E., Conover R.A. Effect of sex type, season, and other factors on *in vitro* establishment of *Carica papaya* L. explants. Journal of the American Society for Horticultural Science 1981; 106:792-794

Litz R.E., O'Hair S.K., Conover R.A. In vitro growth of *Carica papaya* L. cotyledons. Scientia Horticulturae 1983; 19:287-293

Madrigal L.S., Ortiz A.N., Cooke R.D., Fernandez R.H. The dependence of crude papain yields on different collection ('tapping') procedures for papaya latex. Journal of the Science of Food and Agriculture 1980; 31(3):279-285

Magdalita P.M., Adkins S.W., Godwin I.D., Drew R.A. An improved embryo-rescue protocol for a *Carica* interspecific hybrid. Australian Journal of Botany 1996; 44:343-353

Magdalita P.M., Persley D.M., Godwin I.D., Drew R.A., Adkins S.W. Screening *Carica papaya* x *C. cauliflora* hybrids for resistance to papaya ringspot virus-type P. Plant Pathology 1997; 46:837-841

Manshardt R.M. "Papaya." In *Biotechnology in Agriculture No. 8 Biotechnology of Perennial Fruit Crops*, F.A. Hammerschlag and R.E. Litz eds, Alden Press Ltd., Oxford, 1992.

Manshardt R.M., Drew R.A. Biotechnology of papaya. Acta Horticulturae 1998; 461:65-73

Manshardt R., Wenslaff T. Zygotic polyembryony in interspecific hybrids of *Carica papaya* and *C. cauliflora*. Journal of the American Society for Horticultural Science 1989a; 114:684-689

Manshardt R., Wenslaff, T. Interspecific hybridisation of papaya with other *Carica* species. Journal of the American Society for Horticultural Science 1989b; 114:689-694

Miller R.M., Drew R.A. Effect of explant type on proliferation of *Carica papaya*, L. *in vitro*. Plant Cell, Tissue and Organ Culture 1990; 21(1):39-45

Mondal M., Gupta S., Mukherjee B.B. In vitro propagation of shoot buds of *Carica papaya* L. (Caricaceae) var. Honey Dew. Plant Cell Reports 1990; 8:609-612

Monmarson S., Michaux-Ferriere N., Teisson C. Production of high-frequency embryogenic calli from integuments of immature seeds of *Carica Papaya*, L. Journal of Horticultural Science 1995; 70(1):57-64

Moore G., Litz R. Biochemical markers for *Carica papaya C. cauliflora*, and plants from somatic embryos of their hybrid. Journal of the American Society for Horticultural Science 1984; 109:213-218

Murashige T., Skoog F. A revised medium for rapid growth and bioassays with tobacco tissue culture. Physiologia Plantarum 1962; 15:473-497

Pandey R.M., Singh S.P. Field performance of in vitro raised papaya plants. Indian Journal of Horticulture 1988; 45(1-2): 1-7

Pang S.Z., Sanford J.C. Agrobacerium-mediated gene transfer in papaya. Journal of the American Society for Horticultural Science 1988; 113:287-291

Parasnis A.S., Ramakrishna W., Chowdari K.V., Gupta V.S., Ranjekar P.K. "Identification of sex specific markers in papaya." In *Agricultural Biotechnology: Laborataory, Field and Market.* Proceedings of the 4th Asia Pacific Conference on Agricultural Biotechnology, P.J. Larkin ed. 1998; pp 350-352.

Purseglove J.W. "Caricaceae." In *Tropical Crops Dicotyledons* Vol. I. Longmans, Green and Co. Ltd., Bristol, 1968.

Rajeevan M.S., Pandey R.M. Lateral bud culture of papaya (*Carica papaya* L.) for clonal propagation. Plant Cell, Tissue and Organ Culture 1986; 6:181-188

Reuveni O., Shlesinger D.R. Rapid vegetative propagation of papaya plants by cuttings. Acta Horticulturae 1990; 275:301-306

Reuveni O., Shlesinger D.R., Lavi U. *In vitro* clonal propagation of dioecious *Carica papaya*. Plant Cell, Tissue and Organ Culture 1990; 20:41-46

Sharon D., Hillel A., Vainstein A. Applications of DNA fingerprints for identification and genetic analysis of *Carica papaya* and other *Carica* species. Euphytica 1992; 62:119-125

Somsri S., Fletcher R.J., Drew R.A., Jobin M., Lawson W., Graham M.W. Developing molecular markers for sex prediction in papaya. Acta Horticulturae 1998; 461:141-148

Sondur S., Manshardt R., Stiles J. A genetic linkage map of papaya based on randomly amplified polymorphic DNA markers. Theoretical and Applied Genetics 1996; 93:547-553

Stiles J., Lemme C., Sondur S., Morshidi M., Manshardt R. Using randomly amplified polymorphic DNA for evaluating genetic relationships among papaya cultivars. Theoretical and Applied Genetics 1992; 85:697-701

Storey W.B. "Papaya." In *Evolution of Crop Plants*, N.W. Simmonds, ed. Longman Inc., New York, 1976.

Teo C.K.H., Chan I.K. The effects of agar content, nutrient concentration, genotype and light intensity on the *in vitro* rooting of papaya micro-cuttings. Journal of Horticultural Science 1994; 69(2):267-273

Went R.W. The dual effect of auxin on root formation. American Journal of Botany 1939; 26:24-29

Went F.W., Thimann K.V. "Root Formation." In *Phytohormones,* MacMillan, New York, 1937.

Yang J.S., Ye C.A. Plant regeneration from petioles of in vitro regenerated papaya (*Carica papaya* L.) shoots. Botanical Bulletin of Academia Sinica 1992; 33:375-381

Ye K., Huang J., Li B. "Hybrid Papaya Artificial Seed Production for Experimental Field." In *Biotechnology in Agriculture*, C.B. You et al. eds, Kluwer Academic Publishers, Netherlands, 1993

Yie S., Liaw S.I. Plant regeneration from shoot tips and callus of papaya. In Vitro 1977; 13:564-568

Yu T-A., Yeh S-D., Cheng Y-H., Yang J-S. Efficient rooting for establishment of papaya plantlets by micropropagation. Plant Cell, Tissue and Organ Culture 2000; 61:29-35

19. MICROPROPAGATION OF PISTACHIO

AHMET ONAY

Department of Biology, Faculty of Science, University of Dicle, 21280 Diyarbakir, Turkey
(Fax: 90-412-2488039, Email: ahmeto@dicle edu.tr)

CONTENTS

1. INTRODUCTION

1.1. Distribution and biology

Pistachio (*Pistacia vera* L., family Anacardiaceae) is the only edible crop of 11 species in the genus *Pistacia* (Zohary 1952; Özbek & Ayfer 1958; Crane

S.M. Jain and K. Ishii (eds.), Micropropagation of Woody Trees and Fruits, 565–588.

1984). The Pistachio tree is native to western Asia and Asia Minor, from Syria to the Caucasus and Afghanistan (Zohary 1952; Whitehouse 1957). Archaeological evidence in Turkey indicates that nuts were being used for food as early as 7,000 B.C. The tree was introduced into Europe at the beginning of the Christian era (Moldenke & Alma 1952). Subsequently its cultivation spread to other Mediterranean countries at approximately the beginning of the Christian era (Crane & Iwakiri 1981). The tree was first introduced into the United States in 1854 by Charles Mason, who distributed seed for experimental plantings in California (Lemaister 1959). Under favourable conditions the tree grows slowly to a height and spread of 5 to 9 metres, with one or several trunks. A strong apical dominance has also been reported to characterise developmental patterns of vegetative and floral Pistachio organs (Crane & Iwakiri 1985). Shoot extension begins late in March and terminates at the end of April to the middle of May (Crane 1984). Most of the axillary buds differentiate inflorescence buds during April and grown to their ultimate size for the season by late June (Takeda et al. 1979).

Pistachio thrives in areas which have winters cool enough to break bud dormancy and warm, long summers. So the areas which are suitable for Pistachio nut production are limited in the world (Ayfer 1990). They are drought resistant and very tolerant of high summer temperatures, but cannot tolerate excessive dampness and high humidity. Chilling requirements are estimated at 800 to 1,000 hours (Ayfer 1963 1990). The Pistachio is best adapted to the hot, drier regions of the south-east of Turkey, especially in the province of Gaziantep inland areas.

Pistachios are dioecious with male and female flowers on separate trees (Ayfer 1963). Male and female trees must be present for fruit to set, and it is usual for a branch from a male tree to be grafted on a female tree. The small, brownish green flowers are without petals and borne on axillary panicles in early summer. Once flower buds start to appear fewer vegetative buds and more flower buds are produced (Crane & Iwakiri 1981). Vegetative buds are considerably smaller than inflorescence buds and may give rise to lateral branches the following year or they may remain dormant (Crane 1984). Wind carries the pollen from male to the female flowers.

The reddish, wrinkled fruits are borne in heavy clusters somewhat like grapes. Although known as a nut, the fruit of Pistachio is botanically a drupe (Crane & Iwakiri 1981), the edible portion of which is the seed. The oblong kernel is approximately 12 mm in length and 6 mm in diameter and protected by a thin, ivory-coloured, bony shell. Normally the shells split longitudinally when mature along their sutures. The colour of the kernel varies from yellowish through shades of green. Pistachio nuts are very nutritious with high oil, with an average content of about 58% and protein

(19%) contents, and are relatively low in sugars depending on cultivars (Bloch & Brekke 1960). The trees begin bearing fruits in 5 to 8 years, but full bearing is not attained until the 10th or 15th year. Pistachios tend toward biennial bearing, producing a heavy crop one year followed by a light or none the next (Crane & Iwakiri 1981). Production of nuts is also influenced by drought, excessive rain, heat or cold and high winds.

The tree is affected by numerous diseases and pests (Davatchi 1958; Tabatabai 1966). A number of fungi attack the Pistachio. The most serious fungal disease in Turkey is *Septoria pistacina*, which can quickly kill trees of varying age. Most Pistachios are now grafted to *Septoria* resistant *P. terebinthus* rootstocks (Ayfer et al. 1990). The tree also sensitive to the following fungi: *Uromyces terebenthi*, *Phyllsutinia guttata*, *Septoria pistacina*, *Nematospora coryli* and the species of *Aspergillus* and *Penicillium* causing aflatoxin (Turan 1990).

1.2. Industrial use

The major Pistachio producing areas are Iran, California and Turkey and to a lesser extent, Syria, India, China, Greece, Pakistan and elsewhere. *Pistacia vera* L. produces the Pistachio-nuts or Pistachio of commerce which are used in confectionery and flavouring, and some of the other species are used for ornamental planting and as stock on which to graft the commercial species. Pistachio is consumed as a luxury table nut in its countries of origin and elsewhere in the world (Bailey 1947). Pistachio fruits are very delicious and nutritious. The Pistachio seeds contain a high content of oil and protein and most cultivars are relatively low in sugar. In addition, Pistachio leaves, gall, and bark are rich in tannin, which can be used for dying.

1.3. Conventional methods used for Pistachio tree propagation and improvement

Due to cross-pollination, commercial Pistachio nut trees are virtually as variable as wild populations. The breeding strategy is to exploit this genetic variation using seed orchards and controlled crossing. Currently, in Turkey and elsewhere in the world due to difficulties in rooting cuttings Pistachio trees are propagated by grafting or budding mature scions onto seedling rootstocks. Rootstocks for Pistachio are still obtained from seeds, because a successful vegetative propagation method for *Pistacia* species has not yet been found. Rootstocks for Pistachio are of two kinds; wild Pistachio species which are grafted at the place where they happen to grow, and Pistachio seedlings which are used to establish commercial orchards (Ayfer et al. 1990).

1.4. Propagation by cuttings

Currently, rooting of cuttings is not the most common practice of vegetative propagation in Pistachios. However, softwood cuttings of one-year-old *P. chinensis* seedlings treated with 5 mgl^{-1} IBA resulted in 92% rooting, and no improvement was observed even when higher IBA concentrations (10 and 20 mgl^{-1}) were used (Pair & Khatamian 1982). Explants taken from *Pistacia vera* seedlings have been rooted *in vivo* following a quick dip (5 s) in concentrated IBA solutions; 100 and 77.7% rooting being obtained when 500 and 1000 mgl^{-1} IBA were used respectively (Abousalim 1990). As far as adult material is concerned, rooting was attempted *in vitro* using the quick dip method. Optimal rooting achieved was 50% when the cut ends of four-year-old microcuttings were dipped for 10 s in 1000 mgl^{-1} IBA (Abousalim 1990). Rooting of adult hardwood and softwood cuttings of two *Pistacia vera* cultivars and of two clonal rootstocks *P. atlantica* and *P. palestina*) treated with up to 10 mgl^{-1} IBA was unsuccessful (Assaf 1971). Softwood *Pistacia vera* cuttings were successfully rooted under a mist system after very high auxin concentrations were used; 35 mgl^{-1} IBA was found to be optimal and 88% rooting was obtained after six weeks from planting (Al Barazi & Schwabe 1982).

Rooting of adult *Pistacia vera* material has been attempted by several workers (Joley 1960; Joley & Opitz 1971; Sakoury 1976) *in vitro* using a mist system. Unfortunately no more than 5% rooting was achieved. Perhaps the major dilemma with rooting cuttings of Pistachios is the rapid loss in rooting capacity with increasing age of the parent tree. Ironically, when the time is appropriate to select elite mature trees on the bases of their good past performance, propagation by cuttings is the most difficult problem.

1.5. Propagation by budding

Several *Pistacia* species may be used as rootstocks. The rootstock diameter should be large enough to accommodate Pistachio buds which are broad and large as compared with most fruit tree buds (Joley 1979; Woodroof 1979). In the U.S, "Kerman" (pistillate) and "Peters"(staminate) are normally budded onto *P. atlantica* or *P. integerrima* rootstocks (Hall 1975; Joley 1979; Crane and Iwakiri 1981) which have a greater resistance to nematodes and *Verticillum* than *Pistacia vera*. However, in Turkey, Iran and other Middle East countries, in general, *Pistacia vera* is used as a rootstock for new orchards. Seedlings of this species make more lateral roots and thicker stems than the others and they can reach budding size in a shorter time (Ayfer et al. 1990). However, there are growth differences among seedlings, and stock-scion incompatibility may necessitate intergrafting. Seedlings of *P. atlantica* and *P. khinjuk* show a rapid increase in length, which results in thinner seedlings than the others so they reach a budding size later. However, there is no stock-scion incompatibility between cultivated varieties and these rootstocks (Ayfer et al. 1990).

Seedlings of *P. atlantica* and *P. terebinthus* are widely used for commercial production (Joley & Opitz 1971; Woodroof 1979; Hartman & Kester 1983). Top worked cultivars on these two rootstocks quickly outgrow and outyield those grafted onto *Pistacia vera* in spite of the characteristically slow growth in the nursery of the two former rootstocks (Joley 1979). *P. atlantica* and *P. terebinthus* have been reported to be highly susceptible to *Verticillium* (Hartman & Kester 1983). Because of its very slow growth habit, *P. terebinthus* when grafted at the place where it grows, makes dwarf trees and begins to produce early, large and quality fruits. It is generally accepted that *P. terebinthus* is one of the most valuable rootstocks and together with *Pistacia vera* can be a suitable rootstock for intensive cultivation under irrigation conditions (Ayfer et al. 1990). Concerning root knot nematode susceptibility, seedling progeny of *Pistacia vera* x P. *atlantica*, *Pistacia vera* x *P. interregima* have been found to be highly resistant as compared to *Pistacia vera* x *P. terebinthus* seedling, which proved to be the least resistant (Joley & Whitehouse 1953). However, *P. terebinthus* was reported to be tolerant to *Phytophthora* spp (Pontikis 1977).

In Turkish Pistachio-growing regions 'budding' is the most common propagation technique. This technique can only be exercised during a very short period of the year (Kaska et al. 1990). The outlook seems to be that traditional methods of vegetative propagation in Pistachios still suffer from many problems. A potential solution to these problems would be the development of methods for vegetatively propagating the rootstocks. It would also be advantageous, for example, to propagate only preferred rootstocks which impart a low production of non-bearing trees while maintaining the other superior characteristics associated with the rootstocks. Unfortunately, attempts to propagate rootstocks vegetatively by conventional methods of soft and hard wood cuttings have given inconsistent results (Sakoury 1976; Joley & Opitz 1971; Al Barazi & Schwabe 1982). The low efficiency and difficulties in propagating mature trees are the most important problems. Consumption of Pistachio nuts is greater today than ever before and population increases will ensure a rising annual consumption. World production has increased significantly over the past 20 years mainly due to the growing demand in the world markets and the future appears more optimistic for the Pistachio nut. Therefore, there is an urgent need for large numbers of improved, fast growing trees to establish the new Pistachio orchards. The traditional methods used for Pistachio propagation and improvement are apparently not adequate to meet these demands. So, new techniques to supplement the traditional methods must be developed. *In vitro* tissue culture techniques provide a promising and alternative approach to the traditional methods.

2. *IN VITRO* APPROACHES

Tissue culture studies on Pistachio were initiated nearly two decades ago. Three resumes of most of the work carried out with Pistachio trees over the period 1982-2000 were published by Hansman & Owens (1986), Barghchi & Alderson (1989) and, Onay & Jeffree (2000). Cloning by *in vitro* techniques is especially valuable for the propagation of heterozygous and sexually incompatible Pistachio trees. This can be achieved by tissue culture techniques. Rapid asexual multiplication can be achieved by micropropagation through: (1) organogenesis; (2) somatic embryogenesis and (3) micrografting.

2.1. Organogenesis

2.1.1. Summary of past and present work

Table 1 shows that *Pistacia vera* L. has been micropropagated from both juvenile (Barghchi 1982; Alderson & Barghchi 1982; Barghchi & Alderson 1983a,b; Barghchi & Martinelli 1984; Bustamante-Garcia 1984; Barghchi & Alderson 1985; Barghchi 1986b; Abousalim 1990; Yücel et al. 1991; Yang & Lüdders 1993; Parfitt & Almehdi 1984; Dolcet Sanjuan & Claveria 1995; Taskin 1995; Onay 1996; Ghorbel et al. 1997) and adult trees (Barghchi & Martinelli 1984; Barghchi 1986b; Onay 1996; Abousalim 1990; Onay 2000a). Several basal culture media and growth regulators have been used during different growth phases. Clonal propagation through axillary branching has been obtained from young seedlings using shoot tips, apical and axillary buds as well as nodal segments (Barghchi 1982; Barghchi &

Alderson 1983a,b; Bustamante-Garcia 1984; Barghchi & Alderson 1985; Abousalim 1990; Onay 1996). The time required to initiate the cultures varies with the type of explant (Onay 1996), the age of the mother tree (Barghchi 1986b, Abousalim 1990, Onay 1996) and the species (Barghchi & Alderson 1982). Normally two weeks are required for shoot growth and proliferation from juvenile *Pistacia vera* L seedlings (Barghchi & Alderson 1983a,b; Onay 1996). The establishment of explants from more than one-year-old plants was more difficult than those from aseptically germinated seedlings due to browning of the media and tissue and callus growth at the base of the cultured explants (Barghchi & Alderson 1983a,b; Barghchi 1986a,b). The initiation of cultures from adult material usually involves pruning, grafting, BA and GA_3 spray treatments to stimulate new growth of shoots on mature plants (Barghchi & Martinelli 1984; Barghchi & Alderson 1989; Gonzales & Frutos, 1990).

There is only one report on true axillary shoot formation from initial explants of adult material (Onay 2000a). In this investigation, the forced flushing of branch sections in combination with BA was used to promote shoot development. Although adventitious buds are formed on the initial explants, further shoot multiplication can always be obtained from axillary shoots by using shoot tips and nodal bud segments.

Multiple shoots were induced from nodal segments of mature trees of *Pistacia vera* L. on Murashige and Skoog (MS) medium supplemented with benzylaminopurine (BA). Maximum shoot production was obtained from shoot tips taken from *in vitro* proliferated shoots when cultured on solidified MS medium containing 4 mgl^{-1} BA. The multiplication rate was 20 microshoots per explant on the 30th day. Rooting of microshoots was achieved in MS medium supplemented with indole butyric acid (IBA). The rooting percentage for *in vitro* produced shoots is very high, 80%. Roots usually appear after 7-10 days on the rooting medium. Shoots can only be rooted *in vitro*. Acclimatisation of *in vitro* plants is easy and if the quality of the transplants is maintained 50% survival is obtained. Phenotypically stable regenerated plants were established in the greenhouse for adult Pistachio (Onay 2000a).

2.1.2. Collecting material

The time of the year when explants are harvested has been reported to influence their establishment and growth *in vitro* (Barghchi 1986b; Onay 1996). To test the influence of season on culture initiation, shoot tips were

Table 1. *In vitro culture for micropropagation of Pistacia vera L. using organogenesis*

Age	Explant	Medium	PGR (mgl^{-1}) Shoot induction	PGR (mgl^{-1}) Rooting	Culture conditions	Success	Reference
Juvenile, Seedlings (up to 2 years)	Terminal and axillary buds	MS	BA (4) or Kin (2-4)	IBA (2.5)	Agar 0.7% 16 hour light 25 ± 1°C	Multiple shoot bud formation and plantlet regeneration	Barghchi (1982)
Seedlings (up to 2 years)	Terminal and axillary buds	MS	Various conc. BA, NAA, IAA, IBA	IBA	Agar 0.7% 16 hour light 25 ± 1°C	Multiple shoot-bud formation Plantlet regeneration Callus formation	Alderson and Barghchi (1982)
Seedlings (up to 2 years)	Apical tips and nodal buds	MS	Various conc. BA, Kin, NAA, GA_3	IAA, IBA, 2,4-D	Agar 0.7% 16 hour light 25 ± 1°C	Multiple shoot-bud formation Callus formation	Barghchi and Alderson (1983a,b)
Seedlings (up to 2 years)	Apical tips, nodal buds and leaf	Mod. MS	Kin + Aux	IAA, IBA, 2,4-D	Agar 0.7% 16 hour light 25 ± 1°C	Callus and embryoids	Barghchi (1982); Barghchi and Alderson (1983b)
Mature	Nodal/axillary bud segments	MS or WPM	BA (0.3) + IBA (0.05)	-	Agar 0.7% 16 hour light 25 ± 1°C	Shoot growth Callus formation	Barghchi and Martinelli (1984)
Juvenile 2-month-old	Axillary buds	WPM	BA (3) + GA_3 (0.5) + IB (0.001) or 2iP	NAA (1)	Liq. medium or Agar 0.6% 16 hour light 25 °C	Shoot proliferation Multiplication Plantlet regeneration	Bustamante-Garcia MA (1984)
Seedlings (1-2-year-old)	Shoot tips and nodal bud segments	MS	BA (4)	IBA (2.5)	Agar 0.7% 16 hour light 25 ± 1°C	Multiple shoot bud formation Plantlet regeneration	Barghchi & Alderson (1985)
Seedlings	Apical tips or nodal buds	MS	BA (4), Kin (2-4)	IBA (2.5)	Agar 0.8% 16 hour light 25 ± 1°C	Multiple shoot proliferation Single shoot growth Plantlet regeneration	Barghchi (1986a)
Mature	Nodal bud segments or meristem tips	MS or WPM	BA (0.3) + IBA (0.05)	-	Agar 0.7% 16 hour light 25 ± 1°C	Single or multiple shoot growth Callus formation	Barghchi (1986b)

Immature fruits	Mezocarp	MS	2,4-D (4) Kin (2)		Agar 0.7% 16 hour light 26°C	Callus formation	Zaheer et al. (1989)
Seedlings (or mature)	Shoot tips	Mod. MS	BA (0.5-4)	IBA(1-2)	Agar 0.7% 16 hour light 25°C	Multiple shoot formation (or survival of nodal explants) Plant regeneration	Abousalim (1990)
Seedlings (2 week-old)	Nodal buds	MS	Kin (4) + NAA (1)	No hormone	Jeloz 10% 24 hour light 25°C	Multiple shoot Plantlet regeneration	Yucel et al. 1991
Seedlings (1 -2-years-old)	Shoot tips and nodal bud segments	Mod. MS	IBA (0.1) BA (0.5)	IAA(1-2)	Agar 1.5% or Gelrite 1.5% 16 hour light 26 ± 2°C	Plantlet regeneration	Yang and Lüdders (1993)
Seedlings 1-3-year-old	Shoot tips	(DKW)	BA + IBA	IBA (2.5)	Gelrite or Liquid media 16 hour light 25°C	Multiple shoot formation Plantlet regeneration	Parfitt & Almehdi (1994)
Grafted seedlings	Shoot tips	Mod. MS	BA +IBA	NAA	16 hour light 25°C	Plantlet regeneration	Dolcet Sanjuan & Claveria (1995)
Seedlings	Shoot tips and nodal bud segments	Mod. MS	BA (2)	2.5 (2.5)	Agar 0.7% 16 hour light 25 ± 2°C	Multiple shoot formation Plantlet regeneration	Taskin (1995)
Seedlings (1-2-year-old or mature)	Apical tips Nodal bud segments	MS + Gam. Vit.	BA (3)	NAA (1) + IBA (1)	Agar 0.8% 24 hour light 25 ± 2°C	Multiple shoot proliferation or shoot growth and bud formation	Onay (1996)
Seedlings	Meristematic apices	Mod. MS	-	-	-	Vigorous shoots	Ghorbel et al. (1997)
Mature (30 year-old)	Nodal segments	MS + Gam. Vit.	BA (0.5-8)	IBA (1-4)	Agar 0.7% 24 hour light 25 ± 2°C	Multiple shoot proliferation Plantlet regeneration	Onay (2000a)

harvested in mid-June and mid-July 1994, and in early January and early March 1995. All explants were cultured on MS medium supplemented with 1 mgl^{-1} BA. Rates of survival of explants were assessed at the end of a 10-week incubation period.

Table 2. *Effects of time of the year in which explants were harvested*

Date	Total number of explants	% survival
Early-January 1994	30	27
Early-March 1994	30	54
Mid-June 1995	30	84
Mid-July 1995	30	77
χ^2test (3 df)		P < 0.01

The results shown in Table 2 shows that the number of surviving mature explants was affected by the time of the year that the explants were harvested. The differences in the frequencies of explants surviving between the tested dates were highly significant ($P < 0.01$, Table 2) with the highest percentage (84%) obtained from the mid-June 1994 harvested explants. Harvesting of explants in early January resulted in a significantly reduced frequency of survival (27%).

2.1.3. Surface sterilisation

Eradication of fungal and bacterial contaminants is a major problem in the establishment of Pistachio tissue cultures (See. Barghchi 1982; Bustamante-Garcia 1984; Barghchi 1986b; Abousalim 1990; Parfitt & Almehdi 1994; Taskin 1995). None of these authors systematically searched for an optimum procedure and data on the effectiveness of the sterilisation method have not been published except by Abousalim (1990). However, an effective surface sterilisation method for the production of sterile explants from *Pistacia vera* mature seeds or immature kernels, seedlings, and mature meristem tips has been achieved (Onay 1996).

In the case of explants from adult Pistachio trees, the newly formed leafy shoots (2-3 cm long) from the forced buds can be surface sterilised in absolute ethanol for a 2 min followed by a 10 min in 10% H_2O_2 and a 20-min soak in 20% (v/v) NaOCl solution (10-14% available chlorine). The explants were then rinsed 4-5 times in sterile distilled water under aseptic conditions. Buds were then reduced to 5 to 7 mm size and cultured on MS medium.

2.1.4. Culture initiation medium

In general a modified basal Murashige Skoog (1962) medium has been employed in most studies of the culture of *Pistacia* (Barghchi 1982; Barghchi & Alderson 1989; Abousalim 1990; Onay 1996; Onay & Jeffree 2000). In our experiments the most suitable medium for culture initiation was a modified MS medium with Gamborg`s vitamins (M-0404, Sigma Ltd.) containing 200 mgl^{-1} casein hydrolysate and 20 mgl^{-1} l-ascorbic acid, 30 gl^{-1} sucrose and 1 mgl^{-1} BA. The pH was adjusted to 5.7 prior to the addition of 0.7% agar, and the medium was autoclaved at a pressure of 15 PSI for 15 min, and then dispensed into culture vessels. The shoot tips were cultured in Magenta GA-7 culture vessels containing 50 ml medium and sealed with aluminium foil. During establishment and shoot multiplication, the cultures were kept at 25 ± 2°C under continuous light (20 μmol m^{-2} s^{-1}). During culture initiation almost all of the shoot tips or buds from juvenile material develop into shoots and form both axillary and adventitious shoots (Barghchi & Alderson 1983a,b; Onay 1996). However, shoot tips from adult material usually deteriorate at the end of the culture period because of an increase in vitrification and chlorosis (Barghchi & Martinelli 1994; Barghchi 1986c; Abousalim & Mantell 1992; Onay 1996).

More recently a preliminary method for regenerating adventitious shoots from adult material of Pistachio has met increasing success and may prove useful for rapid propagation (Onay 2000a). In this study, terminal lignified stem sections from 30-year-old *Pistacia vera* cultivar `Antep` plants, 3-4 cm long were collected, the base of the fresh green bud or apical tip (Fig 1a) was cut back and the freshly cut ends were immersed in 8.9 mgl^{-1} BA or 9.9 mgl^{-1} IBA solutions for 24 h. Subsequently the cuttings were potted into a sterile 1:1 mixture of sand and soil, and placed in a greenhouse (65 ± 5% R.H.) at 25 °C/24 h photoperiod supplied by mercury fluorescent lamps (25 μmol photons m^{-2} s^{-1}) in order to force the axillary buds to sprout. After 3 weeks, the newly formed leafy shoots (or rejuvenated shoots) from the forced buds were excised and surface sterilised. After 4-5 rinses in sterile distilled water, shoot tips (0.5-0.7 cm) were cultured in Magenta GA-7 vessels containing 50 ml MS medium with Gamborg`s vitamins (M-404, Sigma Ltd.) with 1 mgl^{-1} BA, 200 mgl^{-1} casein hydrolysate, 20 gl^{-1} l-ascorbic acid and 30 gl^{-1} sucrose for axillary shoot initiation.

2.1.5. Shoot multiplication

The best medium for shoot multiplication was the MS medium with Gamborg`s vitamins used with 3 mgl^{-1} BA, 3% sucrose and 0.6% agar at pH

5.7 (Onay 1996). The shoots grown on MS also produced the lowest percentage of calli at the basal end of the explants; this occurred in 15% of the cultures as compared with 50%, 65%, 30% and 20 % cultures for WP, SH, G-5 and R media, respectively (Onay 1996). Significantly different rates of necrotic shoots were obtained with all four media sources at the tested levels. However, few necrotic shoots were produced by cultures on MS medium. Explants on MS produced more and longer shoots than those on WP media and WP supplemented with nitrogen compounds (1250 mgl^{-1} ammonium nitrate and 1900 mgl^{-1} potassium nitrate) present in MS but not present in WP medium. Shoot multiplication took place in Magenta GA-7 culture vessels, containing 50 ml of medium, sealed with aluminium foil. The material was subcultured every fourth week. The culture conditions were as for culture initiation.

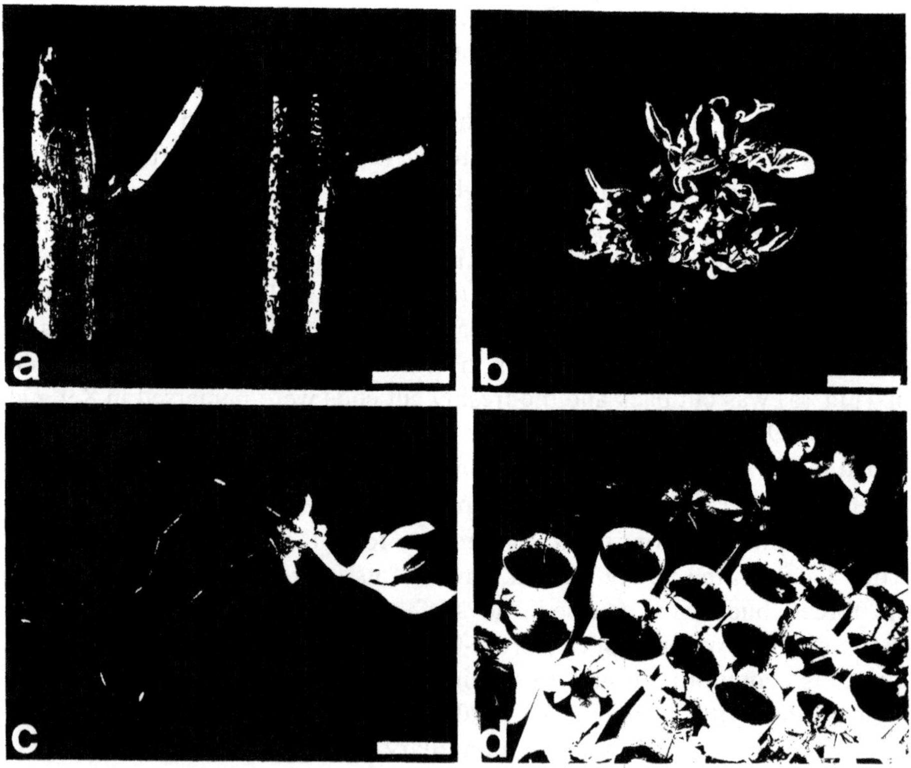

Figures 1 a-d. Stages in micropropagation of Pistachio from nodal segments of mature trees. *(**a**) Fresh green buds, bar: 3.5 mm; (**b**) Multiple shoots, bar : 10 mm; (**c**) Rooted plantlet, bar : 5.0 mm; (**d**) Plantlets in greenhouse, bar : 80 mm.*

For adult material of Pistachio it has reported that it is not possible to base a rapid micropropagation technique initially on axillary shoots alone. However, when low concentrations of cytokinins are used, adventitious shoots can be induced directly from the forced-explants (Onay 2000a). Further multiplication can be based on shoot tips and nodal segments giving a satisfactory multiplication rate of 2-20 new shoots per shoot tip (Fig 1b).

2.1.6. Rooting *in vitro*

Acclimatization of Pistachio microcuttings is generally easy *in vitro*. It is important that shoots used for rooting are in good growing condition. For *in vitro* rooting, healthy elongated shoots (1-3 cm) with 1-2 internodes were selected to obtain a high survival rate during acclimatisation. For root initiation the explants were incubated in Magenta GA-7 vessels. Root elongation took place on MS medium with Gamborg`s vitamins supplemented with 1 mgl^{-1} indole-butyric acid (IBA) and 30 gl^{-1} sucrose. Fully developed plantlets having roots with laterals could be obtained within 30 days (Fig 1c). Rooting ability and percentage increased with subculture and the type of auxin used. Shoots which had undergone more than 2 passages gave a higher rooting response. Magenta GA-vessels were superior to the other culture tubes tested.

2.1.7. Acclimatization

Acclimatization is very important for the future *ex vitro* plant development. After about 7 days on the rooting medium plantlets were washed in running water to remove the agar and transferred to a mixture of peat and perlite or peat and grit 1:1. The plantlets were covered with a Pyrex beaker to maintain a 95% relative humidity for at least two weeks and then the humidity was gradually reduced to 65% before transfer into greenhouse conditions. Nutrient feed (Solinure 7, Fisons) was given to plants undergoing growth on a regular weekly basis. The regenerated plantlets were transferred to plastic containers a containing sterile horticultural substrate moistened with a 0.1% solution of N:P:K (2:1:1) (Solinure 7). The plastic containers, with 1 plantlet each, were placed for at least 4 weeks in a growth room under a 24-h photoperiod (40 $\mu mol\ m^{-2}\ s^{-1}$) with day and night temperatures of 20°C. The relative humidity in the growth room was maintained at the same level (95%) and during the first week the plants were sprayed daily with water. After 2-3 weeks the humidity was gradually reduced to normal (65%) values. With this treatment about 100% survival was obtained.

2.1.8. Field performance

Until now no data were available on nursery and field trials with *in vitro* derived plantlets for Pistachio. Under greenhouse conditions, the viable plantlets set dormant buds, survived over wintering, and appeared phenotypically normal compared to control seedlings under the same conditions (Fig. 1d). Field performance data on these propagules is essential to confirm the feasibility of micropropagation for large scale propagation. If genetic stability of the regenerated plantlets is true, then micropropagation can be considered to be a true clonal propagation method.

2.1.9. Factors affecting culture initiation

2.1.9.1. Phenolics

A serious problem often encountered in the tissue culture of woody species is the presence of phenolic compounds exuded upon excision. Yalpani & Tyman (1983) showed that the Pistachio nut was found to contain a mixture of phenolic acids. The problem of phenolics is encountered during establishment of buds and proliferation of shoots in initial subcultures and this causes inhibition of growth. The addition of antioxidants charcoal and polyvinyl pyrrolidone to the culture medium did not reduce the exudation of phenolics during establishment.

2.1.9.2. Shoot-tip necrosis

Frequently shoot-tip necrosis (STN) resulted from poor ventilation of *in vitro* shoot cultures (Abousalim & Mantell 1994). However, they developed a practical method for alleviating STN by the periodic inversion of culture tubes (once every 7 days) containing liquid medium with 15 mM calcium gluconate.

2.1.9.3. Temperature

The optimum temperature was found to be 25 ± 2°C. At lower temperatures was suppressed and at higher temperature the browning and blackening of shoots was observed.

2.1.9.4. Light

Higher multiplication rates and the healthy vigorous growth of shoots was promoted under a continuous light regime (20 µmol m^{-2} s^{-1}).

2.1.9.5. Culture vessels

With a view to developing a large scale multiplication procedure, various types of culture vessels, such as conical flasks (250 ml), autoclavable

Magenta GA-7 vessels and glass test tubes (12 x 100 mm) (Sigma) etc. have been used. High multiplication rates of shoots were obtained in Magenta GA-7 vessels. In glass test tubes the cultures appeared to produce calli at the base of the shoots.

2.2. Somatic embryogenesis

Although shoot tip culture is the most commonly practised method for micropropagation, somatic embryogenesis is preferred for the production of large numbers of plantlets in fewer steps with a concomitant reduction in labour, time and cost. Recently, protocols (Table 3) have been established for the first stage of a study of somatic embryogenesis in Pistachio (Onay et al. 1995, Taskin et al. 1996; Chatibi et al. 1997; Onay et al. 1997; Onay & Namli 1998; Firat & Onay 1999; Onay 2000b; Onay 2000c; Onay & Jeffree (2000). It is evident from the increasing success of the methods used for somatic embryogenesis in Pistachio that it will gradually replace the shoot tip culture method, being more amenable to automation.

Somatic embryos and embryogenic masses of Pistachio, *Pistacia vera* L. have been successfully encapsulated to form somatic seeds and these are capable of germination (Onay et al. 1996). Both of these approaches can be applied to the preservation of desirable elite Pistachio genotypes and to the management of culture stocks during production.

Indirect somatic embryogenesis can be induced from embryogenic callus, cell suspension cultures or groups of cells of somatic embryos. Any cell which can develop into a somatic embryo is said to posses embryogenic competence. The initiation and further development to form embryogenic masses requires transfer of callus to a different culture regime (Onay et al. 1995). Little information is available about the origin and growth of somatic embryos in culture (Onay & Namli 1998; Onay 2000c).

Histological studies indicate that there seem to be at least two mechanisms which could account for their origin. Somatic embryos can arise from single cells or small cell aggregates by an initial asymmetric division at the epidermal layer of the embryogenic cell mass (ECM). Secondly, somatic embryos can develop from a few small meristematic cells within the ECM.

Histological studies also showed that two distinct regions can be observed in all proliferating ECMs : (1) containing cells with relatively dense cytoplasm, thin walls and vacuoles (meristematic cells) and (2) region of cells with unclear cytoplasm and large vacuoles (Fig 2a).

Table 3. In vitro culture for micropropagation of Pistacia vera L. using somatic embryogenesis

Age	Explant	Medium	PGR (mgl^{-1})	Culture conditions	Success	Reference
Immature	Kernels	MS with Gamborg`s Vit.	BA (1-8)	Agar 0.7% 24 hour light 25 ± 2°C	Embryogenic tissue Somatic embryo Plantlet regeneration	Onay et al. 1995
Immature	Kernels	MS with Gamborg`s Vit	BA (1-8)	Agar 0.7% 24 hour light 25 ± 2°C	Encapsulated embryogenic tissue Synhtetic seeds, Plantlet	Onay et al. 1996
Immature	Kernels	MS with Gamborg`s Vit	BA (0.5-2) ABA (0.5-2)	Liquid 24 hour light 25 ± 2°C	Somatic embryos Plantlet production	Onay et al. 1997
Juvenil seedlings 10-day-old	Leaves	MS	TDZ (0.5-32) or BA (0.5-4)	Liquid 24 hour light 25°C	Callus, Embryogenic tissue Somatic embryo Plantlet regeneration	Onay & Namli (1998)
Immature	Kernels	MS with Gamborg`s Vit	BA (0.5-2) ABA (0.5-2)	Liquid 24 hour light 25 ± 2°C	Somatic embryo germination	Firat & Onay (1999)
Juvenil Seedlings up to 1-year-old	Leaves	MS	TDZ or BA (0.5-4)	Liquid 24 hour light 25°C	Callus, Embryogenic tissue Somatic embryo Plantlet regeneration	Onay & Jeffree (2000)
Immature	Kernels	MS with Gamborg`s Vit	BA (1-16) ABA (0.5-4)	Agar 0.7% 24 hour light 25 ± 2°C	Somatic embryogenesis Plantlet regeneration	Onay et al. 2000
Juvenil, Seedlings up to 1-year-old	Aseptic leaves	MS	TDZ or BA (0.5-4)	Liquid 24 hour light 25°C	Development of somatic embryos from single epidermal cell	Onay (2000b)
Imature	Kernels	MS	BA (1-4)	Agar 7% 24 hour light 25°C	Somatic embryogenesis and Plantlet regeneration	Onay (2000c)
Immature	Cotyledons	MS	NAA (1.5)	Agar 7% 24 hour light 25 ± 2°C	Somatic emryogenesis and Plantlet regeneration	Taskin et al. 1996
Immature	Cotyledons	MS	2,4-D (1-5) + Kin (0.5-2) + NAA(0.1)	-	Somatic emryogenesis	Chatibi et al. 1997

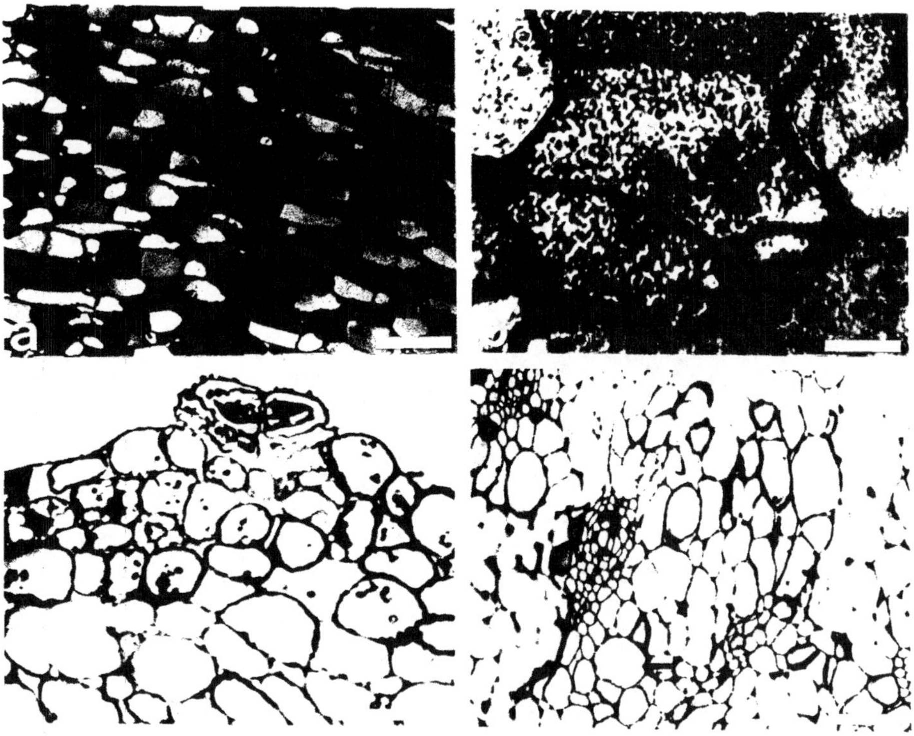

Figures 2 a-d: Histological aspects of somatic embryogenesis in the callus of leaf explants: *(**a**) The original status of ECM at day 0, showing vacuolated and parenchymatous cells filled with food reserves, bar = 100 μm; (**b**) A magnified view of embryogenic cells arising at cut end of ECM after 3 days of incubation on inducing medium. Note the mitotic figure in an embryogenic cell, bar = 50 μm; (**c**) Two-cell stage somatic embryo, bar = 150 μm; (**d**) Globular-stage somatic embryo arising from proliferating meristematic cells within the embryogenic mass, bar = 100 μm.*

A section passing through the ECM after 3 days of culture revealed meristematic activity (Fig 2b). Some of these cells showed signs of proliferation within the epidermal layer (Fig 2c). The embryogenic cells at the cut end of the ECM were also small and compact with densely stained cytoplasm; features associated with early embryogenic ECM development, which gave rise to globular stage somatic embryos (Fig 2d). The epidermal cells of the ECM elongated vertically, and at certain points the developing somatic embryo emerged. This was also seen in histological studies of ECMs examined using scanning electron microscopy (Onay & Namli 1998).

Scanning electron micrographs of embryogenic cultures were unique and could be distinguished from non-embryogenic cultures by their surface structure (Figs 3a, b). Single cell differentiation always developed at the surface of the embryogenic cellular masses (Fig 3c). The scanning electron micrographs also reveal a large number of embryos developed on the surface of explants after 27 days of culture (Fig 3d). Embryos generally developed only a single cotyledon. Embryogenic cells in the callus of leaf explants may be proembryogenic determined cells (PEDC) or induced embryogenic determined cells (IEDC). To extend the expression of embryogenic potential to the explants from mature trees, it is necessary to identify cell or tissue types which are physiologically identical to these juvenile tissues.

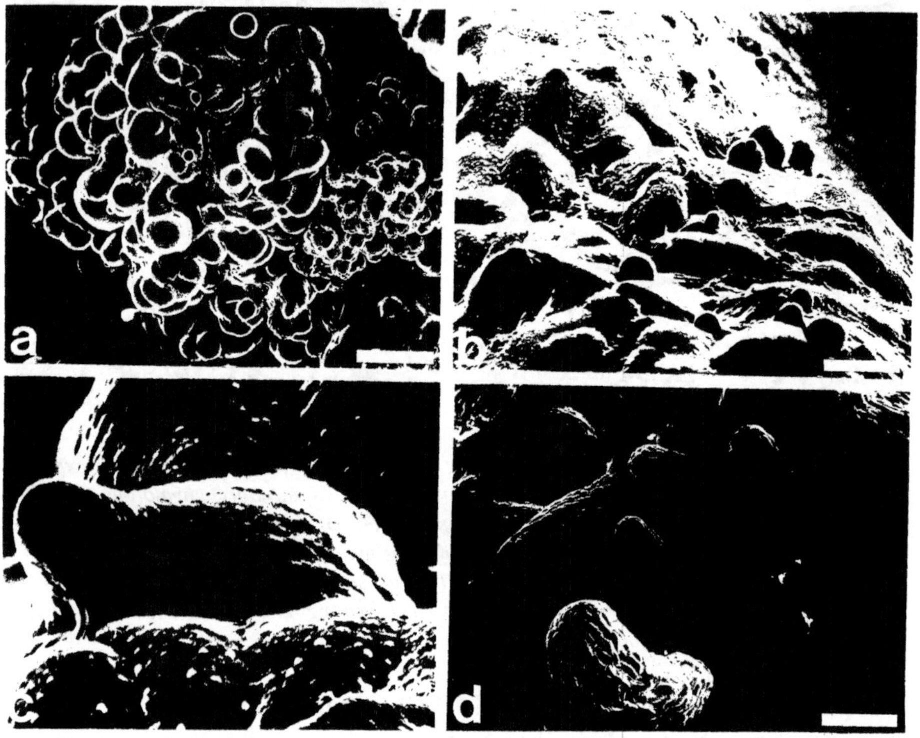

Figures 3 a-d. Scanning electron micrographs of Pistacia vera L. juvenile leaf embryogenic masses raised on MS + 1 mgl^{-1} Thidiazuron which has differentiated somatic embryos on MS + 1 mgl^{-1} BA. ***a***. *Scanning electron micrograph of non-embryogenic masses after 3 days of culture, bar = 10 μm.* ***b***. *Scanning electron micrograph of embryogenic masses after 3 days of culture, bar = 3.6 μm.* ***c***. *Scanning electron micrograph showing single cell differentiation, bar = 1.0 μm.* ***d***. *Scanning electron micrograph showing well-developed somatic embryo on the growth regulator free MS medium after 27 days of culture, bar = 20 μm.*

2.3 : Micrografting

Micrografting, a grafting procedures adapted for tissue cultures is performed either *in vitro* or *in vivo* and is increasingly being used with fruit and nut tree species for a range of purposes (Abousalim & Mantell 1992). These include the production of disease-free scions (Navarro, Roistacher & Murashige 1975), the rejuvenation of mature shoot materials (Hackett 1985) and a study of the graft union (Gebhardt & Goldbach 1988). Pistachio has benefited little from the use of this relatively new approach. The available information on micrografting in this nut species includes the first attempts to rejuvenate mature materials by micrografting *in vitro* mature scion shoot-tips onto juvenile *Pistacia vera* rootstocks (Barghchi & Martinelli, 1984; Barghchi 1986a). However, only very slow growth of the grafted scion was observed and no elongation was obtained despite supplementation of GA_3 to the culture medium. Pistachio micrografting has also been also investigated *in vitro* as well as *in vivo* (Abousalim & Mantell 1992). High levels of graft take were achieved with 10 mm (100%) and 1-3 mm (83-92%) long scions obtained from four-year-old *Pistacia vera* but no rejuvenation was reported when an elite-mature tree was used as a source of scions.

Recently, refinements to *in vitro* micrografting of Pistachio (*Pistacia vera* L. cv. Siirt) have been introduced by Onay et al. (unpublished results). Excised embryos germinated *in vitro* were used as rootstocks. Forced-shoot tips proliferated from 30-year-old trees were used as the source of scions. Grafting inside tubes proved crucial to obtain well developed micrografts *in vitro*. High levels of micrograft take were achieved with 2-4 mm (56.75%) and 4-6 mm (79.25%) long scions obtained from shoot-tips. No response was obtained with scions of < 0.5 mm in length. Slow growth and a lack of axillary shoot development was noticeable on the scions subcultured only once onto the multiplication medium. *In vitro* micrografted plantlets were successfully acclimatized and no problems were encountered with the establishment of micrografted plants *in vivo*.

3. CONCLUSIONS AND PROSPECTS FOR THE FUTURE

Currently, Pistachio plantations is are established from seedlings. Until recently, the seed used was usually obtained randomly rather than from the controlled crossing between clones selected for a breeding programme. The selected clones have usually been established in nursery seed orchards by grafting the elite rootstocks. Since Pistachio is a natural outbreeder much genetic variation exists in any seed populations. Therefore, Pistachio trees grown from seed obtained as a product of open pollination exhibit the continuous variation of almost all characteristics.

Micropropagation techniques, because of the potential they offer for the control of the environment and nutrient conditions, may be available to propagate individual trees vegetatively at a sufficient rate from clones for Pistachio plantations. Within the last two decades, these techniques have found increasing application for the rapid clonal multiplication of desirable Pistachio cultivars. *Pistacia vera* L. has been micropropagated from both juvenile and adult trees.

The results achieved with Pistachio are encouraging since the forced-mature explants have the capacity to form multiple shoots in large numbers (Onay 2000a). Pretreatments of axillary buds with BA prior to culture was effective for inducing multiple bud formation from the mature explants. Possibly the pretreatments confer a physiological status on the explants that is favourable for shoot bud formation. The time required to initiate the cultures varies with the type of explants and the age and physiological condition of the mother trees. Generally 4 weeks are required for bud induction. although adventitious buds are formed on the initial explants, further multiplication can be obtained via axillary shoots using shoot tips and nodal bud segments. Induction of axillary shoots usually requires a low concentration of the cytokinin BA. The rooting percentage for *in vitro* produced shoot tips is very high, 80%. The healthy plantlets were obtained ready to be transferred with a 50% survival rate. Several refinements in the techniques may be necessary. Before scaling up the process, the clonal fidelity of the micropropagated trees must be tested by extensive field trials.

Clonal propagation of Pistachio trees by somatic embryogenesis offers the benefit of producing large number of clonal propagules in bioreactors and rapidly commercial exploitation of somatic embryogenesis in *Pistacia vera* L. The recent achievements of embryogenesis from Pistachio leaf tissues (Onay 1996; Onay & Jeffre 2000; Onay & Namli 1998; Onay 2000c) may be a first step towards the realisation of such methods.. Limitations are imposed by the genotype, the age of the explants and cultural environments. However, somatic embryogenesis has been reported in several types of explant (Onay & Jeffree 2000). Non-synchron of somatic embryos, precocious germination and low frequency of somatic plantlets has limited the application of somatic embryogenesis for Pistachio improvement. Further attempts should be made to synchronise the production of somatic embryos and to initiate a bioreactor process. Besides, synthetic seed technology provides an easy means of handling large number of propagules during transport, and can greatly facilitate automated sowing under green house conditions. Encapsulation (or cryopreservation) of embryogenic masses, somatic embryos, synthetic seeds and regenerated plants has the potential for long term conservation and exchange of germplasm. Also, the basic methodology of alternative techniques such as protoplast culture may be established. Potential applications of protoplast fusion with Pistachio have not been studied. It also merits some priority as a research topic which could lead to further exploitation of rejuvenation effects with Pistachio. In the longer-term, the genetic improvement of this scarce plant in breeding programmes, producing commercial stocks for orchards should be considered.

4. SUMMARY

In this chapter the tissue culture of *Pistacia vera* L. has been reviewed, with a special attention paid to recent results. The vegetative propagation of *Pistacia vera* L., especially from mature trees, is still limited. Micropropagation of *Pistacia vera* L. has recently become better understood and the possibility of including this species among those in mass propagation is now closer to realisation.

5. ACKNOWLEDGEMENTS

I thank Professor Dr. M.M. Yeoman for critically reading the manuscript and Professor Dr. S.M. Jain for providing helpful suggestions.

6. REFERENCES

Abousalim A & Mantell SH (1992) Micrografting of Pistachio (*Pistacia vera* L.cv. `Mateur`). *Plant, Cell, Tissue and Organ Culture* **29**: 231-234

Abousalim A & Mantell SH (1994) A practical method of alleviating shoot-tip necrosis symptoms in *in vitro* shoot culture of *Pistacia vera* cv. `Mateur`. *J of Hort Sci*, **69**: 357-365

Abousalim A (1990) Micropropagation and Micrografting of Pistachio (*P. vera* L. and *Pistacia atlantica* Desf.) PhD Thesis, Department of Horticulture. Wye College, University of London, UK

Al Barazi Z & Schwabe WW (1982) Rooting softwood cutting of adult *Pistacia vera. J Hort Sci*, **57** (2): 247-252

Al Ramadhani RMA (1985) The effect of some growth regulator on the initiation and growth of *Pistacia vera* L. Mosul (Irag) Coll. of Science. Mosul Univ. (Irag).

Alderson PG & Barghchi M (1982) The potential for *Pistacia* propagation through shoot, leaf and callus culture. In: Fujiwara A (ed) Plant tissue culture 1982. Maruzen, Tokyo, pp 739-740

Assaf R (1971) Nouvelle methode de multiplication vegetative de rameaux physiologiquement adultes de pistachier, noyer et pacanier. *Fruits*, **32** (5): 309-319

Ayfer M (1963) Pistachio nut and its problems with special reference to Turkey. Univ. of Ankara Faculty of Agriculture Yearbook, 1963, pp: 189-217

Ayfer M (1990) Past, present and future of Pistachio nut. In: Türkiye 1. Antepfistigi Simpozyumu Bildirileri, 11-12 Eylül 1990- Gaziantep, pp: 14-23

Ayfer M, Okay Y & Erdogan V (1990) Pistachio rootstocks and their propagation. In: Türkiye 1. Antepfistigi Simpozyumu Bildirileri, 11-12 Eylül 1990- Gaziantep, pp: 38-48

Bailey LH (1947) *Pistacia. The standard Encyclopedia of Horticulture*, III, 2648-2650

Barghchi and Alderson (1982) *In vitro* propagation of *Pistacia* species. XXI International congress *in vitro* propagation of *Pistacia* species. Hamburg, p:1244.

Barghchi M & Alderson P. G.(1989) Pistachio (*Pistacia vera* L.). In: Y.P.S. Bajaj (ed.). Biotechnology in agriculture and forestry. vol. 5. Trees II, p.68-98. Springer-Verlag, Berlin, Heidelberg.

Barghchi M & Alderson PG (1983a) *In vitro* propagation of *Pistacia vera* L. from seedling tissues. *J Hort Sci*, **58**: 435-445

Barghchi M & Alderson PG (1983b) *In vitro* propagation of *Pistacia* species. *Acta Hort*, **131**: 49-60

Barghchi M & Alderson PG (1985) *In vitro* propagation of *P. vera* L. and commercial varieties of Ohadi and Kalleghoc. *J Hort Sci*, **60**: 423-440

Barghchi M & Martinelli A (1984) *In vitro* propagation of mature *Pistacia vera* varieties of Kerman (female) and Peter's (male) Pistachio. 41st Easter School Symp. Plant tissue culture and its agricultural applications. Univ. of Nottingham, Abst 75

Barghchi M (1982) *In vitro* propagation of *Pistacia* species. PhD Thesis, Nottingham University, UK

Barghchi M (1986a) *In vitro* micropropagation of *Pistacia* rootstocks. *Proc Int Plant Pro Soc*, **35**: 334-337

Barghchi M (1986b) *In vitro* culture of mature commercial varieties of *Pistacia vera* L. *Proc Int Plant Prop Soc*, **35**: 331-333

Barghchi M (1986c) Control of *in vitro* shoot tip necrosis in Pistachio (*Pistacia vera* L.). In: Plant Pyhsiology Division Biennial Report. p 52. Palmerston North Department of Scientific and Industrial Research (D.S.I.R)

Bloch & Brekke JE (1960) Processsing of Pistachio nuts. *Econ Bot*, **14**:129-144

Bustamante-Garcia MA (1984) Micropropagation and Rejuvenation of *Pistacia* species and the mechanism by which light influences root initiation. PhD Thesis, University of California, Davis, USA

Chatibi A, Kchouk ML & Ghorbel A (1997) Somatic embryogenesis in Pistachio (*Pistacia vera* L.) cv. Mateur. ISHS second international symposium on Pistachios and Almonds Presented on August 24-29 in Pistachio Absracts, in Davis, California, USA

Crane JC & Iwakiri BT (1981) Morphology and Reproduction of Pistachio. *Hort Review*, **13**: 376-393

Crane JC & Iwakiri BT (1985) Vegetative and reproductive apical dominance in Pistachio. *Hort Sci*, **20** (6): 1092-1093

Crane JC (1984) Pistachio Production Problems. *Fruit Var J*, **3**, (38): 74-85

Davatchi AG (1958) Etrude Biologuede la Fauna Entelogique des *Pistacia* sauvages et cultivars. Extr Rev Pathol Veg Entomol Agric France XXX,VII

Dolcet Sanjuan R & Claveria E (1995) Improved shoot-tip micropropagation of *Pistacia vera* L. and the beneficial-effects of methyl jasmonate. *J Amer Society Hort Sci* **120** (6): 938-942

Firat M.Z. & Onay A (1999) Analysis of binominal data obtained from plant-tissue culture studies using generalised linear models. *Turkish Journal of Biology*. Vol. **23**: 261-267

Gebhardt K & Goldbach H (1988) Establishment, graft union characteristics and growth of *Prunus* micrografts. *Physiologia Plantarum*, **72**: 153-159

Ghorbel A, Chatibi A & Kchouk ML (1997) Multivariate analysis of the role of inorganic elements on micropropagation of Pistachio (*Pistacia vera* L.) cv. Mateur. ISHS Second International Symposium on Pistachios and Almonds Presented on August 24-29, 1997 in Davis, California, USA

Gonzales A & Frutos D (1990) *In vitro* culture of *Pistacia vera* L. embryos and aged tree explants. NATO ASI series A. **186**: 335-338

Hackett WP (1985) Juvenility, maturation and rejuvenation in woody plants. *Horticultural Reviews*, **7**: 109-155

Hall J (1975) Propagation of walnuts, almonds and Pistahios in California. *Proc Inter Plant Prop Soc*, **25**: 53-57

Hansman D & Owens y de Novoa C (1986) Micropropagation of temperate nut trees. *Hort Abstr*, No **6**: 403-416

Hartmann HT & Kester DE (1983) Plant propagation: principles and practices. 4th ed. Englewood, New Jersey, USA; Prentice Hall. 727 pp. ISBN 0-13-681007-1

Joley L.E. (1979) Pistachios. In: R.A.Jaynes (ed). Nut tree culture in North America. Northern Nut Growers Assoc, pp163-174, Broken Arrow Road, Hamden, Conn.

Joley LE & Opitz KW (1971) Further experiments with propagation of *Pistacia*. *Proc Inter Plant Prop Soc*, **21**: 67-76

Joley LE & Whitehouse WE (1953) Root knot nematode susceptibility. A factor in the selection of Pistachio nut rootstocks. *Proc Amer Soc Hort Sci*, **61**: 99-102

Joley LE (1960) Experiment with propagation of the genus *Pistacia*. *Proc Inter Prop Soc*, **10**: 287-292

Kaska N, Ak BE & Nikpeyma Y (1990) Application of chip, patch and spring-fall shield budding on different *Pistacia* species. In: Turkiye 1. Antepfistigi Simpozyumu Bildirileri, 11-12 Eylül 1990- Gaziantep, pp: 59-67

Lemaister J (1959) Le Pistacher (Etude bibliographique). *Fruits*, **14**: 57-77

Moldenke H N, Alma L (1952) Plants of Bible. Chronica Botanica

Murashige T, Skoog F (1962) A revised medium for rapid growth and bioassays with tobacco tissue culture. *Physiol Plant*, **15**: 473-497

Navarro L, Roistacher CN & Murashige T (1975) Improvement of shoot tip grafting *in vitro* for virus-free *Citrus*. *J Amr Soc Hort Sci*, **100**: 471-479

Onay A & Namli S (1998) Antepfistiginin (*Pistacia vera* L.) yapraklarinda somatik embriyogenesis. XIV. Ulusal Biyoloji Kongresi, Samsun, 7-10 Eylül 1998, Bitki Fizyolojisi-Bitki Anatomisi ve Hidrobiyoloji Seksiyonu, Cilt 2: 44-53, Samsun

Onay A (1996) *In vitro* organogenesis and embryogenesis of Pistachio, *Pistacia vera*, L. PhD Thesis, University of Edinburgh, UK

Onay A (2000a) Micropropagation of Pistachio from mature trees. *Plant Cell, Tissue and Organ Culture*, **60**: 159-162

Onay A (2000b) Somatic embryogenesis in cultured kernels of Pistachio, *Pistacia vera* L. cv Siirt. Second International Balkan Botanical Congress. Istanbul, May 14-16, 2000, (in press)

Onay A (2000c) Histology of somatic embryo initiation and development in Pistachio (*Pistacia vera* L.). *Turkish Journal of Botany*, **24**: 91-95

Onay A, Firat MZ & Namli O (1997) An improved method for embling production in Pistachio, *Pistacia vera* L. using liquid medium matured somatic embryos. *Turkish Journal of Biology*, **21**, (2): 159-174

Onay A, Jeffree C.E. (2000) Somatic Embryogenesis in Pistachio. In: Somatic embryogenesis in woody plants / edited by S.Mohan Jain, Pramod K. Gupta, Ronald J. Newton (Forestry Sciences). Chapter 10, Section B, Vol: **6**: 361-390.

Kluwer Academic Publishers, The Netherlands

Onay A, Jeffree C.E., Theobald C.,Yeoman M.M. (2000) Analysis of the effects of maturation treatments on the probabilities of somatic embryo germination and embling development in Pistachio, *Pistacia vera L.* using a logistic regression method. *Plant Cell, Tissue and Organ Culture*. Vol **60** No 2: 121-129

Onay A, Jeffree CE & Yeoman MM (1995) Somatic embryogenesis in cultured immature kernels of Pistachio, *Pistacia vera* L. *Plant Cell Reports*, **15**: 192-195

Onay A, Jeffree CE & Yeoman MM (1996) Plant regeneration from encapsulated embryoids and an embryogenic mass of Pistachio. *Plant Cell Reports*, **15**: 723-726

Özbek S & Ayfer M (1958) A hermaphrodite *Pistacia* found in the vicinity of Antep, Turkey. *Proc Amer Hort Sci*, **72**: 240-241

Pair JC & Khatamian H (1982) Propagation and growing of the Chinese Pistachio. *Proc Inter Plant Prop Soc*, **32**: 497-503

Parfitt DE & Almehdi A (1994) Use of high CO_2 atmosphere and medium modifications for the succcesful micropropagation of Pistachio. *Sci Hort*, **56**: 321-329

Pontikis CA (1977) Contribution to studies on the resistance to some hybrids and species of the genus *Pistacia* (as rootstocks) to Phytophthora spp. Nea Agrotiki Epitheorisi, 1:14-16. *Hort Abst*, **47**: 9199

Sakoury K (1976) Propagation of Pistachio by cuttings. MSc Thesis. California State University, Fresno

Tabatabai M (1966) Pistachio and its importance in Iran. Tehran. Publication of the Iranian Forestry Service

Takeda F, J.C. Crane & J Lin (1979) Pistillate flower bud development in Pistachio. *J Amer Soc Hort Sci*, **104**: 229-232

Taskin T (1995) The researches of production of regenerated plants by tissue culture techniques in the *in vitro* conditions from different tissues and organs of *Pistacia vera* L. PhD Thesis, Department of Biology, Graduate school of natural and applied sciences, University of Firat, Turkey

Taskin T, Çobanoglu D, Çolak G, Namli S, Basaran D & Gaspar T (1996) plant regeneration through somatic embryogenesis in *Pistacia* vera. *Soc Bot*, **65**: 59-65

Turan K (1990) Studies on Pistachio diseases and control measures.In: In: Türkiye 1. Antepfistigi Simpozyumu Bildirileri, 11-12 Eylül 1990- Gaziantep, pp: 167-176

Whitehouse WE (1957) The Pistachio nut-A new crop for the Western United States. *Econ Bot*, pp: 281-321

Woodroof JG (1979) Tree nuts, Production, Processing, Products. Vol. II, 2nd Edit. AVI, West port Connecticut

Yalpani M & Tyman JHP (1983) The phenolic acids of *Pistacia vera*. *Phytochemistry*, Vol **22** No 10: 2263-2266

Yang Z & Lüdders P (1993) *In vitro* propagation of Pistachio (*Pistacia vera* L.). *Gartenbauwissenschaft*, **59** (1): 30-34

Yücel S, Onay A, Colak G, Basaran D (1991) The researches of obtaining from *Pistacia vera* L. apical bud and nodal bud by micropropagation. *Acta Hort*, **289**: 167-168

Zaheer A, Nasreen A & Shah FH (1989) Callus formation from the mesocarp tissues of *Pistacia vera* L. *Pak J Sci Ind Res*, Vol **32**, No 8: 549-550

Zohary M (1952) A monographical study of the genus *Pistacia*. *Pal J Bot*, **5**: 187- 228

20. Micropropagation of *Citrus*

Francesco Carimi and Fabio De Pasquale
Istituto di Ricerca per la Genetica degliAgrumi, CNR, Viale delle Scienze, 11, I-90128 Palermo, Italy

S.M. Jain and K. Ishii (eds.), Micropropagation of Woody Trees and Fruits, 589–619.

1. INTRODUCTION

The genus *Citrus* is cultured in more than 100 countries making it one of the most important commercial fruit crops in terms of economic value and human nutrition (Barlass and Skene, 1986). Fresh fruits and juice are the most significant citrus products, but essential oils, pectin, and marmalade, as well as candied and dried rinds, also have commercial value (Barlass and Skene, 1986). In recent years, the importance of *Citrus* for the production of ornamental plants has increased considerably (De Pasquale and Carimi, 1998).

Conventional methods for *Citrus* propagation are based on budwood selection and grafting for scion varieties. Rooted cuttings, or more frequently nucellar seed propagation, are used for preparation of rootstocks (Barlass and Skene, 1986). The importance of the citrus industry and the continuous introduction of new, improved genotypes emphasize the use of modern methods to rapidly propagate new and promising plant material. Many *Citrus* genotypes are apomictic and polyembryonic (Iglesias et al., 1974); in such genotypes, adventitious embryos can arise from the cells of the nucellus of the mother plant and develop together with the zygotic embryos (Carimi et al., 1998a). This event produces a variable percentage of nucellar offspring that is genetically identical to the female parent (Iglesias et al., 1974). Standard commercial citrus rootstocks are propagated by growing open-pollinated seeds, while many rootstocks are highly polyembryonic and reproduce true-to-type via nucellar embryony (Moore, 1986). This method of propagation is not appropriate for genotypes with low levels of polyembryony, since they produce low percentages of nucellar seedlings. In these cases, zygotic seedlings must be identified on the basis of morphology and appropriately eliminated in order to maintain clonal uniformity. In some cases, rootstock seeds are not available, while some potentially interesting rootstocks have low levels of polyembryony and do not produce adequate quantities of nucellar seedlings (Grosser and Chandler, 1986). The propagation of somatic hybrid rootstocks, especially those with trifoliate orange parentage, are amenable to micropropagation (Grosser et al., 2000). In situations where established orchards are being grafted over new cultivars with limited availability, micropropagation techniques could quickly supply budwood. Rapid and cost-effective *in vitro* protocols for propagation of citrus rootstocks would be of great interest in this regard. The production of plants would not be limited by seed supply and more uniform plant material could be produced as a result (Moore, 1986).

2. SANITARY ASPECTS

Among fruit species, *Citrus* crops are most affected by pathogens transmitted through vegetative propagation material (Roistacher, 1991). Since diseases caused by virus and virus-like agents cannot be efficiently controlled by chemical treatments, prevention is the major means of control (Savino and D'Onghia, 1998). Prevention encompasses the application of certification, which includes a set of legal and technical procedures that guarantee the quality of the final propagating plant material derived from selected source plants. These plants must be individually controlled for the presence of virus as well as for their pomological traits (Roistacher, 1993; Savino and D'Onghia, 1998).

Murashige and co-workers have developed an *in vitro* procedure (termed shoot tip grafting) in which an excised apical dome is inserted into an inverted-T incision on a decapitated dark-grown seedling rootstock (Murashige et al., 1972; Navarro et al., 1975). This method allows for the production of *Citrus* plants that are free of virus and virus-like diseases (Roistacher et al., 1976).

Since somatic embryogenesis can be used to eliminate many viral diseases (Bitters et al., 1970; D'Onghia et al., 1997, 2000, 2001), the plant material obtained by somatic embryos regenerated *in vitro* can be used to establish healthy citrus stocks. Somatic embryogenesis from stigma and style culture is a new technique for *Citrus* sanitation and has proven to be highly effective in the complete elimination of psorosis, tristeza, infectious variegation, concave gum, impietratura, cristacortis, exocortis and cachexia (D'Onghia et al., 2000). This technique involves the use of different explant types (thin cell layers of stigma and style or entire organs) excised from citrus flower tissues, which are not ovular in origin and regenerate plantlets genetically identical to the original source (De Pasquale et al., 1994; Carimi et al., 1995, 1998b, 1999; Fiore et al., 2002). For induction of cell proliferation, explants are cultured in Petri dishes containing Murashige and Skoog (MS) (1962) semisolid medium (7 g/L agar) supplemented with 50 g/L sucrose and 13 μM 6-benzylaminopurine (BA). Somatic embryos are formed 2-5 months at the surface of the callus after cultures are initiated. Germinated embryos are transferred to test tubes or Petri dishes containing MS medium and exposed to a 16 hour photoperiod to induce plant development at 27 °C. After 1-2 months of incubation, regenerated plants are grafted on Troyer citrange seedlings growing in soil (De Pasquale et al., 1999). The entire grafted plants are individually inserted into a polyethylene bag containing 5 ml of distilled water, sealed with adhesive tape and incubated for 2 months in a climate chamber at 27 ± 1 °C, under a photosynthetic photon flux of 100 μmol m^{-2} s^{-1}. About 2 months after grafting, the grafted plants are transferred to the screenhouse and exposed to natural day light conditions at 22-27 °C night/day. Holes of increasing size are made in the polyethylene bags to gradually reduce humidity levels and about 4 weeks later the polyethylene bags are removed (De Pasquale et al., 1999). Virological analyses showed that the callus formed on the surface of infected stigmas and styles proved to be highly infective (D'Onghia et al., 1997, 2000, 2001) confirming previous results on callus generated from different explants (Navas Castillo et al., 1995). Nevertheless, all plants regenerated trough somatic embryogenesis from stigma and style culture were free from psorosis, tristeza, infectious variegation, oakleaf pattern complex, cachexia and exocortis (D'Onghia et al., 1997; 2000, 2001) confirming previous results on somatic embryos regenerated from different type of explants (Bitters et al., 1970). In conclusion, the regeneration of somatic embryos from stigma and style cultures can be successfully applied to the production of true-to-type germplasm, free from a number of graft transmissible pathogens. The regenerated plants are able to fruit in a relatively short time (3-4 years from the culture initiation) (D'Onghia et al., 2000). In addition to the establishment of *in vivo* healthy germplasm collections, this technique allows the *in vitro* maintenance of healthy regenerated plants and it appears to have promise as a complementary approach to the conservation of citrus genetic resources in the field (D'Onghia et al., 2000). Moreover, as long as the international demand for *Citrus* is high and continues to rise, *in vitro* conservation of plant materials regenerated by somatic embryogenesis is very promising for the establishment of the safe international exchange of citrus germplasm (D'Onghia et al., 2000).

3. JUVENILITY

During the juvenile phase, flowering does not occur. The transition from the juvenile to the mature phase can be influenced by environmental and genetic factors (Hackett, 1985). In *Citrus*, this transition varies from species to species (Furr et al., 1947) and usually 4-7 years are sufficient to overcome this phase. However, Cervera et al. (1998) reported that sweet oranges may require up to 20 years in order to lose their juvenile character. Early fruit production is very important in *Citrus* and this is one of the reasons for commercial propagation systems of scion varieties based on grafting adult material on rootstocks.

Maturation seems to be responsible for the decline of the regenerative potential of woody plants *in vitro* (Durzan, 1990). In *Citrus*, the best results (in terms of rapid rates of proliferation) are normally obtained using stock material in the juvenile phase as an explant source. In fact, this is the reason why explants derived from juvenile tissues (such as nucellar embryos, somatic embryos or young seedlings) are used as starting material for micropropagation of polyembryonic *Citrus*. Sim et al. (1989) and Cervera et al. (1998) have reported that explants collected from juvenile citrus plants give the best regeneration frequency in plant tissue culture as compared to explants collected from adult plants. Unfortunately, plants regenerated from juvenile tissues usually exhibit strong, undesirable juvenile characters. It has been shown that collecting explants from mature plants provides the means to overcome this problem. Altman and Goren (1974) regenerated shoots from lateral buds collected from a mature plant of sweet orange and Barlass and Skene (1982) regenerated plantlets from nodal segments of mature plants of Cleopatra mandarin, Carrizo citrange, Rangpur lime and sweet orange. Singh et al. (1994) have also regenerated plants from shoot tips collected from mature plants of mandarin and lemon, based on morphological characters (thorniness, growth habit and internode length) the established plants did not appear to be juvenile. Starrantino and Caruso (1988) propagated *in vitro* citrus rootstocks (Flying Dragon, Troyer and Carrizo citrange) using nodal stem segments excised from 4-year-old plants and reported that the multiplication rates remained constant for a long period (5-6 years). The same genotypes propagated *in vitro* from shoot tips maintained a constant multiplication and rooting capacity after three years of subculturing (Starrantino and Caruso, 1987).

D'Onghia et al. (2001) reported that plants regenerated by somatic embryogenesis from style and stigma cultures exhibited initial juvenile growth, characterized by the presence of thorns on stems and branches and the absence of flowers. Nevertheless, within 12 months, the vast majority of the plants developed thornless apical growth. Plants multiplied from thornless portions of vegetation by top grafting onto sour orange rootstock completely lost their juvenile character and began fruiting on the same branches three years after the embryogenic event (D'Onghia et al., 2000; D'Onghia et al., 2001).

Several methods have been described that shorten the juvenile period in *Citrus* (Snowball et al., 1994a; Snowball et al., 1994b; Krajewski and Rabe, 1995). Snowball et al. (1994a) reported that plants pre-grown under greenhouse conditions (24 months) and transferred to natural outside conditions were able to flower within 30 (seedlings of lime, lemon and mandarin) or 48 (seedlings of sweet orange, grapefruit and trifoliate orange) months of germination. Application of paclobutrazol generally increases the probability of the seedlings to flower. It has also been observed that that seedlings of *C. aurantifolia* treated with paclobutrazol (30 mg per plant) flowered within 11 months following germination (Snowball et al., 1994b).

Recently, it has been reported that juvenile seedlings of Troyer citrange (*C. sinensis* X *Poncirus trifoliata*) transformed with *APETALA1* (*AP1*) gene had a very short juvenile phase (Peña et al., 2001). The expression of *AP1*, which promotes flower initiation in *Arabidopsis thaliana*, did not produce any severe developmental abnormalities in *Citrus* and the *35S:AP1* transgenic plants produced fertile flowers and fruits as early as the first year. Furthermore, the zygotic and nucellar seedlings resulting from self-pollination of *35S:AP1* transgenic plants showed a reduction in thorns and flowering in their first spring. A more complete understanding of the effect of *AP1* on the shortening of citrus juvenile phase could help to provide an efficient method of micropropagation for *Citrus* (Peña et al., 2001).

4. PROPAGATION THROUGH SOMATIC EMBRYOGENESIS

The main factors influencing somatic embryogenesis are genotype, composition of the culture medium, and the type/developmental stage of the explant. Generally, explants from juvenile tissues are more inclined towards embryogenesis with respect to those from mature tissues. Many *Citrus* genotypes have been regenerated by somatic embryogenesis from explants derived from different plant tissues. Nevertheless, considerable differences in somatic embryogenic potentials exist. Since many *Citrus* are polyembryonic, and as adventive embryos are generated *in vivo* from the nucellar tissue, several studies on embryogenesis have been performed with this tissue. Mitra and Chaturvedi (1972) reported a direct correlation between the degree of *in vivo* polyembryony and the ability to regenerate somatic embryos *in vitro*.

Somatic embryogenesis, from either ovular tissues or embryos developed from ovules, has been extensively investigated in *Citrus*. Maheshwari and Rangaswamy (1958) first reported *in vitro* regeneration of embryos from nucellus cultures. Subsequently, several reports appeared regarding improvements of somatic embryogenesis protocols. The regeneration of somatic embryos has been obtained from excised nucelli (Rangan et al., 1968), abortive ovules (Bitters et al., 1970), unfertilized ovules (Button and Bornman, 1971), undeveloped ovules (Starrantino and Russo, 1980; Moore, 1985), and isolated nucellar embryos (Litz et al., 1985). However, there are only a few reports on somatic embryogenesis from somatic tissue, which is neither nucellar nor ovular in origin. Hidaka et al. (1981) regenerated diploid plants from anthers of sour orange. Nito and Iwamasa (1990) obtained somatic embryos from cultures derived from satsuma juice vesicles, and Carimi et al. (1994, 1995) developed embryogenic cultures from stigmas and styles of lemon, grapefruit, mandarin, sour orange and sweet orange. Gill et al. (1995) obtained somatic embryos from epicotyl, cotyledon, leaf and root segments of *in vitro* grown nucellar seedlings of mandarin. Recently, embryogenic potential was demonstrated in six different *Citrus* species from cultures of pistil transverse thin cell layer explants (Carimi et al., 1999).

Unfortunately, somatic embryo induction was low from the nucellar tissues of monoembryonic *Citrus* genotypes. Rangan et al. (1968) first reported the direct induction of somatic embryos from nucellus of fertilized ovules of monoembryonic *Citrus*. Button and Kochba (1977), Kobayashi et al. (1981) and Kobayashi et al. (1982) reported that somatic embryogenesis from unfertilized ovules of monoembryonic genotypes was unsuccessful. Carimi et al. (1999) observed that stigma and style transverse thin cell layer explants could optimize the regeneration of somatic embryos in a monoembryonic species such as citron. Usually, the embryogenic calli of *Citrus* can maintain their embryogenic competence for long periods (5-10 years).

Sucrose is the most common carbon source in citrus plant tissue and cell culture media. High concentrations (50 g/L) of sucrose are usually used to induce the formation of embryogenic callus. In some cases, omission of sucrose from the culture medium for a single culture passage greatly stimulated embryogenesis when cultures were subcultured back onto sucrose containing medium (Kochba and Button, 1974). Several studies have investigated the effect of different carbon sources on citrus somatic embryogenesis. Kochba et al. (1978b) reported that it is possible to stimulate somatic embryogenesis in *C. sinensis* by changing the carbon source from sucrose to galactose. Ben-Hayyim and Neumann (1983) reported that glycerol (2% w/v) as the sole carbon source has stimulatory effects on citrus somatic embryogenesis. Jiménez et al. (2001) analyzed the initial changes in the endogenous hormone levels of *Citrus* habituated cultures treated with glycerol or sucrose. It was found that incubating the cells on a medium containing glycerol promotes the expression of somatic embryogenesis; higher levels of auxin and cytokinin in glycerol treated cultures was also observed. Cytokinin BA is essential for the regeneration of citrus somatic embryos. The use of 13.3 μM BA stimulated the embryogenic response from stigmas and styles (De Pasquale et al., 1994), while Moore (1985) found that levels of BA higher than 0.44 μM suppressed embryogenesis in ovules. In some cases, BA in the culture medium was not necessary to induce somatic embryo formation; Starrantino and Russo (1980) obtained somatic embryogenesis from undeveloped ovule of *Citrus* cultured on MS medium supplemented with malt extract (0.5 g/L). Kochba et al (1978a) reported that the addition of low levels of ABA to the culture medium stimulated somatic embryogenesis in cultures from ovular tissues of *C. sinensis*. In *Citrus* cultures, the addition of ethephon (an ethylene donor) to the culture medium can also stimulate somatic embryogenesis in some cases (Kochba et al., 1978a).

A typical procedure for the induction of embryogenic cultures involving the use of undeveloped ovules (Starrantino and Russo, 1980; Moore, 1985; Carimi et al., 1998b) is described below. Open pollinated fruits are harvested 3-9 months after anthesis and washed with water. The skin is peeled off and the fruits are surface-sterilized by immersion for 5 min in ethanol (70% v/v in water) followed by 15 min in 2% (w/v in water) sodium hypochlorite. Without rinsing, the fruits are cut open under sterile conditions. Undeveloped ovules are then dissected and transferred to sterile Petri dishes. Ovule integuments are removed with the aid of a stereo microscope and placed onto MS semisolid medium (7 g/L agar) supplemented with 50 g/L sucrose, 0.5 g/L malt extract in presence or absence of 13 μM BA. The somatic embryos usually appear after 2-4 weeks of culture, but in some cases ovules can differentiate somatic embryos after 3-4 months. Individual somatic embryos (about 2 mm in diameter) are isolated and germination is attempted in MS semisolid medium (7 g/L agar) supplemented with 50 g/L sucrose and 0.5 g/L malt extract. When plantlets are 5-7 cm length, they are washed in distilled sterile water and planted into autoclaved plastic pots containing a 1:1 (v/v) mixture of peat-moss:sand. Plantlets are maintained for 2 months under 95±5% relative humidity in a climate chamber at 27 °C. Later on, plantlets can be transferred to the greenhouse and humidity levels are gradually reduced. The percentage of embryogenic cultures usually is normally between 20 and 70% depending on genetic differences of the material. Usually, this regeneration method does not work with monoembryonic genotypes. In fact, Moore (1985) reported that none of the cultured undeveloped ovules of the 5 monoembryonic cultivars tested became embryogenic. In some cases (grapefruit), the addition of low concentrations (0.062 μM) of 2,2-

dimethylhydrazide (daminozide) greatly increased the induction of somatic embryogenesis from undeveloped ovules (Moore, 1985).

An alternative protocol for the induction of embryogenic cultures involving the use of stigma and style culture (De Pasquale et al., 1994; Carimi et al., 1995, 1998b) is described below. Flowers are collected before opening and surface-sterilized by immersion for 2 min in ethanol (70% v/v in water), 15 min in 2% (w/v in water) sodium hypochlorite, followed by one 5-min rinse in sterile distilled water. Stigma and style explants are excised with a scalpel, and placed vertically onto plastic Petri dishes with the cut surface in contact with the medium. Explants are maintained in MS semisolid medium (7 g/L agar) supplemented with 50 g/L sucrose, 0.5 g/L malt extract and 13 μM BA. Explants and calli are subcultured at 30-day intervals. Usually, the somatic embryos appear after 2-3 months of culture on the surface of the callus, but in some cases somatic embryos are differentiated after 5-7 months of culture. The procedures for germination of somatic embryos and transfer of plantlets to soil are the same described for somatic embryogenesis from undeveloped ovules. The percentage of embryogenic cultures usually is normally between 2 and 58%. In conclusion, the culture from stigmas and styles have been successfully performed by regenerating different genotypes of 14 *Citrus* species (D'Onghia et al., 2000). Among the tested genotypes, only the clementine was not able to regenerate somatic embryos (D'Onghia et al., 2000). A strong influence of the explant type (stigma and style), hormonal combination, and genotype on embryo regeneration was observed. Usually, lemons, sweet oranges (the navel group in particular), and tangerines show the higher regeneration potential, whereas mandarins and grapefruit show a lower percentage of embryo regeneration. In some cases, a different regeneration aptitude was also observed in different genotypes of the same species (D'Onghia et al., 2000).

High somatic embryogenic rates can be achieved in suspension cultures, making this technique ideal for large-scale automated micropropagation of healthy material (Cabasson et al., 1997). Nevertheless, only a limited number of studies have focused on the establishment of protocols for *Citrus* somatic embryo development in liquid culture (Cabasson et al., 1995, 1997; Gavish et al., 1991; Vu et al., 1993). The rate of growth and differentiation of liquid citrus cultures is much slower than other plants (Gavish et al., 1991). In carrot, 14 days after embryogenesis release, embryos are at the torpedo stage, whereas in *Citrus*, during this time frame, only globular somatic embryos are differentiated (Gavish et al., 1991). Cabasson et al. (1995) studied the effect of different carbon source (sucrose and galactose) on somatic embryogenesis in cell suspension cultures of *C. deliciosa*. Embryogenesis was started by filtering cell suspension cultures and collecting the fraction between 100 and 500 μm. The cells were resuspended (4 g/L) in basal Murashige and Tucker (MT) (1969) liquid medium containing sucrose or galactose (0.15 M). Cultures were maintained in a climate chamber on a horizontal rotary shaker (100 rpm) at 27 °C under a 16 h day length (27 μmol m^{-2} s^{-1}). Histocytological analyses revealed proliferation of young globular embryos during the first half of the 18-day culture cycle, independently of the presence of carbon source. However, after the first week of liquid culture on sucrose, suspensions were composed of embryogenic cell clusters that subsequently developed into young globular embryos, and further embryo development failed completely. In contrast, when cells were cultured on galactose, the globular embryos developed during the second week of culture, but they rarely reached the cotyledonary stage when maintained in liquid medium. Regeneration of plantlets was obtained after plating embryos onto solid medium Cabasson et al. (1995). A procedure for the induction of *Citrus* somatic embryogenesis in liquid cultures involving temporary immersion of the cultures (Cabasson et al., 1997) is described below. Cells are routinely subcultured in MT medium (hormone free) supplemented with 0.15 M sucrose and 0.5 g/L malt extract. In order to initiate a subculture cycle, 800 mg of packed cell volume are incubated into 150 mL Erlenmeyer flasks containing 40 mL of liquid medium. Cells are subcultured in fresh medium at 15-day intervals. Somatic embryogenesis is promoted by inoculating 0.1 g of cells per liter of liquid medium. Early globular somatic embryos are visible after 20 days of culture. Subsequently, in order to promote the development of somatic embryos, embryogenic cultures are temporarily immersed (1 min every 4 h) in MT medium supplemented with 0.1 M sorbitol and 0.1 M galactose. Cells and embryos are subcultured in fresh medium at 30-day intervals. More than 60% of somatic embryos reach the cotyledonary stage after 80 days of culture. Somatic embryos at the cotyledonary stage are transferred on a germination medium: MT semisolid medium (8 g/L agar) supplemented with 0.1 M sorbitol, 0.1 M galactose and 1 μM GA_3. In conclusion, this method gives a high number of somatic embryos (6700 *C. deliciosa* somatic embryos per g of fresh weight in 80 days of culture, representing 840 plantlets per gram of fresh weight). However, the somatic embryos regenerated with these temporary immersion cultures have a low germination potential (Cabasson et al., 1997). Cabasson et al. (1997) suggested that further improvements on the regeneration of *Citrus* somatic embryogenesis in liquid cultures must involve more precise definition of the times and frequencies involved in their culture.

5. ORGANOGENESIS

Several reports have appeared on organogenesis from different type of explants of *Citrus* and citrus rootstocks. The morphogenic responses of *Citrus* cultured *in vitro* are influenced by genotype, explant type, and the culture medium. Explants have included shoot tips (Barlass and Skene, 1986), stem sections (Grinblat, 1972; Chaturvedi and Mitra, 1974; Raj Bhansali and Arya, 1979; Barlass and Skene, 1982), root sections (Sauton et al., 1982; Burger and Hackett, 1986; Sim et al., 1989; Bhat et al., 1992), leaf sections (Chaturvedi and Mitra, 1974, Hu and Kong, 1987), stem internodes

(Duran-Vila and Navarro, 1989), epicotyl segments (Edriss and Burger, 1984), and transverse thin cell layer (tTCL) explants excised from stem internodes (Van Le et al., 1999). The regeneration of adventitious shoots has been obtained either directly from the explant or from an intermediate callus phase. The regeneration of *Citrus* and related species by organogenesis has been described in many reports (Table 1). While the literature cited is limited to the last twenty years, the reader may refer to the manuscripts of Skirvin (1981) and Barlass and Skene (1986) for earlier references.

Marin and Duran-Vila (1991) proposed a micropropagation protocol for the conservation of citrus germplasm *in vitro*. This method was based on the establishment of primary cultures in order to recover *in vitro* plants from nodal stem segments (dissected from 1-2-year-old plants) and the maintenance of successive cycles of secondary cultures with the recovery of rooted plantlets in each subculture cycle. The subculture cycle would last 8-12 months and would involve several steps (culture of nodal segments, rooting of shoots, and elongation of rooted shoots). This protocol permitted the long-term maintenance of lemon, Mexican lime, sweet orange and trifoliate oranges; *in vitro* regenerated plants can be maintained for up to 12 months on the same medium before subculture to fresh medium (Marin and Duran-Vila, 1991).

Recently, Van Le et al. (1999) developed a new efficient regeneration method leading to a high frequency of plant regeneration by using tTCL explants excised from the stem internodes of 1-year-old trifoliate orange (*P. trifoliata*). The optimal combination of growth regulators for shoot regeneration was obtained with MS medium supplemented with 10 μM BA and 1 μM N-phenyl-N'-1,2,3-thidiazol-5ylurea (thidiazuron, TDZ) and led to 90% responsive tTCL, forming an average of 37 buds. The highest percentage (100%) of shoot elongation was obtained onto MS medium containing 1 μM gibberellic acid (GA_3), while 60% of shoots formed roots on MS medium amended with 5 μM α-naphthaleneacetic acid (NAA) with 100% survival. The plants regenerated from tTCL had no morphological alterations. This may be due to the direct regeneration of plants from the explant without an intermediate callus phase and a subsequent reduction in somaclonal variation (Van Le et al., 1999). The high efficiency of bud formation could be due to faster transport of nutrients and growth regulators to tTCL explants. Moreover, a higher percentage of responsive cells can be directly exposed to the medium (Van Le et al., 1999).

A typical procedure for plant regeneration involving the use of shoot tips derived from mature plants of *Citrus* (Singh et al., 1994) is described below. Growing shoot tips (5-6 cm long) are collected from plants growing in the field and stripped of leaves; explants are washed with distilled water, surface disinfected for 15 min in 2% (w/v in water) calcium hypochlorite, and rinsed three times with sterile distilled water. Shoot tips (2-3 mm long) are dissected under aseptic conditions and cultured on MS semisolid medium (8 g/L agar) supplemented with 30 g/L sucrose, 4.5 μM BA, 2.3 μM 6-furfurylaminopurine (kinetin, KIN) and 2.2 μM NAA to induce shoot proliferation. After 7 weeks of culture, the regenerated shoots (about 2 cm in length) are cut and transferred to MS medium supplemented with 30 g/L sucrose, 1.1 μM BA, 2.5 μM indole-3-butyric acid (IBA) and 2.7 μM NAA to induce the formation of roots. After 5 weeks, the rooted shoots are transferred to MS inorganics semisolid (10 g/L agar) medium. After 8 weeks, the plantlets are washed with water (to remove agar from the roots), soaked in fungicide, and potted in a sterilized 1:1 (v/v) mixture of sand:soil. The plants are irrigated with MS solution for 1 week and subsequently with water. They are acclimatized at 27±2 °C with a 16 h photoperiod for 3-4 weeks. Polyethylene covers are used to ensure high humidity levels. About 60% of the plants are established in soil. Once established, plants are then transplanted to the field and watered regularly (Singh et al., 1994).

An alternative protocol for plant regeneration involving the use of tTCL explants excised from trifoliate orange stem internodes (Van Le et al., 1999) follows. Twelve month-old plants growing aseptically *in vitro* on MS medium are used as source plant material. tTCL 0.2-0.3 mm thick are excised from the youngest subapical stem internodes and cultured on MS semisolid medium (10 g/L agar) supplemented with 30 g/L sucrose, 10μM BA and 1 μM TDZ. One week after culture, bud primordia arise directly from the surface of the tTCL. The tTCL having clumps of regenerated buds are then transferred to MS medium supplemented with 30 g/L sucrose and 1 μM GA_3 for shoot elongation. When shoots reach 1-2 cm in length (about 4-5 weeks), they are dissected and transferred to MS medium supplemented with 30 g/L sucrose and 5 μM NAA to induce root formation. On this medium, about 60% of shoots generate roots after 4 weeks of incubation. Rooted shoots are transferred to soil in the greenhouse 13 weeks after tTCL culture initiation (Van Le et al., 1999).

Table 1. Regeneration of *Citrus* and related species by organogenesis

Species	Common name	Explant	Medium for callus	Medium for shoot	Medium for root	Reference
C. acida	Wild orange	Epicotyl	MS + 4.5 μM BA	MS + 4.5 μM BA + 5.8 μM GA_3	MS + 5.4 μM NAA	Chakravarty & Goswami, 1999
C. assamensis		Shoot tips		MS+22μM BA	Soirlite	Baruah et al., 1996a-b
C. aurantifolia	Mexican lime	Nodal stem segments		MS + 0.4 μM BA	MS + 16.1 μM NAA	Duran-Vila & Navarro., 1989
C. aurantifolia	Mexican lime	Internodal stem segments		MS + 33.3 μM BA + 5.4 μM NAA	½ MS + 2.7 μM NAA	Pérez-Molphe-Balch & Ochoa-Alejo 1997
C. aurantium	Sour orange	Internodal seedlings stem sections		MT + 22 μM BA + 5.4 μM NAA	½ MT + 5.4 μM NAA	Moore, 1986
C. aurantium	Sour orange	Mature embryos	MT + 9 μM 2,4-D	MT + 44.4 μM BA + 5.4	MT + 5.4 μM NAA	Beloualy, 1991

Species	Common name	Explant	Medium for callus	Medium for shoot	Medium for root	Reference
			+ 22.2 μM BA	μM NAA		
C. grandis	Pummelo	Epicotyl segments		MS + 2.2 μM BA	MS + 2.5 μM IBA	Goh et al., 1995
C. grandis	Pummelo	Root segments		MS + 0.089 μM BA	MS + 2.5 μM IBA	Goh et al., 1995
C. grandis	Pommelo	Shoot-tip		MS + 1.8 μM BA	½ MS + 5.4 μM NAA	Paudyal & Haq, 2000
C. halimii		Hypocotyl segments		MS + 4.4 μM BA	MS + 2.7 μM NAA	Normah et al., 1997
C. indica		Shoot tips		MS + 2.2 μM BA	Soilrite	Baruah et al., 1996a-b
C. jambhiri		Stem and root segments	MS + 0.9 μM KIN + 53.7 μM NAA	½ MS + 22.2 μM BA	½ MS + 5.4 μM NAA	Raman et al., 1992
C. latipes				MS + 2.2 μM BA	Soilrite	Baruah et al., 1996a-b
C. limon	Lemon	Stem and root segments	MS + 0.9 μM KIN + 53.7 μM NAA	½ MS + 22.2 μM BA	½ MS + 5.4 μM NAA	Raman et al., 1992
C. limon	Lemon	Shoot tips		MS + 4.4 μM BA + 4.6 μM KIN + 2.7 μM NAA	MS + 1.1 μM BA + 2.7 μM NAA + 2.5 μM IBA	Singh et al., 1994
C. limon	Lemon	Shoot-tip and node		DKW + 0.76 μM ABA + 8.87 μM BA	DKW + 19.6 μM IBA	Kotsias & Roussos, 2001
C. limon x P. trifoliata	Citremon	Stem segments		MS + 2.2 μM BA + 2.7 μM NAA	MS + 0.8 μM NAA	Mas et al., 1992
C. limonia	Rangpur lime	Nodal and internodal stem segments		MS + 10 μM BA	W + 10 μM NAA	Barlass & Skene, 1982
C. medica	Citron	Nodal stem segments		MS + 4.4 μM BA	MS + 16.1 μM NAA	Duran-Vila & Navarro, 1989
C. mitis	Calamondin	Epicotyl, shoot tip and nodal stem segments		MS + 2.2-4.4 μM BA	½ MS + 4.9 μM IBA	Sim et al., 1989
C. mitis	Calamondin	From root of whole seedlings		MS + 2.2 μM BA	½ MS + 4.9 μM IBA	Sim et al., 1989
C. mitis	Calamondin	Leaf from seedlings		MS + 8.9 μM BA	½ MS + 4.9 μM IBA	Sim et al., 1989

Species	Common name	Explant	Medium for callus	Medium for shoot	Medium for root	Reference
		(whole or segments)				
C. paradisi x P. trifoliata	Swingle citrumelo	Epicotyl stem segments		MT + 150 µM Cumarin	MT + 150 µM Cumarin	Grosser & Chandler, 1986
C. reshni	Cleopatra mandarin	Internodal seedlings stem sections		MT + 22 µM BA + 5.4 µM NAA	½ MT + 5.4 µM NAA	Moore, 1986
C. reticulata	Mandarin	Shoot tips		MS + 4.4 µM BA + 4.6 µM KIN + 2.7 µM NAA	MS + 1.1 µM BA + 2.7 µM NAA + 2.5 µM IBA	Singh et al., 1994
C. reticulata	Mandarin	Internodal stem segments		MS + 33.3 µM BA + 5.4 µM NAA	½ MS + 2.7 µM NAA	Pérez-Molphe-Balch & Ochoa-Alejo 1997
C. sinensis	Sweet orange	Nodal and internodal stem segments		MS + 10 µM BA	W + 10 µM NAA	Barlass & Skene, 1982
C. sinensis	Sweet orange	Nodal stem segments		MS + 4.4 µM BA	MS + 16.1 µM NAA	Duran-Vila & Navarro, 1989
C. sinensis	Sweet orange	Nodal and internodal stem segments		MS + 4.4 µM BA or 13.3 µM BA	MS + 54 µM NAA	Duran-Vila et al., 1992
C. sinensis	Sweet orange	Epicotyl and hypocotyl		MS + 8.89 µM BA + 0.76 µM ABA	Not reported	Maggon & Singh, 1995
C. sinensis	Sweet orange	Nodal stem segments		MS + 2.2 µM BA + 0.5 µM NAA	½ MS + 2.7 µM NAA + 2.5 µM IBA	Tapati et al., 1995
C. sinensis x P. trifoliata	Carrizo-citrange	Shoot tips and nodal sections		MS-KNOP + 22.2 µM BA	MT + 5.4 µM NAA	Kitto & Young, 1981
C. sinensis x P. trifoliata	Carrizo citrange	Nodal and internodal stem segments		MS + 10 µM BA	W + 10 µM NAA	Barlass & Skene, 1982
C. sinensis x P. trifoliata	Troyer citrange	Epicotyl stem segments		MS + 2.2 µM BA + 0.5 µM NAA	MS + 10.7 µM NAA	Edriss & Burger, 1984
C. sinensis x P. trifoliata	Carrizo citrange	Internodal seedlings stem sections		MT + 22 µM BA + 5.4 µM NAA	½ MT + 5.4 µM NAA	Moore, 1986

Species	Common name	Explant	Medium for callus	Medium for shoot	Medium for root	Reference
C. sinensis x P. trifoliata	Carrizo citrange	Shoot-tip		MS + 4.4 μM BA + 2.5 μM IBA + 296 μM AD	MT + 5.4 μM NAA	Starrantino & Caruso, 1987
C. sinensis x P. trifoliata	Troyer citrange	Shoot-tip		MS + 4.4 μM BA + 2.5 μM IBA + 296 μM AD	MT + 5.4 μM NAA	Starrantino & Caruso, 1987
C. sinensis x P. trifoliata	Carrizo citrange	Nodal stem segments		MS + 4.4 μM BA + 2.5 μM IBA + 296 μM AD	MS + 5.4 μM NAA	Starrantino and Caruso, 1988
C. sinensis x P. trifoliata	Troyer citrange	Nodal stem segments		MS + 4.4 μM BA + 2.5 μM IBA + 296 μM AD	MS + 5.4 μM NAA	Starrantino and Caruso, 1988
C. sinensis x P. trifoliata	Carrizo citrange	Mature embryos	MT + 9 μM 2,4-D + 22.2 μM BA	MT + 22.2 μM BA + 5.4 μM NAA	MT + 5.4 μM NAA	Beloualy, 1991
P. trifoliata	Trifoliate orange	Nodal and internodal stem segments		MS + 10 μM BA	W + 10 μM NAA	Barlass and Skene, 1982
P. trifoliata	Flying Dragon	Shoot-tip		MS + 2.2 μM BA + 1.2 μM IBA + 296 μM AD	MS + 5.4 μM NAA	Starrantino and Caruso, 1987
P. trifoliata	Flying Dragon	Nodal stem segments		MS + 2.2 μM BA + 1.2 μM IBA + 296 μM AD	MS + 5.4 μM NAA	Starrantino and Caruso, 1988
P. trifoliata	Trifoliate orange	Mature embryos	MT + 9 μM 2,4-D + 22.2 μM BA	MT + 22.2 μM BA + 5.4 μM NAA	MT + 5.4 μM NAA	Beloualy, 1991
P. trifoliata	Trifoliate orange	Hypocotyl		MS + 44.4 μM BA	½ MS + 0.5-5.0 μM IBA	Harada and Murai, 1996
P. trifoliata	Trifoliate orange	tTCL from stem internodes		MS + 10 μM BA + 1 μM TDZ	MS + 5 μM NAA	Van Le et al., 1999

Abbreviations used in the table:
AD=Adenine; BA = 6-benzylaminopurine; DKW = Driver-Kuniyuki medium; IBA = indole-3-butyric acid; KIN = kinetin; MS = Murashige and Skoog medium; MS-KNOP = medium based on MS microelements and Knop's macroelements and vitamins; NAA = α-naphthaleneacetic acid; TDZ = thidiazuron; tTCL = transverse thin cell layer; W = White medium.

In conclusion, the regeneration of *Citrus* by organogenesis has been described in several reports, although two principal methods of citrus micropropagation have been proposed. The first is based on the following steps: (1) establishment of primary cultures, (2) incubation of explants on proliferating medium (in presence of a cytokinin or a cytokinin plus an auxin) to induce shoot proliferation, (3) incubation of regenerated shoots (1-2 cm) on rooting medium (in presence of an auxin), (4) incubation of rooted shoots on elongation medium (basal medium without plant growth regulators), and (5) establishment of plants on soil (Duran-Vila and Navarro, 1989).

Figure 1. Sweet orange (*Citrus sinensis*) shoot transferred to Jiffy 7 peat block (left). Shoots rooted in Jiffy are transferred in pots (0.5 L) containing a sterile soil. Plant of sweet orange (4-months-old) derived from a shoot rooted directly on Jiffy 7 peat block (right).

The second method is based on the following steps: (1) establishment of primary cultures, (2) incubation of explants on proliferating medium (in the presence of a cytokinin or a cytokinin plus an auxin) to induce shoot proliferation, (3) incubation of shoots-buds on elongation medium (in presence of a cytokinin plus GA_3), (4) incubation of elongated shoots (2 cm) on rooting medium (in presence of an auxin), (5) establishment of plants on soil (Chakravarty and Goswami, 1999; Paudyal and Haq, 2000). An alternative method is based on the following steps: (1) establishment of primary cultures, (2) incubation of explants on proliferating medium (in presence of a cytokinin or a cytokinin plus an auxin) to induce shoot proliferation, (3) incubation of regenerated shoots (1-2 cm) on Jiffy 7® peat pellets under mist (Fig. 1), (5) establishment of plants in the greenhouse (De Pasquale and Carimi, unpublished data). Parthasarathy et al. (1999)

proposed a similar method that involves direct planting of 6- to 8-week-old in-vitro-proliferated shoots (2-2.5 cm long) in sterile soilrite (a single-step rooting and acclimatization of microshoots). This method has been used with success for the micropropagation of different *Citrus* species (Parthasarathy et al., 1999).

6. EXPLANT PREPARATION

Organogenesis in *Citrus* has been achieved from different types of explants, leaf, whole seedlings, shoot tips, nodal and internodal segments, epicotyl and hypocotyl segments, root tips, and root segments (Sim et al., 1989, Duran-Vila and Navarro, 1989). The growth conditions of the stock field-growing plants are very important for the establishment of explants *in vitro*. On the surface of the plant material growing in the field, many microorganisms can be found (in some cases endogenous contamination are also present). These microorganisms can be difficult to eliminate and may also limit the initiation of axenic cultures. If the stock material is collected during the warm humid season (a period that corresponds with a vigorous growth), the bacterial and fungal contamination problems can easily be avoided with a standard procedure for surface-disinfection in *Citrus*. In order to stimulate vigorous plant growth during slow growth periods (during the dry or cold season), the entire plant is pruned and defoliated for new shoot growth. Before the introduction *in vitro*, explants collected from plant cultivated in the field or in the greenhouse are surface-disinfected.

A procedure for citrus surface-disinfection is described below. Stem pieces (10 cm) are stripped of their thorns and leaves, and washed with tap water. Stems are surface-disinfected by immersion for 15 min in 2 % (w/v) sodium hypochlorite, 0.05% (v/v) Tween-20, and rinsed three times in sterile distilled water. Moore (1986) reported that HCl treatment of explants during disinfection significantly reduced contamination of cultures and also proposed an alternative protocol, namely 1 N HCl for 0.5 min, 70% ethanol for 2.5 min, 1.5% sodium hypochlorite (w/v), 0.1% (v/v) Tween-20 for 5 min, followed by 3 rinses with sterile distilled water Moore (1986).

Sterilized explants are cultured horizontally or vertically on culture medium. Usually, nodal stem segments (1 cm in length) are excised longitudinally and the portion containing the bud is cultured with the cut surface in contact with the medium (Marin and Duran-Vila, 1991). Pérez-Molphe-Balch and Ochoa-Alejo (1997) investigated the influence of explant culture position and wounding on efficiency of adventitious shoot formation in Mexican lime and mandarin; an optimal response was obtained when internodal stem segments were cut longitudinally and then placed horizontally with the cut surface in contact with the culture medium. Kitto and Young (1981) reported that shoot tip explants of Rangpur lime, sweet orange, and trifoliate orange do not respond well to culture *in vitro*, whereas Carrizo citrange shoot tip explants proliferated vigorously. They also observed that stimulation of axillary shoots from single nodes (Fig. 2) was the most prolific source of shoots when compared with shoot tip and internodal explants.

Figure 2. Stimulation of axillary shoot from single nodes of sour orange (*Citrus aurantium)*. The nodal stem segment is cut longitudinally and placed horizontally with the cut surface in contact with the culture medium.

Sim et al. (1989) observed a different pattern of adventitious bud formation in calamondin leaves in that the response of the proximal half of the leaf was much higher and faster than that of the distal half. Burger and Hackett (1986) also found a different aptitude of adventitious bud formation in epicotyl and root sections of sweet orange seedlings. In particular, it was observed that in the presence of BA (8.9 μM) and NAA (0.1 μM), epicotyl and root sections nearest the cotyledonary node have the greatest bud-forming potential. Moreira-Dias et al. (2001) studied the adventitious shoot regeneration in epicotyl cuttings of Troyer citrange and found that the number of shoots formed decreased markedly if the explant is dissected far from the cotyledonary node.

El-Morsy and Millet (1996) studied the rhythmic pattern of shoot elongation, with particular attention on the excision time and position of axillary buds on *in vitro* cultures of sour oranges. They reported that axillary buds excised from the top of the growth unit, and therefore those having initiated at the end of the last growth flush, grew more slowly than buds located in the median or basal part (El-Morsy and Millet, 1996).

7. MODES OF CULTURE

7.1 Media composition

The composition of the media used for *in vitro* culture of *Citrus* is usually based on the nutrients and vitamins of the MS medium, although in some cases better results are obtained with MT medium (Murashige and Tucker, 1969). This latter contains the inorganic salts of MS medium in addition to 26.6 µM glycine, 554.9 µM myo-inositol, 40.6 µM nicotinamide, 48.6 µM pyridoxine-HCl and 29.7 µM thiamine-HCl. Kitto and Joung (1981) considered the medium based on Knop's macroelements and vitamins together with the microelements of the MS medium to be the most suitable for propagation of Carrizo citrange. Recently, Kotsias and Roussos (2001) proposed the use of Driver-Kuniyuki medium (DKW) (Driver and Kuniyuki, 1984) for the micropropagation of lemon. In some cases, addition of high levels of vitamins showed favorable effect on *in vitro* growth of *Citrus* (Chakravarty and Goswami, 1999).

Sucrose is usually used as a carbon source at concentrations ranging from 20 to 50 g/L. Kitto and Young (1981) reported that Carrizo citrange shoot proliferation was higher if explants were cultured on medium containing 30 or 40 g/L sucrose as compared to 50 or 60 g/L. Pérez-Molphe-Balch and Ochoa-Alejo (1997) similarly observed that the optimal sucrose concentration for adventitious shoot formation in Mexican lime and mandarin was 30-50 g/L. High sucrose concentration (60 g/L) was deleterious to cultures (Pérez-Molphe-Balch and Ochoa-Alejo, 1997).

In some cases, organogenesis was obtained in liquid medium (Sauton et al., 1982), although a semisolid medium (agar 5-10 g/L) is usually used for *Citrus*. Kitto and Young (1981) reported that agar concentration can influence rooting and that this response increased as the agar concentration was decreased from 20 to 5 g/L. The pH of the medium of 5.6-5.8 is considered to be favorable for growth of *Citrus*; usually the pH is adjusted with KOH or HCl before autoclaving at 121 °C for 15-20 min.

A large number of other compounds may influence the *in vitro* growth of *Citrus*. In some cases, the positive effect of malt extract (0.5-1.5 g/L) on citrus growth has been observed (Moore, 1985; Beloualy, 1991). For example, coumarin (150 µM) can be used to induce both root and shoot formation in Swingle citrumelo epicotyl segments. The addition of this compound to culture media can increase the number of whole plants that recover by at least 5-fold (Grosser and Chandler, 1986). Also, orange juice (10% v/v) can stimulate the growth of citrus callus (Einset, 1978; Duran-Vila and Navarro, 1989).

7.2 Culture conditions

Citrus cultures are usually maintained at 25-28 °C under a 16 h day length with a photosynthetic photon flux density (PPFD) of 30-60 µmol m^{-2} s^{-1} provided by daylight type fluorescent lamps. Duran-Vila et al. (1992) studied the effect of light and temperature on induction and maintenance of callus, bud and shoot regeneration, as well as root formation in sweet oranges. Incubation in the dark or under low PPFD (30 µmol m^{-2} s^{-1}) considerably increased callus induction and growth and also improved the regeneration of shoots and buds from internode segments as well as the recovery of citrus plants regenerated *in vitro* (Duran-Vila et al., 1992). Bud culture, regeneration of buds and shoots from internode explants, and induction of primary callus responded nearly optimal over a fairly wide temperature range (21-30 °C). Growth of established callus cultures was optimal at 30 °C. The highest frequency of root formation was achieved at 27 °C and was more than adequate for all the process studied (Duran-Vila et al., 1992).

Pérez-Molphe-Balch and Ochoa-Alejo (1997) investigated the effect of light on adventitious shoot formation of Mexican lime and mandarin. The best response was obtained when segments were incubated for 21 days in darkness, followed by 29 days under fluorescent light in a 18/8 h light/dark cycle (54 µmol m^{-2} s^{-1} PPFD). However, no statistically significant differences were observed between this treatment and those obtained after constant incubation in a 16/8 h light/dark cycle or 24 h light (Pérez-Molphe-Balch and Ochoa-Alejo, 1997).

Moreira-Dias et al. (2001) studied the effect of different lighting conditions (day-length and photon flux density) on shot regeneration in epicotyl cuttings of Troyer citrange with varying concentrations of BA and NAA. Two optimal medium and illumination conditions for adventitious shoot regeneration were found. The first was an 8 h day-length with a photon flux density of 74 µmol m^{-2} s^{-1} in the presence of 2.2 µM BA, while the second consisted in a 16 h day-length with a photon flux density of 37 µmol m^{-2} s^{-1} in the presence of 22 µM BA (Moreira-Dias et al., 2001).

7.3 Effect of growth regulators

The use of plant growth regulators in citrus tissue culture is of fundamental importance. Many studies have been conducted in order to identify the optimal composition of the medium for callus cultures, bud, shoot and root differentiation, and plant regeneration in *Citrus*. Cytokinins and the auxins are the two most important plant growth regulators utilized to control organogenesis and embryogenesis. Together with auxins, cytokinins also control differentiation of plant cells in culture and in fact, a high cytokinin to auxin ratio promotes shoot production (Buchanan et al., 2000). Gibberellins are usually used to promote shoot elongation in *Citrus*; other hormones, in particular ABA, have been used only on occasion. The hormonal requirements for optimal regeneration of shoots, buds, and roots are different depending on the species under consideration (Duran-Villa and Navarro, 1989). Plant material regenerated from juvenile or mature explants of the same genotype can respond differently to a plant growth regulator. Results on shoot rooting from Carrizo citrange indicated that mature explants require higher levels of auxin (10 μM NAA) as compared to juvenile explants (5 μM NAA) (Barlass and Skene, 1982).

For many *Citrus* species, shoot proliferation medium containing cytokinin alone (Fig. 3) or in combination with an auxin is necessary (Van Le et al., 1999).

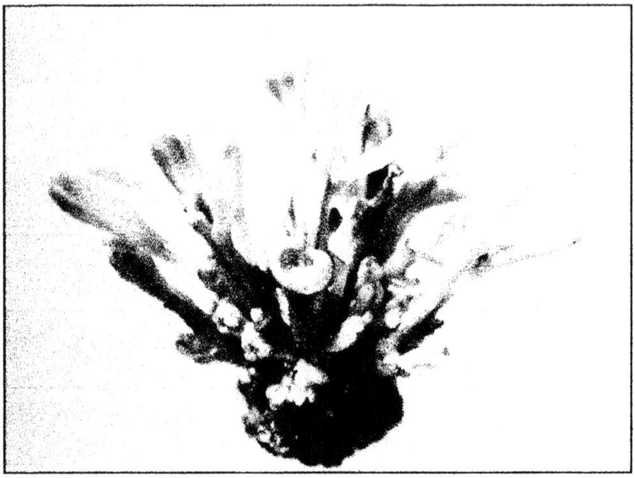

Figure 3. In vitro shoot proliferation of calamondin (*Citrus mitis*) on MS medium supplemented with 13.3 μM BA.

Marin and Duran-Vila (1991) reported that the optimal cytokinin and auxin concentrations for shoot proliferation varies considerably among *Citrus* species, but little variation was observed among cultivars belonging to the same species. The most suitable plant growth regulators seem to be BA and NAA. The promoting effect of BA on shoot proliferation has been reported in *C. aurantifolia* (Pérez-Molphe-Balch and Ochoa-Alejo, 1997), *C. limon* (Duran-Vila and Navarro, 1989), *C. mitis* (Sim et al., 1989), *C. reticulata* (Pérez-Molphe-Balch and Ochoa-Alejo, 1997), and *C. sinensis* (Burger and Hackett, 1986). Duran-Vila and Navarro (1989) observed that when nodal stem segments were cultured in presence of BA, multiple shoots were produced, although only the main shoot developed faster (reaching a size of 1-2 cm); the other buds either developed slowly or were no larger than a few millimeters. These small shoots grew only when the main shoot was removed from the original explant. This phenomenon of apical dominance was evident in sweet orange but absent in citron and lime (Duran-Vila and Navarro, 1989). In mandarin, mexican lime, and trifoliate orange it was observed that TDZ was less active than BA on shoot or bud regeneration (Pérez-Molphe-Balch and Ochoa-Alejo, 1997; Van Le et al.,

1999). Kitto and Young (1981) investigated the effect of different cytokinins on shoot tip proliferation in Carrizo citrange. They reported that only BA induced proliferation, whereas KIN and 6 γ- γ-dimethylallylaminopurine (2iP) had no effect on shoot proliferation. Baruah et al. (1995, 1996a) reported that the effect of BA was superior to kinetin for shoot proliferation of different *Citrus* species. Van Le et al. (1999) reported a synergic effect between a cytokinin (BA) and the cytokinin-like TDZ on *in vitro* bud differentiation in trifoliate orange. The percentage of responsive explants and the number of shoots or buds obtained per explant varied with genotype and culture conditions. In calamondin, all of the explants differentiated 3.7 buds per explant in presence of 66.6 μM BA (Sim et al., 1989). In Carrizo citrange, 3.1 shoots per explant were induced with 22.2 μM BA (Kitto and Young, 1981). In trifoliate orange, 90% of the explants differentiated 37 buds per explant with a combination of 10 μM BA and 1 μM TDZ (Van Le et al., 1999). In sweet orange, all of the explants differentiated 29 shoots and/or buds in presence of 13.3 μM BA (Duran-Vila et al., 1992).

In several *Citrus* genotypes it has been observed that an auxin is not as critical as cytokinin with regards to shoot proliferation (Paudyal and Haq, 2000). In some cases, citrus shoot proliferation was improved by using different auxin concentrations (0.5-5.4 μM NAA) in combination with a cytokinin (Edriss and Burger, 1984; Moore, 1986). The presence of an auxin in the culture medium is generally necessary to promote rooting in *Citrus*. Usually, the incubation of adventitious shoots on NAA or IBA supplemented medium induces the formation of roots in many different *Citrus* species (Fig. 4).

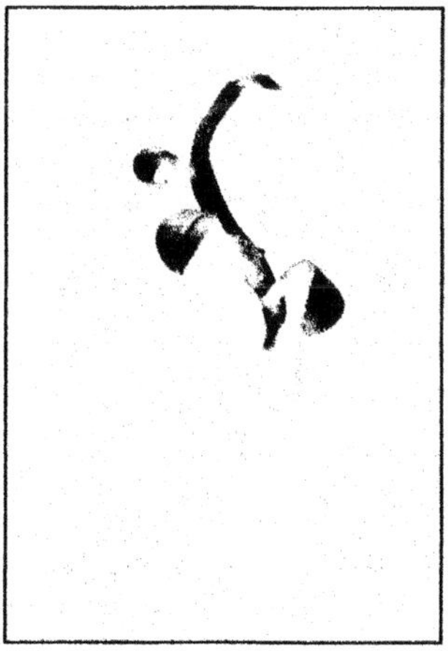

Figure 4. In vitro rooting of lemon (*Citrus limon*) on MS medium supplemented with 2.5 μM IBA.

Kotsias and Roussos (2001) reported that in lemon, NAA gave the lowest rooting percentages as compared to IBA and that the highest rooting percentage was obtained in the presence of 19.6 μM IBA. Nevertheless, Kitto and Young (1981) found that NAA induced higher rooting percentages as compared to IBA or IAA in Carrizo citrange. Normah et al. (1997) reported that NAA induced higher rooting percentages as compared to IBA in *C. halimii*. Singh et al. (1994) reported that a combination of NAA (2.7 μM), IBA (2.5 μM) and BA (1.1 μM) was found to be essential in inducing root formation in lemon and mandarin.

Duran-Vila and Navarro (1989) reported that in *Citrus*, small shoots rooted very poorly and therefore, only shoots with a minimal length of 0.5 cm should be transferred to rooting medium. They also observed that in sweet orange, citron, and lime, root formation was completely inhibited in rooting medium containing BA at concentrations greater than 0.4 μM (Duran-Vila and Navarro, 1989). The percentage of shoots rooting and the time necessary to differentiate roots is dependent on the size of shoots utilized (Chakravarty and Goswami, 1999); usually the optimal shoot length of *Citrus* is about 15-20 mm. In most cases, NAA or IBA at a concentration of 5 μM is sufficient to induce root formation in *Citrus*. Since plantlets maintained in the rooting medium failed to grow further and did not survive when transplanted to soil (Duran-Vila and Navarro, 1989), the shoots that have differentiated roots or root primordia should be transferred from rooting medium to auxin-free medium where shoot and root elongation can proceed.

Maggon and Singh (1995) reported that ABA (0.76 μM) as the sole growth regulator completely inhibited shoot bud regeneration in sweet orange, but in combination with BA (8.89 μM) it strongly promoted shoot differentiation. Another noteworthy effect of ABA was that the bulk of the buds regenerated directly from the epidermal surface of explants without any apparent callus formation, while in presence of BA as the sole growth regulator, shoot buds originated only at the cut ends of explants (Maggon and Singh, 1995). ABA inhibited the further elongation of shoot buds, even after 8-10 weeks after their transfer to the basal medium. The addition of GA_3 (0.01 μM) counteracted the negative effect of ABA and enhanced shoot elongation (Maggon and Singh, 1995). Kotsias and Roussos (2001) also reported that the addition of ABA together with BA enhanced bud formation in lemon with a combination of 0.76 μM ABA and 8.89 μM BA providing the highest number of shoots per explant. On the contrary, treatment with ABA in the culture medium inhibited the shoot elongation, while the addition of GA_3 and silver nitrate (an ethylene action inhibitor) enhanced shoot elongation (Kotsias and Roussos, 2001). The authors reported that the optimal combination of growth regulators for shoot elongation was 8.9 μM BA, 1.4 μM GA_3 and 17.7 μM silver nitrate.

Usually, when a large number of buds are obtained on one explant they remain small and are unable to root (Duran-Vila and Navarro, 1989). In these cases, the addition of GA_3 in the culture medium can promote shoot elongation. Paudyal and Haq (2000) reported that the addition of GA_3 in the shoot-proliferation media during primary culture reduced shoot proliferation. Moreover, the regenerated shoots were long, thin and etiolated. The authors also observed that transfer of shoots after the proliferation phase to 5.8 µM GA_3 and 1.8 µM BA supplemented medium significantly increased shoot elongation; shoots cultured on GA_3 supplemented medium were about twice the length of those cultured in absence of the growth regulator (Paudyal and Haq, 2000). As the growth regulator inhibits the initiation of meristems, as is necessary for shoot elongation once new shoots are differentiated (Omura and Hidaka, 1992; Paudyal and Haq, 2000), GA_3 should be added to the culture medium after the shoot-proliferation phase.

8. ACCLIMATIZATION OF MICROPROPAGATED PLANTS

Only scanty information is available regarding the acclimatization of citrus micropropagated plants. They usually grow *in vitro* under high levels of relative humidity and low levels of light in aseptic conditions on a medium containing high levels of nutrients and sugars that allow heterotrophic growth. Plantlets developed in these conditions usually do not survive transfer from *in vitro* conditions to greenhouse or field environments (Preece and Sutter, 1991). *In vitro* regenerated citrus plants need acclimatization in order to guarantee their survival once transferred to *in vivo* conditions. Well-developed *in vitro* plants of *Citrus* should be gradually exposed to a lower relative humidity and a higher light intensity. The gradual reduction in humidity helps the plant harden and further increases the percentage of plant survival once transferred to soil. If not properly hardened-off once transferred to the greenhouse or to the field environments, micropropagated plants show signs of stress, including the immediate desiccation of the entire plant. A variable percentage of citrus micropropagated plants (0-85%) do not survive the transfer from *in vitro* to *in vivo* conditions (Grosser and Chandler, 1986; Ghorbel et al., 1998; Van Le et al., 1999; Paudyal and Haq, 2000). Hidaka and Kajiura (1989) described a method of acclimatization of micropropagated plantlets of *Citrus* and achieved 60-90% survival rate in a mist house. Parthasarathy et al. (1999) and Hazarika et al. (1998, 1999) proposed a method involving (in a single-step) rooting and acclimatization of citrus microshoots in sterile soilrite with an acclimatization success rate ranging from 90 to 97%.

Micropropagated plants (transferred in sterile soil) are usually maintained in a basal heating bench at 27 °C, a photosynthetic photon flux density of 100 µmol m^{-2} s^{-1} and a high relative humidity (95-98%) for 20 days. The gradual reduction of humidity can be achieved by partially opening the lid of the bench. After 40 days the lid can be completely removed and most plants can survive under green house conditions.

When the regenerated plantlets have been collected from the culture medium, agarized medium has to be completely removed from the plant tissues, as nutrients present in the small pieces of solidified medium can promote the growth of pathogens. To prevent diseases and pests in plants incubated at high relative humidity levels, insecticides (Methomyl, methyl O-(methylcarbamoyl) thiolacetohydroxamate) and fungicides (Fosetyl Al, phosphonic acid monoethyl ester aluminium salt) can be applied at 20-30 day intervals (De Pasquale et al., 1999).

9. GENETIC FIDELITY OF MICROPROPAGATED PLANTS

Plant propagation through micropropagation has many advantages including the potential to produce a large number of plantlets in a short period of time (Rani and Raina, 2000). However, there are several disadvantages to micropropagation, with the foremost being the genetic instability of the resulting plants (Pierik, 1991). The variation associated with *in vitro* regeneration of plants is termed as somaclonal variation (Larkin and Scowcroft, 1981) and it is well known that increases in the length of subculture or number of subcultures enhance the frequency of mutations (Swartz, 1991). For this reason, stock cultures should be renewed periodically. Maggon and Singh (1995) observed that citrus shoot buds originate mainly from the callus formed at the cut surface of explants cultured on a medium containing BA. This approach for shoot regeneration is correlated with an increasing risk of genetic instability (Maggon and Singh, 1995).

Morphological markers have been extensively used in the identification of somaclonal variants, although some genetic changes may not be reflected in the observed phenotypic variation (Rani and Raina, 2000). In same cases, abnormalities observed in off-type plants were noticed only after many years of cultivation in field conditions (Rani and Raina, 2000).

The economic impact of somaclonal variation can be high in *Citrus*, as they have long life cycles. Many *Citrus* species produce fruits 2-3 years after transplantation into soil and investigations on somaclonal variation on morphological traits of the fruits takes long periods of time and becomes rather costly. Moreover, additional long-term studies are needed when nucellar or somatic embryos are used as source material for micropropagation, considering that many observations can not be made until maturity (3-7 years after the embryogenic event).

Unfortunately, extensive studies on the incidence of somaclonal variation in *Citrus* at morphological, cytological, biochemical and molecular levels do not exist. Some researchers have reported having off-types of citrus plants produced during *in vitro* plant regeneration by organogenesis or somatic embryogenesis. Navarro et al. (1985) observed that 29% of plants obtained by somatic embryogenesis of nucelli of monoembryonic *Citrus* presented abnormal phenotypic characteristics. Also, significant genetic variation existed in nucellar seedling populations of *Citrus* (Iglesias et al., 1974, Litz and Jaiswal, 1991). Therefore, it is possible that the mutations, observed by Navarro et al. (1985), were probably already present in the nucelli at the time of explanting (Litz and Jaiswal, 1991). Navarro et al. (1985) also reported that all the plants regenerated by *in vitro* somatic embryogenesis of the nucellus of polyembryonic citrus cultivars were normal. The authors were at a loss to explain the different behavior of monoembryonic and polyembryonic cultivars (Navarro et al., 1985). Starrantino and Caruso (1987) propagated *in vitro* citrus rootstocks using nodal stem segments and observed no change in either morphology or in chromosome number. Duran-Vila et al. (1992) found that none of the plants regenerated from *in vitro* bud cultures or internode stem segments showed phenotypic abnormalities and that no differences, as determined by isoenzyme analysis, were observed among regenerated plants. Chakravarty and Goswami (1999) analyzed plants regenerated from *C. acida* calli of different ages and observed no changes in either morphology or in chromosome number. Moreover, they reported that the plants regenerated *in vitro* were genetically stable.

In conclusion, reports on limited populations of polyembryonic *Citrus* species regenerated by somatic embryogenesis show uniformity, at least with respect to those characteristics examined (Vardi and Spiegel-Roy, 1982; Kobayashi, 1987). More complete investigations on morphology, ploidy level, and leaf isozyme analysis on *Citrus* and citrus hybrids have revealed little or no permanent variations among the plants regenerated through organogenesis or somatic embryogenesis (Gmitter et al., 1992). This technique should not applied to monoembryonic species, as a high percentage of aberrant plants have been obtained when plants were regenerated by somatic embryogenesis from nucelli of these species (Navarro et al., 1985). Moreover, in most of the plants that are recovered from somatic embryos, undesired juvenile characters may appear.

10. CONCLUSION

It is important to point out that the available data on citrus micropropagation normally refers only to experimental conditions. The transfer of micropropagation techniques on a commercial scale will require appropriate, substantial modifications in order to reduce the high cost of production and decrease the risk of variability and production of off-type individuals.

It is difficult to provide information with regard to the time necessary to induce the formation of shoots and root, since in some reports no details were furnished and because inconsistent data were reported. These differences might be due to the fact that genotype, culture medium composition, explant type and age may have a strong effect on *in vitro* citrus growth. Unfortunately, investigations on the performance in the field of micropropagated citrus plants are insufficient.

A critical economic analysis of the benefits of the use of citrus tissue culture versus the traditional propagation systems is not present in the literature. As the process of rooting *in vitro* has been estimated to account for approximately 35 to 75% of the total cost of micropropagation (Preece and Sutter, 1991), preference should be given to those systems that use a single, simultaneous step for both rooting and acclimatization. This would thus substantially decrease the costs of micropropagation.

11. REFERENCES

Altman A and Goren R (1974) Growth and dormancy cycle in citrus bud culture and their hormonal control. Physiol Plant 30:240-245

Barlass M and Skene KGM (1982) *In vitro* plantlet formation from *Citrus* species and hybrids. Scientia Horticulturae 17:333-341

Barlass M and Skene KGM (1986) *Citrus*. In Bajaj YPS (Ed) Biotechnology in Agriculture and Forestry 1. Trees I, pp207-219, SpringerVerlag, Heidelberg

Baruah A, Nagaraju V and Parthasarathy VA (1995) Cytokinin mediated response in shoots of *Citrus grandis* Osbeck. Ann Plant Physiol9:13-16

Baruah A, Nagaraju V and Parthasarathy VA (1996a) Micropropagation of three endangered *Citrus* species. 1. Shoot proliferation in vitro. Ann Plant Physiol 10: 124-128

Baruah A, Nagaraju V and Parthasarathy VA (1996b) Micropropagation of three endangered *Citrus* species. 2. Rooting ex vitro. Ann Plant Physiol 10: 129-132

Beloualy N (1991) Plant regeneration from callus culture of three *Citrus* rootstocks. Plant Cell Tissue Org Cult 24:29-34

Ben-Hayyim G and Neumann H (1983) Stimulatory effect of glycerol on growth and somatic embryogenesis in citrus callus cultures. Z. Pflanzenphysiol. 110:331-337

Bhat SR, Chitralekha P and Chandler KPS (1992) Regeneration of plants from long-term root culture of lime, *Citrus aurantifolia* (Christm.) Swing. Plant Cell Tissue Org Cult 29:19-25

Bitters WP, Murashige T, Rangan TS and Nauer E (1970) Investigations on established virus-free plants through tissue culture. Calif Citrus Nursery Soc 9:27-30

Buchanan BB, Gruissem W and Jones RL (2000) Biochemistry and molecular biology of plants. Amer Soc Plant Physiol, Rockville, Maryland

Burger DW and Hackett WP (1986) Gradients of adventitious bud formation on excised epicotyl and root sections of *Citrus*. Plant Sci 43:229-232

Button J and Bornman CH (1971) Development of nucellar plants from unpollinated and unfertilized ovules of the Washington navel orange *in vitro*. J S Afr Bot 37:127-134

Button J and Kochba J (1977) Tissue culture in the citrus industry. In: Reinert J and Bajaj YPS (Eds) Applied and Fundamental Aspects of Plant Cell, Tissue and Organ Culture (pp 70-92) Springer-Verlag, Heidelberg

Cabasson C, Ollitrault P, Côte FX, Michaux-Ferrière N, Dambier D, Dalnic R and Teisson C (1995) Characteristic of *Citrus* cell cultures during undifferentiated growth on sucrose and somatic embryogenesis on galactose. Physiol Plant 93:464-470

Cabasson C, Alvard D, Dambier D, , Ollitrault P and Teisson C (1997) Improvement of Citrus somatic embryo development by temporary immersion. Plant Cell Tissue Org Cult 50:33-37

Carimi F, De Pasquale F and Crescimanno FG (1994) Somatic embryogenesis from styles of lemon (*Citrus limon*). Plant Cell Tissue Org Cult 37:209-211

Carimi F, De Pasquale F and Crescimanno FG (1995) Somatic embryogenesis in *Citrus* from style culture. Plant Sci 105:81-86

Carimi F, De Pasquale F and Puglia AM (1998a) *In vitro* rescue of zygotic embryos of sour orange, *Citrus aurantium* L., and their detection based on RFLP analysis. Plant Breeding 117:261-266

Carimi F, Tortorici MC, De Pasquale F and Crescimanno FG (1998b) Somatic embryogenesis and plant regeneration from undeveloped ovules and stigma/style explants of sweet orange navel group [*Citrus sinensis* (L.) Osb.]. Plant Cell Tissue Org Cult 54:183-189

Carimi F, De Pasquale F and Crescimanno FG (1999) Somatic embryogenesis and plant regeneration from pistil thin cell layers of *Citrus*. Plant Cell Rep 18:935-940

Cervera M, Juárez J, Navarro A, Pina JA, Duran-Vila N, Navarro L and Peña L (1998) Genetic transformation and regeneration of mature tissues of woody fruit plants bypassing the juvenile stage. Transgenic Res 7:51-59

Chakravarty B and Goswami BC (1999) Plantlet regeneration from long-term callus cultures of *Citrus acida* Roxb. and the uniformity of regenerated plants. Scientia Horticulturae 82:159-169

Chaturvedi HC and Mitra GC (1974) Clonal propagation of *Citrus* from somatic callus cultures. HortScience 9:118-120

De Pasquale F, Carimi F and Crescimanno FG (1994) Somatic embryogenesis from styles of different cultivars of *Citrus limon* (L.) Burm. Aust J Botany 42:587-594

De Pasquale F and Carimi F (1998) Agrumi ornamentali. I Georgofili 5:105-122

De Pasquale F, Giuffrida S and Carimi F (1999) Minigrafting of shoots, roots, inverted roots and somatic embryos for rescue of *in vitro* regenerants of *Citrus*. J Am Soc Hortic Sci 124:152-157

D'Onghia AM, De Pasquale F, Carimi F, Savino V and Crescimanno FG (1997) Somatic embryogenesis from style culture as possible means for virus elimination in *Citrus*. J Phytopathology 145:77-79

D'Onghia AM, Carimi F, De Pasquale F, Djelouah K and Martelli GP (2000) Somatic embryogenesis from style culture: a new technique for the sanitation, conservation, and safe exchange of citrus germplasm. Proc Int Soc Citriculture, In press

D'Onghia AM, Carimi F, De Pasquale F, Djelouah K and Martelli GP (2001) Elimination of citrus psorosis virus by somatic embryogenesis from stigma and style cultures. Plant Pathol 50:266-269

Driver JA and Kuniyuki AH (1984) *In vitro* propagation of paradox walnut rootstock. HortScience 19:507-509

Duran-Vila N and Navarro L (1989) Morphogenesis and tissue cultures of three citrus species. Plant Cell Tissue Org Cult 16:123-133

Duran-Vila N, Gogorcena Y, Ortega V, Ortiz J and Navarro L (1992) Morphogenesis and tissue culture of sweet orange (*Citrus sinensis* (L.) Osb.): Effect of temperature and photosynthetic radiation. Plant Cell Tissue Org Cult 29:11-18

Durzan D (1990) Adult vs. juvenile explants: directed totipotency. In Rodríguez R, Sánchez-Tamés R and Durzan DJ (Eds.) Plant Aging. Basic and applied approaches. NATO ASI Series A: Life Sciences 186:19-25

Edriss MH and Burger DW (1984) *In vitro* propagation of 'Troyer' citrange from epicotyl segments. Scientia Horticulturae 23:159-162

Einset JW (1978) Citrus tissue culture. Stimulation of fruit explant cultures with orange juice. Plant Physiol 62:885-888

El-Morsy AA and Millet B (1996) Rhythmic growth and optimization of micropropagation: the effect of excision time and position of axillary buds on *in vitro* culture of *Citrus aurantium* L. Annals Botany 78:197-202

Fiore S, De Pasquale F, Carimi F and Sajeva M (2002) Effect of 2,4-D and 4-CPPU on somatic embryogenesis from stigma and style transverse thin layers of *Citrus*. Plant Cell Tissue Org Cult 68:57-63

Furr JR, Cooper WC and Reece PC (1947) An investigation on flower formation in adult and juvenile citrus trees. Am J Botany, 34:1-8

Gavish H, Vardi A and Fluhr R (1991) Extracellular proteins and early embryo development in *Citrus* nucellar cell cultures. Physiol Plant 82:606-616

Gill MIS, Singh Z, Dhillon BS and Gosal SS (1995) Somatic embryogenesis and plantlet regeneration in mandarin (*Citrus reticulata* Blanco). Scientia Horticulturae 63:167-174

Gmitter FG, Grosser JW and Moore GA (1992) Citrus. In: Litz RE and Hammerschlag F (Eds.) Biotechnology of perennial fruit crops. CAB International, Oxon, UK pp335-369

Goh CJ, Sim GE, Morales CL and Loh CS (1995) Plantlet regeneration through different morphogenic pathway in pommelo tissue culture. Plant Cell Tissue Org Cult 43:301-303

Ghorbel R, Navarro L and Duran-Vila N (1998) Morphogenesis and regeneration of whole plants of grapefruit (*Citrus paradisi*), sour orange (*C. aurantium*) and alemow (*C. macrophylla*). J Hort Sci & Biotec 73:323-327

Grinblat U (1972) Differentiation of citrus stem *in vitro*. J Am Soc Hort Sci 97:559-603

Grosser JW and Chandler JL (1986) *In vitro* multiplication of Swingle citrumelo rootstock with coumarin. HortScience 21:518-520

Grosser JW, Ollitrault P and Olivares-Fuster O (2000) Somatic hybridization in *Citrus*: an effective tool to facilitate variety improvement. *In vitro* Cell Dev Biol Plant 36:434-449

Hackett WP (1985) Juvenility, maturation, and rejuvenation in woody plants. Horticultural Reviews 7:109-155

Harada H and Murai Y (1996) Clonal propagation of *Poncirus trifoliata* through culture of shoot primordia. J Hort Sci 71:887-892

Hazarika BN, Singh IP, Nagaraju V and Parthasarathy VA (1998) An efficient method of acclimatization of micropropagated plantlets of citrus. Ann Plant Physiol 12: 47-51

Hazarika BN, Nagaraju V and Parthasarathy VA (1999) Acclimatization techniques of citrus plantlets from *in vitro*. Adv Plant-Sci 12: 97-102

Hidaka T, Yamada Y and Shichijo T (1981) Plantlet formation from anthers of *Citrus aurantium* L. Proc Int Soc Citriculture 1:153-155

Hidaka T and Kajiura I (1989) A simple method for acclimatization of *in vitro* plantlets of citrus. Bull Fruit Res Stn 16:19-28

Hu JT and Kong KL (1987) The organogenesis of buds from entire lamina of *Citrus sinensis* in tissue culture and their anatomical observations. J Fruit Science 7:81-84

Iglesias L, Lima H and Simon JP (1974) Isoenzyme identification of zygotic and nucellar seedlings in *Citrus*. J Hered 65:81-84

Jiménez VM, Guevara E, Herrera J and Bangerth F (2001) Endogenous hormone levels in habituated nucellar *Citrus* callus during the initial stages of regeneration. Plant Cell Rep 20:92-100

Kitto SL and Young MJ (1981) *In vitro* propagation of Carrizo citrange. HortScience 16:305-306

Kobayashi S, Ikeda I, Nakatani M (1981) Role of the primordium cell in nucellar embryogenesis in *Citrus*. Proc Int Soc Citriculture 1: 44-48

Kobayashi S, Ikeda I and Nakatani N (1982) Studies on nucellar embryogenesis in *Citrus*. III. On the differences in ability to form embryoids in *in vitro* culture of ovules from poly- and mono-embryonic cultivars. Bull Fruit Tree Res Sta E 4:21-27

Kobayashi S (1987) Uniformity of plants regenerated from orange (*Citrus sinensis* Osb.) protoplasts. Theor Appl Genet 74:10-14

Kochba J and Button J (1974) The stimulation of embryogenesis and embryoid development in habituated ovular callus from the 'Shamouti' orange (*Citrus sinensis*) as affected by tissue age and sucrose concentration. Z Pflanzenphysiol 73:415-421

Kochba J, Spiegel-Roy P, Neumann H and Saad S (1978a) Stimulation on embryogenesis in citrus ovular callus by ABA, ethephon, CCC and alar and its suppression by GA_3. Z Pflanzenphysiol. 89:427-432

Kochba J, Spiegel-Roy P, Saad S and Neumann H (1978b) Stimulation of embryogenesis in *Citrus* tissue culture by galactose. Naturwissenchaften 65:261

Kotsias D and Roussos PA (2001) An investigation on the effect of different plant growth regulating compounds in *in vitro* shoot tip and node culture of lemon seedlings. Scientia Horticulturae 89:115-128

Krajewski AJ and Rabe E (1995) Citrus flowering: a critical evaluation. J Hortic Sci 70:357-374

Larkin PJ and Scowcroft WR (1981) Somaclonal variation – a novel source of variability from cell cultures for plant improvement. Theor Appl Genet 60:197-214

Litz RE, Moore GA and Srinivasan C (1985) *In vitro* system for propagation and improvement of tropical fruits and palm. Horticultural Reviews 7:157-200

Litz RE and Jaiswal VS (1991) Micropropagation of tropical and subtropical fruits. In: Debergh PC and Zimmerman RH (Eds.) Micropropagation. Kluwer Academic Publishers, The Netherlands pp 247-263

Maheshwari P and Rangaswamy NS (1958) Polyembryony and *in vitro* culture of embryos of *Citrus* and *Mangifera*. Ind J Hort 15:275-282

Maggon R and Singh BD (1995) Promotion of adventitious bud regeneration by ABA in combination with BAP in epicotyl and hypocotyl explants of sweet orange (*Citrus sinensis* L. Osbeck). Scientia Horticulturae 63:123-128

Marin ML and Duran-Vila N (1991) Conservation of citrus germplasm *in vitro*. J Am Soc Hort Sci 116:740-746

Mas O, del Valle N, Rios A and Ramos M (1992) *In vitro* propagation of Citremon. Proc Int Soc Citriculture 1:318-320

Mitra GC and Chaturvedi HC (1972) Embryoids and complete plants from unpollinated ovaries and from ovules of *in vivo*-grown emasculated flower buds of *Citrus* spp. Torrey Bot Club 99:184-189

Moore GA (1985) Factors affecting *in vitro* embryogenesis from undeveloped ovules of mature citrus fruits. J Am Soc Hort Sci 110:66-70

Moore GA (1986) *In vitro* propagation of *Citrus* rootstocks. HortScience 21:300-301

Moreira-Dias JM, Molina RV, Guardiola JL and García-Luis A (2001) Daylength and photon flux density influence the growth regulator effects on morphogenesis in epicotyl segments of Troyer citrange. Scientia Horticulturae 87:275-290

Murashige T and Skoog F (1962) A revised medium for rapid growth and bioassays with tobacco tissue cultures. Physiol Plant 15:473-497

Murashige T and Tucker DPH (1969) Growth factors requirements of citrus tissue culture. 1st Int Citrus Symp, 3:1155-1161

Murashige T, Bitters WP, Rangan TS, Nauer EM, Roistacher CN and Holliday PB (1972) A technique of shoot apex grafting and its utilization towards recovering virus-free *Citrus* clones. HortScience 7:118-110

Navarro L, Roistacher CN and Murashige T (1975) Improvement of shoot-tip grafting *in vitro* for virus-free citrus. J Am Soc Hort Sci 100:471-479

Navarro L, Ortiz JM and Juarez J (1985) Aberrant citrus plants obtained by somatic embryogenesis of nucelli cultured *in vitro*. HortScience 20:214-215

Navas Castillo J, Moreno P and Duran-Vila N (1995) *Citrus* psorosis, ringspot, cristacortis and concave gum pathogens are maintained in callus culture. Plant Cell Tissue Org Cult 40:133-137

Nito N and Iwamasa M (1990) *In vitro* plantlet formation from juice vesicle callus of satsuma (*Citrus unshiu* Marc.). Plant Cell Tissue Org Cult 20:137-140

Normah MN, Hamidah S and Ghani FD (1997) Micropropagation of *Citrus halimii* – an endangered species of South-east Asia. Plant Cell Tissue Org Cult 50:225-227

Omura M and Hidaka T (1992) Shoot tip culture of *Citrus*. I. Culture condition. Bull Fruit Tree Res Stat 22:23-26

Parthasarathy VA, Nagaraju V, Hazarika BN and Baruah A (1999) An efficient method of acclimatization of micropropagated plantlets of citrus. Tropical Agriculture 76: 147-149

Paudyal KP and Haq N (2000) *In vitro* propagation of pummelo (*Citrus grandis* L. Osbeck). *In Vitro* Cell Dev Biol Plant 36:511-516

Peña L, Martin-Trillo M, Juárez J, Pina JA, Navarro L and Martinez-Zapater JM (2001) Constitutive expression of Arabidopsis LEAFY or APETALA1 genes in citrus reduces their generation time. Nature Biotechnology, 19:263-267

Pérez-Molphe-Balch E and Ochoa-Alejo N (1997) *In vitro* plant regeneration of mexican lime and mandarin by direct organogenesis. HortScience 32:931-934

Pierik RLM (1991) Commercial aspects of micropropagation. In: Parakash J, Pierik RLM (Eds.) Horticulture – New technologies and applications. Dordrecht, Kluwer Academic, pp141-153

Preece JE and Sutter EG (1991) Acclimatization of micropropagated plants to the greenhouse and the fiel. In: Debergh PC and Zimmerman RH (Eds.) Micropropagation. Kluwer Academic Publishers, The Netherlands pp 71-93

Raj Bhansali R and Arya HC (1979) Organogenesis in *Citrus limetoides* Tanaka (sweet lime) callus culture. Phytomorphology 28:97-100

Raman H, Gosal SS and Brar DS (1992) Plant regeneration from callus cultures of *Citrus limon* and *C. jambhiri*. Crop Improvement 19: 100-103

Rani V and Raina SN (2000) Genetic fidelity of organized meristem-derived micropropagated plants: a critical reappraisal. *In Vitro* Cell Dev Biol Plant 36:319-330

Rangan TS, Murashige T and Bitters WP (1968) *In vitro* initiation of nucellar embryons in monoembryonic citrus. HortScience 3:226-227

Roistacher CN, Navarro L and Murashige T (1976) Recovery of citrus selections free of several viruses, exocortis viroid, and *Spiroplasma citri* by shoot tip grafting *in vitro*. Proc 7th Conf IOCV, Riverside, 786-793

Roistacher CN (1991) Graft-transmissible disease of Citrus. Handbook for detection and diagnosis. FAO, Rome

Roistacher CN (1993) Arguments for establishing a mandatory certification program for Citrus. Proc 4th Conf Int Soc Citrus Nurserymen, South Africa, pp18-34

Sauton A, Mouras A and Lutz A (1982) Plant regeneration from citrus root meristems. J Hort Sci 57:227-231

Savino V and D'Onghia AM (1998) Certification of *Citrus* in Italy. Options Méditerranéennes, Proc Mediterranean Network Certification Citrus, Italy, 21:37-43

Sim GE, Goh CJ and Loh CS (1989) Micropropagation of *Citrus mitis* Blanco-multiple bud formation from shoot and root explants in the presence of 6-benzylaminopurine. Plant Sci 59:203-210

Singh S, Ray BK, Bhattacharyya S and Deka PC (1994) *In vitro* propagation of *Citrus reticulata* Blanco and *Citrus limon* Burm. HortScience 29:214-216

Skirvin RM (1981) Fruit crops. In Conger BV (Ed.) Cloning agricultural plants via *in vitro* techniques. Boca Raton, Florida, CRC Press, pp 51-139

Snowball AM, Halligan EA, Warrington IJ and Mullins MG (1994a) Phase change in citrus: growth and flowering of citrus seedlings from thirteen genetically diverse seedling families. J Hortic Sci 69:141-148

Snowball AM, Warrington IJ, Halligan EA and Mullins MG (1994b) Phase change in citrus: The effect of main stem node number, branch habit and paclobutrazol application on flowering in citrus seedlings. J Hortic Sci 69:149-160

Starrantino A and Russo F (1980) Seedlings from undeveloped ovules of ripe fruits of polyembryonic citrus cultivars. HortScience 15:296-297

Starrantino A and Caruso A (1987) Experiences on the '*in vitro*' propagation of some citrus rootstocks. Acta Horticulturae 212:471-478

Starrantino A and Caruso A (1988) *In vitro* culture for citrus micropropagation. Acta Horticulturae 227:444-446

Swartz HJ (1991) Post culture behavior: genetic and epigenetic effects and related problems. In: Debergh PC and Zimmerman RH (Eds.) Micropropagation. Kluwer Academic Publishers, The Netherlands pp 95-121

Tapati D, Mitra GC, Chatterjee A and Das T (1995) Micropropagation of *Citrus sinensis* var. Mosamby – an important scion. Phytomorphology 45:57-64

Van Le B, Thanh Ha N, Anh Hong LT and Tran Thanh Van K (1999) High frequency shoot regeneration from trifoliate orange (*Poncirus trifoliata* L. Raf.) using the thin cell layer method. Comptes Rendus Acad Sci Paris, Life Sciences 322:11056-1111

Vardi A and Spiegel-Roy P (1982) Plant regeneration from *Citrus* protoplasts: Variability in methodological requirements among cultivars and species. Theor Appl Genet 62:171-176

Vu JCV, Niedz RP and Yelenosky G (1993) Glycerol stimulation of chlorophyll synthesis, embryogenesis and carboxylation and sucrose metabolism enzymes in nucellar callus of 'Hamlin' sweet orange. Plant Cell Tissue Org Cult 33:75-80

21. MICROPROPAGATION OF OLIVE (*Olea europaea* L.)

Maurizio Lambardi

Istituto sulla Propagazione delle Specie Legnose, CNR / National Research Council, via Ponte di Formicola n. 76, 50018 Scandicci (Firenze), Italy. E-mail: lambardi@fi.cnr.it

Eddo Rugini

Dipartimento di Produzione Vegetale, sez. Ortofloroarboricoltura, Università degli Studi della Tuscia, via S. Camillo de Lellis, 01100 Viterbo, Italy. E-mail: rugini@unitus.it

CONTENTS

S.M. Jain and K. Ishii (eds.), Micropropagation of Woody Trees and Fruits, 621–646.

1. INTRODUCTION

1.1. Botany and Geographical Distribution

The olive (*Olea europaea* L.), together with the fig and the date, is one of the oldest fruit trees in the Mediterranean area (Zohary and Spiegel-Roy, 1975). The species, originating in the eastern part of the Mediterranean basin, belongs to the genus *Olea*, which is one of over 30 genera of the Oleaceae family. The basic chromosome number of the genus *Olea* is 2n=46 (x=23). Although the genus comprises some 60 species distributed over most of Africa, the warmest zones of Europe, South Asia, Central America, South Australia and New Zealand, the *O. europaea* is typical of the Mediterranean area. Although several different classifications of *Olea europaea* are reported, it is generally accepted that the Euro-Mediterranean domesticated olive includes a wild type (*O. europaea* subsp. *oleaster*) and a cultivated form (*O. europaea* subsp. *sativa* Hoffm. et Link = *O. europaea* subsp. *europaea*). The oleaster is a shrub producing small bitter-tasting fruit with low oil content. In addition to olive and oleaster, the term "olevaster" is also widely used to refer to domesticated escapes (Lavee, 1996).

The olive is an evergreen, medium-sized, slow-growing long-lived fruit tree, with a characteristic basitonous habit of growth. As a Mediterranean tree, it originates from dry, warm, subtropical climates and, as a consequence, it shows great adaptability to difficult soil conditions (like dry and shallow soils) and to very high temperatures. However, the olive tree performs best in temperate dry climates, without sharp and prolonged drops in temperature (particularly below −10°C) during winter. Plants from vegetatively propagated clones, under traditional cultivation, start to bear fruit 3-4 years after planting and, if properly managed, continue fruiting for hundreds of years.

From the Mediterranean basin, olive trees arrived in Latin America in the 16th century, and were introduced into California by the Franciscan friars in the 18th century. In the last century, the considerable increase in olive oil consumption led to a wide diffusion of the species in many other countries, such as Australia, South Africa, Japan, New Zealand, China, India and Pakistan.

1.2. Economic and Social Importance of Olive

The olive is cultivated over a total world surface area of almost 10 million hectares, on 60% of which it represents the main crop. A rough estimate of the global number of olive trees is over 800 million. The annual yield of olives is estimated at 10 million tons, most of which is used for oil production and less than 10% consumed as table olives. As regards the world trend in the last 30 years, the production and the consumption of olive oil have increased together. It is unlikely that this trend will change in the near future, considering the recent introduction or increase of olive cultivation and olive-oil consumption in countries such as Japan, Australia, China and South Africa. The traditional area of olive cultivation is the Mediterranean basin, which accounts for 95% of the olive orchards of the world, and where, mainly in Spain, Italy and Greece, almost 99% of the world's olive oil and more than 80% of its table olives are produced (IOOC, 2001). It has been estimated that in the year 2008 olive production will have risen by about 3 million tons, as a result of an increase of olive cultivation of 120,000 hectares per year.

Although the production of oil and table olives is by far the most important reason for olive cultivation, the species serves other important functions. In several countries of the Mediterranean basin the olive has been closely associated from ancient times with the traditional landscape, and its presence, together with cypress, is typical of religious sites. In these areas, it is one of the few species with characteristics of high drought resistance, and it is suitable for use in difficult terrains such as calcareous, rocky and sloping soils. Finally, its wood is of great value due to its natural durability and beauty; hence, in some areas (like the South of Italy), it represents a secondary source of profit from olive orchards.

1.3. Traditional *In Vivo* Propagation Methods

Although in the past traditional propagation largely made use of rootable organs and natural formations (ovules, stump suckers, 2- or more-year old portions of branches), at present the olive is almost entirely reproduced by cuttings and graftings (Fabbri *et al.*, 2002).

Since the 60s, mist propagation of leafy stem cuttings is the most sophisticated and widely used technique for producing self-rooted olive trees. Leafy cuttings are obtained from one-year-old vigorous shoots. Cuttings (about 15 cm long, and provided with 2-3 pairs of leaves in their upper part) are treated with ethanol solutions or talcum dispersions of indol-3-butyric acid (IBA), prior to being placed in a bottom-heated bench under mist conditions. This method reduces leaf transpiration and temperature, enabling cuttings to remain turgid throughout root emission. As regards grafting propagation, this technique is nowaday confined to specific areas (mainly Central and Southern Italy) where, traditionally, many olive farmers still prefer grafted to self-rooted plants when establishing olive orchards, particularly in shallow soils (Lambardi, 1999). The most common procedure makes use of scions (4-5 cm long) from one-year old branches, which, in spring, are bark grafted on potted or in-field growing olive seedlings.

In spite of recent advances in nursery technology, several problems still affect conventional vegetative propagation of olive, i.e.:

- cuttings root properly only if collected in two specific periods of the

growing season, i.e., before blossoming (spring), and at the onset of autumn growth;
- adventitious rooting capacity varies greatly among cultivars, though usually greater in oil than table cultivars;
- olive cultivars are made up of a combination of clones ("cultivar-populations"); hence, they can perform very differently in different nurseries, even when all the physiological, agronomical and propagation conditions are similar;
- grafting propagation has been largely abandoned (except in Italian nurseries) because of the high costs of management and the lack of specific clonal rootstocks.

In consequence of the above, much is hoped for from the development and the perfecting of micropropagation procedures, to modernize and improve propagation technology, standardize the characteristics of nursery plants, and enhance their performance when in field. Moreover, *in vitro* techniques allow for (i) rapid propagation of cultivars which are difficult to reproduce by traditional propagation, (ii) production of disease-free plants, (iii) application of bioengineering methods, and (iv) conservation of valuable germplasm.

2. CONVENTIONAL MICROPROPAGATION: AXILLARY BUD DEVELOPMENT AND SHOOT PROLIFERATION

Micropropagation of olive has already been reviewed in the past (e.g., Rugini 1986; Rugini and Fedeli, 1990; Cañas *et al.*, 1992). These reports evidenced that, at the beginning of the 90s, few olive cultivars could be efficiently propagated by axillary bud development and shoot proliferation. Moreover, at that time micropropagation was often initiated using explants from embryos and seedlings (e.g., Bao *et al.*, 1980; García-Berenguer and Durán González, 1990; Cañas *et al.*, 1992), but this approach is of minor interest when selected cultivars or clones must be reproduced. When using explants from adult trees, several problems impeded the development of effective protocols of micropropagation, among which: (i) the heavy oxidation of tissues when explants (shoot tips, meristems) were collected from in-field or greenhouse plants, (ii) the difficulty of getting sterile shoots when nodal explants were used, and (iii) the laboriousness of establishing shoot cultures with some cultivars. Over the last decade, many advances have been made towards the solution of these problems and the optimization of the various steps involved in olive micropropagation. Table 1 concisely summarizes the cultivars for which a complete procedure of micropropagation from mature trees has been reported.

2.1. Initiation of Cultures, Axillary Bud Development and Shoot Proliferation

The initiation of olive micropropagation using buds or nodal segments from adult field-grown plants is difficult and time consuming, mainly because of the rapid oxidation of tissues after plating. Best results are obtained when explants are collected from organs with strong characteristics of juvenility (such as suckers; Rugini and Fontanazza, 1981) or from vigorous upgrowing shoots. A suitable material for the establishment of micropropagation from in-field trees are the ovules (Rama and Pontikis, 1990), which are spheroblasts naturally rich in adventitious buds. Cañas *et al.* (1992) proposed an effective procedure in which uninodal leafy explants were excised from shoots sprouted from ovules cultured in semi-aseptic conditions. Following axillary bud stimulation, conventional shoot proliferation and rooting was obtained. However, no information was given on the state of juvenility of derived plants, which can be expected to be considerable when ovules are utilized as the source of explants.

Explants collected from potted young stock plants, grown in greenhouse, are the ideal material to initiate *in vitro* propagation of olive, particularly when tender apical twigs and nodal segments (Fig. 1A) are excised from vigorous shoots, soon after sprouting (Rugini and Fedeli, 1990). Tissue sterilization before its introduction *in vitro* is a fundamental step in olive micropropagation. Although old procedures also include passages in ethanol solutions (Scaramuzzi and De Gaetano, 1974; Grossoni, 1979; Rugini and Fontanazza, 1981), the present tendency is to avoid the use of ethyl alcohol, which causes tissue dehydration. Treatments with calcium or sodium hypochloride, at different concentrations, alone or in combination with $HgCl_2$, have been proposed (Rugini, 1991; Mencuccini, 1995). Martino *et al.* (1999) obtained best sterilization and survival rates when shoot tips from old field-grown trees of 'Moraiolo' were immersed for 5 min in 0.05-0.1% $HgCl_2$, rinsed in sterile water, treated with 15% sodium hypochloride, and finally rinsed three times in sterile water before plating. Standardi *et al.* (1998) reported that microbe contamination is reduced when a slight water stress is induced on potted stock plants by diminishing the amount they are watered a few days before explant collection.

2.1.1. Medium Formulation

The development of a specific olive medium (OM) for axillary bud stimulation and subsequent shoot multiplication marked an important step forward in the improvement of olive micropropagation. The medium was formulated on the basis of the analysis of the main mineral elements of shoot apices from field plants, during their rapid growth (Rugini, 1984). The major differences between MS (Murashige and Skoog, 1962) and OM

Table 1. Summary of best culture conditions in micropropagation of olive cultivars. Only procedures that were initiated with explants from mature trees are reported (m, modified medium; NR, not reported).

Cultivar	Original explant	SHOOT PROLIFERATION			SHOOT ROOTING			Reference
		Medium	Hormones	Ratio	Medium	Hormones	%	
Arbequina	Micrografted shoots	DKW [1]	BA, 4.4 μM + IBA, 0.05 μM	7-8	½ DKW	IBA, 0.5 μM	57	Revilla *et al.*, 1996
Arbequina	Uninodal explants	OM	zeatin, 10 mg/l	4	OM	IBA, 3 mg/l or NAA, 1 mg/l	NR	Otero and Docampo, 1998
Arbequina, Empeltre, Picual	Uninodal explants	OM	BA, 1 μM + TDZ, 1 μM	2.5-2.8	Compost	IBA, 15 μM + IAA, 10 μM	75	Garcia-Fèrriz *et al.*, 2000
Carolea	Uninodal explants	½ OM	zeatin, 4 mg/l	3	½ OM	NAA, 2 mg/l or IBA, 2.5 mg/l + Putrescine, 160 mg/l	NR	Briccoli Bati *et al.*, 1994
Dolce Agogia	Uninodal explants from suckers	½ MS	zeatin, 10 mg/l + IBA, 0.5 mg/l + GA_3, 0.5 mg/l	6	½ Knop / Heller [2]	NAA, 4 mg/l	84	Rugini and Fontanazza, 1981
Frantoio	Uninodal explants	OM	zeatin, 4 mg/l or 2iP, 4 mg/l	9-10	Various	NAA, 1 mg/l	80	Rugini, 1984
FS-17	Uninodal explants	mOM	zeatin, 2 mg/l or 2iP, 4 mg/l	7.2	½ MS	NAA, 2 mg/l	100	Mencuccini, 1998
Kalamon	Uninodal explants [3]	mOM	zeatin, 10 mg/l + crude extract [4]	6	mOM	NAA, 2 mg/l + crude extract [4]	82	Rama and Pontikis, 1990

Kalamon	Uninodal explants	WPM [5]	BA, 1 mg/l + IBA, 1 mg/l + GA_3, 0.1 mg/l	9.6	WPM	IBA, 2 mg/l	80	Dimassi-Theriou, 1994
Maurino	Three-nodal explants	mMS	zeatin, 0.5 mg/l or TIBA, 0.5 mg/l	6	mMS	K-IBA, 500 mg/l ("dip method")	60	Bartolini *et al.*, 1990
Memecik, Domat	Nodal explants	mOM	BA, 1 mg/l	13	NR			Seyhan and Özzambak, 1994
Meski	Uninodal explants	mOM	zeatin, 1 mg/l	11	mOM	IBA, 1 g/l ("dip method")	100	Chaari-Rkhis *et al.*, 2000
Moraiolo	Shoot apices	OM	zeatin, 4 mg/l or 2iP, 4 mg/l	9-10	Various	NAA, 1 mg/l	80	Rugini, 1984
Nocellara etnea	Uninodal explants	mOM	zeatin, 4 mg/l	5.6	mOM	NAA, 1.5 mg/l	100	Briccoli Bati and Lombardo, 1995
Tunisian cultivars [6]	Uni- or binodal explants	mMS	zeatin, 2 mg/l + kinetin, 1-2 mg/l (or BA, 2 mg/l)	1-8	½ mMS	NAA, 2 mg/l	100	Chaari-Rkhis *et al.*, 1999

(1) From Driver and Kuniyuki (1984).
(2) From Heller (1953).
(3) Uninodal explants from shoots developed from excised ovules.
(4) Crude extract was obtained by fresh tissue maceration and methanol extraction of olive ovules.
(5) From McCown and Lloyd (1981).
(6) Cvs 'Sig de Sfax', 'Chemlali de Sfax', 'Chetoui' and 'Meski'.

medium formulations are: (i) the OM medium is richer in Ca, Mg, S, P, B, Cu and Zn, (ii) it has a slightly different Ca/N ratio (1:11), and (ii) it also contains glutammine as a nitrogen source. The Ca/N ratio, in particular, proved to be an important factor in olive micropropagation, as it can interfere with the maintenance of healthy shoots (Fiorino and Leva, 1986). Among the microelements, the role of boron is controversial: while shoot proliferation of 'Maurino' and 'Nocellara etnea' cultivars was improved by the presence of boron in the culture medium (Leva *et al.*, 1992), its absence doubled the number of shoots per explant of the 'Kalamon' cultivar (Dimassi, 1999). However, the effectiveness of OM formulation for a wide pool of olive cultivars is demonstrated in Table 1, as two-thirds of the procedures were developed using the medium in its original formulation or with modifications.

As regards the carbon source, it has been shown that, in comparison with sucrose, mannitol (one of the major carbohydrates of olive metabolism) increases shoot proliferation, improves the general quality and uniformity of shoot cultures and reduces basal callus formation (Leva *et al.*, 1994).

Cozza *et al.* (1997) reported that the medium formulation can considerably influence the nutritional state, the shoot morphology and the leaf histology of olive micropropagated plantlets ('Nocellara etnea' and 'Nocellara del Belice'). In particular, greater amounts of N, P and Mg, and a reduced internode length of shoots were observed following shoot culture in half-strength OM medium, rather than in a modified (Fiorino and Leva, 1986) half-strength MS medium. Moreover, in half-strength OM medium, vascular elements of leaves were fewer in number and smaller in size, and cell expansion was reduced.

2.1.2. Growth Regulators

Olive is characterized by a strong apical dominance. As a consequence, the proliferation of shoots through the stimulation of axillary buds (a typical approach with the majority of fruit species) is of minor importance. In olive, shoot proliferation is a consequence of "multiplication" of shoots which, at each subculture, is achieved mainly by means of uni- or binodal segmentation of elongated shoots (Fig. 1B and 1C). Cytokinins play a major role in the regulation of this phenomenon. The cytokinin which induces maximum stimulation of axillary buds is zeatin, it is applied in a wide range of concentrations, alone or in combination with other growth regulators. In Table 1, its concentration ranges, when used alone, from 0.5 mg/l (cv 'Maurino'; Bartolini *et al.*, 1990) up to 10 mg/l (cv 'Arbequina'; Otero and Docampo, 1998). However, Mencuccini *et al.* (1997) showed that the proliferating activity of 'Moraiolo' shoot cultures was the same when 2 or 4 mg/l of zeatin (mixed isomers) were applied.

It is commonly stated that the other cytokinins cannot replace zeatin in promoting olive bud stimulation, although some exceptions have been reported (e.g., Seyhan and Özzanbak, 1994; Dimassi-Theriou, 1994; Revilla *et al.*, 1996). In general, a lower proliferative efficiency is reported for 6-γ,γ-dimethylallylaminopurina (2iP; Rugini, 1986; Mencuccini *et al.*, 1997), while 6-benzyladenine (BA), the cytokinin most often used in tissue culture, is not very effective in that it produces short thin shoots and, occasionally, abundant callus proliferation (Rugini, 1991). Various growth regulators – BA, kinetin, thidiazuron (TDZ), gibberellic acid (GA_3), IBA – have been used in different combinations (with or without zeatin) to promote the stimulation of axillary buds (see Table 1), but the results of these reports are always restricted to one or few specific cultivars; hence, they can never be considered for general application. Nevertheless, in consideration of the high cost of zeatin, efforts are currently being made to find cytokinin combinations that can replace, or at least reduce, zeatin use, without losing proliferative potential of shoot cultures. In this sense, the combined use of zeatin with 2iP and TDZ (a cytokinin-like substance) seems promising (Zuccherelli and Zuccherelli, 2000). After proliferation, a passage in a medium containing filter-sterilized GA_3 (20-40 mg/l) induces a rapid elongation of shoots, which are more suitable for subsequent rooting (Rugini *et al.*, 2001).

2.2. Root Induction

Great advances have been made in rooting of micropropagated shoots over the last decade, so that even cultivars "recalcitrant" to cutting propagation (e.g., 'Nocellara etnea') can now be satisfactorily rooted *in vitro* (Briccoli Bati and Lombardo, 1995). Medium formulation and auxin treatment were the first factors to attract the attention of researchers. Table 1 shows that, with only two exceptions ('Dolce Agogia', 'FS-17'), rooting was performed on the same medium of shoot proliferation, generally by reducing to ½ the strength of the original macroelement formulation. As regards the use of growth regulators, both IBA and 1-naphthaleneacetic acid (NAA) have been indicated as most effective for the stimulation of rooting (Fig. 1D). It is often reported that NAA produces a stronger stimulation than IBA, as a lower hormone concentration is required to obtain similar rooting activity (Briccoli Bati *et al.*, 1994; Briccoli Bati and Godino, 1996; Otero and Docampo, 1998). However, Briccoli Bati *et al.* (1999) showed that, in comparison with NAA, IBA was able to induce earlier rhizogenesis of 'Nocellara etnea' shoots, especially when 50-100 ppm H_3PO_4 was added to the rooting medium. Furthermore, they also demonstrated the considerable influence of the proliferation medium on subsequent shoot rooting expression.

Alternatively to the traditional use of rooting media, Bartolini *et al.* (1990) proposed a "dip rooting method", which consisted in dipping the base of 'Maurino' microcuttings for 30 min in a concentrated solution of potassium-salt IBA, before culturing in a hormone-free medium. Similarly,

Rugini and Fedeli (1990) reported high rooting percentages by dipping entire microcuttings, or only their basal parts, in a IBA solution (100-200 mg/l) for 10-20 sec. Although very effective in rooting stimulation, the method has never been reported on since.

Over time, other rooting methods have been proposed to improve micropropagation of difficult-to-root cultivars, i.e., the basal etiolation of shoots and the addition of polyamines to rooting medium. Basal etiolation, performed by black painting of the outside of the jars and by covering the agarized rooting medium with black sterile polycarbonate granules, proved to be effective in enhancing the rootability of 'Moraiolo' shoots (Rugini *et al.*, 1993). Olive rooting is also promoted by the addition of 1 mM putrescine to an auxin-containing rooting medium. Indeed, this treatment increased the rooting percentage and the number of roots per shoot; earlier rooting was induced as well (Rugini *et al.*, 1997). These marked effects are presumably due to a general low content of endogenous polyamines in olive shoots at the end of proliferation, in comparison with other woody species (e.g., walnut and chestnut) which are not affected by putrescine treatment (Rugini *et al.*, 1993). It was proved that putrescine affects early root differentiation similarly to exogenous treatments with H_2O_2, a product of polyamine catabolism. These observations led to the hypothesis that putrescine acts during the root induction phase by means of an increase of peroxidase activity through H_2O_2. In fact, during the first 48 hours of root induction, a drastic reduction of polyamines was observed at the basal part of able-to-root microcuttings, even when cultured in putrescine-containing medium (Rugini *et al.*, 1997).

2.3. Acclimatation and Field Performance of Micropropagated Plants

Acclimatation is another critical point in olive micropropagation, due to the drastic change of climatic conditions (humidity, light intensity, asepsis) which characterizes the passage from the *in vitro* to the *in vivo* environment. This problem is accentuated by the particular histology of leaves from *in vitro* culture, which makes them even more prone to dessiccation, as well as poorly functional in the acquisition of autotrophic condition. It was reported, for instance, that micropropagated plants of 'Nocellara del Belice' and 'Nocellara etnea' had leaves with a notable reduction of both thickness and central vascular bundle diameter when compared with in-field grown plants. Moreover, the epidermal layer showed a thin cuticle, the stomatal density was low, and the palisade tissue was composed of only one layer, instead of the normal 3 to 4 layers. These differences were more accentuated in the 'Nocellara del Belice' cultivar (Cozza *et al.*, 1997). As a consequence, to reduce as much as possible the stress resulting from their transfer to *in vivo* conditions, the plantlets must be potted in appropriate compost substrates (e.g., peat moss, perlite and polystirole granules, 2:2:1; Rugini and Fedeli, 1990), and acclimatated under a transparent plastic film or in fog conditions. Following acclimatation, a one-year hardening period is required before their final plantation in orchard (Mencuccini, 1995). It is worthy of note that micropropagated polyploids (4n) olive plants showed great precociousness in blooming, as they started to bloom during the first year after their transplantation into pots (Rugini, unpublished).

2.4. Genetic and Morphological Stability of Micropropagated Plants

To date, few reports have dealt with the genetic and agronomic characteristics of micropropagated olive trees after in-field plantation. With 'Maurino', the genetic fidelity of *in vitro* propagated trees to the donor plant was proved by PCR-RAPD analysis (Leva *et al.*, 2000). Moreover, trees did not show any difference to self-rooted plants in their architecture and morphology; indeed, they showed superior growth during the second year. Similarly, the conformity to the original germplasm collection of 'Empeltre', 'Arbequina' and 'Picual' plants from micropropagation was clearly proved by PCR-RAPD with the use of only 4 primers (Garcia-Fèrriz *et al.*, 2000). Briccoli Bati *et al.* (2000) observed that, after in-field plantation, micropropagated plants of 'Nocellara etnea' and 'Carolea' did not exhibit juvenile traits and had a short unproductive period, as they started to bear fruit in the second and third growing season, respectively. Acclimatated plants of several other cultivars (such as 'Frantoio', 'Canino', 'Leccino', 'Moraiolo') were able to differentiate flowers even during their growth in pots, while over 2 years of juvenility was observed only in plants micropropagated from sucker explants or from 'Canino' somatic embryos (Rugini, unpublished).

3. NEW APPROACHES TO OLIVE REGENERATION

In the last two decades, great efforts have been made to develop procedures of olive regeneration other than axillary bud stimulation, such as organogenesis from callus culture and somatic embryogenesis. An important reason was to explore nonconventional methods of genetic improvement, such as transformation by *Agrobacterium*, somaclonal variation, protoplast manipulation (reviewed by Rugini *et al.*, 2000) or microprojectile bombardment (Lambardi *et al.*, 1999a). In this context, it is fundamental to apply bioengeneering techniques to *in vitro* culture systems characterized by high morphogenetic potential, in order to get plant regeneration from manipulated cells. Moreover, olive can take advantage of efficient regeneration techniques (such as somatic embryogenesis) in other ways, i.e., (i) by overcoming the problem of non-rootability of some cultivars, which requires laborious grafting procedures, and (ii) with the production of synthetic seeds.

3.1. Shoot Organogenesis

Since adventitious shoots and roots were obtained for the first time from seedling explants (Gilad and Lavee, 1974), few reports have dealt with olive regeneration by shoot organogenesis. Nevertheless, information on several aspects of shoot organogenesis (e.g., regeneration potential of cultivars and tissues, culture conditions, shoot development and rooting) is available.

Direct shoot regeneration was induced on olive hypocotyls in White's medium, supplemented with 0.05 μM NAA and 2.5 μM BA (Bao *et al.*, 1980), as well as from cotyledons of mature seeds (Rugini, 1986). Wang *et al.* (1979) described two histologically different meristematic tissues originating from olive callus, i.e., cellular masses and nodules, which evolved, respectively, into adventitious roots and buds.

High shoot organogenesis was obtained from callus, which had been previously induced on cotyledon segments excised from 'Tanche' and 'Picual' embryos (Cañas and Benbadis, 1988). Initial callus proliferation was produced on OMc medium (an OM medium, modified in the macroelements as from Bourgin and Nitsch, 1967), supplemented with a high auxin (IBA)/cytokinin (2iP or zeatin) ratio. Shoot organogenesis was stimulated when the calli were transferred onto OM medium containing 4 mg/l 2iP. Maximum shoot regeneration was observed in calli from cotyledon segments proximal to the embryo axes rather than the distal ones, suggesting that a gradient of regeneration potential existed from the proximal to the distal region of olive cotyledons. Rooting of isolated shoots was easily obtained on 1 mg/l IBA- or NAA-containing OM medium.

Mencuccini and Rugini (1993) obtained adventitious buds in petioles from *in vitro*-grown shoots of olive (cvs 'Moraiolo', 'Dolce Agogia' and 'Halkidikis'). Attempts to regenerate shoots from leaf discs and midribs, as well as from petioles excised from field or potted plants, were totally unsuccessful. The efficiency of shoot organogenesis was heavily dependent on: (i) the cultivar ('Moraiolo' being the best), (ii) the position of petiole along the shoot (apical nodes being better than basal ones), (iii) the medium/hormone combination, and (iv) the maintenance of darkness during culture. Following rooting, the regenerated plantlets did not show morphological differences in comparison with the micropropagated donor plantlets.

3.2. Somatic Embryogenesis

New procedures of somatic embryogenesis have been reported in recent years for olive (Table 2), providing important information on the influence of a variety of factors. These advances can be regarded as a starting point from which further progress in the technique can be made.

3.2.1. Somatic Embryogenesis from Zygotic Embryo and Seedling Explants

The induction of somatic embryos has been obtained from both immature and mature zygotic embryo explants, with or without callus interposition.

Rugini (1988) was the first to report somatic embryogenesis from immature zygotic embryos of various cultivars ('Dolce Agogia', 'Leccino', 'Frantoio' and 'Moraiolo'), after they were cultured in half-strength MS medium containing 0.5 or 2.5 μM BA. At higher BA concentrations somatic embryogenesis was inhibited. The majority (70%) of somatic embryos originated directly from the immature embryo (often arranged in a crown-like position), with no callus formation. It is interesting to note that embryogenesis occurred only when the zygotic embryos were harvested 75 days after full bloom, while only callus was obtained from those collected before or after this period. The embryogenic capacity could be maintained for at least 2 months following embryo storage at 14-15°C.

The existence of a "window" of embryogenic competence during zygotic embryo development was also reported by Leva *et al.* (1995). They observed that only cotyledonary explants from immature embryos (cvs 'Picholine', 'Frangivento' and 'Frantoio') harvested between 60 and 90 days after anthesis were competent for embryogenic callus induction, following their culture in SH (Schenk and Hildebrandt, 1972) medium containing various combinations of NAA and 2iP. When cotyledons came from earlier (30 days) or later (130 days) collections, no evidence of somatic embryogenesis was observed. Differently, embryo-like structures could be observed in calli, originated from cotyledonary explants, when 'Chalkidikis' zygotic embryos were harvested 126 days after full bloom (Pritsa and Voyiatzis, 1999).

When radicles, distal and proximal cotyledon segments were excised from mature zygotic embryos of wild olive (*Olea europaea* var. *sylvestris*), during the first 21 days of culture on inductive medium, they originated calli that showed a different morphogenetic potential (Orinos and Mitrakos, 1991). Indeed, when they were subcultured on OMc medium containing 2.5 μM IBA, all the three forms of calli underwent a "developmental" stage of rhizogenesis during the first 30 days, followed by a stage during which low-level somatic embryogenesis (in a range of 5 to 7%), but not rhizogenesis, occurred. Moreover, somatic embryogenesis was always more pronounced in calli from radicle explants. The dissimilar morphogenetic expression of calli from different tissues was confirmed also when mature embryo of the cv 'Koroneiki' were used as the source of explants (Mitrakos *et al.*, 1992). Here, both rhizogenesis and somatic embryogenesis were high from radicle calli and low from distal-cotyledon calli, while only high rhizogenesis was promoted in calli from proximal cotyledon segments. It is notable that the

Table 2. Summary of reports on somatic embryogenesis of olive. Culture conditions of "Callus Induction" and "Somatic Embryogenesis Expression" are indicated as: medium formulation, growth regulators, light regime. Only medium formulation and growth regulators are indicated for "Somatic Embryo Development", as the light regime is always 16h (D, darkness; 16h, 16 hour photoperiod; – , hormone-free medium; NR, not reported).

Explants	**Callus Induction**	**Somatic Embryogenesis Expression**	**Somatic Embryo Development**	**Reference**
Immature zygotic embryos [(1)]		½ MS, 0.5-2.5 µM BA, D	OM, –	Rugini, 1988
Cotyledons and radicles from mature embryos [(2)]	OMc, 2.5 µM 2iP + 25 µM IBA, 16h	OMc, 2.5 µM IBA, 16h	½ MS, –	Orinos and Mitrakos, 1991
Radicles from mature zygotic embryos (cv 'Koroneiki')	OMc, 2.5 µM 2iP + + 25 µM IBA, 16h	OMc, – , 16h	NR	Mitrakos *et al.*, 1992
Taproot callus from seedlings (cv 'S. Agostino')	MS, 25 µM NAA + 0.5 µM BA, 16h	½ MS, 0.5 µM NAA + 0.5 µM BA, D	½ MS, 2.5 µM zeatin	Rugini *et al.*, 1995
Cotyledons from immature zygotic embryos [(3)]	SH, 1.25-2.5 µM NAA ± 0.5 µM 2iP, D	SH, – , 16h	SH, –	Leva *et al.*, 1995

Petioles from *in vitro*-growing shoots [4]	½ MS, 30 µM TDZ + 0.54 µM NAA, D	OMc, 0.5 µM 2iP + 0.44 µM BA + 0.25 µM IBA + 0.42 mM cefotaxime, D	Liquid OMc, – [5]	Rugini and Caricato, 1995
Root explants from germinated seeds ('Nabali')	MS, 5 µM 2,4-D [6] + 0.5 µM kinetin + 5 µM NAA, D	MS, 5 µM 2iP + 0.5 µM 2,4-D + 0.5 µM IBA, 16h	MS, –	Shibli *et al.*, 2001

(1) Embryos were excised from drupes of the cvs 'Dolce Agogia', 'Leccino', 'Frantoio' and 'Moraiolo', harvested 75 days after full bloom.
(2) Mature zygotic embryos were excised from drupes of wild olive (*Olea europaea* var. *sylvestris* (Miller) Lehr).
(3) Embryos were excised from drupes of the cvs 'Frangivento', 'Frantoio' and 'Picholine'.
(4) Shoot cultures had been initiated 3-5 years before from mature trees of the cvs 'Canino' and 'Moraiolo', and subcultured every 40 days.
(5) Cyclic embryogenesis was then established by transferring primary somatic embryos onto hormone-free OMc medium, supplemented with 0.1% activated charcoal and solidified with Phytagel.
(6) 2,4-D, 2,4-dichlorophenoxyacetic acid.

high level of somatic embryogenesis (up to 40%) was obtained in radicle calli that, after 14-21 days in induction medium, were subcultured on OMc medium without exogenous growth regulators. The high embryogenic potential of root callus was confirmed in trials with seedlings of the cvs 'S. Agostino' (Rugini *et al.*, 1995) and 'Nabali' (Shibli *et al.*, 2001).

It is well known that the combination and concentration of growth regulators in the medium regulate the different phases of somatic embryogenesis (from callus induction to somatic embryo development and germination). Table 2 shows a few similarities among the various procedures, i.e., (i) a high auxin/cytokinin ratio is required for callus induction, (ii) 2iP is the prevailing cytokinin in this phase (at a concentration of 0.5-2.5 μM), and (iii) a hormone-free medium is generally required for the development and germination of somatic embryos. No common trends can be observed during the expression phase of somatic embryogenesis, as hormone-free media, media supplemented with cytokinins or auxins alone, and media containing combinations of them are reported as well.

Another point of inconsistency concerns the light regimes applied during culture. Both darkness and 16h photoperiod conditions have been reported during both callus induction, and somatic embryogenesis expression, but only in a few cases the selected light regime was the result of specific experiments (Rugini, 1988; Shibli *et al.*, 2001).

3.2.2. Somatic Embryogenesis from Mature Tissue Explants

A cyclical system of somatic embryogenesis from mature tissue of olive was described by Rugini and Caricato (1995), and subsequently indicated as "double regeneration system". The original explants (i.e., petioles from micropropagated shoots of the cvs 'Canino' and 'Moraiolo') produced callus on half-strength MS medium supplemented with 30 μM TDZ, 0.54 μM NAA and 4% sucrose. When the calli were subcultured in the dark on gelled OMc medium supplemented with growth regulators (as reported in Table 2), 1 g/l casein hydrolysate and 3% sucrose, morphogenetic masses were produced. Afterwards, these masses differentiated primary somatic embryos on a filter paper soaked in liquid OMc medium. By subculturing primary somatic embryos on hormone-free OMc medium, with the addition of 0.1% activated charcoal and solidified with phytagel, cycles of secondary somatic embryogenis were established and maintained for years by monthly subculturing. Somatic embryo germination was achieved after their isolation and transfer into shaken liquid OMc medium. Histological observations (Lambardi *et al.*, 1999b) showed that, in this system, secondary somatic embryos originated from the epidermal (or, rarely, the first subepidermal) layer of the primary embryo, mainly from its basal part (Fig. 2A). Subsequent divisions of embryo tissues led to the formation of highly proliferating masses (Fig. 2B). Although it was not possible to ascertain a mono- or multicellular origin of somatic embryos, a very limited number of cells of the primary explant was apparently involved in the formation of embryo primordia (Fig. 2C). In the embryogenic masses, together with a majority of perfect somatic embryos (Fig. 2D), severel other forms could be recognized, such as fused embryos (Fig. 2E), teratomic leaves, claviform structures, embryos with fused cotyledons. These abnormal structures never developed further (Benelli *et al.*, 2001a).

3.3. Synthetic Seeds and Micrografting

Important advances have been made recently in research on the technology of synthetic seeds, so that many different propagules (other than somatic embryos) can now be used for the production of "synseeds", among which apical and axillary buds, shoot tips, adventitious buds, protocorms (Standardi and Piccioni, 1998). In olive, Micheli *et al.* (1998) proposed a synthetic seed technique aiming towards the improvement of conventional micropropagation and the application to germplasm preservation. After encapsulation in an alginate matrix, containing a nutritive medium, apical and nodal buds from micropropagated shoot cultures (cv 'Moraiolo') maintained up to 49% viability and regrew satisfactorily after storage at 4°C for 45 days. However, as rooting was unsatisfactory and restricted only to microcuttings formed from apical bud beads, the technique seems still a long way from being applicable in practice. Recently, olive somatic embryos (cv 'Canino') have also been used for synthetic seed preparation (Lambardi, unpublished). Following encapsulation in calcium-alginate (Fig. 2F) and plating of the beads in a solidified nutrient medium, survival and further development of somatic embryos was achieved.

The feasibility of olive *in vitro* propagation by micrografting has been reported. Revilla *et al.* (1996) induced rejuvenation of mature olive trees (cv 'Arbequina') after cycles of shoot micrografting on *in vitro*-grown seedling rootstocks. Troncoso *et al.* (1999) cleft-micrografted uninodal explants (from *in vitro*-grown 'Cañivano' seedlings) on *in vitro* 'Arbequina' seedlings, prepared with a cut just under the basal pair of leaves. After 60 days, 85% grafting survival was achieved, which then dropped to 67% after plantlet hardening. These results suggest that micrografting should be further explored as a simple and effective method for olive multiplication.

4. *IN VITRO* CONSERVATION OF OLIVE

Almost 4000 cultivars of olive are cultivated in the world, mainly in the Mediterranean basin, although this huge number is partially due to many synonyms among cultivar names (Bartolini *et al.*, 1998). Hence, it is evident that germplasm conservation strategies are of fundamental importance in the preservation of olive genetic variability. As for the majority of woody species, in-field gene banking (clonal orchards) is at present the only option open for olive conservation, even though germplasm kept this way remains vulnerable to loss due to disease, pests, extreme environmental conditions and economic pressures. As reported elsewhere in this book (chapter "Application of Tissue Culture to the Germplasm Conservation of Temperate Broad-Leaf Trees"), an alternative or complementary approach is the preservation of germplasm by exploiting tissue culture technology. Here, advances in olive slow growth storage and cryopreservation are reviewed.

4.1. Slow Growth Storage

Lambardi *et al.* (2000) showed that the maintenance of darkness or light conditions during slow growth storage can markedly influence post-conservation shoot regrowth. Olive shoot cultures (cvs 'Leccino' and 'Frantoio'), subcultured monthly on OM medium supplemented with 10 µM zeatin, were transferred into gas-permeable containers on hormone-free OM medium and then stored at 4°C, under a 8h photoperiod (25 µmol m^{-2} s^{-1} photosynthetically active radiation) or in total darkness. After 4 months 100% survival was achieved with both cultivars (Table 3), but shoots conserved at 4°C under 8h photoperiod were in healthier condition and had a more intense leaf colour. On prolonging storage under 8h photoperiod to 8 months, the percentage of shoot survival dropped drastically to 40% ('Leccino') and 30% ('Frantoio'), while shoot regrowth (after subculture onto fresh proliferation medium at 23°C and 16h photoperiod) was even lower. In contrast, when shoots were stored for 8 months in the dark, 80% ('Frantoio') and 90% ('Leccino') of the shoots regrew satisfactorily and produced new nodes similar to those of unstored shoots.

Gardi *et al.* (2001) stored shoot cultures of 'Ascolana tenera', 'Frantoio' and 'Moraiolo' at 6°C and in the dark. 'Frantoio' was the one that performed best, as 100% survival and regrowth could be obtained after 5 months of conservation, while both the other cultivars maintained this maximum regrowth potential only up to 2 months.

4.2. Cryopreservation

Promising results have recently been obtained with the application of cryogenic techniques to the long-term conservation of olive germplasm. Martinez *et al.* (1999) obtained 30% survival of olive shoot tips (cv 'Arbequina') following the removal of up to 30% of their moisture content, direct immersion into liquid nitrogen (–196°C), and rewarming at room temperature. Lambardi *et al.* (2000) applied a procedure of vitrification and one-step freezing in liquid nitrogen to shoot tips excised from *in vitro*-grown shoot cultures of the cv 'Frantoio'. Following re-warming at 40°C and plating in a regrowth medium, 15% survival rate was achieved, but only from shoot tips which had been obtained from apical buds. The surviving

Table 3. Performance of olive shoot cultures, after storage at 4°C for up to 12 months (LEC, cv 'Leccino'; FRA, cv 'Frantoio'; SE, standard error). From Lambardi *et al.*, 2000.

Duration of shoot storage	Survival (%)		Regrowth (%)[a]		New nodes (x ± SE)[a]	
	LEC	FRA	LEC	FRA	LEC	FRA
8h photoperiod						
4 months	100	100	80	60	2.6 ± 0.4	6.2 ± 1.0
8 months	40	30	15	10	2.0 ± 0.0	1.5 ± 0.5
12 months	0	0	0	0	—	—
Darkness						
4 months	100	100	95	85	3.5 ± 0.6	4.9 ± 0.8
8 months	100	90	90	80	4.3 ± 0.4	4.2 ± 0.5
12 months	66	40	0	0	—	—
Unstored	100	100	100	100	4.5 ± 0.5	5.8 ± 0.7

[a] The percentage of regrowing shoots and the new nodes (i.e., the average number of new nodes per regrowing shoot) were recorded 6 weeks after transferring of cultures at 23°C under a 16h photoperiod.

shoot tips remained green and started regrowth 4 weeks after plating. Differently from the vitrification technique, the application of encapsulation-dehydration procedures was not effective in the protection of either 'Frantoio' (Benelli *et al.*, 2001b), or 'Arbequina' (Martinez *et al.*, 1999) explants during ultra-rapid freezing.

Alternatively to shoot tips, embryogenic cultures of olive proved to be a highly suitable material for cryopreservation using the vitrification approach (Lambardi *et al.*, 2000). The technique was applied to portions of embryogenic masses of the cv 'Canino' (described in 3.2.2.), containing somatic embryo primordia at different stages of development. After a 90-min incubation in a vitrification solution (PVS2; Sakai *et al.*, 1991), 38% of the cryopreserved embryogenic samples survived and soon started to proliferate. Moreover, the recovered embryogenic tissue showed enhanced proliferative and morphogenetic activity. The embryo primordia which were present in the embryogenic tissue before cryopreservation greened when transferred into the light and developed rapidly to the cotyledonary stage (Fig. 2G).

5. CONCLUSIONS AND PROSPECTS

Although several fruit species and rootstocks (e.g., apple, peach, plum, kiwifruit) are today routinely reproduced by conventional micropropagation, the commercial *in vitro* reproduction of olive (one of the most important fruit trees in the Mediterranean area) is still at the beginning. There are two main reasons for this anomaly: first, important olive cultivars are still difficult (e.g., 'Pendolino', 'Frantoio', 'Kalamata') or even recalcitrant ('Leccino', 'Picholine') when an attempt is made to establish them *in vitro* (Zuccherelli and Zuccherelli, 2000); second, the stimulation of axillary buds and the subsequent shoot elongation still require the presence in the medium of high concentrations of zeatin, a costly cytokinin which contributes considerably to raising the final cost of micropropagated plants. However, olive micropropagation is now competitive and, for some cultivars, even more economical than traditional cutting propagation (Rugini *et al.*, 2001). It has recently been esteemated that an 18-month old olive plant of an easy-to-micropropagate cultivar can be sold at the price of 1 Euro (about 1 US Dollar), when the total production of the commercial laboratory is at least 100,000 olive plants, plus another 900,000 plants of different species (Zamboni, 1999). Considering the present cost of a plant from traditional propagation (3-4 Euro), the expansion of micropropagated olive production can be predicted for the near future. In corroboration of this, in the last decade the number of cultivars that can be successfully propagated *in vitro* has risen considerably, thanks to the numerous scientific advances regarding the important role that various medium constituents (e.g., specific macro and microelements, mannitol, cytokinins other than zeatin, auxins, polyamine) can play in the improvement of the various steps (axillary bud stimulation, shoot elongation, adventitious rooting) of micropropagation.

At the same time, much significant progress has also been made in the development of other regeneration systems, in particular somatic embryogenesis which has recently been obtained also from mature tissue of cultivars. Knowledge on important aspects, such as embryogenic potential of tissues, effect of exogenous growth regulators, histological origin and development of somatic embryos, has been acquired, opening the door to a more ends-directed exploitation of the technique for olive propagation and genetic improvement. Finally, the promising results with medium- and long-term conservation of olive explants by slow growth storage and cryopreservation support the idea of an olive *vitro*-bank creation, which could contribute greatly to the safeguarding of olive biodiversity.

6. REFERENCES

Bao Z.H., Ma Y.F., Liu J.F., Wang K.J., Zhang P.F., Ni D.X., Yang W.Q. (1980) Induction of plantlets from the hypocotyl of *Olea europaea* L. *in vitro*. Acta Bot. Sin. 2: 96-97.

Bartolini G., Leva A.R., Benelli A. (1990) Advances in *in vitro* culture of the olive: propagation of cv Maurino. Acta Hortic. 286: 41-44.

Bartolini G., Prevost G., Messeri C. and Carignani G., 1998. Olive Germplasm: Cultivars and World-wide Collecions. FAO, Roma, pp. 462.

Benelli C., Fabbri A., Grassi S., Lambardi M., Rugini E. (2001a) Histology of somatic embryogenesis in mature tissue of olive (*Olea europaea* L.). J. Hortic. Sci. Biotech. 76: 112-119.

Benelli C., De Carlo A., Lambardi M., Lynch P.T. (2001b) Vitrification of shoot tips, nodal segments and embryogenic tissue of olive (*Olea europaea* L.) for germplasm cryopreservation. Acta Hortic. 560: 137-140.

Bourgin J.P., Nitsch J.P. (1967) Obtention de Nicotiana haploides à partir d'étamines cultivées *in vitro*. Ann. Physiol. Veg. 9: 377-383.

Briccoli Bati C., Librandi A., Sirianni T. (1994) *In vitro* propagation of two olive cultivars. Abstracts "VIIIth International Congress of Plant Tissue and cell Culture". Firenze, June 12-17, p. 72.

Briccoli Bati C., Lombardo N. (1995) Propagazione *in vitro* della cv Nocellara etnea. Proceedings "L'Olivicoltura Mediterranea: stato e prospettive della coltura e della ricerca". Rende (Cosenza, Italy), January 26-28, pp. 249-257.

Briccoli Bati C., Godino G. (1996) Rizogenesi *in vitro* di alcune cultivar di olivo meridionali. Proceedings "III Giornate Scientifiche S.O.I.". Erice, Italy, March 10-14, pp. 331-332.

Briccoli Bati C., Fodale A., Mulé R., Trombino T. (1999) Trials to increase *in vitro* rooting of *Olea europaea* L. cuttings. Acta Hortic. 474: 91-94.

Briccoli Bati C., Nuzzo V., Godino G. (2000) Preliminary agronomic evaluation of two olive cultivars obtained by micropropagation. Abstract Book "4th International Symposium on Olive Growing – OLIVE 2000". Valenzano (Bari, Italy), 25-30 September, p. 6/135.

Cañas L.A., Benbadis A. (1988). *In vitro* plant regeneration from cotyledons fragments of the olive tree (*Olea europaea* L.). Plant Sci. 54: 65-74.

Cañas L.A., Avila J., Vicente M., Benbadis A. (1992) Micropropagation of Olive (*Olea europaea* L.). In: Y.P.S. Bajaj (ed) Hightech and Micropropagation I. Biotechnology in Agriculture and Forestry, Vol. 17. Springer, Berlin Heidelberg New York, pp. 493-505.

Chaari-Rkhis A., Trigui A., Drira N. (1999) Micropropagation of Tunisian cultivars olive trees : preliminary results. Acta Hortic. 474 : 79-81.

Chaari-Rkhis A., Maalej M., Chelli Chaabouni A., Drira N. (2000) 'Meski' olive variety propagated by tissue culture. Abstract Book "4th International Symposium on Olive Growing – OLIVE 2000". Valenzano (Bari, Italy), 25-30 September, p. 6/132.

Cozza R., Turco D., Briccoli Bati C., Bitonti M.B . (1997) Influence of growth medium on mineral composition and leaf histology in micropropagated plantlets of Olea europaea . Plant Cell Tiss. Org. Cult. 51: 215-233.

Dimassi-Theriou K. (1994) *In vitro* propagation of cv. Kalamon olives (*Olea europaea sativa* L.). Adv. Hort. Sci. 8: 185-189.

Dimassi K. (1999) Micropropagation studies of the cv. Kalamon olives (*Olea europaea sativa* L.). Acta Hortic. 474: 83-86.

Driver J.A., Kuniyuki A.H. (1984) *In vitro* propagation of Paradox walnut rootstock. HortScience 19 : 507-509.

Fabbri A., Bartolini G., Lambardi M., Kailis S. (2002) Olive Propagation Manual. CSIRO Publ., Collingwood (Australia). *In press*.

Fiorino P., Leva A.R. (1986) Investigations on the micropropagation of the olive (*Olea europaea* L.) and influence of some mineral elements on the proliferation and rooting of explants. Olea 17: 101-104.

García-Berenguer A., Durán González R. (1990) Mineral media for *in vitro* propagation of juvenile 'Picual' microcuttings. Acta Hortic. 286: 61-64.

Garcia-Fèrriz L., Ghorbel R., Ybarra M., Mari A., Belay A., Trujillo I. (2000) Micropropagation of olive mature trees. Acta Hortic. (Proceedings "4th International Symposium on Olive Growing – OLIVE 2000". Valenzano-Bari, Italy, 25-30 September). *In press*.

Gardi T., Micheli M., Piccioni E., Sisani G., Standardi A. (2001) Cold storage of micropropagated shoots of some fruit species. Italus Hortus 8(4): 32-40 (with English abstract).

Gilad F., Lavee S. (1974) Callus formation and organogenesis from various parts of developing olive embryos. Abstract Book "III International Congress on Plant Tissue and Cell Culture". Leicester, p. 87.

Grossoni P. (1979) Problem concerning the *in vitro* culture of *Olea europaea* L. Gior. Bot. Ital. 113: 75-88.

Heller R. (1953) Researches on the mineral nutrition of plant tissue. Ann. Sci. Nat. Bot. Biol. Vég., 11th Ser., 14: 1-223.

IOOC (2001). International Olive Oil Council. In: *http://www.internationaloliveoil.org/*, Economics, Production.

Lambardi M. (1999). Olive growing in Italy: present situation and future prospects. Proceedings "Third Olive Cultural and Scientific Symposium". New Norcia, Australia, 9-11 August, pp. 29-39.

Lambardi M., Amorosi S., Caricato G., Benelli C., Branca C., Rugini E. (1999a) Microprojectile-DNA delivery in somatic embryos of olive (*Olea europaea* L.). Acta Hortic. 474: 505-509.

Lambardi M., Caccavale A., Rugini E., Caricato G. (1999b) Histological observation on somatic embryos of olive (*Olea europaea* L.). Acta Hortic. 474: 67-70.

Lambardi M., Benelli C., De Carlo A., Fabbri A., Grassi S., Lynch P.T. (2000) Medium- and long-term *in vitro* conservation of olive germplasm (*Olea europaea* L.). Acta Hortic. (Proceedings "4th International Symposium on Olive Growing – OLIVE 2000". Valenzano-Bari, Italy, 25-30 September). *In press*.

Lavee S. (1996) Biology and physiology of the olive. In: International Olive Oil Council (ed) World Olive Encyclopaedia. IOOC, Madrid, pp. 59-110.

Leva A..R., Petruccelli R., Goretti R., Panicucci M. (1992) Ruolo di alcuni microelementi e carboidrati nella proliferazione "*in vitro*" di cv di olivo (*Olea europaea* L.). Proceedings of the Congress "Olive Oil Quality". Firenze, 1-3 December, pp. 333-334.

Leva A.R., Petruccelli R., Bartolini G. (1994) Mannitol "*in vitro*" culture of *Olea europaea* L. (cv. Maurino). Acta Hortic. 356: 43-46.

Leva A.R., Muleo R., Petruccelli R. (1995) Long-term somatic embryogenesis from immature olive cotyledons. J. Hortic. Sci. 70: 417-421.

Leva A.R., Muleo R., Petruccelli R., Montagni G. (2000) Field performance of micropropagated olive plants (cv Maurino): morphological and molecular features. Abstract Book "4th International Symposium on Olive Growing – OLIVE 2000". Valenzano (Bari, Italy), 25-30 September, p. 6/80.

Martinez D., Arroyo-Garcia R., Revilla A.M. (1999) Cryoconservation of *in vitro* grown shoot-tips of *Olea europaea* L. var. Arbequina. CryoLetters 20: 29-36.

Martino L., Cuozzo L., Brunori A. (1999) Establishment of meristem tip culture from field-grown olive (*Olea europaea* L.) cv. Moraiolo. Agr. Med. 129: 193-198.

McCown B.H., Lloyd G. (1981) Woody plant medium (WPM). A mineral nutrient formulation for microculture of woody plant species. HortScience 16: 89.

Mencuccini M. (1995) Micropropagazione e miglioramento genetico *in vitro* dell'olivo: stato dell'arte e prospettive future. Frutticoltura 12: 73-82.

Mencuccini M. (1998) Micropropagazione della cv FS-17. Olivo&Olio 6: 35-39.

Mencuccini M., Rugini E. (1993) *In vitro* shoot regeneration from Olive cultivar tissue. Plant Cell Tiss. Org. Cult. 32: 283-288.

Mencuccini M., Micheli M., Standardi A. (1997) Micropropagazione dell'olivo: effetto di alcune citochinine sulla proliferazione. Italus Hortus 4: 32-37.

Micheli M., Mencuccini M., Standardi A. (1998) Encapsulation of *in vitro* proliferated buds of olive. Adv. Hort. Sci. 12: 163-168.

Mitrakos K., Alexaki A., Papadimitriou P. (1992) Dependence of olive morphogenesis on callus origin and age. J. Plant Physiol. 139: 269-273.

Murashige T., Skoog F. (1962) A revised medium for rapid growth and bioassays with tobacco tissue cultures. Physiol. Plant. 15: 473-497.

Orinos T., Mitrakos K. (1991) Rhizogenesis and somatic embryogenesis in calli from wild olive (*Olea europaea* var. *sylvestris* (Miller) Lehr) mature zygotic embryos. Plant Cell Tiss. Org. Cult. 27: 183-187.

Otero M.L., Docampo D.M. (1998) Micropropagation of olive (*Olea europaea* L.) cv. Arbequina from juvenile cuttings. Phyton 63: 133-140.

Pritsa T.S., Voyiatzis D.G. (1999) The *in vitro* morphogenetic capacity of olive embryos, as affected by their developmental stage and the L-arginine and L-glutamine concentration in the nutrient substrate. Acta Hortic. 474: 87-90.

Rama P. and Pontikis C.A. (1990) *In vitro* propagation of olive (*Olea europaea sativa* L.) 'Kalamon'. J. Hort. Sci. 65: 347-353.

Revilla M.A., Pacheco J., Casares A., Rodriguez R. (1996) *In vitro* reinvigoration of mature olive trees (*Olea europaea* L.) through micrografting. *In vitro* Cell. Dev. Biol.-Plant 32: 257-261.

Rugini E. (1984) *In vitro* propagation of some olive (*Olèa europaea sativa* L.) cultivars with different root-ability, and medium development using analytical data from developing shoots and embryos. Sci. Hortic. 24: 123-134.

Rugini E. (1986) Olive (*Olea europaea* L.). In: Y.P.S. Bajaj (ed) Trees I. Biotechnology in Agriculture and Forestry, Vol. 1. Springer, Berlin Heidelberg New York, pp. 253-267.

Rugini E. (1988) Somatic embryogenesis and plant regeneration in olive (*Olea europaea* L.). Plant Cell Tiss. Org. Cult. 14: 207-214.

Rugini E. (1991) Stato dell'arte della coltura *in vitro* dell'olivo e prospettive per il miglioramento genetico. In: S. Sansavini (ed) Biotecnologie e Resistenze Genetiche nelle Piante da frutto. AGROBIOFRUT, Cesena, pp. 132-147.

Rugini E., Fontanazza G. (1981) *In vitro* propagation of 'Dolce Agogia' olive. HortScience 16: 492-493.

Rugini E., Fedeli E. (1990) Olive (*Olea europaea* L.) as an oilseed crop. In: Y.P.S. Bajaj (ed) Legumes and oilseed crops I. Biotechnology in Agriculture and Forestry, Vol. 10. Springer, Berlin Heidelberg New York, pp. 593-641.

Rugini E., Jacoboni A., Luppino M. (1993) Role of basal shoot darkening and exogenous putrescine treatments on *in vitro* rooting and on endogenous polyamine changes in difficult-to-root woody species. Sci. Hortic. 53: 63-72.

Rugini E., Caricato G. (1995) Somatic embryogenesis and plant recovery from mature tissues of olive cultivars (*Olea europaea* L.) "Canino" and "Moraiolo". Plant Cell Rep. 14: 257-260.

Rugini E., Pezza A., Muganu M., Caricato G. (1995) Somatic embryogenesis in olive (*Olea europaea* L.). In: Y.P.S. Bajaj (ed) Somatic Embryogenesis and Synthetic Seed I. Biotechnology in Agriculture and Forestry, Vol. 30. Springer, Berlin Heidelberg New York, pp. 404-414.

Rugini E., Di Francesco G., Muganu M., Astolfi S., Caricato G. (1997) The effects of polyamines and hydrogen peroxide on root formation in olive and the role of polyamines as early marker for rooting ability. In: Altman A. and Waisel A. (eds) Biology of Root Formation and Development. Plenum Press, New York, pp. 65-73.

Rugini E., Biasi R., Muleo R. (2000) Olive (*Olea europaea* var. sativa) transformation. In: S.M. Jain and S.C. Minocha (eds) Molecular Biology of Woody Plants. Vol. 2. Kluwer Ac. Pub., Dordrecht, pp. 245-279.

Rugini E., Biasi R., Muganu M., Pannelli G., Aversa G., Maggini F., Martelli G.P., Zamboni E., Zuccherelli G., Barba M. (2001) L'organizzazione di un moderno vivaismo olivicolo alla base della produzione di piante certificate. Frutticoltura 5: 11-24.

Sakai A., Kobayash S., Oiyama I. (1991) Survival by vitrification of nucellar cells of navel orange (*Citrus sinensis* Osb. var. brasiliensis Tanaka) cooled to –196°C. J. Plant Physiol. 137: 465-470.

Scaramuzzi F., De Gaetano A. (1974) Ricerche sulla rizogenesi di espianti di fusto di *Olea europaea* L., cv Ogliarola di Monopoli, coltivati *in vitro*. Riv. Ortoflorofrut. Ital. 58: 419-436.

Schenk R.U., Hildebrandt A.C. (1972) Medium and techniques for induction and growth of monocotyledons and dicotyledons plant cell cultures. Can. J. Bot. 50: 199-204.

Seyhan S., Özzanbak E. (1994) Shoot multiplication of some olive (*Olea europaea* L.) cultivars. Acta Hortic. 356: 35-38.

Shibli R.A., Shatnawi M., Abu-Ein, Al-Juboory K.H. (2001) Somatic embryogenesis and plant recovery from callus of 'Nabali' Olive (*Olea europaea* L.). Sci. Hortic. 88: 243-256.

Standardi A., Piccioni E. (1998) Recent perspectives on synthetic seeds technology using nonembryogenic *in vitro*-derived explants. Int. J. Plant Sci. 159: 968-978.

Standardi A., Micheli M., Piccioni E. (1998) Propagazione "*in vitro*" dell'olivo: acquisizioni e prospettive. Frutticoltura 7/8: 19-23.

Troncoso A., Liñan J., Cantos M., Acebedo M.M., Rapoport H.F. (1999) Feasibility and anatomical development of an *in vitro* olive cleft-graft. J. Hort. Sci. Biotech. 74: 584-587.

Wang K.J., Zhang P.F., Ni D.X., Zhu X.Z., Yang W.Q., Bao Z.H. (1979) Cytohistological studies on the tissue culture of *Olea europaea* L. Acta Bot. Sin. 21: 225-230.

Zamboni E. (1999). Messa a punto di una tecnica di moltiplicazione su larga scala dell'olivo (*Olea europaea* L.) e sua valutazione economica. Doctoral Dissertation, University of Tuscia, Viterbo (Italy), pp. 95.

Zohary A., Spiegel-Roy P. (1975) Beginning of fruit growing in the Old World. Science 187: 319-327.

Zuccherelli G., Zuccherelli S. (2000) *In vitro* propagation of fifty olive cultivars. Acta Hortic. (Proceedings "4th International Symposium on Olive Growing – OLIVE 2000". Valenzano-Bari, Italy, 25-30 September). *In press*.

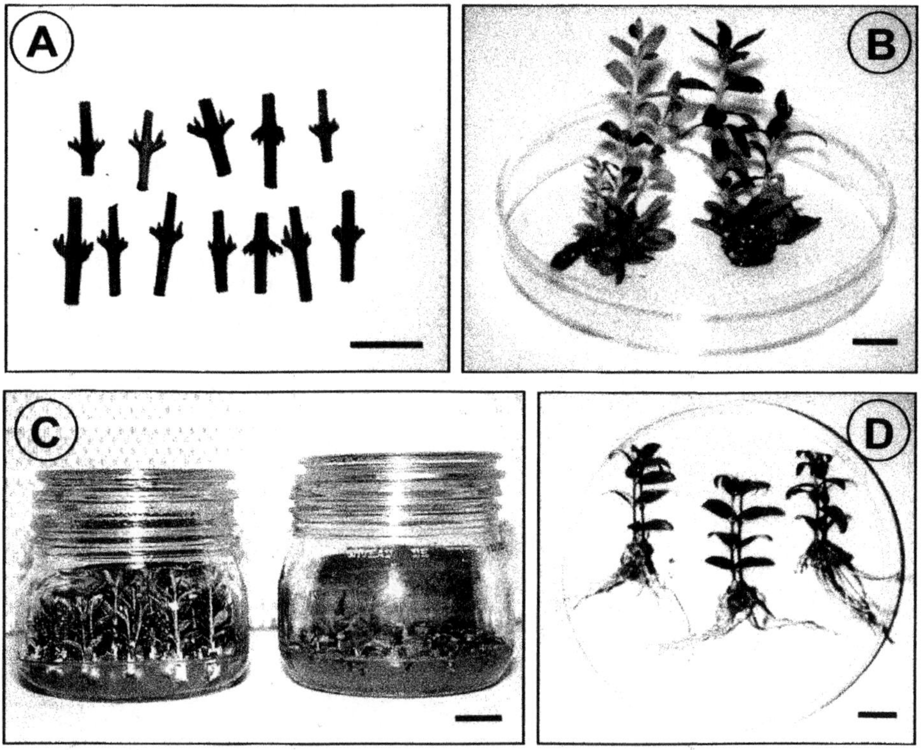

Figure 1. Olive micropropagation (cv 'Frantoio') by axillary bud development. **A**: One-centimeter (on average) nodal segments from tender twigs are used as initial explants (bar, 1 cm). **B**: 'Frantoio' multiplication is obtained on zeatin-containing OM medium, mainly by shoot elongation rather than by shoot proliferation. Callus production at the base of shoots is often observed (bar, 1 cm). **C**: At each subculture, elongated shoots are divided into binodal segments (bar, 2 cm). **D**: Rooting of elongated shoots is easily obtained on IBA-containing OM medium (bar, 1 cm). Lambardi, unpublished.

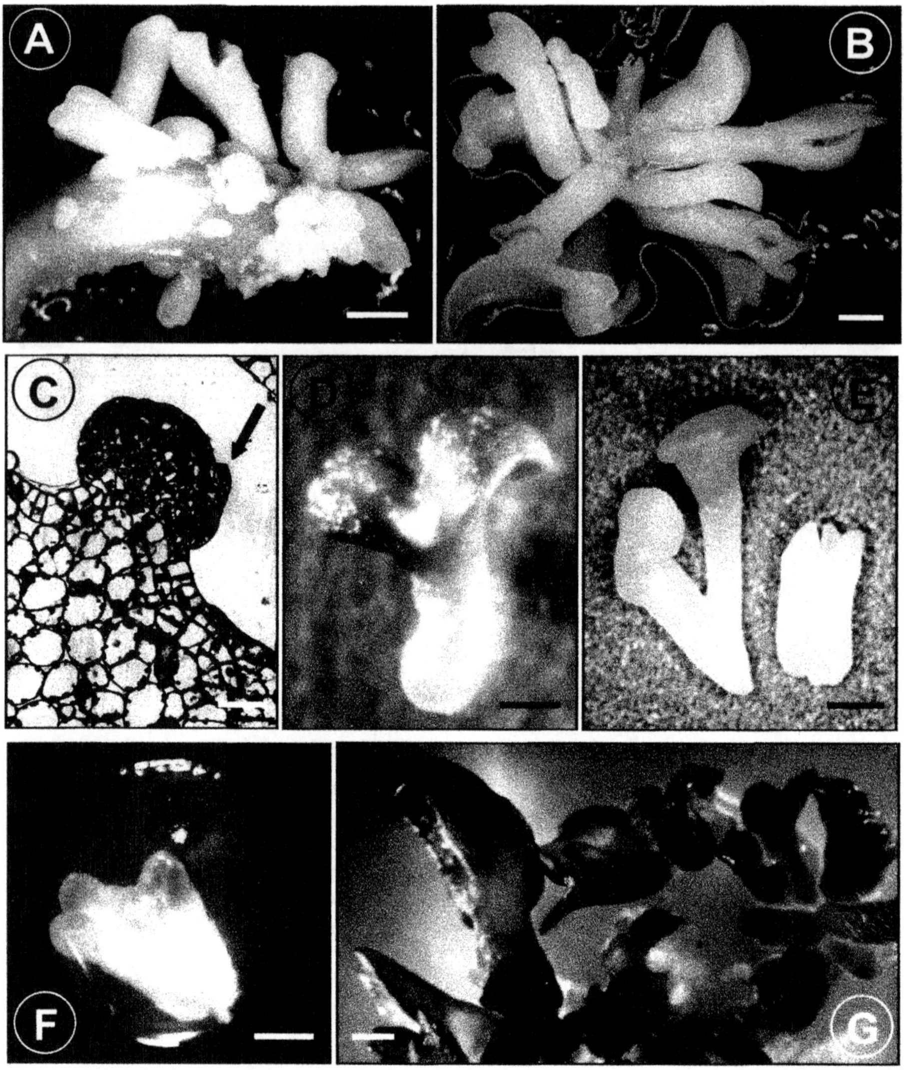

Figure 2. Somatic embryogenesis and cryopreservation of olive (cv 'Canino'). **A**: Many secondary somatic embryos at different stages of development, formed at the radicle end of a primary somatic embryo, with no evidence of interposed callus (bar, 1 mm). **B**: Further embryo development to the cotyledonary stage (bar, 1 mm). **C**: A globular proembryo originated from a few cells of the epidermal layer of a primary somatic embryo. The arrow shows the initial of a cotyledon (bar, 0.1 mm). **D**: An isolated well-formed somatic embryo (bar, 1 mm). **E**: Two examples of fused somatic embryos (bar, 1 mm). **F**: Synthetic seed preparation using olive somatic embryos (bar, 0.5 mm). **G**: Following cryopreservation of embryogenic masses and their plating in the light, many somatic embryos have greened and developed to the cotyledonary stage (bar, 0.5 mm). (C, from Lambardi *et al.*, 1999b; D and E, from Benelli *et al.*, 2001a).

22. MICROPROPAGATION OF KIWIFRUIT (*Actinidia spp.*)

Eddo Rugini and Patricia Gutiérrez-Pesce

Dipartimento di Produzione Vegetale, sez. Ortofloroarboricoltura, Università degli Studi della Tuscia, via S. Camillo de Lellis, 01100 Viterbo, Italy. E-mail: rugini@unitus.it

CONTENTS

S.M. Jain and K. Ishii (eds.), Micropropagation of Woody Trees and Fruits, 647–669.

1.0 INTRODUCTION

The genus '*Actinidia*' name is derived from the Greek "aktis" which means irradiation for its radiate stigma. This name was attributed by a Danish researcher, Nathaniel Wallich, and found plants in Nepal. The common name is kiwi and/or kiwifruit which is accepted world-wide and is attributed to the New Zealand kiwifruit. Other names which are attributed to the fruit of other kiwi growing countries include: 'Chinese gooseberries' in U.S.A.; 'Groseille de China' in France; 'Mao-yantao' or 'Mao-ertao' and 'Yang-tao' or 'Yang-taw', with bristly hair and soft hair respectively in China and 'Actinidia' the plant and 'Kiwi' the fruit, in Italy.

1.1 Genetic base, centre of origin, current distribution and economic importance

The genus *Actinidia* Lindl consists over 110 taxa and 60 species and belongs to the family *Actinidiaceae*. *Actinidia* is widespread within the temperate and subtropical regions of the Asian continent. It is considered native to China, since most of the species are spread in the Southwest of the country. Other species of the *Actinidiaceae* family have been found in the Korea, Japan, Nepal, India, Vietnam and Russia (Ferguson, 1990). In these areas the fruits were collected from the wild, and only recently, several species were introduced for cultivation and selected for the breeding programmes. The cultivated area of kiwifruit in production worldwide is 1,004,537 Mt, with the major producers being Italy (382,300 Mt), New Zealand (265,000 Mt), and Chile (120,000 Mt) (FAO statistical database 2001).

There are a few economically important species which have the same chromosome number but different ploidy levels, these includes species such as: *A. chinensis* (2n=2x=58), *A. deliciosa* (2n=6x=174), *A. arguta* (2n=4x=116), *A. eriantha* (2n=x=58), *A. latifolia* (2n=2x=58), *A. chrysantha* (2n=4x=116) and *A. melanandra* (2n=4x=116). Other species such as *A. kolomikta* (2n=2x=58) and *A. polygama* (2n=2x=58) are also currently being used for genetic improvement. Only *A. deliciosa* (A. Chev.) C.F. Liang et A.R. Fergusson is intensively cultivated worldwide. It is similar to *A. chinensis* Planch. But, until recently, both were considered under the same botanical name *A. chinensis*. Although, both species differed in ploidy level, fruit size, shape, and taste.

The genus *Actinidia* is genetically very diverse (Huang *et al.*, 1999). It is unusual that the commercial production is based almost exclusively on the pistillate cultivar 'Hayward', with two pollinators, 'Matua' and 'Tomuri'. This cultivar was selected from *A. deliciosa* by scientists from New Zealand who imported this species from China around 1930. In New Zealand, the cultivation became economically important from the middle 1960's and only in early 1970's did this species become widely distributed. The other pistillate cultivars like 'Abbott', 'Allison' and 'Bruno', were cultivated together with 'Hayward' and soon lost their popularity in favour of 'Hayward' because cv. Hayward had better shape, size and long storability fruits.

However, this cultivar is not free from defects, these include high risk of sudden disease occurrence, pest infection or frost damage due to its very narrow genetic base, and the dioicism character typical of the *genus*.

In the recent years, few accessions from cv. Hayward have been selected among the offspring of *in vivo* crosses and from somaclonal variation derived from micropropagated plants. They included hairless fruit, and early fruit ripening, comprising the accession Summer 4605® and Summer 3373® having 35 and 50 days earlier ripening than cv Hayward respectively (Ossani, 1994; Summerkiwi® 20010160 and 20010159 patented by Ossani 2001). Accessions with hairless fruit are susceptible to fruit fly (*Ceratitis capitata*) for this reason it has not been included in the cultivation.

Recently, the New Zealand cultivar 'Zesprigold' (Hort16A) obtained from *A. chinensis*, presents interesting characteristics such as, yellow pulp, higher vitamin C content and sweet flavour. The cultivar also has some negative aspects such as, early blossom and shooting, fruits are almost completely hairless and have a mucro that can damage the other fruit during handling. Ripening is contemporary to Hayward. The most suitable pollinators used for this cultivar are "K2" and "K3".

Kiwifruit are characterised by the high amount of vitamin C with *A. deliciosa* containing 140 mg/100 g of pulp and *A. eriantha* over 1000 mg/100 g of pulp. However, like many other fruit it has low vitamin B and carotene contents. Minerals are also high, especially potassium and magnesium as well as copper, zinc, and manganese. Fruits are consumed mainly as fresh within 1-6 months after harvesting. However, a small amount of the total production is used for food industry, such as jams, juices, and syrups and in New Zealand, the residue after processing is used to produce wine (Monastra, 1991).

1.2. Traditional *in vivo* propagation methods

1.2.1 By seed

In vivo propagation by seed is used for obtaining both new varieties as well as varieties for the production of rootstocks. Seeds of kiwi are considered "immature" since they mature only after fruit ripening. The seeds must be extracted from the pulp of very ripened fruits and dried if necessary. To overcome dormancy seeds are incubated at 2-4° C in humid substrates for approximately one month. For uniform germination rates, it is advisable to alternate the temperatures between 10-12° C during the day and 2-4° C during the night for a few weeks (Zuccherelli, 1994).

This system of *in vivo* propagation has several disadvantages: it is time consuming, the genetic variability of the offspring is large and about 50% of offspring are male. Traditionally, the sex of the plant has been known after the first blooming (Zuccherelli & Zuccherelli, 1985). In the recent years, the development of molecular biology techniques has permitted the determination of the sex of the offspring earlier. This comprises the use of molecular markers, determination of polyamines, and isoperoxidase and peroxidase tests (Auxtova *et al.*, 1994; Biasi *et al.*, 1999; Harvey *et al.*, 1997a, b; Hirsch *et al.*, 1997). Furthermore, the use of rootstocks is not necessary because specific rootstocks are not available and in kiwi is easily propagated by cuttings.

1.2.2 By stem cutting

Semi-herbaceous stem cuttings harvested during the spring season are the most widely used means of propagation, while stem cutting of woody branches obtained during winter rest are used less because of the low rooting percentage. Herbaceous branches also give good results provided the environmental conditions prevent dehydration. In each case, the success of rooting is dependent of the nutritional state of the mother plant, season, type, concentration of rooting hormone used (α-Naphthaleneacetic acid (NAA), indole-3-butyric acid (IBA) or indole acetic acid (IAA)), temperature (20-30°C), light intensity (5,000-6,000 lux), and humidity (Zuccherelli, 1994).

1.2.3 By grafting

Graftings are used when the rootstocks are able to positively effect the vegetative and reproductive characteristics of the scion. Since not all selected rootstocks are available this practice is hardly used. If rootstocks only come from seedlings, they may affect negatively the grafting by inducing different characteristics to the scion. Grafting is very easy in kiwifruit with many methods available to use. Spring grafting is carried out both by using the scion from semi-herbaceous branches of the same year or woody twigs from the previous year. Scions from the same year appear to give the best results (Revilla *et al.*, 1992, Zuccherelli, 1994). Budding is usually carried out within the summer season (Zuccherelli, 1994). Grafting can also be used to speed up blooming of the scion of the juvenile plants, by grafting on old kiwi plants.

2.0 MODERN PROPAGATION METHODS OR *IN VITRO* MULTIPLICATION

In vitro multiplication is the most common and modern method for plant propagation, which facilitates the growth and/or maintenance of plant cells, tissues, organs or whole plants on a nutrient culture medium. *In vitro* multiplication methods can be distinguished in two main groups:

1. *Conventional micropropagation methods,* these include the *in vitro* production of shoots by axillary bud stimulation and subsequently the rooting of the neo-formed shoots.
2. *Modern approaches of micropropagation* include shoot organogenesis followed by rooting of the neoformed shoots, somatic embryogenesis and embryo rescue.

2.1. Conventional micropropagation

Conventional micropropagation is effective when *in vivo* vegetative propagation is slow, difficult and laborious. The most frequent applications for conventional micropropagation are clonal propagation for large-scale plant multiplication, germplasm conservation, somaclonal variation and genetic manipulation. Plant tissues used as starting material are shoot tips (Revilla *et al.*, 1992), shoot meristems (Standardi, 1981), whole buds and nodal segments (Velayandom *et al.*, 1985). The first micropropagation protocol for *Actinidia* was proposed by Harada (1975) and was subsequently improved (Wang *et al.*, 1982; Standardi, 1983; Wessels *et al.*, 1984; Monette, 1986a).

In *A. deliciosa*, shoot meristems have been demonstrated to be a suitable starting material for micropropagation, since genetic mutations are limited and virus free plants are obtained (Standardi, 1981). Liquid culture of shoot tips of *A. deliciosa* resulted in a greater increase of fresh weight and number of shoots produced than in agar-solidified medium (Monette, 1986a).

2.1.1 Axillary bud stimulation

Axillary bud stimulation allows the bud to develop under the influence of a relatively high cytokinin concentration which decreases apical dominance by stimulating the basal growth of neoformed buds (Pierik, 1987). A large number of uniform shoots can be produced from a single explant in a short period of time because this method is simpler than other method of micropropagation and the genetic stability is usually preserved.

In *Actinidia* sp., this technique has become the most viable propagation method for large-scale plant production (Revilla *et al.*, 1992, Oliveira *et al.*, 1994; Zuccherelli, 1994) using nodal segments (Velayandom *et al.*, 1985) and/or shoot tips (Monette, 1986a, b; Revilla *et al.*, 1992). However, attention should be given to avoid the propagation of shoots regenerated from basal callus because *Actinidia* sp. regenerate easily from undifferentiated cells and have a high capacity for somatic variation.

2.1.2 Shoot rooting

The methodology for root induction varies, most of them include:- IBA as a potent root inductor: immersion of the basal part of the shoot in high IBA concentrations (50 mg/l) for a few seconds (Gui, 1979; Huang *et al.*, 1983; Jona & Gribaudo, 1983; Standardi, 1981, 1982, 1983; Wessels *et al.*, 1984). Higher IBA concentrations (1 g/l) also give 80-90% rooting (Revilla *et al.*, 1992; Gonzàlez *et al.*, 1990).

An alternative method includes low IBA concentrations (0.3, 1, 2 mg/l) for a long basal shoot immersion period (overnight) which gives high rooting percentages (Revilla & Power, 1988; Harada, 1975; De Paoli *et al.*, 1994). Other auxins have also been used, such as, IAA (Zuccherelli, 1994), NAA (Velayandom *et al.*, 1985) with good results. Pedroso *et al.*, (1992) reported higher rooting percentage when *in vitro* shoots of *A. deliciosa* were directly transferred to rooting substrate in the greenhouse. Furthermore, 24 h treatment with 20 mg/l IBA induced thin and non-functional roots, probably due to the use of growth regulators in the proliferation medium (Aldrufeu *et al.*, 1985).

2.2.0 Modern approaches of proliferation and regeneration

Leaf pieces, leaf petioles, and zygotic embryos can be used to obtain new explants through organogenesis and somatic embryogenesis. Kiwifruit has demonstrated relative amenability to *in vitro* multiplication, both via organogenesis and to a lesser extent via embryogenesis.

2.2.1 Somatic embryogenesis

Embryogenesis is the process of embryo formation. Embryos can develop either from fertilised egg cells, i.e. the zygote, or from asexual or somatic cells, in this case called somatic embryos. In each case, the cells must have an embryogenic competence, which means it is relatively easy to induce somatic embryogenesis from tissue culture. The level of embryogenic competence is generally high in non-meristematic tissues of immature and mature seedlings and floral structures. Direct and indirect somatic embryogenesis has been described as the first one where no callus formation is present, and the second one where embryoids arise from callus cells.

Potential uses of somatic embryos include: clonal propagation, artificial seed production, protoplast culture, embryogenic cell suspension, gene transfer and cryopreservation. Unfortunately, the embryos are subjected to high somaclonal variation, which is a limiting factor for its application. Furthermore, due to the multicellular origin of the somatic embryos, the risk to recover chimeric plants is high. However, it may be useful for genetic improvement through mutation.

As in many other species, the critical factors in kiwifruit somatic embryogenesis are: genotype and explant competence, preconditioning and culture conditions. Several attempts have been made to induce somatic embryogenesis from different tissues of *A. chinensis, A. arguta* and *A. deliciosa.* The first attempt to induce embryogenesis in kiwifruit explants was carried out with *A. chinensis* var. *chinensis* which has demonstrated to be highly responsive to embryogenesis depending of the explants used. In most cases using *A. deliciosa*, only globular stages and embryoids were obtained from stem and root segments (Harada, 1975; Bini, 1979) but these failed to yield plantlets; endosperm tissue developed embryogenic calli, embryoids and regenerate complete plantlets (Gui *et al.*, 1982; Huang *et al.*, 1983; Zenguang *et al.*, 1983). Additionally, Mu *et al.* (1990) did not obtain complete plants with the same tissue of several interspecific crosses of *Actinidia.*

The stamen filaments of *A. deliciosa* plants gave rise to roots and callus with subsequent shoots and somatic embryos differentiation. However, they failed to grow further than the torpedo stage (Brossard-Chriqui & Tripathi, 1984). Anthers of male plants of *A. chinensis* and *A. deliciosa* differentiated the largest number of embryoids from callus tissues derived from the anther wall, which developed into bipolar heart-shaped structures and in some cases they developed into plantlets (Fraser & Harvey, 1986). Oliveira & Pais (1992) obtained successful somatic embryogenesis by preconditioning axillary buds of cv. Hayward. Pre-conditioning treatments were carried out by induction the competent tissue with different stresses before use. This pre-conditioning treatment gave rise to leaves able to differentiate somatic embryos at different stages.

2.2.2.0 Organogenesis

Organogenesis refers to the process whereby tissues and cells of callus may be induced to form shoots or roots and then complete plants. These organs can develop either from an intermediate callus originated from an initial explant (indirect organogenesis) (Krikorian, 1995), or directly from explant tissue (direct organogenesis). This ability of plant tissues may be useful to exploit clonal variation in an attempt to produce stable plants, for its unicellular origin.

2.2.2.1 Explant tissues

Actinidia species demonstrate a remarkable organogenetic property from different types of tissues (Revilla *et al.*, 1992). Organogenesis has been accomplished using roots of *A. chinensis* (Harada, 1975) and *A. deliciosa* cv. Tomuri (Chiariotti *et al.*, 1991), stems (Harada, 1975, Augé & Boccon-Gibot, 1982), shoot meristems (Augé & Boccon-Gibot, 1982), cambial tissue of stems of *in vivo A. chinensis* cv. Tomuri plants and *A. arguta* Pl. (Leva & Magrini, 1986; Barbieri & Morini, 1987), stem-segments of field-grown plants of *A. deliciosa* (Leva & Bertocci, 1988), nodal shoot segments from field-grown and micropropagated plants of *A. deliciosa* cv. Hayward (Rey *et al.*, 1992), hypocotyl and stem segments from seedlings (Uematsu *et al.*, 1991), staminal filaments (Tripathi & Saussay, 1980; Brossard-Chriqui and Tripathi, 1984), endosperm (Gui *et al.*, 1982; Huang *et al.*, 1983), allowing the novel genetic constitution derived from interspecific hybrids (Mu *et al.*, 1990), leaf explants from *in vitro* grown plants (Predieri *et al.*, 1988; Rugini *et al.*, 1991; Janssen and Gardner, 1993), protoplasts (Raquel & Oliveira, 1996), and from fruit galls of *A. polygama* (Tanaka *et al.*, 1997).

Young stem segments were reported as the most suitable for micropropagation, followed by roots. Whereas, leaf blades produced few buds and petioles almost none (Augé *et al.*, 1982). However, Pais *et al.* (1987) found a suitable way to obtain meristem differentiation and later plant development from petiole segments of *in vivo* plants of *A. chinensis.* Plants were vertically inoculated in the medium, while petioles were put horizontally and they gave rise to friable calli suitable for protoplast isolation. The same methodology was used successfully with petioles taken from *in vivo* (Revilla & Power, 1988) and *in vitro* propagated shoots of *A. deliciosa* cv Hayward (Gonzalez *et al*., 1995)

2.2.2.1 Callus, cell suspension culture and protoplasts

Callus can be obtained *in vitro* under certain stimuli and/or *in vivo* after injury. There are some differences between types of callus (*in vivo* /*in vitro*) in structure and physiology. Callus culture can be initiated with cells of any plant species that are plated on a nutrient solid medium.

Cell suspension cultures consist of mixture of cell aggregates, cell clusters and single cells that grow in a nutrient liquid medium. When the culture is carried out on solid medium their growth declines slowly (Gamborg, 1982). Callus and cell culture can be used for regeneration or propagation purposes (organogenesis, somatic embryogenesis), for physiological, biochemical (secondary metabolite production, stress), genetic studies, for inducing somaclonal variation (*in vitro* selection and mutation), and germplasm preservation.

The source of material for callus and cell culture can be any type of explant: juvenile/mature tissues, *in vitro*/*ex vitro* plant material, zygotic/maternal, vascular cambia, storage organs, pericycle of roots cotyledons, leaf mesophylls, provascular tissues, and/or from induced galls such as in *A. polygama* (Sashida *et al*., 1994). Almost all tissues respond to callus induction, while maintenance is more difficult.

The first observations of genetic mutations within globular embryoids derived from calli was made by Harada (1975), Hirsch & Bligny-Fortune (1979) and Gui (1979) & Bini (1979) during the development of a suitable micropropagation protocol for *A. chinensis*.

The first attempt to develop a strategy for leaf mesophyll protoplast culture was carried out by Cossio & Marino (1983) who published quantitative data on direct isolation and purification of protoplasts from *A. deliciosa* cv. Hayward. However, the protoplasts failed to divide. Other attempts have been carried out for plant regeneration from leaf callus-derived protoplasts of *A. chinensis* var. *chinensis* (Tsai & Cai, 1988; Xiao *et al*., 1992). Mii & Ohashi, (1988) reported culture of cell suspension derived protoplasts from *A. chinensis* cv. Abbott, but only a reduced number of callus could regenerate shoots due to the frequent necrosis during the early period of colony growth. In *A. deliciosa* cv Hayward, Oliveira & Pais (1991) obtained normal-type plant regeneration from protoplasts culture derived from *in vitro*-induced callus of *in vivo* grown plants. Contrary to these results, studies on protoplast culture of *A. deliciosa* revealed considerable variation in chromosome number, morphological characters, sex differentiation, and vegetative growth of protoplast-derived plants. This indicated the possibility of producing new breeding resources by protoplast culture in kiwi (Cai *et al*., 1992, 1993).

Successful methodologies for micro callus formation from leaf mesophyll protoplast was developed for several species of *Actinidia* (Xiao & Hirsch, 1996). This research was carried out with 6 species and varieties: *A. arguta* var. *arguta*, *A. arguta* var. *purpurea*, *A. arguta* cv. *Issai*, *A. deliciosa*, *A. kolomikta* and *A. polygama*. Several factors influenced purification, browning, survival and sustained division of the protoplasts such as culture type (solid or liquid), medium ingredients (such as hormones, sugars and amino acids) and genotype.

An improvement and simplification of this technique was recently presented by Zhang *et al.* (1998) which used leaf protoplasts of *in vitro* seedlings of *A. eriantha* and obtained plant regeneration. Protoplast isolation from leaves using this method may also facilitate somatic hybridisation in *Actinidia*. The authors observed not only variation in chromosome number within the protoplasts derived plants but also variation in the number of nucleus per cell.

Leaf explants offer the possibility to produce plants through indirect organogenesis of *A. chinensis* (Ludvova & Ostrolucka, 1998), after intensive proliferation of callus cultures from leaf explants. The morphogenic potential of callus cells was manifested both in organogenesis (adventitious buds) and somatic embryogenesis.

2.2.3 Embryo rescue method

Embryo rescue constitutes the sterile isolation from the seed of both immature and/or mature embryos. These isolated embryos are then placed on an artificial medium with the goal of obtaining a viable plant. Immature embryos are collected from unripe seeds with the aim to avoid embryo abortion that can occur during fruit ripening. Mature embryos are collected from ripe seeds with the aim to avoid inhibition of seed germination from natural crosses and/or to induce germination of embryos derived from hybridisation between sexually incompatible genotypes, which normally do not germinate in nature. Embryo rescue is commonly used to obtain novel hybrid combinations and to transfer the entire cytoplasm or specific cytoplasmic traits.

In *Actinidia*, embryo culture of interspecific hybridisation has been tried in order to avoid embryo and endosperm abortion in some cross combination. The first attempts of embryo rescue carried out by Hirsch *et al.*, (1991) gave poor results after hybridisation between large-fruited species (*A. deliciosa*) and cold resistant ones (*A. arguta, A. kolomikta* and *A. polygama*). While normal seedlings were produced from embryos of normal appearance from the viable hybrids obtained between *A. deliciosa* and *A. arguta* (Hayward and clone 26), and between *A. deliciosa* and *A. eriantha*, (Mu *et al.*, 1991)

Successful results have been obtained in a comparative study on the effectiveness of different techniques and media in rescuing interspecific hybrids of *Actinidia* species and ploidy races. The embryo rescue was the unique tool, which led to hybrid plantlets of 4 interspecific hybridisation crosses. Several embryos were rescued when incubated in the proper media (Hirsh *et al.*, 2001).

3.0 Culture medium and environmental conditions

Morphogenesis of *in vitro* cultured tissues, as well as shoot development, are regulated by plant hormones, specifically auxins and cytokinins. Morphogenesis varies with the type, combination and concentration of growth regulators added to the basal medium. *In vitro* growth also requires the creation of an artificial environment conditions suitable for further growth. The environmental conditions should consider the:- physical form of the medium; pH; humidity and gas atmosphere; light quality and quantity and temperature (Debergh & Read, 1991). Factors such as light, relative humidity and gas exchange has been scarcely studied. However, they can stimulate an abnormal functioning of photosynthesis and poor development of leaf systems (Infante *et al*., 1989a; Pierik, 1988).

Different explant types, and culture media have been used for shoot/embryo induction, regeneration and efficient rooting for different kiwifruit genotypes (Canhoto & Cruz, 1987, Rey *et al*., 1992; Cai *et al*., 1993; Gonzalez *et al*., 1995), apical meristem culture (Standardi, 1981), and shoot tips (Monette, 1986a).
The procedure for plant multiplication involves: callus induction and formation from the initial explant and shoot stimulation and development. During these two steps the required levels of exogenous hormones may be different (Krikorian, 1995). Moreover, this hormonal control was studied mainly by exogenous application of plant growth regulators (PGRs) and only in few cases in relation to endogenous levels (Letham & Palni, 1983). The analysis of the same item in morphogenic areas, explants or genotypes, as well as in non-morphogenic ones is a good way to study the control of morphogenesis in *in vitro* cultures (Tran Thanh Van, 1981).

3.1. Endogenous compounds and PGRs

In the last years more efforts have been focused for understanding the process of uptake PGRs, especially cytokinins. Regarding uptake and metabolism of zeatin, Einset & Silverstone (1987) determined that this compound was rapidly accumulated in roots and stem tissue of *Actinidia arguta* Planc. plants when they were incubated on basal medium supplemented with its precursor N^6-(Δ^2-isopentenyl)adenine.

Centeno *et al*., (1996) studied the relationship between exogenous applied PGRs, endogenous hormonal contents, and *in vitro* morphogenic responses of *in vitro* culture of *A. deliciosa* petioles by comparing endogenous auxin and cytokinin levels with different callogenic and caulogenic abilities. They attributed a major role to auxins in callus formation and maintenance and to cytokinins in the caulogenic responses of kiwi tissues. The most important differences found in hormonal content between organogenic and non-organogenic callus concerned benzyladenine levels. This fact is explained because the stress conditions accompanying *in vitro* culture induced a great change in hormonal content in kiwi tissues (Debergh & Read, 1991). Later, this studied was carried out in order to establish a relation between the levels of active cytokinins and callus induction and initiation. 6-benzyladenine (BA) metabolism was triggered mainly into 9-glycosylation pathway for BA inactivation. Results of this research showed a direct relation between cytokinin content in the explant and the induction and initiation of basal callus in petioles. The different proliferation ability of the petiolar portion examined could be associated with the maturation and differentiation state of the basal and apical cells (Centeno *et al.*, 1998).

The uptake and metabolism of 6-benzyladenine were studied in *A. deliciosa* explants grown in ventilated cultures as a system to avoid hyperhydric shoots when propagation was carried out in liquid medium (Cañal *et al.*, 2000). The authors found that 65% of the initial amount of BA disappeared in the first half hour of culture and then it was converted into 7 different glucosides. Similar results were reported previously in non-ventilated systems, but the amount of BA uptake in the first 30 min was only 17% (Feito *et al.*, 1994; Feito *et al.*, 1995; Moncaleàn *et al.*, 1999). It was demonstrated that 2 glucosides were formed rapidly from BA from the initial step of culture. Cañal *et al.* (2000) explained this difference could be due to a mass flux because of transpiration of explants which have low relative humidity in the ventilated system.

A recent study on the transport and distribution of BA used radioactivity for quantifying in the different portions of the microshoots of *A. deliciosa*, and for identifying its metabolites in apoplast and symplast. The different effect of solid and liquid medium on the transport and distribution of N-6-benzyladenine along the microshoots was evaluated by Feito *et al.* (2001). They found that explants from liquid medium have a higher BA content, its inactivation is slower, its distribution is more effective, and its presence in the meristematic tissue is also greater than in the solid medium.

In a recent study, Moncalean *et al.* (2001) demonstrated that BA not only has an important effect on the different phases of the micropropagation, but also regulates the development of the regenerants. The highest quality shoots in terms of multiplication index, weight and length of shoots, presence of callus, were those obtained from culturing in the presence of 4.4 μM of BA for 1 day. Zeatin seems to be the best in inducing regeneration of shoots from callus, in cv Hayward (Rugini *et al.* 1991) while BAP is used normally for shoot proliferation.

3.2 Light

In some species light is a determining factor for the formation of new organ during the morphogenetic process. In callus culture of *A. deliciosa*, light quality (blue and red) controlled organogenesis, in particular red light, while no differentiation occurred in white light (Muleo and Morini, 1990). The effect of solar light has been also study *in vitro*: dry and fresh matter accumulation on proliferating axillary shoots as well as net photosynthetic capacity was greatest under solar light regimen, while shoot proliferation was the highest under white light. Solar light leaves also had a thicker epidermis than those grown under white and blue light, which can reduce water loss when plant are acclimatised (Infante *et al.*, 1994).

Light does not play a single role in plant organogenesis but strongly interacts with different compounds of the culture medium. The effect of combinations of light intensity, magnesium and sucrose on leaf anatomy, photosynthesis, starch and total sugar accumulation in kiwi shoots grown *in vitro* was study by Dimassi-Theriou & Bosabalidis (1997). They found that an increase of light intensity had a direct effect on size of the leaf mesophyll cells and an inverse effect in the number of chloroplasts and content of starch grains. Total sugar accumulation was favoured by normal light intensity and low light intensity favoured starch grain formation. The photosynthetic capacity greatly increased by raising 3-times magnesium concentration in the culture medium in absence of sucrose.

3.3 Carbon sources or carbohydrates

Sucrose is the most common sugar that is used as source of energy and osmotic stimuli within tissue culture, but other sugars (such as sorbitol, mannitol, glucose, maltose, lactose and raffinose) can be also used for sustaining specific growth requirements in many species. In addition, different stages of morphogenesis and organogenesis require specific nutritive requirements.

Different responses to the carbon sources have been observed in callus culture of *Actinidia*: morphogenesis and growth was enhanced by using sucrose (Huang and Tan, 1988); combinations of sucrose with mannitol or sorbitol enhanced growth and morphogenesis respect to sucrose alone in *A. chinensis* (Leva, 1986), preconditioning of callus with mannitol or sorbitol induced higher regeneration levels than those obtained in non-preconditioning callus on *A. deliciosa* (Leva and Bertocci, 1988). Lowest levels of growth was observed when sugar alcohols were used for more than five cultures in *A. deliciosa*, but they showed a high regenerative potential and production of normal shoots after transferred to sucrose-containing medium (Leva and Muleo, 1993).

Concentration of the sugar in the nutrient medium also play an important role on growth and differentiation from callus as demonstrated by histological examinations carried out by Leva *et al.* (1991). High sucrose and glucose concentration on *A. deliciosa* cv Hayward calli produced high rates of lignification diminishing growth and differentiation of vascular tissue in cambium-like areas.

Sucrose concentration affects photosynthetic capacity in tissue culture of already differentiated explants, particularly when combined with light intensity. In shoot cultures of *A. deliciosa* without sucrose, the total sugars in leaves were found to be 25-30% of the value measured when sucrose was added to the medium and derived from CO_2 fixation by photosynthesis under normal light intensity (Dimassi-Theriou and Bosabalidis, 1997).

The possibility of producing *in vitro* plants in a medium with reduced or omitted sucrose has been investigated in other species by increasing both vessel CO_2 concentration and light intensity (Chaumont & Gudin, 1985; Honjo *et al.*, 1988; Kozai & Iwanami 1988;) and represents an interesting alternative to traditional method in large scale production systems.

3.4. CO_2 supply

Photoautotrophism *in vitro* is strongly limited by low CO_2 concentration in the closed containers during the period with artificial light (Kosai, 1991), so carbon enrichment during the *in vitro* phase was suggested for improving hardening and *in vivo* acclimation in other species by stimulating photosynthetic capability (Ziv, 1986).
The addition to CO_2 to the vessels containing shoots of *A. deliciosa* cv Hayward increased fresh and dry weight, while it did not change the proliferation rate (Predieri *et al.*, 1990). This aptitude is due to a particular ability of kiwi plants to fix CO_2 during the first hours of light exposure (Infante *et al.*, 1989b; Dimassi-Theriou and Bosabalidis, 1997).

On the contrary, in proliferating cultures of *A. deliciosa* cv Tomuri, catabolic CO_2 accumulated during the dark period was reduced during the photoperiod with no differences between CO_2 enriched and non-enriched cultures at the end of the dark period, while oxygen concentration was not reduced by the respiratory process during the dark period. Oxygen accumulation produced by photosynthesis damaged had toxic effect on chloroplasts of green tissues (Righetti *et al.*, 1993).

Other important factor, not always consider during *in vitro* culture, is oxygen concentration in culture vessels. Its concentration is not constant as in open cultures, because the *in vitro* environment is not sufficiently complex to reduce O_2 excess via thermodynamic processes or by O_2 requiring enzymes. The accumulation of O_2 is not completely represented by the O_2 released by photosynthesis since superoxides (O_2^-) and singlet O_2 ($^1O_2$) are also photochemically produced (Badger, 1985). In this case, the process may cause oxidative injury to photosynthetic tissues not scavenged by efficient antioxidant systems (Elstner, 1982; Asada & Takahashi, 1987; Monk *et al*, 1989).

4.0 APPLICATION OF *IN VITRO* CULTURE TECHNIQUES IN *ACTINIDIA* SPP.

4.1. Somaclonal variation and genetic transformation

Genetic modification has been widely reviewed recently by Rugini et al., 2000 and Baldoni and Rugini, 2002. Somaclonal variation occurs as a result of *in vitro* methods and is found in many species, thus creating additional genetic variation without the need to outcross (Larkin *et al.*, 1989). Somaclonal variation is commonly found within callus, cell and protoplasts cultures. The value of somaclonal variation as a tool for plant improvement in morphological and biochemical traits and resistance to biotic and abiotic stress has been described for many fruit plant species (McPheeters *et al.*, 1993). Some of these reports demonstrated the effectiveness of *in vitro* selective and/or screening procedures to obtain different cell lines and plants showing beneficial variations for some characters with the advantage that agronomic useful traits have a higher probability of being retained (Hammerschlag, 1992; Donovan *et al.*,1994; Skirvin *et al.*, 1994).

The ability of kiwifruit to regenerate plants from callus encouraged scientists to use somaclonal variation for improving some agromonic and commercial characteristics such as increased the fruit size, calcareous and drought tolerance.

Somaclonal variation has been observed in different tissues of *Actinidia* after *in vitro* culture. Hirsch *et al.* (1991) studied the occurrence of variation of *in vitro* regenerated plantlets coming from different explants such as stem, leaf, root, ovary, anther filament, mature endosperm and embryos. They found that plants regenerated from mature embryos and endosperm of fruitful male plants and the female cv. 'Hayward' showed great variability in morphological traits. Endosperm-derived plants of *A. chinensis* were produced and planted in China (Gui and Mu, 1982). Differences in flowering, sex expression, fruit quality such as fruit size and shape, seed number, nutrition value (vitamin C and soluble solid content), and chromosome number (from 58 to 146) have also been observed among these clones. The plants were aneuploid with only a few triploids but none of them were parthenocarpic (Gui *et al.*, 1993). Considerable morphological variations regarding vigour and flowering have also been observed in protoplast-derived plants which were micro-grafted on mature rootstocks respect to the self-rooted kiwifruit clones of *A. deliciosa* (Ke *et al.*, 1993). Anatomical modifications in the cell wall of cultured and natural fruit cells have also been observed (Fisher *et al.* 1996). Piccotino *et al.* (1997) showed that micropropagated plants have a smaller size of xylem vessels, marked juvenile traits, elongated leaves and lower quantity of epicuticular waxes, in comparison with woody cuttings.

Results of ploidy manipulation have also been obtained with anther culture (Fraser *et al.*, 1991). The doubling of chromosomes by colchicine and oryzaline treatments in one parthenocarpic trihaploid from cv. Hayward (Chalak & Legave, 1996), and in one staminate cultivar which often produce hermaphrodite flowers (Rugini *et al.* 1996) have been observed. Furthermore, Rugini *et al.* (1994) observed an increase in ploidy level among regenerated plants from callus in both staminate and pistillate cultivars without the use of chemicals.

DNA dodecaploid plants have been detected among somaclones regenerated from leaf-derived callus of cv. Hayward. The stem and leaf characteristics of the pots-grown somaclones appeared to be similar to hexaploid somaclones but the growth rate was slower than that of the hexaploids plants (Boase & Hooping, 1995).

Studies were carried out on the effect of different pH media on the plant regeneration ability of leaves of *A. deliciosa* Chev. cv Tomuri and Hayward in order to obtain some kiwifruit somaclones under selective pressure of increasing pH *in vitro* conditions. Leaf-derived somaclones have shown a high degree of variability when grown *in vitro* on standard culture media. Moreover, culture medium of pH 7 and 7.5 were often effective in selecting variants with the best *in vitro* proliferation and rooting characters (Marino & Battistini, 1990). When these somaclones where cultured at pH $\leq$ 8, they showed a decreased in all characters evaluated except those regenerated at pH 7 for 'Tomuri' and one line at pH 7.5 of 'Hayward'. The selection pressure exerted by the high pH during regeneration appeared to be effective in obtaining variants tolerant to high culture pH (Marino & Bertazza, 1998). When these somaclones were transferred to pots and field with high pH and lime levels, they demonstrated an altered tolerance to lime-induced iron chlorosis (Marino & Bertazza, 1996; Marino *et al.*, 1998).

Muleo *et al* (1996) cultured calli in high concentration of glucose-like synthetic compounds: 3-O-Methylglucose and N-Methyl-D-glucamine. Both compounds induced a similar type of genetic variation in regenerated plants, contrary to NaCl at the same water potential. RAPD analysis demonstrated different frequencies and patterns of polymorphic fragments.

4.2. Germplasm conservation

Kiwifruit germplasm has been conserved by living collections in field plots. However, this is not suitable for long-term preservation because of the extensive maintenance requirements, vulnerability to pests and diseases and climate variability.

4.2.1. Slow growth preservation

An approach for short and medium-term germplasm preservation is the *in vitro* storage techniques by inducing minimum growth (Monette, 1995). Culture conditions are modified (culture in the dark at reduced temperatures) so that the cultures can be stored for up to one year before they need to be transferred into fresh medium (Monette, 1986b). Unfortunately, slow growth induces the accumulation of somaclonally variant cells.

4.2.2 Cryopreservation

Since the maintenance and preservation of an extensive gene pool for *Actinidia* species and hybrids is essential for future breeding, new efforts have been carried out to stop growth completely in cryopreservation, by the conservation of the explants in liquid nitrogen (-196° C). Long term storage was first attempted by Jian and Sun (1989), who reported successful cryopreservation of dormant vegetative stem segments of *A. chinensis,* but only callus re-growth and some regenerated plantlets was achieved after 120 days of storage in liquid nitrogen. No data has yet been reported concerning long-term storage of kiwifruit meristems or shoot tips by cryopreservation.

An optimised encapsulation-dehydration protocol for the cryopreservation of shoot tips from *in vitro* grown *A. arguta* x *A. deliciosa, A. deliciosa* (cvs. Hayward and Tomuri), *A. chinensis*, *A. eriantha* and several diploid hybrids was reported by Bachiri *et al.* (2001). In this method, tolerance to high sucrose concentration during the pre-conditioning and PGR balance in the post-thaw medium were crucial factors. When NAA was replaced by zeatin the recovery of plantlets was higher than with NAA and/or BAP. The balance between BAP and zeatin also appeared to be important. After recovery, plantlets were returned to their usual stock culture medium. The high recovery of plants (65-95%) of various kiwifruit species, cultivars and hybrids of different ploidy, suggests that encapsulation-dehydration is the preferred pre-treatment for long-term germplasm preservation, because no special equipment is required and the procedure is easy to handle (Bachiri *et al.*, 2001).

Wu *et al.* (2001) studied some of the factors that affect the cryopreservation of shoot tips and the genetic stability of recovered plants. They found that one month of cold acclimation of mother plants at 5° C optimised the protocol for cryopreservation, and no differences were noted between *in vitro* plantlets produced from cryo-preserved and control shoot tips.

4.3. Synthetic seeds

Encapsulation of somatic embryos or vegetative propagules can be considered an important application of micropropagation, to improve the success of *in vitro* derived plant to be delivered to the field, greenhouse and/or to contribute to synthetic seed technology. Attention has been paid to encapsulation of somatic embryos of different species (Redenbaugh *et al.*, 1993) while few research has been carried out on the encapsulation of non embryogenic *in vitro* derived vegetative propagules, such as shoot tips or axillary buds of different species (Mathur *et al.*, 1989; Ganapathi *et al.*, 1992; Bapat, 1993; Piccioni & Standardi, 1995).

In *A. deliciosa,* successful results have been obtained by using a sodium alginate matrix to encapsulate micropropagated buds before culture on enriched medium (Piccioni & Standardi, 1995). However, a more extensive research of the nutritive conditions in the regrowth medium and in the capsules is needed because of the nutritional deficiency found. This technique has been recently improved by applying cold treatment to the *in vitro* proliferating mother shoots and by increasing the concentration of sucrose of the encapsulated apical and axillary buds which were able to convert into whole plants (Adriani *et al.*, 2000).

4.4. Production of secondary metabolites within plant cell cultures

Cell suspensions are used for the production of secondary metabolites in preference to callus culture because cultures are grown under standard conditions for short growth cycles. In addition, the suspension cultures are not subjected to the seasonal changes, are less complex in organisation than the entire plant and purified enzymes and active cell free systems can be prepared easily.

In Japan, the fruit galls of *A. polygama* have analgesic and tonic properties. Sashida *et al.*, (1994) studied formation of triterpenoids in callus tissue and a comparison of triterpenoid composition in callus tissue and regenerated and natural plants. From the research eight triterpenoids containing two new ones were isolated from calli culture of *A. polygama.*

5.0 TRANSFER TO GREENHOUSE

This is a critical step after micropropagation due to the high risk of dehydration of the rooted shoots, because the relative humidity in the jars is about 100%, and they have to start the autonomous nutrition. The first step is to wash all the residues of agarised medium from the roots that could facilitate fungal attack. The micropropagated plants are then transplanted. The first sixty-days is the critical period after transplanting and it must to be possible to control the environmental conditions within the greenhouse such as, relative humidity, temperature and light. Plants must be kept in greenhouse for one year prior to transfer to the field (Zuccherelli, 1994).

6.0. TECHNIQUES TO ACCELERATE FLOWERING OF JUVENILE PLANTS

Normally the plants from *in vitro* culture, particularly those derived from organogenesis and/or somatic embryogenesis, are juvenile in growth and development. To accelerate the flowering within the field, the plants were pruned to a height of 30-40 cm. The remaining stems were arced in order to stimulate the growth of the basal buds, from which shoots will develop. The shoots will be pruned again at the same height of 30-40 cm. The procedure is repeated 2-3 times within the same season. Consequently, these plants will flower earlier when transferred to the field. Alternatively, grafting of juvenile scion on old plants will accelerate flowering.

7.0 SUMMARY AND CONCLUSIONS

The largest component of the genetic pool within kiwifruit has not been yet been fully exploited because there is very little demand from the consumer for change in the end-product (Ferguson *et al.*, 1990). For this reason, the world-wide kiwi production industry is dependent on the production of a unique cultivar belonging to the species *A. deliciosa*, which was selected years ago from crosses among a small number of plants. Recently, selection has included the species *A. chinensis* which seems promising because its germplasm base shows variability larger than in *A. deliciosa.*

By developing *in vitro* culture techniques, all genotypes can easily be propagated mainly by axillary bud stimulation. Furthermore, it is cheaper than traditional propagation methods and the plants obtained are uniform and virus free. Somaclonal variation has been observed in different tissues, particularly in callus, cell and protoplast cultures. The ability of this species to regenerate plants from callus has encouraged researchers to use somaclonal variation to improve some agronomic and commercial characteristics. Suitable starting material for *in vitro* culture are shoot tips, collected every season, the best during spring since the growth response is more rapid under *in vitro* conditions and the possibility of contamination is low.

The efficient *in vitro* regeneration systems, which are well established within kiwi, is useful to provide new genotypes, by inducing mutations through *in vitro* selection, somaclonal variation or genetic manipulation through recombinant DNA technology. Organogenesis may be useful to exploit clonal variation to obtain stable plants, for its unicellular origin, and lesser extent for rapid propagation systems because of the risk of genetic variation..

Finally the development of embryo rescue techniques has permitted to overcome difficulties of germination of the hybrid plantlets of 4 interspecific hybridisation crosses, which was very useful to transfer important traits, such as, resistance to frost, high content of vitamin C, from some minor kiwi species to more important and diffused species.

8. REFERENCES

Adriani, M., Piccioni, E., Standardi, A. Effect of different treatments on the conversion of 'Hayward kiwifruit' synthetic seeds to whole plants following encapsulation of in vitro derived buds. New Zealand Journal of Crop and Horticultural Science 2000; 28: 59-67.

Aldrufeu, A., Cols, N., Adillon, J. Aplicaciòn de la técnica de cultivo *in vitro* a la multiplicaciòn de la *Actinidia chinensis* Planch. Porencias y comunicaciones das Primeras jornadas técnicas sobre *Actinidia* (kiwi), Vigo (Pontevedra) Spain. 17-19 Oct. 1985, pp 119-125.

Asada, K., Takahashi, M. Production and scavenging of active oxygen in photosynthesis. In: *Photoinhibition* DJ Kyle, CB Osmond, CJ Arntzen eds. Elsevier Science, Amsterdam 1987, pp 227-287.

Augè, R., Boccon-Gibot, J.B. Les applications de l'horticulture. In: *La culture in vitro.* H Vidalie, ed Tech et Docum Lavoiseir, Paris 1982, 75-102.

Augé, R., Vidalie, H., Ginoux, R. Mise au point de la multuplication de l'*Actinidia chinensis.* E.N.I.T.A. *Angers* 1982: 713-723.

Auxtova, O., Samaj, J., Cholvadova, B., Khanolova, E. Isoperoxidase and isopolyphenoloxidase spectra in male and female tissues of *Actinidia deliciosa* in vitro. Biologia Plantarum 1994; 94: 535-541.

Bachiri, Y., Song, G.Q., Plessis, P., Shoar-Ghaffari, A., Rekab, T., Morisset, C. Routine cryopreservation of kiwifruit (Actinidia spp) germplasm by encapsulation-dehydration importance of plant growth regulators. CryoLetters 2001; 22: 61-74.

Badger, M.R. Photosynthetic oxygen exchange. Annual Review Plant Physiology 1985; 36:27-53.

Baldoni L and Rugini E.. Gene modification in fruit crops, In: *Fruit and Vegetable Biotechnology,* V. Valpuesta (ed). Woodhead Publishing Limited England CRC Press, Boca Raton, Boston New York, Washington, DC, 2002; pp 25-113.

Bapat, V.A. Studies on synthetic seeds of sandalwood (*Santalum album* L.) and mulberry (*Morus indica* L.). In: *Synseeds: applications of synthetic seeds to crop improvement.* Redenbaugh K eds.. CRC press Inc. Boca Raton Fl, USA 1993, pp 381-408.

Barbieri, C., Morini, S. Plant regeneration from *Actinidia* callus cultures. Journal of Horticultural Science 1987; 62:107-109.

Biasi, R., Franceschetti, M., Falasca, G., Altamura, M.M., Bagni, N. Polyamines as markers in sexual reproduction of kiwifruit. In: *Anther and Pollen: Biology, Biotechnology* Clement, C., E. Pacini and J.C Andran eds.. Springer-Verlag, 1999

Bini, G. Moltiplicazione 'in vitro' di *Actinidia chinensis* Pl. Proceedings of the Congress: 'In vitro culture techniques' for large scale propagation of horticultural crops: 211-218. Pistoia October 6, 1979.

Boase, M.R., Hopping, M.E. DNA dodecaploid plants detected among somaclones of *Actinidia deliciosa* var. *deliciosa* cv. Hayward. Plant Cell Reports 1995; 14:319-323.

Brossard-Chriqui, D. Tripathi, B.L. Comparaison des aptitudes morphogénétiques des étamines fertiles ou stériles d'*Actinidia chinensis* cultivées in vitro. Canadian Journal of Botany 1984; 62:1940-1946.

Cai, Q.Q., Qian, Y.Q., Ke, S.Q., He, Z.C., Jiang, R.X., Zhou, Y.L., Ye, Y.P., Hong, S.R., Huang, R.H. Studies on the somaclonal variation of regenerated plants from protoplasts of *A. deliciosa.* Acta Botanica Sinica 1992; 34:822-828.

Cai, Q.Q., Qian, Y.Q., Ke, S.Q., He, Z.C. Regeneration of plants from protoplast of kiwifruit (*Actinidia deliciosa*). In: *Biotechnology in agriculture and forestry*, 23. Bajaj YPS ed.. Springer, Berlin Heidelberg New York, 1993, pp 3-17.

Cañal, M.J., Fernàndez, H., Fernàndez, P., Centeno, M.L., Fernàndez, B. Uptake and metabolism of 6-benzyladenine in ventilated cultures of kiwifruit plants. Plant Growth Regulation 2000; 30:209-214.

Canhoto, J.M., Cruz, G.S. In vitro multiplication of *Actinidia chinensis* Planch. by culture of young leaves. Boletim da Sociedade Broteriana 1987; 2, 60:239-252.

Centeno, M.L., Rodriguez, A.M., Albuerne, M.A., Feito, I., Fernàndez, B. Uptake, distribution and metabolism of 6-benzyladenine and cytokinins contain during callus initiation from *Actinidia deliciosa.* Journal of Plant Physiology 1998; 152:480-486.

Centeno, M.L., Rodrìguez, A., Feito I., Fernàndez, B. Relationship between endogenous auxin and cytokinin levels and morphogenic responses in *Actinidia deliciosa* tissue cultures. Plant Cell Reports 1996; 16:58-62.

Chalak, L., Legave, J.M. Oryzalin combined with adventitious regeneration for an efficient chromosome doubling of trihaploid kiwifruit. Plant Cell Reports 1996; 16: 97-100.

Chaumond, D., Gudin, C. Transition from photomixotrophic to photoautotropic growth of *Asparagus officinalis* in suspension culture. Biomass 1985; 8:41-58.

Chiariotti, A., Caboni, E., Frattarelli, A. Shoot regeneration from in vitro roots of kiwi. Acta Horticulturae1991; 289:97-98.

Cossio, F., Marino G. Indagini preliminari sull'isolamento e la coltura dei protoplasti di *Actinidia chinensis. Rivista di Ortoflorofrutticoltura* It. 1983; 67:455-464.

De Paoli, G., Rossi V., Scozzoli, A. Micropropagazione della piante ortoflorofrutticole. Edagricole, Edizioni Agricole. Bologna, Italy. 1994; pp 196-197.

Debergh, P.C., Read, P.E. Micropropagation. In: *Micropropagation: Technology and Application*, PC Debergh, RH Zimmerman eds. Kluwer Academic Publishers, London 1991; pp 1-13.

Dimassi-Theriou, K. Bosabalidis, A.M. Effects of light, magnesium and sucrose on leaf anatomy, photosynthesis, starch and total sugar accumulation, in kiwi cultured in vitro. Plant Cell Tissue and Organ Culture 1997; 47:127-134.

Donovan, A.M., Ridout M.S., James D.J. Assessment of somaclonal variation in apple. 2. Rooting ability and shoot proliferation in vitro. Journal of Horticultural Science 1994; 69:115-122.

Einset, J.W., Silverstone A. Hydroxylation of N^6-(Δ^2-isopentenyl)adenine to zeatin. Plant Physiology 1987; 84:208-209.

Elstner, E.F. Oxygen activation and oxygen toxicity. Annual Review of Plant Physiology 1982; 33:73 96.

Feito I., Gonzalez, A., Centeno, M.L., Fernandez, B., Rodriguez, A. Title and distribution of benzyladenine in *Actinidia deliciosa* explants cultured in liquid and solid media. Plant Physiology and Biochemistry 2001; 39:909-916.

Feito, I., Rodrìguez, A., Centeno, M.L., Sànchez Tamés, R., Fernàndez, B. Effect of the physical nature of the culture medium on the metabolism of benzyladenine and endogenous cytokinins in *Actinidia deliciosa* tissues cultured in vitro. Physiologia Plantarum 1994; 91:449-453.

Feito, I., Rodrìguez, A., Centeno, M.L., Sànchez Tamés R., Fernàndez B., Effect of applied benzyladenine on endogenous cytokinin content during the early stages of bud development of kiwifruit. Physiologia Plantarum 1995; 95:241-246.

Ferguson, A.R. The genus *Actinidia*. In: *Kiwifruit: Science and Management*. IJ Warrington, GC Weston, eds.., Ray Richards Publisher, Aukland, 1990 pp. 15-35.

Fisher, M., Wegryzn, R.F., Hallett, I.C., Redgwell, R.J. Chemical and structural features of kiwifruit cell walls: Comparison of fruit and suspension-cultured cells. Carbohydrate Research 1996; 295:195-208

Fraser, L.G., Harvey C.F. Somatic embryogenesis from anther-derived callus in two *Actinidia* species. Scientia Horticulturae 1986; 29:335-346

Fraser, L.G., Harvey, C.F., Kent, J. Ploidy manipulation of kiwifruit in tissue culture. Acta Hortculturae 1991; 297: 109-114.

Gamborg O.L. Callus and cell culture. In: *Plant tissue culture methods*. LR Wetter, F Constabel eds.. Plant Biochemistry Section Prairie Regional Lab Saskatoon, Canada, 1982; pp 1-10.

Ganapathi, T.R., Suprasanna, P., Bapat, V.A., Rao, P.S. Propagation of banana through encapsulated shoot tips. Plant Cell Reports 1992; 11:571-575.

Gonzalez, M.V., Rey, M.A., Revilla, M.A., Rodriguez, R. Morphogenetic patterns in kiwi tissue culture and after *Agrobacterium* co-culture. Acta Horticulturae 1990; 282:359-366.

Gonzalez, M.V., Rey, M., Rodriguez, R. Plant regeneration from petioles of kiwifruit microshoots. Horticultural Science 1995; 30:1302-1303.

Gui, Y.L. Induction of callus and regeneration of plantlets in stem segment culture of Chinese gooseberry. Acta Botanica Sinica 1979; 21:339-344.

Gui, Y., Hong, S., Ke, S., Skirvin, R.M. Fruit and vegetative characteristics of endosperm-derived kiwifruit (*Actinidia chinensis* F.) plants. *Euphytica* 1993; 7, 57-62.

Gui, Y.L., Mu, S.K., Xu, T.Y. Studies on morphological differentiation of endosperm plants of Chinese gooseberry in vitro. Acta Botanica Sinica 1982; 24: 216-221.

Hammerschlag, F.A. Somaclonal variation. In: *Biotechnology of perennial fruit crops*. FA Hammerschlag, RE Litz eds.. CAB International Wallingford, UK, 1992; 35-55.

Harada, H. In vitro organ culture of *Actinidia chinensis* Pl as a technique for vegetative multiplication. Journal of the Horticultural Society 1975; 50:81-83.

Harvey, C.F., Fraser L.G., Gill, G.P. Sex determination in *Actinidia*. Acta Horticulturae 1997a; 444:85-88.

Harvey, C.F., Gill, G.P., Fraser, L.G., McNeilage, M.A. Sex determination in *Actinidia*. 1.Sex-linked markers and progeny sex ratio in diploid *A. chinensis*. *Sex*. Plant Report 1997b; 10:149-154.

Hirsch, A.M., Bligny-Fortune D. Organogenesis in tissue cultures of two plants of the genus *Actinidia* (*A. chinensis* and *A. polygama*). Relationship between organogenesis and peroxidase. Comptes rendus Hebd. Des Sea. De l'Ac. Des Sc O. 1979; 288:1159-1162.

Hirsch, A.M., Fortune D., Xiao, X.G., Blanchet, P. Somaclonal variations related to kiwifruit micropropagation, study of fruitful male plants and use of peroxidase as an earlier sex marker. Acta Horticulturae 1991; 297:123-131.

Hirsch, A.M., Fortune, D., Chalak, L., Legave, J.M., Chat, J., Monet, R. Peroxidase test: A test for sex screening in kiwifruit seedlings. Acta Horticulturae 1997; 444:89-95.

Hirsch, A.M., Testolin R., Brown, S., Chat, J., Fortune F.D., Bureau J.M., De Nay, D. Embryo rescue from interspecific crosses in the genus *Actinidia*. Plant Cell Reports 2001; 20:508-516.

Honjo, T., Futaya, Y., Takakura, T. Effect of CO2 enrichment on the growth of *Cymbidium* PLB in vitro. Acta Horticulturae 1988; 230:185-188.

Huang, H., Li, J., Lang P., Jiang, Z. Integrated approaches to germplasm conservation and phylogenetic relationships in *Actinidia*. 4th Int. Symp. on kiwifruit. Santiago de Chile. 11-14 January. 1999; pag 21

Huang, Z., Huangfu, Y., Xu, L. Triploid plantlets of Chinese gooseberry obtained by endosperm culture. Kexue Tongbao 1983; 28:112-117.

Huang, Z.G., Tan, C.Y. Chinese gooseberry, kiwifruit (*Actinidia* spp). In: *Biotecnology in agriculture and forestry*, 6. YPS Bajaj ed. Springer-Verlag, 1988, pp 166-180.

Infante, R., Rotondi, A., Marino, G., Fasolo, F. Solar light effects on growth, net photosynthesis and leaf morphology of in vitro kiwifruit (*Actinidia deliciosa*) cv Hayward. *In vitro* Cell Developemental Biology 1994; 30P:160-163.

Infante, R., Magnanini, E., Righetti, B. CO_2 evolution, photosynthetic rate and multiplication response of *Actinidia deliciosa* in vitro under different radiant flux density. Physiologia Plantarum 1989a; 78:125-133.

Infante, R., Magnanini, E., Righetti, B. The role of light and CO_2 in optimizing the conditions for shoot proliferation of *Actinidia deliciosa* in vitro. Physiologia Plantarum 1989b; 77:191-195.

Janssen, B.J., Gardner, R.C. The use of transient GUS expression to develop an *Agrobacterium*-mediated gene transfer system for kiwifruit. Plant Cell Reports 1993; 13, 28-31.

Jian, L.C., Sun, L.H. Acta Botanica Sinica 1989; 31:66-68.

Jona, R., Gribaudo, I. Comparazione sperimentale di diverse metodologie di micropropagazione dell'Actinidia. In: Incontro Frutticolo sull'Actinidia. Udine, Italy. 1983; pp 545-556.

Ke, S., Cai, Q., Skirvin, R.M. Micrografting speeds growth and fruiting of protoplast-derived clones of kiwifruit (*Actinidia deliciosa*). Journal of Horticultural Science 1993; 68: 837-840

Kosai, T. Photoautothophic micropropagation. In vitro 1991; 279:47-51.

Kozai, T., Iwanami, Y. Effects of CO_2 enrichment and sucrose concentration under high photon fluxes on plantlet growth of carnation (*Dianthus caryophyllus* L.) in tissue culture during the preparation stage. Journal of the Japanese Society of Horticultural Science 1988; 57:279-288.

Krikorian, A.D. Hormones in tissue culture and micropropagation. In: *Plant Hormones: Physiology, Biochemistry and Molecular Biology* PJ Davies ed.. Kluwer Academic Publishers, London, 1995, pp 774-798.

Larkin, P. J., Banks, P.M., Bhati, R., Brettell, S., Davies, P.A., Ryan, S.A., Scowcroft, W. R., Spindler, L.H., Tanner, G.J. From somatic variation to variant plants: mechanisms and applications. Genome 1989; 31:705-711.

Letham, D.S., Palni, M.S. The biosynthesis and metabolism of cytokinins. Annual Review of Plant Physiology 1983; 34:163-197.

Leva, A.R. Test on organogenesis from callus of *Actinidia chinensis*. Acta Horticulturae 1986; 179:883-884.

Leva, A.R., Bartolini, G., Muleo, R., Biricolti, S., Benelli, A. Micromorphological studies on *Actinidia* cv Hayward calli as influenced by different carbohydrates. Acta Horticulturae 1991; 300:381-383.

Leva, A.R., Bertocci F. The effect of sorbitol and mannitol on the morphogenesis of *Actinidia deliciosa* (A. Chev.) Liang and Ferguson callus. Acta Horticulturae 1988; 227:447-449.

Leva, A.R., Magrini P., Fattori di crescita del callo e produzione di germogli in *Actinidia chinensis* Pl. ed *Actinidia arguta* Pl. *Rivista Ortoflorofrutticoltura*. It. 1986; 70:49-56.

Leva, A.R., Muleo, R. Effects of 'media osmotic agents' on the growth and morphogenesis of *Actinidia deliciosa* cv Hayward callus. *In Vitro* Cell Developmental Biology 1993; 29:59-64.

Ludvova, A., Ostrolucka, M.G. Morphogenic processes in callus tissue cultures and de novo regeneration of plants in *Actinidia chinensis* Planch. Acta Societatis Botanicorum Poloniae 1998; 67:217-222.

Marino, G., Battistini, S. Leaf-callus growth, shoot regeneration and somaclonal variation in *Actinidia deliciosa*: Effect of medium pH. Acta Horticulturae 1990; 280, 37-44.

Marino, G., Bertazza, G. Variabilità somaclonale e selezione in vitro per tolleranza al calcare in *Actinidia deliciosa*. Atti del Convegno Nazionale 'La Coltura dell'Actinidia. Faenza 10-12 Ottobre 1996. Italia 1996; 71-77 pp.

Marino, G., Bertazza, G. Selection-pressure effects of medium pH during regeneration on successive performances of leaf-derived 'Tomuri' and 'Hayward' kiwifruit (*Actinidia deliciosa*) somaclones cultured on proliferation culture media with variable pH. Journal of Horticultural Science and Biotechnology 1998; 73:664-669.

Marino, G., Bertazza, G., Buscaroli, C. In vivo growth and tolerance to lime-induced iron chlorosis of leaf-derived cvs. Tomuri and Hayward kiwifruit (*Actinidia deliciosa*) somaclones. Journal of Horticultural Science and Biotechnology 1998; 73:670-675.

Mathur, J., Ahuja, P.S., Lai, N., Mathur, A.K. Micropropagation of *Valeriana wallichii* D. C. using encapsulated apical and axial shoot buds. Plant Science 1989; 60: 111-116.

McPheeters, K.D., Skirvin, R.M., Norton, M.A., Somaclonal variation in tissue culture. In: *Plant cell culture and its applications*. BV Charlwood, , RA Dixon and JM Torres eds.. Edward Arnolld, Sevenoaks, England. 1993

Mii, M. Ohashi, H., Plantlet regeneration from protoplast of kiwifruit, *Actinidia chinensis* Planch. Acta Horticulturae 1988; 230:167-170.

Monastra, F. Actinidia. In: *Frutticoltura speciale*. REDA, Roma, 1991

Moncalean, P., Canal, M.J. Feito, I., Rodrìguez, A., Fernàndez, B. Cytokinins and mineral nutrition in *Actinidia deliciosa* shoots cultured in vitro. Journal of Plant Physiology 1999; 115:606-612.

Moncalean, P., Rodriguez, A., Fernandez, B., In vitro responses of *Actinidia deliciosa* explants to different BA incubation periods. Plant Cell Tissue and Organ Culture 2001; 67:257-266.

Monette, P.L. Micropropagation of kiwifruit using non-axenic shoot tips. Plant Cell Tissue Organ Culture 1986a; 6, 73-82.

Monette, P.L. Cold storage of kiwifruit shoot tips in vitro. Horticultural Science 1986b; 21:1203-1205.

Monette, P.L. Cryopreservation of Plant Germplasm I. In: *Biotechnology in Agriculture and Forestry* 32. YPS Bajaj ed. Pub Springer Verlag, Berlin, 1995; pp 321-331.

Monk, L.S., Fagerstedt, K.V., Crawford, R.M.M. Oxygen toxicity and superoxide dismutase as an antioxidant in physiological stress. Physiology Plantarum 1989; 76:456-459.

Mu, S.K., Fraser, LG., Harvey, C.F. Initiation of callus and regeneration of plantlets from endosperm of *Actinidia* interspecific hybrids. Scienta Horticulturae 1990; 44:107-117.

Mu, X., Tsai D., An H., Wang W. Embryology and embryo rescue of interspecific hybrids in *Actinidia.* Acta Horticulturae 1991; 297: 93-97.

Muleo, R., Bigozzi, L., Boscherini, G., de Agazio, M., Zacchini, M., Bartolini, G., Loreti, F., Leva, A.R. Modifiche nell'habitus di varianti somaclonali rigenerati in mezzi iperosmotici; relazione con le poliammine. Atti del Convegno Nazionale 'La Coltura dell'Actinidia. Faenza 10-12 Ottobre 1996, Italia 77-85 pp.

Muleo, R., Morini, S. Effect of light quality on regeneration from callus of *Actinidia deliciosa.* Acta Horticulturae 1990; 280:155-158.

Oliveira, M.M., Barroso, J.G., Martins, M., Pais, M.S., Genetic transformation in *Actinidia deliciosa* (Kiwifruit). In: *Biotechnology in Agriculture and Forestry, 29.* YPS Bajaj ed. Springer-Verlag, Berlin, Hedelberg, 1994 pp 189-210

Oliveira, M.M., Pais, M.S.S. Plant regeneration from protoplasts of long term-term callus cultures of *Actinidia deliciosa* var deliciosa cv. Hayward (Kiwifruit) Plant Cell Reports 1991; 9:643-646.

Oliveira, M.M., Pais M.S. Somatic embryogenesis in leaves and leaf-derived protoplast of *Actinidia deliciosa* var *deliciosa* cv Hayward (kiwifruit). Plant Cell Reports 1992; 11, 314-317.

Ossani, V. Otto anni di esperienze e tentativi di miglioramento genetico dell'actinidia. Rivista di Frutticoltura. 1994; 2, 59-61.

Pais, M.S., Oliveira, M.M., Barroso, J. Use of petiole segments of *Actinidia chinensis* (kiwi) for plant differentiation and production of friable calli for protoplast isolation. Acta Horticulturae 1987; 212:687-690.

Panatta, G. Valore dietetico-nutrizionale dell'actinidia. Rivista di Frutticoltura. 1990; 10:55-60.

Predieri, S., Mezzetti B., Ranieri, R. Rigenerazione 'in vitro' da foglie di *Actinidia deliciosa* cv Hayward. Rivista Frutticoltura Ortofloricoltura 1988; 50:69-72.

Pedroso M. C.,. Oliveira, M.M., Pais, M.S.S. Micropropagation and simultaneous rooting of *Actinidia deliciosa* var. 'Hayward'. Horticultural Science 1992; 27:443-445.

Piccioni, E., Standardi, A., Encapsulation of micropropagated buds of six woody species. Plant Cell Tissue and Organ Culture 1995; 42:221-226.

Piccotino, D., Massai, R., Dichio B., Nuzzo, V. Morphological and anatomical modifications induced by in vitro propagation of kiwifruit plants. Acta Horticulturae 1997; 444, 127-132.

Pierik, R.L.M. *In Vitro Culture of Higher Plants.* Martinus Nijhoff Publishers, Kluwer, 1987

Pierik, R.L.M. Handicaps for large scale commercial application of micropropagation. Acta Horticulturae 1988; 230:63-71.

Predieri, S., Infante, R., Fasolo, F., Righetti, B. CO_2 enrichment effect on in vitro grown apple and kiwifruit. Acta Horticulturae 1990; 300:107-109.

Raquel, M.H., Oliveira, M.M. Kiwifruit leaf protoplats competent for plant regeneration and direct DNA transfer. Plant Science 1996; 121:107-114.

Redenbaugh, K., Fujii, J.A.A., Slade, D. Hydrated coatings for synthetic seed. In: *Synseeds: applications of synthetic seeds to crop improvement.* K Redenbaugh ed. CRC press Inc. Boca Raton Fl, USA, 1993 pp 35-46.

Revilla, M.A., Power, J.B. Morphogenic potential of long-term callus culture of *Actinidia.* Journal of Horticultural Science 1988; 63:541-545.

Revilla, M.A., Rey, M.A., Gonzalez-Rio, F., Gonzalez, M.V., Diaz-Sala, C., Rodriguez, R. Micropropagation of kiwi (*Actinidia* spp), in: *Biotecnology in Agriculture and Forestry* **18**. YPS Bajaj ed. Springer-Verlag Berlin Heidelberg, 1992; pp 399-423.

Rey, M., Fernandez, M.T., Gonzalez, M.V., Rodriguez, R. Kiwifruit micropropagation through callus shoot-bud induction. *In Vitro* Cellular Developemental Biology 1992; 28:148-152.

Righetti, B., Magnanini, E., Rossi, F., Photosynthetic carbon dioxide uptake and oxygen accumulation during in vitro culture of *Actinidia deliciosa* cv Tomuri. Environmental and Experimental Botany 1993; 33:523-528.
Rugini, E., Muganu, M., Natali, S., Gutierrez-Pesce, P., Starrantino, A. Effetti dei geni *rol* di *Agrobacterium rhizogenes* sulle caratteristiche fenotipiche di alcune specie frutticole. In Atti giornate scientifiche S.O.I. San Benedetto del Tronto. 22-24 Giugno 1994; pp. 53-60.
Rugini, E., Pellegrineschi, A., Mencuccini, M., Mariotti, D. Increase of rooting ability in the woody species kiwi (*Actinidia deliciosa* A. Chev.) by transformation with *Agrobacterium rhizogenes rol* genes. Plant Cell Reports 1991; 6/7:291-295.
Rugini, E., Pilotti, M., Testolin, R., Cipriani, G., Huang,W.G. Uso di biotecnologie nel miglioramento genetico delle actinidie. In: Atti convegno nazionale . La coltura dell'actinidia, Faenza 10-12 Ottobre 1996; pp 61-70.
Rugini E., Muganu M., Gutierrez-Pesce P. Transformation of *Actinidia* spp In: *Molecular Biology of Woody Plants,* Vol 2 (SM Jain and SC Minocha) eds), Kluwer Academic Publishers, 2000; pp 191-243.
Sashida, Y., Ogawa, K., Yamanouchi, T., Tanaka, H., Shoyama Y., Nishioka,I. Triterpenoids from callus tissue of *Actinidia polygama.* Phytochemistry 1994;5:377-380.
Skirvin, R.M., McPheeters K.D., Norton,M., Sources and frequency of somaclonal variation. Horticultural Science 1994; 9:1232-1246.
Standardi, A. Micropropagazione dell'*Actinidia chinensis* Planch. Mediante coltura 'in vitro' di apici meristematici'. Frutticoltura 1981; 43:23-27.
Standardi, A. Effects of repeated subcultures in shoots of *Actinidia chinensis* Pl. In: *Plant Tissue Culture.* A Fujiwara ed. Maruzen, Tokyo, 1982; pp 737-738.
Standardi, A. La 'micropropagazione' nella moltiplicazione dell'actinidia. Frutticoltura 1983; 45:17-22.
Tanaka, H., Shoyama, Y., Sasaki, Y., Sashida, Y., Micropropagation of *Actinidia polygama* from fruit galls. Plant Cell Tissue and Organ Culture 1997; 48:135-137.
Tran Thanh Van, K.M. Control of morphogenesis in in vitro culture. Annual Review of Plant Physiology 1981; 32:291-357.
Tripathi, B.K., Saussay, R. Sur la multiplication vegetative de l'*Actinidia chinensis* Planchon, 'Chinese gooseberry' par culture de racines issues de filets staminaux. Comptes Rendus de l'Academie des Sciences 1980; 291:1067-1069.
Tsai, (Cai) C. K. Plant regeneration from leaf callus protoplasts of *Actinidia chinensis* Plantch var. *chinensis.* Plant Science 1988; 54:231-235.
Uematsu, C., Murase, M., Ichikawae, H., Imamura, J. *Agrobacterium*-mediated transformation and regeneration of kiwi fruit. Plant Cell Report 1991; 10:286-290.
Velayandom, L., Hirsch, A.M., Fortune, D. Tissue culture of nodal stem segments of *Actinidia chinensis* (L.) Planchon, as a method of micropropagation. Comptes Rendus de l'Academie des Sciences III – Sciences de la Vie-Life Sciences 1985; 301:598-600.
Wang, J.X., Li, S.Z., Li, B.W., Ren, E.Y. Propagation of *Actinidia chinensis* by tissue culture. Liaoning Agricultural Sciences 1982; 1:32-34.
Wessels, E., Nel, D.D., Von Staden, D.F.A. In vitro propagation of *Actinidia chinensis* Pl. cultivar Hayward. Deciduous Fruit Grower 1984; 34:453-457.
Wu, Y., Zhao, Y., Engelmann, F., Zhou, M. Cryopresevation of kiwi shoot tips. CryoLetters 2001; 22 :277-284.
Xiao, X.G. Hirsch, A.M. Microcallus formation from leaf mesophyll protoplasts in the genus *Actinidia* Lindl. Plant Cell Report 1996; 15:896-899.
Xiao, Z.A., Shen, D.X., Lin, B.N. Plant regeneration from leaf callus protoplast of *Actinidia chinensis* Planch. Acta Botanica Sinica 1992; 34:736-742.
Zenguang, H., Youli, H., Lyin, X. Kexue Tongbao 1983; 28:112-117.
Zhang, Y.L., Qian, Y.Q., Mu, X.J., Cai, Q.G., Zhou, Y.L., Wei, X.P. Plant regeneration from in vitro cultured seedling leaf protoplasts of *Actinidia eriantha* Benth. Plant Cell Reports 1998; 17:819-821.
Ziv, M. In vitro hardening and acclimatisation of tissue culture plants. In: *Plant tissue and its agricultural application.* LA Withers PG Alderson eds. Butterworths, London, 1986; pp 187-196.
Zuccherelli, G. *L'Actinidia e i nuovi kiwi.* Collana 'Frutticoltura moderna', Edagricole, Bologna. 1994; 360 pp.
Zuccherelli, G., Zuccherelli, G. *L'Actinidia pianta da frutto e da giardino.* Collana di Manuale Tecnici dell'Edagricole, Edagricole, Bologna, It. 1985

23. Micropropagation of tea *(Camellia sinensis* L.)

Tapan Kumar Mondal

Center for Advance Study in Tea Science and Technology, Department of Plantation Crops and Processing, Uttar Banga Krishi Viswavidyalaya, P.O: Pundibari, Dist Cooch Behar, Pin: 736165, West Bengal, India. (Email: mondaltk@yahoo.com).

CONTENTS

S.M. Jain and K. Ishii (eds.), Micropropagation of Woody Trees and Fruits, 671–719.

1. INTRODUCTION

1. 1. Botany and Geographical Distribution

Tea (*Camellia sinensis* L.), of family Theacea is the oldest non-alcoholic caffeine containing beverage crop in the world. Chinese were the first to use tea as a medicinal drink, later as beverage and have been doing so for the past 3000 years (Eden, 1958). Today, tea has become the most popular and cheapest beverage drink world-wide. The cultivated taxa comprise of three main natural hybrids which are: *C. sinensis* (L.) O. Kuntze or China tea, *C. assamica* (Masters) or Assam tea and *C. assamica* sub sp. *lasiocalyx* (Planchon ex Watt.) or Cambod tea or Southern tea. Tea is an evergreen, perennial, cross-pollinated plant and grows naturally as tall as 15 meters. However, under cultivated condition, the bush height 60 - 100 cm is maintained for harvesting the young leaves, which continues even more than 100 years. The flowers are white in colour and born singly or in pairs at the axils. The fruits are green in colour with 2-3 seeds and start bearing within 5-6 years after planting. Leaf is the main criterion by which three types of tea are classified. Briefly, a) Assam type has biggest leaves, b) China type has smallest leaves, and c) Cambod leaves size are in-between of both, Assam and China type.

The genus *Camellia* had 82 species in 1958 (Sealy, 1958) and accounts for more than 325 species in 2000 (Thakor, 2000) which indicates genetical instability and high out-breeding nature of the genus. Presently over 600 cultivated varieties are available world-wide, many of which have unique traits such as high caffeine content, blister blight disease resistant etc. (Mondal *et al.*, 2002d). Tea breeds well with wild relatives and thus tea taxonomists have always been interested to identify such hybrids due to their suspected involvement in the genetic pool of tea. Two particularly interesting taxa are *C. irrawadiensis* and *C. taliensis* whose morphological distribution overlap with that of tea (Banerjee, 1992). It has also been postulated that some desirable traits such as anthocyanin pigmentation or special quality characters of Darjeeling tea might have introduced to tea gene pool from wild species (Wood and Barua, 1958). Tea is diploid (2n=30; basic chromosome number, X=15) and karyotype of chromosome ranges from 1.28μ to 3.44μ (Bezbaruah, 1971). However higher ploidy level such as triploids e.g TV-29, UPASI-3 (2n=45), tetraploids (2n=60), pentaploids (2n=75) and deviation from normal chromosome number (2n$\pm$1 to 29) have also been identified among the natural tea populace (Singh, 1980; Zhan *et al.*, 1987).

South East Asia is the original home of tea. Specifically, primary center of origin of tea is considered the point of intersection of latitude 29°N and longitude 98°E, near the source of Irrawaddy river which is at the confluence of North-East India, North Burma, South-West China and Tibet provinces (Wight, 1959). Tea grows within the latitudinal range of 45°N to 34°S in 52 different countries. Despite occupying only 16.4 % of the total tea growing areas of the world, India ranks first as a producer, consumer and exporter of tea.

1.2. Economic importance and health benefits

Tea was initially used as a medicine and subsequently as beverage and now has proven well to be a future potential as an important raw material in the pharmaceutical industry. Apart from being used as beverage, tea leaves are also used as vegetables such as 'leppet tea' in Burma and 'meing tea' in Thailand. Although the tea oil is used as lubricant but extraction of oil from tea seed is not economically viable (Wealth of India, 1950). Moreover, tea seed cakes also contain saponins. It has a poor value as a fertilizer thus unfit for animal feed on the account of its low nitrogen, phosphorus and potassium contents but can be used successfully in the manufacture of nematocide (Wealth of India, 1950).

Tea leaves have more than 700 chemical constituents, among them, the compounds closely related to human health are flavanoides, amino acids, vitamins, caffeine and polysaccharides. Leaves are rich in Vitamin C and concentration of which are comparable with that of lemon. It is also a good source of Vitamin E and Vitamin K. More over, 'tea drinking' has recently been proven to be associated with cell-mediated immune function of human body. Tea plays an important role in improving beneficial intestine microflora as well as in providing immunity against intestinal disorders and in protecting the cell membranes from oxidative damages. Tea also prevents dental caries due to the presence of fluorine. The role of tea in normalizing blood pressure, lipid depressing activity, prevention of coronary heart diseases and diabetes by reducing the blood-glucose activity is also well established (Chen, 1999). Tea also possesses germicidal and germistatic activities against various gram-positive and gram negative human pathogenic bacteria such as *Vibrio cholera*, *Salmonella*, *Clostridium* etc. (Chen, 1999). Both green and black tea infusions contain a number of antioxidant mainly catechins that have anticarcinogenic, anti-mutagenic and anti-tumor properties. Among the different catechins, epigallocatechin-gallate is the most active component. Several epidemiological studies have proved that tea consumption plays a protective role against human cancer.

1.3. Conventional propagation and breeding

Traditionally tea is propagated either through seeds or vegetative cuttings. Nevertheless, seed-grown plants show a high degree of variability. Moreover, a tea seed loses viability quickly and need to be shown immediately. Therefore, the alternative choice is through vegetative propagation of elite clones. The method of vegetative propagation has been modified over the years and present established method of using cuttings consisting of a single leaf with nodal axillary bud and the associated piece of inter-node of current year growth which was first developed by Tunstall (1931). Subsequently, this method was standardized (Tubbs, 1932) and used as an easy and simple technique for vegetative propagation of tea plant. In recent years, grafting, as an alternative propagation technique has gained considerable popularity as grafted plants show 100% increase of yield with better quality than either of the non-grafted cultivars.

Conventional breeding is done by either natural hybridization or by clonal selection. In natural hybridization, based on the known characters such as yield, quality or diseases resistance etc., two parents are planted side by side in an isolated place and allow them to bear fruits naturally. Subsequently, seeds (F_1) are harvested, raised and planted. If average performance of these plants is better than either parents, then seeds are distributed to growers as hybrid seed or Bi-clonal seed. However, if found any outstanding performer among them, release as clone for that geographical area. Following this method, tea research institutes world-wide have generated the clones for their own region. Instead of two, if more parents are used, seeds are termed as polyclonal seed. The idea is to introduce more variability among the F_1 seeds. Since it is difficult to know about the pedigree of these seeds, (as pollen may come from any male), hence the reproducibility of performance are low. Thus this process is not used presently.

Clonal selection is an age-old practice in tea. Most of the commercial tea gardens were established with seeds, as a result high genetic variability exist among the tea existing population. In many instances elite plants have been identified among the exiting bushes and released as clone. Majority of the tea clones have been developed through selection. However, pedigree of the clone remains unknown.

Although, conventional tea breeding is well established and contributed much for tea improvement over the past several decades, but time consuming and labour intensive. The bottlenecks of conventional breeding are: (1) perennial nature, (2) long gestation periods, (3) high inbreeding depression, (4) self-incompatibility, (5) unavailability of distinct mutant of different biotic and abiotic stress, (6) lack of distinct selection criteria, (7) low success rate of hand pollination, (8) short flowering time (2-3 months), (9) long duration for seed maturation (12-18 months), (10) clonal difference of flowering time and fruit bearing capability of some clones. Although vegetative propagation is an effective method of clonal propagation, yet it is limited by several factors such as: (i) slower rates of propagation, (ii) unavailability of suitable planting material due to winter dormancy, drought in some tea growing area etc., (iii) poor survival rate in nursery due to poor root formation of some clones and (iv) seasonal dependent rooting ability of the cuttings. Therefore, micropropagation technique appears to be ideal choice for the circumvention of the problems related to conventional propagation.

2. Micropropagation

Micropropagation of tea has been reviewed well by several workers in past (Kato, 1989a; Vieitez, *et al.*, 1992; Dood, 1994; Das, 2001). According to the literature, while Forrest, (1969) was pioneer for tissue culture, Kato (1985), initiated a systematic study on micropropagation of tea. Since then till late 80's thrust on tea micropropagation was on increasing the multiplication rate in *in vitro*. However, the emphasis on survival rate including hardening protocol was given only on early 90's. Studies on field performance of micropropagated tea and commercial exploitation started only in the beginning of the new millennium. These studies are summarized in Table-1 and discussed below.

2.1. Explants

Establishment of initial culture in *in vitro* and speed of multiplication depends upon suitable explants, origin and their availability through out the year. Generally, shoot tips and nodal segments with dormant axillary buds of either juvenile or adult origin of current year growth are commonly used explants for tea micropropagation (Vieitez *et al.*, 1992). While Iddagoda *et al.*, (1988) and Jha and Sen, (1992) have used zygotic embryos, immature and mature cotyledons for the induction of adventitious buds, Kato (1985), employed epidermal layers of stem segments, stem segments without epidermal layer, and intact stem segments for the regeneration of shoots and documented that callus from the epidermal layers formed buds more rapidly than the callus from other origins. Flower stalks, stem pieces and leaf petioles have been attempted for the induction of adventitious buds but except callus formation, regeneration could not be achieved (Sarwar, 1985). Regeneration of adventitious shoot buds from leaf explants were also reported in tea (Kato 1996).

2.2. Media composition for initiation and multiplication

Maximum efforts have been made to standardize the media formulation for increasing the multiplication rate of tea micropropagation. The most common basal media used are either full or half-strength MS salts (Murashige and Skoog, 1962). However, there are also few reports on the use of the WPM (Lloyd and McCown, 1980) and Heller's (1953) medium (Table-1). Nakamura (1987a) compared different basal medium namely MS, B_5 medium (Gamborg *et al.*, 1968) and Nitsch and Nitsch, (1969), accordingly concluded that MS medium was the best for tea shoot multiplication. Tahardi and Shu (1992), achieved axillary shoot proliferation on WPM containing various concentration of Thidiazuron (TDZ) within 10-12 weeks. Although several researchers have reported that MS medium were highly suitable for initiation of multiple shoots, use of half-strength MS salts was also reported for multiplication and shoot proliferation in tea (Phukan and Mitra, 1984; Banerjee and Agarwal, 1990; Agarwal *et al.*, 1992). Manipulation of vitamin compositions along with organic and inorganic salts of MS, were found suitable for initiation and multiplication of axillary shoots of tea (Arulpragasam and Latiff, 1986).

Among the different PGRs, addition of 6- benzyladenine (BA) (1-6 mg/l) and indole-3-butyric acid (IBA) (0.01-2.0 mg/l) in the culture medium have suited well for shoot proliferation and growth of axillary buds, both for shoot initiation and subsequent multiplication (Table-1). Kato (1985) succeeded indirect organogenesis via callus phase from three different types of explant, namely epidermal layer of stem segment, intact stem segment and stem segment without epidermal layer on BA (10 mg/l) containing medium. Although 2,4-dichlorophenoxyacetic acid (2,4-D) and α-naphthaleneacetic acid (NAA) were found to induce callus and hence were ineffective for the induction, growth and development of tea shoots (Nakamura, 1988b). Auxin such as NAA together with BA produced either callus or induced 4-5 shoot buds per explant within 8-12 weeks (Phukan and Mitra 1984; Bag *et al.*, 1997). Picloram and 2,4,5-trichlorophenoxyacetic acid (2,4,5-T) have been used successfully for the elongation of tea shoots by several researchers (Arulpragasam and Latiff, 1986; Nakamura, 1987a,b; Iddagoda *et al.*, 1988; Nakamura, 1989a; Jain *et al.*, 1991). Higher concentrations of these auxins (10.0 mg/l) were found to be critical for shoot elongation of 30 different clones (Kuranuki and Shibata, 1993). Importance of indole-3-acetic acid (IAA) (0.1 - 2.0 mg/l) and kinetin (0.21 - 8.0 mg/l) for both induction and multiplication of axillary shoots were emphasized by several researchers (Phukan and Mitra 1984; Sarwar, 1985; Das and Barman, 1988).

The effect of thidiazuron (TDZ) on micropropagation of tea was compared with that of BA using nodal segments from *in vitro* raised seedlings in two basal media namely WPM and MS. It was clearly evident that MS was superior to WPM in all combination. Extremely low concentrations of TDZ (1 pM-100 nM) alone were effective in inducing shoot bud proliferation and maintaining high rates of shoot multiplication on hormone-free media up to 24 subcultures. On the other hand, higher concentrations of BA (1-10 μM) and its continued presence were required to initiate and sustain shoot proliferation. Although, all the explants produced callus on TDZ (5, 10 and 15 μM) in combination with 2,4-D or IBA (5, 10 and 15 μM) in both the basal media, optimal response (98%) of shoot proliferation was observed in MS supplemented with 5μM TDZ and 10 μM NAA, yet TDZ alone was more effective than BA (Table-2). Number of shoots formed was higher in explants initiated on medium containing TDZ when compared with medium supplemented with BA but multiplication rates were more or less similar (i.e. 2-3 times) after each subculture (Table-3). Since very low concentrations of TDZ are used only at the initial phase, the overall higher cost of TDZ is overcome. TDZ thus appeared to be a potent cytokin for tea micropropagation with high proliferation rates.

The significant role of growth adjuvants for tissue culture of tea is well documented. These include coconut milk (Phukan and Mitra, 1984; Sarathchandra *et al.*, 1988; Banerjee and Agarwal, 1990; Agarwal *et al.*, 1992; Jha and Sen 1992; Rajkumar and Ayyapan, 1992), yeast extract (Phukan and Mitra, 1984; Sarwar, 1985; Banerjee and Agarwal, 1990; Phukan and Mitra, 1990), casein acid hydrolysate (Chen and Liao, 1983; Jha and Sen, 1992), serine and glutamine as nitrogen sources (Chen and Liao, 1982) etc. Among the different carbon source, sucrose with the concentration between 15-60 g/l remains most suitable for tea micropropagation. This was reconfirmed by Nakamura (1990), showing that 3-6% sucrose was the best for adventitious bud formation as compared to four different sugars viz. lactose, galactose, glucose or maltose for adventitious bud formation. Therefore, it can be concluded that MS medium containing low cytokinin, especially BA, seems to be the best for micropropagation of tea.

2. 3. Media composition for rooting

Establishment of microshoots in hardening chamber depends upon the efficiency of rooting. In tea, both *in vitro* and *ex vitro* root induction of microshoots are reported. While *in vitro* rooting of tea depends upon concentration and duration of auxin treatment, salt strength of basal medium or physical condition of the cultures, *ex vitro* rooting was influenced by pH of the hardening media, percentage of relative humidity of the hardening chamber.

Reduction of MS salt concentrations to half-strength favored not only root induction and but also root elongation in tea (Kato, 1985; Banerjee and Agarwal, 1990). Generally, IBA (0.5-8 mg/l) was preferred over NAA because roots induced by NAA are shorter and thicker with accompanying calli making subsequent transplanting difficult. Although rooting occurred much later when treated with IBA, but roots were long and fibrous (Nakamura, 1987a; Banerjee and Agarwal, 1990).

Although liquid medium with filter paper bridge was successfully used for *in vitro* rooting in tea (Tian-Ling, 1982; Kato, 1985; Nakamura, 1987b), Jain *et al.*, (1993) found that *ex vitro* rooting was better than all types of *in vitro* rooting methods such as liquid shake culture, agar-solidified medium or filter paper bridges. They achieved 97% rooting in shoots whose cut ends were dipped in IBA (50 mg/l) for 2 hrs before transplanting them into potting mixtures. Another important observation that was made by Banerjee and Agarwal (1990) that low light and a low pH (4.5-4.6) were favorable for root induction in tea. This finding was in agreement with

Nakamura, (1987a), indicating that dark treatment after dipping the shoots in auxin also promoted rooting in tea. Rooting of tea in *in vitro* also depend upon the genotype. Murali *et al.*, (1996) demonstrated that among the 4 different cultivars, highest rooting was achieved with UPASI-26 and UPASI-27 cultivars when compared with UPASI-3 and BSB-1 cultivars after IBA treatment (1 mg/l) for 30 min. Therefore, they concluded that genotypic differences strongly influence rooting ability of tea which require finer tuning of the media optimum for individual cultivar. In general, *in vitro* rooting of tea micro-shoots was achieved either by culturing in media containing low auxin for long time or by high concentration 'auxin-shock' treatment followed by subculture to an auxin-free medium.

2. 4. Hardening and Field transfer

Perhaps the most critical phase of micropropagation is the establishment of *in vitro* plantlets to the greenhouse. Achievement of uniform plant growth and their high survival rate not only demands good greenhouse conditions, but also requires modifications of the internal micro-climate to match the local environmental need. Arulpragasam *et al.*, (1988) transferred 5-8 cm rooted plants to small plastic pots containing fumigated soil and kept in humid chamber for a period of one week to ten days. Plants were gradually acclimatized to field conditions and planted in larger concrete pots. Das *et al.*, (1990) observed better hardening on pre-conditioning plants at low temperature (22°C) and low light intensity (250 μ mol $m^{-2}s^{-1}$) after transferring to soil sleeves. The plants were covered with poly bags and kept in thatch-house for few weeks and gradually exposed to sunshine before transferring to field. Generally, the standard procedure is to transfer the rooted plantlets of tea to potting mixtures containing peat and soil (1:1) under high humidity created by misting or fogging units (Banerjee and Agarwal, 1990; Agarwal *et al.*, 1992; Jha and Sen, 1992; Jain *et al.*, 1993), but use of vermiculite: soil (1:1) is also worked very well (Tian-Ling, 1982; Kato, 1985). Detail study with various parameters such as time of harvesting of micro shoots, shoot size, soil pH, plant growth regulators, CO_2 enrichment and light condition was made by Sharma *et al.*, (1999). Among the different parameters, soil pH coupled with CO_2 enrichment is the most critical factor for hardening to achieve high percentage of survival rate. Following this protocol, 300 plants were transferred to the field.

In our laboratory, microproagated tea shoots of 4-5 cm height were harvested and treated with IBA (500 mg/l) for 30 mints before transferring to Hikkotray containing pre-sterilized sand and cowdung (1:1). This Hikkotrays were then kept in poly tunnel with intermittent watering for 90 days inside of an indigenously developed poly house (90 % survival rate). After which they were transferred to polythene sleeves for 1 year in the same poly house, containing black virgin soil (Rajshekaran and Mohan Kumar 1992; Mondal *et al*., 1998). Following these techniques at Research and Development Department of Tata Tea Ltd, India, more than 45000 plants of 8 different cultivars were transferred to the field and leaves are being harvested regularly to manufacture tea for last several years (Annual report, 1999).

Biological hardening is a new alternative concept for hardening the micropropagated plants. Like in any other plants, micropropagated tea plants often experience high mortality during or following laboratory to land transfer. Apart from various abiotic factors, one major cause of mortality of such 'aseptically' raised plants is their sudden exposure to the soil microbial communities present in the soil. Tissue-cultured plants, at least initially, do not resist sufficiently against the soil microbial attack (Palni *et al.*, 1998). Therefore, microbial culture can be used to over come the 'transient transplant shock' and better development of the plants on the transfer to the soil. Although antagonistic fungus *Trichoderma* and arbuscular mycorrhiza such as VAM and *Piriformospora indica* is most suitable for micropropagated raised woody plants (Singh *et al.*, 2000; Pandey *et al.*, 2002), Pandey *et al.*, (2000), tested two bacteria namely *Bacillus subtilis*, *Pseudomonus corrugata* as mcicrobial inoculants for hardening of micropropagated tea plants prior to transfer in open land. They found that bacterial inoculation enhanced survival rate of 100, 96 and 88% as against 50, 52, and 36% in corresponding control plants in rainy, winter and summer seasons respectively. Rhizoplane and rhizosphere soil analysis showed that major biotic factor responsible for mortality was fungus *(Fusarium oxyporum)*. Bacterial inoculations also resulted in plant growth promotion of tissue culture as well as seed raised plants of tea.

2. 5. Micrografting

Micrografting, a hardening tool of *in vitro* raised shoots have been used in a wide range of plants such as citrus, cherry, kiwifruit, pistachio, stone fruits, apple, grape, *Larix decidua* and *Picea* spp where, *in vitro* raised scions directly transplanted either on *in vitro* raised root stocks under sterile conditions or *in vivo* raised stocks (Banerjee *et al.*, 2000). In tea, Prakash *et al.*, (1999) reported the grafting of *in vitro* raised tea shoots on seedlings at the green house. Microshoots of cultivar Banuri-96 were either grafted on seedlings of the same cultivar (autograft) or on UPASI-9 seedlings (heterograft). Three important parameters such as (i) effect of PGRs, (ii) assessment of compatibility and (iii) effect of age of root-stock as well as season on the graft union were evaluated. A higher amount of foliar development was observed when BA (5mg/l) and NAA (5 mg/l) treatment was given at the graft union for 10 min. Also, significantly higher percentage of survival was observed in autografts as compared to heterografts. Age of the root-stock was found to be the most important factor. The success rate was higher when 4-month-old seedlings were used, than in either the 2- or 12-month-old seedlings.

However, using the same technique, we reduced the hardening period up to 1 year of *in vitro* raised tea shoot when grafted on 3-month-old tea seedling. Our result (Mondal and Parathi, 2002c) differs in two aspects from the earlier report (Prakash *et al.*, 1999). Firstly, we achieved a higher rate of survival (98 %) with out any PGR treatment. Therefore we concluded that PGR play an insignificant role for the graft compatibility. This was an absolute necessary for commercial use of the technique to eliminate the cost of PGR. Secondly, we found that the growth of grafted plant was much higher than ungrafted *in vitro* raised scion or seedling (Table-4). One of the reasons for better growth of microprogated plants than the vegetative counter part, is the higher root volume which help former to absorb more water and nutrients from soil **(Fig.1)**. It is due to the pre-existing tap root system of the seedling. This was not only of great importance to reduce the hardening period of tissue culture plant but better root system will resist the plant during drought periods. Tissue culture tea plant of the same cultivar took 12-18 months time in the hardening phase where as the same cultivar of microgafted one, took 6-8 months, which vitally reduced the hardening time by one cycle. This plays an important role in tea breeding where progress is slow due to the slow growth habit.

2. 6. Field performance of micropropagated plants

Though several reports on micropropagation of tea state that hardened plantlets are transplanted to potting mixture, only a few details of field transfer are available. Till date, no systematic study has been conducted to assess the growth performance of field grown micropropagated raised tea plants.

Sharma *et al.*, (1999) analyzed the nursery performance of 17-month-old micropropagated and vegetatively propagated tea plants of cultivar Banuari-96. No remarkable differences were noticed in the nursery except rooting time reduced by one month in micropropagated shoots. An average height and stem thickness at collar region were better in vegetatively propagated plants; an average number of leaves were double (16 leaves/plant) in micropropagated plants as compared to control (6 leaves/plant). However, reason for such difference was not mentioned in the report but 17month time is too early to reflect the actual performance for this woody perennial.

In a comparative study of 8- and 4-year-old, field-grown micropropagated and vegetatively propagated tea plants, namely "UPASI-9 and TTL-1", overall yields were comparable. While the yield of micropropagated UPASI-9 recorded 3656 kg made tea/ha/annum, vegetatively propagated plants of UPASI-9 yielded 3596 kg made tea/ha/annum after second cycle of pruning (8 years). On the other hand, micropropagated TTL-1 plants registered 2321kg made tea/ha/annum and their vegetative counter parts yielded 2383 kg made tea/ha/annum after first cycle of pruning (4 years) with 10000 plants/ha and 22% out turn. Chemical parameters of made tea such as thearubigin, theaflavin, total soluble solids and total colour evidenced no significant difference between the two types for both the cultivars, which is obvious as both are genetically same and planted in same environment. In tea, chemical parameters depend upon the genetic make-up as well as on environment. Although the different physiological parameters such as photosynthetic rate, chlorophyll content etc., remained the same, two different morphological variations were detected. The number of lateral shoots produced after 'centering' were significantly more in micropropagated raised plants as compare to vegetatively propagated plants. This is probably due to the effects of the various plant growth regulator treatments that the tissue culture raised plants experienced under *in vitro* conditions. Secondly, the root volumes of tissue culture plants were also higher than vegetatively propagated plants. The reason may be micropropagated shoots were treated with IBA (500mg/l) for 30 min to induce rooting which may be responsible for better root development in field.

2. 7. Problems of tea micropropagation

Like other woody perennials, major problems encountered in tea tissue culture are phenolic exudation from explant and microbial contamination in tissue culture medium. We describe attempts to overcome these problems in tea tissue culture.

2.7.1. Explant browning

High phenolic contents, for which tea is valued, are exuded in the culture medium from the cut ends of explants which undergoes enzymatic oxidation to form some toxic compounds. These toxic compounds create problems by lowering down the pH of the tissue culture medium. Sarwar (1985) tried different chemicals such as ascorbic acid, catechol, L-cystein, phloroglucinol, phenyl-thiourea, polyvinyl pyrolidone-10, sodium diethyl dithio-carbonate, sodium flouride and thiourea along with different strengths of MS inorganic salts. He found that highly reduced strengths of MS (1/20) prevented explants browning. While ascorbic acid was used by several researchers (Iddagoda *et al.*, 1988; Agarwal *et al.*, 1992), polyvinyl pyrolidone was recommended by Creze and Beauchesne (1980) to prevent explant browning of *in vitro* cultures. Magnitude of polyphenols exudation varies the different cultivars of tea. Murali *et al.*, 1996, categorized that polyphenol leaching in *in vitro* was highest in Cambod type followed by China type and was least with Assam type. Pandidurai *et al.*, (1996) reported that successive transfer of the explant in to a new culture medium at periodic intervals worked well to prevent explant browning of tea.

2.7.2. Microbial contamination

The success of micropropagation depends on prevention or avoidance of microbial contamination *in vitro* cultures, which can be achieved through proper surface sterilization. In tea often explants are taken from field grown plants that are heavily contaminated by various epiphytic and endophytic organisms, cause severe losses at every stage of micropropagation (Debergh and Vanderschaeghe, 1988). Generally, cultures are pre-screened when explants from field-grown plants are used. Different other strategies have also been employed to overcome these problems in tea. Ogutuga and Northcote (1970a) used 5-15 mm stem segments from the green house-grown tea plant and surface-sterilized with 70% ethanol followed by 7% sodium hypochlorite solution. Kato (1985 and 1989b) took stems with 3-4 leaves from the green house-grown seedlings (cultivar Yabukuta) and sterilized with 7 % calcium hypochlorite solution for 20 min. Arulpragasam and Latiff (1986) snapped off the leaf carefully with petioles as close as possible to the stem without any damage to the axillary buds. The shoots were surface-sterilized in 10-15% clorox solution for 15 min under constant agitation and inoculated in culture medium after rinsing with distilled water. Kuranuki and Shibata (1993) minimized the problems of endogenous bacteria by reducing the size of the explant to an apical meristem and a leaf primodium less than 0.5 mm. According to Nakamura (1989a) rate of contamination depends upon the season of explant collection, which was minimized by collecting the explants at the beginning of first seasonal flush when air temperature was relatively low and the plants were juvenile and vigorous. Haldeman *et al.*, (1987) had shown significant reduction in bacterial and fungal contaminations with 24-h intermediate culture of the traditionally disinfected materials in liquid medium containing benomyl at 1, 2 or 4 g/l and rifampicin at 10,25 or 50 mg/l without any phytotoxic effects. Das and Barman (1988) treated the explants with 1% streptomycin sulfate for 1 min prior to inoculation in order to prevent contamination. However, 0.05-1% mercuric chlorite solutions were also used for surface sterilization of the explants (Rajakumar and Ayyappan, 1992; Rajashekaran and Mohankumar, 1992; Jha and Sen, 1992; Agarwal *et al.*, 1992)

3.0.Alternative approaches to tea micropropagation

In tea, somatic embryogenesis is the only alternative tool to conventional micropropagation as efforts on either by neoformation of adventitious buds or micropropagation though protoplast and suspension culture are less. Since ever the work started on somatic embryogenesis of tea in early 80's, promising results have been achieved till today. The various factors, which govern the somatic embryogenesis of tea, are concisely summarized in Table-5 and described below.

3.1. Somatic embryogenesis

Genetic engineering through *in vitro* manipulation of cell is an additional tool to tea breeding. The prerequisite step for such a technology is an efficient *in vitro* plant regeneration system. Somatic embryogenesis is considered to be the most efficient regeneration system of tea (Jain and Newton, 1990). However, the efficacy of such a system for plant production depends on the efficiency of multiplication and conversion rate of somatic embryos. The advantage of somatic embryogenesis is the development of adventitious embryos from explant tissue without an intervening callus phase which helps in maintaining genetic fidelity (Bano *et al.*, 1991). Therefore, it has a tremendous potential in clonal propagation (Mondal *et al.*, 2001a) and most importantly through genetic transformation of tea (Mondal *et al.*, 1999). It has been successfully used for artificial seed production (Mondal *et al.*, 2000b), cryopreservation for long- term storage of germplasm and some inter-specific hybrid crosses of *Camellia* (Nadamitsu *et al.*, 1986), where immature somatic embryos were rescued, cultured before abortion. It can also be used for producing diseases free plants, androgenic or haploid plant production of tea (Chen and Liao, 1982). Additionally, unlike the micropropagated plants, tap roots of somatic seedling will help them better to combat drought, which is a major abiotic stress problem in tea growing areas of the world.

3.1.1. Induction of somatic embryos

Induction of somatic embryogenesis depends upon several factors such as explants type, physiological stage of the explants and media formulation including plant growth regulators. A judicious selection of all the factors can achieve a good yield of embryo in tea.

3.1.1.1. Explants

Although induction of somatic embryos in tea began with pollen as explants (Chen and Liao, 1983) but mature cotyledons or zygotic embryos were used mostly (Table-5). Use of immature cotyledons (Abraham and Raman, 1986; Nakamura, 1988a; Bano *et al.*, 1991), de-cotylenated embryos (Nakamura, 1985; Paratasilpin, 1990), de-embryonated cotyledons (Rajkumar and Ayyappan, 1992; Ponsamuel *et al.*, 1996), nodal cuttings (Akula and Akula 1999) and juvenile leaves (Sarathchandra *et al.*, 1988) for the production of somatic embryos is also documented. However, there is no report of somatic embryogenesis from *in vitro* grown root of tea.

3.1.1. 2. Physiological stage and genotypic variation

Seeds of different tea varieties mature at different time. Therefore, the physiological stage and genotypic variation are inter-related and hence both play important role for the successful induction of somatic embryos from cotyledons (Vieitez, 1994). However, systematic study on the effect of physiological maturity for different cultivars is scanty. Nakamura (1988a) found that the optimum timing for cotyledon culture in *C. sinensis* was late September to mid-October. This was when seeds were physiologically matured in Shizuoka, Japan. In that report, a high percentage of embryogenesis (45-50%) was observed during this period in contrast to 15% from immature seeds collected during August.

Seed maturity of 4 different Indian tea cultivars was attained at different periods (Table-6). Primary somatic embryos were induced from de-embrocated cotylends of tea on MS + BAP (2mg/l) + IBA (0.2mg/l) (Fig.2 a). Among them, the best time for the induction of somatic embryogenesis for cultivars TG 270/2/B and UPASI-9 was during the monsoons (July-August); embryogenic response was 11.1 % and 50.0 %, respectively. The best time to induce somatic embryogenesis in cultivar Tuckdah-78 was during autumn (September-October); the rate of induction was 44.1%, whereas the highest number of somatic embryo induction was 66.7% in cultivar Kangra Jat during winter (November-December) (Mondal *et al.*, 2000b).

Nakamura (1988a) screened 13 Japanese tea cultivars and observed variation between 0- 50% in somatic embryogenesis induction. Among the cultivars, the best response was reported in "Yabukita". She also reported formation of 28% somatic embryos in *C. sinensis,* against 2% in *C. assamica.* Kato (1996) used Japanese cultivars of *C. sinensis* for embryogenesis from *in vitro* leaves and reconfirmed the highest frequency of embryogenesis in the cultivar "Yubukita".

3.1.1.3.Basal medium and plant growth regulators

Although MS is the most commonly used medium for the induction of somatic embryos in tea (Table-5), WPM (Lloyd and McCown, 1980), and Nitsch and Nitsch, (1969) media have also been used by few researchers. Type, concentration and time of application of different plant growth regulators have also been extensively worked out. In general, a high cytokinin-to-low auxin or low cytokinin alone was found to be necessary for the induction of somatic embryos in tea but reduction or omission of cytokinin in subsequent sub-culturing is also known. BA (0-10mg/l) has been widely used for tea and it's wild relatives but in certain reports, kinetin (0.05 -10 mg/l) has also been used in the induction medium (Wu *et al.*, 1981; Bano *et al.*, 1991; Wachira and Ogado, 1995).

Among the different auxins, IBA (0-2 mg/l) was the most widely used for the somatic embryo induction in tea. However, different concentrations of NAA (Paratasilpin, 1990; Bag *et al.*, 1997; Balasubramanian *et al.*, 2000), 2,4-D (Das and Barman 1988; Bano *et al.*, 1991), IAA (Wu *et al.*, 1981; Sood *et al.*, 1993) were also used to elicit a better response for induction of somatic embryos in *C. sinensis*. Ponsamuel *et al.*, (1996) used some novel auxins, namely, α-napthoxyacetic acid (NOA), tetraphenylboron (TPB), phenylboronic acid (PBOA), etc., for induction of somatic embryos from immature cotyledons of *C. sinensis* cultivar UPASI-10. They found that 1 mM of each TPB and PBOA were capable of inducing somatic embryos. We found that half-strength of MS salts contain NAA (2.5 mg/l) and BA (0.2 mg/l) were the best for induction of somatic embryo of all 4 different cultivars. Among them a highest 66% somatic embryos were formed by cultivar Kangra Jat (Mondal *et al.*, 2000b). Akula *et al.*, (2000) reported addition of betaine (1000 mg/l) and ABA (7.5 mg/l) to the induction medium significantly enhanced the rapid formation of somatic embryos without callusing from mature seeds of tea within two weeks of culture initiation.

3.1.1.4. Growth adjuvants

Although growth adjuvant plays an important role but it seems that in tea their requirement is less, which may be due to fact that tea cotyledons have high degree of inherent embryogenic capacity (Vieitez, 1994) and hence dose not demand any growth adjuvant. While, Arulpragasam *et al.*, (1988) added yeast extract to the induction medium, others have used coconut milk instead (Sarathchandra *et al.*, 1988; Rajkumar and Ayyappan, 1992) for somatic embryo formation. Addition of adenine sulfate (80 mg/l) as an additional nitrogen source in the induction medium for *C. sinensis* cultivar, T-78 dose not improve embryogenesis significantly (Jha *et al.*, 1992).

3. 2. Secondary embryogenesis

Exploitation of embryogenesis through biotechnological approach demands a system of repetitive/recurrent or secondary somatic embryogenesis. The practical implication of secondary embryogenesis is genetic transformations as foreign gene can be introduced into primary embryos of important crops and transgenic plants can be developed as in case of tea (Mondal *et al.*, 2001h). Two different growth patterns for secondary embryogenesis have been reported in tea: (1) somatic embryo-to-somatic embryo, commonly known as repetitive embryogenesis which is more frequent in tea and (2) callus-to-somatic embryo in which the multiplication of somatic embryos depends upon sub-culturing of callus (Vieitez, 1994).However, multiplication rate, synchronized secondary somatic embryo formation and their long-term maintenance of embryogenic capacity have not been studied well.

Abraham and Raman (1986) used a single medium containing BA and NAA for the induction, germination and maintenance of somatic embryos. Kato (1986a) used GA_3 (1 mg/l) for induction of secondary embryos on isolated primary embryos. However, Jha *et al.*, (1992) obtained secondary embryos on cotyledons of primary embryo within 6-8 weeks on MS or B_5 media supplemented with BA (0.1-3 mg/l), IBA (0.1-1 mg/l), IAA (0.1-2 mg/l) and GA_3 (5.0 mg/l). Apart from growth regulators, nutritional effect of nitrate salts of potassium and ammonium, together with different concentrations of sulphate salts of aluminum, potassium, magnesium and ammonium on secondary somatic embryogenesis were investigated (Mondal *et al.*, 2001a). It has been found that potassium nitrate (800 mg/l) and ammonium nitrate (825 mg/l) along with (260 mg/l) potassium sulphate produced maximum number of secondary embryo (i.e. 20-25 secondary embryos per primary embryo in 91.6% responsive explants) (data not shown). Among the different factorial combinations of glutamine, BA and IBA tested, maximum number of synchronous secondary embryos developed on a medium supplemented with BA (2 mg/l), IBA (0.2 mg/l) and L-glutamine (1g/l) (Table-7). Following this media formulation synchronus globular, heart-shaped and torpedo shaped somatic embryo were developed (Fig.2 b, c, d). Furthermore, high embryogenic capacity of this cultivar could be maintained on the above medium for more than four years, highest so far among the published reports.

Akula and Akula (1999) developed a unique system for repetitive embryogenesis in tea. Uniform size of somatic embryo at the globular stage induced on nodal explants of cultivar TRI-2025 was taken. Among different methods tested namely roller drum, liquid medium shaker and temporary immersion system, the highest multiplication rate (> 20 folds) were achieved with later. They continued harvesting of secondary embryos up to 2 years from the original set of experiment by regular sub culturing in every 6 weeks. After 3 cycles, the multiplying embryos showed a tendency to form callus in the same medium, which was over come by subculture in lower concentration of growth regulators for 6 weeks and transferred them to hormone-free basal medium.

3.1.1. 5. Maturation and germination

The success of embryogenesis also depends on high rate of somatic embryo germination and whole plant regeneration which occur through series of events, such as synchronized embryo maturation followed by cotyledon expansion, hypocotyl-root axis elongation and emergence of the radical (Webster *et al.*, 1990). However, it has been noticed that as compare to other woody plants, induction of somatic embryos on tea cotyledon is easier due to inherent capacity but have low germination rates, specially on hormone free culture media (Vieitez 1994). Perhaps, this is due to the lack of physiological understanding of the somatic embryogenic pathway. Two common problems that often encounter with tea somatic embryos are precocious and abnormal germination due to lack of storage of food material, improper balance of endogenous hormone profile coupled with venerability of somatic embryo to desiccation. Therefore judicious supply of external nutrient and plant growth regulators, which will improve the conversion rate of tea somatic embryos, are discussed below.

3.1.1.5.1. Sugars

Sugars serve dual purpose of acting as a carbon source and also as inducers of desiccation tolerance during maturation and germination of somatic embryos (Lecouteux *et al.*, 1993) by acting as an osmotic agent (Tremblay and Tremblay 1995). One way to resist desiccation is by preventing osmotic changes, which can be achieved by addition of an osmotic stabilizer to the culture medium. This is generally accomplished by inclusion of an osmoprotectant such as sucrose. The sucrose concentration in the culture medium and duration of culture treatment effected germination rate of tea somatic embryos (Mondal *et al.*, 2002c). They registered significantly higher germination percentage (35.2 %) at 3% sucrose after 5 weeks of treatment, where as no response was observed at any of the higher sucrose concentrations (6 or 9%) with periods of treatment (4, 5 and 6 weeks). This may due to readily available of reserve carbohydrates during maturation, which is important for subsequent successful germination of zygotic embryos within seeds. Thus it is not surprising that the embryos cultured on 3% sucrose for periods of 5 weeks germinated normally to the extent of 35.2% but declined thereafter.

3.1.1. 5. 2. Desiccation

Although desiccation promotes both germination energy and germination capacity of somatic embryos (Roberts *et al.*, 1990), yet tea somatic embryos were found to be desiccation sensitive and were adversely affected when desiccated. Probably the recalcitrant behavior of tea seed is reflected in the somatic embryos also. In tea, drying and death coupled with poor germination (1.2 -1.6%) of somatic embryos were observed within a week when the globular embryos were desiccated for 3 weeks under high relative humidity created by different chemicals i.e. 60% relative humidity by Mg NO_3, 6 H_2O or 90% relative humidity by $ZnNO_3$, $7H_2O$. Therefore, it is clearly evident that to achieve the successful maturation one should use the osmotic protectant in the culture media to prevent the tea somatic embryo from desiccation.

3.1.1. 5. 3. Plant growth regulators and additives

Among plant growth regulators, abscisic acid (ABA) has been widely used for the maturation of somatic embryos of woody plants particularly because it inhibits precocious germination by conferring desiccation tolerance (Roberts *et al.*, 1990) and promoting accumulation of storage lipids (Avgioglu and Knox, 1989) and proteins (Roberts *et al.*, 1990). Although there is no report till date on the effect of ABA on tea somatic embryo maturation and germination, our results showed that none of the somatic embryos germinated after ABA (0.5, 2.5, 12.5 and 62.5mg/l) treatment on tea somatic embryos for 3, 4, and 5 weeks (Mondal *et al.*, 2002a). However, GA_3 also (0.5-5 mg/l) stimulated germination of somatic embryos of several species and hybrids of *Camellia* including *C. sinensis*. Kato (1986a) found that BA (10 mg/l) and IBA (0.5mg/l) were equally useful for the induction of the concomitant development of shoots and roots in *C. sinensis* embryos. Jha *et al.*, (1992) obtained 20% germination of tea somatic embryos on B_5 medium (Gamborg *et al.*, 1968) containing BA (3 mg/l) and IAA (2mg/l).

Ponsamuel *et al.*, (1996) studied the influence of other growth regulators such as Brassin, synthetic analog of a naturally occurring brassionoide (Maugh, 1981), on somatic embryogenesis and plant regeneration from the immature cotyledonary tissue of cultivated tea (*Camellia sinensis*). They reported about 50% germination of *C. sinensis* somatic embryos on 0.8% agar solidified MS medium containing 1mM brassin. The positive influence of brassin on germination rate may be due to the inherent sensitivity of the tissue to this compound as brassinoide has been reported to be present in tea leaves (Ikekawa, 1991).

The effect of different concentration of maltose and trans-cinamic acid (t-CA) alone or in combination was evaluated to improve the germination efficiency of tea (Mondal *et al.*, 2002a). While none of the untreated somatic embryos germinated, significantly, a low percentage (3-6%) of germination was observed at all concentration of either maltose or t-CA alone (Table-8). On the other hand, normal and significantly high germination (70.6 %) was observed (Fig. 2e) when the embryos were cultured on a medium supplemented with both t-CA (3 mg/l) and maltose (4%) for 4 weeks followed by their transfer to MS medium containing GA_3 (1.5 mg/l). The success achieved with maltose and its superiority over sucrose is probably due to difference in its breakdown products. Since maltose is broken down more slowly than sucrose, it can provide a readily metabolize carbon source (glucose) to the embryos which probably lack any stored carbohydrate. This further reflects on the inefficiency of the developing embryos in accumulating reserves during the embryo maturation phase on the other hand. Cinamic acid is an important compound on a wide variety of metabolic pathways such as favonoids biosynthesis, the phenolics synthesis and most important in the synthesis of malonyl-Co-A, the key precursor of fatty acid synthesis pathway. However, the inability of either maltose or transcinamic acid alone for germination indicate that the requirement of both sugars and fatty acid which is essential during embryo maturation and subsequent germination of tea somatic embryo.

3. 3. *In vivo* embryogenesis of tea

In most reports on somatic embryogenesis, till now emphasis has been given to manipulate the nutrient composition, growth regulators in culture medium, physical conditions of incubation and other stress treatments to induce somatic embryos. However, induction of *in vivo* embryogenesis of tea is possible without using the conventional tissue culture media (Mondal *et al.*, 2001b). In an attempt to germinate the tea seeds under sterile conditions in steel boxes with moist sand for the use of protoplast isolation, we observed the induction of embryogenesis on cotyledon surface of mature tea seeds under *in vivo* conditions at 28°C.

The induction was however, influenced by genotype, time of seed collection and temperature of incubation. Selected regime of temperature (24°C, 28°C and 32°C) for incubating the seeds was examined so as to determine the effect on embryogenesis *in vivo*. At 32°C, no response was evinced in any of the three genotypes irrespective of the time of seed collection. Only a very low germination frequency (1.6 %) was observed in the Kangra Jat seeds. At 24°C a small percentage (4.2%) of Kangra Jat seeds germinated but there was no development of somatic embryos. However, when the seeds were incubated at 28°C both germination and adventive embryos were observed. The percentage of germination was lower in comparison to that of embryo formation (Table-9). However, a considerable amount of seeds were remained unresponsive irrespective of the treatments.

The somatic embryos per seed were also genotype specific. While the maximum number of somatic embryos (15-16 per seed) were observed after 90 days of incubation at 28°C in UPASI-9, that of Kangra Jat and T-78 showed 10-12 and 6-8 such embryos per seed respectively (data not shown). After 28-42 days, the embryos underwent to a typical embryogenic pathway as evident by cup-shaped structures. The maximum number of seeds showing formation of embryogenic response was observed in UPASI-9 compared to two other cultivars. The response in UPASI-9 seeds collected in September was 23.3% while that of T-78 collected in the month of October was 16.5% and Kangra Jat collected in the month of November was 11.7%.

Therefore, we concluded that induction of embryogenic response is not only dependent upon the external supply of nutrients and growth regulators, but also on temperature. At a particular temperature, mature tea seeds are capable of expressing their somatic embryogenic potential even under *in vivo* conditions. Bhatia *et al.*, (1986) observed *in vivo* response of callusing and adventitious shoot formation from de-embryonated *Arachis hypogea* cotyledons inoculated on moist sand and cotton wool in enamel trays. The greater influence of temperature on embryogenesis, forced us to propose a hypothesis that, in *Camellia* the induction of *in vivo* embryogenesis must be regulated by temperature-dependent gene. Although reason for this observation is not clear presently, seeds of *Camellia* appear to have a considerable inherent capacity for embryogenesis. Thus at the right physiological stage, with certain levels of internal hormone and appropriate moisture profile of the substrate with sterile conditions, the tea seeds are able to produce embryos without any exogenous nutrient.

3. 4. Hardening and field transfer

Information on hardening and field transfer of tea somatic embryo derived plants is scanty in the literature. One reason may be establishment of somatic embryo derived plantlets is easy with compare to micropropagated plant due to presence of tap root system in the former. Generally, sand or soil with some additives such as peat or vermiculite under high humid condition is sufficient for initial establishment. Although Wu *et al.*, (1981) were first to transfer the plantlets to soil but it was not mention clearly about the composition of soil mixture and other conditions used by them. Kato (1989a) grew healthy tea somatic seedlings under natural conditions by transferring them in to a mixture of vermiculite: soil (1:1) and they were covered under the plastic. On the other hand, Jha *et al.,* (1992) hardened tea plantlets for 8 weeks in quarter-strength MS salts before transplanting to pots containing a mixture of peat and soil (1:1) and achieved an acclimatization rate of 70%.

Multiple shoots differentiated from the germinated embryos were successfully rooted in autoclaved 3:1 mixture of sand: peat in small pots. The potted shoots were watered and kept under poly-tunnel to prevent them from excessive loss of moisture and to maintain a high relative humidity. After 8 weeks, the small plants had developed root system and were transferred into 10 cm long lay flat polythene sleeves containing a mixture of soil and sand in a ratio of 3:1. The plants were then shifted to a shaded nursery bed (Wachira and Ogada, 1995)

In order to improve rooting, Ponsamuel *et al.,* (1996) treated the plantlets with 1mM indole-3-acctonitrile, 1mM brassin and 10 mM phloroglucinol in liquid MS medium for 15 days and after profuse root proliferation, the plantlets were acclimatized in pots containing vermiculite. Eventually the plants were transplanted in the greenhouse. Akula and Akula (1999) transferred the small plantlets with a strong tap root and 4-6 leaves in to small pots filled with pre-sterilized potting mix (sand: peat: vermiculite, 1:2:1) and kept in to a green house with misting facility, at 80-95% humidity under low light. The new leaves were observed with in 5-6 weeks when they were taken in to bigger pots. Following this procedure they achieved 90-95% survival rate and more than 200 plantlets were transferred to field at Indonesia.

Plantlets with young leaves and stout roots, with a height of 4-5 cm, were taken from culture room to Hikkotrays containing pre-sterilized sand and cow dung (1:1). This Hikkotrays were then kept in poly tunnel with intermittent watering for 60 days in side indigenously developed poly house (90 % survival rate) after which they were transferred to polythene sleeves filled with black virgin soil and kept for further 1 year in the same poly house (Mondal *et al.*, 2001a). Following these techniques, we could produce 3000 somatic seedlings at Research and Development Department of Tata Tea Ltd, India, which will be transferred to the field for their performance in the next planting season.

4. *IN VITRO* CONSERVATION

Conservation of tea germplasm is in alarming situation as tea industry world-wide has undertaken a massive replanting of old plantation and replacing with improved tea vegetatively propagated clones causing a narrow genetic base of tea. Therefore, in-field gene banking (clonal orchards) is at present the only option open for tea germplasm conservation, although germplasm kept this way remains vulnerable to loss due to disease, pests, extreme environmental conditions and economic pressures. An alternative or complementary approach is the preservation of germplasm by exploiting tissue culture technology at a low temperature under *in vitro* condition. Here, the efforts made in tea germplasm conservation are discussed.

4.1. Low temperature storage

The potential use of low temperature storage is numerous for *in vitro* conservation of germplasm, their exchange, hand-pollinated hybrids with reduced seed fertility and genetically modified plants with sterile or unstable genotypes. However, efficacy of storage depends upon efficiency with which synthetic seeds can be stored at low temperature without reducing multiplication rate or germination efficiency. In tea, the first successful plant regeneration from syn-seed was produced by Mondal *et al.*, (2000b). We investigated the effect of low temperature/cold storage (4°C) up to 75 days on germination rate of encapsulated somatic embryos derived from cotyledons of cultivar, Kangra Jat. A maximum germination efficiency of 59 % was achieved with naked somatic embryo kept in germination media in culture room at 25±°C. However germination rate of unencapsulated somatic embryo reduced more rapidly from 59 % to 9.7% with increasing periods of storage time. Similar observation was made earlier in *C. japonica* somatic embryos (Janeiro *et al.*, 1995). It was postulated that this may be due to naked embryo just excised from embryo cluster which was unable to recover "excision shock" while detaching from mother tissue coupled with less capability to resist drying than the embryo clusters. On the other hand, germination efficiency of untreated encapsulated somatic embryo was 34.3%, which was reduced to 7.7% upto 60 days of storage. No germination was recorded either unencapsulated or encapsulated somatic embryo stored for 75 days. The lower germination rate of encapsulated somatic embryos may be related to both oxygen deficiencies and rapid drying in the gel bead (Redenbaugh *et al.*, 1991). The reduction of germination efficiency recovered in alfalfa (Redenbough *et al.*, 1987), *Asparagus cooperi*, (Ghosh and Sen 1994) and *Eucalyptus citrodora* (Muralidharan and Mashcarenhas, 1995) from cold storage was reported earlier. However, a considerable amount of secondary somatic embryogenic response was observed irrespective of the duration of cold treatment. Therefore, we concluded that tea somatic embryos were susceptible to cold. The germinated somatic embryos were grown into plants, successfully hardened and transferred to the field.

It is well known that cold pretreatment of somatic embryos of certain woody perennials is known to promote both normal maturation and germination capacity (Tulecke, 1987). While two months of cold-treated tea somatic embryos at 4°C improved the plantlet conversion ability of secondary embryos of tea significantly (up to 76-100%) as compare to embryonic line, which accounts for 32-65% germination rate in control. The incidence of secondary somatic embryogenesis during germination also decreased after the cold treatment. While cold storage reduced the competence for secondary somatic embryogenesis in *Camellia*, it increased their capacity for conversion in to plantlets (Mondal *et al.*, 2000b).

In order to evaluate the effect of cold storage on bud sprouting response, encapsulated nodal explants were stored under both at culture laboratory condition at (25± 2°C) and at 4°C for varying sets of 15- days interval up to 60 days. Storage of artificial seeds at 4°C did not shown any adverse effect on shoot proliferation efficiency, rather the time taken for initial sprouting was reduced by 15 days as compare to the ones that were stored at 25°C. The bud sprouting efficiency (90.2% to 92%) of both non-encapsulated and encapsulated nodal explants was maintained up to 45 days but declined to 45.9 and 50% when stored up to 60 days (Mondal *et al.*, 2002b). Therefore, we concluded that while somatic embryos are unsuitable for low temperature storage but nodal explants are not only suitable but also sprout bud earlier in tea.

4. 2. Cryopreservation

The preservation of intact seeds and embryonic axes, at ultra-low temperature (-196°C) in liquid nitrogen was attempted in *C. sinensis* (Chaudhury *et al.*, 1990). While the intact seeds could not be cryo-preserved due to their large size and high moister content, the excised embryonic axes could be preserved successfully by desiccating them up to 13% moisture (Chaudhury *et al.*, 1991). The cryopreserved embryonic axes could be revived and germinated by placing them simply on a moisten filter paper for 20-25 days followed by culturing on nutrient medium leading to healthy seedlings of 5-6 cm height. In the following year, Kuranuki and Yashida (1991) reported that the excised embryonic axes were not only able to tolerate the frost but also desiccation. Kurunuki and Sakai, (1995) cold-hardened the 2-months old *in vitro* grown plantlets at 10°C by culturing on half-strength MS salts and subsequently cryopreserved at 5°C. However, details of cold recovery were not mentioned in their report.

5. Somaclonal and gametoclonal variation

Although somaclonal variation (Scowscroft and Larkin, 1982) is undesirable, stable and desirable somaclones can be micropropagated particularly for perennials. Till recently, no effort was made to exploit the somaclonal variation in tea. The reasons may be, tea being cross-pollinated, variations exist among natural population and most of the tea breeder world-wide were engaged to identify the superior variant. Secondly, in tea, number of tissue culture plants, which have been transferred to the field, may not represent a sizable amount to study the somaclonal or gametoclonal variants. Recently, Rajkumar *et al.*, (2001) undertook a detail study about somaclonal variation of tea. The plants were derived from somatic embryo, which were induced on cotyledonary tissues of clone UPASI-10. They identified 5 somaclonal variants in the field on the basis of morphological, physiological and biochemical parameters. Interestingly based on the 'membrane stability index' study they concluded the somaclones would be tolerant to low temperature stress and thus suitable to grow at high altitude. However, DNA marker study would have thrown more light on the homogeneity of variants and nature of heritability.

There were few preliminary attempts earlier the work of Rajkumar *et al.*, (2001) to identify the somaclonal variant among the micropropagated tea. A wide range of chromosomal variation was observed in callus derived from leaf and cotyledon explants of tea (Das, 1992). The study revealed that more than 60% cells were diploid in 8-week-old calli with remaining cells being of triploids, tetraploids or aneuploids with more than 60 chromosomes. However, Jha *et al.*, (1992) reported that embryo derived plants had chromosome number of 2n=30, suggesting an apparent genetic uniformity.

Variation in plants regenerated from gametophytic tissue has been reported in some cases due to uncovering 'residual heterozygosity' (Evans and Sharp 1986). Variation in the chromosome number of gametes or gametophytic tissue plays an important role in gametoclonal variation. This is evident from a range of aneuploids and mixoploids recovered from anther cultures of wheat, maize and sexual hybrid of wheat and triticle (Hu, 1983). In tea, Chen and Liao (1993) examined the chromosome number of anther-derived plantlets. They demonstrated that 9 out of the 12 such plantlets were aneuploids and most of them had chromosome number of 2n= 18-22. Further, morphological variations were not mentioned.

6. Detection of genetic fidelity among *in vitro* raised plants by RAPD analysis

The most important part of any *in vitro* propagation system is mass multiplication of plantlets that are phenotypically uniform and genetically akin to the mother plant, otherwise the advantage of desirable characters of elite supreme clones will not be achieved. Several approaches have been applied for identifying variants among micropropagated plants. They are phenotypic variation (Vuylsteke *et al.*, 1988), karyotypic analysis of metaphase chromosomes (Jha *et al.*, 1992), and biochemical analysis (Damaco *et al.*, 1996). Importantly, a major disadvantage of these techniques is the limited numbers of informative markers and dependence upon environmental or developmental process (Rani *et al.*, 1995). Besides these limitations, the above approaches are not fully suitable for detecting DNA sequence polymorphisms of *in vitro* raised plants. On the other hand, RAPD (Random amplified polymorphic DNA) has been used advantageously for a number of crop species to detect genetic diversity among micropropagated plants (Isabel *et al.*, 1993; Rani *et al.*, 1995; Damasco *et al.*, 1996). We applied RAPD analysis with 50 primers to investigate genetic variability of *in vitro* raised tea plants of cultivar T-78 (Mondal and Chand, 2001f). Among them, 39 primers developed 197 monomorphic bands among all the plants. The primer SC 10-12 produced highest 8 and whereas primer SC 10-57 produced only a single of 1 monomorphic band. Twenty four out of 221 bands produced by remaining 11 primers (SC 10-15, SC 10-56, SC 10-71, SC 10-83, OPA-12, OPA-15, OPA-18, OPM-06, OPM-09, OPM-12 and OPM-18) were polymorphic for four micropropagated tea plants. These 24 loci were monomorphic for the remaining plants. The polymorphic amplification products of 11 primers that were present in the 4 plants compared to remaining 14 plants by conducting a second DNA extraction from respective plants. In a separate experiment, survey with these 11 primers revealed monomorphism of the amplified DNA fragments among 5 vegetatively propagated plants of the same cultivar (T-78). Therefore, we concluded that variation occurred as a result of mutation during micropropagation. Interestingly, the marker profiles among these 4 plants were identical with all 11 primers suggesting a complete homogeneity among them.

Various studies indicate that *in vitro* culture produces abundant cytological anomalies affecting both structural and numerical chromosome constitutions, which can be correlated with many phenotypic abnormalities (Groose and Bingham, 1984; Evans and Sharp, 1986). However, occurrence of variability among tissue culture plants have been investigated extensively and reported in oil palm (Corley *et al.*, 1986), blackberry (Harry *et al.*, 1983), coleus (Marcotrigiano *et al.*, 1990), strawberry (Moore *et al.*, 1991) and plantains (Vuylsteke *et al.*, 1998). Various reasons have been assigned to the occurrence of variability among *in vitro*-propagated plants. Smith (1998) reviewed the factors contributing to this variation and divided these into two, intrinsic factors, which largely depends on the genetic stability of the explant and an extrinsic factor depending on culture media and particular growth regulators. Different types of cultured tissue produced different genetic instability. They have been ranked in the order of low to high from micropropagation of isolated shoot tips and meristems, adventitious shoots, somatic embryos, organogenesis from callus and cells from Protoplast (Scowcroft, 1984). Such variants at times have a negative effect, affecting the high commercial value of the crop such as tea. Although, Hammerschlag (1992) suggested that somaclonal variation among tissue-cultured plants can be either eliminated or minimized if special efforts are made to distinguish between axillary and adventitious shoots that are produced during in vitro propagation and roughing out the adventitious shoots but is not always possible to monitor fully in case of commercial tissue culture laboratory wages are given on the basis of number of plant inoculated per day time. In that situation some tissue culture plants can be randomly selected for RAPD analysis using the polymorphic primers.

7. CONCLUSIONS AND PROSPECTS

With the information available so far it can be concluded that while extensive work has been done on micropropagation of tea but challenge remain on commercial application of micropropagation of tea. A wide range of media composition, explants, plant growth regulators, varietal difference along with physiological stage of the explants have been studied. While MS is the most preferred basal medium, BA is mostly used as PGR. Important aspects which remain uninvestigated are: (1) field behavior of micopropagated tea specially after the 2nd round of pruning which is indicative for stabilized-yield, (2) low temperature long-term storage of tea shoot,

(3) lower down the cost of production of micropropagated plant using bioreactor technology coupled with low cost medium and improved hardening procedure.

It is noteworthy to mention here that since ever the work of Forrest, (1969), micropropagation of tea has not been commercially exploited. Very few micropropagated tea plants are on the field. Akula and Akula (1999) have rightly pointed out excellent reasons. The reason are loss of juvenility of *in vitro* cultures that hinder long term production, lack of tap root system that makes micropropagated tea more vulnerable to drought. But the major point is the higher cost of production of micropropagated tea plants. In a conservative estimation, the cost of production of one micropropagated tea plant is seven times higher in compare to vegetatively propagated plant which has been reduced much in recent years than earlier with the advancement of hardening protocol such as carbon enrichment, biological hardening, micrografting and specially designed hardening chamber etc. that leads to higher survival rate. However where the breeding works have been taken extensively that generates high demand for large planting material of superior clones including transgenic tea to supply within a short period of time to the growers, micropropagation started seeing silver line for commercial exploitation as in case of Tata Tea Ltd, India.

Somatic embryogenesis is an inherent character of tea cotyledon, so much so that, it has been produced *in vivo* without any tissue culture media. There are excellent reports on induction, maturation, germination and long-term multiplication of somatic embryos. Transgenic teas were produced by co-cultivation of somatic embryos with *Agrobacterium tumifaciences* (Mondal *et al.*, 2001h). However, except few, somatic embryogenesis in tea has been developed only from cotyledon. Therefore, effort should be directed towards: (1) evaluation of other type of explants such as *in vitro* leaf, *in vitro* roots etc for induction of somatic embryo, (2) investigating the unicellular or multicelluar origin of tea somatic embryo which is not clear presently, (3) detection of biochemical changes in different stages of development and their co-relation for the requirement of growth regulator in the culture medium, (4) induction of indirect somatic embryogenesis and development of suspension culture of embryogenic cells, (5) investigating the genetic fidelity among the somatic embryo derived plantlets with molecular markers, and (6) field performance of somatic embryo-derived plants.

Being highly cross-pollinated, tea out-breeds freely with wild relatives and as result plenty of variation exist among cultivated gene pool, some of which are extremely important for future breeding. Therefore it is essential to conserve this variability in a very systematic manner worldwide, as tea industry all over the world has taken massive replanting of old seedling population. Thus a protocol which facilitate the exchange of germplasm among the scientific community and long-term cryopreservation of tea germplasm is extremely essential, which need to be studied extensively.

8. REFERENCES

Abraham G. C., Raman K. (1986) Somatic embryogenesis in tissue culture of immature cotyledons of tea (*Camellia sinensis)*. In: Somers D.A., Gengenbach B.G., Biesboor D.D., Hackett W.P., Green C.E. (eds) 6th Int. Congr.on Plant Tissue and Cell Culture. Univ. Minnesota, Minneapolis. pp. 294.

Agarwal B., Singh U., Banerjee M. (1992) *In vitro* clonal propagation of tea (*Camellia sinensis* (L.) O. Kuntze). Plant Cell Tissue Org. Cult. 30: 1-5.

Akula A., Dodd W. A. (1998) Direct somatic embryogenesis in a selected tea clone, 'TRI-2025' *(Camellia sinens*is (L).O. Kuntze) from nodal segment. Plant Cell Rep. 17: 804-809.

Akula A., Akula C. (1999) Somatic embryogenesis in tea (*Camellia sinensis* (L) O Kuntze. In: Jain S.M., Gupta P.K., Newton R. J. (eds) Somatic embryogenesis in woody plants, Vol. 5. Kluwer Academic Publishers, The Greate Britain, pp. 239-259.

Akula A., Akula C., Bateson M. (2000) Betaine a novel candidate for rapid induction of somatic embryogenesis in tea (*Camelia sinensis* (L.) O.Kuntze). Plant Growth Regu. 30: 241-246.

Annual Report (1999). Research and Development Department, Tata Tea Ltd, Munnar, India

Arulpragasam P.V., Latiff R., (1986) Studies on the tissue culture on tea (*Camellia sinensis* (L.) O. Kuntze). 1. Development of a culture method for the multiplication of shoots. Sri Lank J. Tea Sci. 55: 44-47.

Arulpragasam P.V., Latiff R., Seneviratne P. (1988) Studies on the tissue culture of tea (*Camellia sinensis* (L.) O. Kuntze). 3. Regeneration of plants from cotyledon callus. Sri Lank J. Tea Sci. 57: 20-23.

Avgioglu A., Knox R. B. (1989) Storage lipid accumulation by zygotic and somatic embryo in culture. Ann. Bot. 63: 409-412.

Bag N., Palni L.M.S., Nandi S. K. (1997) Mass propagation of tea using tissue culture methods. Physiol. Mol. Biol. Plants. 3: 99-103.

Balasubramanian S., Marimuthu S., Rajkumar R., Haridas V. (2000) Somatic embryogenesis and multiple shoot induction in *Camellia sinensis* (L.) O. Kuntze. J. Plant Crops. 28: 44-49.

Banerjee A. K., Agrawal D. C., Nalawade S. M., Krishnamurty K. V. (2000) Recovery *of in vitro* cotton shoots through micrografting. Curr. Sci. 78: 623-626.

Banerjee B. (1992) Botanical classification of tea. In: Wilson K.C., Clifford M.N. (eds) Tea cultivation to consumption. Chapman and Hall, London, pp. 25-51.

Banerjee M., Agarwal B. (1990) *In vitro* rooting of tea, *Camellia sinensis* (L.) O. Kuntze. Ind. J. Exp. Biol. 28: 936-939.

Bano Z., Rajaratnam S., Mohanty B. D. (1991) Somatic embryogenesis in cotyledon culture of tea (*Thea sinensis* L.) J. Hort. Sci. 66: 465-470.

Beretta D., Vanoli M., Eccher T. (1987) The influence of glucose, vitamins and IBA on rooting of *Camellia* shoots *in vitro*. Abstr. Symp. on Vegetative Propagation of Woody Species, Pisa, Italy p 105.

Bezbaruah H. P. (1971) Cytological investigation in the family theaceae-I. Chromosome numbers in some *Camellia* species and allied genera. Carylogia. 24: 421-426.

Bhatia C.R., Murty G.S.S., Mouli C., Kale D. M. (1986) Nuclear techniques and *in vitro* culture for plant improvement. In: Proceedings of an International sympossium on nuclear techniques and *in vitro* culture for plant improvement. IAEE, FAO and UN, Vienna, 19-23 August. pp. 419-427.

Chaudhaury R., Lakhanpal S., Chandel K. P. S. (1990) Germination and desiccation tolerance of tea (*Camellia sinensis* (L.) O. Kuntze) seeds and feasibility of cryopreservation. Sri Lank J Tea Sci. 59: 89-94.

Chaudhury R., Radhamani J., Chandel K.P.S. (1991) Preliminary observation in the cryo-preservation of desiccated embryonic axes of tea (*Camellia sinensis*) L.O.Kuntze) seeds for genetic conservation. Cryo-letters 12:31-36.

Chen Z. (1999) Pharmalogical functions of tea. In: Jain N.K. (ed) Global advances in tea Science, Aravali Books International (P) Ltd, New Delhi, pp.333-358.

Chen Z., Liao H. (1982) Obtaining plantlet through anther culture of tea plants. Zhongguo chaye. 4:6-7.

Chen Z., Liao H. (1983) A success in bringing out tea plants from the anthers. China Tea 5:6-7.

Corley R.H.V., Lee C.H., Wong, C.V. (1986) Abnormal flower development in oil palm clones. Planter 62: 233-240.

Creze J., Beauchesne M. G. (1980) *Camellia* cultivation *in vitro*. Int. *Camellia* J. 12: 31-34.

Damasco O.P., Godwin I.D., Smith M.K., Adkins S.W. (1996) Gibberellic acid detection of dwarf off-types in micropropagated Cavendish bananas. Aus. J. Expt. Agric. 36 : 237-241.

Das S. C. (1992) Non-conventional techniques of regenerating polyploids in tea. Proc. 31st Tocklai Conf., TRA, Jorhat; 1992: 26-30.

Das S. C. (2001) Tea. In: Parthasarathy VA., Bose TK., Deka PC., Das P., Mitra SK., Mohandas S. Biotechnology of Horticultural Crops, Vol.1. (eds) Naya Prokash., India, pp. 526-546.

Das S.C., Barman T. S. (1988) Current state and future potential of tissue culture in tea. Proc. 30th Tocklai Conf. TRA Jorhat, pp. 90-94.

Das S.C., Barman T.S., Singh R. (1990) Proc. Int. Conf. Tea Res. Assoc. Global Perspective, January11-12, 1990, TRA, Calcutta, pp. 19-22.

Debergh P.C., Vanderschaeghe A. M. (1988) Some symptoms indicating the presence of bacterial contaminants in plant tissue culture. Acta Hort. 255: 77-81.

Dood A. W. (1994). Tissue culure of tea (*Camellia sinensis* (L.) O.Kuntze) – a review. Int. J. Trop Argic. XII (3 & 4) 212-247.

Eden T. (1958) The development of tea culture. In: Eden T. (ed) Tea, Longman, London pp. 1-4.

Evans D.A., Sharp W.R. (1986) Somaclonal and gametoclonal variation: In: Evans D.A., Sharp W.R., Ammirato P.V. (eds) Hand Book of plant cell culture, Vol.4. Technique and applications, Macmillan Publishing Company, New York. pp. 97-132.

Florin B., Tessereau H., Leconteux C., Didier C., Petiard V. (1993) Long term preservation of somatic embryos. In: Redenbaugh K (ed) Synseeds: application of synthetic seeds to crop improvement (pp133-162) CRC press, USA.

Forrest G.I. (1969) Studies on the polyphenol metabolism of tissue culture derived from the tea plant (*C. sinensis* L.) Biochem. J. 113: 765-772.

Gamborg O., Miller R., Ojima K. (1968) Nutrient requirements of suspension cultures of soyabean root cells. Exp. Cell. Res. 50: 157-158.

Ghosh B., Sen S. (1994) Plant regeneration from alginate encapsulated somatic embryos of *Asperagus cooperi* baker. Plant Cell Rep. 13, 381-385.

Groose R.W., Bingham E.T. (1984) Variation in plants regenerated from tissue culture of tetraploid alfalfa heterozygous for several traits. Crop Sci. 24: 655-658.

Haldeman JH., Thomas R.L., Mckamy D.L. (1987) Use of benomyl and rifampicin for *in vitro* shoot tip culture of *Camellia sinensis* and *Camellia japonica*. Hort. Sci. 22: 306-307.

Hammerschlag F. A. (1992) Somaclonal variation: In: Hammerschlag, F. A., Lit R.E. (eds) Biotechnology of perennial fruit crops. C.A.B International, pp. 35-55.

Haridas V., Balasaravanan T., Rajkumar R., Marimuthu S. (2000) Factor influencing somatic embryogenesis in *Camellia sinensis* (L.) O.Kuntze. In: Muraleedharan A., Rajkumar R. (eds) Recent Advances in Plantation Crops Research. Allied Publishers Ltd, India, pp. 31-35.

Harry J.S., Gene J.G., Zimmerman R.H. (1983) Field performance and phenotypic stability of tissue culture propagated thornless blackberries. J. Amer. Soc. Hort. Sci. 108: 285-290; 1983.

Heller R. (1953) Recherches sur la nutrition minerale des tissus vegetaux cultives *in vitro* Annales des Sciences Naturelles (Botanique) Biologie Vegetale. 14: 1-223.

Hu H. (1983) Genetic stability and variability of pollen derived plants . In: Sen S.K., Giles K.L. (eds) Plant cell culture in crop improvement. Plenum Press, New York , pp. 145-157.

Iddagoda N., Kataeva N.N., Butenko R. G. (1988) *In vitro* clonal micropropagation of tea (*Camellia sinensis* L.) 1. Defining the optimum condition for culturing by means of a mathematical design techniques. Ind. J. Plant Physiol. 31: 1-10.

Ikekawa N. (1991) In: Patterson G.W., Nes W.D. (eds) Physiology and biochemistry of sterols American Oil chemists Society. Champaign. Vol. II. pp. 347-360.

Isabel N.L., Tremblay M.M., Tremblay F.M., Bousquet. J. (1993) RAPD as an aid to evalute the genetic integrity of somatic embryogenesis derived population of *Picea mariana* (Mill) B.S.P. Theor. Appl. Genet. 86:81-87

Jain S.M., Newton R. J. (1990) Prospects of biotechnology for tea improvement. Proc. Indian. Natl. Sci. Acad. 6:441-448.

Jain SM., Das S.C., Barman T.S. (1991) Induction of roots from regenerated shoots of tea (*Camellia sinensis* L.). Acta Hort. 289:339-340.

Jain S.M., Das S.C., Barman T.S. (1993) Enhancement of root induction from *in vitro* regenerated shoots of tea (*Camellia sinensis* L.) Proc. Ind. Natl. Sci. Acad. 59:623-628.

Jha T.B. Sen S.K. (1992) Micropropagation of an elite Darjeeling tea clone. Plant Cell Rep. 11:101-104.

Jha T.B., Jha S., Sen S.K. (1992) Somatic embryogenesis from immature cotyledon of an elite Darjeeling tea clone. Plant Sci. 84:209-213.

Kato M. (1982) Results of organ culture on *Camellia japonica* and *C. sinensis*. Jpn. J. Breed. 32: 267-277.

Janeiro, L.V., Ballester, A. and Vieitez, A.M., 1995. Effect of cold storage on somatic embryogenesis systems of *Camellia*. J. Hort. Science, 70 (4): 665-672.

Kato M. (1985) Regeneration of plantlets from tea stem callus. Jpn. J. Breed. 35: 317-322.

Kato M. (1986a) Micropropagation through cotyledon culture in *Camellia japonica* L. and *Camellia sinensis* L. Jpn. J. Breed. 36:31-38.

Kato M. (1989a) *Camellia sinensis* L. (Tea): *In vitro* regeneration. In: Bajaj Y.S.P (ed) Biotechnology in Agriculture and Forestry, Vol. 7. Medicinal and Aromatic Plants Vol. II. Springer-Verlag, Berlin, Heidelberg, pp. 83-98.

Kato M. (1989b) Polyploids of *Camellia* through culture of somatic embryos. Hort. Sci., 24: 1023-1025.

Kato M. (1996) Somatic embryogenesis from immature leaves of *in vitro* grown tea shoots. Plant Cell Rep. 15:920-923.

Kuranuki Y., Sakai A. (1995) Cryo-preservation of *in vitro* grown shoot tips of tea (*Camellia sinensis*) by vitrification. Cryo-letters. 16:345-352.

Kuranuki Y., Shibata M. (1992) Effect of concentration of plant growth regulators on the shoot apex culture of tea plant. Bull. Shizuoka Tea Expt. Station. 16:1-6.

Kuranuki Y., Shibata M. (1993) Improvement of medium components for *in vitro* cuttings of tea plant. 2. Optimum concentration of plant growth regulators. J. Tea Sci. 77: 39-45.

Kuranuki Y., Yoshida S. (1991) Cryopreservation of tea seeds and excised embryonic axes in liquid nitrogen. Proc. Int. Symp. Tea Sci. Shizuoka, Japan. pp 419-420.

Lecouteux C.G., Lai F.M., Mckersie B.D. (1993) Maturation of alfalfa (*Medicago sativa* L.) somatic embryos by abscisic acid, sucrose and chilling stress. Plant Science. 94:207-213.

Llyod G., McCown B. (1980) Commercially feasible micropropagation of mountain laurel, *Kalmia lalifolia* by use of shoot tip culture. Comb. Proc. Int. Pl. Prop. Soc. 30:421-427.

Marcatrigiano M., Boyle T.H., Morgan P.A., Ambach, K.L. (1990) Leaf color variants from coleus shoots cultures. J. Amer. Soc. Hort. Sci. 115: 681-696.

Maugh T. H. (1981) The natural occurring brassionoide in the plant species. Science. 212: 33-34.

Mondal TK., Bhattacharya A., Sood A., Ahuja P.S. (1998) Micropropagation of tea using thidiazuran. Plant Growth Regu. 26: 57-61.

Mondal T.K., Bhattachraya A., Sood A., Ahuja P.S. (1999) An efficient protocol for somatic embryogenesis and its use in developing transgenic tea (*Camellia sinensis (L)* O. Kuntze) for field transfer. In: Altman A., Ziv M., Izhar S. (eds) Plant Biotechnology and *In Vitro* Biology in 21st Century. Kluwer Academic Publishers, Dordrecht, The Netherlands, pp. 101-104.

Mondal TK., Bhattacharya A., Sood A., Ahuja P.S. (2000b) Factor effecting induction and storage of encapsulated tea (*Camellia sinensis* L. O.Kuntze) somatic embryos. Tea. 21: 92-100.

Mondal T.K., Bhattacharya A., Ahuja P.S. (2001a) Induction of synchronous secondary embryogenesis of Tea (*Camellia sinensis*). J. Plant Physiol. 158: 945-951.

Mondal TK., Bhattachray A., Sharma M., Ahuja P.S. (2001b) Induction of *in vivo* somatic embryogenesis in tea (*Camellia sinensis*) cotyledons. Curr. Sci. 81: 101-104.

Mondal T.K., Chand P.K. (2001f) Detection of genetic instability among the miocropropagated tea (*Camellia sinensis)* Plants. In Vitro Cell and Dev Biol. 37:1-5.

Mondal TK., Bhattacharya A., Ahuja P.S., Chand P.K. (2001h) Factor effecting *Agrobacterium tumefaciens* mediated transformation of tea (*Camellia sinensis* (L).O.Kuntze) Plant Cell Rep. 20:712-720.

Mondal TK., Bhattachrya A., Sood A., Ahuja P.S. (2002a) Higher conservation frequency of somatic embryos for improved transformation efficiency in tea (*Camellia sinensis* (L.) O. Kuntze) J. Plant Physiol. (In press).

Mondal T.K., Bhattacharya A., Sood A., Ahuja P.S. (2002b) Cold treatment and storage of synthetic seed of tea (*Camellia sinensis).* Curr. Sci. (Accepted).

Mondal T.K., Parathi R. (2002c) Micrografting - A technique to shorten the hardening time of *in vitro* raised tea plants. J Hort and Biotech. (Communicated)

Mondal. T.K., Bhattacharya A., Laxmikumaran L. (2002d). Recent advance of Camellia Biotechnology. Plant Cell Tissu Org Cult. (Communicated).

Moore P. P., Robbins J.A., Sjulin T.M. (1991) Field performance of "Olympus" strawberry subclones. Hort. Sci. 26:192-194.

Murali K.S., Pandidurai V., Manivel L., Rajkumar R. (1996) Clonal variation in multiplication of tea through tissue culture. J. Plant Crops. 24: 517-522.

Muralidharan E.M., Mascarenhas A.F. (1995) Somatic embryogenisis in *Eucalyptus*. In: Jain S., Gupta P.K., Newton R. (eds) Somatic embryogenesis in woody plant. Vol. 2. Angiosperm. Kluwer Academic Publishers. Dordrecht. The Netherlands, pp. 23-40.

Murashige T., Skoog F. (1962) A revised medium for rapid growth and bioassay with tobacco tissue cultures. Physiol Plant. 15: 473-497.

Nadamitsu S., Andoh Y., Kondo K., Segawa M. (1986) Interspecific hybrids between *Camellia vietnamensis* and *C. chrysantha* by cotyledon culture. Jpn. J. Breed. 36: 309-313.

Nakamura Y. (1990) Effect of sugar on formation of adventitious buds and growth of axillary buds in tissue culture of tea. Bull. Shizuoka Tea Expt. Station. 15: 1-5.

Nakamura Y. (1985) Effect of origin of explants on differentiation of root and its varietal difference in tissue culture of tea plant. Tea Res. J. 62: 1-8.

Nakamura Y. (1987a) Shoot tip culture of tea cultivar Yabukita. Tea Res. J. 65: 1-7.

Nakamura Y. (1987b) *In vitro* rapid plantlet culture from axillary buds of tea plant (*C. sinensis* (L.) O. Kuntze). Bull. Shizuoka Tea Expt. Station. 13: 23-27.

Nakamura Y. (1988a) Efficient differentiation of adventitious embryos from cotyledon culture of *Camellia sinensis* and other *Camellia* species. Tea Res. J. 67: 1-12.

Nakamura Y. (1988b) Effects of the kinds of auxins on callus induction and root differentiation from stem segment culture of *Camellia sinensis* (L.) O. Kuntze. Tea. Res. J. 68: 1-7.

Nakamura Y. (1989a) Differentiation of adventitious buds and its varietal difference in stem segment culture of *Camellia sinensis* (L.) O. Kuntze. Tea Res. J. 70: 41-49.

Nitsch J.P., Nitsch C. (1969) Haploid plants from pollen grains. Science. 163: 185.

Ogutuga D.B.A., Northcote D.H. (1970a) Caffeine formation in tea callus tissue. J. Expt Bot. 21: 258-273.

Palni L.M.S., Bag N., Nadeem M., Tamata S., Vyas P., Bisht MS., Purohit V.K., Kumar A., Nandi S.K., Pandey A., Purohit A.N. (1998). Micropropagation: Conservation through tissue culture of selected Himalayan Plants. In: Research for Mountain Development: Some initiatives and accomplishments. Nainital: Gyanodaya Prakashan, pp431-452.

Pandey A., Bag N., Chandra, B., Palni L.M.S. (2002) Biological hardening: a promising technology for tissue culture industry. In: Role of plant tissue culture in biodiversity conservation and economic development. Nainital: Gyanodaya Prakashan, pp565-577.

Pandey A., Palni L.M.S., Bag N. (2000). Biological hardening of tissue culture raised tea plants through rhizosphere bacteria. Biotech. Letter. 22: 1087-1091.

Pandidurai V., Murali K.S., Manivel L., Rajkumar R. (1996) Factors affecting *in vitro* shoot multiplication and root regeneration in tea. J. Plantation Crops 24 (Suppl): 603-609.

Paratasilpin T. (1990) Comparative studies on somatic embryogenesis in *Camellia sinensis* var. *sinensis* and *C. sinensis* var. *assamica* (Mast.) Pierre. J. Sci. Soc. Thailand. 16: 23-41.

Phukan M.K., Mitra G.C. (1984) Regeneration of tea shoots from nodal explants in tissue culture. Curr. Sci. 53: 874-876.

Phukan M.K., Mitra G.C. (1990) Nutrient requirements for growth and multiplication of tea plants *in vitro*. Bangladesh J. Bot. 19: 65-71.

Ponsamuel J., Samson NP., Ganeshan P.S., Satyaprakash V., Abrahan G.C. (1996) Somatic embryogenesis and plant regeneration from the immature cotyledonary tissues of cultivated tea (*Camellia sinensis* (L.) O. Kuntze). Plant Cell Rep. 16: 210-214.

Prakash O., Sood A., Sharma M., Ahuja P.S. (1999) Grafting micropropagated tea (*Camellia sinensis* (L.) O. Kuntze) shoots on tea seedling- a new approach to tea propagation. Plant Cell Rep. 18: 137-142.

Rajkumar R., Ayyappan P. (1992) Somatic embryogenesis from cotyledonary explants of *Camellia sinensis* (L.) O. Kuntze. The Planters Chronic. May. 227-229.

Rajkumar R., Ayyappan P. (1992) Micrppropagation of *Camellia sinensis* (L.) O kuntze. J Plant Crops 20: 252-256.

Rajkumar R., Balasusaravanam S., Jayakumar D., Haridas V., Marimuthu S. (2001). Physiological and biochemical feathers of field grown somaclonal variants of Tea. UPASI tea research foundation Bulletin No 54, 73-81.

Rajkumar R., Ayyappan P. (1992) Micropropagation of *Camellia sinensis* (L.) O.Kuntze. J Plant. Crops 20: 252-256.

Rajashekaran P., Mohankumar P. (1992) Rapid micropropagation of tea (*Camellia* spp). J Plant Crops. 20:248-251

Rani V., Ajay P., Raina S.N. (1995) Random amplified polymorphic DNA (RAPD) markers for genetic analysis in micropropagated plants of *Populus deltoides* Marsh. Plant Cell Rep. 14 : 459-462.

Redenbaugh K., Fujii J.A., Slade D. (1991) Synthetic seed technology. In: Vasil K.I. (ed) Scale-up and automation in plant propagation. Cell culture and somatic cell genetics of plants, Vol. 8. Academic Press Inc., New York, pp. 35-74.

Redenbaugh K., Slade D., Viss P., Fujii J.A. (1987) Encapsulation of somatic embryos in synthetic seed coats. Hort Sci 22, 803-809.

Roberts D.R., Sutton B.C.S., Flinn B.S. (1990) Synchronous and high frequency germination of interior spruce somatic embryo following partial drying at high relative humidity. Can. J. Bot. 68: 1086-1093.

Sandal I., Bhattacharya A., Ahuja P.S. (2001) An efficient liquid culture system for tea shoot proliferation. Plant Cell, Tissue Organ Culture. 65: 75-80.

Sarathchandra TM., Upali P.D., Wijeweardena R.G.A. (1988) Studies on the tissue culture of tea (*Camellia sinensis* (L.) O. Kuntze) 4. Somatic embryogenesis in stem and leaf callus cultures. Sri Lank. J. Tea Sci. 52: 50-54.

Sarwar M. (1985) Callus formation from explanted organs of tea (*Camellia sinensis* L.). J. Tea Sci. 54: 18-22.

Scowcroft W.R. (1984) Genetic variability in tissue culture: Impact on germplasm conservation and utilization. Report on the IBPGR, Rome, p.152.

Scowcroft W.R., Larkin P.J. (1982). Somaclonal variation – a new option for plant improvement. In: Vasil I.K., Scowcroft W.R., Frey K.J. (eds) Plant improvement and somatic cell genetics, Academic Press, New York, pp. 157-178.

Sealy J.R. (1958) A revision of the genus *Camellia.* R. Hortic Soc. London.

Sharma M., Sood A., Nagar P.K., Prakash O., Ahuja P.S. (1999) Direct rooting and hardening of tea microshoots in the field. Plant Cell Tissue Org Cult. 58: 111-118.

Shibata M., Kuranuki Y. (1993) Improvement of medium components for *in vitro* cuttings of tea plant. 1. Effects of concentration of some components of MS medium and comparison between MS medium and woody plant medium. J. Tea Sci. 77: 39-45.

Singh A., Sharma J., Rexer K-H., Varma A. (2000) Plant productivity determinants beyond minerals, water and light: *Piriformospora indica-* A revolutionary plant growth promoting fungus. Curr. Sci. 79 (11): 1548-1554.

Singh I.D. (1980) Non conventional approaches to the breeding of tea in north-east India. Two and a bud. 27: 3-6.

Smith M.K. (1998). A review of factors influencing the genetic stability of micropropagated banana fruits. Australian J. Expt. Agric. 43: 219-223.

Sood A., Palni L.M.S., Sharma M., Rao D.V., Chand G., Jain N.K. (1993) Micropropagation of tea using cotyledon culture and encapsulated somatic embryos. J. Plant. Crops. 21: 295-300.

Tahardi J.S., Shu W. (1992) Commercialization of clonal micropropagation of superior tea genotypes using tissue culture technology USAID/CDR Network meeting on tea crop biotech Costa Rica.

Thakor, B.H. (2000). A re-examination of the phylogenetis relationships within the genus *Camellia.* Int Camellia J. 2000. 130-134.

Tian-Ling. (1982) Regeneration of plantlets in cultures of immature cotyledons and young embryos of *Camellia oleifera* Abel. Acta. Biol. Exp. Sin., 15: 393-403.

Tremblay M. M., Tremblay F.M. (1995) Maturation of black spruce somatic embryos: sucrose hydrolysis and resulting osmotic pressure of the medium. Plant Cell Tiss. Org. Cult. 42: 39-46.

Tubbs F. R. (1932). A note on vegetative propagation of tea. Tea Quarterly, 5, 154-156.

Tulecke W. (1987) Somatic embryogenesis in woody perennials. In: Bonga J.M., Durzan D.J. (eds) Cell and Tissue Culture in Forestry. Vol. 2. Martinus Nijhoff Publishers, Dordrecht, pp. 61-69.

Tunstall A. C. (1931) A note of propagation of green shoot cuttings. Quarterly J. ITA, 49-51.

Vieitez A.M. (1994) Somatic embryogenesis in *Camellia* spp. In: Jain S.M., Gupta PK., Newton RJ. (eds) Somatic Embryogenesis in Woody Plants, Kluwer Academic Publishers, The Netherlands, pp. 235-276.

Vieitez A.M., Vieitez M.L., Ballester A., Vieitez E. (1992) Micropropagation of *Camellia* spp. In: Bajaj Y.P.S. (ed) Biotechnology in Agriculture and Forestry, Vol. 19. High Tech and Micropropagation III. Springer-Verlag, Berlin Heidelberg. pp 361-387.

Vuylsteke D., Swennen R., Wilson G.F., Langhe E.D. (1988) Phenotypic variation among *in vitro* propagated platain (*Musa* sp. Cultivar 'AAB') Sci. Hort. 36: 79-80.

Wachira F., Ogado J. (1995) *In vitro* regeneration of *Camellia sinensis* (L.) O. Kuntze by somatic embryo. Plant Cell Rep. 14: 463-466.

Wealth of India. (1950) A directory of Indian raw materials and industrial products. In: Bhatnagar S.S. Chairman, Editorial Committee, Council of Scientific and Industrial Res, Vol. II. New Delhi, India, pp. 26-51.

Webster F.B., Roberts D.R., Mclnnis S.M., Sutton B.C.S. (1990) Propagation of interior spruce by somatic embryogenesis. Can. J. For. Res. 20: 1757-1762.

Wight W. (1959) Nomenclature and classification of tea plant. Nature. 183: 1726-1728.
Wood D.J., Barua D.N. (1958) Species hybrids of tea. Nature. 181: 1674-1675.
Wu C.T., Huang T.K., Chen G.R., Chen S.Y. (1981) A review on the tissue culture of tea plants and on the utilization of callus derived plantlets In: Rao A.N. (ed) Tissue culture of Economically Important plants, Proc. Costed. Symp Singapore, pp.104-106.
Yan M.Q., Ping C. (1983) Studies on development of embryoids from the culture cotyledons of *Thea sinensis* L. Sci. Silv. Sin. 19: 25-29.
Zhan Z., Ke N., Chen B. (1987) The cytology of tea clonal cultivars fujian shuixian and their infertile mechanism. Proc of Intl Tea quality. Human Health Symp, China, pp. 46.

Table1 Micropropagation of tea.

Species/ cultivar	Explant	Medium				Novelty	Special treatment	Response	Response time	Hardening/ Field transfer	Reference
		Initiation	Multiplication	Regener-ation/ organo-genesis	Rooting						
C. sinensis, TV1	Terminal bud and nodal stem segment of field grown plant	MS+IAA(0.25)+ BA(4.0)+ CW(10%)+YE (200)	1/2 MS+IAA (0.50) +BA(4.0)+ CW(10%)+YE (200)	-	1/2 MS+ ascorbic acid (2)+IBA (7.0gm/l)+ Sucrose(4%)	Hardening and rooting of micropopagated shoots was major problem	A pre-treatment of shoots with an aqueous solution of IBA(100) greatly increased rooting	Shoot proliferation and rooting (60%)	4-5 weeks for proli feration and 4 weeks of rooting	Hardening was done in peat:soil (1:1) mixture and kept in greenhouse with 85-90% humidity	Agarwal *et al.*, 1992
C. sinensis (L.) O.Kuntze cv.TRI 2025, CY9 and PK2	Shoot (4-5cm)tip and nodal segment of mature plant	MS+modified vit+IBA(0.01)+ GA_3 (0.1)+BAP(1) for TRI 2025 and MS+modified vit+IAA(0.1)+ BA(1) for others	MS+modified vit+IBA(0.01) + GA_3 (0.1)+BAP(2) for TRI 2025 and MS+ modified vit+ IAA(0.1)+ BA (2)	-	-	-	-	Axillary shoot proliferation (2-4)	4 weeks	-	Arulpra-Gasam and Latiff, 1986
C. sinensis (L.) O. Kuntze	Nodal segment	MS+BAP(1.12)+ NAA(0.2)	MS+BAP(2.25) + IBA(0.2)	-	½ MS+NAA (3.6)	Rooting observed after withdrawal from auxin containing medium with in 60 days	-	Axillary shoot multiplicatio n and rooting	10 weeks	Rooted plants were subsequently transferred to community pot with plastic covering	Bag *et al.*, 1997
C. sinensis, TV1	Shoot tip and nodal segment (0.7-1cm) of mature field grown plant.	MS+YE(200)+CM (10%)+IAA(0.5)+ BAP(4).	1/2MS+CM (10%)+IAA (0.25) +BAP (4)	-	½ MS +modified vit+IBA (7)	-	A pre-treatment of the shoots for 30 min. with aqueous solution of IBA (100) was helpful for induction of rootings. Highly acidic soil pH (4.5-4.6) favoured rooting	Rooting (60%) of *in vitro* grown shoot	25-30 days	Potting mixture was sterile peat and soil 1:1 and irrigated with 1/2 MS. After 30 days directly transfer to same potting mixture in greenhouse (with 85% RH)	Banerjee and Agarwal, 1990
C. saluenensis x *C. japonica*	Shoot culture	-	-	-	½ MS	-	A pre-treatment with IBA (1-2g/L) for 15 min before placing rooting medium increase rooting	Root induction and plantlet regeneration	-	-	Beretta *et al.*, 1987
C. sinensis L. cv. Kolkhida, kymyn, [illegible]	Zygotic embryo (1-2mm)	1/2MS+ Modified vit+GA_3 (0.1)+ 3% sucrose	1/2MS+Modified vit+ BAP(2)+ IAA (0.2)+ [illegible]	-	½MS + IBA (0.5-2)	-	-	Axillary bud proliferation (6-8), rooting	8 weeks and 3-4 weeks	Transfer to the pot but media not mentioned	Iddagoda *et al.*,1988

			+2% sucrose						tively		
C. sinensis L.	*In vitro* micro shoot	MS+BA(6)+NAA (0.5)+GA_3(2)	MS+BA(6)+NAA (0.5)+GA_3(2)	-	1) Shoots were treated with IBA (50) before putting in soil mixture 2) Liquid medium 1/2MS+ IBA (5) in dark 3) Filter paper bridge 1/2MS+0.5 IBA 4) Solidified agar with MS+. IBA (0.5)	Work was emphasized on rooting of microshoots	-	Rooting 97% for medium 1, 50% for medium 2 and 2.44% for medium 3	2-3 weeks	Soil:peat (1:1) with light humidity	Jain *et al.*, 1993 and Jain *et al.*, 1991
C. sinensis (L.)O. Kuntze, T-78	Shoot tips and cotyledonary nodes from germinated seedling .Nodal segment of field grown plant.	1/2 MS+CW (10%) +CH(1g/l)+BA(5) +IBA(0.1)	1/2 MS+CW (10%)+CH(1g/l+ BA(1)+IBA (0.1) shoot tips and nodal segment	-	MS+IBA (100) for 10 days and then transfer to MS	Direct organogenesis from hypocotyl segment failed	Treatment with IBA (500) for 30-40 min induced rooting. Cytological analysis done to see genetic uniformity	Axillary bud proliferation. 8 and 35 from shoot tip and cotyledonary nodes respectively, *in vitro* rooting (80-90%)	18 weeks, 10 days respec-tively	Potting mixture was peat and soil (1:1).60% survival of plantlets	Jha and Sen, 1992
C. sinensis cv. Yabukita	Epidermal layer, Intact stem segment, stem segment without epidermal layer	MS+ IBA(4)+ BAP(2) from all three explants	-	MS+ IBA (0.5) +BAP (10)	1/2 MS+ IBA(0.5)	Regeneration ability depends upon explant used	In rooting, filter paper bridge with liquid medium was used	Organo-genesis (20% from epidermal layer)	4 weeks for callus and 8 weeks for organo-genesis	Rooted plant after six weeks transplanted to pot containing vermiculite and soil	Kato, 1985
C. sinensis with 30 different clones.	Shoot apex	MS+BAP(1)+IBA (1)+GA_3 (0.01)	-	-	-	-	-	Shoot proliferation	-	-	Kuranuki and Shibata 1992
C. sinensis with 30 different clones	*In vitro* cuttings	-	1/2 MS+BAP (1)+IBA(0.01)+ GA_3 (10)	-	-	-	-	Shoot proliferation	7 weeks	-	Kuranuki and Shibata, 1993
C. sinensis cv. China hybrid	Nodal segments from *in vitro* grown seedlings	MS+ TDZ (1.1) + NAA (1.86)	MS	-	Treatment with IBA (500) for 30 mints	A novel cytokinin TDZ was evaluated for the first time	-	Axillary shoot development	24 weeks for shoot proliferatio n	Pot mixture was soil:river bed sand: FYM:: 9:1:1	Mondal *et al.*, 1998
C. sinensis	Nodal segments from *in vitro* grown seedlings	MS+ TDZ (1.1) + NAA (1.86)	MS	-	Treatment with IBA (500) for 30 mints	Artificial seed of tea was produced for the first time	Sodium alginate encapsulated beads were stored at 4°C	Cold treatment shorten the bud sprouting	-	Pot mixture was soil:river bed sand: FYM:: 9:1:1	Mondal *et al.*, 2002b

								time by 15 days			
C. sinensis with 23 cultivars.	Stem segment	MS+IAA (.01-1)	MS+IAA (0.01-1)+GA_3 (1-5)	-	-	Histological observation of adventitious buds showed that the buds differentiated only from the small lumps of callus developing on the cutting surface. Vertical difference was observed for adventitious bud formation	-	Adventitious bud formation (21.2%) from callus	-	-	Nakamura, 1989a
C. sinensis cv. Yubukita	Axillary bud	MS+BA(1)+GA_3 (1)	MS+BA(1)+GA_3 (1)	-	1/2MS+IBA (3) or MS+NAA (1 and 10)	Liquid and solid media with same composition does not differ the growth	Dark incubation was better for axillary bud growth and rooting	Axillary shoot proliferation, rooting	-	-	Nakamura, 1987a.
C. sinensis (L.) cv. Yabukita	Shoot tips of mature tea bush	MS, Gamborg (B_5), Nitsch and Nitsch (1969)	MS+BA(1)+GA_3 (5)	-	1/2MS + IBA(3)	Shoot tip were precultured to avoid bacterial contamination	Liquid medium with filter paper bridge were used for rooting	Shoot multiplication, rooting	10 weeks and 7 weeks respectively	-	Nakamura, 1987b.
C. sinensis	-	-	1/2MS+IBA(0.1) +BA(1)+GA_3(5)	1/2MS+ IAA(1)+ GA_3(5)	-	Out of five sugar glucose and galactose with sucrose (3-6%) help in differentiation of adventitious bud formation. For growth of axillary bud maltose and lactose was better with 1-3% sucrose	-	Callus, differentiation of roots, differentiation of adventitious buds and axillary shoot proliferation.	-	-	Nakamura, 1990
C. sinensis cv. Yubakita	Stem segment of *in vitro* shoot.	1/2MS+IBA(0.1) or 1/2MS+2ip(30)	1/2MS+ IBA(0.1)+ BAP(1)+ GA_3 (5)	-	-	Different auxins and cytokins has been tried for multiplication of shoot	-	Proliferation of shoot.	One year	-	Nakamura, 1990.
C. sinensis UPASI-3	Nodal explant	½ MS	½ MS +BAP (3)	-	¼ MS(liquid) +IBA (3)	-	-	Shoot multiplication and rooting (60%)	-	Rooted plants were transfer to black soil:sand (1:1) as potting mix.	Pandidurai *et al.*, 1996
C. sinensis	Nodal segment of *in vitro* shoots	1/2MS+Modified Vit+IAA(1)+Kn (3)+YE(0.15%)+C[illegible]	1/2MS+Modified Vit+IAA(1)+Kn (3)+YE(0.15%)+ [illegible]	-	1/2MS+IBA (8)	-	Salt concentration modified [illegible]	Shoot multiplication and rooting [illegible]	16-24 weeks and 2-4 weeks [illegible]	-	Phukan and Mitra, 1990

		W(10%)	CW(10%)					(60-70%)	respec-tively		
C. sinensis (L.) O. Kuntze	Shoot tip, nodal segment of field grown plant	MS+0.1% YE +10%CM + NAA (2) + BAP(6) for indirect regeneration. MS+ IAA (2)+ Kn (8) for direct regeneration.	For indirect regeneration 1/2M +IAA (1)+Kn (3)	-	-	-	-	4-5 shoots bud from NAA +BAP treated callus and 3-4 shoot bud from IAA+Kn treated explant	8-12 weeks and 8 weeks respec-tively	-	Phukan and Mitra., 1984
C. sinensis cv. Banuri-96	Nodal segments from field grown plants	½ MS + 8% agar	½ MS + IAA (0.2) + BAP (2.5)	-	Grafted on seedlings of tea plants	Micro-grafting method for hardening was described	BAP (5) + NAA (5)	83.3% graft union.	6-7 weeks for new foliage dvelopment	83.3 % survival in the field	Prakash *et al.*, 1999
C. sinensis UPASI-3, UPASI-9	Nodal segments	MS+ BAP(5)+CM (10%)	MS + GA_3 (1) + BAP(0.5)	-	MS + IAA(1)	-	-	Shoot proliferation and rooting (18%)	60 shoot/explant with in 8 weeks	-	Rajkumar and Ayyappan (1992)
C. sinensis UPASI-9	Nodal explant (1-1.5 cm)	MS + BAP (0.5)	MS + BAP (5)	-	-	-	-	Axillary shoot multiplication	20-25 axillary bud with in 20-24 weeks	Rooting done on black soil:sand (1:1) as potting mix. And following this procedure 45000 tissue culture plants of 8 cultivars was transferred to fields (see text)	Rajshekaran and Kumar, 1992
C. sinensis (TV-1, T-78, Upasi-9 and Kangra Jat	Nodal segments with slight callusing at the base from *in vitro* raised cultures	Liquid MS static cultures + TDZ (1.1)	Same medium	-	IBA ( ) treatment for 30 minutes before placing onto sand in closed jars under *in vitro* conditions			88% rooting		88% survival in the green house	Sandal *et al.*, 2000
C. sinensis L.	Flower stalks, leaf pieces, petioles, immature cotyledons, stem pieces and shoot tips	MS+2,4-D(4.42), NAA (3.72) +BA(2.25), Kn (2.15)	MS+2,4-D (10)+YE (0.2%)	Buds with 2-3 leaves were initiated on MS+BA (2.25)	Explant browning were checked by lower concentration of MS salt 1/20th	Growth of buds occurred when cultures were incubated in darks for first 10 days	-	Callus and shoot-bud regeneration (50%)	-	-	Sarwar, 1985
C. sinensis cv. Banuri-96	Single node cuttings of *in vitro* shoots	½ MS + 8% agar	½ MS + IAA (0.2) + BAP (2.5)	-	-	A specially designed hardening chamber along with optimal conditions	IBA treatment (500) for 30 minutes was necessary for	71.6 % root induction and 73% survival the	18 weeks for rooting	Over 300 plants have been transferred to the field.	Sharma *et al.*, 1999

						was described.	rooting.	field			
C. sinensis	*In vitro* cuttings	MS+BAP(1)+IBA (0.01)+GA_3(5)	-	-	-	-	-	Shoot proliferation	-	-	Shibata and Kuranuki, 1993
C. sinensis, TRI 2025	Axillary buds	Modified WPM+ BAP(2.25)+IBA (1.01)+TDZ(0.01)	Modified WPM+ BAP(2.25)+IBA (1.01)+TDZ(.01)	-	-	-	-	Shoot proliferation	10-12 weeks	-	Tahardi and Shu,1992
Figures in parenthesis denote concentration (mg/l)											

Table 2 Effect of TDZ/BAP in MS/WPM basal media on induction of bud break in nodal explants.

Concentration	**Percentage of bud break**			
	MS		WPM	
	TDZ	**BAP**	**TDZ**	**BAP**
0	-	-	-	-
100μM	94.0 ±1.4 CF	-	94.7±1.1 CF	-
10μM	95.8 ±1.2 CF	33.8 ±1.0 SP	95.5 ±1.1 CF	1.5 ± 0.2 SP
1μM	96.7 ±0.8 CF	12.8 ±0.5 SP	96.0 ±1.0 CF	1.3 ±0.2 SP
100 nM	54.5 ±1.2 SP	-	43.5 ±0.9 SP	-
10 nM	49.2 ±0.7 SP	-	38.5 ±0.7 SP	-
1nM	49.0 ±0.9 SP	-	39.0 ±0.7 SP	-
100pM	29.0 ±0.9 SP	-	28.7 ±0.8 SP	-
10pM	27.5 ±0.6 SP	-	28.7 ±0.8 SP	-
1pM	28.2 ±0.8 SP	-	28.3 ±0.72 SP	-

SP = Shoot proliferation; CF= Callus formation; - = No response

Table 3 Comparative performance of explants initiated on TDZ or BAP containing media during successive culture passage.

Culture passage	No.of shoots/explants		Shoot length (mm)		Shoot diameter (mm)		No.of internodes		Internode length (mm)	
	MS*	MS+BAP	MS*	MS+BAP	MS*	MS+BAP	MS*	MS+BAP	MS*	MS+BAP
2	13.00±1.74	11.33± 0.42	30.83±6.67	22.33±2.73	2.25±0.27	1.20±0.24	7.33±0.42	5.20±0.58	0.36±0.07	0.42±0.13
3	37.17±3.77	23.33±0.80	60.00±3.87	40.00±1.18	2.25±0.11	2.25±0.11	10.50±0.50	7.33±0.42	0.56±0.09	0.98±0.07
6	44.17±1.08	30.33±1.20	66.17±2.17	44.20±1.74	2.30±0.11	2.70±0.12	11.0±0.60	9.70±.21	0.5±0.03	1.15±0.07

***Explants initiated on media containing TDZ were sub-cultured on to a hormone-free medium whereas explants initiated on media containing BAP were maintained on the same medium**

Data (X ± SE) pooled from three independent experiments each with three replicates

Table-4 Description of different graft combination and their performance.

Graft combination		Survival percentage	Height (cm).
Scion	Stock (Seedling with parental combination)		
TTL-4	UPASI-9 X C-6017	91.5	48.6 ± 1.23
TTL-4	Non-grafted (Control)	82.0	25.2± 0.67
TTL-5	B/4/142 X TRI-2026	95.0	47.3± 1.11
TTL-5	Non-grafted (Control)	87.5	35.5± 0.97
N/17/17	B/4/142 X TRI-2026	92.0	66.1± 1.10
N/17/17	Non-grafted (Control)	85.0	17.0± 0.45
RD/1/101	UPASI-9 X C-6017	94.8	45.2± 0.84
RD/1/101	Non-grafted (Control)	83.0	36.0± 1.23

Table 5 Somatic embryogenesis in tea.

Species/Cultivar	Explant	Medium				Response time and % of response	Special treatment/Problem /Novelty	Hardening status	Reference
		Induction	Maturation	Germination	Multiplication				
C. sinensis	Immature cotyledon	MS+BA+IBA	-	MS+BA+ NAA	-	Direct and indirect embryo-genesis and plantlet regeneration	-	-	Abraham and Raman, 1986
C. sinensis TRI 2025	Nodal segment	MS+BAP(0.5)+ GA_3 (3)	-	-	MS basal with a 50% dilution of macro salts	60% conversion rate	-	90% survival rate in greenhouse	Akula and Dood,1998
C. sinensis	Mature seed	½ MS macro+ Full micro MS+ adenine hemisulphate (100)+ glutamine (100)	½ MS macro+ Full micro MS+ adenine hemisulphate (100)+ glutamine (100)	½ MS macro+ Full micro MS+ adenine hemisulphate (100)+ glutamine (100)	-	15-20% germination	-	-	Akula *et al.*, 2000
C. sinensis (L.) O. Kuntze., Bi-clonal seed stock.	Cotyledon pieces	Modified MS+BAP (10)+YE (2)	-	Modified MS+BAP (10) with 2% sucrose	-	-	Incubation for 6-8 weeks in liquid culture medium was necessary for embryo formation. Embryos were formed through callus phase	4-5 leaves stage germinated embryo were transferred to plastic pot containing a mixture of sterile soil and sand (1:1). After 2 weeks they were transferred to soil	Arulpragasam *et al.*, 1988
C. sinensis	Cotyleon segment	MS+BAP (1.13)+ NAA(22.7)	MS+BAP (1.13)+ NAA (22.7)	MS+BAP (2.26)+NAA (22.7) +GA_3(0.12)	-	60% plantlet was achieved	-	The rooted plants were subsequently transferred to community pots	Bag *et al.*, 1997
C. sinensis	Mature cotyledon, leaf	MS + BAP (2) + 2,4-D (5)	-	MS + BAP(2) + NAA (3)	-	14% from mature cotyledon, 2% from leaf for induction and 12% germination	-	-	Balasubramanian *et al.*, 2000
Thea sinensis (L.), B-61	Cotyledon slice	MS+K(0.05)+2,4-D (0.5)	1/2MS+Kn (0.05)+activa-ted charcol (0.2%)+	1/2MS+Kn (0.05)+activa-ted charcol (0.2%)+	-	40% germination of somatic embryo	For germination high light intensity (5000 lux) was used. Irregular germination was	-	Bano *et al.*, 1991

			glucose (1.5%)	glucose (1.5%)			problem		
C. sinensis	Immature cotyledon	Modified MS + BAP (3) + NAA (2)	-	MS + BAP (5)	-	-	-	-	Haridas *et al.*, 2000
C. sinensis, T-78	Cotyledon	MS+BAP(10)+IBA (0.5)+adenine (80)	MS+BAP(10)+ IBA(0.5)+ adenine (80)	B_5+BAP(3)+ IAA(2)	-	70% survival in acclimatization	Authors reported secondary embryogenesis	Plantlets were transferred to potting mixture containing peat:soil (1:1)	Jha *et al.*, 1992
C. sinensis, Yabukita	Cotyledon segments	MS+BAP (4) +IBA (2)	-	MS+BA(10)+ IBA (0.5)	-	Direct embryogenesis and plant regeneration	-	-	Kato, 1982 and Kato,1986a
C. sinensis, Yabukita, Sayamamidori, Benikaori, Akane.	Immature leaves of *in vitro* grown shoots	Liquid MS+2,4-D (0.5)	MS+BAP(10)+ IBA(0.5) or MS+BAP (4)+IBA(2) in presence of GA_3	MS+BAP(10)+ IBA(0.5)or MS+BAP (4)+IBA(2) in presence of GA_3	-	-	Younger leaves were better than older leaves to produce somatic embryo	-	Kato,1996
C. sinensis, Kangra Jat, UPASI-9, T-78 and TG 270/2/B	Mature de-embryonated cotyledon	MS + BAP (2) + IBA (0.2)	-	4 weeks culture on Modified MS + trans-cinamic acid (3) + maltose (4%) followed by their transfer to MS + GA_3 (1.5Modified)	-	Maximum 66.7 % induction was recorded by Kangra Jat	Artificial seed was produced and found tea synthetic seeds are cold susceptible and hence can not be stored for long time at 4°C	-	Mondal *et al.*, 2000b
C. sinensis	Mature de-embryonated cotyledon	MS + BAP (2) + IBA (0.2)	Modified MS + BAP (2) + IBA (0.2) + Glutamine (1) + K_2SO_4	Modified MS + BAP (2) + IBA (0.2) + Glutamine (1)	-	70% conversion rate	Secondary embryos were induced which developed and germinated synchronously	Following this procedure 3000 plants have been transferred to fields at Tata Tea Ltd, India (see text)	Mondal *et al.*, 2001a
C. sinensis	Primary somatic embryo	MS + BAP (2) + IBA (0.2)	Modified MS + BAP (2) + IBA (0.2) + Glutamine (1) + K_2SO_4	4 weeks culture on Modified MS + trans-cinamic acid (3) + maltose (4%) followed by their transfer to MS + GA_3 (1.5Modified)	Modified MS + BAP (2) + IBA (0.2) + Glutamine (1)	70% conversion rate	Embryogenic capacity was maintained for more than 4 years	Plantlets (4-5 cm height) were transferred to green house in hikkotray containing sand: virgin soil (1:1)	Mondal *et al.*, 2002a
C. sinensis, Kangra Jat, T-78, UPASI-9	Mature seeds	Moist sterile sand at 28°C	-	-	-	23 % induction rate	First report for production of in vivo	-	Mondal *et al.*, 2001b

							somatic embryo in any woody species including tea.		
C. sinensis	De-cotyledonated embryo	Nitsch and Nitsch medium (1969)	Nitsch and Nitsch (1969) medium+N-Z-amino type-A (0.1%)	Same medium with 2.6 time addition of N-Z amino type-A.	-	-	Kinetin and BAP were not required for shoot growth but 4-pyridylura had positive effect on the formation of multiple shoots	-	Nakamura, 1985
C. sinensis with 13 cultivars, *C. japonica 3 cultivars., C. sasanqua, C. brevistela, C. nokoensis, C. japonica* (cv. Kosyougatu)x*C. granthamiana.*	Half sliced cotyledon	MS+BA (1-5)	-	-	-	Late September to mid October was best time for induction of somatic embryo	Maximum 93% adventive embryos were formed in *C. brevistela*	-	Nakamura, 1988a
C. sinensis and *C. assamica*	De-embryonated cotyledon segments	B_5+NAA(0.1)+BAP (2)	B_5+NAA (0.1)+BAP(2)	B_5+NAA (0.1)+BAP (2)	-	China tea (28%), Assam tea (2%) germination of somatic embryo	-	-	Paratasilpin, 1990
C. sinensis, UPASI-10	De-embryonated immature cotyledon	MS+PBOA (1μM)+ BA (0.5μM) with 6% agar	MS+Brassin (1μM) with 0.8%. agar	MS+Brassin (1μM) with 0.08% agar	-	50% germination	Shoots were treated with 1μM IAN+1μM brassin +10μM phloroglucinol in liquid MS for 15 days for rooting	Plantlets were transferred in vermiculite in greenhouse. After 200 days plants were transferred to red acidic soil	Ponsamuel *et al.*,1996
C. sinensis, UPASI-10	Cotyledon, cotyledon without zygotic embryo, cotyledon slice	Modified MS+BAP(0.25)	MS+BAP (3-5)+CW (10%)+GA_3 (0.25-1)	MS+BAP (3-5) +CW (10%)+ GA_3 (0.25 -1)	-	80-90% germination of somatic embryo	-	-	Rajkumar and Ayyappan, 1992
C. sinensis, TRI 202.	Nodal segment and leaf	Vacin and Went medium (1949)+ CW (15%) MS+IBA(0.1)+BAP (1) for nodal	-	-	Callus, embryo-like structure	-	-	-	Sarathchandra *et al.*, 1988

		segment MS+ 2,4-D(1)+Kn(1) for embryo like structure.							
C. sinensis, Kangra Jat	Cotyledon slice.	1/2MS+BAP(2)+IAA (0.2)	-	1/2MS+BAP (2)+IAA(0.2)+ GA_3 (0.2)	-	-	Hypertrophied structure under proper media composition might gave somatic embryos	-	Sood *et al.*, 1993
C. sinensis	Cotyledon slice from in *vitro* raised plantlets	Modified MS	Modified MS	Modified MS +K(1.8)or BAP(1)	-	53-60% germination	Auxin treatment led to callusing	-	Wachira and Ogado, 1995
C. sinensis , Chyi-Men and Pyng-Shoei.	Mature cotyledon	MS+K(10) + IAA (1)	-	-	-	Indirect embryogenesis and plant regeneration.	-	-	Wu *et al.*, 1981
C. sinensis	Mature cotyledon	-	-	-	-	Embryo genesis	-	-	Yan and Ping, 1983

Figures in parenthesis without assigned units denote concentrations in mg/l

Table-6. Effect of time of seed collection on induction of somatic embryogenesis in different tea cultivars*

Cultivars	Variety Type		Time of seed collection						
			Spring Mar-April	Summer May-June	Monsoon July-Aug	Autumn Sept-Oct	Winter Nov-Dec	Late Winter Jan-Feb	Duration of response (Days)
TG 270/2/B	Assam	% response Developmental Stage	- Immature	- Intermed	11.1 ± 1.4 Mature	- Aged	- Aged	- Aged	60
UPASI-9	Assam	% response Developmental Stage	- Immature	- Immature	50.0±0.7	-	-	-	60
T-78	China	% response Developmental Stage	- Immature	- Immature	- mature	44.1±1.7 Aged	11.1± 1 Aged	- Aged	120
Kangra Jat	China	% response Developmental Stage	-	- Immature	- Immature	16.7± 1.3 Intermed	66.7± 0.9 mature	- Aged	120

* Data (X + SE) pooled from three independent experiment, each with three replicates

Table 7 Effect of different growth regulators and adjuvants on secondary embryo development.

BAP (mg/l)	IBA (mg/l)*															
	0				0.1				0.2				0.3			
	L-glutamine (gm/l)															
	0	0.5	1. 0	1.5	0	0.5	1. 0	1. 5	0	0.5	1. 0	1.5	0	0.5	1. 0	1.5
0	24±1.2	25±1.16	36±1.72	28±0.68	37±0.90	43±0.92	47±0.23	42±1.62	42±0.70	51±0.72	57±1.80	53±1.67	40±1.20	41±1.28	43±0.29	47±2.20
1	23±1.5	24±0.13	28±0.22	28±1.67	42±1.02	68±3.12	72±0.67	67±0.17	68±1.90	81±2.20	87±1.12	86±0.73	68±0.71	75±1.68	72±0.34	74±2.09
2	37±0.6	42±2.17	42±3.00	39±0.73	68±0.70	72±0.44	86±2.23	80±0.46	72±1.90	89±1.20	91±0.67	86±0.39	72±0.90	82±1.23	85±1.98	79±2.67
3	28±1.2	37±1.62	40±1.72	39±1.36	68±1.15	75±1.18	82±3.35	76±0.88	71±1.20	82±3.90	85±0.40	84±2.29	70±0.61	70±1.12	75±1.39	70±0.22
4	28±0.7	30±1.12	32±1.93	27±1.32	51±0.70	62±1.23	60±0.68	67±0.77	63±1.20	64±0.29	67±2.19	65±0.30	61±1.23	62±1.12	67±0.78	65±2.11

Data (X ± SE) (* % explant produce secondary embryos) pooled from three independent experiments each with three replicates

Table 8 Effect of maltose and trans-cinamic acid on germination and maturation of tea somatic embryo.

Culture treatment (weeks)	Trans-cinamic acid (mg/l) 0% Maltose (mg/l)				1.5 % Maltose (mg/l)				3% Maltose (mg/l)				6% Maltose (mg/l)			
	0	3	4	5	0	3	4	5	0	3	4	5	0	3	4	5
0	0.71^{w}	0.71^{w}	0.71^{w}	0.71^{w}	$1.83^{s\text{-}u}$	$1.81^{s\text{-}u}$	$1.9^{s\text{-}u}$	$2.05^{s\text{-}u}$	$1.86^{s\text{-}u}$	$2.29^{s\text{-}u}$	$1.8^{s\text{-}u}$	$2.62^{s\text{-}u}$	$1.81^{s\text{-}u}$	$1.81^{s\text{-}u}$	$2.49^{s\text{-}u}$	$2.01^{s\text{-}u}$
	(nil)	(nil)	(nil)	(nil)	(3.0)	(3.0)	(4.0)	(4.0)	(3.2)	(5.0)	(3.0)	(6.4)	(3.0)	(3.0)	(6.2)	(4.4)
3	0.71^{w}	$2.68^{s\text{-}u}$	4.0k-i	3.88l-p	$1.75^{s\text{-}u}$	$4.18^{l\text{-}n}$	$4.64^{g\text{-}i}$	$4.40^{j\text{-}i}$	$2.14^{s\text{-}u}$	$5.39^{d\text{-}h}$	$6.10^{b\text{-}d}$	$5.28^{d\text{-}l}$	$2.01^{s\text{-}u}$	$5.23^{j\text{-}i}$	$5.52^{c\text{-}d}$	$5.51^{c\text{-}d}$
	(nil)	(7.2)	(20.0)	(14.6)	(3.0)	(17.0)	(21.2)	(19.0)	(4.6)	(28.6)	(36.8)	(27.4)	(4.0)	(27.0)	(30.0)	(30.0)
4	0.71^{w}	$3.93^{l\text{-}o}$	$4.80^{f\text{-}i}$	$4.66^{g\text{-}i}$	$2.02^{s\text{-}u}$	$5.55^{c\text{-}d}$	$6.71^{b\text{-}c}$	$6.15^{b\text{-}d}$	$1.82^{s\text{-}u}$	7.78^{a}	8.43^{a}	7.87^{a}	$1.91^{s\text{-}u}$	2.06s-u	$4.80^{f\text{-}d}$	$4.78^{f\text{-}d}$
	(nil)	(15.0)	(23.8)	(21.2)	(3.6)	(30.4)	(44.6)	(37.4)	(3.0)	(60.0)	(70.6)	(61.4)	(3.4)	(22.6)	(22.4)	(17.8)
5	0.71w	$4.01^{k\text{-}n}$	$4.32^{l\text{-}m}$	$3.64^{l\text{-}p}$	$1.98^{s\text{-}u}$	$4.75^{f\text{-}d}$	$5.53^{c\text{-}d}$	$4.52^{h\text{-}i}$	$2.15^{s\text{-}u}$	$6.37^{b\text{-}c}$	6.81^{b}	$6.18^{b\text{-}d}$	$2.01^{s\text{-}u}$	$5.24^{e\text{-}i}$	$5.52^{c\text{-}d}$	$5.51^{c\text{-}d}$
	(nil)	(15.6)	(18.4)	(13.0)	(4.0)	(22.2)	(30.2)	(20.0)	(4.6)	(40.2)	(46.0)	(37.8)	(5.4)	(7.2)	(11.2)	(10.2)

Data ($X \pm SE$) pooled from three independent experiments, each with three replicates

Table 9 Effect of temperature, genotypes and time of seed collection on *in vivo* formation of somatic embryos in tea seed*.

Clones	Month of seed collection	Control		Incubation temperature					
				24°C		28°C		32°C	
		G	SE	G	SE	G	SE	G	SE
Kangra Jat	Sept.	-	-	-	-	-	-	-	-
	Oct.	45± 0.67	-	-	-	-	-	-	-
	Nov.	95±1.34	-	4.2±1.9	-	9.2±1.2	11.7±0.69	1.6±1.84	-
T-78	Sept.	45±1.74	-	-	-	3.2±0.67	5.0±1.6	-	-
	Oct.	97±0.23	-	1.7±0.18	-	-	16.5±0.84	-	-
	Nov.	27±0.27	-	-	-	-	-	-	-
UPASI-9	Sept.	98±1.05	-	-	-	11.2±1.03	23.3±0.59	-	-
	Oct.	47±0.55	-	-	-	-	-	-	-
	Nov.	-	-	-	-	-	-	-	-

*Data (X±SE) pooled from three consecutive independent years (1994-95, 1995-96 and 1996-97) experiments. G = Germination frequency, SE = Somatic embryo formation

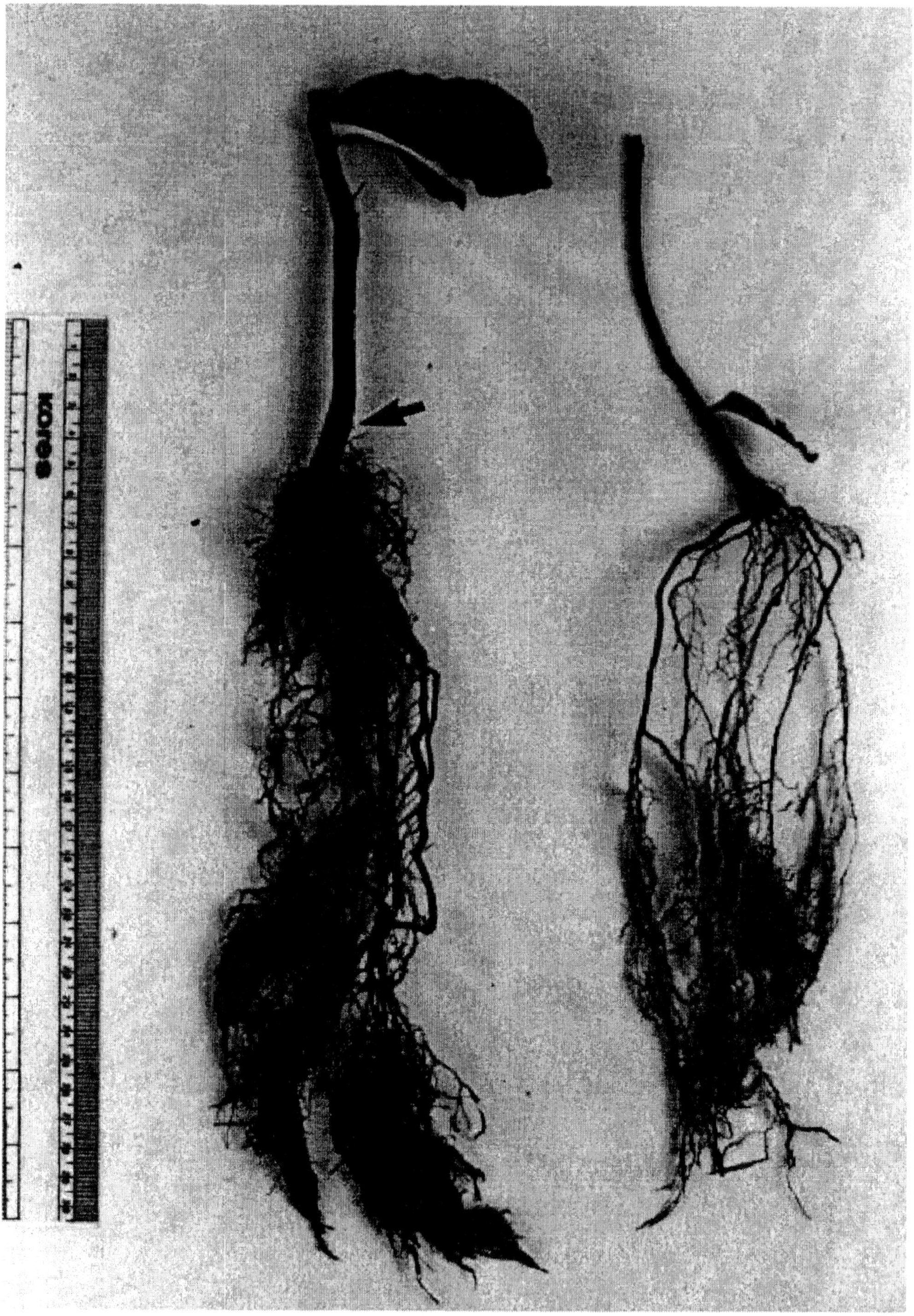

Fig.1. Roots of tea plants, cultivar "TTL-4" of 12 months old. Left: Micrografted plants (arrow indicate the mark of grafting), right: Ungrafted plants.

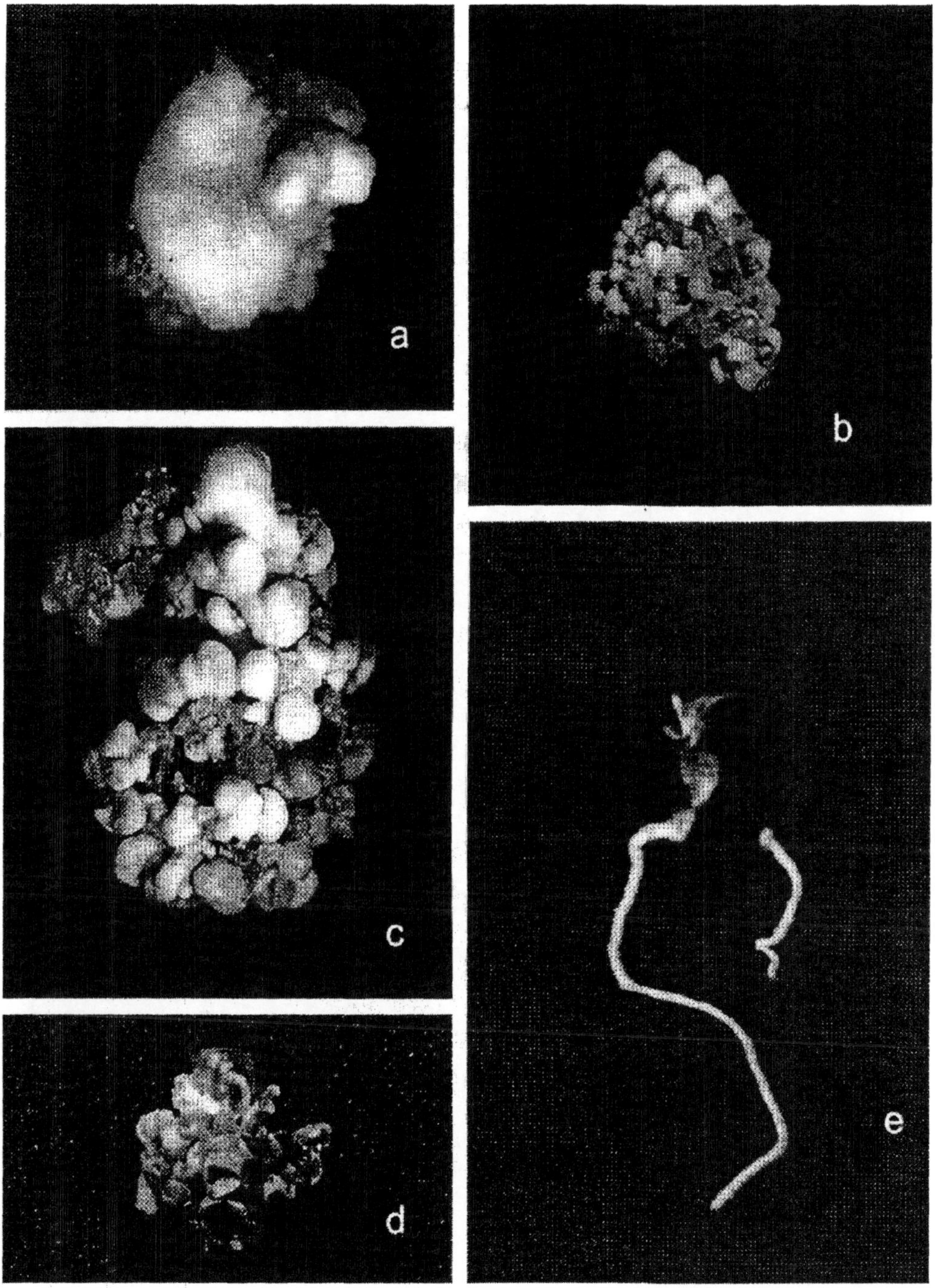

Fig.2. a. Induction of primary somatic embryos on de-embryonated cotyledon of tea. b. Synchronous globular secondary embryos on medium supplemented with potassium sulphate. c. Synchronous, secondary, heart-shaped embryos on medium supplemented with potassium sulphate, d. Synchronous, secondary, torpedo-shaped embryos on medium supplemented with potassium sulphate, e. Germination of somatic embryos.

24. MICROPROPAGATION OF LITCHI (*Litchi chinensis* Sonn.)

Neera Bhalla-Sarin*, U.S. Prasad#, A. S. Kantharajah and S. Mohan Jain*****

*School of Life Sciences, Jawaharlal Nehru University, New Delhi-110067, India
#Department of Botany, T.M. Bhagalpur University, Bhagalpur, India
** Centre for Horticulture and Plant Sciences, University of Western Sydney, Locked Bag 1797, Penrith South DC, NSW 1797, Australia.

*** FAO/IAEA Joint Division, International Atomic Energy Agency, Box 100, A-1400, Wagramer strasse 5, Vienna, Austria. E-mail: s.m.jain@iaea.org

CHAPTER CONTENTS

1. INTRODUCTION

Litchi (*Litchi chinensis* Sonn.) is one of the precious and economically important fruit crops of the world. The species belongs to the family Sapindaceae, which comprises about 2000 species of tropical and subtropical trees, shrubs, and vines that have been classified into 140 genera (Chapman, 1984). Litchi has been shown to possess variable diploid chromosome numbers where 2n = 28, 30, 32. The variation in chromosome number is thought to be because the modern species had more than one

S.M. Jain and K. Ishii (eds.), Micropropagation of Woody Trees and Fruits, 721–731.

wild progenitor. Litchi fruit contain high levels of ascorbic acid ranging from 40 to 80 mg/100 g of flesh, making it an excellent source of vitamin C (Cavaletto, 1980). The edible portion of fruit is a thick juicy aril having high moisture content (77 to 83 %), low fat (less than 1 %) and high sugar (14 to 23° Brix) and phosphorus contents. The fruit is an excellent thirst quencher and serves as a tonic for brain, heart and liver. In addition to medicinal uses, litchi fruit and their starchy seeds are used in brewing sweets and fragrant wine.

1.1. Botanical aspect

The Litchi tree is an evergreen, round-topped tree of about 10-13 m in height. It is a mycorrhizal tree. The foliage comprises 2-4 pairs of leaflets, which are 3-6 inches long, coriaceous, elliptic-oblong to lanceolate, shortly acute, glabrous and shining above and glauscescent beneath. Flowers are polygamous, regular, small and inconspicuous having greenish-white or yellowish colour. Flowers are born in profusion on terminal panicles of leafless or leafy branches, having small valvate sepals without petals. There are eight stamens with hairy filaments, which are incurved in bud but straight erect and far exerted later. The ovary is bilobed, compressed and silky, usually only one lobe develops into a fruit. The stigma is bilobed. According to their hermaphrodite nature, flowers are classified into three classes viz. type I, type II, type III. Type I and type III flowers are male while, type II flowers function as a fruit-bearing female. Most of the flowers are of type III and only 20 % of the flowers are of fruit producing type II females. Fruit grow in loose bunches of 2-20. Each fruit is oval or round in shape, about 3 to 4 cm in diameter. Ripe fruit possess a dry, brittle, and tubercule pericarp. Colour of the ripe fruit is rose-red to deep red, which is likened to a large strawberry in appearance. Colour changes to dull-brown as the fruit dries. Botanically the fruit is a dry nut but possession of a white translucent fleshy and juicy aril (edible part), developing from the funicle and surrounding the large seed makes it fleshy. The aril has firm texture with sub-acidic flavour. On drying, the aril shrinks away from the thin outer shell (the pericarp or rind) and remains as a rather tough structure around the seed. The flavour is that of a raisin that bear no resemblance to that of the fresh delicious fruit.

1.2. Origin and Distribution

Litchi originated in southern China where it has been cultivated since antiquity (Huang and Xu, 1983). Distribution of litchi into other tropical and subtropical countries occurred during the sixteenth and seventeenth centuries. Litchi reached Burma (now Myanmar) and eastern India by the end of the 17th century or shortly thereafter. Although, it may have reached India through Assam (Eastern India), there it is known to be of later occurrence. From regions of Burma and India, it spread to the rest of the tropical and sub-tropical world (Goto, 1960; Liang, 1981). Exotic litchi fruit have attracted attention worldwide, including Taiwan, Malaysia, Mauritius, Burma, West Indies, South Africa, Australia, Israel and U.S.A.

1.3. Climatic Requirement

Litchi is native to subtropical regions where cool, dry winters and warm, wet and humid summers prevail. Litchi can be grown on a variety of soil types having good drainage. Most favourable soil types are neutral to slightly alkaline with lime, alluvial silty loams of clays or laterite soils on the hills. Litchi plants require ample rainfall and high temperature for satisfactory growth. However, low temperatures severely affect young plants. Temperature affects flowering and anthesis. The optimum temperature for flower initiation has been observed to be 15 to 20 / 20 to 15 °C (day/night) while pollination is optimum at 19-22 °C (Menzel and Simpson, 1991). Hot, dry conditions result in poor fruit set.

1.4. Breeding

Wide genetic variability has been found in different cultivars of litchi grown in China and India. These cultivars have high potential for genetic improvement of litchi through hybridisation and selection. However, hybridisation of two parents by artificial pollination is not suitable for litchi due to their highly heterozygous nature. Selection within open pollinated accessions is common method for developing new cultivars with improved characters. The improvement of desirable fruit and tree characteristics of litchi were described by Galan Sauco (1989). Fruit characters including large fruit size, high proportion of edible aril portion, acceptable sugar/acid ratio, small or shrivelled seeds, bright colour, firmness and resistance to diseases are the important considerations for new cultivar development. Similarly, tree characters like high yield, regular bearing, resistance to water stress, wind, soil salinity, diseases and insects, and adaptation to different climatic conditions are highly desirable traits for the development of improved cultivars. Intergeneric hybridisation between litchi and longan has been demonstrated (McConchie *et al.*, 1994), using litchi as the female parent. The hybrid plants were similar to the female parents but with smaller leaves.

1.5. Production

The total production of litchi fruit is 350, 000 tonnes worldwide. China is the largest producer of litchi in the world, followed by India (Underhill *et al.*, 1997). The most important litchi-producing region is on the banks of Pearl River in South China. The total area of cultivation of litchi is 38,000 hectares with an annual production of 84,000 tonnes. In India, Bihar is the most important litchi growing state, which contributes 77 % of total litchi production in the country. Cultivation is spread over

22,503 hectares with an annual production of 27000 tonnes. Litchi is mostly harvested during May-June and yield is around 80-100 kg per tree per year in India.

1.6. Propagation

Litchi is highly heterozygous because it is open pollinated. Usually it is not propagated from seed because seedlings are genetically diverse and often characterised by a long juvenile period (5-12 years and in some cases even up to 25 years), irregular bearing and poor fruit quality (Loebel, 1976; Pandey and Sharma, 1989). Many cultivars have abortive, non-viable seeds while in some cultivars the seeds are short-lived, e.g. remain viable only for 3-4 days; and are extremely sensitive to changes in moisture content (Menzel, 1985; Ray and Sharma, 1985, 1987). The short-life span of litchi seed is presumably responsible for the slow spread of litchi from China to the rest of the world (Nijjar, 1981). Litchi seeds have justifiably been classified as recalcitrant (Hanson, 1984; Chin *et al.*, 1984) as they rapidly lose their germination capacity in the open environment (Kumari-Singh and Prasad, 1991; Xia *et al.*, 1992 a, b; Prasad *et al.*, 1996). Seeds readily germinate when sown soon after separation from whole fruit but delay in sowing for a day or two is detrimental to their viability. Seed viability was extended by keeping them in polyethylene bags (Prasad *et al.*, 1996). However, the causes of loss of viability under moisture stress are not understood. Therefore, seeds are planted for the selection and breeding purposes or for rootstocks. Development of new cultivars is a slow process because it may take up to 10 years or even more before the trees begin to bear fruit (Hamilton and Yee, 1970; Menzel, 1983).

Litchi can be propagated through various asexual means such as grafting, stem cutting, air layering or marcottage and budding (Menzel, 1985; Pandey and Sharma, 1989; Ram and Mujumdar, 1981). These methods are useful in raising "true-to-type" plants and desired cultivars are properly maintained and become easily available in large numbers. However, such propagation methods involve depletion of branches in great numbers from the mother trees (Ray and Sharma, 1987). Although, conventional vegetative propagation methods are slow and inefficient (Chapman, 1984), these methods are used under horticultural practices in many countries. Several researchers have used auxins, indole butyric acid (IBA) and α-naphthalene acetic acid (NAA) to promote root initiation and better root development. Genotype, physiological condition and wood type of the parent tree in addition to environmental factors are important considerations for successful development of a root system by stem cuttings (Pandey and Sharma, 1989).

Grafting and budding are not in vogue primarily because of the non-availability of vigorously growing healthy seedlings of litchi for use as root-stocks. Therefore, alternative methods for raising the chosen elite cultivars must be attempted. The alternative approach to vegetative propagation is *in vitro* culture and regeneration of litchi plants for mass-scale propagation of desirable genotypes.

2. *IN VITRO* PROPAGATION

Techniques for plant cell, tissue and organ culture have created many opportunities for plant development. These methods that allow rapid multiplication of true to parental type plants *in vitro*, popularly called 'micropropagation', have several advantages over conventional vegetative propagation of economically important crops. Many woody species are propagated commercially by axillary bud proliferation as it typically results in an average tenfold increase in shoot number per monthly culture passage. It is feasible to produce 10^6 propagules or plants starting from a single explant. Axillary buds are preformed meristems, which can be treated with hormones to break dormancy and produce shoot branches. The shoots are separated and rooted to produce plantlets. The shoots can also be used as propagules for further propagation. Although, several protocols have been reported for many economically important fruit crops, litchi has lagged behind due to its recalcitrant nature. Nevertheless, little success has been achieved by *in vitro* clonal propagation of litchi. The problems faced and the successes in attempting *in vitro* propagation of litchi are discussed.

2.1 Axillary bud development and shoot proliferation

Application of tissue culture for improvement of litchi has been reviewed previously (Kantharajah *et al.*, 1997). The review reported that although there have been many attempts to multiply litchi by *in vitro* cloning, knowledge of the process is scarce. Secretion of polyphenols by the tissue and heavy contamination of explants, hindered development of suitable and reliable methods for *in vitro* culture (Kantharajah *et al.*, 1997). Yu (1991) reported that exudation of polyphenolics and contamination depended on explant age, and explant treatment as well as growing conditions of the donor plant, respectively. Sterilization of litchi explants requires special attention. Neera Bhalla-Sarin (personnel communication) group developed a method, which involves wiping the explants using 5% (v/v) ethanol, followed by washing in running tap water for 2hrs for pre-surface sterilization. The explants are then soaked in bavistin (1% w/v) for 45-60 min followed by three rinses with sterile double distilled water. Next, the explants are treated with 0.1% (w/v) mercuric chloride for 5 min. followed by thorough washing with sterile water. Finally the explants are dipped in 0.8% (v/v) sodium hypochlorite solution twice for three minutes each and then washed thrice with sterile water. Bavistin (0.1%) included in the medium for culturing explants helps to maintain aseptic cultures. Other fungicides like thiabendazole (TBZ) or fenbendazole (FBZ) were not effective in controlling the endogenous fungus.

Among the antibiotics tested (gentamycin, tetracycline and streptomycin) gentamycin (25mg/l) was suitable for controlling microbial contamination in litchi explants (Neera Bhalla Sarin, unpublished results). An equally rigorous protocol is also followed to avoid the problem of phenolic exudation by explants in the medium. Antioxidants like ascorbic and citric acid at a concentration of 100ppm each along with polyvinyl pyrrolidone (PVP, 2%) were added to MS medium in order to control browning of the explant and medium. MS medium supplemented with activated charcoal (1%) was effective to a limited extent. Another method, which involved the use of liquid medium in which a filter paper bridge supporting the explant was submerged, helped in escaping the adverse effect of polyphenolic compounds.

Das et al. (1999) suggested two methods of shoot multiplication. The first involved shoot bud initiation from seeds directly germinated on filter-paper bridge submerged in liquid MS medium supplemented with 20 mgl^{-1} BA. In the second method (*in planta*) BA (100 micro gram on alternate days) was administrated on the axillary bud regions, of plants grown and maintained under sterile soil (vermiculite) conditions. The cytokinin treatment *in planta* consisted of placing sterile filter paper moistened with sterile distilled water over the node and adding different concentrations of BA. Both methods of multiple shoot induction were effective for the five genotypes of litchi that were tested. While the *in planta* method yielded about eight shoot buds from each treated node in eight weeks, the first method resulted in as many as 27.5±8.6 shoot buds from the cotyledonary node within four weeks. Chandra and Padaria (1999) cultured shoot buds of litchi cv. Seedless on MS medium supplemented with 0.2 mg BA + 0.1 mg IAA+ 0.5 mg GA_3/litre) and obtained shoot differentiation and growth.

2.2 Adventitious shoot bud development

Besides the reports on axillary bud development, some success has also been reported in litchi on adventitious shoot bud development. Yu et al. (1991) studied the effect of various factors on tissue culture of litchi. They used shoot segments (0.8 cm long) from cv. Liudong1 and reported a contamination rate of 100% in test material taken after 10 continuous rainy days and 20% after 15 continuous sunny days. Highest regeneration rate in the shoot segments was observed when cultured on half strength MS medium fortified with 1.0 mg BA, 0.2 mg IAA and 0.6 mg GA_3/litre. The regenerated segments were grafted on stocks cultured *in vitro* (seedling age 15-20 days). New shoots were produced after 13 days, which developed into complete plants with roots, stem and leaves after 20 days.

Kantharajah et al. (1992) succeeded in developing adventitious shoot buds *in vitro* from embryogenic shoots of litchi. They reported that pretreatment of embryos with

100 mg^{-1} BA in MS liquid medium for three hours was optimum for adventitious shoot formation. Subsequent transfer to MS semi solid medium containing thidizuran (TDZ 1.0 mg^{-1}) resulted in the maximum frequency of 15 shoots/explant in cv. Bengal.

2.3 Rooting

Adventitious and axillary shoots developed in cultures in a medium containing cytokinin generally lack roots. To obtain complete plants, regenerated shoots must be transferred to rooting medium, which is different from the shoot regeneration medium. In litchi, rooting of *in vitro* regenerated shoots has been generally carried out on auxin containing medium. Kantharajah *et al.* (1992) showed that 65 % of adventitious shoots formed roots on MS medium supplemented with 0.5 mgl^{-1} NAA and activated charcoal and successfully established rooted shoots in the green house. Pulse treatment with 25 mgl^{-1} NAA for 15 minutes has been used to induce roots on regenerated shoots (Das *et al.*, 1999). The treated shoots were elongated and rooted directly in vermiculite.

2.4 Somatic embryogenesis

Somatic embryogenesis is the process by which somatic cells develop through the stages of embryogeny to give whole plants without the fusion of gametes. The potential application of somatic embryogenesis lies in its power to achieve high rates of multiplication of plants. Unlike axillary or adventitious shoots, somatic embryos are complete propagules, carrying a pre-formed radicle and hence they bypass the stage of root induction. These and several other advantages have made somatic embryogenesis a suitable method for multiplication of plants. Somatic embryogenesis has also been attempted in litchi. Zhou et al. (1993) reported that MS medium supplemented with 2.0 mgl^{-1} 2,4-D + 0.2 mgl^{-1} BA + 0.1 mgl^{-1} NAA was ideal for embryogenic callus formation from immature embryos (20-, 30- and 50-day-old) of 4 varieties of litchi, although rapidly growing calli were also obtained on MS medium containing 4.0 or 6.0 mgl^{-1} 2,4-D + 0.1 mgl^{-1} BA + 0.1 mgl^{-1} NAA. Somatic embryo development and complete plantlets were obtained on MS medium with low concentration of NAA and IBA. Kuang et al. (1996) found that 20 to 30 day-old embryos cultured on MS medium containing 2,4-D produced embryogenic calli. Somatic embryos were promoted when embryogenic callus was cultured on medium with low concentrations of NAA and IBA. They also found different types of somatic embryos including seven abnormal types, which may have been due to chromosomal variability. In contrast, Liao and Ma (1998) reported that a five year-old litchi plant derived by somatic embryogenesis showed no difference in flowering, fruit set or fruit characteristics from the original plant. This report demonstrated that litchi plants could be propagated *in vitro* without genetic abnormalities.

Liao and Ma (1998) carried out a thorough investigation of somatic embryogenesis and plantlet regeneration in litchi cv. Yuherbau. Secondary embryos appeared in large numbers on the surface of immature primary somatic embryos in a culture medium containing 0.05 mg NAA, 0.05 mg 2ip (isopentenyl adenosine), 0.2 mg ABA / litre. The amount of sucrose in the medium was increased to 45 g/litre. The secondary embryos appeared either directly from cotyledons, hypocotyls and roots of somatic embryos or indirectly from embryo derived callus. The immature somatic embryos gave rise to more number of secondary embryos as compared to mature somatic embryos. On an average 50-70 secondary embryos could be produced from one somatic embryo after four weeks of culture. These secondary embryos, initiated from the epidermal and subepidermal cells of somatic embryos, developed into plantlets, which were successfully transplanted to pots and later into the orchard. Somatic embryogenesis has the potential to greatly speed up the propagation of litchi, and to provide significant economic advantage. As the conventional methods are long-term and laborious, somatic embryogenesis offers a method for propagating large numbers of uniform plants of litchi in less time and possibly at less cost than from marcottage and other vegetative propagation methods used. After trails to identify which embryogenic cell lines yield superior trees, these lines can be used to generate numerous plants with desirable traits, thus providing large genetic gain.

3. APPLICATIONS

3.1. Artificial seeds

The time is emerging when nurseries will have the option to use somatic embryos as "artificial seeds" (Redenbaugh, 1990). These are prepared by encapsulating or desiccating the tissue culture derived embryos in a protective coating such as that obtained from sodium alginate (a common food thickener extracted from brown algae) because of its low toxicity to the embryo. Automation of encapsulation can further speed up the process. The artificial seeds would be directly sown in the field, thus eliminating the steps of hardening and transplantation. Artificial seed technology if perfected could prove very useful for the propagation of "true to the parent" type plants of litchi.

3.2. Protoplast culture

Protoplasts are cells without cell walls. Because the cell walls are absent, protoplasts of two species or genera can be fused to produce somatic hybrids. Kantharajah (1998) reported that callus tissues obtained from embryonic shoots of litchi yielded viable protoplasts, which could subsequently undergo cell wall formation and cell division. However, regenerants were not obtained from microcalli derived from protoplasts. Yu *et al.* (1996) found that protoplasts isolated from embryogenic cell suspensions of litchi with a solution containing CPW salts, 11 % mannitol, 0.8 % cellulose R-10, and 0.4 % macerozyme R-10, produced callus and subsequently somatic embryos when transferred to a solid medium. However, the plantlets were obtained by culturing somatic embryos on different media for 2-5 months. Yu *et al.* (2000) reported that protoplasts isolated from embryogenic suspension cultures and maintained by alternative culture in liquid and on solid media containing silver thiosulphate, yielded microcalli and subsequently somatic embryos and plants. Lai *et al.* (2001) demonstrated that protoplasts isolated from embryogenic callus of litchi and longan cultures can be fused by electrofusion with up to 20 % frequency and microcalli could be formed from fusion products. Thus a lot of potential exists for litchi improvement and micropropagation using the technique of protoplast culture.

3.3. Cryopreservation

The work of Nag and Street (1973) stimulated an international effort to grow numerous plants *in vitro* and preserve them by cryopreservation, which involves storage at about -320°F (-196°C). Embryogenic cultures as well as other propagules can be maintained in cryogenic storage while their field trails are being carried out. Choudhury et al. (1998) showed that litchi axes with a moisture content of 11.5-18.3% from seeds harvested 60-80 days after anthesis survived in liquid nitrogen at -196°C. This technique needs to be perfected for litchi so that clonal material obtained can be preserved, transported, exported and grown as and when required.

4. CONCLUSIONS AND PERSPECTIVES

Conventional methods used for litchi propagation are extremely slow and inefficient. *In vitro* techniques offer a great a potential for micropropagation of litchi. A beginning has already been made in this direction as is obvious from the handful of successful reports. However, to commercialise litchi micropropagation technique more intensive efforts are needed. Some elite varieties of litchi in which the seeds of aborted are termed 'chicken tongues'. Consumers prefer them because they have 20% more flesh as compared to the normal varieties. The desirable characters of micropropagated plants include seedlessness, greater flesh and increased shelf life of the fruit. Therefore, the future need is to identify the elite litchi trees, which are heavy bearers as well as possess all the advantages of the fruit as listed and clonally multiply these using the *in vitro* techniques. Molecular markers could be used for identification of such elite trees in future. More work is needed to induce somatic embryogenesis, which can be used for clonal propagation and induced mutations. The efforts should be made to increase somatic embryo germination rates for making this technology feasible to commercial application. Somatic embryogenic cultures could be cryopreserved for long-term storage and also for the prevention of somaclonal variation.

Refinement of techniques of micropropagation would prove to be a boon not only for the tissue culturists but also for the horticulture industry. This is one of the horticultural crops, which needs more inputs in terms of research and experimentation to achieve the desired goals.

5.0. REFERENCE

Cavaletto C. G., 1980. Lychee. In: S. Nagy and P. E. Shaw (Editors), Tropical and Subtropical Fruits. AVI Publishing Inc., Westport, CT.

Chandra R. and Padaria J. C., 1999. Litchi shoot bud culture for micropropagation. J. Appl. Hort. Lucknow, 1(1): 38-40.

Chapman, K. R., 1984. Sapindaceae. In: P. E. Page (Editor), Tropical Tree Fruits for Australia. Queensland Department of Primary Industries, Queensland, Australia, pp. 179-197.

Chaudhury R., Malik S. K., Marzalina M., Khoo K. C., Jayanthi N., Tsan F. Y. and Krishnapillay B., 1998. Cryopreservation and in vitro technology for conservation of recalcitrant seeds of some tree species. IUFRO Seed Symposium.

Chin H. F., Hor Y. L. and Lassim M. B., 1984. Identification of recalcitrant seeds. Seed Sci. & Tech., 12: 429-436.

Das D. K., Prakash N. S., Neera Bhalla S., Bhalla Sarin N. and Kishor P. B. K., 1999a. Multiple-shoot induction in *Litchi chinensis* - a tropical tree species. Proc. of Plant tissue culture and biotechnology: emerging trends, Hyderabad, India, 29-31 january, 1997: 166-170.

Das D. K., Shiva-Prakash N. and Bhalla Sarin N., 1999b. Multiple shoot induction and plant regeneration in litchi (*Litchi chinensis* Sonn.). Plant Cell Rep., 18: 691-695.

Fu, L. F., and Tang, T. Y. 1983. Plantlets formed from litchi pollen. Plant Physiol. Commun. Zhiwu Shenglixue Tongxun No. 2, 28-29.

Galan Sauco V., 1989. Litchi Cultivation (in Spanish). FAO Plant Production and Protection Paper No.83. FAO, Rome, Italy, pp. 136.

Goto Y. B., 1960. Lychee and its processing. Pacific Region Food Conf., 1: 15-23.

Hamilton R. A. and Yee W., 1970. Lychee cultivars in Hawaii. Proc. Tropical Region American Soc. Hort. Sci., 14: 7-12.

Hanson J., 1984. The storage of seeds of recalcitrant tree fruits. In: G. Allen (Editor), Crop Genet. resource conserv. evaluation. Unwin, London, pp. 53-62.

Huang, H., and Xu, J. 1983. The development patterns of fruit tissues and their correlative relationships in *Litchi chinensis* Sonn. Scient. Hortic. 19, 335-342.

Kantharajah A. S., Ebert G. and Ludders P., 1997. In vitro techniques for crop improvement in lychee - a review. Erwerbsobstbau, 39(4): 116-118.

Kantharajah A. S., McConchie C. A. and Dodd W. A., 1992. In vitro embryo culture and induction of multiple shoots in lychee (*Litchi chinensis* Sonn.). Ann. Bot., 70(2): 153-156.

Kantharajah A. S. and Dodd W. A., 1998. Developments in tissue culture for crop improvement of lychee (*Litchi chinensis* Sonn.). Procc.of IX International Congress on Plant Tissue and Cell Culture. June 14-19, 1998, Jerusalem, Israel, pp. 86.

Kuang Z., Zhou L., Ma X., Chen J., Cai J., Kuang Z. S., Zhou L. N., Ma X. J., Chen J. Q. and Cai J., 1996. Study on the types of embryoid in tissue culture of *Litchi chinensis* Sonn. J. Fruit Sci., 13(Supplement): 25-28.

Kumari-Singh A. and Prasad U. S., 1991. Dehydration pattern and viability loss in seeds of two cultivars of litchi (*Litchi chinensis* Sonn.). Seed Res., 19: 41-43.

Lai Z. X. and Chen Z. G., 2001. Electrofusion of protoplasts between longan and litchi. Journal of Fujian Agricultural University, 30(3): 347-352.

Liang J., 1981. The litchi, origin, utillization and development of its culture. J. Agric. Trinidad Bot. Appl., 28: 259-270.

Liao Y. W. and Ma S. S., 1998. Adventitious embryogenesis of *Litchi chinensis*. J. Chinese Soc. Hortic. Sci., 44(1): 29-40.

Loebel R., 1976. The litchi. Trop. Fruit Bull. of N. S. W. Deptt. of Agric., 6: 1-12.

McConchie C. A., Vithanage V. and Batten D. J., 1994. Intergeneric hybridisation between litchi (*Litchi chinensis* Sonn.) and longan (*Dimocarpus longan* Lour.). Annals of Botany, 74(2): 111-118.

Menzel C. M., 1983. The control of floral initiation in lychee: a review. Scien. Hort., 21: 201-215.

Menzel C. M., 1985. Propagation of lychee: a review. Scien. Hort., 25: 31-48.

Menzel C. M. and Simpson D. R., 1991. Effects of temperature and leaf water stress on panicle and flower development of litchi (*Litchi chinensis* Sonn.). J. Hortic. Sci., 66: 335-344.

Nag, K. K. and Street H. E., 1973. Carrot embryogenesis from frozen cultured cells. Nature, 245: 270-272.

Nijjar G. R., 1981. Litchi cultivation. Punjab Agric. Univ., Ludhiana, India, 1-3 pp.

Pandey R. M. and Sharma H. C., 1989. The litchi. ICAR, New Delhi, India, 1-79 pp.

Prasad J. S., Kumar R., Mishra M., Singh A. K. and Prasad U. S., 1996. Characterstics of litchi seed germination. Hort. Sci., 31: 1187-1189.

Ram M. and Majundar P. K., 1981. A note on propagation of litchi (*Litchi chinensis* Sonn.) by stolling. Sci. Cult., 47: 332-334.

Ray P. K. and Sharma S. B., 1985. Viability of *Litchi chinensis* seed when stored in air and water. J. Agric. Sci. Camb., 104: 247-248.

Ray P. K. and Sharma S. B., 1987. Growth, maturity, germination and storage of litchi seeds. Scient. Hortic., 33(3-4): 213-221.

Redenbaugh K., 1990. Application of artificial seed to tropical crops. Hort. Sci., 25

Underhill S. J. R., Coates L. M. and Saks Y., 1997. Litchi. In: S. K. Mitra (Editor), Post Harvest Physiology and Storage of Tropical fruits. CAB International, Wallingford, UK, pp. 191-208.

Xia Q. H., Chen R. J. and Fu J. R., 1992a. Effects of desication temprature and other factors on the germination of lychee (*Litchi chinensis* Sonn.) and longan (*Euphoria Longon* Setud.) seeds. Seed Sci. & Tech., 20: 119-127.

Xia Q. H., Chen R. J. and Fu J. R., 1992b. Moist storage of lychee (*Litchi chinensis* Sonn.) and Longan (*Euphoria longon* Setud.) seeds. Seed Sci. & Tech., 20: 269-279.

Yu Y. B., 1991. Study on some factors in tissue culture of lychee (*Litchi chinensis*). Fujian Agric. Sci. Tech., No. 5: 17-18.

Yu, C., Chen, Z., Lu, L., Lin, J., Yu, C. H., Chen, Z. G., Lu, L. X., and Lin, J. W. 2000. Somatic embryogenesis and plant regeneration from litchi protoplasts isolated from embryogenic suspensions. Plant Cell Tissue Org. Cult., 61, 51-58.

Yu, C., Chen, Z., Lu, L., Yu, C. H., Chen, Z. G., and Lu, L. X. 1996. Regeneration of plantlets from protoplasts culture of litchi. J. Fujian Agric. Univ., 25, 386.

Zhou L. N., Kuang Z. S., Ma X. J., Chen J. Q., Cao J., Tang X. J. and Ling X., 1993. Preliminary studies on culture of immature embryos and embryogenesis of somatic cells of litchi (*Litchi chinensis* Sonn.). Guangdong Agric. Sci., No. 5: 14-15.

SECTION D

25. MICROPROPAGATION OF WOODY PLANTS USING BIOREACTOR

Paek Kee Yoeup and Debasis Chakrabarty
Research Center for the Development of Advanced Horticultural Technology, Chungbuk National University, Cheongju 361-763, South Korea

Chapter Contents

S.M. Jain and K. Ishii (eds.), Micropropagation of Woody Trees and Fruits, 735–756.

1. INTRODUCTION

The use of plant tissue culture techniques has revolutionized the plant propagation and now plants are commercially micropropagated. In recent years, many micropropagation companies world over have emerged. However, few companies are generating good profits. As a result micropropagation industry has not grown appreciably. High skilled labour cost in developed countries is the limiting factor for the development of this industry. Commercial laboratories so far, have been using the procedures developed at the research level in which skilled operators manipulate explants and sub-culture of shoots in laminer hood. The methods limit the total number of propagules a person can handle. This restricts its application to plants in which a high unit cost of production can be economically justified (Christie et al., 1995; Cervelli and Senaratna, 1995). In addition, current method of tissue culture is difficult to automate. Many significant biological and cultural problems must be solved before the technique can be amenable to automation.

Bioreactor technology is quite sophisticated for microbial cell culture and plant cell cultures for secondary metabolite production (Lee, 1997; Yu, 2000; Gao et al., 2000). Bioreactors with computer control systems offer theoretically various advantages due to possibilities of automation, saving labour and reduction of production costs (Preil, 1991, Sharma, 1992, Christie et al., 1995, Honda et al., 2001; Paek et al., 2001) Since, the quality and performance of the end product, micropropagated plants, are likely to be affected by bioreactor design and conditions (Takayama and Akita, 1994; Sajc et al., 2000). Nonetheless, modified bioreactors would reduce manual labour and provide optimum growth conditions by controlling chemical and physical environments.

Woody plants may be defined as trees and shrubs whose stems live for a number of years and increase in diameter each year by addition of woody tissue (Ray et al., 1983). Research with micropropagation of woody plants has proceeded at a tremendous pace, and even though they are considered as a difficult group to work with (Dunstan et al., 1995). The application of bioreactors in woody plant propagation is still at infancy stage. Recent progress is restricted to only a small number of plant species. It demonstrates the feasibility of this technology and for scientific progress. We believe that the investment of research efforts into these group of plants certainly will benefit both the investigator and the broader scientific community; open up a bright future for the modernization of agriculture. Thus the present aim of this chapter is to identify the problems related to large-scale woody plant micropropagation and to summarize the most recent developments in bioreactor design.

2. BIOREACTOR DESIGN

Usually micropropagation can be achieved in three different ways: 1) direct organogenesis via shoot tips or axillary buds, 2) by indirect organogenesis through callus, and 3) by somatic embryogenesis. Organogenesis via shoot tips or axillary buds is considered to be the most widely used commercial methods of regeneration. However, the procedures are labour intensive and more specialized due to many steps with manual manipulations involved and the low plant multiplication rates. Gradual acclimatization of plants to the greenhouse and then to the field are also needed. The possibility for micropropagation through somatic embryogenesis provides a valuable alternative to propagation system via organogenesis (Denchev et al., 1992). This is amenable to a higher degree of automation and holds much promise for the mass propagation of plants at low cost because very large number of somatic embryos can be produced in a short period of time in a limited volume of medium (Preil, 1991; Paek et al., 2001; Ibaraki and Kurata, 2001). This should help towards making such a system economically viable for many woody species.

The improvement of knowledge of the biological, physical and engineering components of bioreactors system for the growth of plant cells made it possible to grow plant cells in bioreactors for the growth of regenerable cells and tissues in liquid media in bioreactors. The use of bioreactor for plant propagation was first reported in 1981 for Begonia propagation (Takayama and Misawa, 1981). Since then it has proved applicable to many plant species including shoots, bulbs, microtubers, corms and somatic embryos (Son et al., 1999; Kim, 1999; Park et al., 2000 Lian, 2001; Paek et al., 2001).

Bioreactors are vessel, filled with liquid medium, for culturing the shoots and embryos in completely submerged conditions. The culture conditions and the characteristics of culture in bioreactor are somewhat different from agar culture. The use of liquid media in a bioreactor has advantages as : (i) large number of plantlets can be easily produced and also easy to scale up; (ii) handling of cultures such as inoculation or harvest is easy, saving labour and time; and (iii) forced aeration stimulates the growth rate to produce biomass (Sharma, 1992; Heyerdahl et al., 1995; Paek et al 2001).

The construction of bioreactor vessel designed for shoot and embryo cultures requires optimal mixing of oxygen, nutrients and culture without severe shear stress. Various types of bioreactors have been developed and have potential for use in plant shoot and embryo cultures (Takayama and Akita, 1994; Heyerdahl et al., 1995; Sajc et al., 2000; Honda et al., 2001; Paek et al., 2001). They are fundamentally classified based on agitation methods and vessel construction into two categories: 1) mechanically agitated bioreactors (aeration agitation bioreactor, rotating drum bioreactor, spin filter bioreactor) 2) pneumatically agitated and non-agitated bioreactors (simple aeration bioreactor, bubble column bioreactor, air-lift bioreactor, balloon type bubble bioreactor, overlay aeration bioreactors).

2.1. Mechanically Agitated Bioreactors

2.1.1. Aeration-agitation Bioreactors

These are often referred as stirred tank bioreactors (STR)s and use impellers such as turbines, screws, paddles and helical ribbons (Fig. 1). Although STRs are valuable for suspension cultures, they normally generate high shear forces. However, recent advances in STRs seem to have overcome this disadvantage. The helical ribbon impeller was efficient for high density suspension culture of plants and for somatic embryo production (Archambault et al., 1994). In studies using bioreactors for the production of somatic embryos, STRs were often used, but little research has been carried out to compare the different types of bioreactors.

2.1.2. Rotating Drum Bioreactor

This consists of a drum-shaped container mounted on rollers for support and rotation. The drum is rotated at 2-6 rpm to minimize the shear stress. A reactor vessel having no baffle plate can be used for plantlet formation through shoot and embryo culture (Takayama and Akita, 1994).

2.1.3 Spin Filter Bioreactor

In this bioreactor, the rotating filter mixes the suspension cultures and simultaneously removes spent medium and adds fresh medium. The spinning filter agitates the medium without generating the shear observed with blade stirrers. In order to maintain continuous callus cell proliferation and embryo development on a large-scale, two spin filter bioreactors were configured in series. A more efficient strategy would be operation of the first stage bioreactor (cell proliferation) as a continuous culture. Continuous culture provides more homogeneous and constant culture conditions for the cells and maintains an actively growing cell population available for embryo development at all times. The second stage bioreactor results in the periodic production of embryos. The embryo development could be controlled by way of nutrient requirements and air components. The production of plants via somatic embryogenesis has been recognized as the most suitable for mass propagation using spin filter bioreactor (Honda et al., 2001).

2.2. Pneumatically Agitated and Non-agitated Bioreactors

Air-driven bioreactors are simple in design, and are equipped with a sparger at the bottom of the reactor vessel, which agitates the medium and also supplies oxygen (simple aeration bioreactor, bubble column bioreactor). In some cases culture vessels are equipped with draft tubes (air-lift bioreactor).

2.2.1. Air-lift

They are similar to STRs (without an impeller). The air-lift bioreactor overcomes two disadvantages of the stirred tank bioreactor, the high energy requirement for agitation and high shear force, by creating a rapid, low shear movement. Efficient mixing in the liquid phase is done with air bubbles, using internal or external recirculation loops. In order to sparge air into bioreactor, compressed air is generally passed through a filter with small pores. Relatively low shear rate, due to low relative velocity between the liquid and bubbles in cocurrent flow is particularly desirable for shear sensitive cell cultures. Marvituna and Buyukalaca (1996) compared the performance of a STR and an air-lift bioreactor and concluded that the magnetic STR was better for *Capsicum annuum* somatic embryo production. In alfalfa, an air-lift bioreactor also gave slightly higher yields than a flask or a STR with propeller type of impeller (Stuart et al., 1987).

2.2.2. Simple Aeration Bioreactor and Bubble Column Bioreactor

Like an air-lift bioreactor, in a bubble column the bubble creates less shear stress. The major difference between the air-lift and bubble column bioreactor is the provision of a circulation system and hydrodynamic behavior. It is useful for plant organ cultures especially for propagation of various species through tissue cultures of shoots, bulbs, corms and tubers (Takayama 1991). Moreover, the division of a bubble column into segments, and installation of multiple spargers increases the mass transfer (Buitelaar et al., 1991).

The disadvantages of airlift and bubble column bioreactors are: a) foaming induced by large volumes of air, b) a tendency of cells to be thrown out of the solution by bubbles or air, and c) cell growth on the wall of the vessel (in the foam). The quantity of cells contained within this region represents a significant quantity of total biomass. The phenomenon of foaming and wall growth is due to diameters of the vessel and the top of the section are the same. This problem has been solved by designing improved models of air-lift and/or bubble column bioreactors that have larger top section diameter and/or balloon type bioreactor (Paek et al., 2001).

2.2.3. Balloon Type Bubble Bioreactor (BTBB)

The lay out of BTBB is shown in Figure 2. This consists of a balloon shaped container with a cap on the top section. In the bottom of the reactor a 'Y' or a 'T'-shaped tube was attached as a inoculum and harvest ports. By using a concentric tube for cell lifting at the air flow part of the vessel bottom, foaming was drastically reduced. In addition, the cap on the upper portion of the bioreactor provided space to install extra ports for the measurements of pH and dissolved oxygen. These bioreactors are reliable for micropropagation including somatic embryo production (Son et al., 1999; Kim, 1999; Lian, 2001; Paek et al., 2001).

2.2.4. Overlay Aeration Bioreactor

Aeration is performed by blowing the air against the surface of the medium and sometimes combined with gentle agitation. But the oxygen transfer coefficient K_La (h^{-1}) without agitation is quite low (less than 1). The use of this type of bioreactor has not been successful with plant tissue cultures (Ishibashi et al., 1987).

2.2.5. Automated Temporary Immersion Bioreactor

Temporary immersion system, commercially designated as 'RITA', has been described by Teisson and Alvard (1995). The bioreactor system consists of two containers; one for growing plants and a storage tank for liquid medium. The two containers were connected by silicone and glass tubes. Air pressure from an air compressor pushed the medium from one container to the other to immerse the plants completely. The air flow was reversed to withdraw the medium from the culture container. In each case, the air flow was sterilized by passing through 0.2 mm hydrophobic filters. Electronic timers controlled the frequency and length of the immersion period. Three way solenoid valves provided on/off operation. This system has been successfully used with several plant species including pear (Damiano et al., 2000), pineapple (Escalona et al., 1999) and *coffea arabica* (Etienne et al., 1997a).

3. APPLICATION OF BIOREACTORS FOR MASS PROPAGATION OF WOODY PLANTS

3.1. Micropropagation Using Shoot Tip or Axillary Bud Proliferation

Automation of organogenesis in bioreactors is a possible way to reduce costs of micropropagation. However, very few reports are available on woody plant micropropagation via shoot tip or axillary bud proliferation. Bioreactors containing liquid media are used for large-scale growth of various tissues. However, liquid medium environment is detrimental to many of woody plant species. Hyperhydricity frequently occurs when tissues are grown in the liquid media (Christie et al., 1995). Submersion of some tissues readily induces hyperhydricity in some cases. To avoid the problem of hyperhydricity in liquid media several different approaches have been made. Firstly, growth retardants and osmotic agents (Ziv and Ariel, 1991; Escalona et al., 1999) have been included in the medium. Secondly, environmental control systems have been used to alter the physical environment around the tissues for the production of less abnormal plantlets e.g., more aeration or bubbles in the liquid and thirdly, new designs of bioreactors and automated liquid culture systems have been redesigned especially for plantlet growth (Walker, 1995).

3.1.1 Temporary immersion system

Teisson and Alvard (1995) described a temporary immersion (TI) system for plant propagation called 'RITA' which has been successfully used for micropropagation of several plant species including some woody plants such as pear (Damiano et al., 2000) and pineapple (Escalona et al., 1999), allowing significant increase in multiplication rate and automation. The advantages of this technique are: (a) reduction of hyperhydricity in tissues, possibly due to the atmosphere in the culture bottle that is renewed continuously, (b) improve economics, as minimal operator handling is required, (c) reduction of the contamination levels. The TI culture combines the advantages of solid and liquid culture. Solid culture allows aeration but do not provide full contact with nutrient media. Liquid culture permits an efficient nutrient uptake. Escalona et al. (1999) tested the effectiveness of TI and compared it with micropropagation using solid and liquid culture in pineapple. They observed that multiplication rate increased by 300 and 400%, over that obtained from liquid medium and conventional solid medium, respectively. This protocol reduce the production cost by 20% per pineapple plant as compared to the conventional method. Similarly, Damiano et al., (2000) established a TI system for *in vitro* culture of wild pear (*Pyrus communis* var. Pyraster L). The immersion period length was very critical. When the period of contact of the explant with the medium was too short, the multiplication rate was poor. The explant showed necrosis and hyperhydricity, especially when explant was exposed to the medium for a long time. Attempts to apply this technique to more automated systems such as bioreactors are also now in progress.

3.1.2. Ebb and Flood system

Recently we developed a novel type ebb & flood bioreactor system for mass propagation of apple root stock M9 EMLA (unpublished). The lay out of the ebb & flood bioreactor system is shown in Fig 3. The principle equipment in an ebb & flood bioreactor is the same as those of BTBB. However in order to avoid the complete submersion of explants in the liquid medium, a supporting net was used to hold the plant material. In this system, medium is pumped from a storage tank into culture vessels. A series of channels help to supply nutrient solution evenly to the plant materials, resulting in uniform growth. The medium remains in the vessel for a few minutes, after that it is drained back to the storage tank for reuse. The drainage process is controlled by a solinoid valve, at 4-8 hour interval depending on plant species and explants.

Nodal cuttings (1-1.5 cm) from *in vitro* grown plantlets of apple rootstock M9 EMLA was used as the source for bioreactor culture. Two types of BTBB (immersion and periodic immersion liquid culture using ebb and flood) were used in this experiment. 40 nodal explants were transferred to 5 litre BTBB with MS liquid medium supplemented with 30 g L^{-1} sucrose, 0.3 mg.L^{-1} IBA and 1 mg. L^{-1} BA. The pH of the medium was adjusted to 5.8 before autoclaving at 121°C and 1.2 kg cm $^{-2}$ pressure for 30 min (2 litre medium per bioreactor). In the immersion type bioreactor, nodal cuttings were touching the liquid media during the whole period where as the ebb and flood was programmed to immerse the nodal cuttings in medium for 15 min every 8 h. The volume of input air for immersion culture was adjusted to 0.1 vvm (volume/culture volume, min). Bioreactors were maintained at 25°C in 16 h photoperiod (80 µmol m^{-2} s^{-1}). The multiplication rate, fresh weight and hyperhydricity were determined after 40 days of culture. Although the mutiplication rate was highest in immersion culture, a large number of hyperhydric plantlets (29.83%) were produced. In case of ebb & flood system, hyperhydricity was reduced (17.11%) as compared to immersion system but still it was a problem. So for complete elemination of hyperhydricity, we supplied compressed air inside the bioreactor chamber to reduce the humidity (Figure 4a). This approach significantly reduced the hyperhydricity (8.24%) during bioreactor culture of apple plantlets (Unpublished). Plantlets regenerated from bioreactor culture were successfully transferred to hydrophonic culture for *ex vitro* rooting and acclimatization (Figure 4b and c). Similarly an automated bioreactor functioning on the same principle was designed for large-scale grape root stock 5BB propagation (Shim 2002).

Although micropropagation of several woody species have been reported using conventional solid culture methods, so far only few can successfully be propagated in bioreactors. This is mainly due to lack of systematic, factorial experiments in bioreactors, where many of the interactions between biological and physical parameters can be revealed. Both scientists and manufacturers of bioreactors demand high product quality. This requires in depth knowledge about the process, reliable measurement techniques and good control tools. In bioreactor culture, research concerning improvement of the physical and chemical environments, such as increased number of air exchanges, increase PPF and CO2, are necessary for practical use of this technique.

3.2. Somatic Embryo Production in Bioreactor

Somatic embryogenesis is a distinctive process, describing the production of bipolar structures normally capable of germination without separate shoot and root induction phase. Somatic embryogenesis also offers a potential system for large-scale plant propagation in automated bioreactor. Conventional micropropagation requires intensive labour and quite often, it limits commercial application. Automation can reduce the labour requirements, which is linked to scale up of cultures. Somatic embryos could be easier to handle since they are relatively small and uniform in size. They do not require the cutting of plantlets into segments and transferring of segments onto new media during proliferation. In addition, somatic embryos have a potential for long-term storage through cryopreservation or desiccation, which facilitates flexibility in scheduling of production and transportation and, therefore, fits large-scale production (Ibaraki and Kurata, 2001).

Biologically, as well as technically, embryogenesis is a very complex process. The formation of embryos from somatic cells was first described in carrot by Steward et al.,(1958) and since then somatic embryogenesis has been reported in scores of species (Jain et al., 1995,1999,2000). Somatic embryogenesis with woody plants dates back to the 1960s (Konar and Oberoi, 1965) and to date somatic embryogenesis has been reported for approximately 150 woody species and related hybrids (Dunstan et al., 1995; Jain et al., 1995,1999,2000). Although there have been some reports on somatic embryo production in bioreactors (Table 1), it does not seem to be a well-established technique. The problems of somatic embryo production in a bioreactor are inconsistent results related to bioreactor types and optimization of culture parameters, including dissolved oxygen concentration, aeration rate and agitation rate. In some cases, embryogenic tissues have a lower rate of final plantlet development after bioreactor culture when compared with solid medium culture. It should also be noted that a successful bioreactor system for one species might not be applicable to other species.

Embryogenic tissue can be grown for the multiplication of somatic embryos in a bioreactor. However, maturation of embryos in bioreactors is limited to a few model systems such as Coffee (Etienne et al., 1997a), Sandelwood (Das et al., 1999) and some conifer species (Ingram and Mavituna, 2000). Development beyond the proembryo stage has not been achieved in liquid medium for forestry species, and embryos must be matured on solid medium. Scaling up the multiplication of somatic embryos has been more successful in flasks in some cases, and avoids the potential catastrophe of losing millions of embryos through contamination.

3.2.1. Coffee

A somatic embryogenesis procedure using a temporary immersion bioreactor (The RITA bioreactor, CIRAD, Montpellier, France) was recently developed for *coffea arabica,* enabling a mass and virtually synchronous production of germinated somatic embryos, without the need for selection before acclimatization (Etienne et al., 1997a). Conditions for showing them directly on to horticultural soil *ex vitro* conditions have also been developed (Etienne et al., 1999). This methodology is currently being applied on a large-scale in Central America. However, despite an increased synchronization in the development of the embryos, they still show morphological heterogeneity in the bioreactor. Also heterogeneity is seen in plant development in the nursery. They also showed that the embryo development stage (torpedo, germinated or plantlet) directly affected the efficiency of both acclimatization and further conversion into plantlets. Recently Etienne et al., (2002) used 1 litre bioreactor to obtain approximately 800 *Coffea arabica* somatic embryos, 86% of which reached the 'germinated' stage but with morphological heterogeneity. They concluded that the heterogeneity of somatic embryos in a bioreactor affected both the plant conversion efficiency in soil and the plant growth in nursery, where it mainly resulted in retarded growth, primarily in plantlets derived from the somatic embryos with small cotyledons.

3.2.2. Sandalwood

Bapat et al., (1990) reported the successful production of sandalwood somatic embryos using a stirred tank bioreactor. Conversion of non-embryogenic cells to embryogenic cells was achieved in a 7-litre bioreactor. Pro-embryo cells were grown in a 1-litre bioreactor for maturation. A small magnetic Teflon bar was used to stirr the medium uniformly in the vessel with minimum damage to the somatic embryos. Viable plantlets were regenerated from mature embryos upon transfer of the embryos from bioreactor to the solid medium. Recently Das et al., (1999) reported a mass propagation method for *Santalum album* L. through *in vitro* cultivation in liquid media in a 3.5 litre air-lift bioreactor. Embryogenesis, maturation and germination were performed in the same bioreactor by the removal of old medium and input of fresh medium without any subculturing and about 3000 somatic seedlings were obtained, 59.3% of them were abnormal, in 6 weeks time, per litre of medium. About 70% of the somatic seedlings grew to healthy plantlets and tolerated hardening trials (Das et al., 1999).

3.2.3. Siberian ginseng

Paek et al., (2001) reported an efficient protocol for the production of somatic embryo of *Acanthopanax koreanum* in bioreactor. Ten grams of induced embryogenic determined cells (IEDC) were inoculated into a 1000 ml Erlenmeyer flask containing 500 ml culture medium and cultured for 3 weeks. Then the cultured cells were transferred into a 20 litre BTBB and 14 litre fresh weight of

IEDC was harvested after 6 weeks. By inoculating 0.2 ml suspended IEDCs into a plastic petridish containing 20 ml semi solid medium, more than 5000 germinated somatic embryos were obtained. By using the same protocol, more than 500000 somatic embryos of thornless *Aralia elata* at different developmental stages, were harvested from a 10 litre BTBB after 6 weeks of culture. The somatic embryos regenerated plantlets (somatic seedlings) with well developed aerial and root system, showed vigorous vegetative growth in greenhouse (Figure 5a-c). Recently Kim and Kim (2001) developed an efficient mass production of Siberian ginseng somatic embryos in a bioreactor where torpedo shaped somatic embryos were transferred to 5-10 litre air-lift bioreactors and cultured for 10-15 days, which resulted in plant regeneration from somatic embryos (Figure 4d and e). Recently we developed a protocol for large-scale production of Siberian ginseng somatic embryos in a 500 litre BTBB (Figure 4g and f). By inoculating 3.5 kg IEDC, 60 kg mature somatic embryos (5.7 kg dry weight) were harvested from 500 litre BTBB (Unpublished). This protocol is currently being applied on a large-scale commercial production of secondary metabolites from mature somatic embryos of Siberian ginseng in Korea (Microplants Co., Ltd. Hanwha Chemical R & D Center, 6 Shinsung-Dong, Daejon, South Korea and CBN biotech, Chungbuk National University, Cheongju, South Korea).

3.2.4. Pepper

Marvituna and Buyukalaca (1996) studied *Capsicum annuum* embryogenesis in a bioreactor, from maintenance to embryo maturation, as a series of drain-and-fill batches, keeping embryogenic cultures in the reactor at all times. They compared two types of bioreactor, an air lift and a magnetically stirred 'hanging stirred bar', each with 1.8 litre working volumes, and different mixing/aeration rates. The performance of the magnetically stirred bioreactor at 0.45 vvm aeration and 80 rpm mixing rate was better in terms of percentage embryogenesis.

Similarly Nishihira et al., (1998) tested somatic embryo production of *Aralia cordata* using an air lift or rotating drum bioreactors and suggested that a 100 ml Erlenmeyer flask was more efficient than bioreactors. However, 95% normal somatic embryos were obtained in bioreactors as compared to 65% normal somatic embryos in Erlenmeyer flask.

3.2.5. Conifers

Several conifer species are able to produce somatic embryo on solid medium (Timmis, 1998). There are very few reports of somatic embryo development and maturation from liquid cultures. The maturation of embryonal suspensor masses (ESM) scarcely occurs in submerged liquid culture. Therefore, STR and air lift bioreactors are unsuccessful in embryo production. Perfusion bioreactors using absorbent pads (Attree et al., 1994; Gupta and Timmis 1999) and the regular immersion bioreactor (Paques et al., 1995) are effective for the embryo maturation of these species. The bioreactor cultures of conifer somatic embryos was reported with *P. mariana* and *P. glauca-engelmannii* (Tautorus et al., 1992), *P. glauca* (Attree et al., 1994), *P. abies* (Paques et al., 1995), *P. stichensis* (Moorhouse et al., 1996; Ingram and Mavituna, 2000) and *Pseudotsuga mrnziesii* (Gupta and Timmis, 1999).

Attree et al., (1994) produced vigorous somatic embryos of *P. glauca* in a bioreactor and frequency of conversion of normal plantlets (on support systems soaked with liquid medium) was much higher when compared with somatic embryos produced on agar-solidified medium and zygotic embryos from seeds. Tautorus et al., (1992) reported that a greater growth rate of *P. mariana* and *P. glauca-engelmannii* embryogenic suspensions was achieved in an air lift bioreactor as compared to the mechanically agitated vessel but the maturation of somatic embryos has been restricted to solid gelled medium (Tautorus et al., 1994).

Paques et al., (1995) succeeded in cotyledonary embryo production of *P. abies* using a bioreactor with polyurethane layers that were placed vertically in liquid maturation medium. After 6 weeks of culture on a maturation medium, the cotyledonary embryos were harvested from the ESMs.

Successful production of somatic embryos of Douglas fir in a bioreactor with pads has been reported by Gupta and Timmis (1999). The medium was supplied semicontinuously from the lower surface of the pads to the developing embryos on the top and higher yields of quality embryos were obtained using this system.

Ingram and Mavituna (2000) studied the quantitative data on growth of aggregated and non-aggregated embryogenic suspensions of *P. stichensis* in shake flasks, air lift, bubble, stirred tank and hanging stirrer bar bioreactors and compared the maturation frequencies of somatic embryos derived from different proliferation regimes using both solid medium and submerged culture. The number of mature embryos produced in the bubble bioreactor was significantly higher than those produced from stirred tank and hanging stirred bar bioreactors with both solid medium and submerged culture.

3.2.6. Other Woody Species

Besides above reported results, several automated systems for somatic embryogenesis based on the addition and removal of liquid media (temporary immersion) has been developed and successfully extended to some woody species such as *Hevea brasilensis* (Etienne et al., 1997b), *Citrus* (Cabasson et al., 1997) and tea (Akula et al., 2000). This technique could have significant commercial implications for large-scale production with considerably reduced production costs.

However, somatic embryogenesis based systems have not been widely commercialized. Many commercial limitations have been imposed as a result of comparisons to true seed production. Many somatic embryos in culture showed considerable variation in their size and shape, in the number and nature of the cotyledons, and in the timing and efficiency of their maturity and germination. Synchronization of the bioreactor product is a challenge for plant system, especially it is difficult to engineer production of somatic embryo such that all units have achieved the same stage of maturation at harvest. Embryo sorting, separating and harvesting have been automated by a number of researchers (Hamalainen et al., 1993; Harrell et al., 1993; Kurata, 1994). These systems have used image analysis to detect the shape and quality of embryos and various types of mechanical devices for separating and harvesting. In every case specialized software and equipment can be developed for each species because of the shape and growth habit of the somatic embryos and many of the differences between systems result from the different physiologies that the equipment has to process. In addition to the biology of somatic embryogenesis systems, market constraints have inhibited successful commercialization. Redenbaugh et al., (1991) noted that for most crops in which somatic embryogenesis system exist, the per-unit seed value is too low and somatic embryo is too high to compete with seed. In case where the per-unit seed value is high, low conversion frequencies of somatic embryos are cited as the primary obstacle to successful crop establishment. Because of these limitations, large-scale production of somatic embryos are currently more of a concept than a reality.

4. Conclusion

The advantages of using bioreactor for mass propagation are large-scale potential for plant propagation, rigorous control of culture parameters and scheduling, virus and pathogen free plant propagules, and probable low cost per unit. But these benefits must be balanced against several unique challenges associated with bioreactor technology, including large capital investment (equipment costs), occasional somaclonal variation when propagules are cultivated in liquid suspensions, subsequent difficulties in acclimatization of a liquid cultured propagules to the *ex vitro* environment, synchronization of the somatic embryo development, contamination (if contamination is introduced into a large batch of propagules, the cost and lost time can be devastating) and reproducibility of mass propagation bioreactor system. However there are still undergoing considerable research and development as demonstrated by the significant contribution in current plant tissue culture literature and conferences. It is expected that in future bioreactor technology may revolutionize the economics of many crops in terms of expediting delivery of products of breeding.

5. Acknowledgements

The Korean Science and Engineering Foundation (KOSEF) supported this work with the assistance from the Research Center for the Development of Advanced Horticultural Technology at Chungbuk National University, Cheongju, 361-763, Korea.

6. References

Akula A., Becker D., Bateson M. High yielding repetitive somatic embryogenesis and plant recovery in a selected tea clone, 'TRI-2025', by temporary immersion. Plant Cell Rep. 2000; 19:1140-1145.

Archambault J., Williams R. D., Lavoie L., Pepin M. F., Chavarie C. Production of somatic embryos in a helical ribbon impeller bioreactor. Biotechnol. Bioeng. 1994; 44: 930-943.

Attree S. M., Pomeroy M. K., Fowke L. C. Production of vigorous, desiccation tolerant white spruce (*Picea glauca* [Moench.] Voss.) synthetic seeds in a reactor. Plant Cell Rep. 1994; 13: 601-606.

Bapat V. A., Fulzele D. P., Heble M. R., Rao P. S. Production of sandalwood somatic embryos in bioreactors. Curr. Sci 1990; 59: 746-748.

Buitelaar R. N., Langenhoff A. A. N., Heidstra R., Tramper J. Growth and theophene production by hairy root cultures of *Tagetes patula* in various two liquid phase bioreactors. Enzyme Microbiol. Tech. 1991; 13: 487-494.

Cabasson C., Alvard D., Dambeir D., Ollitrault P., Teisson C. Improvement of *Citrus* somatic embryo development by temporary immersion. Plant Cell Tiss. Org. Cult. 1997; 50:33-37.

Cervelli R., Senaratna T. Economic aspects of somatic embryogenesis. In: *Automation and environmental control in plant tissue culture*. Christie J. A., Kozai T., Smith M. L., eds. Kluwer Academic Publishers, Dordrecht. 1995.

Christie J. A., Kozai T., Takayama S. Automation in plant tissue culture- general introduction and overview. In: *Automation and environmental control in plant tissue culture*. Christie J. A., Kozai T., Smith M. L., eds. Kluwer Academic Publishers, Dordrecht. 1995.

Damiano C., Caboni E., Frattarelii A., Giorgioni M., Liberali M., Lauri P., Angeli S. D. Innovative micropropagation of temperate fruit trees: the case of pear. Acta Hortic. 2000; 530: 181-185.

Das S., Das S., Pal S., Ponde N. R., Gupta S. D., Mujib A., Sahoo S. S., Dey S. A novel process for rapid mass propagation of *Santalum album* L. in liquid media and bioreactor. Acta Hortic. 1999; 502: 281-288.

Denchev P. D., Kuklin A. I., Scragg A. H. Somatic embryo production in bioreactors. J. Biotech. 1992; 26: 99-109.

Dunstan D. I., Tautorus T. E., Thorpe T. A. Somatic embryogenesis in woody plants. In: *In vitro* embryogenesis in plants. Thorpe T. E., ed. Kluwer Academic Publishers, Dordrecht. 1995.

Escalona M., Lorenzo J. C., Gonzalez B., Daquinta M., Gonzalez J. L., Desjardins, Y., Borroto, C. G. Pineapple (*Ananas comosus* L. Merr) micropropagation in temporary immersion systems. Plant Cell Rep. 1999; 18:743-748.

Etienne D. B., Bertrand B., Schlonvoigt A., Ettienne H. The morphological variability within a population of coffee somatic embryos produced in a bioreactor affects the regeneration and the development of plants in the nursery. Plant Cell Tiss. Org. Cult. 2002; 68: 153-162.

Etienne D. B., Bertrand B., Vasquez N., Etienne H. Direct sowing of *Coffea arabica* somatic embryos mass produced in a bioreactor and regeneration of plants. Plant Cell Rep. 1999; 19:111-117.

P. H., Olsen O. A. S., Eide A. K. H. Engineering aspects of plant propagation in bioreactors. In: *Automation and environmental control in plant tissue culture*. Christie J. A., Kozai T., Smith M. L., eds. Kluwer Academic Publishers, Dordrecht. 1995.

Honda H., Liu C., Kobayashi T. "Large scale Plant propagation". In *Advances in Biochemical Engineering/Biotechnology*, Vol. 72, Scheper T., ed. Springer-Verlag Berlin Heidelberg, 2001.

Ibaraki Y., Kurata K. Automation of somatic embryo production. Plant Cell Tiss. Org. Cult. 2001; 65: 179-199.

Ingram B., Mavituna F. Effect of bioreactor configuration on the growth and maturation of *Picea sitchensis* somatic embryo cultures. Plant Cell Tiss. Org. Cult. 2000; 61: 87-96.

Ishibashi T., Odawara Y., Fujiwara K., Ookuma M. New oxygen supplying method for animal cell reactors. Hitachi Hyouron. 1987; 69: 301-305 (in Japanese).

Jain S. M., Gupta P. K. and Newton R. J (Eds.). *Somatic embryogenesis in woody plant.* Volumes 1-3 , 1995; Kluwer Academic Publishers, The Netherlands.

Jain S. M., Gupta P. K. and Newton R. J (Eds.). *Somatic embryogenesis in woody plants,* Volumes 4-5 , 1999; Kluwer Academic Publishers, The Netherlands.

Jain S. M., Gupta P. K. and Newton R. J (Eds.). *Somatic embryogenesis in woody plants,* Volume 6 , 2000; Kluwer Academic Publishers, The Netherlands.

Kim J. W., Kim H. S. Mass production of Siberian ginseng (*Eleutherococcus senticosus*) somatic embryos by cell culturing. Oriental Pharmacy Exp. Med. 2001; 1:34-38.

Kim Y. S. *Development of large scale process for mass production of lily bulblets by bioreactor system.* 1999; M. Sc thesis. Chungbuk National University, Korea.

Konar R. N., Oberoi Y. P., *In vitro* development of embryoids on the cotyledons of *Biota orientalis.* Phytomorphology. 1965; 15: 137.

Kurata K. Image analysis for embryogenesis. In: *Automation and environmental control in plant tissue culture.* Christie J. A., Kozai T., Smith M. L., eds. Kluwer Academic Publishers, Dordrecht. 1995.

Lee Y. H. *Design of bioreactor system for small and pilot scale cultivation of plant cells.* 1997; M. Sc. thesis. Chungbuk National University, Korea.

Lian M. L. *Several Factors affecting bulblet production in vitro and establishment of mass production system of Lilium Hybrid using bioreactor.* 2001; Ph. D thesis. Chungbuk National University, Korea.

Mavituna F., Buyukalaca S. Large-scale production of pepper somatic embryos using bioreactors. Acta Hortic. 1995; 412: 58-63.

Mavituna F., Buyukalaca S. Somatic embryogenesis of pepper in bioreactors: a study of bioreactor type and oxygen uptake rates. Applied Micro. Biotech. 1996; 46: 327-333.

Moorhouse S. D., Wilson G., Hennerty M. J., Selby C., Antsaoir S. M. A plant cell bioreactor with medium perfusion for control of somatic embryogenesis in liquid cell suspensions. Plant Growth Regul. 1996; 20: 53-56.

Nishihira T., Hayashi Y., Matsumoto K. Somatic embryo induction from cell suspension culture of *Aralia cordata* using bioreactors. J. Jap. Soc. Hort. Sci. 1998; 67: 87-92.

Paek, K. Y., Hahn E. J., Son S. H. Application of bioreactors for large- scale micropropagation systems of plants. *In vitro* Cell. Dev. Biol.-Plant. 2001; 37: 149-157.

Paques M., Bercetche J., Palada M. Prospects and limits of somatic embryogenesis of *Picea abies.* In: *Somatic embryogenesis in Woody Plants.* Vol. 1. Jain S., Gupta P., Newton R. eds. Kluwer Academic Publishers, Dordrecht. 1995.

Park S. Y., Murthy H. N., Paek K. Y. Mass multiplication of protocorm like bodies using bioreactor system and subsequent plant regeneration in *Phalaenopsis.* Plant Cell Tiss. Org. Cult. 2000; 63: 67-72.

Preil W. Application of bioreactors in plant propagation. In: Micropropagation. Debergh P. C., Zimmerman R. H. eds. Kluwe Academic Publishers. 1991.

Ray P. M., Steeves T. A., Fultz S. A. *Botany.* Saunders College Publishing, New York, 1983.

Redenbaugh K., Fujii J., Slade D., Viss P., Kissler M. Artificial seeds- encapsulated somatic embryos. In: Biotechnology in Agriculture and Forestry. Y.P.S. Bajaj (Ed.) 1991; 17: 395-416.

Sajc L., Grubisic D., Novakovic G. V. Bioreactors for plant engineering: an outlook for further research. Biochem. Eng. J. 2000; 4: 89-99.

Sharma N. K. Automation in plant tissue culture: problems and prospects. Curr. Sci. 1992; 7:507-512.

Shim S. W. *Environmental conditions affecting in vitro growth and acclimatization of '5BB' grape rootstock.* 2002; Ph. D thesis. Chungbuk National University, Korea

Son S. H., Choi S. M., Kwon S. R., Lee Y. H., Paek K. Y. Large Scale culture of plant cell and tissue by bioreactor system. J. Plant Biotech. 1999; 1: 1-8.

Steward F. C., Mapes M. O., Mears K. Growth and organized development of cultured cells. II. Organization in cultures grown from freely suspended cells. Am. J. Bot. 1958; 45: 705.

Stuart D. A., Strickland S. G., Walker K. A. Bioreactor production of alfalfa somatic embryos. HortScience. 1987; 22: 800-803.

Takayama S., Akita M. The types of bioreactors used for shoots and embryos. Plant Cell Tiss. Org. Cult. 1994; 39: 147-156.

Takayama S., Misawa M. Mass propagation of *Begonia hiemalis* plantlets by shake culture. Plant Cell Physiol. 1981; 22: 461-467.

Takayama S. Mass propagation of plants through shake and bioreactor culture techniques. In: Y.P.S. Bajaj (Ed.). *Biotechnology in Agriculture and Forestry*, Vol. 17 (pp 495-515). Springer-Verlag, Berlin.

Tautorus T. E., Dunstan D. I. Scale-up of embryogenic plant suspension cultures in bioreactors. In: Jain S. M., Gupta P. K. and Newton R. J (Eds.). *Somatic embryogenesis in woody plant.* Volumes 1-3 , 1995; Kluwer Academic Publishers, The Netherlands.

Tautorus T. E., Lulsdorf M. M., Kikcio S. I., Dunstan D. I. Bioreactor culture of *Picea mariana* Mill. (Black spruce) and the species complex *Picea glauca-engelmannii* (interior spruce) somatic embryos. App. Microbiol. Biotechnol. 1992; 38: 46-51.

Tautorus T. E., Lulsdorf M. M., Kikcio S. I., Dunstan D. I. Nutrient utilization during bioreactor culture, and regeneration of somatic embryo cultures of *Picea mariana* and *Picea glauca-engelmannii. In vitro* Cell Dev. Biol. 1994; 30P: 58-63.

Teisson C., Alvard D. A new concept of plant *in vitro* cultivation liquid medium: temporary immersion. In: *Current issues in plant molecular and cellular biology*. Terzi M., Cella R., Falavigna A. eds. Kluwer Academic Publishers, Dordrecht.

Timmis R. Bioprocessing for tree production in the forest industry: conifer somatic embryogenesis. Biotech. Progress. 1998; 14: 156-166.

Walker P. N. System analysis and engineering. In: *Automation and environmental control in plant tissue culture*. Christie J. A., Kozai T., Smith M. L., eds. Kluwer Academic Publishers, Dordrecht. 1995.

Yu K. W. *Production of the useful metabolites through bioreactor culture of Korean ginseng (Panax ginseng* C. A. Meyer). 2000; Ph. D thesis. Chungbuk National University, Korea

Ziv M., Ariel T. Bud proliferation and plant regeneration in liquid cultured Philodendron treated with ancymidol and paclobutrazol. J. Plant Growth Reg. 1991; 10: 53-57.

Table 1. Somatic embryo production by bioreactor

Plant species	Reactor type	References
Aralia cordata	Air-lift, rotating drum and bubble column	Nishihira et al., 1998
Capsicum annum	Air-lift	Marvituna & Buyukalaca, 1996
	Air-lift, magnetically stirred hanging stirred bar	Marvituna & Buyukalaca, 1995
Coffea arabica	Temporary immersion bioreactor	Etienne-Barry et al., 1997
	Temporary immersion bioreactor	Etienne-Barry et al., 1999
	Temporary immersion bioreactor	Etienne-Barry et al., 2002
Camellia sinensis	Temporary immersion bioreactor	Akula et al., 2000
Eleutherococcus senticosus	Air lift	Kim & Kim, 2001
Santalum album	Air lift	Bapat et al., 1990
	Air lift	Das et al., 1999
Picea glauca	Perfusion bioreactor with a flat absorbent pad for supporting cells	Attree et al., 1994
P. abies	Regular immersion bioreactor with polyurethane layers.	Paques at al., 1995
P. sitchensis	STR with polypropylene capillary membrane for aeration tubeing	Moorhouse et al., 1996
	Stirred tank hanging stirred bar	Ingram & Mavituna, 2000
P. mariana		
P. glauca-engelnannii	Mechanically stirred bioreactor	Tautorus et al., 1992
Pseudotsuga menziesii	Semi continuous perfusion bioreactor with pads for supporting cells.	Gupta & Timmis, 1999

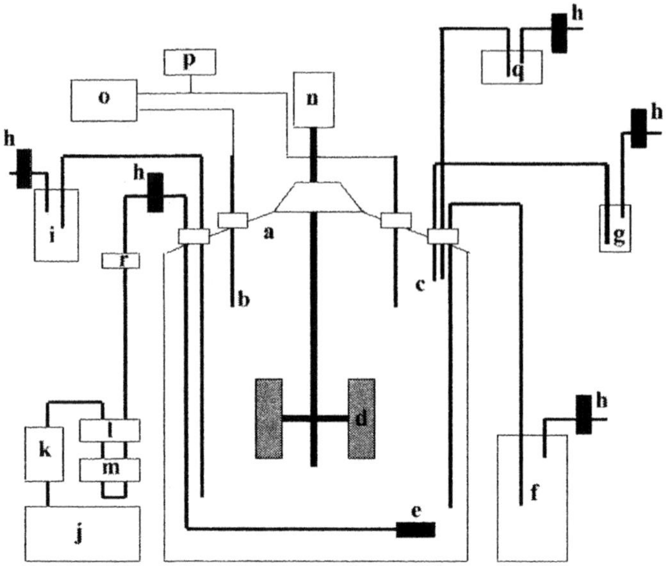

Figure 1. Layout of stirred tank bioreactor. a, Wheaton vessel for 20 or 50 liters; b, dissolved oxygen probe; c, pH probe; d, impeller; e, sparger; f, inoculation and harvest vessel; g, vet trap; h, membrane filter; i;, sampling port; j, air compressor; k, filtering system. L, cooling system; m, heating system; n, top driven motor; o, controller; p, computer; q, anti-foam port; r, air flow meter.

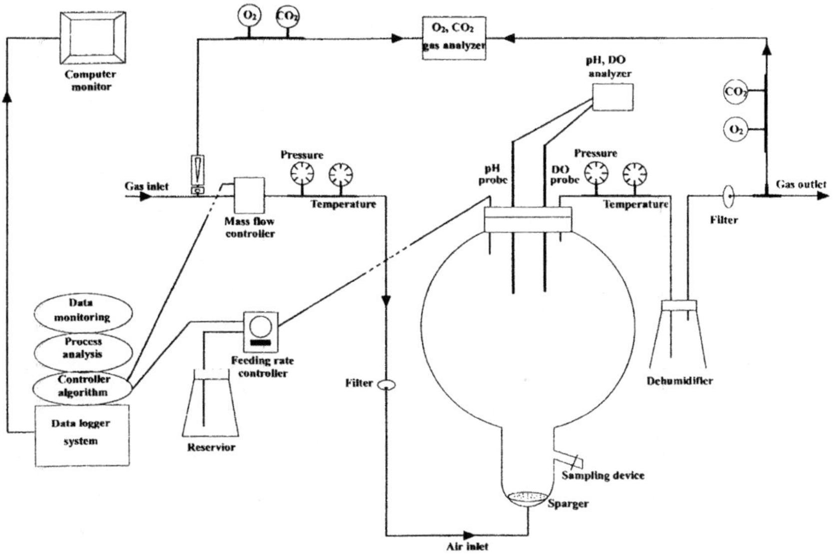

Figure 2. Configuration of a balloon-type bubble bioreactor system.

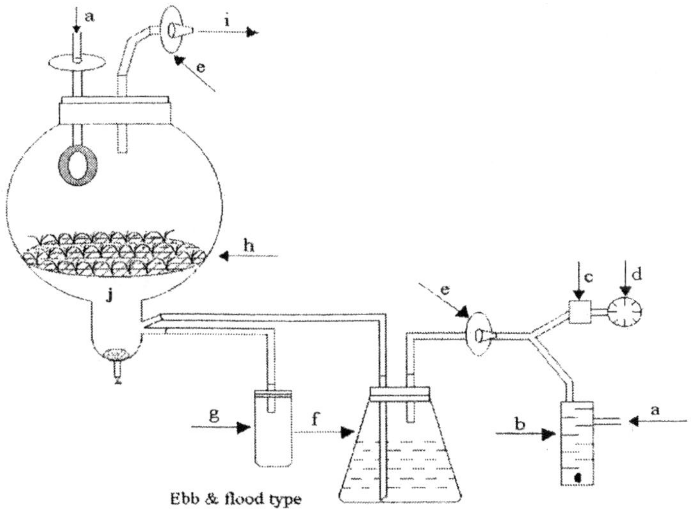

Figure 3. Layout of an ebb and flood bioreactor system (a) air inlet; (b) air flow meter; (c) timer; (d) solenoid valve; (e) membrane filter; (f) medium reservoir; (g) sampling port; (h) plantlets; (i) air outlet; and (j) supporter (net).

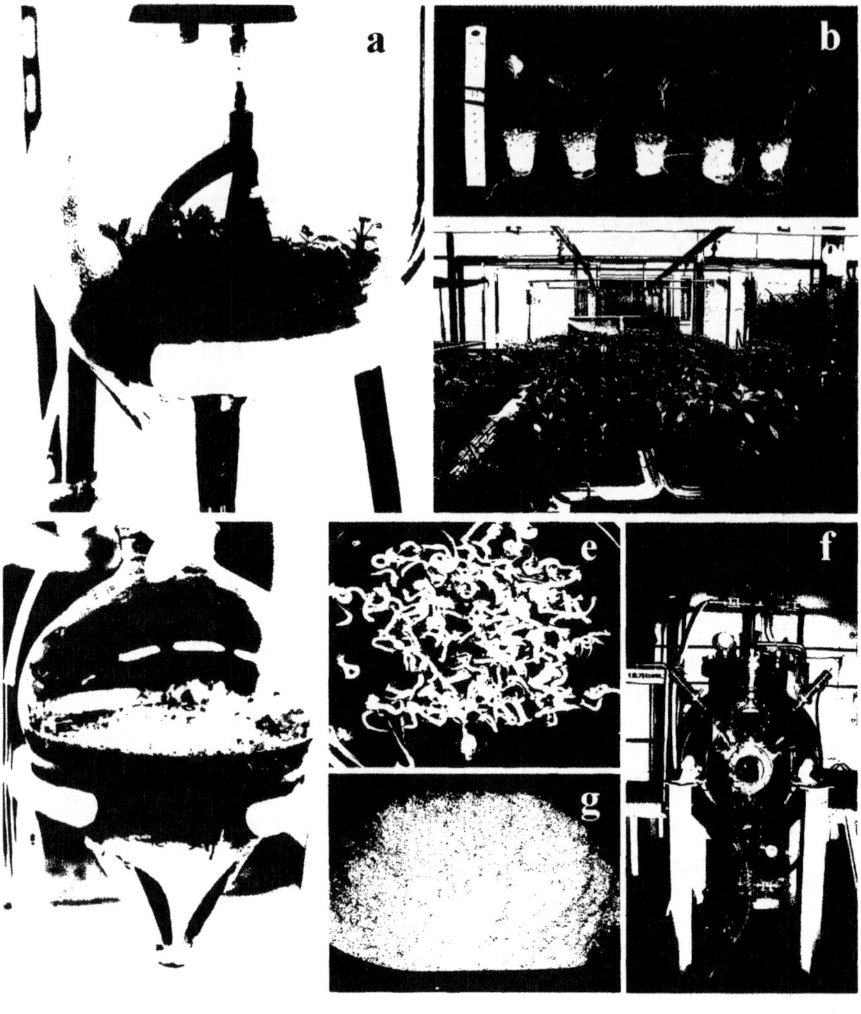

Figure 4. Micropropagation system using bioreactors. a, Apple plantlets in a BTBB after 40 days of culture. b, rooting of in vitro regenerated apple plantlets under ex vitro condition. c, large scale production of apple plantlets using bioreactor system. d, Mass propagation of Siberian ginseng somatic embryos in 5 litre BTBB from embryogenic cells. e, germinated somatic embryos showing cotyledon and root after 30 days of culture in a BTBB. f, Pilot-scale BTBB (500 litre) for commercial production of Siberian ginseng somatic embryos. g, harvesting of Siberian ginseng somatic embryos from 500 litre BTBB after 30 days of culture.

Figure 5. Micropropagation of thornless Aralia alata using bioreactor system. a, selected clone of thornless Aralia alata. b, somatic embryos harvested from 10 litre BTBB. c, plantlets regenerated from somatic embryos grown under green house condition.

26. PHOTOAUTOTROPHIC MICRO-PROPAGATION OF WOODY AND TROPICAL PLANTS

Toyoki Kozai[1] and Quynh Thi Nguyen[2]
[1] *Chiba University, Matsudo, Chiba 271-8510, Japan*
[2] *Institute of Tropical Biology, NCST-VN, 1 Mac Dinh Chi Street, Hochiminh City, Vietnam*

1. INTRODUCTION

A growing human population, global warming, and loss of biodiversity are having a tremendous impact on our environment. The world human population increased substantially during the last ten years and was predicted to be 8 billion by the year 2020. This means that during the next 20 years population will increase by 2 billion (ISAAA, 2001).

In Asian, African and South American countries where the latest annual human population growth is nearly 3%, environmental pollution and shortages of food, natural resources and fossil fuel may become more serious in the next several decades. On the other hand, the forests in these countries have been harvested at a faster rate than they are being regenerated, either naturally or artificially. It is obvious that an increase in human population is responsible for wilderness and forest destruction in the past decades.

The harvested wood material is mostly utilized as pulp, paper, timber, furniture etc.; whereas a significant portion of hardwoods is used as fuel (Thorpe et al., 1991; Kozai et al., 2000). In addition, the rapid spread of pests, diseases and disastrous fires or floods may destroy the very existence of certain tree species.

According to a survey by FAO (1999), the forestland in the tropical regions was decreasing at an annual rate of 0.7% during the period 1990-1995, resulting in a yearly net decrease in forest area of 12.6 million hectares. Desertification has significantly decreased biomass of woody plants. The destruction of forests and continuous use of coal and oil have led to a steady increase in CO_2 concentration levels in the atmospheric air, which is partly causing the recent global warming. Such serious damage of the world's environment is threatening the existence of our earth.

The reforestation of denuded forest area is needed for environmental protection and conservation, for lowering the atmospheric CO_2 concentration and for stabilizing of local and global climates. It has been predicted that the

S.M. Jain and K. Ishii (eds.), Micropropagation of Woody Trees and Fruits, 757–781.

demand for quality transplants will increase in the future decades. For reforestation of vast deforested areas in the tropics, there is an urgent need for 25-40 billions of transplants annually (FAO, 1999). However, the volume of transplants such as rooted cuttings supplied by traditional vegetative propagation methods cannot meet the demand of transplants in the 21st century (Lapitan, 1999). The immediate need for mass production of quality transplants emphasizes the use of micropropagation methods.

Transplant production based on micropropagation methods has advantages over transplant production using seeds or cuttings with respect to genetic and phenotypic uniformity, enormous multiplication rate and scheduled-year-round production of disease-free or pathogen-free transplants (Aitken-Christie et al., 1995). However, the reduction of production costs for micropropagation of many plant species, especially woody transplants, was not as rapid as expected. The reasons for the high production cost of plants *in vitro* by conventional, heterotrophic or photomixotrophic micropropagation are: (1) a significant loss of plants *in vitro* by microbial contamination at the multiplication stage, (2) poor growth and development of plants *in vitro*, (3) a significant percentage of morphological and physiological disorders which are partly due to the presence of plant growth regulators and/or unfavorable environmental conditions *in vitro* (4) poor rooting partly due to callus formation at the plant base and thus, (5) excessive loss of these plants after transplanting in the greenhouse or field, (6) high labor cost, (7) high energy cost for lighting, cooling, sterilization, etc., (8) a significant cost for culture medium components and vessels, and (9) over- or underproduction of plants *in vitro*.

2. REASONS FOR LOW GROWTH AND DEVELOPMENT RATES OF PLANTS *IN VITRO* UNDER HETEROTROPHIC OR PHOTOMIXOTROPHIC CONDITION

The environment *in vitro* in conventional micropropagation is generally characterized by high relative humidity (RH); low light intensity or photosynthetic photon flux (PPF); large variation in diurnal CO_2 concentration; constant air temperature; presence of sugar, vitamins, plant growth regulators and toxic substances such as phenols in the medium; absence of microorganism; and accumulation of ethylene and other volatile gases in the vessel headspace.

In conventional micropropagation, the amount of CO_2 uptake or net photosynthetic rate of plants *in vitro* is lower than plants in greenhouse or field controls (Donnelly et al., 1984). The net photosynthesis rate of birch plants regenerated *in vitro* was only one-third of those grown in the greenhouse, indicating the lack of full development of photosynthetic competency of plants *in vitro* (Smith et al. 1986). The low net photosynthetic rate of plants *in vitro* is attributed to the low RuBPcase activity (Grout, 1988), which was probably due to the high sucrose concentration in leaves of plants *in vitro* (Hdider and Desjardins, 1994).

On the other hand, CO_2 concentration in the culture vessels containing chlorophyllous or leafy plants *in vitro* decreases from a range of 3000 to 9000 $\mu mol\ mol^{-1}$ in the dark period to lower than 100 $\mu mol\ mol^{-1}$ in the photoperiod in conventional micropropagation (Kurata and Kozai, 1992). The low CO_2 concentration in the vessel during most of the photoperiod is partly due to a low ventilation rate of airtight vessels and a limited quantity of CO_2 in small culture vessels. The low CO_2 concentration inhibits the photosynthetic activity of plants *in vitro* and forces them to develop a heterotrophic or photomixotrophic growth by absorbing sugar from the culture medium as their main carbon source.

At a CO_2 concentration around the CO_2 compensation point (50-70 $\mu mol\ mol^{-1}$), plants *in vitro* requires only a minimum PPF, because a higher PPF would not promote photosynthesis of plants *in vitro* (Kozai, 1988).

Using airtight vessels can prevent microbial contamination in *in vitro* cultures. However, the use of such culture vessels having a limited air exchange with the ambient atmosphere creates stagnant air, high RH and water condensation on the inner surface of the vessel. They are partly responsible for physiological and morphological disorders such as the lack of epicuticular wax formation (Brainerd et al., 1981) or hyperhydration of plants grown *in vitro* (Aitken-Christie et al., 1995). Hyperhydricity of plants *in vitro* is characterized by high water content of the leaf and stem tissues, thin palisade layers, and increase in the mesophyll air space of plants *in vitro* (Brainerd et al., 1981; Short et al., 1987).

The reduction of RH in the culture vessel is known to reduce plant hyperhydricity. Under a low RH in the culture vessel, plants *in vitro* form normal wax, and promote the stomatal functioning. Thus, the acclimatization in the *ex vitro* stage may be shortened or eliminated.

In conventional micropropagation, ethylene accumulates at a high level up to 2 or 3 $\mu mol\ mol^{-1}$ due to the airtight culture vessel (De Proft et al., 1985; Jackson et al., 1991). The accumulation of ethylene has an adverse effect on plant development such as the decrease in leaf expansion and shoot length (Jackson et al., 1991). New shoot regeneration is also inhibited by a high ethylene concentration in the sealed vessel (Biddington, 1992).

Development of new micropropagation systems for the control of *in vitro* environment is necessary to overcome the major shortcomings as described earlier for large-scale production of high quality plants *in vitro*. A great number of quality transplants can be produced at low cost with photoautotrophic (no sugar in the culture medium) micropropagation.

For with liquid medium containing plant growth regulators, genetic fidelity is often problematic, and only a portion (often, 5-10 %) of germinated somatic embryos can actually be used as transplants. Also, the field performance of plants regenerated *in vitro* via calli and/or using bioreactors needs to be examined carefully. On the other hand, most (often, 90-95 %) of plants regenerated *in vitro* from leafy single node cuttings under normal environmental conditions are considered to be genetically stable and can actually be used as transplants.

3. COST REDUCTION BY PHOTOAUTOTROPHIC MICROPROPAGATION

In photoautotrophic micropropagation, plants *in vitro* are propagated by using relatively small leafy or chlorophyllous explants on a sugar-free medium under pathogen-free and favorable *in vitro* environmental conditions, which enable *in vitro* plants to photosynthesize and produce their own carbohydrates for growth. A photoautotrophic micropropagation system with the improvement of *in vitro* aerial and root zone environments has significantly enhanced the growth of plants *in vitro*, increased the multiplication rate and thus, shortened the multiplication period of plants *in vitro* (Kozai et al., 1988).

When plants *in vitro* are grown on sugar-free medium in a ventilated vessel under pathogen-free condition, loss of plants *in vitro* due to microbial contamination can be considerably reduced (microbes multiply rapidly only on the sugar-containing medium in most cases). On the other hand, plants *in vitro* on the sugar-free medium need increased light intensity and CO_2 concentration in the vessel during the photoperiod to promote their photosynthesis. Thus, the production costs of micropropagated plants can be reduced significantly if the increase in CO_2 concentration during the photoperiod and decrease in relative humidity in the vessel containing sugar-free medium can be achieved at a low cost.

Reduction of relative humidity and ethylene concentration in the vessel will promote the transpiration and the mineral uptake of plants *in vitro*, and improve physiological and morphological characteristics of plants *in vitro*, respectively (Kozai et al., 1995). In addition, plants *in vitro* are proved to grow better when gelling agent such as agar or Gelrite is replaced by air porous supporting material such as vermiculite or cellulose fibers, which improves rooting of plants *in vitro* by increasing oxygen or nutrient availability in the root zone environment.

The increase in CO_2 concentration and the decrease in relative humidity and ethylene concentration in the vessel can be achieved either by natural or forced ventilation. In the natural ventilation method, CO_2 in the vessel can be increased by enhancing the air diffusion (CO_2 and water vapor) through gas-permeable filters, attached to the lid or sidewalls of the culture vessel. The CO_2 concentration in the vessel can also be increased by increasing the CO_2 concentration in the culture room (Kozai et al., 1987).

In the forced ventilation method, a particular gas mixture is flushed directly to the vessel with an air pump (Fujiwara et al., 1988). The advantage of this method is that a large culture vessel which can accommodate thousands of *in vitro* plants can be used for commercial mass propagation of *in vitro* plants (Xiao et al., 2000). The important factor in both methods is the rate of ventilation. The air circulation in the vessel increases with the increase in vessel ventilation rate and, thus, enhances the diffusion coefficient of the air in the vessel.

4. GROWTH OF PLANTS *IN VITRO* USING SMALL CULTURE VESSELS UNDER DIFFERENT SUGAR CONCENTRATION, CO_2 CONCENTRATION, PPF AND SUPPORTING MATERIAL CONDITIONS

Experimental results on the photoautotrophic micropropagation of woody and tropical plants using relatively small vessels like Magenta GA-7 with an air volume of 300-400 ml are shown in this section. For photoautotrophic micropropagation, microporous gas-permeable filters are attached to the lid or sidewalls of the culture vessel. The use of these filter discs increases the natural ventilation rate of the culture vessel. The number of air exchanges (defined as hourly ventilation rate divided by the vessel volume) of a Magenta-type vessel is about 0.2 h^{-1}. The number of air exchanges of Magenta-type vessels attached by one, two or three microporous gas filter discs (10 mm in diameter each) with an average microporous diameter of 0.5 μm is about 2, 3, and 4 h^{-1}, respectively (Kozai et al., 1995). Thus, one filter disc attached to the lid of the conventional vessel without the disc increases the vessel ventilation rate about 5 times. The method of measuring the number of air exchanges of the culture vessel is described in Aitken-Christie et al. (1995). By attaching one, two and three gas filter discs to the lid of the Magenta-type vessel, the CO_2 concentration during the photoperiod in the vessel containing plants increases to about 150, 200 and 250 $\mu mol\ mol^{-1}$, respectively, when the CO_2 concentration in the culture room is kept at 350-400 $\mu mol\ mol^{-1}$, as in the standard atmospheric air. Physical relationships among the ventilation rate, CO_2 concentrations in the vessel and the culture room, photosynthetic characteristics of plants *in vitro*, etc., are described using mathematical equations in Jeong et al. (1993), Aitken-Christie et al. (1995) and Kozai et al. (1995). If a CO_2 concentration in the vessel higher than 350 $\mu mol\ mol^{-1}$ is needed during the photoperiod, CO_2 concentration in the culture room needs to be increased to about 1,000 μmol mol^{-1} or higher, using a CO_2 controller, which is widely used for CO_2 enrichment for greenhouse crops. Gas permeable filter discs are currently rather expensive and its high price is restricting the commercialization of photoautotrophic micropropagation using small vessels attached with the gas permeable filter discs. Reduction in price of the filter discs is expected for wider commercialization of photoautotrophic micropropagation using small vessels.

4.1. Pine

Kubota (1994) demonstrated the photoautotrophic growth of radiata pine (*Pinus radiata*) shoots cultured *in vitro* for 49 days in a 500 ml vessel with 4.9 air exchanges per hour under 160 μmol m^{-2} s^{-1} PPF without CO_2 enrichment. Rockwool multi-blocks were used as a supporting material with a liquid medium containing basal components of a modified LP (Aitken-Christie et al., 1988) medium. Dry weight of shoots cultured photoautotrophically was 24% greater than that of shoots cultured conventionally (0.1 h^{-1} air exchanges of the vessel, 60 μmol m^{-2} s^{-1} PPF, and gelled medium containing 30 g l^{-1} sucrose). Root formation did not occur in both treatments during the culture period. Under CO_2 enriched (850 μmol mol^{-1}) conditions, radiata pine shoots were cultured photoautotrophically in a gelled medium with different basal compositions (Aitken-Christie et al., 1992). An addition of activated charcoal into the medium enhanced shoot elongation and fresh weight increment after 12 weeks of culture. However, no roots were formed during the culture period. As shown in other species, a combination of fibrous supporting materials and CO_2 enrichment may be beneficial for growth and rooting of radiata pine, but this hypothesis has not yet been examined.

4.2. Eucalyptus

Kirdmanee (1995) studied the growth of eucalyptus (*Eucalyptus camaldulensis*) shoots (2.2 cm long) cultured photoautotrophically *in vitro* for 6 weeks in Magenta-type vessels (air volume of 370 ml) under CO_2-nonenriched (400 μmol mol^{-1} in the culture room) or CO_2-enriched (1200 μmol mol^{-1} in the culture room) conditions. Each vessel contained one of four different types of supporting materials: agar, Gelrite, plastic net or vermiculite. Gas permeable filter discs (Milli-Seal, Millipore, Tokyo, Japan) were attached to the sidewalls of the vessel. CO_2 enrichment significantly increased the growth (dry weight and number of primary roots) of plants on all types of supporting materials (Fig. 1a).

The highest growth was obtained in the vermiculite, followed by the plastic net, Gelrite, and agar (in descending order) under either CO_2-nonenriched or CO_2-enriched conditions (Kirdmanee et al., 1995). The growth of plants *ex vitro* was highest and the percent damaged leaves/roots was lowest in the vermiculite under the CO_2-enriched condition. An extensive root system with many secondary roots was produced *in vitro* in the vermiculite (Fig. 1b).

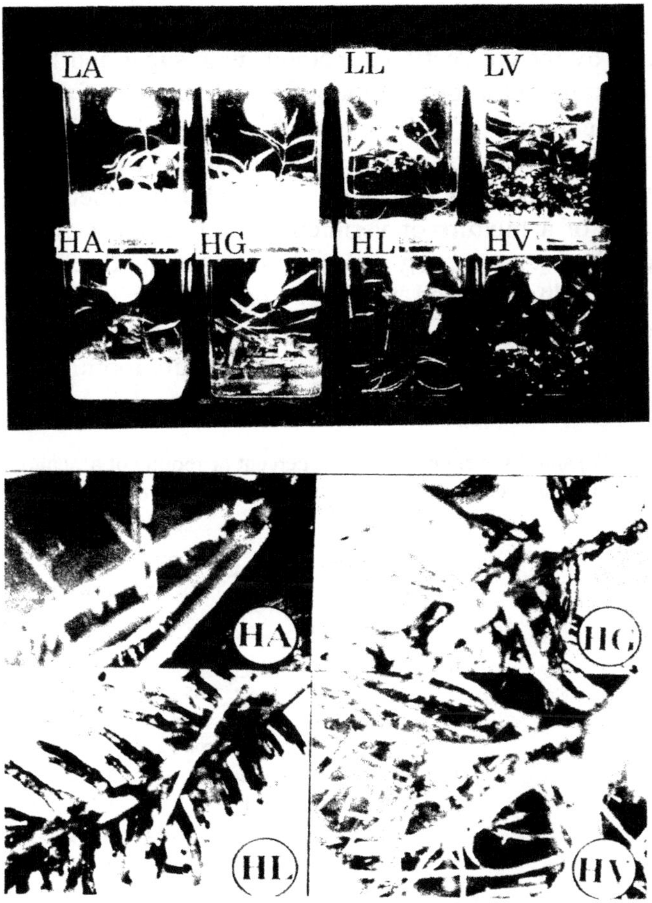

Figure 1. Photoautotrophic growth of eucalyptus (*Eucalyptus camaldulensis*) for 6 weeks (By courtesy of C. Kirdmanee). Eucalyptus plants *in vitro* under CO_2 non-enriched (lower) and CO_2 enriched (upper) conditions, using 4 different supporting materials (a). Roots of eucalyptus plants under CO_2 enriched conditions (b). In the treatment legends, H denotes the CO_2 enriched conditions; A, G, L and V denote agar, Gelrite, plastic net, vermiculite, respectively.

4.3. Acacia

Growth *in vitro* and survival percentage *ex vitro* of acacia (*Acacia mangium*) plants were examined, using apical shoots cultured either photomixotrophically (IBA added into a sugar-containing medium) or photoautotrophically (without IBA and sugar in the medium) (Ermayanti et al., 1999). The combination of high CO_2 concentration (1500 μmol mol^{-1}) in the culture room (CO_2 enrichment) and high natural ventilation rate (number of air exchanges: 6.7 h^{-1}) of the vessel increased the fresh and dry weights of plants *in vitro*, while the presence/absence of sugar and plant growth regulators did not have any significant effects (Table 1). Moreover, acacia plants cultured by the conventional method (with sugar and IBA in the agar-containing medium and a low number of air exchanges of the culture vessel under CO_2 concentration of 400 μmol mol^{-1} in the culture room) showed the lowest fresh and dry weights, and no root formation (Table 1). The rooting of plants *in vitro* was especially enhanced in Florialite treatments, a porous solid cube consisting of a mixture of vermiculite and cellulose fibers (Nisshinbo Ind. Inc., Japan), as a supporting material. Net photosynthetic rate was higher throughout the culture period in the treatment with sugar-free Florialite medium and a high number of air exchanges of the culture vessel under CO_2 concentration of about 1500 μmol mol^{-1} in the culture room.

Table 1. Fresh (FW) and dry weight (DW), and percent of rooting of acacia plants cultured *in vitro* for 28 days (Ermayanti et al., 1999). Means ± SD are shown.

Treatment					FW (mg/plant)	DW (mg/plant)	% rooting
Sucrose conc. (g l^{-1})	Growth regulator	CO_2 conc. (μmol mol^{-1})	No. of air exchanges (h^{-1})	Substrate			
30	Yes^z	1500	6.7	Florialite	222 ± 89	40 ± 21	94
30	No	1500	6.7	Florialite	155 ± 21	32 ± 1	81
0	Yes	1500	6.7	Florialite	410 ± 145	62 ± 24	100
0	No	1500	6.7	Florialite	298 ± 4	41 ± 14	82
30	Yes	400	0.7	Florialite	152 ± 15	22 ± 2	38
30	No	400	0.7	Florialite	116^y	17^y	75
0	Yes	400	0.7	Florialite	110 ± 33	18 ± 4	82
0	No	400	0.7	Florialite	139 ± 38	20 ± 4	46
30 (control)	No	400	0.7	Agar	100^y	14^y	0
ANALYSIS OF VARIANCE^x							
Sucrose concentration					NS	NS	NS
Growth regulator					NS	NS	NS
CO_2 enrichment and No. of air exchanges (h^{-1})					**	**	NS

^{z} Medium contained 1 mg l^{-1} IBA. ^{y} Only one replication for treatment with 30 g l^{-1} sucrose, without growth regulator and CO_2 of 400 mmol mol^{-1}. ^{x}ANOVA (Analysis of variance) was applied for 9 treatments (except for the control treatment). NS, nonsignificant; **, significant at $p \leq 0.01$.

4.4. Paulownia

Single nodal cuttings of *Paulownia fortunei* plants with two leaves for each cutting were cultured *in vitro* without plant growth regulator in Magenta-type vessels under PPF of 120 µmol m^{-2} s^{-1} and CO_2 concentration of 450 µmol mol^{-1} (Nguyen and Kozai, 2001d). On day 30, paulownia plants grew more vigorously under a photoautotrophic (or sugar-free) condition than under a photomixotrophic (20 g l^{-1} sucrose) condition, especially when vermiculite or Florialite was used as supporting materials in place of gelled medium such as agar or Gelrite. There was no significant difference in the multiplication ratio (number of cuttings for the next subculture per explant) when plants grew *in vitro* under the photoautotrophic or photomixotrophic condition on different kinds of supporting material (agar, Gelrite, sand, vermiculite and Florialite). Paulownia plants cultured *in vitro* photoautotrophically without plant growth regulator in 800-ml Phytacon columnar-type vessels (Sigma Co., USA) demonstrated a significant increase in the dry weight and shoot length under longer photoperiod (16-h d^{-1} vs. 10 h d^{-1}) and greater number of air exchanges (2.9 h^{-1} vs. 1.3 h^{-1}).

When paulownia plants *in vitro* were cultured photoautotrophically in a vermiculite-based medium for 28 days, high CO_2 concentration (1600 µmol mol^{-1}) and high PPF (250 µmol m^{-2} s^{-1}) significantly promoted their growth (Fig. 2a). High CO_2 concentration and high PPF enhanced the development of lateral roots and normal vascular systems. Paulownia plants cultured under the CO_2-enriched and high PPF condition in the *in vitro* stage continued increasing their growth significantly during 15 days in the *ex vitro* stage (Fig. 2b).

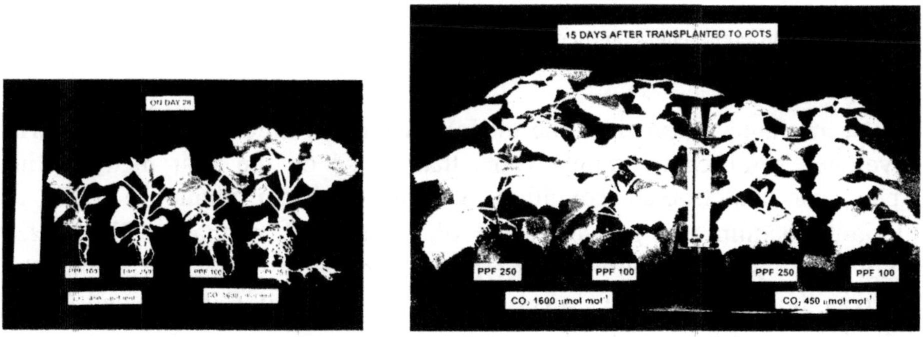

Figure 2. Photoautotrophic growth of paulownia plants *in vitro* as affected by high/low CO_2 concentration and PPF (Nguyen and Kozai, 2001d): on day 28 in the *in vitro* stage (a), 15 days after planted to pots (b).

4.5. Neem

Single node cuttings of neem (*Azadirachta indica*) were cultured on agar or Florialite medium with presence (20 g l^{-1}) or absence of sucrose in Magenta-type vessels (air volume of 370 ml) having low (0.15 h^{-1}) or high (3.5 h^{-1}) number of air exchanges at CO_2 concentration of 450 μmol mol^{-1} and PPF of 100 μmol m^{-2} s^{-1} (Nguyen and Kozai, 2001b). Dry weight and leaf fresh weight of neem plants *in vitro* were significantly greater when cultured without sugar and high number of air exchanges of the vessel than those with sugar and low number of air exchanges (Fig. 3).

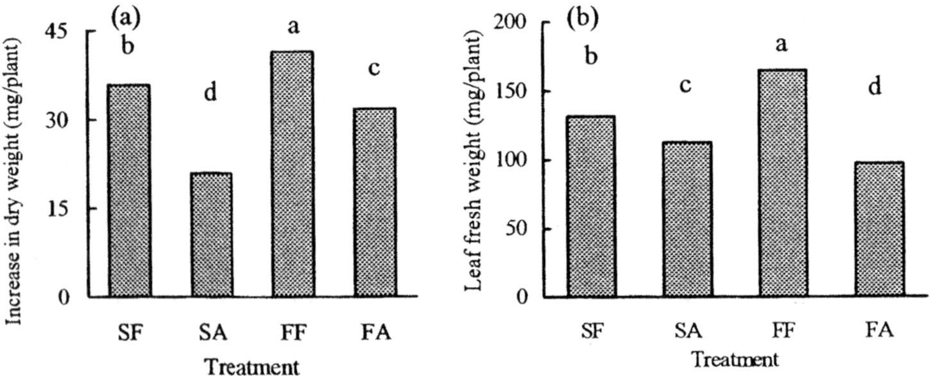

Figure 3. Increased dry weight (a) and leaf fresh weight (b) of neem plants *in vitro* on day 28 (Nguyen and Kozai, 2001b). In the treatment legends, S and F at the left denote the presence or absence of sucrose, respectively. F and A at the right denote Florialite or agar medium, respectively. Different letters on each column show a significant difference at the 5% level by LSD test.

Neem plants cultured on the medium with Florialite supporting material developed significantly longer roots than those on the agar medium did. Under photoautotrophic condition, growth of neem plants *in vitro* on Florialite medium was greater at PPF of 70 μmol m^{-2} s^{-1} than at PPF of 230 μmol m^{-2} s^{-1}. At higher PPF, shoot elongation of neem plants was suppressed, whereas the root elongation was promoted significantly (Nguyen and Kozai, 2001a).

4.6. Coffee

Nguyen et al. (1999a) examined the growth of single node cuttings of *Coffea arabusta* plants cultured *in vitro* at different sugar concentrations (0 or 20 g l^{-1}), types of supporting materials (agar or Florialite) and vessel ventilation rates or number of air exchanges (0.2 or 2.3 h^{-1}) at PPF of 200 $\mu mol\ m^{-2}\ s^{-1}$. Fresh weight, shoot length, root length and leaf area of plants cultured on Florialite with sugar-free nutrient solution under high number of air exchanges per hour were greater than those cultured on sugar-containing media with agar or Florialite. At the end of a 40-day culture period, calli were formed at the shoot base of plants grown *in vitro* on sugar-containing medium with agar as a supporting material (Fig. 4). On the other hand, calli were not formed at the shoot base of plants grown *in vitro* on sugar-free medium with Florialite as a supporting material

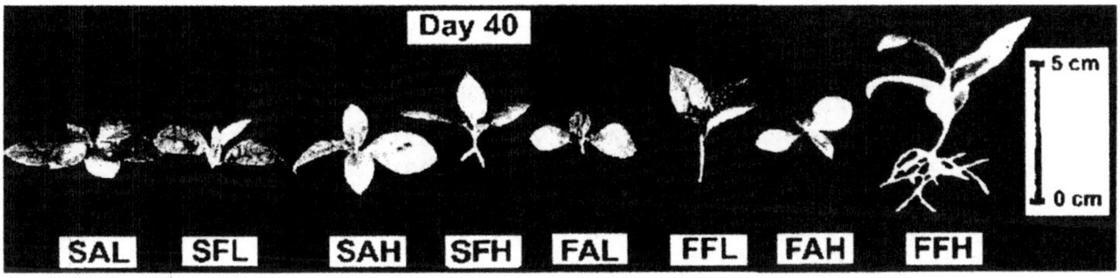

Figure 4. *Coffea arabusta* plants in vitro on day 40 as affected by presence/absence of sugar in the culture media, types of supporting materials (agar and Florialite) and high/low number of air exchanges of the vessel (Kozai et al., 2000).

Our study on photosynthesis and related physiological parameters of coffee plants demonstrated that the CO_2 saturation point of *Coffea arabusta* plantlets was high (4500 to 5000 $\mu mol\ mol^{-1}$) but the light (PPF) saturation point was not as high as those of other plant species (Nguyen et al., 1999b). The net photosynthetic rate of coffee plants *in vitro* was significantly increased when these plants were cultured on Florialite with sugar-free nutrient solution under high vessel ventilation rate per plant (Nguyen et al., 1999a).

Coffee (*C. arabusta*) somatic embryos at cotyledonary and germination stages could be cultured photoautotrophically and did not show any symptom of hyperhydricity in their leaves as compared with those cultured photomixotrophically (Afreen et al., 2001). Pretreatment of somatic embryos at the cotyledonary stage under increased PPF conditions enhanced their photoautotrophic growth in the following culture period (Afreen et al., 2001). These findings support the hypothesis that the growth of somatic embryos at cotyledonary stage can be enhanced significantly by controlling the environmental conditions for promoting photosynthesis.

4.7. Mangosteen

Ermayanti et al. (1999) demonstrated the photoautotrophic growth of mangosteen (*Garcinia mangostana*) by culturing shoots excised from aseptically germinated seedlings on vermiculite or agar medium with the presence (30 g l^{-1}) or absence of sugar and plant growth regulators (10 mg l^{-1} of 2-ip and 1 mg l^{-1} of IBA) in the culture vessel with low (0.1 h^{-1}) or high (4.4 h^{-1}) number of air exchanges at a PPF of 110 μmol m^{-2} s^{-1} under high CO_2 concentration (1300 μmol mol^{-1} in the culture room) (Table 2).

Table 2. Dry weight (DW), percent (%) rooting, number of leaves (NoL) of mangosteen plants cultured for 30 days (Ermayanti et al., 1999). Means ± SD are shown [8].

Treatment					DW	%	NoL
Sucrose conc. (g l^{-1})	Growth regulators	CO_2 conc. (μmol mol^{-1})	No. of air exchanges (h^{-1})	Substrate	(g/plant)	rooting	
30	Yes^z	1300	4.4	Vermiculite	0.30 ± 0.13	40	6 ± 1.0$^{NS\,y}$
30	No	1300	4.4	Vermiculite	0.23 ± 0.18	20	4 ± 0.9^{NS}
0	Yes	1300	4.4	Vermiculite	0.23 ± 0.10	20	6 ± 0.9*
0	No	1300	4.4	Vermiculite	0.20 ± 0.11	40	4 ± 0.9^{NS}
30 (control)	Yes	400	0.1	Agar	0.21 ± 0.11	0	5 ± 0.9
ANALYSIS OF VARIANCE^x							
Sucrose concentration					NS	NS	NS
Growth regulator					NS	NS	**

^{z} Medium contained 10 mg l^{-1} 2-ip and 1 mg l^{-1} IBA.
^{y} NS, *: Nonsignificantly or significantly different from the control treatment at $p \leq 0.05$ according to *t*-test, respectively.
^{x} ANOVA was applied for 4 treatments (except for the control treatment). NS, nonsignificant; **, significant at $p \leq 0.01$.

Dry weight of mangosteen plants on day 30 was not significantly different among treatments. Addition of the growth regulator in the medium increased the number of leaves (Table 2). Twenty to forty percent of the shoots exhibited root induction in the treatments with a high number of air exchanges and vermiculite as supporting material, either with or without sugar/plant growth regulator. On the other hand, in the control treatment with a low number of air exchanges and sugar containing agar medium, no rooting was observed. The CO_2 concentration was the lowest, and the net photosynthetic rate per leaf area in the control treatment was about ten percent of that in the treatments with a high number of air exchanges (Table 2).

4.8. Banana

A study on the photoautotrophy of banana (*Musa* spp.) plants *in vitro* at the rooting stage showed that there was no significant difference in growth between plants cultured either on sugar-free medium or on sugar-containing (20 g l^{-1} sucrose) medium in the same light/dark period condition (16 h/8 h d^{-1}, 12 h/12h d^{-1} or 8 h/16h d^{-1}). Dry weight was about 3 times greater at a photoperiod of 16 h d^{-1} than at 8 h d^{-1}. The shoot/root ratio, on the other hand, was smaller at longer photoperiods, showing that roots develop as well as shoots, especially on sugar-free medium (Nguyen et al., 1999c).

When cultured photoautotrophically *in vitro,* increasing PPF from 50 to 100 µmol m^{-2} s^{-1} and increasing the ventilation rate nearly doubled the fresh weight, and significantly increased a number of unfolded leaves per plant (Nguyen et al., 1999c). There was no significant difference in increased dry weight and number of unfolded leaves of banana plants when PPF was increased from 100 to 200 µmol m^{-2} s^{-1}, even at a high ventilation rate.

Nguyen and Kozai (2001c) demonstrated the growth of banana (*Musa* spp.) shoots *in vitro* at the multiplication stage under photomixotrophic (with 30 g l^{-1} sucrose and 5 mg l^{-1} BA containing medium and 0.2 h^{-1} number of air exchanges of culture vessels) and photoautotrophic (sugar free medium and 3.9 h^{-1} number of air exchanges) conditions for 28 days. The shoot fresh and dry weights on day 14 in photoautotrophic conditions were significantly greater at a PPF of 200 µmol m^{-2} s^{-1} than at a PPF of 100 µmol m^{-2} s^{-1}. The net photosynthetic rate of photoautotrophic banana shoots was significantly greater than that of photomixotrophic shoots. Banana shoots under photoautotrophic conditions had a larger number of unfolded leaves and a greater leaf area than photomixotrophic plants by days 14 and 28 at CO_2 concentrations of both 475 and 1340 µmol mol^{-1} (Nguyen and Kozai , 2001c).

5. AN INTERPRETATION OF THE GROWTH PROMOTION AND QUALITY IMPROVEMENT IN PHOTOAUTOTROPHIC MICROPROPAGATION

The growth of chlorophyllous explants (shoots and leafy nodal cuttings) of most of the woody plant species examined above is greater under photoautotrophic conditions than under photomixotrophic conditions, in the presence of increased CO_2 concentration in the culture vessel during the photoperiod (which is achieved by the use of gas permeable filters in combination with CO_2 enrichment in the culture room). The increase in PPF is also required in most cases. Increases in CO_2 concentration and PPF promoted photosynthesis and thus carbohydrate accumulation of plants (Aitken-Christie et al., 1995; Jeong et al., 1995; Kozai et al., 1995). The use of fibrous or porous supporting materials in place of agar has been shown to be beneficial under high CO_2 concentration and PPF conditions for enhancing root formation and function. These materials are especially effective in promoting the formation of lateral roots and normal vascular systems, which thereby benefits the overall growth and quality of the woody plants.

In addition, an increase in the number of air exchanges of the culture vessel enhances the air movement around the plants growing in the culture vessel. Then, it, in turn, promotes the diffusion of CO_2 and water vapor around plants, resulting in the promotion of photosynthesis and transpiration of plants in vitro (Kitaya et al., 1997).

The increase in the number of air exchanges decreases the relative humidity in the culture vessel from nearly 100% to 85-90% (Aitken-Christie et al., 1995). This decrease in relative humidity in combination with the enhanced air movement in the culture vessel increases the transpiration rate of *in vitro* plants significantly, and thus increases water and nutrient uptake of in-vitro plants (Aitken-Christie et al., 1995).

It should be also noted that calli are not formed at the base of the shoot when grown on sugar-free medium with porous supporting material, which is an advantage of woody plant micropropagation. Physiological mechanism of this phenomenon needs to be studied in the future.

6. FORCED VENTILATION SYSTEMS USING LARGE CULTURE VESSELS AND THEIR APPLICATION

As described above, the CO_2 concentration inside a culture vessel during the photoperiod can be increased and the relative humidity and ethylene concentration can be reduced by attaching gas-permeable filter discs to the lid or the sidewalls of the culture vessel, or by increasing the gaseous concentrations in the culture room. This method of using gas permeable filter discs is a simple way to effectively enhance the growth of plants *in vitro*. However, the CO_2 concentration and other gaseous concentrations in the culture vessel with natural ventilation are interrelated with a number of factors such as the metabolic activity of the plants *in vitro*, the plant size and leaf area, the number of air exchanges of the culture vessel and the culture room environment. Thus, the gaseous concentrations in the culture vessel with natural ventilation are often unpredictable and uncontrollable. Furthermore, it is difficult to provide a high number of air exchanges for a large culture vessel.

Forced ventilation is a method involving the use of mechanical force generated by an air pump or an air compressor to flush a particular gas mixture directly through the culture vessel (often through microporous filters to prevent microorganisms from entering the vessel). With this system, the gaseous composition (CO_2, water vapor, etc.) of incoming air and ventilation rate and/or air current speed in the culture vessel can be controlled relatively precisely by use of a needle valve, a mass flow controller or an air pump with an inverter (Aitken-Christie et al., 1995; Jeong et al., 1995).

Photoautotrophic micropropagation makes it possible to use large culture vessels with minimum risk of microbial contamination. In photoautotrophic micropropagation using a large culture vessel with forced ventilation, the labor cost could be reduced by nearly fifty percent as compared with that in conventional, photomixotrophic micropropagation (Xiao et al., 2000). With a proper control of the gaseous composition in culture vessels with forced ventilation, the growth of plants *in vitro* can be promoted significantly compared with culture vessels with natural ventilation.

A large culture vessel with a nutrient supply unit makes it possible to measure and control the pH, composition and volume of nutrient solution in the culture vessel. The control of plant growth rate is relatively easy in such vessels with a nutrient control unit (Zobayed et al., 2000a).

Fujiwara et al. (1988) developed a large culture vessel with a forced ventilation system for enhancing the photoautotrophic growth of strawberry (*Fragaria* x *ananassa* Duch.) explants and/or plants *in vitro* during the rooting and acclimatization stages, where CO_2 gas was mixed with air and pumped into the vessel. This was a kind of aseptic micro-hydroponic system with a nutrient solution control system.

In 1992, Kubota and Kozai described the photoautotrophic growth of potato (*Solanum tuberosum* L.) plants *in vitro* using a polycarbonate vessel with forced ventilation, containing a multi-cell tray with rockwool cubes. The net photosynthetic rate of potato plants was significantly greater than plants grown in conventional (small) culture vessels with natural ventilation.

Heo and Kozai (1997) demonstrated a forced ventilation micropropagation system using a culture vessel (air volume of 12 l) containing a multi-cell tray widely used for plug seedling production. The cells were filled with sterilized vermiculite or cellulose plugs. The photoautotrophic growth of sweetpotato plants cultured *in vitro* with this system was several times greater than the photomixotrophic growth of plants cultured in small culture vessels containing sugar and with natural ventilation (Fig. 5a & b).

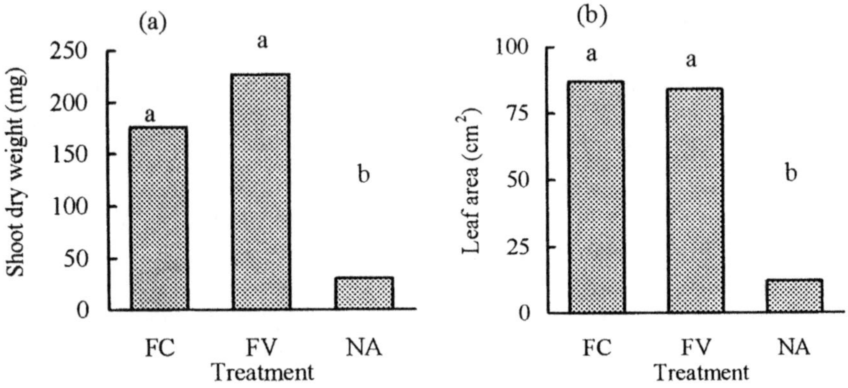

Figure 5. Shoot dry weight (a) and leaf area (b) of sweetpotato plantlets on day 22 as affected by ventilation methods (Heo and Kozai, 1997). In the treatment legends, F and N denote forced and natural ventilations, respectively; C, V and A denote cellulose plug, vermiculite and agar, respectively.

However, in both of the forced ventilation systems mentioned above, the growth in the culture vessel was not uniform. The plants were larger near the air inlet and comparatively smaller near the air outlet.

Zobayed et al. (1999) developed large culture vessels with air distribution pipes for forced ventilation with a major aim to provide an air current pattern that enables uniform distribution of CO_2 concentration and relative humidity as well as air current speed in the culture vessel, and consequently, the uniform growth of sweetpotato plants *in vitro*. The feasibility of forced ventilation systems has been tested for eucalyptus and sweetpotato (Zobayed et al., 2000a) (Fig. 8).

Compared with plants cultured *in vitro* cultured photomixotrophically, the plants cultured *in vitro* photoautotrophically in the large vessel with forced ventilation had a higher net photosynthetic rate, normal stomatal closing and opening, and significantly higher epicuticular leaf-wax content (Zobayed et al., 2000b). Plants cultured in the large vessel with forced ventilation were considered acclimatized *in vitro* and their transpiration rates and percent water loss remained lower than those of conventional plants when transplanted to *ex vitro* conditions (Zobayed et al., 2001a; 2001b).

Coffea arabusta plants *in vitro* cultured in a forced ventilation system had significantly higher fresh and dry weights, shoot length and leaf area tan those cultured with natural ventilation (Nguyen et al., 2001a). The study included 40 days in the *in vitro* stage and 10 days in the *ex vitro* stage (Fig. 6). Mean fresh and dry weights, leaf area, shoot and root lengths and net photosynthetic rate per plantlet were significantly greater in forced high rate treatments compared with those in natural and forced low rate treatments. PPF had a distinct effect on shoot length suppression and root elongation of coffee plantlets in forced high rate treatments. With the forced ventilation method in photoautotrophic micropropagation, the CO_2 concentration inside the culture box could be easily adjusted to promote the net photosynthetic rate of plants.

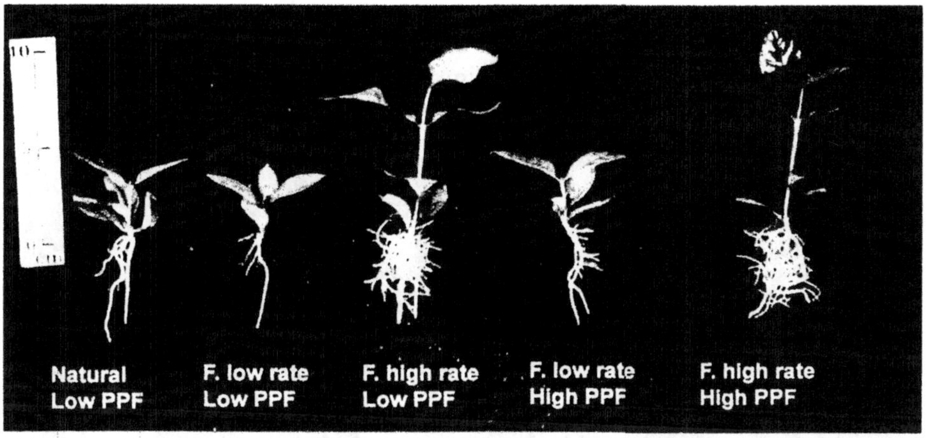

Figure 6. Coffee (*C. arabusta*) plants on day 10 of the *ex vitro* stage as affected by ventilation methods in the *in vitro* stage. In the treatment legends, Natural, Low rate and High rate denote natural and forced low and high rate ventilations (Nguyen et al., 2001a).

The growth of shoots and roots of coffee (*C. arabusta*) somatic embryos cultured photoautotrophically in a large vessel were much greater with forced ventilation than with natural ventilation (Afreen et al., 2002). The percent survival was 98% for plants *ex vitro* derived from somatic embryos grown in the large vessel (volume approx. 3 l) with a forced ventilation system, while it was only 30 % for those grown in the Magenta-type vessel system with natural ventilation. The high percent survival of the photoautotrophically grown coffee plants was thought to be due to their development of normal stomatal functioning with wax formation on leaves and photosynthetic organs. This achievement may lead to a wider application of somatic embryos for producing woody transplants in clonal forestry.

A forced ventilation system for photoautotrophic growth of paulownia (*Paulownia fortunei*) using a large vessel (volume approx. 17 l) was also developed (Nguyen and Kozai, 2001a). The CO_2 uptake rate increased with the increase in the airflow rate over the culture period of 28 days. The growth of single-node leafy cuttings of paulownia was significantly higher under photoautotrophic, forced ventilation conditions than under photomixotrophic, natural ventilation conditions.

Nguyen et al. (2001d) examined the growth of yam (*Dioscorea alata*) leafy cuttings in a forced ventilation system with large vessels. The system was operated under a CO_2 concentration of 550 ± 50 μmol mol^{-1} and a PPF of 150 μmol m^{-2} s^{-1}. On day 20, increased fresh and dry weights, shoot length and leaf area of yam plants *in vitro* under photoautotrophic forced ventilation conditions were significantly greater than those of plants grown under photomixotrophic, natural ventilation conditions with the same PPF and CO_2 concentration of the culture room. After yam plants that were produced under photoautotrophic, forced ventilation conditions were transferred to the *ex vitro* stage, they continued growing significantly faster than plants produced under photomixotrophic, natural ventilation conditions (Fig. 7).

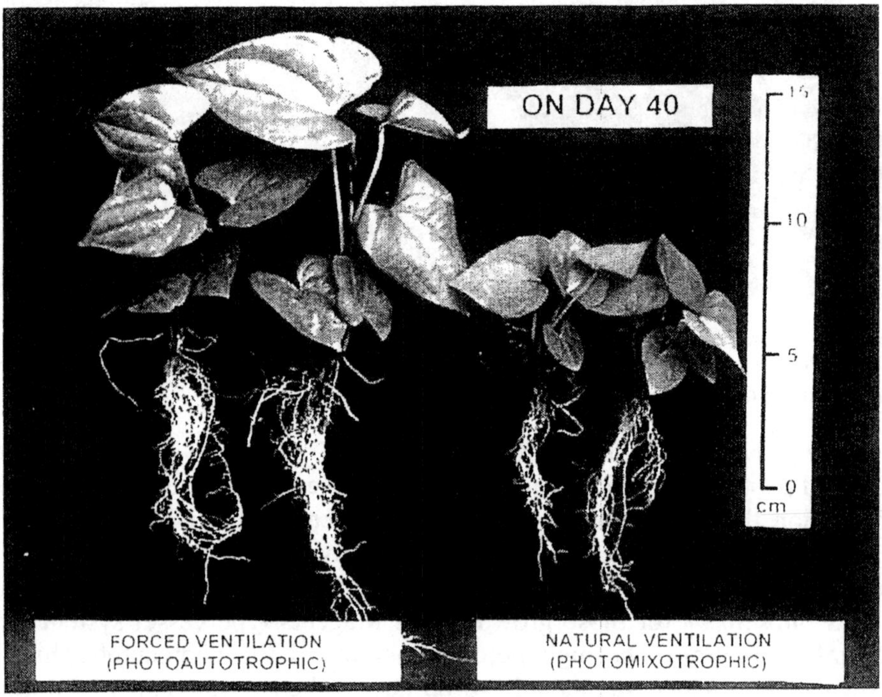

Figure 7. Effect of different types of ventilation during the in vitro stage on the subsequent growth of yam (*Dioscorea alata*) plants. Measurements were made 20 days after the plants were transplanted to pots *ex vitro* (Nguyen et al., 2001d).

Figure 8. Scaled-up photoautotrophic culture vessel (picture taken after removal of the lid). The vessel was 610 mm long, 310 mm wide and 105 mm high (volume approx. 20 l) (Zobayed et al., 2000).

7. SCALED-UP MICROPROPAGATION SYSTEM BY USE OF A CLEAN CULTURE ROOM

A scaled-up forced ventilation micropropagation system can be further developed into a clean culture room, considering itself as a large culture vessel containing plants grown directly in sterile multicelled trays without any covers, which are placed on shelves with a common headspace (Kozai et al., 2000). This kind of photoautotrophic micropropagation system can also be used as a transplant production system using small cuttings under artificial light and pathogen-free conditions or simply a closed transplant production system (Kozai, 1998; Kubota and Chun, 2000).

In research and development for commercialization of such a closed transplant production system, considerable reduction in electricity consumption for lighting and air conditioning has been achieved at Chiba University, Japan using a prototype system. In Japan, electric energy consumption per transplant has been estimated to be 0.7 MJ (0.2 kWh) when a plug tray with seventy-two cells are used (Ohyama and Kozai., 1998). This corresponds to 2-3 Japanese Yen (1.5-2.0 US cents) per transplant, which may be acceptable with view of the fact that the current market price for a micropropagated plant is 50 to 100 Japanese Yen.

Since the cost of electricity is proportional to the total area of the plug trays or the number of plug trays, the electricity cost per transplant decreases with increasing number of cells or transplants per tray. The electric cost can be halved if a plug tray with one-hundred-fifty is used instead of the tray with 72 cells. Remarkable reduction in electricity consumption can be expected in the near future by improving the lighting and air-conditioning systems (Nishimura et al., 2001).

To supply a large number of transplants needed for reforestation (20 to 40 billions of plants per year), large-scale photoautotrophic micropropagation systems with fully automated environmental control and transportation systems may be feasible and may help to compensate partly the worldwide loss of forest and natural resources.

A pilot-scale model of a closed transplant production system, having an automated photoautotrophic micropropagation system was developed at the Matsudo campus of Chiba University, Japan in 2000 (Chun and Kozai, 2000). This system has a floor area of 500 m^2, and consists of a sterilized area and a non-sterilized area. In order to keep the plants pathogen-free, no person is permitted to enter the culture room in this system for handling the trays with plants or for environmental control. Thus, tray transportation and environmental control in the culture room are fully automated. Although this system can produce pathogen-free transplants of any species, it is presently being used to produce virus-free transplants of sweetpotato (*Ipomoea batatas* (L.) Lam).

Kozai et al. (unpublished) also developed another type of the closed system for commercial production of value-added plug transplants for horticultural plants, which consists of a structure covered with thermally insulated opaque walls, multi-layered shelves with fluorescent lamps, home-use air conditioners, and supply units of CO_2 and nutrient solution. This type is developed for production of transplants from seeds. Thus, the degree of asepsis is not required to be very high. This system can also be used for propagation and transplant production of woody and tropical plants using cuttings, which are not completely aseptic but pathogen-free.

In this system, ventilation and heat transmission through walls were reduced as much as possible (1) to minimize the annual cooling and heating costs, (2) to minimize the amount of water for irrigation (Because of the minimized ventilation, net water loss from the closed system is minimized.), (3) to minimize the amount of CO_2 needed for promoting photosynthesis, (4) to prevent insects/pathogens entering into the system, (5) to keep the optimal environment inside the system, and (6) to minimize the discharge of nutrient-rich water to the outside environment.

Electric energy consumption per transplant was 0.7 MJ (0.2 kWh) and its electricity cost was lower than 3 Japanese Yen when tomato transplants with 2-3 true leaves were produced in 128-cell plug trays. The estimated market price of a transplant was 25-30 Japanese Yen. Eighty percent of the total electric energy consumption was for lighting, 17 % for cooling, 0 % for heating and 3 % for others equipment such as an air circulation fan and nutrient solution circulation pump. No power was consumed for heating. When the walls and floor of the closed system are insulated well using thermal insulators with 100-150 mm thick, a heating unit and thus heating cost are not required even during the winter in a cold region. Also, no condensation occurs on the inside surface of walls when the walls are insulated.

Ninety-five percent of the water vapor evapotranspirated from the plug trays was collected as liquid water at the cooling coils of air conditioners and it was reused for irrigation. Relative humidity inside the closed system is kept at 60-70 % during the photoperiod and 90 % during the dark period. Ninety-five percent of CO_2 supplied to the system was fixed as dry matter by the transplants.

The tomato transplant production period was shortened to 17 days, compared with 20-40 days in a greenhouse. Similar results were obtained for eggplant, sweetpotato and Chinese cabbage transplants. In general, the use of the closed system made it possible accurately control the growth, development, vigorousness and uniformity of the transplants. For commercial transplant production of some horticultural crops, the closed system is more economical than a greenhouse. It can produce higher quality transplants using less water, agrochemicals, CO_2 and labor than the greenhouse, and thuscause less environmental pollution.

8. CONCLUDING REMARKS

In the near future, photoautotrophic or sugar-free micropropagation will be commonly used, especially in developing countries, for producing a large number of genetically superior and pathogen-free transplants at low production cost. In fact, the photoautotrophic micropropagation system has been commercialized widely in Kunming, China (Xiao et al., 2000) and partly in Vietnam. At the same time, the photoautotrophic micropropagation system will be adopted on a large scale as a useful means for solving the global problems of environmental conservation, food production and bio-resource production in the 21st century.

Photoautotrophic micropropagation can be done either by using small, conventional culture vessels with gas permeable filters for enhancing natural ventilation or large culture vessels with air pumps for forced ventilation.

The concept of photoautotrophic micropropagation can be further expanded into the idea of 'closed transplant production system'. With a closed transplant production system, the quality of transplants can be further improved; the consumption of artificial energy (electric energy and/or fossil fuels) and other resources (labor, facilities, supplies, etc.) can be decreased; and the release of environmental pollutants can be minimized.

ACKNOWLEDGEMENT

The authors would like to acknowledge Drs. Chalermpol Kirdmanee, Jeongwook Heo, Tri Muji Ermayanti and Sayed M.A. Zobayed for granting permission to use their tables and photographs.

REFERENCES

Afreen F., Zobayed S.M.A., Kozai T. 2001. Mass-propagation of coffee from photoautotrophic somatic embryos. In: Morohoshi N., Komamine A. (eds.) Molecular Breedings of Woody Plants. Elsevier Science B.V., Amsterdam, The Netherlands,

Afreen F., Zobayed S.M.A., Kubota C., Kozai T. 2002. Photoautotrophic growth of *Coffea arabusta* somatic embryos. (submitted).

Aitken-Christie J., Singh A.P., Davies H. 1988. Multiplication of meristematic tissue: a new tissue culture system for radiata pine. In: J.W. Hanover and D.E. Keathley (eds.) Genetic Manipulation of Woody Plants. Plenum Publishing Co., New York, 413-432.

Aitken-Christie J., Davies H., Holland L., Kubota C., Fujiwara K. 1992. Effect of nutrient media composition on sugar-free growth and chlorophyll fluorescence of *Pinus radiata* shoots in vitro. Acta Hort. 319:125-130.

Aitken-Christie, J., Kozai, T., Smith M.A.L. 1995. Automation and Environmental Control in Plant Tissue Culture. Kluwer Academic Publishers, Dordrecht, The Netherlands, 574p.

Biddington N.L., 1992. The influence of ethylene in plant tissue culture. Plant Growth Regulation 11:173-187.

Brainerd K.E., Fuchigami L.H., Kwiatkowski S., Clark C.S. 1981. Leaf anatomy and water stress of aseptically cultured 'Pixy' plum grown under different environments. HortScience 16:173-175.

Chun C., Kozai T. 2000. Closed transplant production system at Chiba university. In: Kubota C. & Chun C. (eds.) Transplant Production in the 21st Century. Kluwer Academic Publishers, Dordrecht, The Netherlands, 20–27.

De Proft M.P., Maene L.J., Debergh P.C. 1985. Carbon dioxide and ethylene evolution in the culture atmosphere of Magnolia cultured in vitro. Physiol. Plant. 65:375-179.

Donnelly D.J., Vidaver W.E., 1984. Pigment content and gas exchange of red raspberry in vitro and ex vitro. J. Am. Soc. Hort. Sci. 109:177-181.

Ermayanti T.M., Imelda M., Tajuddin T., Kubota C., Kozai T. 1999. Growth promotion by controlling in vitro environment in micropropagation of tropical plant species. Proc. The Tokyo Intl. Forum on Conservation and Sustainable Use of Tropical Bioresources. NEDO and IBA, Tokyo, 10-25.

FAO. 1999. State of the world's forests. Food and Agriculture Organization of the United Nations. 149p.

Fujiwara K., Kozai T., Watanabe I. 1988. Development of a photoautotrophic tissue culture system for shoots and/or plantlets at rooting and acclimatization stages. Acta Hort. 230: 153-158.

Grout B.W.W. 1988. Photosynthesis of regenerated plantlets in vitro, and the stresses of transplanting. Acta Hort. 230:129-135.

Heo J., Kozai T. 1997. Photoautotrophic growth enhancement of in vitro plug plants using a forced ventilation micropropagation system. Abstr. Autumn Meeting of Japan Society of Horticultural Science, 298-299.

Heo J., Wilson S.B., Kozai T. 2001. A forced ventilation micropropagation system for photoautotrophic production of sweetpotato plug plantlets in a scaled-up culture vessel: I. Growth and uniformity. HortTech. 11:90-94.

Hdider C., Desjardins Y. 1994. Effects of sucrose on photosynthesis and phosphoenolpyruvate carboxylase activity of in vitro cultured strawberry plantlets. Plant Cell Tiss. Org. Cult. 36: 27–33.

ISAAA. 2001. GM crops and the environment. Global Knowledge Center on Crop Biotechnology, SE*Asia*Center, Manila, Philippines, April 2001.

Jackson M.B., Abbott A.J., Belcher A.R., Hall K.C., Butler R., Camerson J. 1991. Ventilation in plant tissue cultures and effects of poor aeration on ethylene and carbon dioxide accumulation, oxygen depletion and explant development. Ann. Bot. 67: 229-237.

Jeong B.R., Fujiwara K., Kozai T. 1993. Carbon dioxide enrichment in autotrophic micropropagation: Methods and advantages. HortTech. 3:332-334.

Jeong B.R., Fujiwara K., Kozai T. 1995. Environmental control and photoautotrophic micropropagation. Horticultural Reviews 17: 125-172.

Kirdmanee, C. 1995. Environmental control and its effect in *Eucalyptus* micropropagation under photoautotrophic conditions. Ph.D. Dissertation, Chiba University, 172p.

Kirdmanee C., Kitaya, Y., Kozai T. 1995. Effects of CO_2 enrichment and supporting material in vitro on photoautotrophic growth of *Eucalyptus* plantlets in vitro and ex vitro. In Vitro Cell Dev. Biol.-Plant 31: 144-149.

Kitaya Y., Ohmura Y., Kozai T., Kubota C. 1997. Visualization and analysis of air currents on plant tissue culture vessels. Environ. Control in Biol. 35(2): 139-141. (*in Japanese with English abstract, figures and table*)

Kozai T., Oki H., Fujiwara K. 1987. Effects of CO_2 enrichment and sucrose concentration under high photosynthetic photon fluxes on growth of tissue-cultured *Cymbidium* plantlets during the preparation stage. Symp. Florizel on Plant Micropropagation in Hort. Ind., 1987, 135-141.

Kozai T. 1988. Autotrophic (sugar free) tissue culture for promoting the growth of plantlets *in vitro* and for reducing biological contamination. In: T.W. Flegel, G. Tharun and Y. Yuthavonh (eds.) Intl. Symp. Applic. Biotech. Small Ind. Dev. Developing Countries. 21-24 Sept. 1988, Bangkok, Thailand, 39-51.

Kozai T. 1998. Transplant production under artificial light in closed systems – Quality improvement of seedlings and plantlets by environmental control. Proc. of the 3rd Asian Crop Science Conference, Taichung, Taiwan, 296-308.

Kozai T., Koyama Y., Watanabe I. 1988. Multiplication and rooting of potato plantlets in vitro with sugar free medium under high photosynthetic photon flux. Acta Hort. 230:121-127.

Kozai T. 1991. Controlled Environments in conventional and automated micropropagation. In: I. Vasil (ed.) Cell Culture and Somatic Cell Genetics of Plants. Academic Press, New York, N.Y., U.S.A., Vol. 8, 213-230.

Kozai T., Zimmerman R.H., Kitaya Y., Fujiwara K. (eds.) 1995. Environmental Effects and their Control in Plant Tissue Culture. Acta Hort. 393, 230p.

Kozai T., Kubota C., Zobayed S., Nguyen Q.T., Afreen-Zobayed F. 2000. Developing a mass propagation system for woody plants. In: Watanabe K., Komamine A. (eds.) Challenge to the Crisis of the Earth's Biosphere in the 21st Century. Landes Biosphere, Texas, USA, 289-302.

Kubota C., Kozai T. 1992. Growth and net photosynthetic rate of *Solanum tuberosum* in vitro under forced and natural ventilation. HortScience 27: 1312-1314.

Kubota C. 1994. Growth regulation in plant micropropagation by controlling in-vitro physical environment. Ph.D. Dissertation, Chiba University, 161p.

Kubota C., Chun C (eds.). 2000. Transplant Production in the 21st Century, Kluwer Academic Publishers, Dordrecht, The Netherlands, 290p.

Kurata K., Kozai T. 1992. Transplant Production Systems. Kluwer Academic Publishers, Dordrecht, The Netherlands, 299p.

Lapitan G. P. 1999. Developing tissue culture as an alternative propagation technique for the production of high quality seedlings for reforestation. Proc. of the 7th Intl. Workshop of Bio-Refor, Manila, Philippines, 1998.

Nguyen Q.T., Kozai T., Nguyen U.V. 1999a. Effects of sucrose concentration, supporting material and number of air exchanges of the vessel on the growth of in vitro coffee plantlets. Plant Cell Tiss. Org. Cult. 58:51-57.

Nguyen Q.T., Kozai T., Niu G., Nguyen U.V. 1999b. Photosynthetic characteristics of coffee (Coffea arabusta) plantlets in vitro in response to different CO_2 concentrations and light intensities. Plant Cell Tiss. Org. Cult. 55:133-139.

Nguyen Q.T., Nguyen L.K., Nguyen U.V., Kozai T. 1999c. Photoautotrophic micropropagation of tropical plants. In: Altman A., Ziv M., Izhar S. (eds.) Plant Biotechnology and In Vitro Biology in the 21st Century. Kluwer Academic Publishers, Dordrecht, The Netherlands, 659-662.

Nguyen Q.T., Kozai T., Heo J., Thai D.X. 2001a. Photoautotrophic growth response of in vitro cultured coffee plantlets to ventilation methods and photosynthetic photon fluxes under carbon dioxide enriched condition. Plant Cell Tiss. Org. Cult. 66:217-225.

Nguyen Q.T., Kozai T. 2001b. Photoautotrophic micropropagation of tropical and subtropical woody plants. In: Morohoshi N., Komamine A. (eds.) Molecular Breedings of Woody Plants. Elsevier Science B.V., Amsterdam, The Netherlands, 335-344.

Nguyen Q.T., Kozai T. 2001c. Growth of in vitro banana (*Musa* spp.) shoots under photomixotrophic and photoautotrophic conditions. In Vitro Cell. Dev. Biol.-Plant. 824-829.

Nguyen Q.T., Le H.T., Thai D.X., Kozai T. 2001d. Growth enhancement of in vitro yam (*Dioscorea alata*) plantlets under photoautotrophic condition using a forced ventilation system. Proc. The 12th Symp. of the International Society for Tropical Root Crops (ISTRC), September 10-16, 2000, Tsukuba, Japan. (*in press*)

Nishimura M, Kozai T., Kubota C., Chun C. 2001. Analysis of electric energy consumption and its cost for a closed-type transplant production system. J. Soc. High. Tech. Agr. 13:204-209. (*in Japanese with English abstract, figures and tables*)

Ohyama K., Kozai T. 1998. Estimating electric energy consumption and its cost in a transplant production factory with artificial lighting: A case study. J. Soc. High. Tech. Agr. 10:96-107. (*in Japanese with English abstract, figures and tables*)

Short K.C., Warburton J., Roberts A.V. 1987. In vitro hardening of cauliflower and chrysanthemum plantlets to humidity. Acta Hort. 212:329-334.

Smith M.A.L., Palta J.P., McCown B.H. 1986. Comparative anatomy and physiology of microcultured, seedling, and greenhouse-grown asian white birch. J. Am. Soc. Hort. Sci. 111:437-42.

Thorpe T.A., Harry I.S., Kumar P.P. 1991. Application of micropropagation to forestry. In: Debergh P.C., Zimmerman R.H. (eds.) Micropropagation - Technology and Application. Kluwer Academic Publishers, Dordrecht, The Netherlands, 311-336.

Xiao Y., Zhao J., Kozai T. 2000. Practical sugar-free micropropagation system using large vessels with forced ventilation. In: Kubota C., Chun C. (eds.) Transplant Production in the 21st Century. Kluwer Academic Publishers, Dordrecht, The Netherlands, 266-273.

Zobayed S.M.A., Afreen-Zobayed F., Kubota C., Kozai, T. 2000a. Mass propagation of *Eucalyptus camaldulensis* in a scaled-up vessel under in vitro photoautotrophic condition. Annals Bot. 85:587-592.

Zobayed S.M.A., Afreen-Zobayed F., Kubota C., Kozai T. 2000b. Water control and survival of *Ipomoea batatas* grown photoautotrophically under forced ventilation and photomixotrophically under natural ventilation. Annals Bot. 86:603-610.

Zobayed S.M.A., Afreen F., Kubota C., Kozai T. 2001a. Large-scale photoautotrophic micropropagation in a scale-up vessel. In: Morohoshi N., Komamine A. (eds.) Molecular Breedings of Woody Plants. Elsevier Science B.V., Amsterdam, The Netherlands.

Zobayed S.M.A., F. Afreen and T. Kozai. 2001b. Physiology of *Eucalyptus* plantlets grown photoautotrophically in a scaled-up vessel. In Vitro Cell. Dev. Biol. - Plant 37: 807-813.

27. THIN CELL LAYER (TCL) MORPHOGENESIS AS A POWERFUL TOOL IN WOODY PLANT AND FRUIT CROP MICROPROPAGATION AND BIOTECHNOLOGY, FLORAL GENETICS AND GENETIC TRANSFORMATION

Duong Tan Nhut[1], Jaime A. Teixeira da Silva[2], Bui Van Le[3] and K. Tran Thanh Van[4]

[1]*Institute of Biology in Dalat, 116 Xo Viet Nghe Tinh, Dalat, Lam Dong, Vietnam*
Email: nhutduong@hotmail.com
[2]*University of Lisbon, Faculty of Science (FCUL), Department of Applied Plant Biology, C-2 Campo Grande, 1749-016 Lisbon, Portugal*
[3]*Vietnam National University-Ho Chi Minh City, University of Natural Sciences, 227 Nguyen Van Cu Str., Dist. 5, Ho Chi Minh City, Vietnam*
[4]*Institute of Plant Biotechnology, VMR 8168-Paris-Sud University, 91405 Orsay Cedex, France*

Contents

S.M. Jain and K. Ishii (eds.), Micropropagation of Woody Trees and Fruits, 783–814.

Key words

Agrobacterium; biotechnology; embryogenesis; floriculture; flower identity; genetic transformation; *in vitro* flowering; morphogenesis; thin cell layer; tissue culture

Abbreviations

BA	N^6-benzyladenine or
BAP	Benzylaminopurine
CPPU	Forchlofenuron (N-(2-chloro-4-pyridyl)-N-phenylurea)
IBA	Indole-3-butyric acid
NAA	α-Napthalene acetic acid
PGR	Plant Growth Regulator
SAM	Shoot Apical Meristem
TCL	Thin Cell Layer (tTCL = transverse TCL; lTCL = longitudinal TCL)
TDZ	Thidiazuron (N-phenyl-N'-1,2,3-thidiazol-5-ylurea)
2,4-D	2,4-dichlorophenoxyacetic acid

1. INTRODUCTION

The rapid evolution of advanced molecular techniques that the scientific community has witnessed over the past two decades has come to effectively complement conventional breeding, and in many cases, speed it up. The use of biotechnology – largely as new tissue culture and cell biology methods and gene transfer techniques – has not only come to shorten the breeding process, but has also allowed scientists to overcome some of the substantial agronomic and environmental problems that have not been solved using conventional genetic methods.

Traditionally somaclonal variation may be a source of variability, the downside to this phenomenon being the incapacity to control its evolution or its outcome. Genetic and phenotypic variation, especially in the woody plant and fruit crop sectors are not desirable characteristics, especially where uniformity is required. The manipulation and strict control of *in vitro* cultures and their development is the key to the efficient production of cell, organ or tissue cultures and will pave the way for how crops will be utilized and genetically altered in the future. Thin cell layers (TCLs) comprise a system by which the morphogenetic and developmental pathways of specific organs – derived from other specific or non-specific cells, tissues or organs – may be clearly directed and controlled. Moreover, it allows for the study of cytological, physiological, biochemical and molecular changes occurring in a particular morphogenetic program. This strict regulation of the morphogenetic pathways will allow for, *inter alia*, the controlled production of somatic embryos and their subsequent use as synthetic seed, or as mass propagation units. It would also enhance the production capacity of secondary metabolites and pharmaceuticals through transgenic organ cultures, such as those produced by *Agrobacterium rhizogenes*, or by other autotrophic bioreactor plant cultures. The efficiency of genetic transformation is clearly enhanced as a result of cell and/or tissue specificity of gene insertion, and the subsequent successful and controlled regeneration of transformed tissue. Moreover, the use of TCLs allows for the potential production of *in vitro* flowers (independent of or in conjunction with photoperiod, vernalization and/or other environmental cues), and can be used as an explant source, or potentially as a new, long-term ornamental propagule, eliminating the problem of post-harvest deterioration.

The global food and forestry sectors are in a constant search for a higher quality of life for new and innovative products to suit the needs of populations. Lower and more competitive prices and a greater consumer demand (for quantity and quality) has created new challenges and placed some stress on these sectors to produce new products that suit the needs of the consumer. In order to meet this demand, tissue culture and molecular biology techniques are proving useful in the production of woody and fruit crops with novel characteristics (resistance to various pests and diseases, longer shelf life, increased secondary metabolites, cold, frost and salinity tolerance).

TCL technology is a solution to many of the above issues currently hindering the efficient progress of woody plant and fruit crop improvement, since it addresses the issue of plant breeding at the first stage of the problem: regeneration. Since the regeneration of specific organs may be effectively manipulated through the use of TCLs, in conjunction with specific controlled *in vitro* conditions and exogenously applied plant growth regulators, many problems hindering the improvement of *in vitro* plant systems are potentially removed. We will demonstrate, through this comprehensive review, the truth of this claim, and that the application and success of this system in woody plant and fruit crops is widespread. The possibilities of this tool for crop improvement are endless, and go tightly hand-in-hand with molecular and genetic engineering tools. Moreover this system provides a means of mass propagation of a species of interest, and has thus profound potential economic spin-offs.

2. TCL TECHNOLOGY: CONCEPT AND PROCEDURE

Regeneration of plant species has primarily focused on the methodology, but an understanding of the mechanisms that control this process require either a developmental mutant or a model system. Plants have the remarkable ability of reprogramming their already programmed ontogenic program of differentiation, and of mitotically building new morphogenetic patterns, structures (callus, roots, shoots, somatic embryos or flowers) and functions. In the case of shoots and somatic embryos, new individuals can be formed without the need for meiosis and sexual reproduction. The first phase of somatic embryogenesis and organogenesis involves the mitosis of a cell or group of cells, resulting in either: a) friable callus (large and highly vacuolated), b) epidermized nodules comprised of meristematic cells from which the embryos or shoot/root meristems arise, c) direct somatic embryos or d) direct shoot, root or flower primordial meristems. With callus, some cells remain small, becoming isodiametric and meristematic, forming an epidermal structure. The control of meristem, organ or embryo formation from callus lies in the balance between cell division and cell extension. Development within plants is temporally and spatially controlled in a network of different units: organs, tissues and cells. If there is a change in this network, then inhibition at the organism level may occur. The concept of the TCL method is to isolate cells from this inhibition network and to reprogram them by manipulating the *in vitro* conditions. Many of the enzymes or polysaccharides released during excision and wounding of plant tissues are necessary for inducing plant growth and development (Tran Thanh Van and Mutaftschiew 1990). The rationale behind the use of few cells within a TCL is that they are in close contact with wounded cells (the sites for the synthesis of new cell wall constituents and release of oligosaccharides) and nutrients and other factors placed within the medium to "control" morphogenesis. They are thus more medium-dependent. The use of larger explants (such as conventional stem or leaf fragments), in contrast, shows a strong polarity in the response to the medium, and the fact that they may have a greater level of endogenous substances, including PGRs, makes them more medium-independant.

TCL explants are of a small size, and are excised either a) longitudinally (lTCL), being thus composed of a few tissue types or b) transversally (tTCL), thus composed of several tissue types, but which are normally too small to separate. The latter are sometimes referred to as thin sections. There are still no strict definitions for each, but rather broad size guidelines exist: lTCLs are 0.5-1mm wide, 5-10mm long; tTCLs are 0.1-5mm thick. Micro TCLs, despite being a few micrometers thick (using a microtome) (Lee-Stadlemann et al. 1989), may easily become necrotic and be contaminated. The ideal medium conditions suitable for survival of TCLs are species-dependent, and require testing of all conditions of the *in vitro* culture system (including PGRs, nutrients, light, osmolarity and temperature, *inter alia* (Tran Thanh Van 1980).

3. MODEL TCL SYSTEMS

3.1. Tobacco (*Nicotiana tabacum*)

Tobacco is one of the most well studied plants, alongside *Arabidopsis thaliana* and *Antirrhinum majus*. It is often utilized as an unrelated control plant where other species are concerned, is an economically important crop, and is the plant in which most extensive TCL studies exist. Four morphogenetic programs: direct flower formation, direct root formation, direct bud formation, and callus without organogenesis, were induced by varying the concentration of carbohydrates and growth substances as shown in Table 1 (Tran Thanh Van 1980, 1981, Tran Thanh Van and Trinh 1986). lTCLs from tobacco (*Nicotiana tabacum* Wisc.38) were excised from floral ramifications and cultured on solid MS medium. Glucose or sucrose, Kinetin and indole butyric acid (IBA) were added at the concentration shown in Table 1. Cultures were incubated in continuous light of 60 µmol $m^{-2}s^{-1}$, at 24-27°C and 50% humidity. A shift from the floral state to the vegetative state was possible by the addition of cell wall oligosaccharides to the medium, suggesting that oligosaccharides released by PGR-treatment of TCLs (Tran Thanh Van et al. 1985a) or by pH (Cousson et al. 1989, Tran Thanh Van and Mutaftschiew 1990) can act as signaling molecules.

Histological studies of morphogenetic programs have shown that all organs and callus arises from the subepidermal layer, and are visible after an average of 8-10 days from the initialization of culture. Flower buds are visible from 12-21 days after culture initiation and form well-developed anthers with fertile pollen (Tran Thanh Van and Dien 1975, Tran Thanh Van and Chlyah 1976, Tran Thanh Van and Trinh 1986). Pollination of *de novo* flowers resulted in fruit with viable seed (Trinh et al. 1987).

Table 1. Control of morphogenetic program by variation of carbohydrate, IBA or Kinetin in *Nicotiana tabacum* Wisc.38 thin cell layers.

Program	Carbohydrate	PGRs (µM)	Light	#/TCL	DT (days)
Flower	Glucose 30	1 IBA+1 K	L	30-50	12-21
Vegetative	Sucrose 30	1 IBA+10K	L	500-800	10-12
Root	Sucrose 10	10 IBA+0.1K	L/D	10-20	16-18
Callus	Sucrose 30	3 IBA+0.1 K	L	-	8-10

DT= Differentiation time, L=light, D=dark

Other factors controlling morphogenesis in TCLs could be carried out by controlling light, sugar and oligosaccharide concentration, ionic composition of the culture medium and pH (Cousson and Tran Thanh Van 1983, 1992, Tran Thanh Van et al. 1985, Cousson et al. 1989). Furthermore, the auxin/cytokinin ratio in tobacco TCLs has been shown to apply for the control of *de novo* bud and or root formation in many species (Klimaszewska and Keller 1985, Pélissier et al. 1990, Becher et al. 1992, Golds et al. 1993, Okole and Schulz 1996, Ahn et al. 1996, Gendy et al. 1996, Bui Van Le et al. 1997, 1998a, 1998b, 1999a, 1999b).

The flower program can only be induced on TCLs excised from floral branches and not from the base of the stem (Tran Thanh Van 1973a, 1973b).

The suggestion that the TCLs are pre-induced to form flowers is not substantiated by the activation of genes involved in floral differentiation. This, together with the fact that other organ programs can be induced from the same TCL source suggests that the genes necessary for floral induction and differentiation are activated *in vitro*, in response to *in vitro* medium conditions and additives.

3.2. Lily (*Lilium longiflorum*)

Lilies are fast becoming one of the most important bulbous crop species globally. The traditional asexual propagation of *Lilium* spp. by bulb scales (Stimart and Ascher 1978) as well as the lack of efficient micropropagation systems for species within the *Lilium* genus (Priyadarshi and Sen 1992) has prompted members within this group to extensively study TCL as a tool and solution for these shortcomings (reviewed in Nhut et al. 2001d). Due to the extensive nature of TCL studies on lily, and with the successful manipulation of all morphogenetic programs, it is also used here as a model system. In an attempt to regenerate different parts of Easter lily (*L. longiflorum* Thumb., 2n=2x=24) the effect of tTCL explant source (receptacle, stem node and internode, pseudo-bulblet – defined as a compact bulb without visible bulb scales – and leaf) together with different factors such as sucrose concentration, explant position, activated charcoal (AC) and PGRs on the mass propagation of *Lilium* was studied (Nhut et al 2001b). The ideal tTCL explant size for bulblet formation is of 0.3-3 mm thickness. For all the studies on bulblet formation using TCL methods, a MS/2 medium supplemented with 1-2.7 μM α-Napthalene acetic acid (NAA) or 10 μM IBA and sucrose at concentrations from 20-30 g/l was used for the rooting of shoots, bulblets and pseudo-bulblets. Plantlets obtained in the light developed well on this medium and were subsequently transferred to the greenhouse, with a 90-100% survival rate.

Leaf explant culture Young leaf explant thin cell layers (0.3 mm) were excised and cultured in MS medium supplemented with 3% sucrose and 2 μM BA combined with 6 μM NAA (Nhut 1998). Results showed that young *Lilium* leaf explants were efficient for shoot regeneration. Pseudo-bulblets formed on tTCL leaf explants, with more bulbs forming on the adaxial surface of leaf sections. A gradient of bulblet formation in the leaf was observed. Other studies on bulblet formation from leaf sections of *Lilium* have also been reported (Stenberg 1977, Niimi 1986, Gerrits and De Klerk, 1995).

Pseudo-bulblet culture Shoot regeneration and subsequent plant production have been achieved in TCL pseudo-bulblet explants of *in vitro L. longiflorum* using forchlofenuron (CPPU; N-(2-chloro-4-pyridyl)N-phenylurea). All bulblet explants were harvested from stem nodes of *in vitro*-grown plants. They were excised transversally (0.3 mm) and placed on an MS medium with 1-3% sucrose and supplemented with 1-3 μM CPPU.

Induction of buds and shoot proliferation occurred under light conditions with a 16h-photoperiod at a photon fluence rate of 100 μmol $m^{-2}s^{-1}$. A maximum of 15 bulblets can be obtained from one pseudo-bulblet TCL explant. The study on the effect of sucrose concentration in combination with CPPU on shoot regeneration showed that at low sucrose concentrations (2-4%) shoots were obtained, but at high concentrations (6-9%) bulblet formation in tTCL explants resulted in a decreased number of regenerated plants (Bui et al. 1999a, Nhut 2001a, Nhut et al. 2001f). These plantlets were rooted on MS with 10 μM IBA and hardened-off in the greenhouse and were used in micropropagation in *Lilium*.

Receptacle culture Closed flower buds of *L. longiflorum*, 8 cm in length, were used in this experiment. tTCL explants (2-3 mm), prepared from the receptacle and ovary of a flower stalk – denuded of flower, bracts and perianth – were cultured on MS media containing 3% sucrose, 2 μM N^6-benzyladenine (BA) and 5 μM NAA (optimal condition) to induce bud proliferation in light. Buds appeared in receptacle tTCL explants within 3 weeks of culture, but not in the ovary or flower stalks (Nhut et al. 2000a, Nhut 2001h). These buds continued to develop into bulblets. After 6 weeks culture the maximum proliferation response was scored by the number of bulblets formed per tTCL (an average of 41 shoots formed directly on the surface of receptacle sections). Bulblets were transferred to MS/2 medium without phytohormones to develop into shoots, before transfer to rooting medium. Receptacle culture of *Lilium* flower stalks was shown to be a suitable method for the successful micropropagation of *Lilium* by TCL.

Stem node culture Stem node (2-3 mm) tTCL sections from plantlets derived from shoot tips of dormant bulbs were cultured on MS/2 medium with 20 g/l sucrose, 0.8% agar and 1μM BA (Nhut 1998). After 4 weeks culture pseudo-bulblets formed on each cultured stem node tTCL (an average of 2 pseudo-bulblets per tTCL). The pseudo-bulblets continued to develop after transfer onto medium with a high BA concentration (2.3 μM), forming an average of 6 pseudo-bulblets per tTCL. The pseudo-bulblets developed into the flowering stage without dormancy being observed (Nhut 2001b).

Young stem culture tTCL square epidermal layers (4x4 mm, 1 mm thick) of tip-off young stems (stems denuded of apical shoot tips) from dormant bulbs were cultured on MS/2 medium with 20 g/l sucrose, 8 g/l agar and 1 mg/l AC (Nhut et al. 2001b). An average of 4 bulblets was formed per tTCL after 4 weeks culture. Shoots did not form in AC-free medium, indicating that AC has the same effect as a cytokinin-like hormone on the development of tTCL explants. A gradient potential of bulblet formation from the tip-off stem (every 1 mm along the stem length) was also observed. The optimal sucrose concentration was shown to be 20-40 g/l for *L. longiflorum* bulblet formation. In other studies (Nhut 2001c, 2001e), similarly prepared transverse young stem sections of *L. longiflorum* were shown to form different organs when explants were exposed to different PGRs. They formed bulblets when cultured on MS with 5.4 µM NAA, roots with 2.2 µM 2,4-dichlorophenoxyacetic acid (2,4-D), shoots with 1.0 µM IBA, plantlets (shoots and roots) with 2.0 µM IBA, PLBs with 1.1 µM thidiazuron (TDZ; N-phenyl-N'-1,2,3-thidiazol-5-ylurea) and somatic embryos and embryogenic callus with 5.4 µM and 0.4, 1.1 and 2.2 µM of NAA and TDZ, respectively.

Somatic embryogenesis Somatic embryogenesis has been achieved in tTCLs of *in vitro L. longiflorum* pseudo-bulblet explants using TDZ and NAA. All pseudo-bulblet explants were harvested from stem nodes of *in vitro*-grown plants. They were excised transversally (0.5 mm) and placed on MS/2 medium containing 3% sucrose, NAA and TDZ (Nhut et al. 2000, Nhut 2001g). Induction of somatic embryogenesis occurred after 6 weeks under light and somatic embryos developed shoots and roots on MS/2 PGR-free medium with 1-3% sucrose.

4. APPLIED TCL SYSTEMS

4.1. Bamboo (*Bambusa spp. and Dendrocalamus spp.*)

This popular woody monocot is a landscape ornamental and is usually vegetatively propagated by division of culms, but has been used for inducing somatic embryos by the TCL method. tTCLs from young leaves sectioned through a bud from the base to the apex were cultured on medium with 18 µM 2.4-D, forming, after five weeks in the dark, friable and mucilaginous callus or compact and epidermized embryonic nodules, depending on the horizontal and vertical gradient: the inner leaf tTCLs and the tTCLs of the basal part of this leaf were the most responsive for somatic embryogenesis (Jullien and Tran Thanh Van 1994). The pale yellow nodules were similar to the embryonic nodules differentiated on *Iris pallida* leaf tTCLs. Flavonoids were related to auxin transport while polysaccharide mucilaginous secretion was activated by 2,4-D in embryonic maize cells (Everett et al. 1985), and often associated with somatic embryogenesis of *Dactylis glomerata* (Hahne et al. 1988) and *Pennisetum americanum* (Vasil and Vasil 1982). The embryonic nodules of *Bambusa* formed ELSs that failed to develop any further. Their protein pattern presented two bands that were absent in friable callus and in plant root samples. Somatic embryos of *Dactylis glomerata* showed 11 specific polypeptides (Hahne et al. 1998). Somatic embryogenesis with subsequent plant development and in *vitro* flowering were obtained in *Bambusa glaucescens*, *Dendrocalamus giganteus* and *D. strictus* after a callus phase initiated on excised zygotic embryos or nodal segments of 10-day-old *in vitro*-grown seedlings (Rout and Das 1994).

Semi thin TCLs (3-5 mm) of 2-year-old *D. brandisii* Kruz taken from single node stem segments were micropropagated on MS with BAP, with an increase in shoot cluster formation with an increase in BAP concentration, with optimal shoot formation obtained at 10 µM of both BAP and coumarin, a reaction similar to that found in *B. tulda* (Saxena 1990). Rooting could then be successfully induced in shoots by placing on MS with 10 μM of both IBA and coumarin. This method has also been successfully applied to *B. glaucescens* 'Golden Goddess' (Jullien and Tran Thanh Van 1994), *D. asper* (Arya et al 1999) and to *D. strictus* (Saxena and Dhawan 1999).

4.2. Banana (*Musa* sp.)

This tropical fruit tree is a beautiful leafy home and kitchen garden plant and an attractive ornamental, especially the red banana varieties. TCL or micro-cross sections (50-500 μm cut by microtome) of banana and plantain leaf base were used for callus and shoot bud induction. As in *Populus* cross-sections, 400 μm was the optimum thickness: an average of 15 plantlets could be obtained, compared to 4-5 from standard protocols (shoot meristem, cell or protoplast culture; Okole and Schulz, 1996), making the TCL system an attractive alternative because of its direct and high frequency bud differentiation. With thinner micro-cross sections (50-300 μm), 1-2 shoots formed, perhaps due to the smaller number of intact cells and/or the composition of the medium, especially the mineral concentration that must be adapted to the TCL method.

4.3. Cassava (*Manihot esculenta*)

This plant is usually a tuber crop that is the staple food diet for millions of people in tropical and subtropical South America, Africa and Asia. The variegated form of this species is, however, a popular and attractive leafy landscape plant. Immature leaves from 15 genotypes, 3-6 mm in size were excised and cultured in the presence of different 2,4-D concentrations. Somatic embryos developed directly on the explants with 4-6 mg/l 2,4-D. Secondary embryogenesis was induced and long term culture established and maintained up to 18 months (Szabados et al. 1987). Plantlets developed from primary and secondary embryos after the addition of 0.1 mg/l BA, 1 mg/l GA_3 and lowering the 2,4-D concentration to 0.01 mg/l. Differences in response were reported with genotypes: usually up to 15-20 plantlets developed on one explant or 2 or 3 for the most reluctant genotype.

4.4. Citrus (*Citrus* spp. and *Ponciris trifoliata*)

Citrus is the most widely grown fruit crop worldwide. Among this species several citrus serve as ornamental trees, such as orange, lemon, and mandarin. Trifoliate orange or *Poncirus trifoliata*, a rootstock species, is also used as an ornamental plant due to its attractive leaves and flowers. Using traditional methods to multiply this species requires a long time due to the length of regeneration. Several reports show that *Citrus* species can be regenerated by hypocotyl organogenesis (Edriss and Burger 1984, Burger and Hackett 1986) or when shoot tips are used but the efficiency of micropropagation is low (Kitto and Young 1981, Sim et al. 1989): in *Citrus mitis*, an average of only 3.7 buds per shoot tip explant were formed with 1 mg/l BA (Sim et al. 1989). The lTCL epicotyl explants (1x5 mm subepidermal tissue) were excised from donor plants in commercial 1-2 year olds after sterilization and culture with different combinations of BA (1-50 μM) and TDZ (1-10 μM) to determine the optimal condition for the *in vitro* organogenesis of *P. trifoliata*. A higher percentage of responsive lTCLs and a higher frequency of bud regeneration were obtained with a BA and TDZ combination: at 10 μM BA and 1 μM TDZ, 90% of lTCLs were responsive, with 37 buds per lTCL being obtained after 4 weeks culture. Plantlets developed on 1 μM GA_3 medium and rooted on 5 μM NAA. With this technique more than 40 shoots per lTCL can be multiplied. In separate studies, callus, somatic embryo and plant regeration morphogenetic programmes were controlled in six *Citrus* species (*C. deliciosa*, *C. limon*, *C. madurensis*, *C. medica*, *C. tardiva* and *C. sinensis*) from pistil tTCL cultures (Carimi et al. 1999), but showed different somatic embryogenic frequencies depending on the pistil origin: stigma>ovary>style. All somatic embryogenesis was achieved on MS medium supplemented with either BA or malt extract. The TCL method can be successfully applied to other *Citrus* species improving the micropropagation rate, as has been seen in *C. aurantium*.

4.5. Coconut palm (*Cocos nucifera*)

Trees of this species have great importance in terms of their fruit and many by-products, but have an equally weighty value as landscape plants. Gupta et al. (1984) regenerated coconut plants by using zygotic and somatic embryos. Thin transverse slices (0.2-0.5 mm thick) from different sized young inflorescences were also used to induce nodular structures that later developed into shoot-like structures, but smaller explants have a better survival rate than larger ones (Blake 1983, Sugimura and Salvana 1989).

4.6. Conifers (*Pseudotsuga menziesii*, *Sequoiadendron* spp)

Many trees within this group are highly valuable and frequently utilized landscape plants. For most woody species and especially conifers, true-to-type cloning of mature trees requires rejuvenation by grafting or serial cutting. Longitudinal TCLs excised from *Pseudotsuga menziesti*, *Sequoiadendron giganteum* and *Sequoia sempervirens* (California Redwood) adult elite tree shoot tips regenerated shoots without a callus phase. In *Sequoia*, fertile cones were obtained from tTCL rejuvenated shoot internodes that developed spontaneously from the base of the main stem (Tran Thanh van et al. 1985b). Monteuuis (1991) showed that meristems excised from a 100-year-old *S. giganteum* shoot during bud break cultured *in vitro* regenerated juvenile lines that remained juvenile during 3 years.

4.7. Garcinia / Kokum (*Garcinia mangostana*)

This is a food flavourant as well as a medicinal plant used for reducing cholesterol and controlling appetite. High frequency of direct shoot bud regeneration was obtained on thin transverse 3 mm sections made across the leaf vein (Goh et al. 1994). The shoot buds developed at the distal pole of each thin section cultured in the presence of 20 µM BA, forming roots and plantlets after transfer to IBA. The number of shoots was 50 times higher on thin transverse sections than on half leaf explants.

4.8. Monterey pine (*Pinus radiata*)

Pine trees are popular landscape and garden ornamental trees, and many species have important essential oils and chemical constituents. Cotyledon sections (3-5 mm long) formed new shoot primordia from single cells of the first subepidermal cell layer 3 days after culture with 25 µM BA (Villalobo et al. 1985). Organogenesis occurred more intensely on the border of each cotyledon tTCL where it is thinner. Shoots were obtained after 3 weeks culture. The structures organized directly from meristematic centers similar to those described for the monolayer of *Torenia* (Chlyah et al. 1975).

4.9 Paulownia (*Paulownia fortunei*)

Paulownia is a tree with growing ornamental and landscape value with widespread utilization. *Paulownia* sp. has been previously propagated by conventional practices (Burger et al 1985, Rao et al 1996, Bergmann and Moon 1997). tTCLs (1 mm thick) from disinfected young stem segments (Fig. 1A) from 1-year old trees could form shoots (Fig. 1B) at an optimal concentration of 5 µM BAP with 0.1 µM NAA (12 buds/tTCL). These shoots could root easily on PGR-free MS, 1/2 MS or MS with 1 µM IBA (Fig. 1C), with a 100% acclimatization rate (Fig. 1D). This effective protocol has allowed for the efficient commercial micropropagation of *Paulownia* (Fig. 1E).

4.10. Poplar (*Populus spp.)*

This is a woody ornamental tree often used in landscape design. Associated with general tissue culture recalcitrance of woody species, the use of the TCL system in the *in vitro* multiplication of this species brings hope to others that were equally recalcitrant. Shoot regeneration frequency was compared between: a) thin section or "micro-cross sections" (100, 200, 300, 400 or 500 µm) made through leaf mid-veins, and b) different widths: <1 mm, 1-2 mm or >3 mm for 400 µm sections. They were cultured on medium containing 0.2 mg/l BA and 0.01 mg/l NAA (Lee-Stadlemann et al. 1989). Results showed that thin sections were more effective (25 times greater) than 1 cm explants, the optimal thickness being 300, 400 or 500 µm. Thin sections produced buds even when they were randomly oriented on the medium. In contrast to larger explants, the initial explant orientation (pole proximal/distal or abaxial/ adaxial) was not critical. The efficiency of bud regeneration on micro cross-sections 1-2 mm long was 3 times higher than 1 mm long ones, demonstrating the importance of explant size.

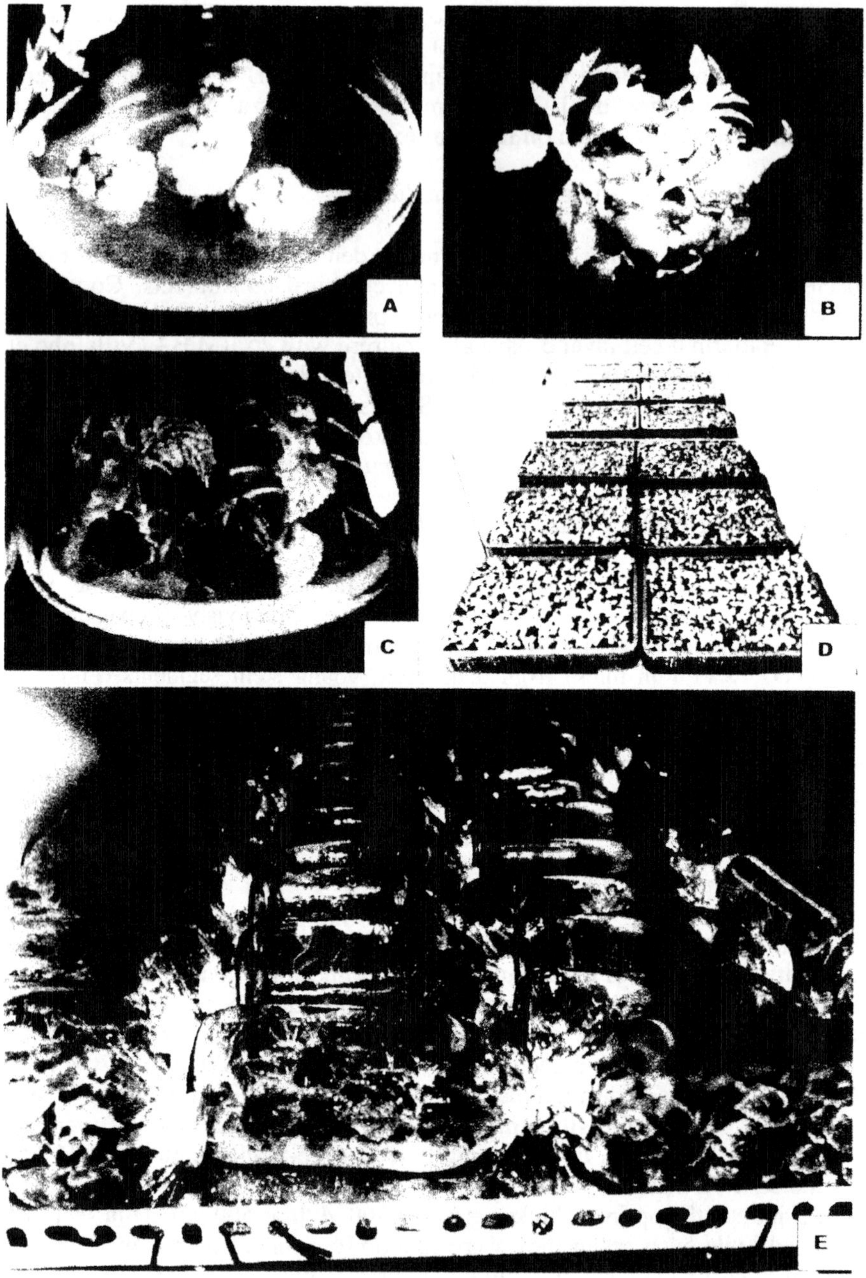
A
B
C
D
E

4.11. Rose (*Rosa spp.)*

Rose is one of the four major cut flower species commercially explored worldwide. Micropropagated plants prove to be superior to budded plants in both yield and flower quality. Although somatic embryogenesis has the potential to be a very efficient method for rose micropropagation there are only few studies on successful somatic embryogenesis of rose from leaves and immature leaves (Hsia and Korban 1996), stem segments (Rout et al. 1991), immature seeds (Kunitake et al. 1993), petioles or roots (Marchant et al. 1996), all of which used micropropagated tissue culture material as the explant source. Explants were commonly cultured on full or half MS, supplemented with 0.05-5 mg/l 2,4-D or NAA usually in combination with BAP, Zeatin or Kinetin. Preincubation at a high (100 μM) 2,4-D concentration increased the frequency of both organogenic and embryogenic callus from *Rosa hybrida* leaf explants (Hsia and Korban 1996). Sucrose at 2-3% was used as the sole carbon source in most cases. However, replacement of sucrose by galactose or fructose increased somatic embryogenesis from leaf explants of some *R. hybrida* cultivars (Hsia and Korban 1996). Breaking of bud dormancy in roses is important for rapid multiplication of roses using grafting. TCLs were excised longitudinally from dormant buds from *R. hybrida* 'Baccara' floral stalks and cultured on medium with 10 μM BAP and 3 μM GA_3. After one week on culture medium, bud breaking and development began, with more than 7 buds per lTCL being obtained after 4 weeks culture. The synergetic effect of dormancy break and bud development was observed in the presence of BAP and GA_3. This TCL protocol could be applied to the rapid multiplication of rose species used for grafting onto rootstocks.

4.12. Tomato (*Lycopersicon esculentum*)

Many bushy dwarfs are used as ornamental house and pot plants, and even as hanging basket plants, in addition to their normal culinary use. TCL technology was applied to studies pertaining to the molecular basis of host-parasite interactions (in particular nematodes) *in vitro* (Radin and Eisenback 1991). Tomato peduncle epidermal TCLs were induced to form roots when cultured on MS with Kinetin and IAA, and rooted cultures could then be utilized in separate experiments with *Meloidogyne incognita* inoculation.

5. GENETIC DETERMINANTS IN VEGETATIVE DIFFERENTIATION, MORPHOGENESIS AND EMBRYOGENESIS RELATED TO THE USE OF TCLs

The analysis of mechanisms that control somatic embryogenesis or organogenesis in plants can be via the identification of developmental mutants or by the development of model systems in which the morphogenetic program can be strictly controlled (through the manipulation of *in vitro* biotic and abiotic conditions). Typical zygotic and somatic embryos in higher plants differentiate into an apical meristem and a basal meristem, the former giving rise to the shoot meristem, the latter into a root meristem. In *Nasturtium*, the conversion of a root meristem into a shoot meristem is possible, by the simple *in vitro* application of Kinetin (Aeschbacher et al. 1994), or of root meristems into somatic embryos (Sticklen 1991) to cortical cell TCLs. Root organogenesis is controlled by the *RAC* gene in *Arabidopsis* (Muller et al. 1985), the *RTCS* gene in maize (Hetz et al. 1996) and the *MORTAL* gene in white clover (White et al. 1998), while shoot organogenesis is controlled by the *CKI* family of genes (Kakimoto 1996). One mm thick tTCLs from soybean cotyledonary nodes cultured on 1 μM BA produced multiple shoot clusters, and when used as a pretreatment for *Sorghum*, *Panax ginseng* and *Digitaria* seedlings, it enhanced somatic embryogenesis (Ahn et al. 1996). Interestingly, tTCLs of twice the thickness initiated the floral program when exposed to a higher (20 μM) concentration of BA. Such shifts from the vegetative to the floral state may be accounted for by differing rates of cell division, and cytokinins (Tran Thanh Van 1991, Binns 1994, auxins and oligosaccharides (Tran Thanh Van et al. 1985, Lerouge et al. 1990, Marfa et al. 1991) play an important role in determining the differentiation of flowers or vegetative buds in several plant species. Laux et al (1996) found that the *WUS* gene is responsible for *Arabidopsis* shoot apical meristem (SAM) maintenance, necessary for the self-perpetuation of the SAM, which in turn is essential for the repetitive initiation of shoot structures during plant development. The presence of the *PkMADS* or of *StMADS* in *Paulownia kawakamii* and *Solanum tuberosum*, respectively, are required for shoot bud induction in excised leaf cultures (Kumar and Prakash 2000). Since protein synthesis, energy stores, DNA sysnthesis and auxin transport are all very important in the early stages of SAM determination, an understanding of these mechanisms with studies using TCL explants (Kim and Ernst 1994) will allow for the isolation and characterization of genes and their functions involved in the determination and development of these tissues.

6. TCL IN FLORAL DIFFERENTIATION AND *IN VITRO* FLOWERING

The SAM is an autonomous unit that gives rise to the shoot, but is also a group of undifferentiated and dividing cells that are patterned by information sent out from the surrounding differentiated cells through an elaborate and versatile communication network (Rinne and van der Schoot 1998). The ability to develop flowers defines a plant's reproductive success. During vegetative growth the SAM establishes the shoot architecture by giving rise to a reiterative series of primordia that develop as leaves and/or lateral vegetative shoots (branches) with their own apical meristem. The *CLV1* gene creates a balance between maintenance and organogenesis in the SAM. Following the switch to reproductive growth – termed floral induction – most or all of the SAM give rise to flowers, under appropriate, environmental stimulative conditions. In the past few years, the mechanisms that define the timing of floral program activation have been elucidated, as have been the genes that determine the fate of meristems, and those that define floral initiation process (FLIP genes) and development (reviewed in Weigel 1995, Levy and Dean 1998, Piñero and Coupland 1998, Pidkowich et al. 1999), form and shape (or floral organ identity (FOIG) genes in *Arabidopsis*, *Antirrhinum* and *Pisum*). The information to date strongly suggests that the general rules governing floral development in *Arabidopsis* are applicable to all dicots, and hence have massive applications in the understanding of flowering in floricultural crops or the control of flowering in ornamental crops. It has been shown that not only does the *LEC1* gene control somatic embryogenesis, as is the presence of this gene the only pre-requisite for this organogenic pathway (Lotan et al. 1998). Recently, a *MYB/TTG/GL2* gene cassette has been shown to control and regulate the placement and fate of specialized epidermal cells in the root, hypocotyls and leaf (Benfey 1999). Organ fate in cells derived from the SAM i.e. meristematic cells, are characterized by KNOX genes, including the *STM* gene that regulates the formation of the vegetative meristem (Lenhard and Laux 1999), while the *CUC1* and *CUC2* genes are required to separate organ primordia in both shoot and floral meristems (Aida et al. 1997). These KNOX genes specify axes of symmetry/asymmetry (Hudson 1999). The process of differentiation is controlled by a *SRD* group of genes, which are temperature-sensitive (Sugiyama 1999).

AG in *Arabidopsis* and *PFG* in *Petunia* (Immink et al. 1999) are required for floral meristem determinacy (i.e. mutants have indeterminate growth; Yanofsky et al. 1990), while the *CLF* gene is its negative regulator (Goodrich et al. 1997). The *CEN* gene (Bradley et al. 1996) is fundamental in the formation of a floral meristem with indeterminate growth, while the *TFL* genes code for its growth counterpart. Floral symmetry is controlled by the *CYC* and *DICH* gene in *Antirrhinum* (Luo et al. 1996), with homology to the *TB1* gene in maize (Doebley et al. 1995), and its negative control by the *DIV* gene (Almeida et al. 1997). Floral mutants (*PIN*, *EFL*, *TFL*) pertaining directly to the timing (control of early or late) of flowering have been identified (Okada et al. 1991, Shannon and Meeks-Wagner 1991, Zagotta et al. 1992, reviewed in Koornneef et al. 1998), while juvenility has been attributed to the *EMF* gene (Sung et al. 1992). *PIN* mutants exhibited a phenotype of no meristematic structure, similar to that observed when wild-type TCLs were cultured in the presence of polar auxin transport inhibitors morphactin or NPA, while *emf* mutants exhibited an endogenous level and IAA polar transport level at 8% and 10% of the wild type, respectively (Okada et al. 1991). Cell wall oligosaccharides are released upon treatment with PGRs, as seen in tobacco TCL systems, and these are hypothesized to be cell-to-cell communication networks that share the auxin polar transport (Tran Thanh Van 1973a, Tran Thanh Van et al. 1985, 1990, Mutafstchiew et al. 1987). The transition from vegetative to floral state is generally coordinated by environmental cues such as daylength (photoperiod), light quality and temperature. Light perception influences floral initiation, and this is regulated by *PHYB* in red light photoreceptors

(Reed et al. 1993), *CRY1* (Bagnall et al. 1996) or *CRY2* (Koornneef et al. 1991, Guo et al. 1998) in blue light photoreceptors, in *Arabidopsis*. The *GI* (Fowler et al. 1999, Park et al. 1999), *TOC1* (Millar et al. 1995) and *FKF1* (Nelson et al. 2000) genes couple flowering time to the circadian clock, and control flowering in response to daylength. *CO* (Simon et al. 1996), *FCA* (Macknight et al. 1997) and *GI*, *FT* and *FWA* (Kornneef et al. 1998) mediate the response to photoperiod under long days, and both the circadian clock and light receptors are involved in this response. *FLC* plays a central role in the vernalization response, and inhibits flowering (Michaels and Amasino 1999), as does *FLF* (Sheldon et al. 1999) and *FRI* (Clarke and Dean 1994). TCLs were used inexperiments aimed at substituting the vernalization stimulus for flowering in *Geum urbanum* by prolonged cold treatment of initial explants, or by treating SAMs with Kinetin and gibberellins, stimulus for flowering in *G. urbanum* by prolonged cold treatment of initial explants, or by treating SAMs with Kinetin and gibberellins, activating cell division in the corpus and medullary zones (Tran Thanh Van 1985).

With respect to sex determination, the presence of the *SLM* genes, independent of the gender of the plant, indicates that the mechanism underlying sex expression is downstream of FOIGs (reviewed in Juarez and Banks 1998). A number of genes and their alleles that are differentially expressed in male or female buds of *Silene latifolia* have been identified (e.g. *MEN-9*), but not yet fully characterized (Robertson et al. 1997). *SK1* and the *TS* family of genes in maize causes efemination of plants, while the *TRA* and *FEM1* genes in *Ceratopteris richardii* account for femaleness and maleness, respectively (Juarez and Banks 1998), regulated by *MAN-1* and *HER* genes. Following floral initiation, floral whorl and organ identity and shape genes are many, regulating the control of these processes. These genes have been extensively reviewed in *A. majus* (reviewed in Coen 1996) and *A. thaliana* (reviewed in Meyerowitz 1998).

Even though many floral initiation and development genes have been elucidated at an *in planta* level, the use of TCL systems to understand the expression of these genes may be of vital importance, since there exists the capacity to manipulate all environmental parameters within the medium, while changing only one at a time. Moreover, once the function of these FLIP or KNOX genes has been fully elucidated, one of the potential practical applications is their activation or control in specific cellular and tissue programs, which can be strictly controlled in TCL systems. The use of TCLs provides a convenient system for the study of these FOIGs, since it is simple, reducing the variable background due to interactions between organ/tissue/cell and environment to a minimum, and since the transport of nutrients and metabolites, PGRs and signal perception often found in whole plants is eliminated. The TCL system thus provides the researcher with a system whereby a specific pattern of morphological development can be used to test the presence of genes supposedly involved with that program. However, the TCL system is limited in that it is not always possible to convert from one mophogenetic program to another, as these may be controlled *in planta* by factors that are unknown or cannot be controlled *in vitro*, such as florigens (Bernier 1988) or antiflorigens (or repressors of the transition from vegetative to floral state e.g. *SVP* (Hartmann et al. 2000), *LFY*, *CAL*, *AP1*, *TFL1* and *FUL* (Ferrandiz et al. 2000) in *Arabidopsis* and *PFG* in *Petunia* (Immink et al. 1999). Often the pre-existence of meristems or meristematic tissue is vital for the TCL system to function, in which case activation of cell division in the corpus zone is a prerequisite for floral differentiation, and may be induced by exogenously-applied cytokinins. The expression of the genes are expressed *in planta*, and represent expression or inhibition following a special or temporal sequence preimposed during ontogenetic processes from the embryogenic phase to the juvenile, mature and reproductive phases. The use of *in vitro* TCL systems in which different morphogenetic patterns (somatic embryo, root, vegetative shoot, flower) can be programmed directly and separately from differentiated cells (and not from embryogenic or meristematic cells) strengthens the function of floral and organ identity genes, which are observed *in planta*.

The use of TCLs has proved to be successful in the induction of *in vitro* flowers of tobacco (Tran Thanh Van, 1973a), *G urbanum* (Tran Thanh Van 1985) and *Petunia* sp. (Mulin and Tran Thanh Van 1989) but floral initiation could only be observed when cotyledonary nodes were utilized as initial explants in these species. Even though *in vitro* flowering has been induced in other species, the origin of floral meristem formation is tissue-specific, and these systems can be further refined if TCLs were to be utilized. In the case of tobacco, a day-neutral species, the key factors for floral differentiation *in vitro* from floral branch TCL explants are: light intensity and quality, sugar concentration; relative concentration of NAA or IBA versus cytokinins (especially kinetin). This control is lost in photoperiodic *Nicotiana* species, but restored once natural photoperiods are applied to *in vitro* cultures. Of interest, the rhizogenic program is overwritten when the use of *Agrobacterium rhizogenes* (with Ri plasmids) produce hairy roots (Tran Thanh Van 1990).

7. TCL AS A TOOL FOR GENETIC ENGINEERING AND CROP IMPROVEMENT THROUGH TRANSFORMATION

Genetic engineering is a tool that, hand-in-hand with an understanding of the genetic and molecular functioning of cells is used to achieve genetic transformation. In no way does it substitute traditional plant breeding, but rather complements it. Plant transformation is a core technology in the genetic engineering of plants, and is normally composed of three processes: 1) the introduction of genes into plant cells; 2) selection of transgenic cells and 3) regeneration of transgenic plants. However the limiting factor has been the third step. Without successful regeneration, in a controlled and defined manner, leading to the formation of organs or plants that are genetically and physiologically normal, there will be no success of stable transgene expression. One of the greatest hindrances to successful (systemic distribution of a transgene) is the formation of chimeras or even escapes on selective medium. Often initial explants utilized in the regeneration procedure are excessively large, and the capacity of both transformed and untransformed cells (following the gene introduction method) to regenerate is the same, especially if the selection level is low, or if the regenerating shoots arise from the surface of cells not in contact with the selection medium, or from within the explant, where the selective agent has not had time to diffuse through the explant, and exert its selective pressure. The utilization of TCLs can eliminate the presence of untransformed cells, and exposes all cells on the thin layer to the selective medium. Only cells harbouring the selector gene within their genome proliferate on the selective medium, thus eliminating the formation of chimeric tissues, especially from organs derived from single cells. Success of transformation varies widely between ornamental species, and has been achieved primarily by *Agrobacterium*-mediated gene transfer, and to a lesser extent by direct gene transfer (particle bombardment, electroporation, electrophoresis, silicon fibres, magnetite particles, or protoplast manipulation). The reports on the use of a TCL system as an initial receptor explant for gene transfer are few, but those that exist demonstrate the effectiveness of introducing a gene into an explant with defined cellular structure and with a controlled regeneration programme, allowing for the formation of non-chimeric transgenic plants.

Transgenic tobacco plants transformed with the *rol*B gene have more pronounced rhizogenesis and flowering, and appear to be involved in the promotion of meristem formation (Altamura et al. 1994).

Using traditional breeding methods, breeders have been able to create new varieties that have desirable traits, such as yield, colour, shape, plant architecture, vase life and resistance against pest and diseases. One of the major goals of breeders in some ornamental species is to obtain blue flowers. The first application of gene technology to modify flower colour led to creation of an orange pelargonindin-production petunia variety by expressing the dihydroflavanol-4-reductase gene (*dfr*) from maize in the suitable petunia line (Meyer et al. 1987, Oud et al. 1995). Recent advances in gene isolation and transfer between species have led to the emergence of plants with novel aesthetic properties, such as the isolation of a flavonoid 3′, 5′ hydroxylase (*hyd*) gene from *Petunia* (Holton et al. 1993) and its introduction into pelargonidin or cyanidin-producing rose cultivars to produce blue rose varieties (Holton and Cornish 1995). By using chalcone synthase gene (*chs*) antisense, white flowers were obtained in gerbera (Elomaa et al. 1993) and pink chrysanthemum (Courtney-Gutterson et al. 1994).

Thus genetic biotechnology is becoming an even more important tool in achieving such novelties, but efficiency of regeneration is a prerequisite for success in genetic transformation, and for the rapid maintenance of the modified phenotype through micropropagation. Furthermore, an understanding of the basic mechanisms of organogenesis and or somatic embryogenesis is needed to select the morphogenetic program (vegetative bud, somatic embryogenesis, callus or roots) and other factors important to obtain high frequencies of genetic transformation. The effect of transformation also depends on cells that are competent for both regeneration and transformation (Creemers-Molenaar et al. 1994). Tobacco (cv. *Samsun*) lTCLs excised from floral pedicels and cultured to induce vegetative buds (Tran Thanh Van 1980) were used to study cell competence for regeneration and transformation by biolistics. lTCLs were precultured at 0, 3, 5, 7 or 15 days in medium for bud induction and then bombarded using the particle inflow gun for DNA delivery (Finer et al. 1992). GUS transient expression showed that GUS expression increases with explant pre-culture time. Ten days pre-culture is the optimal period for obtaining genetic transformants. Similarly, other ornamental species may be regenerated and genetically transformed by using the TCL method, as in tTCL *Gladiolus* explants at the pre-mature bud stage (15 days on medium with 3 μM TDZ) by using biolistic transformation. Thus the pre-culture period necessary for induced cell competent for plant regeneration from TCL explants is an important factor for transgenic plant formation. Moreover TCL explants have an increased surface when infected or exposed to *Agrobacterium* or biolistic transformation.

TCLs have been used as an explant source for the study of transformation, with transgenic plants obtained (Charest et al. 1988, Trinh et al. 1987, Tran Thanh Van et al. 1990, Bui Van Le et al. 1998b). In *Brassica napus* ssp. olifera 'Westar' tTCLs a high frequency of transformation (40-50%) was obtained when tTCLs were *Agro*infected for 30 s and co-cultivated for 2 days before the addition of an antibiotic (Charest et al. 1988). In *Nicotiana plumbaginifolia* tTCLs, transformants were obtained on a 100 mg/l kanamycin selective medium after co-cultivation with *A. tumefaciens* for 2 days (Trinh et al. 1987). Transgenic plants with the *bar* gene were confirmed by Southern analysis, revealing 3 basta-resistant lines when *Digitaria sanguinalis* tTCLs were biolistically transformed (Bui Van Le et al. 1998c).

In addition to the technique used for transforming ornamental species (Robinson et al. 1993; Mol et al. 1995; Burchi et al. 1996), the development of a transformation method using TCL explants to induce a morphogenetic program may further increase this system's desirability.

8. CONCLUSIONS

All patterns of morphogenesis displayed by a plant (callus, root, shoot, flower, somatic embryo) can be induced either separately or in combination, each with a well-defined reversible/irreversible phase. Plant tissue culture and molecular biotechnology are at an important crossroad where one or the other cannot solely be utilized to achieve the successful analysis of *in vitro* or *in planta* physiological and genetic systems. Both have powerful techniques that permit the understanding of, *inter alia*, mechanisms that control processes such as transgene expression, *in vitro* flowering and morphogenesis. TCL systems have been used extensively as a tool that allows for the understanding of these processes, and opens the way for new research that may further elucidate certain physiological and genetic pathways and processes, which, in plant tissue culture and molecular biotechnology, still remain a paradigm. The TCL system is a simplified system that requires only a small amount of plant material and medium volume (circ. 15 µl per TCL), and provides a good system for the study of fundamental and applied aspects of regeneration and transformation. The TCL system has been effectively utilized to study organogenesis and embryogenesis in ornamental, floricultural and medicinal species, in which all organs have been induced *in vitro*.

Acknowledgements: The authors wish to thank Professor Oluf L. Gamborg for critically reading the manuscript.

REFERENCES

Aeschbacher, R.A., Schiefelbein, J.W. and Benfey, P.N. (1994) The genetic and molecular basis of root development, *Ann. Rev. Plant Physiol. Plant Mol. Biol* **45**, 25-45.

Ahn, I.O., Bui Van Le, Gendy, C. and and Tran Thanh Van, K. (1996) Direct somatic embryogenesis through thin cell layer culture in *Panax ginseng*, *Plant Cell Tiss.Org. Cult* **45**, 237-243.

Aida, M., Ishida, T., Fukaki, H., Fujisawa, H. and Tasaka, M. (1997) Genes involved in organ separation in *Arabidopsis*: an analysis of the *cup-shaped cotyledon* mutant, *Plant Cell* **9**, 841-857.

Almeida, J., Rocheta, M. and Galego, L. (1997) Genetic control of flower shape in *Antirrhinum majus*, *Development* **124**, 1387-1392.

Altamura, M.M., Capitani, F., Gazza, L., Capone, I. And Costatino, P. (1994) The plant oncogene *rol*B stimulates the formation of flower and root meristemoids in tobacco thin cell layers, *New Phytol* **126**, 283-293.

Anderson, N.O. (1987) Reclassification of the genus *Chrysanthemum*, *Hort. Sci* **22**, 313.

Anderson, W.C. (1984) A revised tissue culture medium for shoot multiplication of rhododendron, *J. Amer. Soc. Hort. Sci* **109**, 343-347.

Arya, S., Sharma, S., Kaur, R. and Arya, L.D. (1999) Micropropagation of Dendrocalamus asper by shoot proliferation using seeds, *Plant Cell Rep* 18, 879-882.

Bagnall, D.J., King, R.W. and Hangarter, R.P. (1996) Blue-light promotion of flowering is absent in *hy4* mutants of *Arabidopsis*, *Planta* **200**, 278-280.

Bajaj, Y.B.S., Sidhu, M.M.S. and Gill, A.P.S. (1992) Micropropagation of *Gladiolus*, In: Bajaj, Y.P.S. (ed) *Biotechnology in agriculture and forestry vol. 20, High-Tech and Micropropagation IV*, Springer-Verlag Berlin, pp 135-143.

Becher, T., Haberland, A. and Koop, H.U. (1992) Callus formation and plant regeneration in standard and microexplants from seedling of barley (*Hordeum vulgare* L.), *Plant Cell Rep* **11**, 39-43.

Benfey, P.N. (1999) Is the shoot a root with a view?, *Curr. Opin. Plant Biol* **2**, 39-43.

Bergmann, B.A. and Moon, H-K. (1997). *In Vitro* adventitious shoot production in *Paulownia*, *Plant Cell Rep* 16, 315-319.

Bernier, G. (1988) The control of floral evocation and morphogenesis, *Ann. Rev. Plant Physiol* **45**, 175-219.

Bhattacharya, P., Dey, S., Das, N. and Bhattacharya, B.C. (1990) Rapid mass propagation of *Chrysanthemum morifolium* by callus derived from stem and leaf explants, *Plant Cell Rep* **9**: 439-442.

Bilkey, P.C. and Cocking, E.C. (1981) Increased plant vigor by *in vitro* propagation of *Saintpaulia ionantha* Wendl. from sub-epidermal tissue, *Hort. Sci* **16**, 643-644.

Bilkey, P.C., McCown, B.H. and Hildebrand, A.C. (1978) Micropropagation of African violet from petiole cross-sections, *Hort. Sci* **13**, 37-38.

Binns, A.N. (1994) Cytokinin accumulation and action: biochemical, genetic and molecular approaches, *Ann. Rev. Plant Physiol. Plant Mol. Biol* **45**, 173-196.

Blake, J. (1983) Tissue culture propagation of coconut, date and oil palm, In: J.H. Dodds (ed) *Tissue culture of trees*, Croom Helm, London, pp 29-50.

Bradley, D., Carpenter, R., Copsey, L., Vincent, C., Rothstein, S. and Coen, E. (1996) Control of inflorescence architecture in *Antirrhinum*, *Nature* **379**, 791-797.

Bui, V.L., Hang Phuong, N.T., Anh Hong, L.T. and Tran Thanh Van, K. (1999b) High frequency shoot regeneration from *Rhynchostylis gigantea* (Orchidaceae) using thin cell layer, *Plant Growth Reg* **28**, 179-185

Bui, V.L., Jeanneau, M., Nghieng Thao, D.M., Vidal, J. and Tran Thanh Van, K. (1998b) Rapid regeneration of whole plants in large crabgrass (*Digitaria sanguinalis* L.) using thin cell layer culture, *Plant Cell Rep* **18**, 166-172.

Bui, V. L., Nghieng Thao, D.M., Gendy, C., Vidal, J. and Tran Thanh Van, K. (1997) Somatic embryogenesis on thin cell layers of a C_4 species, *Digitaria sanguinalis* (L.) Scop, *Plant Cell. Tiss. Org. Cult* **49**, 201-208.

Bui, V.L., Nghieng Thao, D.M., Gendy, C., Vidal, J. and Tran Thanh Van, K. (1998c) Transformation of a C_4 monocot, the *Digitaria sanguinalis* (Large crabgrass) using somatic embryogenesis induced on thin cell layers. Abst p191, 9th Int Congr Plant Tiss Cell. Cult, Jerusalem, Israel, 14-19 June.

Bui, V. L., Nghieng Thao, D.M., Jeanneau, M., Sadik, S., Tu, S., Vidal, J. and Tran Thanh Van, K. (1998a) Rapid plant regneration of a C_4 dicot species: *Amaranthus edulis*, *Plant Sci* **132**, 45-54.

Bui, V.L., Nhut, D.T. and Tran Thanh Van, K. (1999a) Plant production via shoot regeneration from thin cell layer pseudo-bulblet explants of *Lilium longiflorum in vitro*, *C. R. Acad. Sci, Paris* **322**, 303-310.

Burchi, G., Mercuri, A., Benedetti, L. and Giovanini, A. (1996) Transformation methods applicable to ornamental plants, *Plant Tiss. Cult. Biotech* **2**, 94-104.

Burger, D.W. and Hackett, W.P. (1986) Gradients of adventitious bud formation on excised epicotyl and root sections of *Citrus*, *Plant Sci* **43**, 229-232.

Burger, D.W., Liu, L. and Wu, L. (1985). Rapid micropropagation of *Paulownia tomentosa*, *Hort. Sci* 20, 750-751.

Bush, S.R., Earle, E.D. and Langhans, R.W. (1976) Plantlets from petals segments, petal epidermis, and shoot tip of the periclinal chimera, *Chrysanthemum morifolium* 'Indianapolis', *Amer. J. Bot* **63**, 729-737.

Carimi, F., De Pasquale, F. and Crescimanno, F.G. (1999) Somatic embryogenesis and plant regeneration from pistil thin cell layers of *Citrus*, *Plant Cell Rep* **18**, 935-940.

Cassells, A.C. and Plunkett, A. (1984) Production and growth analysis of plants from leaf cutting and from tissue cultures of disks from mature leaves and young axenic leaves of African violet (*Saintpaulia ionantha* Wendl), *Sci. Hort* **23**, 361-369.

Cassells, A.C., Plunkett, A. and Kelleher, D. (1986) Sreening of *Saintpaulia ionantha* Wendl. cultivars for caulogenetic potential based on the *in vitro* response of young axenic leaves on auxin and cytokinin factoral media, *Sci. Hort* **30**, 151-157.

Chang, C. and Chang, C.W. (1998) Plant regeneration from callus culture of *Cymbidium ensifolium* var. misericors, *Plant Cell Rep* **17**, 251-255.

Charest, P.J., Holbrook, L.A., Gabard, J., Iyer, V.N. and Miki, B.I. (1988) *Agrobacterium*-mediated transformation of thin cell layer explants from *Brassica napus* L, *Theor. Appl. Genet* **75**, 438-445.

Chen, Y. and Piluek, C. (1995) Effects of thidiazuron and N^6-benzylaminopurine on shoot regeneration of *Phalaenopsis*, *Plant Growth Reg* **16**, 99-101.

Chlyah, A. and Tran Thanh Van, K. (1974) Histological changes epidermal and subepidermal cell layers of *Begonia* rex induced to form *de novo* unicellular hairs, buds and roots, *Bot. Gaz* **145**, 55-59.

Chlyah, A. and Tran Thanh Van, M. (1975) Differential reactivity in epidermal cells of *Begonia* rex excised and growth *in vitro*, *Physiol. Plant* **35**, 16-20.

Chouard, P. and Aghion, D. (1961) Modalité de la formation des bourgeons floraux sur des culture de segments de tigre de tabac, *C.R. Acad. Sci. Paris*, **252**, 3864-3866.

Chu, C.Y., Knight, S.L. and Smith, M.A.L. (1993) Effect of liquid culture on the growth and development of miniature rose (*Rosa chinensis* Jasq 'Minima'), *Plant Cell Tiss. Org. Cult* **32**, 329-334.

Clarke, J.H. and Dean, C. (1994) Mapping *FRI*, a locus controlling flowering time and vernalization response in *Arabidopsis thaliana*, *Mol. Gen. Genet* **242**, 81-89.

Coen, E.S. (1996) Floral symmetry, *EMBO J* **16**, 6777-6788.

Compton, M.E. and Veilleux, R.E. (1992) Thin Cell Layer morphogenesis, *Hort. Rev* **14**, 239-264.

Courtney-Gutterson, N., Napoli, C., Lemieux, C., Morgan, A., Firoozabady, E. and Robinson, K.E.P. (1994) Modification of flower color in florist's chrysanthemum: Production of a white-flowering variety through molecular genetics, *Biotechnology* **12**, 268-271.

Cousson, A. and Tran Thanh Van, K. (1983) Light- and sugar-mediated control of direct de novo flower differentiation from tobacco thin cell layers, *Plant Physiol* **72**, 33-36.

Cousson, A. and Tran Thanh Van, K. (1992) Influence of ionic composition of the culture medium on *de novo* flower formation in tobacco thin cell layers, *Can. J. Bot* **71**, 506-511.

Cousson, A., Toubart, P. and Tran Thanh Van, K. (1989) Control of morphogenetic pathway in thin cell layers of tobacco by pH, *Can. J. Bot* **67**, 650-654.

Creemers-Molenaar, J., Harkkert, J.C., Van Staveren, M.J. and Gilissen, L.J.W. (1994) Histology of the morphogenic response in thin cell layer explants from vegetative tobacco plants, *Ann. Bot* **73**, 547-555.

Criley, R.A. and Broschat, T.K. (1992) *Heliconia*: Botany and horticulture of a new floral crop. *Hort. Rev* **14**, 1-56.

Detrez, C., Tetu, T., Sangwan, R.S. and Sangwan-Norrel, B.S. (1988) Direct organogenesis from petiole and thin cell layer explants in sugar beet cultured *in vitro*, *J. Exp. Bot* **39**, 917-926.

Dever, L.V., Blackwell, R.D., Fullwood, N.J., Lacuesta, M., Leegood, R.C., Onck, L.A., Pearson, M. and Lea, P.J. (1995) The isolation and characterization of mutants of the C_4 photosynthetic pathway, *J. Exp. Bot* **39**, 917-926.

Dillen, W. and Buysens, S. (1989) A simple technique to overcome vitrification in *Gypsophila paniculata* L, *Plant Cell Tiss. Org. Cult* **19**, 181-188.

Doebley, J., Stec, A. and Hibbard, L. (1997) The evolution of apical dominance in maize, *Nature* **386**, 485-488.

Edriss, M.H. and Burger, D.W. (1984) *In vitro* propagation of 'Troyer' citrange from epicotyl segments, *Sci. Hort* **23**, 159-162.

Elomaa, P., Honkanen, J., Puska, R., Seppänen, P., Helariutta, Y., Koes, R., Gerats, A.G.M., Mol, J.N.M. and Stuitje, A.R. (1993) *Agrobacterium*-mediated transfer of antisense chalcone synthase cDNA to *Gerbera hybrida* inhibits flower pigmentation, *Biotechnology* **11**, 508-511.

Ernst, R. (1994) Effect of thidiazuron on *in vitro* propagation of *Phalaenopsis* and *Doritaenopsis* (Orchidaceae), *Plant Cell Tiss. Org. Cult* **39**, 273-275.

Everett, N.P., Wach, M.J. and Ashworth, D.J. (1985) Biochemical markers of embryogenesis in tissue culture of the maize inbred B73, *Plant Sci* **41**, 133-140.

Faure, O. (1989) Embryons somatiques de *Vitis rupestris* et embryos zygotiques de *Vitis* sp.: morphologie, histologie, histochimie et dévеloppment, *Can. J. Bot* **68**, 2305-2315.

Fellman, C.C., Read, P.E. and Hosier, M.A. (1987) Effect of thidiazuron and CPPU on meristem formation and shoot proliferation, *Hort. Sci* **22**, 1197-1200.

Ferrandiz, C., Gu, Q., Martienssen, R. and Yanofsky, M.F. (2000) Redundant regulation of meristem identity and plant architecture by *FRUITFULL*, *APETALA1* and *CAULIFLOWER*, *Development* **127**, 725-734.

Finer, J.J., Vain, P., Jone, M.W. and McMullen, M.D. (1992) Development of the particle inflow gun for DNA delivery to plant cells, *Plant Cell Rep* **11**, 323-328.

Flores, H.E. and Teutonico, R.A. (1986) Amaranths (*Amaranthus* spp.): potential grain and vegetable crops, In: Y.P.S. Bajaj (ed) *Biotechnology in agriculture and forestry 2, crop I*, Springer-Verlag, Tokyo, pp: 568-578.

Fowler, S., Lee, K., Onouchi, H., Samach, A., Richardson, K., Morris, B., Coupland, G. and Putterill, J. (1999) *GIGANTEA*: a circadian cloack-controlled gene that regulates photoperiodic flowering in *Arabidopsis* and encodes a protein with several possible membrane-spanning domains, *EMBO J* **18**, 4679-4688.

Gendy, C., Séne, M., Bui Van Le, Vidal, J. and Tran Thanh Van, K. (1996) Somatic embryogenesis and plant regeneration in *Sorghum bicolor* (L.) Moench, *Plant Cell Rep* **15**, 900-904.

Gerrits, M.M. and De Klerk, G.J. (1995) Dry-matter partioning between bulbs and leaves in plantlets of *Lilium speciosum* regenerated *in vitro*, *Acta Bot. Neerl* **41**, 461-468.

Gill, R., Gerrath, J. and Saxena, P.K. (1992) High-frequency direct embryogenesis in thin layer cultures of hybrid seed geranium (*Pelargonium*), *Can. J. Bot* **71**, 408-413.

Goh, C-J., Lakshmanan, P. and Loh, C-S. (1994) High frequency direct shoot bud regeneration from excised leaves of mangosteen (*Garcinia mangostana* L.), *Plant Sci* **101**, 173-180.

Goh, C-J., Nathan, M.J. and Kumar, P.P. (1995) Direct organogenesis and induction of morphogenic callus through thin section culture of *Heliconia psittacorum*, *Sci. Hort* **62**, 113-120.

Golds, T.J., Babczinsky, S. and Koop, H-U. (1993) Regeneration from intact and sectioned immature embryos of barley (*Hordeum vulgare* L.): The scutellum exhibits an apico-basal gradient of embryogenic capacity, *Plant Sci* **90**, 211-218.

Goodrich, J., Puangsomlee, P., Martin, M., Long, D., Meyerowitz, E.M. and Coupland, G. (1997) A polycomb-group gene regulates homeotic gene expression in *Arabidopsis*, *Nature* **386**, 44-51.

Gozu, Y., Yokoyama, M., Nakamura, M., Namba, R., Yomogida, K., Yanagi, M. and Nakamura, S. (1993) *In vitro* propagation of *Iris pallida*, *Plant Cell Rep* **13**, 12-16.

Gray, D.J. and Mortensen, J.A. (1987) Initiation and maintenance of long term somatic embryogenesis from anthers and ovaries of *Vitis longii* 'Microsperma', *Plant Cell Tiss. Org. Cult* **9**, 73-80.

Guo, H., Yang, H., Mockler, T.C. and Lin, C. (1998) Regulation of flowering time by *Arabidopsis* photoreceptors, *Science* **279**, 1360-1363.

Gupta, P.K., Kendurka, S.V., Kulkarni, V.M., Shigurka, M.V. and Mascarenhas, A.F. (1984) Somatic embryogenesis and plants from zygotic embryos of coconut (*Cocos nucifera* L.) *in vitro*, *Plant Cell Rep* **3**, 222-225.

Hahne, G., Mayer, J.E. and Lorz, H. (1988) Embryogenic and callus-specific proteins in somatic embryogenesis of the grass, *Dactylis glomerata* L, *Plant Sci* **55**, 267-279.

Hartmann, U., Hoehmann, S., Nettesheim, K., Wisman, E., Saedler, H. and Huijser, P. (2000) Molecular cloning of *SVP*: a negative regulator of the floral transition in *Arabidopsis*, *Plant J* **21**, 351-360.

Hetz, W., Hochholdinger, F., Schwall, M. and Feix, G. (1996) Isolation and characterization of *rtcs*, a maize mutant deficient in the formation of nodal roots, *Plant J* **10**, 845-857.

Holton, T.A. and Cornish, E.C. (1995) Genetic and biochemistry of anthocyanin biosynthesis, *Plant Cell* 7, 1071-1083.

Holton, T.A., Brugliera, F., Lester, D.R., Tanaka, Y., Hyland, C.D., Menting, J.G.T., Lu, C.Y., Farcy, E., Stevenson, T.W. and Cornish, E.C. (1993) Cloning and expression of cytochrome P450 genes controlling flower colour, *Nature* **366**, 276-279.

Hosokawa, K., Nakano, M., Oikawa, Y. and Yamamura, S. (1996) Adventitious shoot regeneration from leaf, stem and root explants of commercial cultivars *Gentiana*, *Plant Cell Rep* **15**, 578-581.

Hosokawa, K., Oikawa, Y. and Yamamura, S. (1998) Mass propagation of ornamental gentian in liquid medium, *Plant Cell Rep* **17**, 747-751.

Hsia, C.N. & Korban, S.S. (1996) Factors affecting *in vitro* establishment and shoot proliferation of *Rosa hybrida* L. and *Rosa chinensis minima*, *In Vitro Cell. Dev. Biol* **32**, 217-222.

Hsia, C.N. & Korban, S.S. (1997) The influence of cytokinins and ionic strength of Anderson's medium on shoot establishment and proliferation of evergreen azalea, *Euphytica* **93**, 11-17.

Hudson, A. (1999) Axioms and axes in leaf formation? *Curr. Opin. Plant Biol* **2**, 56-60.

Huetteman, C.A. and Preece, J.E. (1993) Thidiazuron a potent cytokinin for woody plant tissue culture, *Plant Cell Tiss. Org. Cult* **33**, 105-119.

Immink, R.G.H., Hannapel, D.J., Ferrario, S., Busscher, M., Franken, J., Lookeren, C.M.M. and Angenent, G.C. (1999). A petunia MADS box gene involved in the transition from vegetative to reproductive development, *Development* **126**, 5117-5126.

Juarez, C. and Banks, J-A. (1998) Sex determination in plants, *Curr. Opin. Plant Biol* **1**, 68-72.

Jullien, F. and Tran Thanh Van, K. (1994) Micropropagation and embryoid formation from young leaves of *Bambusa glaucescens* "Golden Goddess", *Plant Sci* **98**, 199-207.

Jungnickel, F. and Zaid, S. (1992) Micropropagation of African Violets (*Saintpaulia* spp. and cvs.), In: Bajaj YPS (ed), *Biotechnology in agriculture and forestry vol. 20, high-tech. and micropropagation IV*, Springer-Verlag Berlin. pp 357-395.

Jéhan, H., Courtois, D., Ehrets, E., Lerch, K. and Pétard, V. (1994) Plant regeneration of *Iris pallida* Lam. and *Iris germanica* L. via somatic embryogenesis from leaves, apices and young flowers, *Plant Cell Rep* **13**, 671-675.

Kakimoto, T. (1996) CKI1, a histidine kinase homolog implicated in cytokinin signal transduction, *Science* **274**, 982-985.

Kamo, K. (1994) Effect of phytohormone on plant regeneration from callus of gladiolus cultivar 'Jenny Lee', *In Vitro Cell. Dev. Biol*, **30**, 26-31.

Karp, A. (1995) Somaclonal variation as a tool for crop improvement, *Euphytica* **85**, 295-302.

Kaul, V., Miller, R.M., Hutchinson, J.F. and Richards, D. (1990) Shoot regeneration from stem and leaf explants of *Dendranthema grandiflora* Tzvelev (syn. *Chrysanthemum morifolium* Ramat.), *Plant Cell Tiss. Org. Cult* **21**, 21-30.

Khalid, N., Davey, M.R. and Power, J.B. (1989) An assessment of somaclonal variation in *Chrysanthemum morifolium*: The generation of plants of potential commercial value, *Sci. Hort* **38**, 287-294.

Kim, K.N. and Ernst, S.G. (1994) Effects of inhibitors on phenocritical events of *in vitro* shoot organogenesis in tobacco thin cell layers, *Plant Sci* **103**, 59-66.

Kitto, S.L. and Young, M.J. (1981) *In vitro* propagation of Carrizo citrange, *Hort. Sci* **16**, 305-306.

Klimaszewska, K. and Keller, W.A. (1985) High frequency plant regeneration from thin cell layer explants of *Brassica napus*, *Plant Cell Tiss. Org. Cult* **4**, 193-197.

Koornneef, M., Alonso-Blanco, C., Peeters, A.J.M. and Soppe, W. (1998) Genetic control of flowering time in *Arabidopsis*, *Ann. Rev. Plant Physiol. Plant Mol. Biol* **49**, 345-370.

Koornneef, M., Manhart, C.J. and van den Veen, J.H. (1991) A genetic and physiological analysis of late flowering mutants in *Arabidopsis thaliana*, *Mol. Gen. Gen* **229**, 57-66.

Kumar, P.P. and Prakash, A.P. (2000) A novel MADS-box gene is required for shoot bud induction in excised leaf cultures of *Paulownia kawakamii*, 2000 World Congress on In Vitro Biology. P-1018: 43-A.

Kunitake, H., Imamizo, H. and Mii, M. (1993) Somatic embryogenesis and plant regeneration from immature seed-derived calli of rugosa rose (*Rose rugosa* Thumb.), *Plant Sci* **90**, 187-194.

Lakshmanan, P., Loh, C-S. and Goh, C-J. (1995) An *in vitro* method for rapid regeneration of a monopodial orchid hybrid *Aranda* Deborah using thin section culture, *Plant Cell Rep* **14**, 510-514.

Laux, T., Mayer, K.F.X., Berger, J. and Juergen, G. (1996) The *WUSCHEL* gene is required for shoot and floral meristem integrity in *Arabidopsis*, *Development* **122**, 87-96.

Ledger, S.E., Deroles, S.C. and Given, N.K. (1991) Regeneration and *Agrobacterium*-mediated transformation of crhysanthemum, *Plant Cell Rep* **10**, 195-199.

Lee, T., Huang, M.E.E. and Pua, E-C. (1997) High frequence shoot regeneraion from leaf disc explants of garland chrysanthemum (*Chrysanthemum coronarium* L.) *in vitro*, *Plant Sci* **126**, 219-226.

Lee-Stadlemann, O.Y., Lee, S., Hackett, W.P. and Read, P.E. (1989) The formation of adventitous buds *in vitro* on micro-sections of hybrid *Populus* leaf mid veins, *Plant Sci* **61**, 263-272.

Lenhard, M and Laux, T. (1999) Shoot meristem formation and maintenance, *Curr. Opin. Plant Biol* **2**, 44-50.

Lerouge, P., Roche, P., Faucher, C., Fabienne, M., Truchet, G., Prom, J.C. and Der, J. (1990) Symbiotic host specificity of Rhizobium meliloti is determined by a sulphated and acylated glucosamine oligosaccharide signal, *Nature* **344**, 781-784.

Levy, Y.Y. and Dean, C. (1998) The transition to flowering, *Plant Cell* **10**, 1973-1989.

Li, Y., Zhang, H. and Han, B-W. (1996) Changes of endogenous hormone contents during floral bud and vegetative bud differentiation in thin cell layer culture of *Cichorium intybus* L. explant, *Acta Bot. Sin* **38**, 131-135.

Lineberger, R.D. and Druckenbrod, M. (1985) Chimeral nature of the pin-wheel flowering African violets (*Saintpsaulia, Gesneriaceae*), *Amer. J. Bot* **72**, 1204-1212.

Lotan, T., Ohto, M-A., Yee, K.M., West, M.A.L., Lo, R., Kwong, R.W., Yamagishi, K., Fischer, R.L., Goldberg, R.B. and Harada, J.J. (1998) *Arabidopsis LEAFY COTYLEDON1* is sufficient to induce embryo development in vegetative cells, *Cell* **93**, 1195-1205.

Lu, C.Y. (1993) The use of thidiazuron in tissue culture. *In Vitro Cell. Dev. Biol* **29**, 92-96.

Lu, C.Y., Nugent, G. and Wardley, T. (1990) Efficient, direct plant regeneration from stem segments of chrysanthemum (*Chrysanthemum morifolium* Ramat. Cv. Royal Purple), *Plant Cell Rep* **8**, 733-736.

Luo, D., Carpenter, R., Vincent, C., Copsey, L. and Coen, E. (1996) Origin of floral asymmetry in *Antirrhinum*, *Nature* **383**, 794-799.

Macknight, R., Bancroft, I., Page, T., Lister, C., Schmidt, R., Love, K., Westphal, L., Murphy, G., Sherson, S. and Cobbett, C. (1997) *FCA*, a gene controlling flowering time in *Arabidopsis*, encodes a protein containing RNA-binding domains, *Cell* **89**, 737-745.

Madsen, M.H., Nauerby, B., Frederiksen, C.G. and Wyndaele, R. (1998) Regeneration of pea (*Pisum sativum* L.) by the thin cell layer nodal system: Influence of explant culture media on rooting and plantlet formation, *Acta Agric. Scandi. Sect. B Soil Plant Sci* **48**, 58-64.

Malaure, R.S., Barclay, G., Power, J.B. and Davey, M.R. (1991) The production of novel plants from florets of *Chrysanthemum morifolium* using tissue culture. 1. Shoot regeneration from ray florets and somaclonal variation exhibited by the regenerated plants, *J. Plant Physiol* **139**, 8-13.

Marchant, R., Davey, M.R., Lucas, J.A. and Power, J.B. (1996) Somatic embryogenesis and plant regeneration in Floribunda rose (*Rose hybrida* L.) cvs. Trumpeter and Glad Tidings, *Plant Sci* **120**, 95-105.

Marfa, V., Gollin, D., Eberhardt, S., Mohnen, D., Darvill, A. and Albersheim, P. (1991) Oligogalacturonides are able to induce flowers to form on tobacco explants, *Plant J* **1**, 217-225.

Marga, F., Vebret, L. and Morvan, H. (1997) Agar fractions could protect apple shoots cultured in liquid media against hyperhydricity, *Plant Cell Tiss. Org. Cult* **49**, 1-5.

Margara, J. and Piollat, M-T. (1982) Influence de la composition minérale et glucide du milieu sur l'organogenèse *in vitro* à partir de pétales de *Begonia* x *elatior*, *C.R. Acad. Sci. Paris* **294**, 545-548.

May, R.A. and Trigiano, R.N. (1991) Somatic embryogenesis and plant regeneration from leaves of *Dendranthema grandiflora*. *J. Amer. Soc. Hort. Sci* **116**, 366-371.

Meyer, P., Heidmann, I., Forkmann, G. and Saedler, H (1987) A new petunia flower colour generated by transformation of a mutant with a maize gene, *Nature* **330**, 677-678.

Meyerowitz, E.M. (1998) Genetic and molecular mechanisms of pattern formation in *Arabidopsis* flower development, *J. Plant Res.* **111**, 233-242.

Michaels, S.D. and Amasino, R.M. (1999) *FLOWERING LOCUS C* encodes a novel MADS domain protein that acts as a repressor of flowering, *Plant Cell* **11**, 949-956.

Millar, A.J. and Kay, S.A. (1991) Circadian control of *cab* gene transcription and mRNA accumulation in *Arabidopsis*, *Plant Cell* **3**, 541-550.

Millar, A.J., Carré, I.A., Strayer, C.A., Chua, N.H. and Kay, S.A. (1995) Circadian clock mutants in *Arabidopsis* identified by luciferase imaging, *Science* **267**, 1161-1163.

Mok, M.C., Mok, D.W.S., Armstrong, D.J., Shudo, K., Isogai, Y. and Okamoto, T. (1982) Cytokinin activity of N-phenyl-N'-1,2,3-thidiazol-5-ylurea (Thidiazuron), *Phytochemistry*, **21**, 1509-1511.

Mol, J.N.M., Holton, T.A. and Koes, R.E. (1995) Floriculture: genetic engineering of commercial traits, *Trends Biotech* **13**, 350-355.

Momcilovic I., Grubisic, D. and Neskovic, M. (1997) Micropropagation of four Gentiana species (*G. lutea, G. cruciata, G.purpurea,* and *G. aucolis*), *Plant Cell Tiss. Org. Cult* **49**, 141-144.

Monteuuis, O. (1991) Rejuvenation of a 100-year *Sequoiadendron giganteum* through *in vitro* meristem culture, I. Organogenic and morphological arguments, *Physiol. Plant* **81**, 111-115.

Mulin, M. and Tran Thanh Van, K. (1989) Obtention of *in vitro* flowers from thin epidermal cell layers of *Petunia* hybrida (Hort.), *Plant Sci* **62**, 113-121.

Muller, J-F., Goujaud, J. and Caboche, M. (1985) Isolation *in vitro* of napthaleneacetic acid-tolerant mutants of *Nicotiana tabacum,* which are impaired in root morphogenesis, *Mol. Gen. Genet* **199**, 194-200.

Murashige, T. and Skoog, F. (1962) A revised medium for rapid growth and bioassay with tobacco tissue cultures, *Physiol. Plant* **15**, 473-497.

Murthy, B.N.S., Murch, S.J. and Saxena, P.K. (1998) Thidiazuron: a potent regulation of *in vitro* plant morphogenesis, *In Vitro Cell. Dev. Biol* **34**, 267-275.

Murthy, B.N.S., Singh, R.P. and Saxena, P.K. (1996) Induction of high-frequency somatic embryogenesis in geranium (*Pelargonium* x *hortorum* Bailey cv Ringo Rose) cotyledonary cultures, *Plant Cell Rep* **15**, 423-426.

Mutafstchiev, S., Cousson, A. and Tran Thanh Van, K. (1987) Modulation of cell growth and differentiation by pH and oligosaccharides, in British Plant Growth Regulator Group, Monograph 16.

Nakano, M., Hoshino, Y., Mii, M. (1994) Adventitious shoot regeneration from cultured petal explants of carnation, *Plant Cell Tiss. Org. Cult* **36**, 15-19.

Nauerby, B., Madsen, M., Christiansen, J. and Wyndaele, R. (1991) A rapid and efficient regeneration system for pea (*Pisum sativum*), suitable for transformation, *Plant Cell Rep* **9**, 676-679.

Nayak, N.R. Rath, S.P. and Patnaik, S. (1997) *In vitro* propagation of three ephityic cymbidium, *Aloifolium* (L.) SW, *Dendrobium aphyllum* (Roxb) Fisch and *Dendrobium moschatum* (Buch-ham) SW through thidiazuron-induced high frequency shoot proliferation, *Sci. Hort* **71**, 416-426.

Nelson, D.C., Lasswell, J., Rogg, L.E., Cohen, M.A. and Bartel, B. (200) *FKF1*, a clock-controlled gene that regulates the transition to flowering in *Arabidopsis*, *Cell* **101**, 331-340.

Nhut, D.T. (1998) Micropropagation of lily (*Lilium longiflorum*) via *in vitro* stem node and pseudo-bulblet culture, *Plant Cell Rep* **117**, 913-6.

Nhut, D.T., Bui, V.L., Fukai, S., Tanaka, M. and Tran Thanh Van, K. (2000) Direct somatic embryogenesis through pseudo-bulblet thin cell layer of *Lilium longiflorum*, *Abstract of 2000 World Congress on In Vitro Biology*, San Diego, California, p. 46. A.

Nhut, D.T., Bui, V.L., Tanaka, M. and Tran Thanh Van, K. (2001a) Shoot induction and plant regeneration from receptacle tissue of *Lilium longiflorum*, *Sci. Hort* **87**, 131-8.

Nhut, D.T., Bui, V.L., Fukai, S., Tanaka, M. and Tran Thanh Van, K. (2001b) Effects of activated charcoal, explant size, explant position and sucrose concentration on plant and shoot regeneration of *Lilium longiflorum* via young stem culture, *Plant Growth Reg* **33**, 59-65.

Nhut, D.T., Bui, V.L. and Tran Thanh Van, K. (2001c) Manipulation of the morphogenetic pathways of *Lilium longiflorum* transverse thin cell layer explants by auxin and cytokinin, *In Vitro Cell. Dev. Biol* **37**, 44-9.

Nhut, D.T., Bui, V.L., da Silva, J.A.T. and Aswath, C.R. (2001d) Thin cell layer culture system in *Lilium*: Regeneration and Transformation Perspectives, *In Vitro Cell. Dev. Biol* (In press).

Nhut, D.T., Huong, N.T.D., Bui, V.L., da Silva, J.A.T., Fukai, S. and Tanaka, M. (2001e) The changes in shoot regeneration potential of protocorm-like bodies derived from *Lilium longiflorum* young stem explants exposed to medium volume, pH, light intensity and sucrose concentration pretreatment, *J. Hort. Sci. Biotech* (Summited paper).

Nhut, D.T., Bui, V.L., Minh, N.T., da Silva, J.A.T., Fukai, S., Tanaka, M. and Tran Thanh Van, K. (2001f) Direct somatic embryogenesis through pseudo-bulblet transverse thin cell layers of *Lilium longiflorum*, *Plant Sci* (Summited paper).

Nhut, D.T. (2001g) Effect of explant age on direct somatic embryogenesis by culturing young stem transverse thin cell layers of *Lilium longiflorum*, *Sci. Hort* (Summited paper).

Nhut, D.T. and da Silva, J.A.T. (2001h) The control of *in vitro* direct main stem formation of *Lilium longiflorum* derived from receptacle culture and micropropagation by using *in vitro* stem nodes, *J. Plant Physiol* (Summited paper).

Nielsen, J.M. Hansen, J. and Brandt, K. (1995) Synergism of thidiazuron and benzyladeninne in axillary shoot formation depends on the sequence of application in *Miscanthus* x *ogiformis* 'Giganteus', *Plant Cell Tiss. Org. Cult* **41**, 165-170.

Niimi, Y. (1986) Application of leaf segment culture to *in vitro* bulblets production of six *Lilium* species, *Acta Bot* **35**: 189-194.

Ohki, S. (1994) Scanning electron microscopy of shoot differentiation *in vitro* from leaf explants of the African violet, *Plant Cell Tiss. Org. Cult* **36**, 157-162.

Okada, K., Ueda, J., Komaki, M.K., Bell, C.J. and Shimura, Y. (1991) Requirement of the auxin polar transport system in early stages of *Arabidopsis* floral bud formation, *Plant Cell* **3**, 677-684.

Okole, B.N. and Schulz, F.A. (1996) Micro-sections of banana and plantains (*Musa* spp) morphogenesis and regeneration of callus and shoot buds, *Plant Sci* **116**, 185-195.

Oud, J.S.N., Schneiders, H., Kool, A.J. and van Grinsven, M.Q.J.M. (1995) Breeding of transgenic orange *Petunia* hybride varieties, *Euphytica* **85**, 403-409.

Park, D.H., Somers, D.E., Kim, Y.S., Choy, Y.H., Lim, H.K., Soh, M.S., Kim, H.J., Kay, S.A. and Nam, H.G. (1999) Control of circadian rhythms and photoperiodic flowering by the *Arabidopsis GIGANTEA* gene, *Science* **285**, 1579-1582.

Peary, J.S., Lineberger, R.D., Malinich, T.J. and Wertz, K. (1988) Stability of leaf variegation in *Saintpaulia ionantha* during *in vitro* propagation and during chimeral separation of a pinwheel flowering form, *Am. J. Bot* **75**, 603-608.

Pedroso, C.M. and Pais, M.S. (1995a) Explant region-specific embryogenic competence and plant recovery in *Camellia japonica*, *In Vitro Cell. Dev. Biol* **31**, 8-14.

Pedroso, C.M. and Pais, M.S. (1995b) Factors controlling somatic embryogenesis, *Plant Cell Tiss. Org. Cult* **43**, 147-154.

Pidkowich, M.S., Klenz, J.E. and Haughn, G.W. (1999) The making of a flower: control of floral meristem identity in *Arabidopsis*, *Trends Plant Sci* **4**, 64-70.

Piñero, M. and Coupland, G. (1998) The control of flowering time and floral identity in *Arabidopsis*, *Plant Physiol* **117**, 1-8.

Priyadarshi, S. and Sen, S. (1992) A revised scheme for mass propagation of Easter lily, *Plant Cell Tiss. Org. Cult* **30**, 193-197.

Pélissier, B., Bouchefra, O., Pépin, R. and Freyssinet, G. (1990) Production of isolated somatic embryos from sunflower thin cell layers, *Plant Cell Rep* **9**, 47-50.

Radin, D.N. and Eisenback, J.D. (1991) Infection of cultured thin cell layer roots of *Lycopersicon esculentum* by *Meloidogyne incognita*, *J. Nematology* **23**, 441-445.

Rao, C.D., Goh, C-J. and Kumar, P.P. (1996). High frequency adventitious shoot regeneration from excised leaves of *Paulownia spp* cultured *in vitro*, *Plant Cell Rep* 16, 204-209.

Reed, J.W., Nagpal, P., Poole, D.S., Furuya, M. and Chory, J. (1993) Mutations in the gene for the red/far-red light receptor phytochrome B alter cell elongation and physiological responses throughout *Arabidopsis* development, *Plant Cell* **5**, 147-157.

Remotti, P.C. and Löffler, H.J.M. (1995) Callus induction and plant regeneration from gladiolus. *Plant Cell Tiss. Org. Cult* **42**,171-178.

Reuther, G. (1977) Embryoid differenzierungsmuster im kallus der gattungen *Iris* und *Asparagus*, *Ber. Deutsh. Bot. Ges* **90**, 417-437.

Rinne, P.L.H. and van der Schoot, C. (1998) Symplastic fields in the tunica of the shoot apical meristem coordinate morphogenetic events, *Development* **125**, 1477-1485.

Robertson, S., Li, Y., Scutt, C., Willis, M. and Gilmartin, P. (1997) Spatial expression dynamics of *Men-9* delineate the thrird whorl in male and female flowers of dioecious *Silene latifolia*, *Plant J* **12**, 155-168.

Robinson, K.E.P. and Firoozabady, E. (1993) Transformation of floriculture crops, *Sci. Hort* **55**, 83-99.

Rout, G.R. and Das, P. (1994) Somatic embryogenesis and *in vitro* flowering of three bamboos, *Plant Cell Rep* **13**, 683-686.

Rout, G.R., Debata, B.K. and Das, P. (1991) Somatic embryogenesis in callus culture of *Rosa hybrida* L. cv Landora, *Plant Cell Tiss. Org. Cult* **27**, 65-69.

Samson, I., Hamama, L. and Letouze, R. (1999) Rôle du 3F-ip (4,4,4 trifluoroisopentenyladenine) et du CPPU (N-(2,chloropyridyl)4-N'phénylurée), molécules de synthèse à activité cytokinine, sur l'initiation et la multiplication de protocormes de régénération de *Phalaenopsis*, in *Biotechnologies, amélioration des plantes et sécurité alimentaire*, Univ. francophones, Ed. Estem, Paris. pp. 553-555.

Saxena, S. (1990) *In Vitro* propagation of the bamboo (*Bambusa tulda* Roxb) through shoot proliferation, *Plant Cell Rep* 9, 431-434.

Saxena, S. and Dhawan, V. (1999) Regeneration and large-scale propagation of bamboo (*Dendrocalamus strictus* Nees) through somatic embryogenesis, *Plant Cell Rep* **18**, 438-443.

Schricke, H., Joulain, D. and Tran Thanh Van, K. (1988) *Iris pallida*: une source d'irones en culture *in vitro*, in Proc. Symp. European IAPTC, *Bioproduction de métabolites par culture de cellules végétales*, Paris, France.

Shannon, S. and Meeks-Wagner, D.R. (1991) A mutation in the *Arabidopsis TFL1* gene affects inflorescence meristem development, *Plant Cell* **3**, 877-892.

Sheldon, C.C., Burn, J.E., Perez, P.P., Metzger, J., Edwards, J.A., Peacock, W.J. and Dennis, E.S. (1999) The *FLF* MADS box gene: a repressor of flowering in *Arabidopsis* regulated by vernalization and methylation, *Plant Cell* **11**, 445-458.

Shetty, K., Carpenter, T.L., Curtis, O.F. and Potter, T.L. (1996) Reduction of hyperhydricity in tissue cultures of organ (*Origanum vulgare*) by extracellular polysaccharide isolated from *Pseudomonas* spp, *Plant Sci* **120**, 175-183.

Sim, G.E., Goh, C.J. and Loh, C.S. (1989) Micropropagation of *Citrus mitis* blanco-multiple bud formation from shoots and roots explants in the presence of 6-benzylaminopurine, *Plant Sci* **59**, 203-210.

Simon, R., Igeño, M.I. and Coupland, G. (1996) Activation of floral meristem identity genes in

Arabidopsis, *Nature* **384**, 59-62.

Start, N.D. and Cumming, B.G. (1976) *In vitro* propagation of *Saintpaulia ionantha* Wendl., *Hort. Sci* **11**, 204-206.

Stefaniak, B. (1994) Somatic embryogenesis and plant regeneration of Gladiolus (*Gladiolus hort*), *Plant Cell Rep* **13**, 386-389.

Stenberg, N.E., Chen, C.H. and Ross, J.G. (1977) Regeneration of plantlets from leaf cultures of *Lilium longiflorum* Thunb. *Proc. SD Acad. Sci* **56**: 152-158.

Sticklen, M.B. (1991) Direct somatic embryogenesis and fertile plants from rice root cultures, *J. Plant Physiol.* **138**, 577-580.

Stimart, D.P. and Ascher, P.D. (1978) Tissue culture of bulb scale section for asexual propagation of *Lilium longiflorum* Thumb, *J. Amer. Soc. Hort. Sci* **103**, 1982-1984.

Sugimura, Y. and Salvana, M.J. (1989) Induction and growth of callus derived from rachilla explants of young inflorescences of coconut palm, *Can. J. Bot* **67**, 272-274.

Sugiyama, M. (1999) Organogenesis *in vitro*, *Curr. Opin. Plant Biol* **2**, 61-64.

Sung, Z.R., Belachew, A., Shunong, B. and Bertrand-Garcia, R. (1992) EMF, an *Arabidopsis* gene required for vegetative shoot development, *Science* **258**, 1645-1647.

Szabados, L., Hoyos, R. and Roca, W. (1987) *In vitro* somatic embryogenesis and plant regeneration of cassava, *Plant Cell Rep* **6**, 248-251.

Takahashi, S., Shodu, K., Okamoto, T., Yamada, T. and Isogai Y. (1978) Cytokinin activity of N-phenyl-N'-(4-pyridyl)urea derivatives, *Phytochemistry* **17**, 1201-1207.

Takayama S and Misawa M (1981) Mass propagation of *Begonia* x *hiemalis* plantlets by shake culture, *Plant Cell Physiol* **22**, 461-467.

Takayama, S. (1989) *Begonia*. In: P. Ammirato et al (eds) *Handbook of plant cell culture*, Vol 5, Mc Grow Hill Inc. USA, pp. 253-282.

Takayama, S. and Misawa, M. (1983) A scheme for mass propagation of *Lilium in vitro*, *Sci. Hort* **18**, 353-362.

Tran Thanh Van, M. (1973a) *In vitro* control of *de novo* flower, bud, root and callus differentiation from excised epidermal tissues, *Nature* **246**, 44-45.

Tran Thanh Van, M. (1973b) Direct flower neoformation from superficial tissue of small explant of *Nicotiana tabacum*, *Planta* **115**, 87-92.

Tran Thanh Van, M. (1974a) Growth and flowering of *Cymbidium* buds normally inhibited by apical dominance, *J. Amer. Soc. Hort. Sci* **99**, 450-453.

Tran Thanh Van, M. (1974b) Methods of acceleration of growth and flowering in a few species of orchids, *Amer. Orchid. Soc. Bull* **43**, 699-707.

Tran Thanh Van, K. and Dien, N.T. (1975) Étude au niveau cellulaire de la differenciation *in vitro* et *de novo* de bourgeons vegetatifs, de racines, ou de cal à partir de couches minces de cellules de type-epidermique de *Nicotiana tabacum* Wisc. 38, *Can. J. Bot* **53**, 553-550.

Tran Thanh Van, K. and Chlyah, A. (1976) Differenciation de bouton floraux, de bourgeons végétatifs, de raciness et de cals à partir de l'assis sous-épidermiques de ramification florales de *Nicotiana tabacum* L. Wisc.38. Etude infrastructurale, *Can. J. Bot* **54**, 1979-1996.

Tran Thanh Van, K. (1980) Control of morphogenesis by inherent and exogenously applied factors in thin cell layers, *Intl. Rev. Cytol* **32**, 291-311.

Tran Thanh Van, K. (1981) Control of morphogenesis, *Ann. Rev. Plant Physiol* **32**, 291-311.

Tran Thanh Van, K., Toubart, P., Cousson, A., Darvill, A.G., Gollin, D.J., Chelf, P. and Albersheim, P. (1985a) Manipulation of the morphogenetic pathways of tobacco explants by oligosaccharins, *Nature* **314**, 615-617.

Tran Thanh Van, K., Yilmaz, A. and Trinh, H.T. (1985b) How to programme *in vitro* different morphogenetic expression in some conifers, In: J.M. Bonga and D.J. Durzan (eds.) *Cell and tissue culture in forestry*, Martinus Nijhoff Publishers, Dordrecht.

Tran Thanh Van, K. and Trinh, T.H. (1986) Fundamental and applied aspects of differentiation *in vitro* and *in vivo*, In: D.A. Evans, W.R. Sharp and P.V. Ammirato (eds) *Handbook of plant cell culture, techniques and application*, Vol 4, McMillan Publishing Company, New York, pp. 316-335.

Tran Thanh Van, K. and Mutaftschiev, S. (1990) Signals influencing cell elongation, cell enlargement, cell division and morphogenesis in Nijkam HJJ, van der Plaas LHW and Aartij J (eds.) *Progress in Plant Cellular and Molecular Biology*, Kluwer Academic, pp 514-519.

Tran Thanh Van, K., Richard, L. and Gendy, C.A. (1990) An experimental model for the analysis of plant/ cell differentiation: thin cell layer. Concept, strategy, methods, records and potential, In: D. Durzan and R. Rodriguez (eds.) *NATO Biotechnology Series*. Plenum press, New York, pp 215-224.

Tran Thanh Van, K. and Bui Van Le (2000) Curent status of thin cell layer method for the induction of organogenesis or somatic embryogenesis, In: *Somatic embryogenesis in woody plants*, Vol 6, S.J. Mohan, P.K. Gupta and R.J. Newton (eds), Kluwer Academic Publishers, Dordrecht, pp 51-92.

Trinh, T.H., Mante, S., Pua, E-C. and Chua, N-H. (1987) Rapid production of transgenic flowering shoots and F1 progeny from *Nicotiana plumbaginifolia* epidermal peels, *Biotechnology* **5**, 1081-1084.

Urban, L.A., Sherman, J.M., Moyer, J.W. and Daub, M.E. (1994) High frequency shoot regeneration and *Agrobacterium*-mediated transformation of chrysanthemum (*Dendranthema grandiflora*), *Plant Sci* **98**, 69-79.

Vajrabhaya, M. and Vajrabhaya, T. (1970) Culture of *Rhynchostylis gigantea* a monopodial orchid, *Amer. Orchid Soc. Bull* **39**, 907-910

Vasil, V. and Vasil, I.K. (1982) Characterization of an embryogenic cell suspension culture derived from cultured inflorescences of *Pennisetum americanum* (pearl millet, Gramineae), *Amer. J. Bot* **69**, 1441-1449.

Villalobo, Y.M., Yeung, E.C. and Thorpe, T.A. (1985) Origin of adventitious shoot formation in excised *Radiata* pine cotyledons culture *in vitro*, *Can. J. Bot* **63**, 2172-2176.

Vissier, C., Qureshi, J.A., Gill, R. and Saxena, P.K. (1992) Morphoregulatory role of TDZ substitution of auxin and cytokinin requirement for the induction of somatic embryogenesis in *geranium* hypocotyl culture, *Plant Physiol* **99**, 1704-1707.

Weigel, D. 1995. The genetics of flower development: from floral induction to ovule morphogenesis, *Ann. Rev. Genet* **29**, 19-39.

White, D.W.R., Woodfield, D.R. and Caradus, J.R. (1998) *Mortal*: a mutant of white clover defective in nodal root development, *Plant Physiol* **116**, 913-921.

Winkelmann, T. and Grunewald, J. (1995) Genotypic for protoplast regeneration in *Saintpaulia ionantha* (H. Wendl.), *Plant Cell Rep* **14**, 704-707.

Xu, L.Q. (1984) The promotive effects of ginseng on rapid clonal *in vitro* propagation of African violet (*Saintpaulia ionantha* Wendl.), *Acta Bot. Sin* **26**, 489-498.

Yanofsky, M.F., Ma, H., Bowman, J.L., Drews, G.N., Feldmann, K.A. and Meyerowitz, E.M. (1990) The protein encoded by the *Arabidopsis* homeotic gene *agamous* resembles transcription factors, *Nature* **346**, 35-39.

Yepes, L.M., Mittak, V., Pang, S-Z., Gonsalves, C., Slightom, J. and Gonsalves, D. (1995)

Biolistic transformation of chrysanthemum with the nucleocapsid gene of tomato spotted wilt virus, *Plant Cell Rep* **14**, 694-698.

Zagotta, M.T., Shannon, S., Jacobs, C. and Meeks-Wagner, D.R. (1992) Early-flowering mutants of *Arabidopsis thaliana*, *Austr. J. Genet* **19**, 411-418.

28. APPLICATION OF TISSUE CULTURE TO THE GERMPLASM CONSERVATION OF TEMPERATE BROAD-LEAF TREES

Maurizio Lambardi and Anna De Carlo

Istituto sulla Propagazione delle Specie Legnose, CNR / National Research Council, via Ponte di Formicola n. 76, 50018 Scandicci (Firenze), Italy. E-mail: lambardi@fi.cnr.it

CONTENTS

S.M. Jain and K. Ishii (eds.), Micropropagation of Woody Trees and Fruits, 815–840.

1. INTRODUCTION

The recovery and conservation of a broad genetic pool of wild and domesticated species are essential for the maintenance of plant biodiversity. Traditional *ex-situ* conservation systems (i.e., seed and field collections), alongside *in-situ* conservation, make a fundamental contribution to the preservation of plant genetic resources. Up to now these approaches have been the only reliable option for the long-term germplasm preservation of woody species, although they have several drawbacks. Numerous forest Angiosperms (e.g., *Acer* spp., *Quercus* spp., chestnut, horsechestnut, many tropical species) have non-orthodox (recalcitrant) seeds that quickly loose their viability or are subject to a rapid decrease in germinability during storage (Pence, 1995). Moreover, recalcitrant seeds do not tolerate the level of dessication that is commonly required to prolong conservation (Roberts, 1973). As regards temperate fruit trees, the majority of the species belonging to this group are reproduced by vegetative propagation, thus requiring the conservation in clonal orchards of huge numbers of accessions, including old and newly selected cultivars, local varieties and wild material. Table 1, for instance, gives a clear idea of the efforts required today for the preservation of the European *Prunus* germplasm, kept in the field repositories of 21 countries.

Although it plays a pre-eminent role in assisting woody plant conservation programmes, the maintenance of clonal orchards today faces several problems and risks, among which:

- the large areas of land that are required, due to the high number of accessions generally preserved in each collection and the necessity of keeping a minimum of two individuals per accession;
- high running costs, mainly for manual labour (pruning, weed and pest management, irrigation, etc.);
- orchards are prone to environmental stresses (e.g., heavy frosts, flooding) which can seriously threaten the survival of non-tolerant species;
- the hazards of pests and diseases, for which a periodic and careful monitoring of trees is essential to keep them healthy and genetically stable.

The last point has to be taken into particular consideration when planning the establishment of clonal collections. In recent years, for instance, the "Sharka" disease, caused by the Plum Pox Virus (PPV), has spread extensively in plum orchards (Fisher and Mannel, 1994), and there are reports of in-field germplasm repositories which have been contaminated as well. It should be noted that no therapy is effective in preventing or eradicating the PPV; therefore, the infected plum trees must be destroyed in order to avoid further contamination, with dramatic

Table 1. Summary of *Prunus* field collections in 21 European countries (from Gass *et al.*, 1996).

Species	**Total accessions** [(1)]	**Original accessions** [(2)]
Almond	> 1.500	> 600
Apricot	> 4.600	> 1.700
Cherry (sour, sweet, related spp.)	> 9.200	> 4.700
Peach	> 660	> 1.600
Plum and prune	>11.300	> 6.700
Interspecific hybrids	> 2.600	> 1.800
Wild related spp.	> 1.000	> 700

[(1)] The number refers to the total of preserved accessions (i.e., the same accession could be present in the clonal collections of different countries).
[(2)] The number refers to indigenous old cultivars and wild material, originating in each specific country (i.e., each accession is unique).

consequences when the contaminated orchard is a field genebank.

1.1. Potential of *In Vitro* Conservation Techniques

Today, biotechnology offers important options for the collection, molecular characterization, disease indexing, pathogen elimination, propagation, documentation, preservation and exchanging of disease-free plant genetic resources (Benson, 1999). In this context, the application of tissue culture technology to the preservation of plant genetic resources has greatly evolved in recent years. For a considerable number of species, it can already be considered a valid alternative to *ex situ* conservation, for others it is rapidly gaining importance (Ashmore, 1997).

Micropropagation (i.e., the procedure of shoot multiplication by axillary or adventitious shoot proliferation) can be considered the first approach towards *in vitro* germplasm preservation. In this system, the cultured shoots are frequently subcultured in fresh media every 3-4 weeks (for the majority of woody species), in order to renew the gaseous atmosphere of the vessels and to keep the cultures in healthy condition. The risk of somaclonal variation, which can compromise the genetic fidelity of the multiplied material (e.g., Bonga and von Aderkas, 1992), increases together with the number of subcultures, particularly after the second year of subculture. Hence, micropropagation, besides being a costly method of plant conservation, can be considered only for short-term preservation of germplasm.

Slow growth storage (also called "minimal growth storage", "cold or low temperature storage", "conservation at above freezing temperature") has repeatedly proved to be a convenient option for medium-term conservation of woody plants. The technique makes a significant extension to the interval between subcultures possible, thus reducing the costs of stock culture maintenance, as well as the risks of contamination during subculturing. Experience with many herbaceous and woody plants has shown that the storage period before the regrowth of material can be prolonged to 2 or even more years, although 12 months between subcultures is a more realistic goal for the majority of species (Grout, 1995).

Cryopreservation is the only sound alternative to clonal orchards for the long-term storage of germplasm of vegetatively propagated trees. At a cryogenic temperature, the rate of chemical and biophysical reactions is so slow that the biological growth and development of the frozen organ/tissue are hampered, while, if a proper technology of ultra-rapid freezing is applied, cell survival is not affected. Hence, in theory, germplasm cryopreservation could be considered unlimited in terms of time.

This report is a review of progress on medium- and long-term conservation of temperate broad-leaf trees by slow growth storage and cryopreservation. It is important to emphasize that the *in vitro* technique does not aim to replace the traditional *in situ* and *ex situ* approaches to plant germplasm preservation. Rather, it should be regarded as complementary to them, in order to develop a multi-option *modus operandi* for the conservation and use of genebanks, to provide a real guarantee against accidental loss of plant genetic resources.

2. ORGANS AND TISSUES FROM *IN VITRO* CULTURE UTILIZED FOR CONSERVATION

Recent advances in tissue culture technology have greatly increased the variety of organs and tissues that have been tested for medium- and long-term storage, among which shoots, nodal segments, shoot tips, meristems, callus, cell and pollen cultures, zygotic and somatic embryos (e.g., Bajaj, 1995). Two categories of explants are considered in this report for germplasm conservation of temperate broad-leaf trees, i.e. (i) the differentiated tissues, which are fundamental for the preservation of vegetatively propagated plants (e.g., fruit trees and plantation crops), and (ii) the embryogenic cultures, a powerful morphogenetic system, suitable for preservation of seed-propagated species and for utilization in genetic transformation. As regards cryopreservation of recalcitrant seeds and excised zygotic embryos, specific reports can be referred to (Pence, 1995; Marzalina and Krishnapillay, 1999).

2.1. Differentiated Explants (Shoots, Nodal Segments, Shoot Tips)

Differentiated explants, used for conservation *in vitro*, are obtained from typical micropropagated cultures, mainly during active stages of proliferation. Shoot cultures, originating from axillary buds, must be used when the aim is the preservation of true-to-type material. When adventitious shoots develop from callus at the base of shoot clumps, the maintenance of genetic fidelity is not guaranteed (Gould, 1986).

Shoot culture explants are used for slow growth storage of woody species. In general, shoots of various sizes and containing different numbers of nodes/leaves are stored. The size ranges from a minimum of 1 cm - e.g., in *Pyrus* spp. (Wanas *et al.*, 1986; Tahtamouni and Shibli, 1999) – up to a maximum of 3-cm long poplar shoots, bearing about 6 leaflets (Hausman *et al.*, 1994). However, three-shoot tufts (smaller than 1 cm) of apple rootstocks, subcultured in fresh medium and immediately stored at 4°C, proved to recover from storage better than 1-cm single microcuttings (apical or middle part) and to produce a greater number of new shoots (Orlikowska, 1992).

Nodal segments of shoots have occasionally been used for slow growth storage of *Morus alba* (Sharma and Thorpe, 1990), *Castanea sativa* x *C. crenata* and *Quercus* spp. (Janeiro *et al.*, 1995). However, in both the chestnut and oak species, no difference was noted between nodal segments and shoot tips (1-cm long) as regards their regrowth following cold storage.

Shoot tips are by far the most widely used explants for cryopreservation of woody species because genetic fidelity to the donor plants is readily maintained. Shoot tips are obtained from apical or axillary buds, excised from *in vitro*-grown shoot cultures. Under sterile conditions, the bud is manipulated in order to obtain the apical meristem, surrounded by some of the original leaf primordia and leaflets (Fig. 1A). As regards the final size of the shoot tip, a minimum of 0.3 mm long was used for *Malus domestica* (Damiano *et al.*, 1998). However, for the majority of broad-leaf woody species, longer explants (1-2 mm, on average) have proved to be most suitable for cryopreservation. Although it can be presumed that the maintenance of some leaf primordia and leaflets has an influence on survival of cryostored explants (e.g., reducing the risk of meristem dessication during excission, interferring with the penetration of cryoprotectants in the inner part on the shoot tip, protecting the meristem from damage during repeated manipulations of the explant), no specific information is available on this aspect. As a consequence, the number of young leaves which are kept is variable, ranging from 2 (*Morus* spp.; Niino *et al.*, 1992b) to 4 pairs (*Eucalyptus grandis* x *E. camaldulensis*; Blakesley and Kiernan, 2001).

2.2. Embryogenic Cultures

Cryopreservation of embryogenic cultures and somatic embryos could be another important means of long-term preservation of valuable germplasm of woody plants. In seed-propagated species, cryopreservation of embryogenic cultures can play an important role in the conservation of plants with recalcitrant (non-orthodox) seeds, as well as of plants characterized by frequent years with insufficient seed production. Moreover, when a procedure of synthetic seed production is available, cryopreservation would lengthen the storage period of the alginate-coated somatic embryos (Bajaj, 1995). As regards vegetatively propagated species, for which the maintenance of the genetic fidelity of cultivars and clones is required, embryogenic cultures are not recommended for germplasm preservation, since the appearance of somaclonal variation cannot be excluded. Nevertheless, cryopreservation allows the maintenance of valuable embryogenic lines utilized in bioengineering, e.g., avoiding the loss of embryogenic potential due to repeated subculturing or allowing the storage of transgenic material while field trials are ongoing.

Cryopreservation of embryogenic cultures of conifers is already technologically advanced (Cyr, 2000), and over 5000 genotypes of 14 conifer species are cryostored in one facility in British Columbia alone (Reed, 2001). However, its application to broad-leaf tree preservation is still limited. Indeed, since cryopreservation was first tried with nucellar cells of navel orange (Sakai *et al*., 1990, 1991), it has been applied to the conservation of isolated alginate-coated somatic embryos of walnut (de Boucaud *et al.,* 1994), olive (Shibli and Al-Juboory, 2000) and ash (Tonon *et al*., 2001), as well as embryogenic cultures of horsechestnut (Lambardi *et al*., 1999), olive (Benelli *et al*., 2001) and ash (Tonon *et al*., 2001).

3. MEDIUM-TERM CONSERVATION BY SLOW GROWTH STORAGE

Slowing the growth rate of *in vitro* stock cultures to a minimum aims to reduce significantly the number of subcultures per year, without affecting explant viability, health and capacity of regrowth under normal growing conditions.

3.1. Temperature and Light Conditions

Typically, the reduction of explant growth is pursued by lowering the environmental temperature of cultures below their active growing temperature (Table 2). Reports on cold storage of woody species show that the temperatures applied ranged from –3°C (in *Prunus* rootstocks; Marino *et al*., 1985) to a maximum of +12°C (in *Vitis* spp.; Galzy and Compan, 1988).

However, almost 60% of species in Table 2 were stored at a temperature between 2°C and 5°C.

Different light conditions can influence low temperature storage. Although the maintenance of stored cultures in total darkness is more common, different combinations of photoperiod and light intensity have occasionally been utilized to induce better conservation conditions. In Table 2, considerable differences can be seen as for both photoperiod (from 8h to 24h), and light intensities (from 1.3 µE m^{-2} s^{-1} up to a maximum of 25 µE m^{-2} s^{-1}).

3.2. Growth Regulators and Growth Retardants

The effectiveness of supplementing plant growth regulators in culture media during cold storage appears to be controversial. The majority of researchers stored shoot cultures in the same media used during proliferation, which usually contain a combination of growth regulators (mainly cytokinins and auxins). Only for a limited number of species have hormone-free media been utilized during conservation at low temperatures, or a comparison between hormone-free and hormone-containing media been made. A hormone-free MS (Murashige and Skoog, 1962) medium gave 100% *Pyrus syriaca* shoot survival after short-term preservation at 8°C (Tahtamouni and Shibli, 1999). Olive shoots could be stored up to 8 months at 4°C on a hormone-free OM (Rugini, 1986) medium, and a high rate (90%) of regrowth after transfer to normal culture conditions was achieved (Lambardi *et al.*, 2000a). On the other hand, the survival of apple rootstocks, stored at 4°C, was negatively affected when a modified MS medium, without 6-benzyladenine (BA), was used (Orlikowska, 1992). Similarly, the addition of cytokinins in the storage medium is indispensable for both short- and long-term preservation of poplar. Indeed, shoot explants of *Populus tremula* x *P. tremuloides* required MS medium containing 3.9 µM 2-isopentenyladenine, to be stored 3 months in the dark at 4°C (Hausman *et al.*, 1994), while the conservation of shoot cultures of *P. alba* x *P. grandidentata* could be prolonged to up to 5 years under similar conditions (25% shoot recovery) only when MS medium was supplemented with 1.33 µM BA (Son *et al.*, 1991).

Although occasionally proposed, the use of growth retardants (chlorocholine chloride, abscisic acid) or osmotically active compounds, such as mannitol, is not advisable when true-to-type material is required, because it can lead to the appearance of somaclonal variation (Ashmore, 1997). In accordance with this general opinion, mannitol treatments (at concentrations of 2-4%) greatly reduced the survival and the recovery of pear shoots stored at 4°C and, when supplemented in combination with sucrose, induced many very small, abnormal shoots that didn't survive (Wanas *et al.*, 1986).

Table 2. Slow growth storage of temperate broad-leaf trees: summary of storage conditions and survival after maximum time of conservation (NR, not reported).

Species	Temperature (°C)	Light conditions[1]	Months	Survival (%)	Reference
Actinidia deliciosa	8	Darkness	12	100	Monette, 1995
Castanea sativa	4	8h (25 $\mu E\ m^{-2}s^{-1}$)	4	50	Lambardi *et al.*, 2001
Castanea sativa x *C. crenata*	2	16h (1.3 $\mu E\ m^{-2}s^{-1}$)	12	96	Janeiro *et al.*, 1995
Corylus spp.	4	Darkness	30	NR	Reed, 1999
Diospyros kaki	10	Darkness	7	NR	Fukui *et al.*, 1994
Eucalyptus grandis	10	24h (4 $\mu E\ m^{-2}s^{-1}$)	6	70	Watt *et al.*, 2000
Malus spp.	4	16h	28	100	Wilkins *et al.*, 1988
Malus spp.	2-4	15h (10 $\mu E\ m^{-2}s^{-1}$)	NR	NR	Reed, 1999
Malus spp.	6	Darkness	6	100	Gardi *et al.*, 2001
Malus domestica	1	Darkness	12	100	Lundergan and Janick, 1979
Malus punica	4	Darkness	18	95	Negri *et al.*, 2000
Malus rootstocks	4	Darkness	12	100	Orlikowska, 1991, 1992
Morus alba	4	Darkness	6	80	Sharma and Thorpe, 1990
Olea europaea	4	Darkness	8	90	Lambardi *et al.*, 2000a
Olea europaea	6	Darkness	5	100	Gardi *et al.*, 2001
Populus alba x *P. grandidentata*	4	Darkness	60	25	Son *et al.*, 1991
Populus tremula x *P.tremuloides*	4	Darkness	3	100	Hausman *et al.*, 1994
Prunus spp.	4	16h	15	100	Wilkins *et al.*, 1988
Prunus armeniaca	3	Darkness	6	100	Pérez-Tornero *et al.*, 1999
Prunus avium	2	16h (1.3 $\mu E\ m^{-2}s^{-1}$)	12	100	Janeiro *et al.*, 1995
Prunus persica	0	16h (3.5 $\mu E\ m^{-2}s^{-1}$)	8	80	Dorion *et al.*, 1994

Prunus persica x *P. amygdalus*	0	16h (3.5 μE $m^{-2}s^{-1}$)	10	92	Dorion *et al.*, 1994
Prunus rootstocks	-3	Darkness	10	100	Marino *et al.*, 1985
Pyrus spp.	4	Darkness	55	NR	Reed, 1999
Pyrus communis	4	16h (20 μE $m^{-2}s^{-1}$)	18	79	Wanas *et al.*, 1986
Pyrus pashia	4	16h	12	100	Wilkins *et al.*,1988
Pyrus pyrifolia	1	Darkness	20	100	Moriguchi, 1995
Pyrus syriaca	8	Darkness	3	100	Tahtamouni and Shibli, 1999
Quercus petraea	2	16h (1.3 μE $m^{-2}s^{-1}$)	12	79	Janeiro *et al.*, 1995
Quercus robur	2	16h (1.3 μE $m^{-2}s^{-1}$)	12	100	Janeiro *et al.*, 1995
Quercus suber	5	Darkness	24	50	Romano and Martins-L., 1999
Vitis ruprestris	12	14h (12 μE $m^{-2}s^{-1}$)	11	73	Galzy and Compan, 1988
Vitis vinifera	12	14h (12 μE $m^{-2}s^{-1}$)	11	93	Galzy and Compan, 1988

(1) In brackets: light intensity applied with the photoperiod (when specified).

4. LONG-TERM CONSERVATION BY CRYOPRESERVATION

The term cryopreservation refers to the storage of specimens in liquid nitrogen, at ultra-low temperature (–196°C). Cryopreservation can be applied to a wide range of organs and tissues from *in vitro*-grown cultures, including shoot tips, embryogenic tissue and somatic embryos. Traditionally, the cryogenic technique has used slow-cooling methods, in which optimal cell dehydration is achieved by a controlled freezing rate of explants to about –40°C before their immersion in liquid nitrogen. This approach, although successful in preventing damage by intra-cellular ice crystallization, is laborious and expensive, due to the cost of the controlled-rate freezing apparatus. Recent refinement of technology has allowed the development of new, simple and effective procedures, commonly known as "vitrification-based techniques" (Engelmann, 1997). The term "vitrification" refers to the physical process of transition of an aqueous solution into an amorphous and glassy (i.e., non-crystalline) state during ultra-rapid freezing. When the vitrification process concerns the cell cytosol, intracellular ice crystal formation is hampered; thus, the tissue remains viable and responsive to reintroduction to standard culture conditions.

The main advantages of the vitrification technique are:

- the explants are directly plunged into liquid nitrogen (one step freezing), a characteristic which simplifies procedures considerably in comparison with the traditional slow-cooling approach;
- it is effective in preserving the integrity of complex organs (like shoot-tips), i.e., preventing partial destruction of meristems and avoiding callusing. Hence, a normal regrowth of the preserved explants after re-warming is assured (Sakai, 2000);
- protocols of cryopreservation by vitrification are highly repetitive.

Two different techniques have been proposed to induce vitrification during ultra-rapid freezing: (i) by loading the explants in a concentrated vitrification solution, and (ii) by partial dehydration of alginate-coated explants in sterile air flow or on silica gel dessiccant (encapsulation-dehydration). The two procedures were applied to woody species for the first time almost silmultaneously (Sakai *et al.*, 1990; Dereuddre *et al.*, 1990). Since then, there has been an increase in the number of woody plants that have been successfully cryopreserved. Table 3 lists the temperate broad-leaf trees for which a successful protocol has been reported, and it refers to one-step freezing of specimens in liquid nitrogen.

4.1. Loading with a Vitrification Solution

The key step in this procedure is the pre-treatment of explants with highly concentrated vitrification solutions, which induces osmotic dehydration of cells. When the cryovials (containing the explants still loaded in the solution) are then directly plunged into liquid nitrogen, a complete vitrification of cytosol and suspending solution occurs during ultra-rapid freezing. The application of this cryogenic technique to woody species became very popular after the PVS2 (Plant Vitrification Solution n. 2) was proposed for cryopreservation of *Citrus sinensis* nucellar cells (Sakai *et al.*, 1991). The solution is a mixture of cryoprotectants, i.e., 30% glycerol, 15% ethylene glycol, and 15% dimethylsulfoxide (DMSO) in MS liquid medium containing 0.4 M sucrose. To date, the PVS2 solution has been utilized in one-step freezing procedures with species belonging to 9 different Genera of temperate broad-leaf trees. This vitrification solution is highly effective in protecting explants from damage caused by ultra-rapid freezing (Fig. 1B). Indeed, survivals higher than 50% were reported for all the tested species, with the only exceptions being *Pyrus syriaca* shoot tips (Tahtamouni and Shibli, 1999) and *Olea europaea* embryogenic samples (Benelli *et al.*, 2001). Highest shoot-tip survival rates were in *Prunus jamasakura* (97%; Niino *et al.*, 1997) and *Populus alba* (90%; Lambardi *et al.*, 2000b. Fig. 1C).

When loading explants with the vitrification solution, survival after cryopreservation can be greatly influenced by the duration of treatment, which must be long enough to ensure sufficient cell dehydration, without any cytotoxic effect. In Table 3, when the solution was applied at 25°C, exposure time ranged from 20 (Matsumoto *et al.*, 2001) to 120 min (Niino *et al.*, 1997). Alternatively, a chilled solution (0°C) was also utilized to minimize the risk of toxicity (Caccavale *et al.*, 1998).

Although loading explants in PVS2 plays a pivotal role in inducing cell vitrification, some steps preceding the PVS2 treatment can contribute to optimizing the storage technique, such as:

- the hardening of stock cultures at low temperatures (generally, 4-5°C) and 8h photoperiod. The majority of the researchers consider a 3-week-cold hardening sufficient to acclimatize shoots before bud excision; however, a 45-day period was required to harden stock cultures from seven cultivars of cherry and sweet cherries (Niino *et al.*, 1997). Stock plants of *Pyrus syriaca* were hardened at 4°C in total darkness (Tahtamouni and Shibli, 1999);
- the hardening of shoot tips (after excision) and embryogenic cultures at 4°C, for a period ranging from 24 (in *Malus* spp.; Wu *et al.*, 1999) to 96 hours (in *Aesculus hippocastanum*; Lambardi *et al.*, 1999). In combination with this treatment, the pre-dehydration of shoot tips in media containing high sucrose concentrations (0.3-0.7 M) has frequently

improved cell survival during ultra-rapid freezing (Niino *et al.*, 1992a, 1992b; Niino *et al.*, 1997; Channuntapipat *et al.*, 2000; Matsumoto and Sakai, 2000; Matsumoto *et al.*, 2001). However, sucrose concentrations higher than 0.09 M were deleterious to shoot tip survival of white poplar (Lambardi *et al.*, 2000b) and plum (De Carlo *et al.*, 2000);

- the loading of shoot tips with cryoprotectants. A common mixture contains 2 M glycerol and 0.4 M sucrose (Matsumoto *et al.*, 1994). This treatment (20-30 min) is used occasionally before the loading of explants with PVS2.

4.1.1. Re-warming

Following storage in liquid nitrogen, transition of the vitrification solution from a vitreous to a crystalline phase can occur during warming of specimens for recrystallization (McFarlane and Forsyth, 1987; Grout, 1995), causing considerable damage to the tissues and compromising the regrowth of meristems after plating. A rapid warming in a waterbath can avoid recrystallization, and ensure a proper recovery of vitrified material. However, how fast this "rapid warming" should be is still controversial; indeed, warming temperatures of 20, 25, 30, 38 and 40°C have been proposed for temperate broad-leaf trees (see Table 3). In *Populus* spp., following cryopreservation, shoot tips were warmed at different temperatures (30 to 50°C). While *P. alba* shoot tips showed best survival when warmed at 40°C, negligible differences were recorded with *P. canescens*. However, for both species, a majority of healthy, well-developing shoots originated, in time, from shoot tips re-warmed at 40°C (Lambardi, 2002).

4.2. Encapsulation-Dehydration

This technique involves initially encapsulation of the explants in calcium alginate beads (Sakai *et al.*, 2000). Then, the complex explant/bead undergoes a double-step process for the removal of freezable water within the tissue: (i) incubation in a concentrated (up to 1 M) sucrose solution (osmotic dehydration), and (ii) air-drying treatment, accomplished by exposure to silica gel (inside a closed vessel), or to airflow of a laminar-flow hood. Following dehydration, the beads are directly plunged into liquid nitrogen. The advantages of the encapsulation-dehydration technique are: (a) the easy manipulation and the protection of delicate explants (shoot tips, embryos, embryogenic tissue), (b) better control of explant dehydration, and (c) the stimulation of shoot-tip regrowth after cryopreservation (Fig. 1D).

To date, the encapsulation-dehydration approach has been successfully tried with shoot tips from 9 different Genera of temperate broad-leaf woody plants. Maximum survival (91%) was reported for apple (*Malus* spp.), after the encapsulated shoot tips were dehydrated by culturing the beads in medium with 1 M sucrose for 1 day, followed by exposure to airflow for up to 4 hours (Zhao *et al.*, 1999a). It is noteworthy that 50% of the species, cryopreserved following an encapsulation-dehydration procedure, had survival rates of 80% or more.

Cryopreservation by encapsulation-dehydration was applied also to somatic embryos of ash (Tonon *et al.*, 2001). In this procedure, isolated embryos were initially encapsulated in a matrix of calcium alginate (Fig. 2A); the beads were then dehydrated in gelled medium containing 0.5 M sucrose for 24 hours and exposed to silica gel for 1 hour, prior to being directly plunged into liquid nitrogen.

4.2.1. Encapsulation-Vitrification

Recently, an "encapsulation-vitrification" technique has been developed. The procedure combines the encapsulation of explants with the use of a vitrification mixture. This technique has been successfully applied to the shoot-tip cryopreservation of apple (Paul *et al.*, 2000) and plum (De Carlo *et al.*, 2000), as well as isolated somatic embryos of olive, cv Nabali (Shibli and Al-Juboory, 2000), and embryogenic culture of ash (Fig. 2B; Tonon *et al.*, 2001). In ash, following cryopreservation, 27% of the embryogenic samples resumed proliferation. Twelve weeks after re-warming and plating, the embryogenic callus grew out of the alginate matrix and *de-novo* formed somatic embryos started to appear (Fig. 2C).

4.3. Other Procedures

During the last ten years cryopreservation by vitrification/one step freezing in liquid nitrogen has been by far the most widely used technique for temperate broad-leaf trees. However, the use of slow-rate freezing procedures has also been reported (Table 4). In this traditional approach to cryopreservation, following a pre-treatment with cryoprotectants, the explants are slow-cooled to –40°C (in general, at a rate of 0.5-1°C/min), prior to being plunged into liquid nitrogen (two-step freezing). Some researchers used the PVS2 vitrification solution followed by a two-step freezing of explants (de Boucaud and Brison, 1995; Brison *et al.*, 1995; Helliot and de Boucaud, 1997). In comparison with the vitrification method, shoot tip survival rate was highly variable with slow-cooling procedures, from 34% (*Juglans regia*; de Boucaud and Brison, 1995) to 92% (*Malus* spp.; Zhao *et al.*, 1999b). Here, the "droplet freezing method" was for the first (and to date the only) time applied to a woody plant. The peculiarity of this procedure consists in the distribution of shoot tips in 5µl droplets of a cryoprotectant mixture (0.3 M sucrose and 15% DMSO), which are then enclosed on sterile pieces of aluminium foil for slow-cooling to –40°C. As regards embryogenic cultures, a cryopreservation protocol was proposed for walnut somatic embryos (de Boucaud *et al.*, 1994). The somatic embryos, at the early globular stage, were encapsulated in alginate beads, dehydrated under laminar-air flow and stored in liquid nitrogen following either a one- or a two-step freezing procedure. Slow freezing (–1°C/min) gave a higher survival percentage (54%) and stimulated proliferation of new somatic embryos.

A very simple method was described for olive (*Olea europaea*) shoot-tip cryopreservation by Martinez *et al.* (1999). It involved dehydration of explants up to 30% moisture content in sterile air of a laminar flow cabinet, after which they were directly plunged into liquid nitrogen. A maximum of 30% survival rate was recorded after warming at room temperature.

A method of cryopreservation of dormant-vegetative buds was applied for long-term conservation of apple (Forsline *et al.*, 1998), based on the original procedure described by Sakai (1960). Here, winter-collected scions from 64 *Malus* accessions were desiccated in a freezer at 4°C and very slowly cooled (1°C per hour) to –30°C, where they remained for 24 hours prior to transfer to the vapour phase of liquid nitrogen (-165°C). After retrieval from storage and grafting onto rootstocks, 16% to 100% of plant regrowth was obtained. The technique is reviewed elsewhere (Niino, 2000).

5. MAINTENANCE OF GENETIC FIDELITY DURING CONSERVATION

The risk of genetic instability has always been a cause of concern when using tissue culture for plant regeneration. Indeed, numerous forms and manifestations of "somaclonal variation" (Larkin and Scowcroft, 1981) have been described over time, as a result of the introduction, manipulation and regeneration of plants *in vitro*. Obviously, this problem concerns also the conservation of germplasm at low and ultra-low temperatures, although the peculiarities of this technology (reduced or almost blocked metabolism of cells, little manipulation of explants, low frequency of subcultures) greatly reduce the risks of variability.

To date, no clear evidence of genetic alteration due to slow growth storage has been produced, although reports going deep into this aspect are few in number. With reference to broad-leaf trees, plants regenerated from cold-stored shoots of grape (*Vitis* spp.) never showed any phenotypic difference to unstored plants, and the cytological analysis proved that the chromosome number was consistent with the diploid condition (Barlass and Skene, 1983). Phenotypic variations in the colour and growth pattern of hybrid poplar (*Populus alba* x *P. grandidentata*) shoots were observed after conservation at 4°C for more than 5 years (Son *et al.*, 1991). In particular, an unusual appearance of albino plants was observed. However, no evidence was reported on the nuclear or cytoplasmic origin of these alterations.

As regards cryopreservation, genetic stability could be affected by chemical reactions, such as free radical formation and molecular damage due to ionising radiation (Grout, 1990), which can still occur at –196°C. Moreover, common use of DMSO as cryoprotectant (alone or in combination with other compounds, at concentrations up to 10%) generate additional risks of genetic or epigenetic changes. However, several cytological, biochemical and molecular methods are now available to monitor the genetic stability of cryopreserved germplasm (Harding, 1999). Early reports on molecular marker applications to woody plant material from ultra-low freezing confirm the reliability of the cryogenic technique in preserving genetic stability. In Conifers, for instance, the genetic fidelity of embryogenic cultures regrown after cryopreservation has already been reported (Cyr *et al.*, 1994; Park *et al.*, 1994; De Verno *et al.*, 1999). On the other hand, the genetic changes evidenced by an embryogenic line of *Abies cephalonica* following cryopreservation were attributed to the pre-treatment with DMSO. Cryostorage *per se* did not induce genetic variability, but rather, it attenuated the mutagenic effect of DMSO by eliminating a high proportion of cells bearing genetic alterations (Aronen *et al.*, 1999).

As regards broad-leaf woody species, to date few reports have investigated the genetic fidelity of cryopreserved embryogenic tissue (Marin *et al.*, 1993; Olivares Fuster *et al.*, 2000) or shoot tips (Helliot *et al.*, 1998). Although the genetic stability of material after brief periods of cryostorage was proved in all these reports, further validations after longer periods of storage are opportune before long-term preservation of tree biodiversity is undertaken.

6. CONCLUSIONS

Recent advances in tissue culture have paved the way to propagation, genetic improvement and conservation of woody plants. Progress of *in vitro* conservation technology, for instance, has been impressive over the past decade, and the possibility of storing valuable germplasm at low and cryogenic temperatures could now contribute markedly to the preservation of plant biodiversity. Slow growth storage is routinely used in many laboratories for medium-term storage of several temperate broad-leaf trees, while cryopreservation procedures are now available for long-term germplasm preservation, in support of *ex situ* traditional approaches (seed and in-field genebanks). Working with different species, diversities among optimal procedures (e.g., temperatures and light conditions during slow growth storage; cryoprotectant mixtures, vitrification and freezing methods, re-warming temperatures in cryopreservation) are a peculiarity of these techniques, and can be ascribed to the intrinsic characteristics of the stored material, regarding genotype, physiological conditions, and type of explant. Research should move in the direction of standardizing the procedures as much as possible and, at the same time, of improving their versatility. In this sense, the development of other vitrification mixtures, or the further exploration of new and promising techniques, such as encapsulation-vitrification, should be among the main aims of cryopreservation research.

7. REFERENCES

Aronen T.S., Krajnakova J., Häggman H.M., Ryynänen L.A. (1999) Genetic fidelity of cryopreserved embryogenic cultures of open-pollinated *Abies cephalonica*. Plant Sci. 142: 163-172.

Ashmore S.E. (1997) Status Report on the Development and Application of *In vitro* Techniques for the Conservation and Use of Plant Genetic Resources. International Plant Genetic Resources Institute, Rome, xii+67 pp.

Bachiri Y., Song P., Shoar G.A., Rekab T., Morissete C. (2001) Routine cryopreservation of kiwifruit (*Actinidia* spp.) germoplasm by incapsulation-dehydration: importance of plant growth regulators. CryoLetters 22: 61-74.

Bajaj Y.P.S. (1995) Cryopreservation of plant cell, tissue, and organ culture for the conservation of germplasm and biodiversity. In: Y.P.S. Bajaj (ed) Cryopreservation of Plant Germplasm I. Biotechnology in Agriculture and Forestry, Vol. 32. Springer, Berlin Heidelberg New York, pp. 3-28.

Barlass M., Skene K.G.M. (1983) Long-term storage of grape *in vitro*. FAO/IBPGR Plant Gen. Res. Newsl. 53: 19-21.

Benelli C., De Carlo A., Lambardi M., Lynch P.T. (2001) Vitrification of shoot tips, nodal segments and embryogenic tissue of olive (*Olea europaea* L.) for germplasm cryopreservation. Acta Hortic. 560: 137-140.

Benson E.E. (1999) Cryopreservation. In: E.E. Benson (ed), Plant Conservation Biotechnology. Taylor & Francis, London, pp. 83-95.

Blakesley D., Pask N., Henshaw G.G., Fay M.F. (1996) Biotechnology and the conservation of forest genetic resources: *in vitro* strategies and cryopreservation. Plant Growth Regul. 20: 11-16.

Blakesley D., Kiernan R.J. (2001) Cryopreservation of axillary buds of a *Eucalyptus grandis* x *Eucalyptus camaldulensis* hybrid. CryoLetters 22: 13-18.

Bonga J.M., von Aderkas P. (1992) Genetic modification. In : J.M. Bonga and P. von Aderkas (eds) *In Vitro* Culture of Trees. Kluwer Ac. Publ., Dordrecht, The Netherlands, pp. 156-165.

Brison M., de Boucaud M.T., Dosba F. (1995) Cryopreservation of *in vitro* grown shoot tips of two interspecific *Prunus* rootstocks. Plant Sci. 105: 235-242.

Caccavale A., Lambardi M., Fabbri A. (1998) Cryopreservation of woody plants by axillary bud vitrification: a first approach with poplar. Acta Hortic. 457: 79-83.

Chunnuntapipat C., Collins G., Bertozzi T., Sedgley M. (2000) Cryopreservation of *in vitro* almond shoot tips by vitrification. J. Hort. Sci. & Biotech. 75 (2): 228-232.

Cyr D.R., Lazaroff W.R., Grimes G.Q., Quan T.D., Bethune T.D., Dunstan D.I., Roberts D.R. (1994) Cryopreservation of interior spruce (*Picea glauca engelmanni* complex) embryogenic cultures. Plant Cell Rep. 13: 574-577.

Cyr D.R. (2000) Cryopreservation: roles in clonal propagation and germplasm conservation of conifers. In: F. Engelmann and H. Takagi (eds), Cryopreservation of Tropical Plant Germplasm. International Plant Genetic Resources Institute, Rome, pp. 261-268.

Damiano C., Engelmann F., Frattarelli A., Shatnawi M.A. (1998) Crioconservazione di apici di melo. Proceed "IV° Convegno Nazionale Biodiversità". Alghero, Italy, September 8-11, pp. 417-420.

de Boucaud M.T., Brison M., Negrier P. (1994) Cryopreservation of walnut somatic embryos. CryoLetters 15: 151-160.

de Boucaud M.T., Brison M. (1995) Cryopreservation of germoplasm of walnut (*Juglans* species). In: Y.P.S. Bajaj (ed) Cryopreservation of Plant Germplsm I. Biotechnology in Agriculture and Forestry, Vol. 32. Springer, Berlin Heidelberg New York, pp. 129-147.

De Carlo A., Benelli C., Lambardi M. (2000) Development of a shoot-tip vitrification protocol and comparison with encapsulation-based procedures for plum (*Pruns domestica* L.) cryopreservation. CryoLetters 21: 215-222.

De Verno L.L., Park Y.S., Bonga J.M., Barret J.D. (1999) Somaclonal variation in cryopreserved embryogenic clones of white spruce [*Picea Glauca* (Moench)Voss.]. Plant Cell Rep. 18: 948-953.

Dereuddre J., Scottez C., Arnaud Y., Duron M. (1990) Effects of cold hardening on cryopreservation of axillary pear (*Pyrus communis* L. cv Beurre Hardy) shoot tips of *in vitro* plantlets. C. R. Acad. Sci. (Paris) 310: 265-272.

Dorion N., Regnard J.L., Serpette I., Bigot C. (1994) Effects of temperature and hypoxic atmosphere on preservation and futher development of *in vitro* shoots of peach ("Armiking") and peach x almond hybrid (GF 677). Sci. Hort. 57: 201-213.

Engelmann F. (1997) Importance of dessication for cryopreservation of recalcitrant seed and vegetatively propagated apices. Plant Gen. Res. Newsl. 112: 9-18.

Fisher M., Mannel R. (1994) Evaluation of spontaneous plum pox virus infection of plum and prune varieties. Acta Hortic. 359: 131-135.

Forsline P.L., Towill L.E., Waddell W., Stushnoff C., Lamboy W.F., McFerson J.R. (1998) Recovery and longevity of cryopreserved dormant apple buds. J. Amer. Soc. Hort. Sci. 123 (3): 365-70.

Fukui H., Ohba H., Nakamura M. (1994) Low temperature of *in vitro* shoots of japanese persimmon (*Diospyros kaki*). Comb. Proc. Int. Plant Prop. Soc. Vol. 44, pp. 245-248.

Galzy R., Compan D. (1988) Growth and nutrition of grapevine during *in vitro* long-term storage. Plant Cell Tiss. Org. Cult. 13: 229-237.

Gardi T., Micheli M., Piccioni E., Sisani G., Standardi A. (2001) Cold storage of micropropagated shoots of some fruit species. Italus Hortus 8(4): 32-40 (with English abstract).

Gass T., Tobutt K.R., Zanetto A. (1996) Report of the Working Group on *Prunus*. Fifth meeting. Menemen-Izmir, Turkey, Febbrary 1-3. IPGRI, Rome, pp. 1-70.

Gould A.R. (1986) Factors controlling generation of variability *in vitro*. In: I.K. Vasil (ed), Cell Culture and Somatic Cell Genetics of Plants, Vol. 3. Plant Regeneration and Genetic Variability. Academic Press Inc., Orlando, pp. 549-612.

Grout B. (1990) *In vitro* conservation of germplasm. In: S.S. Bhojwani (ed), Developments in Crop Science, Vol. 19. Plant Tissue Culture: Applications and Limitations. Elsevier, The Netherlands, pp. 394-410.

Grout B. (1995) Introduction to the *in vitro* preservation of plant cells, tissue and organs. In: B. Grout (ed), Genetic Preservation of Plant Cells *In Vitro*. Springer, Berlin Heidelberg New York, pp. 1-20.

Harding K. (1999) Stability assessments of conserved plant germplasm. In: E.E. Benson (ed), Plant Conservation Biotechnology. Taylor & Francis, London, pp. 97-107.

Hausman J.F., Neys O., Kevers C., Gaspar T. (1994) Effects of *in vitro* storage at 4°C on survival and proliferation of poplar shoots. Plant Cell Tiss. Org. Cult. 38: 65-67.

Helliot B., de Boucaud M.T. (1997) Effect of various parameters on the survival of cryopreserved *Prunus* ferlenain *in vitro* plantlets shoot tips. CryoLetters 18: 133-142.

Helliot B., Dirlewanger E., de Boucaud M.T. (1998) Genetic conformity analysis of *in vitro* grown *Prunus* Ferlenain shoots after cryopreservation by two different procedures. Abstracts "IX International Congress on Plant Tissue and Cell Culture". Jerusalem, Israel, June 14-19, p. 151.

Janeiro L.V., Vieitez A.M., Ballester A. (1995) Cold storage of *in vitro* cultures of wild cherry, chestnut and oak. Ann. Sci. For. 52: 287-293.

Lambardi M., Capuana M., De Carlo A., Tonon G. (1999) Cryopreservation of forest species by vitrification of organ and tissue from *in vitro* culture. In: S. Espinel and E. Ritter (eds), Proceed. Inter. Congress "Applications of Biotechnology to Forest Genetics. Biofor-99". Vitoria-Gasteiz, Spain, September 22-25, pp. 313-320.

Lambardi M., Benelli C., De Carlo A., Fabbri A., Grassi S., Lynch P.T. (2000a) Medium- and long-term *in vitro* conservation of olive germoplasm (*Olea europaea* L.). Acta Hortic. Proceed. "4th International Symposium on Olive Growing - OLIVE 2000". Valenzano (Bari), Italy, September 25-30. *In press*.

Lambardi M., Fabbri A., Caccavale A. (2000b) Cryopreservation of white poplar (*Populus alba* L.) by vitrification of *in vitro*-grown shoot tips. Plant Cell Rep. 19: 213-218.

Lambardi M., Benelli C., De Paoli G., Battistini A. (2001). Biotecnologie per la conservazione del castagno. Proceed. "Convegno Nazionale Castagno 2001". Marradi (Florence), Italy, October 25-27 (with English abstracts). *In press*.

Lambardi M. (2002) Cryopreservation of Germplasm of Populus (Poplar) Species. In: L. Towill (ed) Cryopreservation of Plant Germplasm II. Biotechnology in Agriculture and Forestry, Vol. 50. Springer, Berlin Heidelberg, pp. 269-286.

Larkin P.J., Scowcroft W.R. (1981) Somaclonal variation – a novel source of variability from cell cultures for plant improvement. Theor. Appl. Genet. 60: 197-214.

Lundergan C., Janick J. (1979) Low temperature storage of *in vitro* apple shoots. HortSci. 14 (4): 514.

McFarlane D.R., Forsyth M. (1987) Devitrification and recrystallisation of glass-forming aqueous solutions. In: D.E. Pegg and A.M. Karow (eds), The Biophysics of Organ Cryopreservation. Plenum Press, New York, pp. 237-264.

Marin M.L., Gorgorcena Y., Ortiz J., Duran-Vila N. (1993) Recovery of whole plants of sweet orange from somatic embryos subjected to freezing thawing treatments. Plant Cell Tiss. Org. Cult. 34: 27-33.

Marino G., Rosati P., Sagrati F. (1985) Storage of *in vitro* cultures of *Prunus* rootstocks. Plant Cell Tiss. Org. Cult. 5: 73-78.

Martinez D., Arroyo-Garcia R., Revilla A.M. (1999) Cryoconservation of *in vitro* grown shoot-tips of *Olea europaea* L. var. Arbequina. CryoLetters 20: 29-36.

Marzalina M., Krishnapillay B. (1999) Recalcitrant seed biotechnology application to rain forest conservation. In: E.E. Benson (ed), Plant Conservation Biotechnology. Taylor & Francis, London, pp. 265-276.

Matsumoto T., Sakai A., Yamada K. (1994) Cryopreservation of in vitro-grown apical meristems of wasabi (*Wasabia japonica*) by vitrification and subsequent high plant regeneration. Plant Cell Rep. 13: 442-446.

Matsumoto T., Sakai A. (2000) Cryopreservation of grape *in vitro*-cultured axillary shoot-tips by three-step vitrification. In: F. Engelmann and H. Takagi (eds), Cryopreservation of Tropical Plant Germplasm. International Plant Genetic Resources Institute, Rome, pp. 424-428.

Matsumoto T., Mochida K., Itamura H., Sakai A. (2001) Cryopreservation of persimmon (*Diospyros kaki* Thunb.) by vitrification of dormant shoot tips. Plan Cell Rep. 20: 398-402.

Monette P.L. (1995) Conservation of germplasm of kiwifruit (*Actinidia* species). In: Y.P.S. Bajaj (ed), Biotechnology in Agriculture and Forestry, Vol. 32. Cryopreservation of Plant Germplasm I. Springer, Berlin Heidelberg New York, pp. 321-331.

Moriguchi T. (1995) Cryopreservation and minimum growth storage of pear (*Pyrus* species). In: Y.P.S. Bajaj (ed), Biotechnology in Agriculture and Forestry, Vol. 32. Cryopreservation of Plant Germplasm I. Springer, Berlin Heidelberg New York, pp. 114-128.

Murashige T., Skoog F. (1962) A revised medium for rapid growth and bioassay with tobacco tissue cultures. Physiol. Plant. 15: 473-497.

Negri V., Tosti N., Standardi A. (2000). Slow-growth storage of single node shoots of apple genotypes. Plant Cell Tiss. Org. Cult. 62: 159-162.

Niino T., Sakai A. (1992) Cryopreservation of alginate-coated *in vitro* shoot tips of apple, pear and mulberry. Plant Sci. 87: 199-206.
Niino T., Sakai A., Yakuwa H., Nojiri K. (1992a) Cryopresevation of *in vitro*-grown shoot tips of apple and pear by vitrification. Plant Cell Tiss. Org. Cult. 28: 261-266.
Niino T., Sakai A., Enomoto S., Magosi J., Kato S. (1992b) Cryopreservation of *in vitro*-grown shoot tips of mulberry by vitrification. CryoLetters 13: 303-312.
Niino T., Tashiro K., Suzuki M., Ohuchi S., Magoshi J., Akihama T. (1997) Cryopreservation of *in vitro* shoot tips of cherry and sweet cherry by one-step vitrification. Sci. Hort. 70: 155-163.
Niino T. (2000) Cryopreservation of deciduous fruits and mulberry trees. In: M.K. Razda and E.C. Cocking (eds), Conservation of Plant Genetic Resources *In vitro*. Vol. 2: Applications and Limitations. Enfield Science Publ. Inc., New Hampshire, pp. 195-223.
Olivares Fuster O., Asíns M.J., Duran-Vila N., Navarro L. (2000) Cryopreserved callus, a source of protoplasts for citrus improvement. J. Horticultural Science Biotechnology 75 (6): 635-640.
Orlikowska T. (1991) Effect of *in vitro* storage at 4°C on surviving and proliferation of apple rootstocks. Acta Hortic. 289: 251-253.
Orlikowska T. (1992) Effects of *in vitro* storage at 4°C on survival and proliferation of two apple rootstocks. Plant Cell Tiss. Org. Cult. 31: 1-7.
Park Y.S., Pond S.E., Bonga J.M. (1994) Somatic embryogenesis in white spruce (*Picea glauca*): genetic control in somatic embryos exposed to storage, maturation treatments, germination, and cryopreservation. Theor. Appl. Genet. 89: 742-750.
Paul H., Daigny G., Sangwan-Norreel B.S. (2000) Cryopreservation of apple (*Malus* x *domestica* Borkh.) shoot tips following encapsulation-dehydration or encapsulation-vitrification. Plant Cell Rep. 19: 768-774.
Pence V.C. (1995) Cryopreservation of recalcitrant seeds. In: Y.P.S. Bajaj (ed), Biotechnology in Agriculture and Forestry, vol. 32. Cryopreservation of Plant Germplsm I. Springer, Berlin Heidelberg New York, pp. 29-50.
Pérez-Tornero O., Ortín-Párraga F., Egea J., Burgos L. (1999) Medium-term storage of apricot shoot tips *in vitro* by minimal growth method. HortSci. 34 (7): 1277-1278.
Reed B.M., Denoma J., Luo J., Chang Y., Towill L. (1998) Cryopreservation and long-term storage of pear germplasm. *In vitro* Cell Dev. Biol. – Plant, 34: 256-260.
Reed B.M. (1999) *In vitro* conservation of temperate tree fruit and nut crops. In: E.E. Benson (ed), Plant Conservation Biotechnology. Taylor & Francis, London, pp. 139-153.
Reed B.M. (2001) Implementing cryogenic storage of clonally propagated plants. CryoLetters 22: 97-104.
Roberts H.F. (1973) Predicting the viability of seeds. Seed Science and Technology 1: 499-514.
Romano A., Martins-Loução M.A. (1999) *In vitro* cold storage of cork oak cultures. Plant Cell Tiss. Org. Cult. 59: 155-157.
Rugini E. (1986) Olive (*Olea europea* L.). In: Y.P.S. Bajaj (ed), Biotecnology in Agriculture and Forestry. Vol. 1: Tress I. Springer, Berlin Heidelberg New York, pp. 253-266.
Sakai A. (1960) Survival of the twigs of woody plants at –196°C. Nature 185: 393-394.
Sakai A., Kobayash S., Oiyama I. (1990) Cryopreservation of nucellar cells of navel orange (*Citrus sinensis* Osb. var. brasiliensis Tanaka) by vitrification. Plant Cell Rep. 9: 30-33.
Sakai A., Kobayash S., Oiyama I. (1991) Survival by vitrification of nucellar cells of navel orange (*Citrus sinensis* Osb. var. brasiliensis Tanaka) cooled to –196°C. J. Plant Physiol. 137: 465-470.
Sakai A. (2000) Development of cryopreservation techniques. In: F. Engelmann and H. Takagi (eds), Cryopreservation of Tropical Plant Germplasm. International Plant Genetic Resources Institute, Rome, pp. 1-7.
Sakai A., Matsumoto T., Hirai D., Niino T. (2000) Newly developed encapsulation-dehydratation protocol for plant cryopreservation. CryoLetters 21: 53-62.

Scottez C., Chevreau E., Godard N., Arnaud Y., Duron M., Dereuddre J. (1992) Cryopreservation of cold-acclimatated shoot tips of pear *in vitro* cultures after encapsulation-dehydration. Cryobiology 29: 691-700.

Sharma K.K., Thorpe T. A. (1990) *In* vitro propagation of mulberry. Sci. Hort. 42: 307-320.

Shatnawi M.A., Engelmann F., Frattarelli A., Damaino C. (1999) Cryopreservation of apices of *in vitro* plantlets of almond (*Prunus dulcis* Mill.). CryoLetters 20: 13-20.

Shibli R.A., Al-Juboory K.H. (2000) Cryopreservation of 'Nabali' olive (*Olea europaea* L.) somatic embryos by encapsulation-dehydration and encapsulation-vitrification. CryoLetters 21: 357-366.

Son S.H., Chun Y.W., Hall R.B. (1991) Cold storage of *in vitro* cultures of hybrid poplar shoots (*Populus alba* L. x *P. grandidentata* Mich). Plant Cell Tiss. Org. Cult. 27:161-168.

Tahtamouni R.W., Shibli R.A. (1999) Preservation at low temperature and cryopreservation in wild pear (*Pyrus syriaca*). Adv. Hort. Sci. 13: 156-160.

Tonon G., Lambardi M., De Carlo A., Rossi C. (2001) Crioconservazione di linee embriogeniche di *Fraxinus angustifolia* Whal. Proceed. "VI Convegno Nazionale Biodiversità". Valenzano (Bari), Italy, September 6-7 (with English abstracts). *In press.*

Wanas W.H., Callow J.A., Withers L.A. (1986) Growth limitation for the conservation of pear genotypes. In: Whiters and Alderson (eds), Plant Tissue Culture and its Agricultural Application. London, Butterworths pp. 285-290.

Wang Q., Tanne E., Arav A., Gafny R. (2000) Cryopreservation of *in vitro*-grown shoot tips of grapevine by encapsulation-dehydration. Plant Cell Tiss. Org. Cult. 63:41-46.

Watt M.P., Thokoane N.L., Mycock D., Blakeway F. (2000) *In vitro* storage of *Eucalyptus grandis* germplasm under minimal growth conditions. Plant Cell Tiss. Org. Cult. 61: 161-164.

Wenhuang M., Sanada T. (1995) Experiment of the cryopreservation of apple shoots *in vitro*. Acta Hortic. 403: 59-62.

Wilkins C.P., Newbury H.J., Dodds J.H. (1988) Tissue culture conservation of fruit trees. FAO/IBPGR Plant Gen. Res. Newsl. 73/74: 9-20.

Wu Y., Engelmann F., Zaho Y., Zhou M., Chen S. (1999) Cryopreservation of apple shoot tips: importance of cryopreservation technique. CryoLetters 20: 121-130.

Wu Y., Zhao Y., Engelmann F., Zhou M. (2001) Cryopreservation of kiwi shoot tips. CryoLetters 22: 277-284.

Zhao Y., Wu Y., Engelmann F., Zhou M., Zhang D., Chen S. (1999a) Cryopreservation of apple shoot tips by encapsulation-dehydration: effect of preculture, dehydration and freezing procedure on shoot regeneration. CryoLetters 20: 103-108.

Zhao Y., Wu Y., Engelmann F., Zhou M., Chen S. (1999b) Cryopreservation of apple in vitro shoot tips the droplet freezing method. CryoLetters 20: 109-112

Zhao Y., Wu Y., Engelmann F., Zhou M. (2001) Cryopreservation of axillary buds of grape (*Vitis vinifera*) *in vitro* plantlets. CryoLetters 22: 321-328.

Table 3. Summary of reports on cryopreservation of temperate broad-leaf trees by vitrification/one step freezing in LN. Cryogenic conditions producing maximum survival of explants are reported. (SG, dehydratation on silical gel; LF, dehydratation in airflow; NR, not reported).

Species	Cryopreserved explants	Vitrification procedures	Re-warming °C (min)	Survival (%)	Reference
Loading with vitrification solution					
Aesculus hippocastanum	Embryogenic culture	PVS2, 90 min at 0°C → LN	40 (1)	75	Lambardi *et al.*, 1999
Diospyros kaki	Shoot-tips [(1)]	PVS2, 20 min at 25°C → LN	40	89	Matsumoto *et al.*, 2001
Malus spp.	Shoot-tips	PVS3 [(2)], 80 min at 25°C → LN	25 (2-3)	88	Wu *et al.*, 1999
Malus domestica	Shoot-tips	PVS2, 80 min at 25°C → LN	25	80	Niino *et al.*, 1992a
Morus spp.	Shoot tips	PVS2, 90 min at 25°C → LN	25	81	Niino *et al.*, 1992b
Olea europaea	Embryogenic culture	PVS2, 90 min at 0°C → LN	40	38	Benelli *et al.*, 2001
Populus alba	Shoot tips	PVS2, 60 min at 0°C → LN	40 (1)	90	Lambardi *et al.*, 2000b
Populus canescens	Shoot tips	PVS2, 60 min at 0°C → LN	40 (1)	54	Lambardi, 2002
Prunus spp.	Shoot tips	PVS2, 120 min at 25°C → LN	30	80	Niino *et al.*, 1997
Prunus domestica	Shoot tips	PVS2, 90 min at 0°C → LN	40 (1)	57	De Carlo *et al.*, 2000
Prunus dulcis	Shoot tips	PVS2, 45 min at 25°C → LN	30	88	Chunnuntapipat *et al.*, 2000
Prunus dulcis x *P. persica*	Shoot tips	PVS2, 60 min at 25°C → LN	30	73	Chunnuntapipat *et al.*, 2000

Prunus jamasakura	Shoot tips	PVS2, 105 min at 25°C → LN	30	97	Niino *et al.*, 1997
Pyrus spp.	Shoot tips	PVS2, 90 min at 25°C → LN	25	72	Niino *et al.*, 1992a
Pyrus syriaca	Shoot tips	PVS2 (60%), 20 min → LN	25 (3-5)	30	Tahtamouni and Shibli, 1999
Vitis vinifera	Shoot tips	PVS2, 30 min + 50 min (3) → LN	40	85	Matsumoto and Sakai, 2000
Encapsulation - dehydration					
Actinidia spp.	Shoot tips	SG, 4 h → LN	20	85	Bachiri *et al.*, 2001
Fraxinus angustifolia	Somatic embryos	LF, 1 h → LN	40 (3)	31	Tonon *et al.*, 2001
Fraxinus angustifolia	Embryogenic culture	Encapsulation-vitrification (4) → LN	40 (2)	27	Tonon *et al.*, 2001
Malus spp.	Shoot tips	NR	25 (2-3)	86	Wu *et al.*, 1999
Malus spp.	Shoot tips	LF, 4 h → LN	25	91	Zhao *et al.*, 1999a
Malus domestica	Shoot tips	SG, 7 h → LN	25	80	Niino and Sakai, 1992
Malus domestica	Shoot tips	SG, 14 h → LN	NR	72	Damiano *et al.*, 1998
Malus domestica	Shoot tips	LF, 6 h → LN	25 (15)	88	Paul *et al.*, 2000
Malus domestica	Shoot-tips	Encapsulation-vitrification (5) → LN	25	77	Paul *et al.*, 2000
Morus bomyicis	Shoot tips	LF, 7 h → LN	25	65	Niino and Sakai, 1992
Olea europaea	Somatic embryos (6)	Encapsulation-vitrification (6)	38 (2)	58	Shibli and Al-Juboory, 2000
Prunus domestica	Shoot tips	Encapsulation-vitrification (7) → LN	40 (1)	48	De Carlo *et al.*, 2000

Prunus dulcis	Apices	SG, 9 h → LN	25 (3-4)	62	Shatnawi *et al.*, 1999
Pyrus spp.	Shoot tips	LF, 7 h → LN	25	70	Niino and Sakai, 1992
Pyrus communis	Shoot tips	LF, 3 h → LN	25	80	Dereuddre *et al.*, 1990
Pyrus communis	Shoot tips	LF, 4 h → LN	25	80	Scottez *et al.*, 1992
Pyrus syriaca	Shoot tips	LF, 5 h → LN	25 (3-5)	40	Tahtamouni and Shibli, 1999
Salix hybrid	Shoot tips	LF, 4.5 h → LN	40	83	Blakesley *et al.*, 1996
Vitis spp.	Shoot tips	LF, 7 h → LN	40	60	Wang *et al.*, 2000

(1) Shoot tips were directly excised from dormant axillary buds collected from trees in field.
(2) PVS3, 50% glycerol (w/v) and 50% sucrose (w/v).
(3) Shoot tips were treated with 50% PVS2 (30 min), followed by 100% PVS2 (50 min).
(4) Bead vitrification was induced with PVS2 (90 min at 0°C).
(5) Vitrification solution consisted of 180% (w/v) sucrose and 120% (w/v) ethylene glycol.
(6) Somatic embryogenesis obtained from germinated seeds. Bead vitrification was induced with PVS2 (3 h at 0°C).
(7) Bead vitrification was induced with PVS2 (3 h at 0°C).

Table 4. Summary of reports on cryopreservation of temperate broad-leaf trees by a two-step freezing in LN (slow cooling). Cryogenic conditions producing maximum survival of explants are reported. (SG, dehydratation on silical gel; LF, dehydratation in airflow).

Species	**Cryopreserved explants**	**Vitrification procedures**	**Re-warming °C (min)**	**Survival (%)**	**Reference**
Actinidia deliciosa	Shoot tips	SG, 4 h → 0.2°C/min up to – 40°C	25 (2-3)	56	Wu *et al.*, 2001
Eucalyptus grandis x *E. camaldulensis*	Shoot tips	LF, 5 h → 0.5°C/min up to – 40°C	38	49	Blakesley and Kiernan, 2001
Juglans regia	Somatic embryos	LF, 3 h → 1°C/min up to – 40°C	40 (1)	54	de Boucaud *et al.*, 1994
Juglans regia	Shoot tips	PVS2, 20 min at 0°C → 0.5°C/min up to – 40°C	40	34	de Boucaud and Brison, 1995
Malus spp.	Shoot tips	15% DMSO + 0.3M sucrose, 30 min at 25°C → 0.2 °C/min up to – 40°C [(1)]	25	92	Zhao *et al.*, 1999b
Malus domestica	Shoot tips	10% DMSO + 3% sucr. → 0.5°C/min up to – 40°C	38 (2)	90	Wenhuang and Sanada, 1995
Prunus Ferlenain	Shoot tips	PVSm[(2)], 45 min at 4°C → 5°C/min up to – 40°C	40	60	Helliot and de Boucaud 1997
Prunus rootstocks	Shoot tips	PVSm[(2)], 25 min at 25°C → 1°C/min up to – 40°C	40 (1)	75	Brison *et al.*, 1995
Pyrus spp.	Shoot tips	5% DMSO, 48 h → PGD [(3)], 30 min at 4°C → 0.1°C/min up to – 40°C	45 (1)	80	Reed *et al.*, 1998
Vitis vinifera	Shoot tips	SG, 5 h → 0.2°C/min up to – 40°C	25 (2-3)	40	Zhao *et al.*, 2001

[(1)] "Droplet freezing method".
[(2)] PVSm, 12.5 % DMSO, 15% ethylen glycol, 25% glycerol, 3% polyethylene glycol 8000 and 0.4 M sucrose.
[(3)] PGD, polyethylene glycol (M, 8000), glucose and DMSO (10% each) in liquid medium.

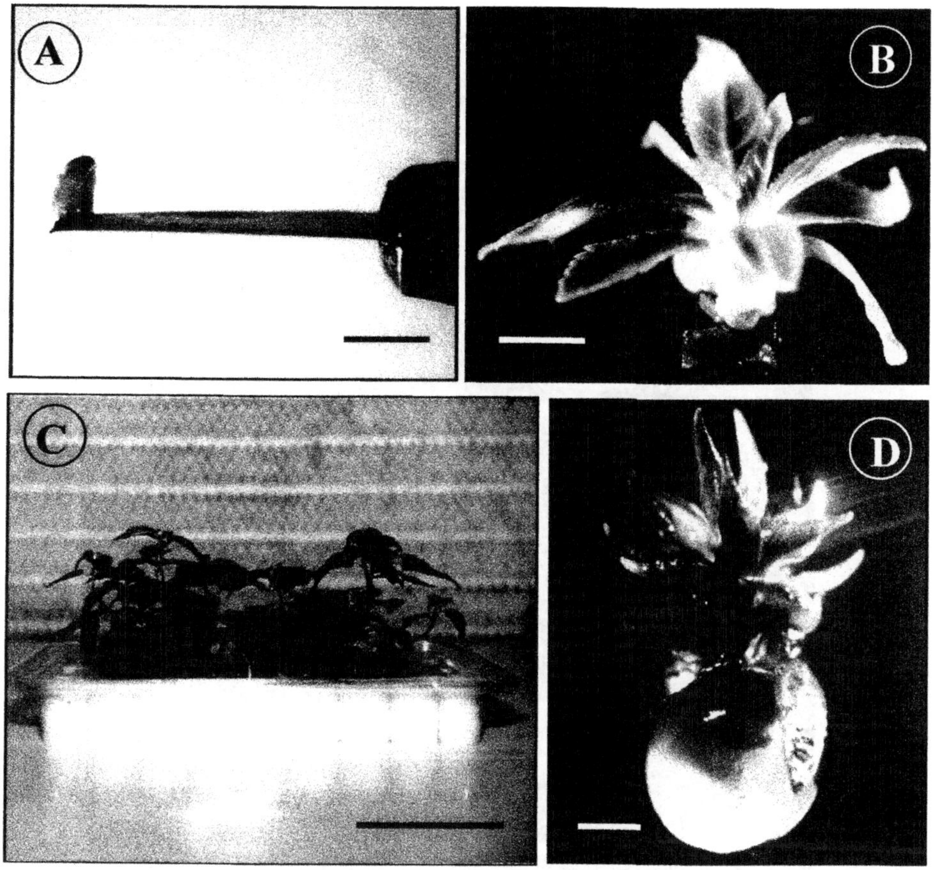

Figure 1. Cryopreservation of shoot tips of temperate broad-leaf species. **A:** a shoot tip of white poplar (*Populus alba*) before being used in cryopreservation trials (bar, 1 mm). **B:** a survived shoot tip of plum (*Prunus domestica*), 3 weeks after cryopreservation (procedure: loading with PVS2; bar, 1 mm). **C:** acclimatized shoots of white poplar from cryopreservation (procedure: loading with PVS2; bar, 5 cm). **D:** shoot tip of plum, growing from the Ca-alginate bead after cryopreservation (procedure: encapsulation-vitrification; bar, 1mm).

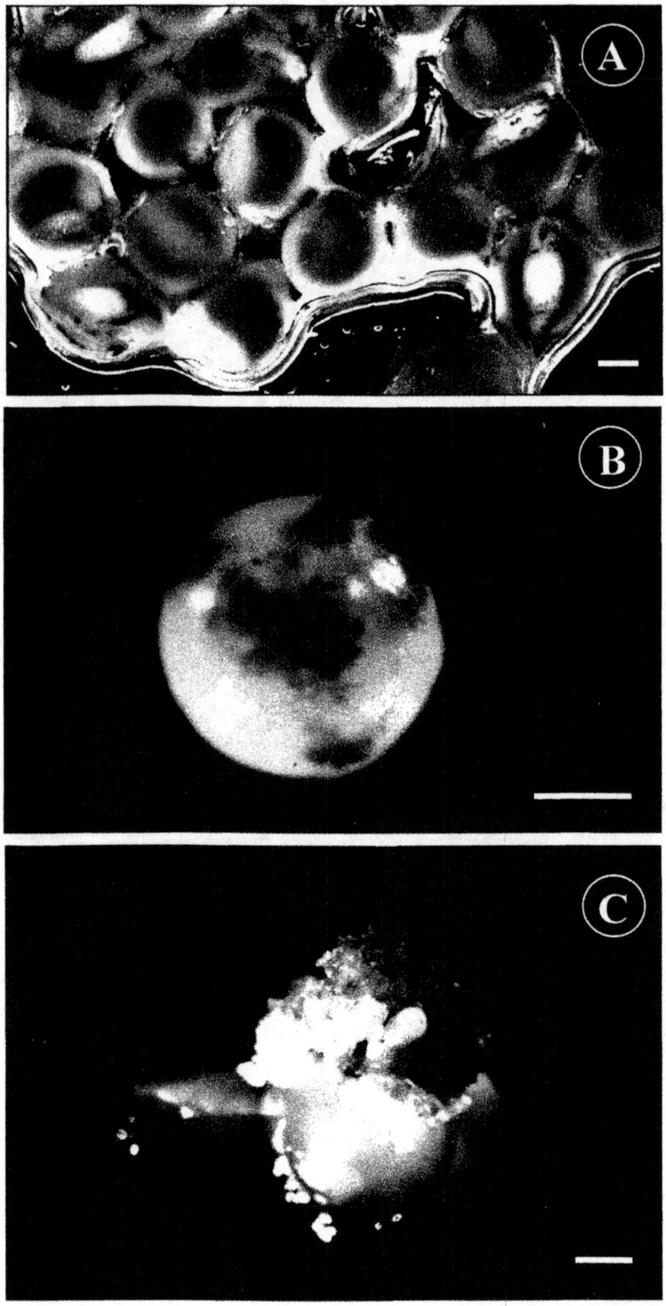

Figure 2. Cryopreservation of somatic embryos and embryogenic callus of ash (*Fraxinus angustifolia*). **A**: Ca-alginate coated somatic embryos before cryopreservation (bar, 1 mm). **B**: a bead containing a portion of embryogenic callus, just after recovery from cryopreservation (bar, 1 mm). **C**: Embryogenic callus coming out of the bead 12 weeks after cryopreservation (procedure: encapsulation-vitrification). A *de-novo* formed somatic embryo is present (from Tonon *et al.*, 2001; bar, 1 mm).